Probability Theory

A. Rényi

Member of the Hungarian Academy of Sciences
Professor of Mathematics at the Eötvös Loránd University, Budapest
Director of the Mathematical Institute
of the Hungarian Academy of Sciences, Budapest

Dover Publications, Inc.
Mineola, New York

Bibliographical Note

This Dover edition, first published in 2007, is an unabridged republication of the work published jointly by North-Holland Publishing Company, Amsterdam, and Akadémiai Kiadó, Publishing House of the Hungarian Academy of Sciences, Budapest, in 1970. This book is the enlarged version of *Wahrscheinlichkeitsrechnung* (VEB Deutscher Verlag der Wissenschaften, Berlin, 1962), *Valószinüségszámitás* (Tankönyvkiadó, Budapest, 1996), and *Calcul des probabilités* (Dunod, Paris, 1966). The English translation is by László Vekerdi.

Library of Congress Cataloging-in-Publication Data

Rényi, Alfréd.
    [Wahrscheinlichkeitsrechnung. English]
    Probability theory / A. Rényi
    p. cm.
    This Dover edition, first published in 2007, is an unabridged republication of the work published jointly by North-Holland Publishing Company, Amsterdam, and Akadémiai Kiadó, Publishing House of the Hungarian Academy of Sciences, Budapest, in 1970. This book is the enlarged version of Wahrscheinlichkeitsrechnung ... [et al.].
    Includes bibliographical references and index.
    ISBN-13: 978-0-486-45867-0
    ISBN-10: 0-486-45867-9
    1. Probabilities. 2. Information theory. I. Title.

QA273.R4313 2007
519.2—dc22

2006053472

Printed in Canada
45867905   2024
www.doverpublications.com

PREFACE

One of the latest works of Alfréd Rényi is presented to the reader in this volume. Before his sudden death on the 1st February, 1970, he corrected the first proof of the book, but he had no longer time for the final proof-reading* and writing the preface he had planned.

This preface is, therefore, a brief memorial to a great mathematician, mentioning a few features of Alfréd Rényi's professional career.

Professor Rényi lectured on probability theory at various universities throughout an uninterrupted series of years, from 1948 till his untimely death. His academic career started at the University of Debrecen and was continued at the University of Budapest where he was professor of the Chair of Theory of Probability. In the meantime he was invited lecturer for shorter or longer terms in several scientific centres of the world. Thus he was visiting professor at Stanford University, Michigan State University, the University of Erlangen, and the University of North Carolina.

Besides his teaching activities, Professor Rényi was director of the Mathematical Institute of the Hungarian Academy of Sciences for one and half decade. Under his direction the Institute developed into an important research centre of the science of mathematics.

He participated in the editorial work of a number of journals. He was the editor of Studia Scientiarum Mathematicarum Hungarica and a member of the Editorial Board of: Acta Mathematica, Annales Sci. Math., Publicationes Math., Matematikai Lapok, Zeitschrift für Wahrscheinlichkeitstheorie, Journal of Applied Probability, Journal of Combinatorial Analysis, Information and Control.

The careful reader will certainly note how the long teaching experience and keen interest in research are amalgamated in the present book. The material of Professor Rényi's courses on probability theory was first published in the form of lecture notes. It appeared as a book in Hungarian in 1954, and completely revised in German translation in 1962. The latter book was the basis of a new Hungarian edition in 1965 and the French

---

* This was done by Mr. P. Bártfai, Mrs. A. Földes and Mrs. L. Rejtő.

translation published in 1966. In the new Hungarian edition the author omitted some theoretical chapters of the German text, inserting new ones dealing with recent results and modern practical methods. The present book contains the complete texts of the Hungarian and German versions and is completed with some additional new material. The presentation of a number of well-chosen problems and exercises will certainly be regarded as a valuable feature of the book; some of them − following the traditions of the Hungarian Mathematical Competitions − have been selected from the material of original publications.

In his lectures and books, Alfréd Rényi always strived at arousing interest in recent results of research, besides presenting the fundamental text-book material of probability theory. Accordingly, he often wrote of problems in which he was just engaged. In the present book the reader will also find many particularities which do not occur in other text-books dealing with the same field. These problems have been selected mostly from among research topics pursued by the author and his school; they are presented with the aim of bringing within the scope of the beginner the spirit of the living and rapidly developing present-day mathematics.

Pál Révész

CONTENTS

CHAPTER I

ALGEBRAS OF EVENTS

§ 1. Fundamental relations

Probability theory deals with events occurring in connection with random mass-phenomena. As it is an abstract mathematical theory, the concept of events is to be dealt with abstractly too; i.e. relations between events are to be characterized axiomatically. For this reason, we consider first of all in this Chapter the "algebras of events". Indeed, relations between events have a primarily logical character: one may assign to every event a proposition stating its occurrence. Thus logical relations between propositions correspond to the relations between events. The algebraic structure of the set of events turns out to be a *Boolean algebra*. Algebras of events as a basis of probability theory were first considered by V. I. Glivenko [3] (cf. also A. N. Kolmogorov [9]).

As stated above, events are to be characterized as abstract concepts. We shall define an algebra of events as a set of events connected with one and the same "experiment", taken in the widest sense of this word. There belongs to every experiment a set of possible outcomes; for every event of the algebra corresponding to the experiment one must be able to decide for each possible outcome whether the event occurred or not.

Let the events $A, B, C, \ldots$ be elements of the same algebra of events. Two events both either occurring or non-occurring at the same time for every outcome of the experiment are said to be identical. The fact that the events A and B are identical is denoted by $A = B$.

The non-occurrence of an event A is itself an event, denoted by $\bar{A}$ and called the *event complementary* to A. From this definition it follows that

$$\bar{\bar{A}} = A. \tag{1}$$

In the realm of logic this corresponds to the proposition that a statement doubly negated coincides with the statement itself.

If A and B are two events of the same algebra of events, we may ask whether they did occur both. Let our experiment be for instance the firing on a target. By a vertical and a horizontal line we subdivide the target into four equal parts. Let event A be a hit in the upper half of the target; event

B one in the right side of it. In this case the statement "A and B occurred both" means the fact that the hit lies in the right upper quadrant of the target (Fig. 1).

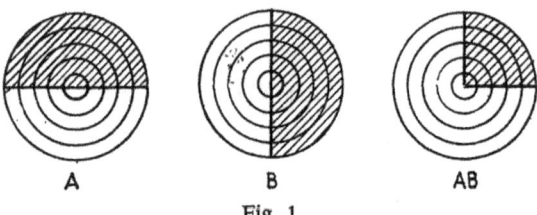

Fig. 1

Event C, occurring if and only if both events A and B occur, is said to be the *product* of events A and B; we write $C = AB$. Thus we have defined an operation, namely the multiplication of events. Let us now see what the properties of this operation are. First, since AB clearly does not depend on the order of A and B, we have the commutative law

$$AB = BA. \tag{2}$$

Also obviously,

$$AA = A, \tag{3}$$

i.e. every event A is idempotent with respect to multiplication. The definition of the product of events may be extended to more than two factors. $A(BC)$ occurs, by definition, if and only if the events A and BC occur; that is, if the events A, B, and C all occur. Evidently, $(AB)C$ has the same meaning. Thus we have the associative law for multiplication:

$$A(BC) = (AB)C. \tag{4}$$

Instead of $A(BC)$ therefore we can write simply ABC. Clearly, the event AB can occur only if A and B do not exclude each other. If A and B are mutually exclusive, AB is an impossible event. It is useful to consider the impossible event as an event too. It will be denoted by O. The fact, that A and B are mutually exclusive, is thus expressed by $AB = O$. Since an event and the complementary event obviously exclude each other, we have

$$A\bar{A} = O. \tag{5}$$

If A and B are two events of an algebra of events, one may ask whether at least one of the events A and B did occur. Let A denote the event that the hit lies in the upper half of the target and B the event that it lies in the right half; the statement, that at least one of the events A and B occurred, means then that the hit does not lie in the left lower quadrant of the target (Fig. 2).

The event occurring exactly when at least one of the events A and B occurs, is said to be the *sum* of A and B and is denoted by $A + B$. It is easy to see

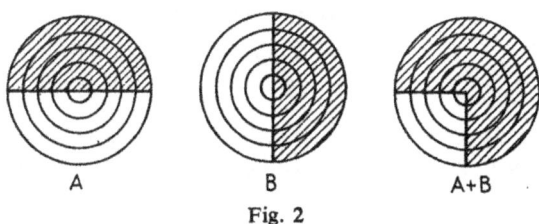

A   B   A+B

Fig. 2

that

$$A + B = B + A \tag{6}$$

(commutative law of addition) and also that

$$A + (B + C) = (A + B) + C \tag{7}$$

(associative law of addition). The definition of the sum is readily extended to the case of more than two events.

The event $A + B$ occurs thus precisely, if A or B occurs; the word "or", however, does not mean in this connection that A and B exclude one another. Thus for instance, in our repeatedly considered example the meaning of $A + B$ is the statement that the hit lies either in the upper half of the target (this is now the event A) or in the right lower quadrant (event $\bar{A}B$). Therefore we have the relation

$$A + B = A + \bar{A}B, \tag{8}$$

where the two terms on the right hand side are now mutually exclusive. By applying relation (8), every sum of events can be transformed in such a way that the terms of the sum become pairwise mutually exclusive.

Clearly the formula

$$A + A = A \tag{9}$$

is valid. Further we see that the event $A + \bar{A}$ certainly occurs; thus by introducing the notation I for the "sure event" we have

$$A + \bar{A} = I. \tag{10}$$

We agree further that

$$\bar{I} = O, \quad \bar{O} = I, \tag{11}$$

i.e. that the event complementary to the sure event is the impossible event O and conversely.

Evidently, the following relations are also valid:

$$AO = O, \tag{12}$$

$$A + O = A, \tag{13}$$

$$AI = A, \tag{14}$$

$$A + I = I. \tag{15}$$

In order to be able to carry out unrestrictedly all operations in the algebra of events, we need some further basic relations. First of all, does the distributive law hold for the addition and multiplication in the algebra of events? Now $A(B + C)$ occurs, by definition, exactly if A occurs and B or C occurs. This, however, means precisely that either A and B occur or A and C occur, i.e. that $AB + AC$ occurs. Therefore we have

$$A(B + C) = AB + AC. \tag{16}$$

From the distributive law follows the so-called "law of inclusion"

$$A + AB = A; \tag{17}$$

since from (14), (16) and (15) we have

$$A + AB = AI + AB = A(I + B) = AI = A.$$

Clearly, rule (17) can be verified directly as well; the direct verification is, however, clumsy for some complicated relations, while by applying the formal rules of operation one can readily get a formal proof. This is the reason why the algebra of events is useful; therefore it is advisable to obtain a certain practice in such formal proofs.

The distributive laws can be extended (just like in ordinary algebra) to more than two terms. In the algebra of events there exists, however, still another distributive law:

$$A + BC = (A + B)(A + C). \tag{18}$$

The validity of (18) is readily seen: $A + BC$ occurs exactly, if A occurs or B and C occur; if A occurs, both factors of the product on the right hand side occur and the same is true if B and C occur, but in no other case.

This consideration being somewhat more difficult as the preceding ones, it is of interest to show how (18) is implied by the already deduced rules of the algebra of events. Indeed, because of (2), (3), (16) and (17) we have

$$(A + B)(A + C) = A + AB + AC + BC = A + BC,$$

which is what we had to show.

Next we prove some further important relations:

$$\overline{AB} = \bar{A} + \bar{B}, \tag{19}$$

$$\overline{A + B} = \bar{A}\bar{B}. \tag{20}$$

The event $\overline{AB}$ occurs exactly, if AB does not occur, hence if the events A and B do not both occur; $\bar{A} + \bar{B}$ occurs exactly, if A or B (or both) do not occur. These two propositions evidently state the same thing; thus (19) is valid. Formula (20) can be proved in the same way.

As to the rules of operation valid for the addition and multiplication of events, we see that both have the same properties (commutativity, associativity, idempotency of every element) and that the relations between the two kinds of rules of operation are symmetrical. Formulas (16) and (18) are obtained from each other — by interchanging everywhere the signs of multiplication and addition. Such formulas are called *dual* to one another. Thus for instance the relations

$$A + AB = A \text{ and } A(A + B) = A$$

are dual to one another. Clearly, there exist relations which are, because of their symmetry, selfdual; e.g. the relation

$$(A + B)(A + C)(B + C) = AB + AC + BC.$$

For sake of brevity we write sometimes $\prod_{k=1}^{n} A_k$ instead of $A_1 A_2 \ldots A_n$ and $\sum_{k=1}^{n} A_k$ instead of $A_1 + A_2 + \ldots + A_n$.

§ 2. Some further operations and relations

Subtraction is defined in the algebra of events by the formula

$$B - A = B\bar{A}. \tag{1}$$

With respect to subtraction the following rules hold:

$$A(B - C) = AB - AC, \ AB - C = (A - C)(B - C); \tag{2}$$

they are the two distributive laws of subtraction. Using the subtraction, the complementary event may be written in the form

$$\bar{A} = I - A. \tag{3}$$

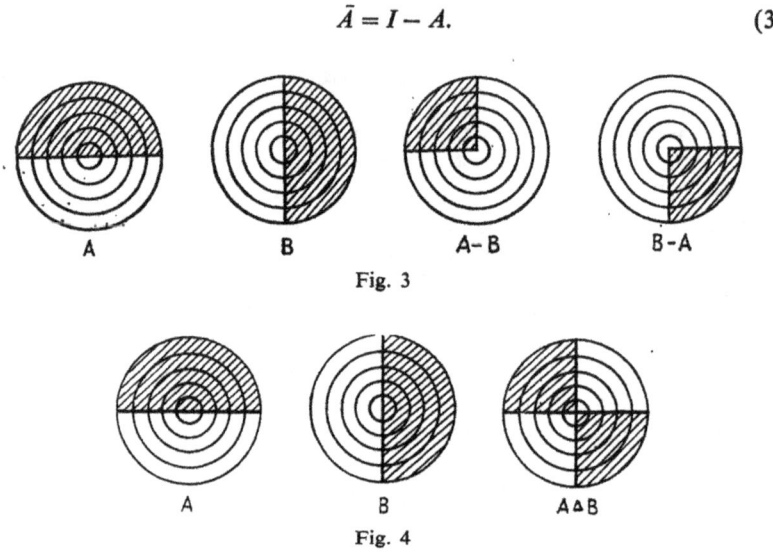

Fig. 3

Fig. 4

The subtraction does not satisfy all the rules of operation known from ordinary algebra. Thus for instance $(A - B) + B$ is in general not equal to A; further $A + (B - C)$ is not always identical to $(A + B) - C$. Hence, if in relations between events there figures the sign of subtraction too, the brackets are not to be omitted without any consideration. There are, however, cases when this omission is allowed, e.g.

$$A - (B + C) = (A - B) - C. \tag{4}$$

The event $A - B$ occurs exactly if A does and B does not occur; in the same way, $B - A$ occurs if B does but A does not occur. The meaning of the expression $(A - B) + (B - A)$ is therefore not O, but the event which consists of the occurrence of one and only one of the events A and B. It is reasonable to introduce for this event a new symbol. We put

$$(A - B) + (B - A) = A \varDelta B. \tag{5}$$

The operation denoted by $\varDelta$ is called the *symmetric difference* of the events A and B (cf. Figs 3 and 4). It fulfils the following rules of operation, derived readily from the already known rules:

$$A \Delta A = O \qquad A \Delta B = B \Delta A \qquad\qquad A(B \Delta C) = AB \Delta AC$$
$$A \Delta O = A \qquad A \Delta B = (A + B) - AB \quad B - A \; = AB \Delta B$$
$$A \Delta I = \bar{A} \quad A + B = (A \Delta B) \Delta AB \qquad B - A \; = (A \Delta B)B \qquad (6)$$

Finally, we mention still another relation between events. If the occurrence of an event A always entails the occurrence of the event B, then we say that the event A implies the event B. We denote this fact by the symbol $\subseteq$. Therefore $A \subseteq B$ means that from the occurrence of the event A always follows the occurrence of the event B.

The following relations hold:

1. $O \subseteq A$.
2. $A \subseteq I$.
3. $A \subseteq A$.
4. $A \subseteq B$ and $B \subseteq C$ imply $A \subseteq C$.
5. $A \subseteq B$ and $B \subseteq A$ imply $A = B$.
6. $A \subseteq A + B$.
7. $AB \subseteq A$.
8. $A \subseteq B$ implies $A = AB$.
9. $A \subseteq C$ and $B \subseteq C$ imply $A + B \subseteq C$.
10. $C \subseteq A$ and $C \subseteq B$ imply $C \subseteq AB$.
11. $A \subseteq B$ implies $B = A + B\bar{A}$.
12. $A \subseteq B$ implies $\bar{B} \subseteq \bar{A}$.
13. $A \subseteq B$ implies $AC \subseteq BC$.
14. $A \subseteq B$ implies $A + C \subseteq B + C$.
15. $AB = O$ and $C \subseteq A$ imply $BC = O$.

It is easy to show that the meaning of $A = AB$ as well as of $B = A + B\bar{A}$ is the same as of $A \subseteq B$; these relations could have served as well for the definition of the relation $\subseteq$. Indeed, if the relation $B = A + B\bar{A}$ is valid, then the occurrence of A implies the occurrence of B. If we have further $A = AB$, then the occurrence of A implies the occurrence of B, since

$$B = BI = B(A + \bar{A}) = BA + B\bar{A} = A + B\bar{A}.$$

From this it follows that for the validity of the relation $A \subseteq B$ the validity of one of the relations $A = AB$ and $B = A + B\bar{A}$ is necessary and sufficient. The latter relation can be stated in the following form: For the validity of $A \subseteq B$ a necessary and sufficient condition is the existence of a C such that $AC = O$ and $B = A + C$; indeed, from this it follows directly $C = B\bar{A}$.

We introduce the following important concept. A system $A_1, A_2, \ldots, A_n$ is called a *complete system of events*, if the relations $A_k \neq O$ $(k = 1, 2, \ldots, n)$;

$$A_j A_k = O \quad \text{for} \quad j \neq k \quad \text{and} \quad A_1 + A_2 + \ldots + A_n = I$$

are valid. For instance $\{A, \bar{A}\}$ is a complete system of events, provided that $A \neq O$ and $A \neq I$.

§ 3. Axiomatical development of the algebra of events

In the preceding paragraphs, we introduced certain operations for events and discussed the rules for these operations. Now we have to make a further abstraction. A set $\mathscr{A}$, of arbitrary elements $A, B, C, \ldots$ is said to be a *Boolean algebra*, if the following conditions are fulfilled: Given any two elements A and B of $\mathscr{A}$, there exists exactly one element of $\mathscr{A}$ called the *product* of A and B and denoted by AB and exactly one element of $\mathscr{A}$ called the *sum* of A and B and denoted by $A + B$;[1] further there corresponds to every element A of $\mathscr{A}$ exactly one element $\bar{A}$ of $\mathscr{A}$. Let there exist two special elements of the set $\mathscr{A}$, namely O and I. Let the elements of the Boolean algebra fulfil the relations obtained in the preceding paragraph; the following *axioms* are therefore assumed to hold:

$$AA = A \tag{1.1}$$
$$AB = BA \tag{1.2}$$
$$A(BC) = (AB)C \tag{1.3}$$
$$A + A = A \tag{2.1}$$
$$A + B = B + A \tag{2.2}$$
$$A + (B + C) = (A + B) + C \tag{2.3}$$
$$A(B + C) = AB + AC \tag{3.1}$$
$$A + BC = (A + B)(A + C) \tag{3.2}$$
$$A\bar{A} = O \tag{4.1}$$
$$A + \bar{A} = I \tag{4.2}$$
$$AI = A \tag{5.1}$$
$$A + O = A \tag{5.2}$$
$$AO = O \tag{5.3}$$
$$A + I = I \tag{5.4}$$

[1] The notations $A \cap B$ and $A \cup B$ are used often instead of AB and $A + B$, respectively.

It is to be noted that these axioms are not all mutually independent; thus for instance (3.2) can be deduced from the others. It is, however, not our aim to examine here which axioms could be omitted from the system.

The totality of the outcomes of an experiment forms a Boolean algebra, if we understand by the product AB of two events A, B the joint occurrence of both events and by the sum $A + B$ of two events the occurrence of at least one of the two events; further, if we denote by $\bar{A}$ the event complementary to A and by O and I the impossible and the sure events, respectively. Indeed, the above 14 axioms are fulfilled in this case. More generally, every subset of the set of the outcomes of an experiment is a Boolean algebra if it contains the sure event, further for every event A its complementary event $\bar{A}$ and for every A and B the events AB and $A + B$.

Clearly, one can find other Boolean algebras as well. Thus for instance, the totality of the subsets of a set H is also a Boolean algebra. We define the sum of two sets as the union of the two sets and their product as the intersection of the two sets. Let I mean the set H itself and O the empty set, further $\bar{A}$ the set complementary to A with respect to H and thus $B - A$ the set complementary to A with respect to B. A direct verification of each axiom shows that this system is indeed a Boolean algebra.

There exists a close connection between Boolean algebras of events and algebras of sets. In our example of the target this connection is clearly visible. This analogy between a Boolean algebra of sets and an algebra of events has an important role in the calculus of probability.

In order to obtain a Boolean algebra, it is not necessary to consider all subsets of a set. A collection T of the subsets of a set H is said to be an *algebra of sets*, if the addition can be always carried out in it, if H itself belongs to T and for a set A its complementary set $\bar{A} = H - A$ belongs to T as well; i.e. if the following conditions are satisfied:[1]

1. $H \in T$.

2. $A \in T$, $B \in T$ implies $A + B \in T$.

3. $A \in T$ implies $\bar{A} \in T$.

The collection of all subsets of a set H is said to be a *complete algebra of sets*. A complete algebra of sets is always a Boolean algebra. Indeed, it is easy to see that the validity of $AB \in T$ follows from $A \in T$ and $B \in T$ by the conditions 1, 2 and 3, since $AB = \overline{\bar{A} + \bar{B}}$. The above 14 axioms are evidently fulfilled.

---

[1] The notation $a \in M$ means here and in the following that a belongs to the set M; $a \notin M$ means that a does not belong to the set M.

§ 4. On the structure of finite algebras of events

An event A is said to be a *compound event*, if it can be represented as the sum of two events which are both different from A:

$$A = B + C, \ B \neq A, \ C \neq A.$$

(The condition $B \neq A$, $C \neq A$ is necessary, to exclude the trivial representations $A = A + O$ and $A = A + A$, which are valid for every A.) Events which do not permit any such representation are said to be *elementary events*. Compound events may be obtained in a number of ways, but elementary events can occur in only one manner. In order to illustrate this by an example, let A denote the event of throwing 10 in a game with two dice. This is a compound event; indeed the number 10 can be obtained by throwing with both dice 5 as well as by getting 6 on one of the dice and 4 on the other. The latter event is again a compound event, since we can have 6 on the first die and 4 on the second and conversely. If A means the result 12 with two dice, then A is an elementary event, since it can be realized only by casting 6 with each die.

If A is an elementary event, then from $B \subseteq A$ follows either $B = O$ or $B = A$. Since $A \subseteq A$ always holds, we denote the fact that $B \subseteq A$ with $B \neq A$ by the symbol $B \subset A$. Clearly, from $B \subset A$ follows $B \subseteq A$, but the converse does not hold. By using this notation the definition of the elementary event may be formulated as follows: The event $A \neq O$ is an elementary event, if and only if there exists no $B(B \neq O)$ such that $B \subset A$.[1] Indeed, if $B \subset A$ is valid for some $B \neq O$, then from relation (11) of § 2 follows $A = B + A\bar{B}$ where B and $A\bar{B}$ are distinct from A; namely $B \neq A$ follows from the assumption $B \subset A$, while from $A\bar{B} = A$ would follow, because of $B \subset A$ and thus $B = AB$, the equation $B = AB = (A\bar{B})B = AO = O$, which contradicts our assumption.

We give here a further characterisation of elementary events: $A \neq O$ is an elementary event, if for an arbitrary event B either $AB = O$ or $AB = A$. Otherwise namely there would exist a decomposition

$$A = AB + A\bar{B}$$

of A, where $AB \neq A$ and $AB \neq O$; the converse is proved readily too.

Next we prove the following

THEOREM 1. *In an algebra consisting of a finite number of events every event can be represented as a sum of elementary events. This representation is unique except for the order of the terms.*

---

[1] The impossible event O is not considered to be an elementary event.

In order to prove the theorem, we need two lemmas:

LEMMA 1. *The product of two distinct elementary events is O.*

For every A_1 and A_2 we obviously have $A_1A_2 \subseteq A_1$. In particular, if A_1 and A_2 are elementary events, we have either $A_1A_2 = O$ or $A_1A_2 = A_1$. But the latter is impossible since it would imply $A_1 \subseteq A_2$, which cannot hold because of $A_1 \neq O$ and $A_1 \neq A_2$.

LEMMA 2. *For every compound event B of an algebra of events with a finite number of events, there exists an elementary event A such that $A \subset B$ holds.*

Since B is a compound event, there exists an $A_1 \neq O$ such that $A_1 \subset B$. If A_1 is elementary, our statement is proved; if, however, it is not elementary, then there exists an $A_2 \subset A_1$. If A_2 is elementary, we have found the elementary event required; if not, there exists again a new decomposition, etc. The procedure must end after a finite number of steps because of the finiteness of the number of events. Therefore A_n must be elementary for a certain number n.

PROOF OF THEOREM 1. Let B be a compound event. According to Lemma 2 there exists then an elementary event A_1 such that $A_1 \subset B$, i.e. $B = A_1 + B_1$. If B_1 is elementary, the first statement of our theorem is proved; if B_1 is not elementary, we obtain, by applying repeatedly Lemma 2, a representation $B_1 = A_2 + B_2$, where A_2 is elementary; if B_2 is compound, the procedure is to be continued. Thus we obtain a representation of B as a sum of elementary events:

$$B = A_1 + A_2 + \ldots + A_r \tag{1}$$

since the number of events is finite. It is evident, from this proof, that all the A_i-s are distinct. If not already known, this could easily be shown because of the rule $A + A = A$ and the commutativity of the addition. (The deduction used above to prove the representability of B as a sum of elementary events is nothing else as the so-called "descente infinie" known from number theory.) It remains still to prove the uniqueness of representation (1). If there would exist two essentially different representations

$$B = A_1 + A_2 + \ldots + A_r = A_1' + A_2' + \ldots + A_s' \tag{2}$$

such that for instance $A_1 \neq A_j'$ ($j = 1, 2, \ldots, s$), then multiplication of both sides of (2) by A_1 would yield, by Lemma 1, $A_1 = O$, i.e. a contradiction. Herewith Theorem 1 is completely proved. The representation (1) is called the *canonical representation* of the event B.

THEOREM 2. *The number of events of a finite algebra of events is necessarily a power of 2.*

PROOF. Let n denote the number of distinct elementary events of a finite algebra of events. Every event of the algebra can be expressed as a sum of a certain number of elementary events; this number r can be one of the numbers $0, 1, \ldots, n$, if the impossible event is included. If r is fixed, the number of the events which can be expressed as a sum of r distinct elementary events is equal to the number of possible selections of exactly r from the n elementary events $A_1, A_2, \ldots, A_n$; that is equal to $\binom{n}{r}$. Thus the number of elements of the whole algebra of events is $\sum_{r=0}^{n} \binom{n}{r}$, which is equal to 2^n.

It follows from Theorem 1 that the sure event I can be represented as the sum of all elementary events

$$I = A_1 + A_2 + \ldots + A_n.$$

Thus always one and only one of the elementary events $A_1, A_2, \ldots, A_n$ occurs. The elementary events form a complete system of events.

Consider now as an example the algebra of events which consists of the possible outcomes of a game with two dice. Clearly, the number of elementary events is 36; let us denote them by A_{ij} $(i, j = 1, 2, \ldots, 6)$ where A_{ij} means that the result for the first die is i, that for the second, j. According to Theorem 2 the number of events of this algebra of events is $2^{36} = 68\ 719\ 476\ 736$. It would thus be impossible to discuss all cases.

We choose therefore another example, namely the tossing a coin twice. The possible elementary events are: 1. first head, second head as well (let A_{11} denote this case); 2. first head, next tail, denoted by A_{12}; 3. first tail, next head, denoted by A_{21}; 4. first tail, second also tail, notation: A_{22}. The number of all possible events is $2^4 = 16$. These are: I, O, the four elementary events, further $A_{11} + A_{12}$, $A_{11} + A_{21}$, $A_{11} + A_{22}$, $A_{12} + A_{21}$, $A_{12} + A_{22}$, $A_{21} + A_{22}$, and besides these the four events $\bar{A}_{11}$, $\bar{A}_{12}$, $\bar{A}_{21}$, $\bar{A}_{22}$ complementary to the four elementary events.

By the canonical representation we have thus obtained a complete description of this finite algebra of events. Now also the rules of operation obtain a new sense. Theorem 1 namely points to a connection which shall lead us to Kolmogorov's theory of probability. A compound event, which is the sum of elementary events, can be characterized uniquely by the set of terms of this sum. In this way one can assign to every event a set, namely the set of the elementary events whose sum is the canonical representation of the event. Let A' denote the collection of elementary events which form the event A and similarly let B' denote the collection of the elementary events from which event B is composed. One can show that the collection of elementary events from which the event AB is composed is the intersection $A'B'$ of A' and B'; further that the collection of elementary events from which the event $A + B$ is composed is equal to the union $A' + B'$ of the sets A' and B'. In this assignment of events to sets, the elementary events themselves correspond to the sets having only one element. Obviously the empty set corresponds to the impossible event. To the sure event corresponds the set of all possible elementary events (with respect to the same experiment); this set will be denoted by H and will be called the *sample space*. Further, it is easy to show that to the complementary event $\bar{A}$ corresponds the complementary set of A' with respect to H.

In the following paragraph we shall show that to every algebra of events corresponds an algebra of the subsets of a set H such that there corresponds to the event $A + B$ the union of the sets belonging to A and B and to the product AB the intersection of the sets belonging to A and B, finally to the complementary event $\bar{A}$ the complementary set with respect to H of the set belonging to A. In other words, *one can find to every algebra of events an algebra of sets which is isomorphic to it.*

The proof of this theorem, due to Stone [1], is not at all simple; it will be accomplished in the next paragraph; on first reading it can be omitted, since Stone's theorem will not be used in what follows. We give the proof only in order to show that the basic assumption of Kolmogorov's theory, i.e. that events can always be represented by sets, does not restrict the generality in any way.

In the case of a finite algebra of events this fact was already established by means of Theorem 1. Here we even have a uniquely determined event corresponding to every subset of the sample space.

The theory of Boolean algebras is a particular case of the theory of more general structures called *lattices* (cf. e.g. G. Birkhoff [1]).

§ 5. Representation of algebras of events by algebras of sets

In this paragraph we prove a theorem due to Stone, mentioned in § 4.[1]

THEOREM. *There can be associated with every algebra of events an algebra of sets isomorphic to it.*

PROOF. Let $\mathscr{A}$ be an algebra of events; let $A, B, C, \ldots$ denote its elements. Consider a subset α of $\mathscr{A}$ having the following properties:

1. The event O is not contained in α.
2. From $A \in \alpha$ and $B \in \alpha$ follows $AB \in \alpha$.

---

[1] The proof given here is due to O. Frink [1].

3. Among the sets satisfying conditions 1 and 2 α is maximal in the following sense: there exists no set β satisfying conditions 1 and 2 and containing α as a proper subset.

The sets α fulfilling these three conditions are briefly called crowds of events.[1]

It is easy to show the following: If α is a crowd of events, we have

a) $I \in \alpha$.
b) $A \in \alpha$ implies $\bar{A} \notin \alpha$ and conversely.
c) If $A + B \in \alpha$, then A or B belongs to α.

Now we prove the property of the crowds of events stated in

LEMMA 1. $AB \in \alpha$ implies $A \in \alpha$ (and clearly, because of the symmetry of the formula, it implies $B \in \alpha$ as well).

PROOF. Suppose that $AB \in \alpha$, $A \notin \alpha$. Let β be the union of α and all events AC, where C runs through all elements of α; α is a proper subset of β, since we have $I \in \alpha$ and thus, by assumption, $A = AI \in \beta$ and $A \notin \alpha$.

The set β fulfils the conditions 1 and 2. First we prove that β satisfies condition 1, and hence that $O \notin \beta$. Indeed, because of $AB \in \alpha$, we have $AB \neq O$, hence $A \neq O$; further, since $(AC)B = (AB)C$ and $(AB)C \in \alpha$, we have for $C \in \alpha$ the relation $AC \neq O$. From this our first statement follows. It remains to show that $D \in \beta$ and $E \in \beta$ imply the relation $DE \in \beta$. Now we have either $D \in \alpha$, $E \in \alpha$ or $D \in \alpha$, $E \notin \alpha$ (or conversely) or else $D \notin \alpha$, $E \notin \alpha$. If $D \in \alpha$, $E \in \alpha$, then, by our assumption, we have $DE \in \alpha$ and thus certainly $DE \in \beta$. If $D \in \alpha$, $E \notin \alpha$, then there exists a $C \in \alpha$ such that $E = AC$, hence $DE = A(CD)$; since $CD \in \alpha$, we have $DE \in \beta$. In the case $D \notin \alpha$, $E \notin \alpha$, there exist two elements C_1 and C_2 of α such that $D = AC_1$ and $E = AC_2$. Then $DE = A(C_1 C_2)$, and since $C_1 C_2 \in \alpha$, we have $DE \in \beta$.

But this contradicts the assumption that α is maximal. Lemma 1 is therefore proved.

Next we prove

LEMMA 2. Every event A $(A \neq O)$ of an algebra of events $\mathcal{A}$ belongs at least to one crowd of events α.

---

[1] In lattice theory such systems are called *ultrafilters*: ultrafilters are commonly characterized as sets complementary to prime ideals. A nonempty subset β of a Boolean algebra $\mathcal{A}$ is called a prime ideal, if the following conditions are fulfilled. 1. $A \in \beta$ and $B \in \beta$ imply $A + B \in \beta$. 2. $A \in \beta$ and $B \in \mathcal{A}$ imply $AB \in \beta$. 3. If $AB \in \beta$, then $A \in \beta$ or $B \in \beta$ (or both). Cf. e.g. G. Aumann [1].

This lemma is the consequence of a general set theoretical theorem of Hausdorff; we have, however, to define some notions in order to state it. A set $\mathscr{R}$ is said to be *partially ordered,* if an ordering relation, denoted by $<$, is defined for certain pairs of its elements; if $a < b$, we say that the element a precedes the element b.

The relation $<$ is required to fulfil the following conditions:

1. For no element does $a < a$ hold.
2. $a < b$ and $b < c$ imply $a < c$.

The relation $<$ is therefore irreflexive and transitive. A subset $\mathscr{L}$ of a partially ordered set $\mathscr{R}$ is called a *chain,* if for any two elements a and b of $\mathscr{L}$ either $a < b$ or $b < a$ holds. An ordered subset $\mathscr{L}$ of $\mathscr{R}$ is thus said to be a chain. A chain $\mathscr{L}$ is said to be a *maximal chain* in $\mathscr{R}$, if there does not exist any element of $\mathscr{R}$ such that by adjoining it to $\mathscr{L}$ the so obtained subset would still remain a chain.

We are now ready to state the above-mentioned lemma.

LEMMA 3. (Hausdorff). *If $\mathscr{R}$ is a partially ordered set, then every chain $\mathscr{L}$ of $\mathscr{R}$ is a subset of a maximal chain.*[1]

Let us return to the proof of Lemma 2. Let $\mathscr{R}$ denote the set of those systems of events β in the algebra of events $\mathscr{A}$ which fulfil conditions 1 and 2 of the crowds of events. If $\beta < \gamma$ means that β is a proper subset of γ, $\mathscr{R}$ is a partially ordered set. If $A \neq O$, the set $\beta = (A)$ consisting only of the element A evidently fulfils conditions 1 and 2. According to Lemma 3 there exists a maximal chain containing $\beta = (A)$ as a subset. Let α denote the union of the subsets γ belonging to this chain. Clearly, α is a crowd of events, since it is the union of sets β fulfilling the rules 1 and 2 defining the crowds of events. Therefore no element of the chain contains the event O and thus α does not contain O either. Further if B_1 and B_2 belong to α, they belong to a subset β_1 respectively a subset β_2 of α. Since either $\beta_1 < \beta_2$ or the contrary must hold, β_1 and β_2 belong both to β_1 or to β_2 and the same holds for $B_1 B_2$. Thus $B_1 B_2$ belongs to α as well. Further we see that α cannot be extended. This is a consequence of the requirement that the chain be a maximal chain. Lemma 2 is thus proved.

Now we can construct to every algebra of events a field of sets isomorphic to it. Let $\mathscr{H}$ be the set of all crowds of events α of the algebra of events $\mathscr{A}$. We assign to every event A of $\mathscr{A}$ the subset $\mathscr{R}_A$ of $\mathscr{H}$ consisting of all crowds of events α containing the event A. The set $\mathscr{R}_A$ will be called the *representa-*

---

[1] As to the proof of Lemma 3 cf. e.g. F. Hausdorff [1] or O. Frink [2].

tive of the event A. As $A \neq O$, $\mathscr{R}_A$ is, by Lemma 2, a nonempty set. We associate to the event O the empty set. The system consisting of the empty set and of all $\mathscr{R}_A$-s shall be denoted by $\mathscr{F}$.

We prove the following relations:

$$\mathscr{R}_A \mathscr{R}_B = \mathscr{R}_{AB}, \tag{1}$$

$$\overline{\mathscr{R}}_A = \mathscr{R}_{\bar{A}}, \tag{2}$$

$$\mathscr{R}_A + \mathscr{R}_B = \mathscr{R}_{A+B}. \tag{3}$$

Further it will be proved that the correspondence $A \to \mathscr{R}_A$ is one-to-one; thus $A \neq B$ implies $\mathscr{R}_A \neq \mathscr{R}_B$. Hence the algebra of sets is isomorphic to the algebra of events $\mathscr{A}$. (We understand of course by $\mathscr{R}_A \mathscr{R}_B$ the intersection of $\mathscr{R}_A$ and $\mathscr{R}_B$ and by $\mathscr{R}_A + \mathscr{R}_B$ the union of both sets. $\overline{\mathscr{R}}_A$ denotes the complementary set with respect to $\mathscr{H}$, and $AB, A + B, \bar{A}$ the corresponding operations in the algebra of events $\mathscr{A}$.)

The relation (1) can be proved as follows. A crowd of events α belonging to both of $\mathscr{R}_A$ and $\mathscr{R}_B$ contains both A and B and thus AB as well. Conversely, if AB belongs to α, then by Lemma 1 $A \in \alpha$ and $B \in \alpha$ and thus α belongs to $\mathscr{R}_A$ and to $\mathscr{R}_B$ hence also to $\mathscr{R}_A \mathscr{R}_B$.

Proof of (2). If the crowd of events α does not belong to the set $\mathscr{R}_A$, it must contain an event B such that $AB = O$; otherwise namely $AB \neq O$ for every element B of α and α could be extended by adjoining A and every product of the form AB ($B \in \alpha$). This leads, just like in the proof of Lemma 1, to a result which contradicts the assumption that α is a maximal chain. From $AB = O$ it follows that $B = AB + \bar{A}B = \bar{A}B$. Hence $\bar{A}$ belongs, according to Lemma 1, to α. Conversely, if $\bar{A}$ belongs to α, A cannot belong to it since $A\bar{A} = O$. From this it follows that $\overline{\mathscr{R}}_A$ consists exactly of the crowds of events containing the event $\bar{A}$, hence $\overline{\mathscr{R}}_A = \mathscr{R}_{\bar{A}}$.

Relation (3) is a direct consequence of relations (1) and (2). Indeed we have

$$\mathscr{R}_{A+B} = \mathscr{R}_{\overline{\bar{A}\bar{B}}} = \overline{\mathscr{R}_{\bar{A}\bar{B}}} = \overline{\mathscr{R}_{\bar{A}} \mathscr{R}_{\bar{B}}} = \overline{\mathscr{R}}_{\bar{A}} + \overline{\mathscr{R}}_{\bar{B}} = \mathscr{R}_A + \mathscr{R}_B.$$

Thus we have proved that $\mathscr{F}$ is an algebra of sets. In order to show that $\mathscr{F}$ is isomorphic to the algebra of events $\mathscr{A}$, it still remains to prove that the correspondence $A \to \mathscr{R}_A$ is one-to-one. If $A \neq B$, we have $A \Delta B \neq O$. Hence at least one of the relations $A\bar{B} \neq O$ and $\bar{A}B \neq O$ is valid as well. Suppose that $\bar{A}B \neq O$. Because of (1) $\mathscr{R}_{\bar{A}B} = \mathscr{R}_{\bar{A}} \mathscr{R}_B$. Hence every crowd of events belonging to $\mathscr{R}_{\bar{A}B}$ belongs to $\mathscr{R}_{\bar{A}}$ and also to $\mathscr{R}_B$, hence it belongs to $\mathscr{R}_B$ and *does not belong* to $\mathscr{R}_A$. Thus we proved the existence of crowds of events which belong to $\mathscr{R}_B$ but do not belong to $\mathscr{R}_A$. Hence $\mathscr{R}_B$ and $\mathscr{R}_A$

cannot coincide. Herewith the theorem is proved. In this proof we have used as regards $\mathscr{A}$ only its property of being a Boolean algebra, therefore we may formulate the theorem just proved in a more general manner: *There exists to every Boolean algebra an algebra of sets isomorphic to it.*

§ 6. Exercises

1. Prove

a) $\overline{AB + CD} = (\bar{A} + \bar{B})(\bar{C} + \bar{D})$,

b) $(A + B)(A + \bar{B}) + (\bar{A} + B)(\bar{A} + \bar{B}) = I$,

c) $(A + B)(A + \bar{B})(\bar{A} + B)(\bar{A} + \bar{B}) = O$,

d) $(A + B)(A + C)(B + C) = AB + AC + BC$,

e) $A - BC = (A - B) + (A - C)$,

f) $A - (B + C) = (A - B) - C$,

g) $(A - B) + C = [(A + C) - B] + BC$,

h) $(A - B) - (C - D) = [A - (B + C)] + (AD - B)$,

i) $A - \{A - [B - (B - C)]\} = ABC$,

j) $ABC + ABD + ACD + BCD =$
$= (A + B)(A + C)(A + D)(B + C)(B + D)(C + D)$,

k) $A + B + C = (A - B) + (B - C) + (C - A) + ABC$,

l) $A \, \varDelta \, (B \, \varDelta \, C) = (A \, \varDelta \, B) \, \varDelta \, C$,

m) $(A + \bar{B}) \, \varDelta \, (\bar{A} + B) = A \, \varDelta \, B$,

n) $A\bar{B} \, \varDelta \, B\bar{A} = A \, \varDelta \, B$.

o) Prove the relations enumerated in § 2 (6) for the symmetric difference.

p) The relation $(A + B) - B = A$ does not hold in general. Under what conditions is it valid?

q) Prove that $A \, \varDelta \, B = C \, \varDelta \, D$ implies $A \, \varDelta \, C = B \, \varDelta \, D$.

Hint. If $A \, \varDelta \, B = C \, \varDelta \, D$ and A, B, C, D are subsets of the same set, then every point of this set belongs to an even number (0, 2, or 4) of the sets A, B, C, D and $A \, \varDelta \, C = B \, \varDelta \, D$ means the same.

r) Prove that the elements of an arbitrary algebra of events form an Abelian group with respect to the symmetric difference as the group operation.

2. The elements of a Boolean algebra form a ring with respect to the operations of symmetric difference and multiplication. The zero element is O, the unit element I.

3. In a finite algebra of events containing n elementary events one can give several complete systems of events. Complete systems of events differing only in the order of the terms are to be considered as identical. Let T_n denote the number of the different complete systems of events.

a) Prove that $T_1 = 1$, $T_2 = 2$, $T_3 = 5$, $T_4 = 15$, $T_5 = 52$, $T_6 = 203$.

b) Prove the recursion formula

$$T_{n+1} = 1 + \sum_{k=1}^{n} \binom{n}{k} T_k$$

and show that $T_{10} = 115\ 975$.

c) Prove

$$1 + \sum_{k=1}^{\infty} \frac{T_k}{k!} x^k = e^{(e^x - 1)}.$$

4. Let γ_n denote the number of complete systems of events consisting of three elements in a finite algebra of events consisting of n elementary events. Show that

$$\gamma_n = \frac{3^{n-1} + 1}{2} - 2^{n-1}.$$

5. Prove the relation

$$T_n = \frac{1}{e} \sum_{k=1}^{\infty} \frac{k^n}{k!}$$

(T_n means the same as in Exercise 3).

6. Let Q_n denote the number of complete systems of events in an algebra with n elementary events, such that every event is the sum of an odd number of different elementary events. Prove that

$$Q_1 = 1, \; Q_2 = 1, \; Q_3 = 2, \; Q_4 = 5, \; Q_5 = 12, \; Q_6 = 37,$$

further that

$$1 + \sum_{n=1}^{\infty} \frac{Q_n}{n!} x^n = e^{shx}.$$

7. We can construct from the events A, B, and C by repeated addition and multiplication eighteen, in general different, events namely $A, B, C, AB, AC, BC, A + B, B + C, C + A, A + BC, B + AC, C + AB, AB + AC, AB + BC, AC + BC, ABC, A + B + C, AB + AC + BC$. (The phrase "in general different" means here that no two of these events are identical for all possible choices of the events A, B, C.) Prove that from 4 events one can construct 166, from 5 events 7579 and from 6 events 7 828 352 events in this way. (No general formula is known for the number of events which can be formed from n events.)

8. The divisors of an arbitrary square-free[1] number N form a Boolean algebra' if the operations are defined as follows: We understand by the "sum" of two divisors of N their least common multiple, by their "product" their greatest common divisor; d being a divisor of N we understand by $\bar{d}$ the number $\dfrac{N}{d}$; the number 1 serves as O and the number N as I.

9. Verify that for the example of Exercise 8, our Theorem 1 is the same as the well-known theorem on the unique representability of (square-free) integers as a product of prime numbers.

10. The numbers $0, 1, \ldots, 2^n-1$ form a Boolean algebra if the rules of operation are defined as follows: Represent these numbers in the binary system. We understand by the "product" of two numbers the number obtained by multiplying the corre-

---

[1] A number is said to be square-free if it is not divisible by any square number except by 1. The square-freeness of N is required only to ensure the existence of the complementary element.

sponding digits of both numbers place for place; by the sum the number obtained
by adding the digits place for place and by replacing everywhere the digit 2 obtained
in the course of addition by 1.

11. Let A, B, C denote electric relays or networks of relays. Any two of these
may be connected in series or in parallel. Two such networks which are either closed
both (allowing current to pass) or both open (not allowing current to pass) are considered as equivalent. Let $A + B$ denote that A and B are coupled in parallel, AB
that they are coupled in series. Let $\bar{A}$ denote a network always closed if A is open
and conversely. Let O denote a network allowing no current to pass and I a network
always closed. Prove that all axioms of Boolean algebras are fulfilled.[1]

Hint. Relation $(A + B)C = AC + BC$ has for instance the meaning that it comes
to the same thing to connect first A and B in parallel and couple the network so obtained
with C in series or to couple first A and C in series, then B and C in series and then
the two systems so obtained in parallel. Both systems are equivalent to each other in
the sense that they either both allow to pass the current or both do not. A similar
consideration holds for the other distributive law. Both distributive laws are illustrated
in Fig. 5.

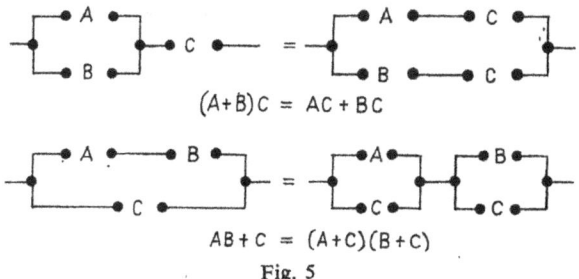

$$(A+B)C = AC + BC$$

$$AB + C = (A+C)(B+C)$$

Fig. 5

12. A domain of the plane is said to be *convex* if it contains for any two of its
points the segment connecting these points as well. We understand by the "sum"
of two convex domains the least convex domain containing both, by their "product"
their intersection which, evidently, is convex as well. Let further I denote the entire
plane and O the empty set. The addition and multiplication fulfil axioms (1.1)–(2.3);
the distributive laws are, however, not valid and the complement $\bar{A}$ is not defined.

13. Let us understand by a linear form a point, a line or a plane of the 3-dimensional
affine space, further let the empty set and the entire 3-dimensional space be called
linear forms too. We define as the sum of a finite number of linear forms the least linear
form containing their set theoretical union; let their product be their (set theoretical)
intersection, which is evidently a linear form too. Prove the same propositions as in
Exercise 12.

[1] This example shows how Boolean algebra can be applied in the theory of networks and why it is of great importance in communication theory and in the construction of computers (cf. e.g. M. A. Gavrilov [1]).

14. Let $A_1, A_2, \ldots, A_n$ be arbitrary events. Form all products of these events containing k distinct factors, and let S_k be the sum of all products. Let P_k be the product of all events representable as a sum of k distinct terms of $A_1, A_2, \ldots, A_n$. Prove the relation

$$S_k = P_{n-k+1} \qquad (k = 1, 2, \ldots, n)$$

in a formal way by applying the rules of operation of Boolean algebras and verify it directly too (cf. the generalization of Exercise 1d).

Hint. S_k has the meaning that among the events $A_1, A_2, \ldots, A_n$ there are at least k which occur and the meaning of P_{n-k+1} is that among these same events there are no $n - k + 1$ which do not occur; these two statements are equivalent.

15. Let H be a set and T a set of certain subsets of H. T is an algebra of sets, if and only if the following conditions are fulfilled:

a) The set H belongs to T.
b) Whenever A and B belong to T, $A - B$ belongs to T as well.

16. Show that condition b) of the preceding exercise cannot be replaced by b') "whenever A and B belong to T, $A \triangle B$ belongs to T as well".

17. If the conditions

α) $A \in T$ implies $\bar{A} \in T$,
β) $A \in T$ and $B \in T$ imply $AB \in T$

are postulated instead of conditions a) and b) of Exercise 15 and T is nonempty, then T is an algebra of sets.

18. Condition β) of the preceding exercise may be replaced by

β') "$A \in T$ and $B \in T$ imply $A + B \in T$."

19. Prove that in Exercise 18 the proposition cannot be replaced by

β") "$A \in T, B \in T$, and $AB = O$ imply $A + B \in T$."

Hint. Let H be a finite set of the elements a, b, c, d and let T consist of the following subsets of H: $\{a, b\}; \{c, d\}; \{a, c\}; \{b, d\}; O; H$.

20. We call a nonempty $\mathcal{R}$ of subsets of a set H that contains with two sets A and B also $A + B$ and $A - B$, a *ring of sets*. A ring of sets $\mathcal{R}$ is thus an algebra of sets if and only if H belongs to $\mathcal{R}$. Prove that a nonempty system of sets containing with A and B also AB and $A - B$ is not necessarily a ring of sets. Show that the condition "with two sets A and B, $A + B$ and AB belong as well to S" is not sufficient for S to be a ring of sets either.

CHAPTER II

PROBABILITY

§ 1. Aim and scope of the theory of probability

There exist in nature several phenomena fitting into a deterministic scheme: given a complex K of circumstances, a certain event A necessarily occurs. On the other hand, there are a number of phenomena in the sciences as well as in our everyday life which cannot be described by such schemes. It is characteristic for such phenomena that given the complex K of the circumstances, the event A may or may not occur. Such events are called *random events* and such schemes are said to be *stochastic schemes*. Consider for instance a radioactive atom during a certain time interval: the atom may or may not decay during the time of observation. The instant of the decay depends on processes in the nucleus which are, however, neither known to us nor observable.

We can see from this example the need of studying stochastic schemes. Very often it is entirely impossible (at least at the present state of matters) to consider all relevant circumstances. But in many practical problems this is not at all necessary. An event A may be a random event with respect to a complex of circumstances and at the same time it may be completely determined with respect to another, more comprehensive complex of circumstances. The randomness or determinedness of an event is a matter of fact: it depends only on whether the given complex of circumstances K does or does not determine the course of the phenomena (that is, the occurrence or the non-occurrence of the event A). But the choice of the complex of circumstances K depends on us, and we have a certain freedom to choose it within the limits of possibilities.

Regarding random mass-phenomena we can sketch their outlines, in spite of their random character. Consider for instance radioactive disintegrations. Each radioactive substance continues to decay according to a well determined "rate"; we can predict, what percentage of the substance will disintegrate during a given time interval. The disintegration follows an exponential law (cf. Ch. III) characterized by the half-life period. (The half-life period is the time interval during which half of the radioactive substance disintegrates. In the case of radium this is some 1600 years.) The exponential law is a typical "probability law". This law is confirmed by the observations with the same accuracy as most of the "deterministic" laws of nature. The radioactive disintegration is thus a mass phenomenon described, as to its regularity, by the theory of probability.

As seen in the above example, phenomena described by a stochastic scheme are also subject to natural laws. But in these cases the complex of the considered circumstances does not determine the exact course of the events; it determines a probability law, giving a bird's view of the outcome.

Probability theory aims at the study of random mass-phenomena, this explains its great practical importance. Indeed we encounter random mass-phenomena in nearly all fields of science, industry, and everyday life.

Almost every "deterministic" scheme of the sciences turns out to be stochastic at a closer examination. The laws of Boyle, Mariotte, and Gay-Lussac for instance are usually considered to be deterministic laws. But the pressure of the gas is caused by the impacts of the molecules of the gas on the walls of the container. The mean pressure of the gas is determined by the number and the velocity of the molecules hitting the wall of the container per time unit. In fact, the pressure of the gas shows small fluctuations, which may, however, be neglected in case of greater gas-masses. As another example consider the chemical reaction of two substances A and B in a watery solution. As it is well known, the velocity of the reaction is in every instant proportional to the product of the concentrations of A and B. This law is commonly considered as a causal one, but in reality the situation is as follows. The atoms (respectively the ions) of the two substances move freely in the solution. The average number of the "encounters" of an ion of substance A with an ion of substance B is proportional to the product of their concentrations; hence this law turns out to be essentially a stochastic one too.

The development of modern science makes it often necessary to examine small fluctuations in phenomena dealt with earlier only in their outlines and considered at that level as causal. In the following, we shall find several occasions to illustrate these principal questions with concrete examples.

§ 2. The notion of probability

Consider an experiment where the circumstances regarded do not uniquely determine the outcome of the experiment, but leave more possibilities. Let A be one of these. If we repeat the experiment under invariant conditions, A will occur at some of these experiments, while in the other experiments $\bar{A}$ will occur. If among n experiments the event A occurred exactly k times, then k is called the *frequency* and $\dfrac{k}{n}$ the *relative frequency* of the event

A in the given sequence of experiments. Generally, the relative frequency of a random event is not constant in different sequences of experiments. Consider as an example screws produced by an automatic machine. Let A denote the event that the screw does not fit the requirements, i.e. that it

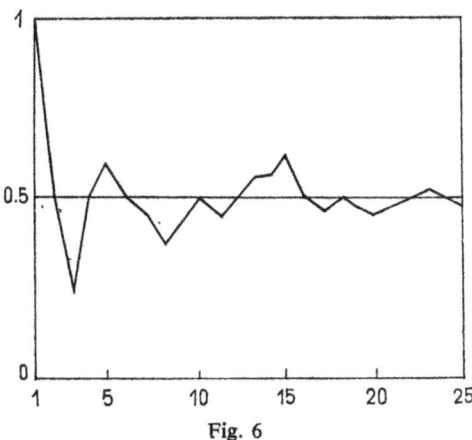

Fig. 6

is defective. The frequency of the defective items will be in general different in every series, for instance in the first series 3 in 100 screws, in the second 5, etc. Under constant circumstances of production the percentage of the defective items fluctuates about a certain value. After a change in the circumstances of production this value may be different, but about the new value there will again be fluctuations.

The mentioned fluctuations of the relative frequency may be observed in the simple experiment of coin tossing by observing for instance the relative frequency of the heads. If H denotes head and T tail, we may obtain in a sequence of 25 tossings the following outcome:

HTTHHTTTHHTHHHTTTHTTHHHTT.

Figure 6 represents the fluctuations of the relative frequency.

Anyone may perform such experiments in a few minutes. The order of heads and tails will every time be different but the general picture will remain essentially similar to the above: the relative frequency will always fluctuate about the value $\frac{1}{2}$.

Figure 7 represents the outcome of an experiment of 400 tossings. As early as in the eighteenth century large sequences of tossings were observed. Buffon for instance performed an experiment of 4040 tossings. The outcome was 2048 times head, hence the relative frequency was 0.5069. In the begin-

ning of our century K. Pearson obtained from 24 000 tossings the value
0.5005 for the relative frequency.

There are thus random events showing a certain stability of the relative
frequency, i.e. the latter fluctuates about a well-determined value and the

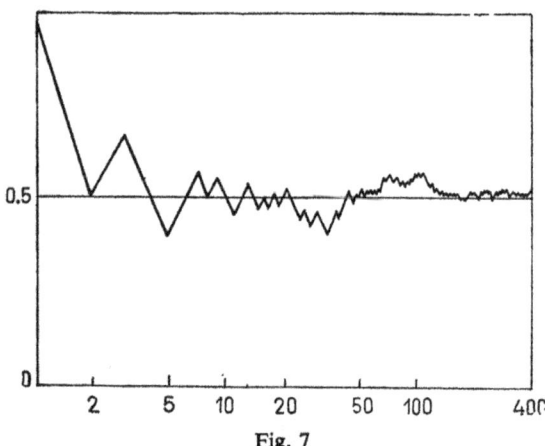

Fig. 7

more trials are performed, the smaller are, generally, the fluctuations. The
number, about which the relative frequency of an event fluctuates, is called
the *probability* of the event in question. Thus, for instance, the probability
of "heads" (supposed the coin is regular) is equal to $\dfrac{1}{2}$.

Consider now another example. In throws of regular die of homogeneous
substance the relative frequency of any one of the faces 1, 2, 3, 4, 5, 6 fluctuates about $\dfrac{1}{6}$, i.e. the probability of each number is equal to $\dfrac{1}{6}$, if,
however, the die is deformed, e.g. by curtailing one of its faces, these
probabilities will be different.

A further example is the following: there is a well-determined probability
that a certain atom of radioactive substance disintegrates during a given
time interval t. That is, in repeated observation of atoms during a time interval t we find that the number of atoms decaying during this time interval
fluctuates about a well-determined value. This value is according to the
observations $1 - c^{-\lambda t}$, where λ is a positive constant depending on the radioactive substance. (E.g. in case of radium, if the time is measured in seconds,
$\lambda = 1.38 \cdot 10^{-11}$.) Later on we shall give a theoretical foundation of this
law, i.e. we shall deduce it from simple assumptions.

Comparing our examples of coin tossing and radioactive disintegration we see that in the first a coin is tossed many a times successively, while in the second a great number of simultaneous events are observed. This difference is, however, unessential. Indeed, instead of tossing one coin successively, we could toss a great number of similar coins simultaneously. The only essential thing from the point of view of probability theory is the (at least conceptually) unrestricted repeatability of the observation under the same circumstances.

To sum up: If the relative frequency of a random event fluctuates about a well-determined number, the latter is called the *probability* of the event in question. The probability of an event may change with the circumstances of the experiment. The only way to decide whether the relative frequency has or does not have the property of stability and to determine the value about which the statistical fluctuations occur, is empirical observation.

Thus the probability is considered to be a value independent of the observer. It gives the approximate value of the relative frequency of an event in a long sequence of experiments.

As the probability of a random event is an objective value which does not depend on the observer, one can approximately calculate the long run behaviour of such events, and the calculations can be empirically verified.

In everyday life we often make subjective judgements concerning the probability of a random event. The mathematical theory of probability, however, does not deal with these subjective judgements but with objective probabilities. These objective probabilities are to be "measured" just like physical quantities. Probability theory and mathematical statistics have their own methods to "measure" probabilities. These methods are mostly indirect and at the end are always based upon the observation of relative frequencies. But at a conceptual level we must sharply distinguish the probability, which is a fixed number, from the relative frequency depending on chance.

§ 3. Probability algebras

The theory developed in this book is due to A. N. Kolmogorov. It is the basis of modern probability theory.

In this theory one starts from the assumption that to all possible (or at least all considered) events in an experiment (or with other words: to all elements of an algebra of events) a numerical value is assigned: the probability of the event in question. If we perform an experiment n times and find k occurrences of the event A, then we have, because of the relation $0 \leq k \leq$ $\leq n$, anyhow $0 \leq \dfrac{k}{n} \leq 1$; that is, the relative frequency of an event is always a number lying between zero and one. Evidently, the probability of an event must therefore lie between zero and one as well. It is further clear that the relative frequency of the sure event is equal to one and that of the impossible event is equal to zero. Hence also the probability of the "sure" event must be equal to one and that of the "impossible" event equal to zero. If A and B are two possible outcomes of the same experiment which mutually exclude each other and if in n performances of the same experiment the event A occurred k_A times and the event B k_B times, then clearly the event $A + B$ occurred $k_A + k_B$ times. Hence, denoting by f_A, f_B and f_{A+B} the relative frequencies of A, B, and $A + B$, respectively, we have:

$$f_{A+B} = f_A + f_B.$$

In other words, the relative frequency of the sum of two mutually exclusive events is always equal to the sum of the relative frequencies of these events. Hence also the probability of the sum of two mutually exclusive events must be equal to the sum of the probabilities of the events. We therefore take the following axioms:

α) *To each element A of an algebra of events a non-negative real number $P(A)$ is assigned, called the probability of the event A.*

β) *The probability of the sure event is equal to one, that is $P(I) = 1$.*

γ) *If $AB = O$, then $P(A + B) = P(A) + P(B)$.*

An algebra of events $\mathscr{A}$ in which to every element A a number $P(A)$ is assigned satisfying Axioms α), β), γ), will be called a *probability algebra*.

Let us first see some consequences of the axioms:

THEOREM 1. *If $B \subseteq A$, then $P(B) \leq P(A)$.*

PROOF. From $B \subseteq A$ it follows $A = B + C$, where $C = A\bar{B}$ and hence $BC = O$. Thus by Axiom γ) we have

$$P(A) = P(B) + P(C).$$

Since because of Axiom α) $P(C) \geq 0$, Theorem 1 follows immediately. It can also be deduced directly from the relation between probability and relative frequency. Indeed, if the occurrence of event B implies the occurrence of event A (i.e. if $B \subseteq A$), then in any sequence of experiments the event A occurs at least as often as the event B.

Since $A \subseteq I$ for every $A \in \mathscr{A}$, it follows from Theorem 1 that

$$P(A) \leq 1 \text{ for every } A.$$

Another important consequence of the axioms is the possibility to find the probability of the contrary event $\bar{A}$ from the probability of the event A. Indeed, we have $A + \bar{A} = I$ and $A\bar{A} = O$, and thus by Axiom $\gamma)$ $P(A) + P(\bar{A}) = P(I) = 1$. Herewith we proved

THEOREM 2. *For any event A the relation*

$$P(A) + P(\bar{A}) = 1$$

holds.

Because of $O = \bar{I}$, it follows from Theorem 2

$$P(O) = 1 - P(I) = 1 - 1 = 0;$$

i.e. the probability of the impossible event is equal to 0.

Theorem 2 can be deduced directly from the conceptual definition of probability. Indeed, if in a sequence consisting of n experiments the event A occurs k times, then $\bar{A}$ occurs exactly $(n - k)$ times. Hence for the relative frequencies f_A respectively $f_{\bar{A}}$ we have

$$f_A + f_{\bar{A}} = 1.$$

Axiom $\gamma)$ states that the probability of the sum of two mutually exclusive events is equal to the sum of the probabilities of the two events. This leads immediately to

THEOREM 3. *If the events $A_1, A_2, \ldots, A_n$ are pairwise exclusive, i.e. if $A_i A_j = O$ for $i \neq j$, then*

$$P(A_1 + A_2 + \ldots + A_n) = P(A_1) + P(A_2) + \ldots + P(A_n).$$

The proof proceeds by mathematical induction. Our theorem is valid for $n = 2$. Suppose that it is proved for $n - 1$. Since clearly $(A_1 + A_2) A_k = O$ $(k = 3, 4, \ldots, n)$, we have because of the induction assumption

$$P(A_1 + \ldots + A_n) = P(A_1 + A_2) + P(A_3) + \ldots + P(A_n) =$$
$$= P(A_1) + P(A_2) + \ldots + P(A_n).$$

From Theorem 3 follows, as a generalization of Theorem 2,

THEOREM 4. *If the events $A_1, A_2, \ldots, A_n$ form a complete system (cf. I, § 2), then*

$$P(A_1) + P(A_2) + \ldots + P(A_n) = 1.$$

Indeed, by assumption

$$A_1 + A_2 + \ldots + A_n = I \quad \text{and} \quad A_i A_j = O$$

for $i \neq j$, further $P(I) = 1$; hence Theorem 4 follows immediately from Theorem 3.

As a further important consequence of the axioms we show how to calculate the probability of the sum of two events A and B, if we drop the restriction that A and B should exclude each other.

THEOREM 5. *Let A and B be arbitrary events. We have then*

$$P(A + B) = P(A) + P(B) - P(AB).$$

PROOF. $A + B$ can be represented as a sum of two mutually exclusive events. We have namely

$$A + B = A + \bar{A}B.$$

Hence by Axiom γ)

$$P(A + B) = P(A) + P(\bar{A}B). \tag{1}$$

On the other hand, we have $B = AB + \bar{A}B$ and $AB \cdot \bar{A}B = O$. From these it follows

$$P(B) = P(AB) + P(\bar{A}B). \tag{2}$$

Subtraction of Equation (2) from (1) leads to our statement. In particular, if we have $AB = O$ our theorem will be reduced to the assertion of Axiom γ).

We shall need the following simple theorems:

THEOREM 6. *If $A \subseteq B$, then*

$$P(B - A) = P(B) - P(A).$$

PROOF. We have, by assumption, $B = A + (B - A)$ and $A(B - A) = O$, hence Theorem 6 follows from Axiom γ).

Clearly the relation $P(B - A) = P(B) - P(A)$ does not hold in general. It is, however, easy to obtain

THEOREM 7. $P(B - A) = P(B) - P(AB)$.

PROOF. We have $(B - A) + AB = B$ and $(B - A)AB = O$; from this our theorem follows because of Axiom γ).

Furthermore we have

THEOREM 8. $P(A \Delta B) = P(A) + P(B) - 2P(AB)$.

PROOF. Because of

$$A \Delta B = (A - B) + (B - A)$$

and

$$(A - B)(B - A) = O$$

we find

$$P(A\Delta B) = P(A - B) + P(B - A),$$

and Theorem 8 follows from Theorem 7.

We proved Theorem 3 by repeated application of Axiom γ). In the same manner, we can obtain by repeated application of Theorem 5 a formula for the probability of the sum of an arbitrary number of events. In particular, we have

$$P(A + B + C) = P(A) + P(B) + P(C) - P(AB) - P(BC) - \\ - P(CA) + P(ABC).$$

More generally we have

THEOREM 9. *Let $A_1, A_2, \ldots, A_n$ be arbitrary events of a probability algebra. Then we have*

$$P(A_1 + A_2 + \ldots + A_n) = \sum_{k=1}^{n} (-1)^{k-1} S_k^{(n)},$$

where

$$S_k^{(n)} = \sum_{1 \le i_1 < i_2 < \ldots < i_k \le n} P(A_{i_1} A_{i_2} \ldots A_{i_k}).$$

In the summation $(i_1, i_2, \ldots, i_k)$ run through all combinations, k at a time, of the numbers $1, 2, \ldots, n$, repetitions not allowed.

Theorem 9 is a particular case of the following more general theorem:

THEOREM 10 (Ch. Jordan). *Let $V_r^{(n)}$ denote the probability of the occurrence of exactly r among the events $A_1, A_2, \ldots, A_n$. Then we have*

$$V_r^{(n)} = \sum_{k=0}^{n-r} (-1)^k \binom{r+k}{k} S_{r+k}^{(n)} \qquad (r = 0, 1, \ldots, n), \tag{3}$$

where $S_0^{(n)} = 1$ and

$$S_l^{(n)} = \sum_{1 \le i_1 < i_2 < \ldots < i_l \le n} P(A_{i_1} A_{i_2} \ldots A_{i_l}) \quad \text{for } l = 1, 2, \ldots \tag{4}$$

and the summation is to be extended over all combinations of the numbers $1, 2, \ldots, n$, l at a time, repetitions not allowed.

We shall prove Theorem 10 by a general principle which can be used for the proof of many similar identities.[1]

---

[1] Cf. A. Rényi [23], [35].

We need some preparatory definitions and remarks. Let $A_1, A_2, \ldots, A_n$ be arbitrary events. There exists a "least algebra of events" $\mathcal{A}$ containing the events $A_1, A_2, \ldots, A_n$. Clearly, $\mathcal{A}$ consists of 2^{2^n} elements and is generated by the 2^n elementary events $\omega = A_{i_1} A_{i_2} \ldots A_{i_k} \bar{A}_{j_1} \ldots \bar{A}_{j_{n-k}}$, where $(i_1, i_2, \ldots, i_k)$ is an arbitrary combination of the numbers $1, 2, \ldots, n$; k at a time $(k = 0, 1, \ldots, n)$ and $j_1, j_2, \ldots, j_{n-k}$ are those of the numbers $1, 2, \ldots, n$ which do not figure among $i_1, i_2, \ldots, i_k$.[1]

We shall call every event belonging to $\mathcal{A}$ "an event expressible in terms of $A_1, A_2, \ldots, A_n$". Every B from $\mathcal{A}$ is namely a sum of certain elementary events ω. Thus B can be expressed by application of the basic operations of the algebra of events to $A_1, A_2, \ldots, A_n$. In other words: Every event $B \in \mathcal{A}$ is a *function* of the events $A_1, \ldots, A_n$. The functional relation between an event $B \in \mathcal{A}$ and the events $A_1, A_2, \ldots, A_n$ representing B does not depend on the specific choice of the events $A_1, A_2, \ldots, A_n$. In particular it does not depend on the probabilities of the events $A_1, A_2, \ldots, A_n$. The event $A_{i_1} A_{i_2} \ldots A_{i_r}$, or the event E_r that exactly r of the events $A_1, A_2, \ldots, A_n$ occur, are for instance events expressible in terms of $A_1, A_2, \ldots, A_n$.

The following theorem contains the principle mentioned above:

THEOREM 11. *Let $A_1, A_2, \ldots, A_n$ be n arbitrary events and $\mathcal{A}$ the algebra of events of all events expressible in terms of the events A_k. Let $c_1, c_2, \ldots, c_m$ be real numbers and $B_1, B_2, \ldots, B_m$ a sequence of events such that $B_k \in \mathcal{A}$ $(k = 1, 2, \ldots, m)$. The inequality*

$$\sum_{k=1}^{m} c_k P(B_k) \geq 0 \tag{5}$$

holds for an arbitrary probability algebra obtained by assigning probabilities to the elements of $\mathcal{A}$, if and only if it holds for those probability algebras, in which the sequence of numbers $P(A_k)$ $(k = 1, 2, \ldots, n)$ consists only of zeros and ones.

PROOF. Since every B_k is the sum of certain elementary events ω, (5) is equivalent to the inequality

$$\sum \lambda_\omega P(\omega) \geq 0, \tag{6}$$

where the real numbers λ_ω depend only on the numbers c_k and on the functional dependence of the events B_k on the events A_j, but do not depend on the numerical values $P(A_j)$. The summation is over all the 2^n elementary

---

[1] The events A_k $(k = 1, \ldots, n)$ are here considered as variables (indefinite events); thus there are no relations assumed between the A_k-s.

events of $\mathcal{A}$. Indeed we have

$$\lambda_\omega = \sum_{\omega \subseteq B_k} c_k,$$

where the summation is over such values of k for which ω figures in the representation of B_k. Since the nonnegative numbers $P(\omega)$ are submitted to the only condition $\Sigma P(\omega) = 1$, (6) holds, in general, if and only if all numbers λ_ω are nonnegative. But when the sequence of numbers $P(A_j)$ $(j = 1, 2, \ldots, n)$ consists of nothing but zeros and ones, one and only one of the elementary events ω has probability 1 and all the others have probability 0. Thus the proposition that $\lambda_\omega \geq 0$ for all ω is equivalent to the proposition that (6) is valid whenever the sequence of numbers $P(A_k)$ consists of zeros and ones only. Theorem 11 is now proved.

From Theorem 11 follows immediately

THEOREM 12. *If $A_1, A_2, \ldots, A_n$ are arbitrary events and $B_1, B_2, \ldots, B_m$ are certain events expressible by $A_1, A_2, \ldots, A_n$, then the relation*

$$\sum_{k=1}^m c_k P(B_k) = 0 \tag{7}$$

holds in every probability algebra, if and only if it holds in all cases when the sequence of numbers $P(A_k)$ consists of zeros and ones only.

PROOF. Apply Theorem 11 in turn to the inequalities $\sum_{k=1}^m c_k P(B_k) \geq 0$ and $\sum_{k=1}^m (-c_k) P(B_k) \geq 0$. Theorem 12 follows immediately.

Now we can prove Theorem 10 (and thus also Theorem 9). If l from the numbers $P(A_k)$ are equal to 1 and the remaining $n - l$ are equal to 0 ($l = 0, 1, 2, \ldots, n$), then (3) is reduced to the identity

$$\sum_{k=0}^{n-r} (-1)^k \binom{r+k}{k} \binom{l}{r+k} = \begin{cases} 1, & \text{if } l = r, \\ 0, & \text{if } l \neq r. \end{cases} \tag{8}$$

For $l < r$ all terms of the left hand side of (8) are equal to 0, for $l = r$ only the term $k = 0$ is distinct from zero, namely 1 and for $l > r$ the sum can be transformed as follows:

$$\sum_{k=0}^{n-r} (-1)^k \binom{r+k}{k} \binom{l}{r+k} =$$
$$= \binom{l}{r} \left(\sum_{k=0}^{l-r} (-1)^k \binom{l-r}{k} \right) = \binom{l}{r} (1-1)^{l-r} = 0.$$

Let us mention as another application of Theorem 11 the inequality of M. Fréchet and A. J. Gumbel (cf. M. Fréchet [2]).

THEOREM 13 (M. Fréchet).

$$\frac{S_{r+1}^{(n)}}{\binom{n}{r+1}} \le \frac{S_r^{(n)}}{\binom{n}{r}} \quad (r = 0, 1, \ldots, n-1), \tag{9}$$

where the same notations are used as in Theorem 9.

THEOREM 14 (A. J. Gumbel).

$$\frac{\binom{n}{r+1} - S_{r+1}^{(n)}}{\binom{n-1}{r}} \le \frac{\binom{n}{r} - S_r^{(n)}}{\binom{n-1}{r-1}} \quad (r = 1, 2, \ldots n-1), \tag{10}$$

where the same notations are used as in Theorem 9.

PROOF of Theorems 13 and 14. Apply Theorem 11. If l of the numbers $P(A_k)$ are equal to 1, and the others are equal to 0, then (9) will be reduced to the trivial inequality

$$\frac{\binom{l}{r+1}}{\binom{n}{r+1}} \le \frac{\binom{l}{r}}{\binom{n}{r}},$$

and Theorem 14 to the likewise trivial inequality

$$\frac{\binom{n}{r+1} - \binom{l}{r+1}}{\binom{n-1}{r}} \le \frac{\binom{n}{r} - \binom{l}{r}}{\binom{n-1}{r-1}}.$$

§ 4. Finite probability algebras

If the set of events of a probability algebra is finite, we have shown in Chapter I the possibility of representing these events by the class of all subsets of a finite set Ω. Let Ω consist of N elements, denoted by $\omega_1, \omega_2, \ldots, \omega_N$.

The probability $P(A)$ of any event A is then uniquely determined by the values of P for the sets which consist of exactly one element. Let $\{\omega_i\}$ be the set consisting of ω_i only and let further be $P(\{\omega_i\}) = p_i$ ($i = 1, 2, \ldots, N$). Then we have for each event A

$$P(A) = \sum_{\omega_i \in A} p_i.$$

Since $P(\Omega) = 1$, the (nonnegative) numbers p_i must obey the condition $\sum_{i=1}^{N} p_i = 1$.

An important particular case is obtained if all the numbers p_i are equal to each other, that is, if they are equal to $\frac{1}{N}$. These special probability algebras are called *classical probability algebras*, since the classical calculus of probability exclusively dealt with these algebras.

At the early stages of development of probability theory one wished to reduce the solution of any kind of problems to this case. But this often turned out to be either too artificial or unnecessarily involved. Since, however, in the games of chance (tossing of a coin, games of dice, roulette, card-games, etc.) the probabilities can be determined in this manner indeed, and since many problems of science and technology may be reduced to the study of classical probability algebras, it is worthwhile to deal with them separately.

In the case of classical probability algebras we have

$$P(A) = \frac{K}{N}$$

where K denotes the number of the elements of A. Thus we arrive at the "classical definition" of probability: The *probability* of an event is equal to the quotient of the number of the favorable cases and of the total number of all possible cases, provided these cases are all equally probable.

Today the "classical definition" of probability is not considered as a definition any more, but only as a way to calculate probabilities, applicable in case of classical probability algebras, i.e. finite probability algebras whose elementary events have, for certain reasons (e.g. symmetry-properties), the same probability.

§ 5. Probabilities and combinatorics

In classical probability algebras the probabilities are determined by combinatorial methods. In what follows we shall give some examples.

Example 1a. A person having N keys in his pocket wishes to open his apartment. He takes one key after the other from his pocket at random and tries to open the door. What is the probability that he finds the right key at the k-th trial? Suppose that the $N!$ possible sequences of the keys all have the same probability. In this case the answer is very simple indeed: N elements have $(N - 1)!$ permutations with a fixed element occupying the k-th place. The probability in question is therefore $\dfrac{(N - 1)!}{N!} = \dfrac{1}{N}$; that is, the probability of finding the right key at the first, second,..., N-th trial, respectively, is always $\dfrac{1}{N}$. If the keys are on the same key-ring and if the same key may be tried more than once, the answer is different and will be dealt with later (cf. Ch. III, § 3, 7.).

Example 1b. An urn contains M red and $N - M$ white balls. Balls are drawn from the urn one after the other without replacement. What is the probability of obtaining the first red ball at the k-th drawing? In order to answer the question we have to determine the total number of all permutations of M red and $N-M$ white balls having for their first $k-1$ balls white ones and for the k-th a red ball. The first $k-1$ balls may be chosen in $\begin{pmatrix} N-M \\ k-1 \end{pmatrix}$ (different) ways from $N - M$ white balls; furthermore, these can be arranged in $(k - 1)!$ different orders. The red ball on the k-th place can be chosen in M different ways and the remaining places can be filled up in $(N - k)!$ different ways. Hence the probability in question — provided that all the $N!$ permutations are equally probable — is given by

$$P_k = \frac{1}{N!} \begin{pmatrix} N - M \\ k - 1 \end{pmatrix} (k - 1)! \, M(N - k)!.$$

Obviously, the special case $M = 1$ is equivalent to Example 1a.

In order to make the calculations more easy, P_k may be written in the following form:

$$P_k = \frac{M}{N - k + 1} \prod_{j=1}^{k-1} \left(1 - \frac{M}{N - j + 1} \right).$$

If N and M are large in comparison to k, and $\dfrac{M}{N}$ is denoted by p $(0 < p < 1)$, then we have approximately $P_k \approx (1 - p)^{k-1} p = \pi_k$. Indeed, if N and M tend to infinity while $p = \dfrac{M}{N}$ remains constant, then P_k tends to the expression π_k.

Example 2. Consider now the following problem. An urn contains N balls, of which M are red and $N - M$ white, $1 \leq M < N$. From the urn n balls are drawn. What is the probability of obtaining k red and $n - k$ white balls?

A new formulation of the same example discloses the practical importance of the problem. In the serial production of machine parts a series of N items contains M rejects. What is the probability, by taking a random sample of n elements, that this sample will contain k rejects?

Solution. A sample of n elements may be chosen from N elements in $\binom{N}{n}$ different ways. Suppose that every such combination is equally probable. Then the probability of every combination is $\binom{N}{n}^{-1}$. Therefore we have only to count, how many combinations contain just k of the rejects. There can be chosen k elements from M in $\binom{M}{k}$ different ways and $n - k$ elements from $N - M$ in $\binom{N - M}{n - k}$ different ways. Therefore the probability in question is:

$$P_k = \frac{\binom{M}{k}\binom{N - M}{n - k}}{\binom{N}{n}}.$$

Example 3. The Maxwell–Boltzmann, Bose–Einstein, and Fermi–Dirac statistics.

We start from the following simple combinatorial problem: How many different ways are there in which n objects can be placed into N cells? Every object can be placed into N different cells, hence there are N possibilities for each object and the number of the possibilities of the different arrangements is N^n. If we assume that the probability of every arrangement is the same, i.e. N^{-n}, then we obtain the probability of a certain cell being occupied by exactly k objects, if we count the possibilities in question. The k objects occupying the cell can be chosen in $\binom{n}{k}$ different ways, and the number of the possibilities to arrange the remaining $n - k$ objects into $N-1$ cells is $(N - 1)^{n-k}$, so the number of possibilities in question will be $\binom{n}{k}(N - 1)^{n-k}$.

The probability in question is

$$W_k = \binom{n}{k} \left(\frac{1}{N}\right)^k \left(1 - \frac{1}{N}\right)^{n-k}.$$

Such "problems of arrangements" are of paramount importance in statistical mechanics. There it is usual to examine the arrangements of certain kinds of particles (molecules, photons, electrons, etc.) in the "phase space". The meaning of this is the following: If every particle can be characterized by K data, then there corresponds to the state of the particle in question a point of the phase space, having for coordinates the data characterizing the particle. In subdividing the phase space into k-dimensional parallelepipeds (cells) the physical system can be described approximately by giving the number of particles in each cell. The assumption that all arrangements have equal probabilities leads to the so-called *Maxwell–Boltzmann statistics*. This can be applied for instance in statistical mechanics to systems of the molecules of a gas. But in the case of photons, electrons, and other elementary particles we must proceed in a different way. For systems of photons for instance, the following model was found to be valid: in distributing n objects into N cells two arrangements containing in each of the cells the same number of objects are not to be considered as distinct. That is, the objects are to be considered as not distinguishable and thus only arrangements having different numbers of objects in the cells can be distinguished from each other. This assumption leads to the so called *Bose–Einstein statistics*.

Next we calculate the number of possible arrangements under Bose–Einstein statistics. This problem is the same as the following question: In how many ways can n shillings be distributed among N persons? (Of course, the number of shillings obtained is of interest, the individuality of the coins being irrelevant.) This number is equal to the number of combinations of N things, n at a time, repetitions allowed, i.e. to $\binom{N+n-1}{n}$. Another solution of our problem is the following: Let to the n objects correspond n collinear points. Let these n points be subdivided by $N-1$ separating lines. Every configuration thus obtained corresponds to one possible arrangement. Every two consecutive lines signify one cell and the number of the points lying between two consecutive lines represents the number of objects in the corresponding cell. If there are no points between two consecutive lines, the corresponding cell is empty. Figure 8 gives a possible arrangement of eight objects into six cells; here the first cell contains one, the second two objects, the third cell is empty, the fourth contains three objects, the fifth is empty, and in the sixth are two objects.

The number of possibilities in question is thus obtained by dividing the number of permutations of the n points and $N-1$ lines by the number obtained by permuting only the points among themselves and the lines among themselves; it is equal to $\binom{N+n-1}{n}$. Hence under the Bose–Einstein hypothesis the probability of each arrangement is $\binom{N+n-1}{n}^{-1}$.

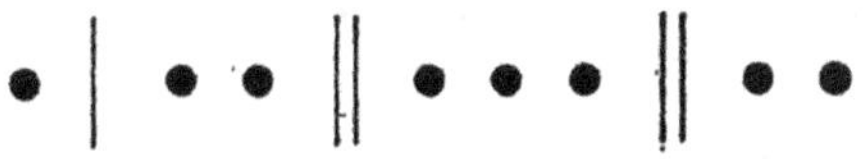

Fig. 8

Now we shall calculate under Bose–Einstein statistics the probability of exactly k particles being in a given cell. For this it suffices to determine the number of possibilities having in the cell in question exactly k particles. This, however, is equal to the number of possibilities of putting $n - k$ particles into the remaining $N - 1$ cells, hence to $\binom{N+n-k-2}{n-k}$; thus the probability is

$$\frac{\binom{N+n-k-2}{n-k}}{\binom{N+n-1}{n}}.$$

We have dealt with Bose–Einstein statistics not only because of their important physical applications. We also wished to make clear that the determination of equally probable cases is not a question of pure logic; experience is involved in it too. This example shows further that a hypothesis relative to equally probable cases cannot always be checked directly. Often we must rely on experimental verification of the consequences of the hypothesis in question.

The Bose–Einstein statistics have no general validity. It does not apply for instance in the case of electrons, where the so-called *Fermi–Dirac statistics* are appropriate. They are obtained by joining to the principle of indistinguishability of the Bose–Einstein statistics the requirement that every cell may be occupied by at most one particle (Pauli's principle). Hence the number of distinct arrangements is equal to the number of the possibilities of choosing n from N elements, i.e. to $\binom{N}{n}$. The probability of a cell being occupied by one particle (more than one is out of question) is

$$\frac{\binom{N-1}{n-1}}{\binom{N}{n}} = \frac{n}{N}.$$

The Fermi–Dirac statistics gives good agreement with experiments in the case of electrons, protons, and neutrons.

§ 6. Kolmogorov probability spaces

The theory discussed up to now can only deal with the most elementary problems of probabilities; those involving an infinite number of possible events are not covered by it. To deal with them we need Kolmogorov's theory, which will now be discussed.

In Kolmogorov's probability theory we assume that there is given an algebra of sets, isomorphic to the algebra of events dealt with. This assumption, as we have seen, does not restrict the generality. We assume further that this algebra of sets contains not only the sum of any two sets belonging to it but also the sum of denumerably many sets belonging to the algebra of sets. Algebras of sets with this property are called *σ-algebras* or *Borel algebras*.

In Kolmogorov's theory we therefore assume the following axioms:

I. *Let there be given a nonempty set* Ω. *The elements of* Ω *are said to be elementary events and are denoted by* ω.

II. *Let be specified an algebra of sets* $\mathscr{A}$ *of the subsets of* Ω; *the sets A of* $\mathscr{A}$ *are called events.*

III. $\mathscr{A}$ *is a σ-algebra, that is*[1]

$$A_k \in \mathscr{A} \quad (k = 1, 2, \ldots) \Rightarrow \sum_{k=1}^{\infty} A_k \in \mathscr{A}.$$

From the Axioms I-III follows immediately that if $A_k \in \mathscr{A}$ $(k = 1, 2, \ldots)$, then also $\prod_{k=1}^{\infty} A_k \in \mathscr{A}$.

The following axioms prescribe the properties of probabilities:

IV. *To each element A of* $\mathscr{A}$ *is assigned a nonnegative real number* $P(A)$, *called the probability of the event A.*

[1] Here and in what follows the sign $\Rightarrow$ stands for the (logical) implication.

V. $P(\Omega) = 1$.

VI. *If $A_1, A_2, \ldots, A_n, \ldots$ is a finite or a denumerably infinite sequence of pairwise disjoint sets belonging to $\mathscr{A}$, then*

$$P(A_1 + A_2 + \ldots + A_n + \ldots) = P(A_1) + P(A_2) + \ldots + P(A_n) + \ldots.$$

Requirement VI is called the σ-*additivity* (or *complete additivity*) of the set function $P(A)$.

A σ-algebra $\mathscr{A}$ of subsets of a set Ω on which a set function $P(A)$ is defined such that Axioms I–VI are fulfilled will be called a *probability space in the sense of Kolmogorov* and will be denoted by $[\Omega, \mathscr{A}, P]$.

Theorems proved in the previous paragraph hold clearly for Kolmogorov probability spaces too, as the Axioms $\alpha), \beta)$, and $\gamma)$ correspond to Axioms IV, V, and VI respectively. Axiom VI, however, requires more than the Axiom $\gamma)$ of probability algebras, since it assumes the additivity of $P(A)$ not only for finitely many, but also for denumerably many pairwise disjoint sets belonging to the σ-algebra $\mathscr{A}$.

Every finite probability algebra is a Kolmogorov probability space, since an additive set function on a finite algebra of sets is trivially σ-additive.

The empty set is denoted by O. Obviously, we have always $P(O) = 0$ (cf. the note to Theorem 2 of § 3).

Apart from finite probability algebras the most simple probability fields are those in which the space Ω consists of denumerably many elements.

Let Ω be a denumerable set, with elements $\omega_1, \omega_2, \ldots, \omega_n, \ldots$; let $\mathscr{A}$ consist of all subsets of Ω. Let the set containing the only element ω_n be denoted by $\{\omega_n\}$; let further be $P(\{\omega_n\}) = p_n$ $(n = 1, 2, \ldots)$. In order that $[\Omega, \mathscr{A}, P]$ be a Kolmogorov probability space the conditions $p_n \geq 0$ and $\sum_{n=1}^{\infty} p_n = 1$ must, according to Axioms IV–VI, be satisfied. Further if A is an arbitrary subset of Ω, then by Axiom VI we have

$$P(A) = \sum_{\omega_k \in A} p_k.$$

Conversely, if the above conditions are fulfilled, $[\Omega, \mathscr{A}, P]$ is in fact a Kolmogorov probability space. Thus we have also proved the consistency of Kolmogorov's axioms. The σ-additivity of P is readily seen, since in any convergent series of nonnegative terms the terms can be rearranged and bracketed in whatever order.

§ 7. The extension of rings of sets, algebras of sets and measures

In this paragraph we shall discuss some results of set theory and measure theory, used in probability theory. We shall not aim at completeness. We assume that the reader is familiar with the fundamentals of measure theory and of the theory of functions of a real variable. Accordingly, proofs are merely sketched or even omitted, especially if dealing with often-used considerations of the theory of functions of a real variable.[1]

We have seen already in Chapter I that every algebra of events is isomorphic to an algebra of sets. It is always assumed in Kolmogorov's theory that the sets assigned to the elements of the algebra of events form a σ-algebra. Hence the algebra of sets constructed to the algebra of events must be extended into a σ-algebra — if it is not already a σ-algebra itself. This extension is always possible, even in the case of a ring of sets.

A system $\mathscr{R}$ of subsets of a set Ω is called a *ring of sets* if

$$A \in \mathscr{R} \text{ and } B \in \mathscr{R} \Rightarrow A + B \in \mathscr{R} \text{ and } A - B \in \mathscr{R}.$$

A ring of sets is said to be a *Borel ring of sets* or σ-*ring*, if

$$A_n \in \mathscr{R} \, (n = 1, 2, \ldots) \Rightarrow \sum_{n=1}^{\infty} A_n \in \mathscr{R}.$$

The ring of sets $\mathscr{R}$ is an algebra of sets iff[2] the set Ω belongs to $\mathscr{R}$. In fact, an algebra of sets $\mathscr{A}$ can be characterized as a system of subsets of a set Ω having the following properties:

  I. $A \in \mathscr{A}$ and $B \in \mathscr{A} \Rightarrow A - B \in \mathscr{A}$.
 II. $A \in \mathscr{A}$ and $B \in \mathscr{A} \Rightarrow A + B \in \mathscr{A}$.
III. $\Omega \in \mathscr{A}$.

This is obvious, as I and III imply that whenever A belongs to $\mathscr{A}$ so does $\Omega - A = \bar{A}$ and thus conditions 1, 2, and 3 of § 3 Chapter I are fulfilled. Conversely, it follows from these conditions that whenever $A \in \mathscr{A}$ and $B \in \mathscr{A}$ hold, so do $\bar{A} \in \mathscr{A}$ and $A - B = \overline{\bar{A} + B} \in \mathscr{A}$ as well. Hence the conditions I, II, III are equivalent to the conditions 1, 2, 3 of Chapter I. We have now the following theorem:

THEOREM 1. *Let Ω be any set and $\mathscr{R}$ a ring consisting of subsets of Ω. There exists then a uniquely determined σ-ring (or Borel ring) $\mathscr{B}(\mathscr{R})$ with the following properties: $\mathscr{B}(\mathscr{R})$ contains $\mathscr{R}$ and an arbitrary σ-ring $\mathscr{R}'$ containing $\mathscr{R}$ contains $\mathscr{B}(\mathscr{R})$ as well. In other words, $\mathscr{B}(\mathscr{R})$ is the least σ-ring containing $\mathscr{R}$.*

PROOF. Obviously, there exists a σ-ring $\mathscr{R}'$ containing $\mathscr{R}$. Such is, for instance, the collection of all subsets of Ω. Let now $\mathscr{B}(\mathscr{R})$ be the intersection

---

[1] More particularly see e.g. P. R. Halmos [1] or V. I. Smirnov [2].
[2] "iff" stands here and in what follows as usual for "if and only if".

of all σ-rings containing $\mathscr{R}$. Evidently, all statements of Theorem 1 are fulfilled by $\mathscr{B}(\mathscr{R})$.

The assumption in Kolmogorov's theory that the sets assigned to the events form a σ-algebra is, according to our theorem, no essential restriction of the generality, since every algebra of sets may be extended by a suitable extension into a σ-algebra.

Let us now consider another example. Let R denote the set of real numbers. Let $\mathscr{R}$ be the collection of those subsets of the real axis which can be represented as sums of a finite number of disjoint half-open intervals (closed to the left and open to the right). Obviously, $\mathscr{R}$ is a ring of sets, though it is not a σ-ring. By Theorem 1 there exists a least σ-ring $\mathscr{B}(\mathscr{R})$ containing $\mathscr{R}$. It is easy to see that $\mathscr{B}(\mathscr{R})$ is in our case a σ-algebra as well. The subsets of R belonging to $\mathscr{B}(\mathscr{R})$ are called *Borel sets*.

Let us now assume that a nonnegative and completely additive set function $P(A)$ is defined on the sets of an algebra of sets (which is not a σ-algebra), further that $P(\Omega) = 1$. We have seen that every algebra of sets $\mathscr{A}$ may be extended into a σ-algebra. But is it possible (and if so how), to extend the definition of P to the elements of $\mathscr{B}(\mathscr{A})$ while preserving the nonnegativity and the σ-additivity?

A nonnegative, completely additive set function defined over a ring of sets is called a *measure*. Our question therefore, in a more general formulation, is whether a measure defined over a ring $\mathscr{A}$ can be extended to the least σ-ring $\mathscr{B}(\mathscr{A})$ containing $\mathscr{A}$.

For our purposes it suffices to consider σ-*finite* measures. A measure μ defined over a ring $\mathscr{R}$ is said to be σ-*finite*, if for every set $A \in \mathscr{R}$ there exists a sequence $A_n \in \mathscr{R}$ $(n = 1, 2, \ldots)$ such that every set A_n has a finite measure $\mu(A_n) < +\infty$ and $A \subseteq \sum_{n=1}^{\infty} A_n$. The following theorem can now be asserted:

THEOREM 2. *If* $\mu(A)$ *is a σ-finite measure defined over a ring of sets $\mathscr{R}$, there exists a uniquely determined σ-finite measure* $\bar{\mu}(A)$ *defined over the extended ring* $\mathscr{B}(\mathscr{R})$ *such that for every* $A \in \mathscr{R}$ *one has* $\mu(A) = \bar{\mu}(A)$.

PROOF. In order to construct $\bar{\mu}$ let us first define a set function μ^* in the following manner. Let $\mu^*(A)$ for any subset of A of Ω be the lower bound of all sums $\sum_{n=1}^{\infty} \mu(A_n)$, where the A_n belong to $\mathscr{R}$ and their union contains A; that is

$$\mu^*(A) = \inf_{A \subseteq \sum_{n=1}^{\infty} A_n} \left(\sum_{n=1}^{\infty} \mu(A_n) \right).$$

It is easy to verify that $\mu^*(A)$ has the following properties:

a) $\mu^*(A) \geq 0$;

b) if $A \in \mathscr{R}$, then $\mu^*(A) = \mu(A)$;

c) if $A \subseteq \sum_{n=1}^{\infty} A_n$, then $\mu^*(A) \leq \sum_{n=1}^{\infty} \mu^*(A_n)$;

$\mu^*(A)$ is called the *outer measure* of the set A. A set A is said to be measurable, if every subset B of Ω satisfies the following relation:

$$\mu^*(B) = \mu^*(AB) + \mu^*(\bar{A}B),$$

where $\bar{A}$ is the set complementary to A with respect to Ω.

Let $\mathscr{R}^*$ be the collection of all sets measurable in this sense. $\bar{\mu}(A)$ is defined on $\mathscr{R}^*$ by the equality $\bar{\mu}(A) = \mu^*(A)$. It is not difficult to prove that

1. $\mathscr{R}^*$ is a σ-algebra containing $\mathscr{R}$ (and thus $\mathscr{B}(\mathscr{R})$ as well);
2. $\bar{\mu}(A)$ is a measure on $\mathscr{R}^*$, i.e. $\bar{\mu}(A)$ is completely additive.

From these statements it follows that $\bar{\mu}(A)$ satisfies the requirements of Theorem 2.

It can be shown that this extension of μ to $\mathscr{B}(\mathscr{R})$ is unique. The measure $\bar{\mu}$ so obtained has, as it is readily seen from the construction, the following further property: If $A \in \mathscr{R}^*$ and $\bar{\mu}(A) = 0$, then from $B \subseteq A$ follows $B \in \mathscr{R}^*$ and thus $\bar{\mu}(B) = 0$. Measures having this property are called *complete* measures. Thus every measure derived from an outer measure in the above described way is complete.

Let us now consider the following example: Let R be the real axis and let $F(x)$ be a nondecreasing function continuous from the left satisfying the relations

$$\lim_{x \to -\infty} F(x) = 0 \qquad \lim_{x \to +\infty} F(x) = 1.$$

Let the ring $\mathscr{R}$ be the collection of all sets consisting of a finite number of intervals closed to the left and open to the right. $\mu(A)$ will be defined as follows: If A consists of the half-open disjoint intervals $[a_k, b_k)$, $a_1 < b_1 < < a_2 < b_2 < \ldots < a_r < b_r$, let then be

$$\mu(A) = \sum_{k=1}^{r} \left(F(b_k) - F(a_k) \right). \tag{1}$$

It is easy to see that $\mu(A)$ is a measure on $\mathscr{R}$, hence $\mu(A) \geq 0$ and if $A_n \in \mathscr{R}$ $(n = 1, 2, \ldots)$ and the A_n are pairwise disjoint, while $\sum_{n=1}^{\infty} A_n = A \in \mathscr{R}$, then

$$\mu(A) = \sum_{n=1}^{\infty} \mu(A_n).$$

To prove this we first prove another important general theorem:

THEOREM 3. *A nonnegative additive set function $\mu(A)$ defined on a ring of sets $\mathscr{R}$ is a measure on $\mathscr{R}$ iff for every sequence of sets $B_n \in \mathscr{R}$ such that $B_{n+1} \subseteq$ $\subseteq B_n$, $\mu(B_n) < + \infty$ $(n = 1, 2, \ldots)$ and $\prod_{n=1}^{\infty} B_n = O$ (i.e. for every decreasing sequence of sets B_n having the empty set as their intersection) the relation*

$$\lim_{n \to \infty} \mu(B_n) = 0 \qquad (2)$$

holds.

The proof is simple. Indeed, if (2) is fulfilled and

$$A_n \in \mathscr{R}, \sum_{n=1}^{\infty} A_n \in \mathscr{R},$$

while $A_n A_m = O$ for $n \neq m$, then we have for every n

$$\mu\left(\sum_{k=1}^{\infty} A_k\right) = \sum_{k=1}^{n-1} \mu(A_k) + \mu\left(\sum_{k=n}^{\infty} A_k\right);$$

thus from the fact that the sets $B_n = \sum_{k=n}^{\infty} A_k$ satisfy the conditions $B_n \in \mathscr{R}$, $B_{n+1} \subseteq B_n$, $\prod_{n=1}^{\infty} B_n = O$ it follows that $\lim \mu(B_n) = 0$; thus

$$\mu\left(\sum_{k=1}^{\infty} A_k\right) = \sum_{k=1}^{\infty} \mu(A_k),$$

i.e. μ is completely additive. Conversely, if μ is completely additive, then whenever $B_n \in \mathscr{R}, B_{n+1} \subseteq B_n, \prod_{n=1}^{\infty} B_n = O$ hold, one has $B_1 = \sum_{n=1}^{\infty} (B_n - B_{n+1})$ where $B_n - B_{n+1} \in \mathscr{R}$ and $(B_n - B_{n+1})(B_m - B_{m+1}) = O$ $(n \neq m)$. Therefore we have

$$\lim_{n \to \infty} \mu(B_n) = \lim_{n \to \infty} \sum_{k=n}^{\infty} \mu(B_k - B_{k+1}) = 0.$$

Now we shall prove that the set function μ defined by (1) satisfies the conditions of Theorem 3. Let therefore be $B_n \in \mathscr{R}$, $B_{n+1} \subseteq B_n$ $(n = 1, 2, \ldots)$ and $\prod_{n=1}^{\infty} B_n = O$. Then $0 \leq \mu(B_{n+1}) \leq \mu(B_n)$ $(n = 1, 2, \ldots)$, thus $\lim_{n \to \infty} \mu(B_n) = c$ does exist and $\mu(B_n) \geq c$ $(n = 1, 2, \ldots)$. We shall show that the assumption $c > 0$ leads to a contradiction. In order to prove this we construct from the set B_1 which consists of the intervals $[a_{1i}, b_{1i})$ another set $B_1' \subset B_1$ consisting of a finite number of closed intervals obtained from every one of the intervals $[a_{1i}, b_{1i})$ by removing a small open subinterval (b_{1i}', b_{1i}) $(a_{1i} < b_{1i}' < b_{1i})$ such that the points b_{1i}' be points of continuity of $F(x)$ and the value of the sum $\sum(F(b_{1i}) - F(b_{1i}'))$ be at most $\frac{c}{4}$. This procedure is repeated with the set $B_1' B_2$ (which may contain closed intervals as well besides the half-open intervals) such that the sum of the increments of the function $F(x)$ belonging to the removed intervals be at most $\frac{c}{8}$. In this way we obtain a set B_2' which consists of a finite number of closed intervals. By continuing the procedure, we obtain a sequence of sets B_n' having the following properties: a) B_n' consists of finitely many closed intervals; b) $B_{n+1}' \subseteq B_n'$; c) $\prod_{n=1}^{\infty} B_n' = O$; d) the sum $\sum_{k} (F(b_{nk}') - F(a_{nk}))$ of the increments of the function $F(x)$, taken over the intervals $[a_{nk}, b_{nk}']$ forming B_n', is at least equal to $\frac{c}{2}$.

These properties are, however, contradictory since from a), b), c) follows the existence of a number N such that, for every $n \geq N$, B_n' is the empty set (indeed, from $B_n' = B_1' B_2' \ldots B_n' \neq O$ for every n, the sets B_n' being closed, the relation $\prod_{n=1}^{\infty} B_n' \neq O$ would follow). But this contradicts d), and thus we proved our statement that the set function μ defined by (1) is a measure on $\mathscr{R}$.

According to Theorem 2 the definition of the measure μ can be extended to all Borel subsets of the real axis. Thus we obtained on these sets a measure μ such that for $A = [a, b)$ the relation $\bar{\mu}(A) = F(b) - F(a)$ is valid.

Especially, if

$$F(x) = \begin{cases} 0 & \text{for} \quad x < 0, \\ x & \text{for} \quad 0 \leq x \leq 1, \\ 1 & \text{for} \quad 1 < x, \end{cases}$$

then the above procedure assigns to every subinterval $[a, b)$ of the interval $[0, 1]$ the value $b - a$.

We have seen that $\bar{\mu}$ is a complete measure determined on a σ-ring $\mathscr{R}^*$, which contains the σ-ring $\mathscr{B}(\mathscr{R})$. If $F(x)$ has the special form mentioned above, this measure is just the ordinary Lebesgue measure defined on the interval $[0, 1]$. Any measure μ constructed by means of a function $F(x)$ satisfying the above conditions is called a *Lebesgue–Stieltjes measure* defined on the real axis.

The same construction can be applied in cases of more than one dimension.

Let $F(x_1, x_2, \ldots, x_n)$ be a function of the real variables $x_1, x_2, \ldots, x_n$ having the following properties:

1. $F(x_1, x_2, \ldots, x_n)$ is in any one of its variables a non-decreasing function continuous from the left.

2. $\lim\limits_{x_k \to -\infty} F(x_1, x_2, \ldots, x_n) = 0$, $(k = 1, 2, \ldots, n)$ and $\lim F(x_1, x_2, \ldots, x_n) =$

$= 1$, if every x_k $(k = 1, 2, \ldots, n)$ tends to $+\infty$.

In order to formulate the third condition let us introduce the following notation: Let $\Delta_h^{(k)}$ $(k = 1, 2, \ldots, n)$ denote the operation of taking the difference with respect to the variable x_k and with difference h; i.e. for a function $G(x_1, x_2, \ldots, x_n)$ we put

$$\Delta_h^{(k)} G(x_1, x_2, \ldots, x_n) = G(x_1, x_2, \ldots, x_k + h, \ldots, x_n) - $$
$$- G(x_1, x_2, \ldots, x_n).$$

Now we can formulate the third condition:

3. For any numbers $h_k \geq 0$ and any real values x_k $(k = 1, 2, \ldots, n)$ the relation

$$\Delta_{h_1}^{(1)} \Delta_{h_2}^{(2)} \ldots \Delta_{h_n}^{(n)} F(x_1, x_2, \ldots, x_n) \geq 0$$

should hold.

Let I be an n-dimensional interval consisting of the points $(x_1, x_2, \ldots, x_n)$ of the n-dimensional space satisfying the inequalities $a_k \leq x_k < b_k$. Let $h_k = b_k - a_k$ and

$$\mu(I) = \Delta_{h_1}^{(1)} \Delta_{h_2}^{(2)} \ldots \Delta_{h_n}^{(n)} F(a_1, a_2, \ldots, a_n). \tag{3}$$

Let $\mathscr{R}$ be the set of all subsets A of the n-dimensional space which can be represented as the union of finitely many pairwise disjoint intervals $I_1, I_2, \ldots, I_r$. For $A = \sum\limits_{k=1}^{r} I_k$ we put

$$\mu(A) = \sum\limits_{k=1}^{r} \mu(I_k). \tag{4}$$

It is readily seen by the same consideration as in the one-dimensional case that the set function μ defined by (3) and (4) is a measure on the ring of sets $\mathscr{R}$; thus μ can be extended to the σ-algebra $\mathscr{B}(\mathscr{R})$ formed by all Borel sets of the n-dimensional space.

Especially, if

$$F(x_1, x_2, \ldots, x_n) = \begin{cases} 0 \text{ for } \min x_i < 0, \\ \prod\limits_{k=1}^{n} \min(x_k, 1) \text{ for } x_k \geq 0 \quad (k = 1, 2, \ldots, n), \end{cases}$$

then the extension of the set function $\mu(A)$ defined above leads to the ordinary n-dimensional Lebesgue-measure defined on the n-dimensional cube $0 \le x_k \le 1$ $(k = 1, 2, \ldots, n)$.

§ 8. Conditional probabilities

In the preceding paragraphs we introduced probabilities by means of the relative frequencies. Accordingly, in order to introduce the notion of conditional probability we shall examine first conditional relative frequencies. If an event B occurs exactly n times in N trials and if among these n trials the event A occurs k times together with the event B, then the quotient $\dfrac{k}{n}$ is called the conditional relative frequency of the event A with respect to the condition B. The conditional relative frequency of an event A with respect to the condition B in a sequence of trials is therefore equal to the simple relative frequency of the event A in a subsequence of the sequence of trials in question; this subsequence contains only those trials of the original sequence in which the event B occurred. If f_B denotes the relative frequency of B in the whole sequence of trials, then $f_B = \dfrac{n}{N}$ defining similarly f_{AB} we have $f_{AB} = \dfrac{k}{N}$. Finally, if $f_{A|B}$ denotes the conditional relative frequency of A with respect to the condition B, then, by definition, $f_{A|B} = \dfrac{k}{n}$. Thus

$$f_{A|B} = \frac{f_{AB}}{f_B}.$$

Since f_{AB} fluctuates around $P(AB)$ and f_B around $P(B)$, the conditional relative frequency $f_{A|B}$ will fluctuate for $P(B) > 0$ around $\dfrac{P(AB)}{P(B)}$. This number $\dfrac{P(AB)}{P(B)}$ shall be called the *conditional probability* of the event A with respect to the condition B; it is assumed that $P(B) > 0$. The notation for the conditional probability is $P(A|B)$; thus we put

$$P(A|B) = \frac{P(AB)}{P(B)}. \tag{1}$$

By means of formula (1) the conditional probability of any event A of a probability algebra with respect to any condition B can be calculated, provided that $P(B) > 0$. If $P(B) = 0$, formula (1) has no sense; the conditional probability $P(A \mid B)$ is thus defined only for $P(B) > 0$.[1] Formula (1) may be expressed in words by saying that the conditional probability of an event A with respect to the condition B is nothing else than the ratio of the probability of the joint occurrence of A and B and the probability of B.

Equality (1) is (in contradiction to the standpoint of many older text-books) neither a theorem nor an axiom; it is the definition of conditional probability.[2] But this definition is not arbitrary; it is a logical consequence of the concept of probability as the number about which the value of the relative frequency fluctuates.

In the older literature of probability theory as well as in some vulgarizations of modern physics one finds often the misleading formulation that the probability of an event A changes because of the observation of the occurrence of an event B. It is, however, obvious that $P(A \mid B)$ and $P(A)$ do not differ because the occurrence of the event B was observed, but because of the adjunction of the occurrence of event B to the originally given complex of conditions.

Let us now state some examples.

Example 1. In the task of pebble-screening one may ask, what part of the pebbles is small enough to pass through a sieve S_A, i.e. what is the probability of a pebble chosen at random to pass through the sieve S_A. Let this event be denoted by A. Assume now that the pebble was already sieved through another sieve S_B, and the pebbles which did not pass through the sieve S_B were separated. What is the probability that a pebble chosen at random from those sieved through the sieve S_B will pass through the sieve S_A as well? Let B denote the event that a pebble passes through S_B, the probability of this event let be denoted by $P(B)$. Let further AB denote the event that a pebble passes through both S_B and S_A, and $P(AB)$ the corresponding probability. Then the probability that a pebble chosen at random from those which passed S_B will pass S_A as well is, according to the above,

$$P(A \mid B) = \frac{P(AB)}{P(B)}.$$

Example 2. Two dice are thrown, a red one and a white one. What is the probability of obtaining two sixes, provided that the white die showed a six?

---

[1] We give a more general definition in Chapter IV.

[2] In view of certain applications it is advisable to generalize the system of axioms of probability theory in such a way that the notion of conditional probability is the fundamental notion (cf. § 11 of this Chapter).

This conditional probability is by definition

$$\frac{\dfrac{1}{36}}{\dfrac{1}{6}} = \frac{1}{6}.$$

From a sheer mathematical point of view the conditional probability $P(A \mid B)$ may be considered as a new probability measure. Indeed, let Ω be an arbitrary set, $\mathscr{A}$ a σ-algebra of the subsets of Ω, and P a probability measure (i.e. a nonnegative, completely additive set function satisfying $P(\Omega) = 1$). Further let B be a fixed element of $\mathscr{A}$ such that $P(B) > 0$. Then $P(A \mid B)$ is a probability measure on $\mathscr{A}$ as well, i.e. $P(A \mid B)$ satisfies Axioms IV, V, and VI. Indeed, by introducing the notation $P^*(A) = P(A \mid B)$ we have $P^*(A) \geq 0$ for $A \in \mathscr{A}$, $P^*(\Omega) = \dfrac{P(\Omega B)}{P(B)} = \dfrac{P(B)}{P(B)} = 1$, further if $A_n \in \mathscr{A}$ and $A_n A_m = 0$ for $n \neq m$, then

$$P^*\left(\sum_{n=1}^{\infty} A_n\right) = \frac{P\left(B \sum_{n=1}^{\infty} A_n\right)}{P(B)} = \frac{\sum_{n=1}^{\infty} P(A_n B)}{P(B)} = \sum_{n=1}^{\infty} P^*(A_n).$$

Hence $[\Omega, \mathscr{A}, P(A \mid B)]$ is again a Kolmogorov probability space. Thus all theorems, proved for ordinary probabilities, remain valid, if the probability of every event is replaced by the conditional probability of the same event relative to some fixed event B (of positive probability).

If A and B are two events of positive probability one can consider besides the conditional probability of A relative to B also the conditional probability of B relative to A.

From the definition it follows readily that

$$P(B \mid A) = \frac{P(A \mid B) P(B)}{P(A)}, \tag{2}$$

hence $P(B \mid A)$ can be expressed by means of $P(A \mid B)$, $P(A)$, and $P(B)$. One can write (2) in the following form, equivalent to it:

$$\frac{P(B \mid A)}{P(B)} = \frac{P(A \mid B)}{P(A)}. \tag{3}$$

Formula (1) can be generalized as follows: If $A_1, A_2, \ldots, A_n$ are arbitrary events such that $P(A_1 A_2 \ldots A_{n-1}) > 0$, we have

$$P(A_1 A_2 \ldots A_n) = P(A_1) \, P(A_2 \mid A_1) \, P(A_3 \mid A_1 A_2) \ldots P(A_n \mid A_1 A_2 \ldots A_{n-1}). \tag{4}$$

This formula is immediately verified by expressing the conditional probabilities on the right hand side of (4) by means of (1).

§ 9. The independence of events

Let A and B be two events of a probability algebra; assume that $P(A) > 0$, and $P(B) > 0$. In the preceding paragraph the conditional probability $P(A \mid B)$ was defined. Generally it is different from $P(A)$. If, however, it is not, i.e. if

$$P(A \mid B) = P(A) \tag{1}$$

then we say that *A is independent of B*. If A is independent of B, then B is independent of A as well; indeed, by Formulas (2) and (3) of the preceding paragraph

$$P(B \mid A) = P(B). \tag{1'}$$

It is therefore permissible to say that A and B are independent of each other. From Formula (1) of § 8 follows readily a definition of independence of two events that is symmetrical in A and B. Indeed, because of the independence just defined we have

$$P(AB) = P(A)\,P(B). \tag{2}$$

A and B being independent, (2) is valid; conversely, if (2) holds and $P(A)$, $P(B)$ are both positive, then (1) and (1') hold as well, thus A and B are independent. Hence (2) is the necessary and sufficient condition of the independence, thus it may serve as a definition, either. Old textbooks of probability theory used to call relation (2) the *product rule of probabilities*. However, according to the interpretation followed in this book (2) is not a theorem but the definition of independence. (Since we take Formula (2) as the definition of independence, any event A with $P(A) = 0$ or $P(A) = 1$ is independent of every event B.)

If A and B are independent, A and $\bar{B}$ are independent as well. Namely from (2) it follows that

$$P(A\bar{B}) = P(A) - P(AB) = P(A) - P(A)P(B) = P(A)P(\bar{B}).$$

Therefore the independence of A and B implies the independence of A and $\bar{B}$ and, similarly, that of $\bar{A}$ and B, further of $\bar{A}$ and $\bar{B}$.

The independence of two complete systems of events is defined in the following manner: The complete systems of events $(A_1, A_2, \ldots, A_m)$ and $(B_1, B_2, \ldots, B_n)$ are said to be independent, if the relations

$$P(A_j B_k) = P(A_j)P(B_k) \qquad (j = 1, 2, \ldots, m; \; k = 1, 2, \ldots, n) \tag{3}$$

are valid for them. It is easy to see that from the $m \cdot n$ conditions figuring in
(3) every one containing A_m or B_n can be omitted. If the remaining $mn -$
$- (m + n - 1) = (m - 1)(n - 1)$ conditions are fulfilled, the omitted
ones are necessarily fulfilled too, as is seen from the relations

$$\sum_{k=1}^{n} P(A_j B_k) = P(A_j) \quad (j = 1, 2, \ldots, m) \tag{4}$$

and

$$\sum_{j=1}^{m} P(A_j B_k) = P(B_k) \quad (k = 1, 2, \ldots, n). \tag{5}$$

Indeed, if (3) is fulfilled for a given j and $k = 1, 2, \ldots, n - 1$, then from
(4) it follows that (3) is fulfilled for $k = n$ as well. Similarly, whenever (3)
holds for some k and for $j = 1, 2, \ldots, m - 1$, it holds for $j = m$ too.

If $m = n = 2$, we get again the result proved earlier that three of the
four conditions

$$P(AB) = P(A)P(B),$$
$$P(\bar{A}B) = P(\bar{A})P(B),$$
$$P(A\bar{B}) = P(A)P(\bar{B}),$$
$$P(\bar{A}\bar{B}) = P(\bar{A})P(\bar{B})$$

are superfluous, since the validity of one implies necessarily the validity of
the remaining three. Thus the independence of the events A and B is equivalent to the independence of the complete systems of events $(A, \bar{A})$ and
$(B, \bar{B})$. This follows also from the relation

$$P(AB) - P(A)P(B) = -(P(\bar{A}B) - P(\bar{A})P(B)) \tag{6}$$

valid for any two events A and B.

Example 1. For two tosses of a coin, four different outcomes are possible:
head–head, tail–head, head–tail, and tail–tail. Suppose that these possibilities
are all equally probable; then each will have the probability $\frac{1}{4}$. We obtain
the same result by the concept of independence, assuming that head and
tail are equally probable at both tosses $\left(\text{both having probability } \frac{1}{2}\right)$ and that
the two tosses are independent from each other. Thus the probability of
each possibility is

$$\frac{1}{2} \cdot \frac{1}{2} = \frac{1}{4}.$$

Let us now extend the concept of independence to more than two events. If A, B, and C are pairwise independent (i.e. A and B, A and C, B and C are independent) events of the same probability algebra, the non-existence of any dependence between the events A, B, and C does not follow. This may be seen from the following example.

Let us throw two dice; let A denote the event of obtaining an even number with the first die, B the event of throwing an odd number with the second, finally C the event of throwing either both even or both odd numbers. Then

$$P(A) = P(B) = P(C) = \frac{1}{2}.$$

further

$$P(AB) = P(AC) = P(BC) = \frac{1}{4}.$$

The events A, B, and C are therefore pairwise independent. Nevertheless,

$$P(ABC) = 0,$$

thus

$$P((AB)C) \neq P(AB) \cdot P(C),$$

i.e. AB is not independent from C.

We shall say that A, B, and C are *completely independent*, if they are pairwise independent and each of them is independent of the product of the remaining two. Thus A, B and C are completely independent, if the relations

$$P(AB) = P(A)P(B),$$
$$P(AC) = P(A)P(C),$$
$$P(BC) = P(B)P(C),$$
$$P(ABC) = P(A)P(B)P(C)$$

are valid. The first three of these relations express the pairwise independence of A, B, and C, the fourth the fact that each of the events is independent of the product of the remaining two. Indeed, from the first three conditions we have:

$$P(AB)P(C) = P(AC)P(B) = P(BC)P(A) = P(A)P(B)P(C).$$

The (complete) independence of more than three events may be defined in a similar manner. The events $A_1, A_2, \ldots, A_n$ are said to be *completely independent*, if for any $k = 2, 3, \ldots, n$ the relation

$$P(A_{i_1} A_{i_2} \ldots A_{i_k}) = P(A_{i_1}) P(A_{i_2}) \ldots P(A_{i_k}) \tag{7}$$

is valid for any combination $(i_1, i_2, \ldots, i_k)$ from the numbers $1, 2, \ldots, n$. Since from n objects one can choose k objects in $\binom{n}{k}$ ways, (7) consists of $2^n - n - 1$ conditions. In what follows, by saying for more than two events that they are independent we shall mean that they are completely independent in the sense just defined. If only pairwise independence is meant this will be stated explicitly. The independence of more than two complete systems of events can be defined in a similar manner.

Combinatorial methods for the calculation of probabilities have already been mentioned. They rested upon the assumption of the equiprobability of certain events. By means of the concept of independence, however, this assumption may often be reduced to more simple assumptions. Besides the simplification of the assumptions, this reduction has the advantage that the checking of the practical validity of our assumptions becomes sometimes more easy.

Example 2. Sampling without replacement. An urn contains n different objects, numbered somehow from 1 to n. We draw one after the other k items without replacement. What is the probability that we obtain a given combination of k elements? Clearly the number of possible combinations is $\binom{n}{k}$.

It was supposed that all combinations are equally probable, the probability looked for is thus $\binom{n}{k}^{-1}$.

This result may also be obtained from the following simpler assumption: *at every drawing the conditional probability of drawing any object still in the urn is the same.* Here the probability that a given combination occurs in a given order is $\dfrac{1}{n} \cdot \dfrac{1}{n-1} \cdots \dfrac{1}{n-k+1}$. Namely, at the first drawing there are in the urn n objects, the probability of choosing any one is $\dfrac{1}{n}$; at the second drawing the conditional probability of choosing any one of the $n-1$ objects which are still in the urn is $\dfrac{1}{n-1}$, etc. Since the elements of the combination in question may be chosen from the urn in $k!$ different orders, the obtained result must be multiplied by $k!$ and thus we get that the probability of drawing a combination of k arbitrary elements is

$$\frac{k!}{n(n-1)\ldots(n-k+1)} = \binom{n}{k}^{-1}.$$

Example 3. Sampling with replacement. An urn contains N balls, namely M red and $N - M$ white. Let $\dfrac{M}{N} = p$.

What is the probability that we obtain in n drawings k times a red ball, if the chosen ball is always replaced into the urn and the balls are again well mixed.

In every drawing the probability of choosing a certain ball is equal to $\dfrac{1}{N}$ and the outcomes of the individual drawings are independent. Hence the probability asked for is

$$W_k = \binom{n}{k} p^k (1 - p)^{n-k}. \tag{8}$$

Indeed, let A_i denote the event of choosing a red ball at the i-th drawing $(i = 1, 2, \ldots, n)$. These events are, because of the replacement, independent of each other. The probability that at the i_1-th, i_2-th, $\ldots$, i_k-th drawing a red and at all the other $(j_1$-th, j_2-th, $\ldots, j_{n-k}$-th) drawings a white ball will be chosen is nothing else than the probability of the event

$$A_{i_1} A_{i_2} \ldots A_{i_k} \bar{A}_{j_1} \bar{A}_{j_2} \ldots \bar{A}_{j_{n-k}}.$$

As the events $A_{i_1}, A_{i_2}, \ldots, A_{i_k}, A_{j_1}, A_{j_2}, \ldots, A_{j_{n-k}}$ are completely independent and $P(A_i) = p$, $P(\bar{A}_j) = 1 - p$, we get

$$P(A_{i_1} \ldots A_{i_k} \bar{A}_{j_1} \ldots \bar{A}_{j_{n-k}}) = p^k (1 - p)^{n-k}.$$

Since the order is irrelevant and only the number of red balls drawn is of interest, the value so obtained must still be multiplied by the number of the possible orderings, i.e. by $\binom{n}{k}$. Thus we obtain (8).

This result can immediately be generalized for experiments with more than two possible outcomes. Let the possible outcomes in every experiment be $A^{(1)}, A^{(2)}, \ldots, A^{(r)}$; let their probabilities be denoted by $P(A^{(h)}) = p_h$ $(h = 1, 2, \ldots, r)$. Of course we have $\sum_{h=1}^{r} p_h = 1$. Assume that in repeated performance of the experiment the outcomes of the individual experiments are independent of each other. Then the probability that in n repetitions of the experiment event $A^{(1)}$ occurs k_1 times, event $A^{(2)}$ k_2 times, $\ldots$, event $A^{(r)}$ k_r times, is

$$W_{k_1 k_2 \ldots k_r} = \frac{n!}{k_1! \, k_2! \ldots k_r!} \, p_1^{k_1} p_2^{k_2} \ldots p_r^{k_r}, \tag{9}$$

where $\sum_{h=1}^{r} k_h = n$. For $r = 2$ Formula (9) reduces to (8).

§ 10. "Geometric" probabilities

Let Ω be a measurable subset of the n-dimensional Euclidean space with positive, finite Lebesgue measure. Let further $\mathscr{A}$ be the set of all measurable subsets of Ω and $\mu(A)$ the n-dimensional Lebesgue measure of the measurable set A. Let $P(A)$ be defined by

$$P(A) = \frac{\mu(A)}{\mu(\Omega)}. \tag{1}$$

It is easy to see from the results of § 7 that $[\Omega, \mathscr{A}, P]$ is a Kolmogorov probability space. In this probability space probabilities may be obtained by geometric determination of measures. Probabilities were thus calculated already in the Eighteenth Century.[1]

Some simple examples will be presented here.

Example 1. In shooting at a square target we assume that every shot hits the target (i.e. we consider only shots with this property). Let the probability that the bullet hits a given part of the target be proportional to the area of the part in question. What is the probability that the hit lies in the part A? Clearly we only have to determine the factor of proportionality. If Ω denotes the entire target, the probability belonging to it must be equal to 1. Hence

$$P(A) = \frac{\mu(A)}{\mu(\Omega)}$$

where $\mu(\Omega)$ denotes the area of the entire target and $\mu(A)$ that of A. Thus for instance the probability of hitting the left lower quadrant of the target is equal to $\frac{1}{4}$.

As seen from this example, not every subset of the sample space can be considered as an event. Indeed, one cannot assign an event to every subset of the target, since the "area", as it is well known, cannot be defined for every subset such that it is completely additive and that the areas of congruent figures are equal.

In general, the distribution of probability is said to be *uniform*, if the probability that an object situated at random lies in a subset can be obtained according to the definition (1) from a geometric measure μ invariant under displacement (e.g. volume, area, length of arc, etc.).

Example 2. A man forgot to wind up his watch and thus it stopped. What is the probability that the minute hand stopped between 3 and 6? Suppose

---

[1] Of course instead of Lebesgue measure the notion of the area (and volume) of elementary geometry was applied.

the probability that the minute hand stops on a given arc of the circumference of the face of the watch is proportional to the length of the arc in question. Then the probability asked for will be equal to the quotient of the length of the arc in question, and the whole circumference of the face; i.e. in our case to $\frac{1}{4}$.

In the above two examples the determination of the probabilities was reduced to the determination of the area or of the length of the arc in certain geometric configurations. Though this method is intuitively very convincing it is nevertheless a very special method. Before applying it to further examples, let us see its relation to the already described combinatorial method. This relation is most evident in Example 2. If we neglect the fractions of the minutes and are looking for the probability that the minute hand stops between the zeroth and the first, the first and second, . . ., the k-th and $k + 1$-th minute ($k = 0, 1, \ldots, 59$), then we have a sample space consisting of 60 elementary events; the probability of every event is the same, viz. $\frac{1}{60}$. In the case of the example of the target let us assume, for sake of simplicity, that the sides of the square target are 1 m long. Let us subdivide the target into n^2 congruent little squares with sides parallel to the sides of the target. The probability that a hit lies in a set which can be obtained as the union of a certain number of the little squares is obtained by dividing the number of the little squares in question through n^2. Thus we see that geometric probabilities can be approximately determined by a combinatorial method. We must not, however, restrict ourselves to some fixed n in the subdivision, for then we could not obtain the probability of a hit lying in a domain limited by a general curve. If the mentioned subdivision is performed for every n however large, then the probability of measurable sets, or to be more precise, of every domain having an area in the sense of Jordan, can be calculated by means of limits. For this calculation we have to consider the quotient $\frac{k_n}{n^2}$ where k_n means the number of small squares lying in the domain if the large square is subdivided into congruent small squares and we have to determine the limit of $\frac{k_n}{n^2}$ for $n \to \infty$.

Probabilities obtained in a combinatorial way (without passing to the limit) are always rational numbers; geometric probabilities, however, may assume any value between 0 and 1. Thus for instance the probability that the hit lies in the circle inscribed into the square target is equal to $\frac{\pi}{4}$.

The reduction of the calculation of geometric probabilities to combinatorial considerations is nowadays of historical interest only. Namely it was thought for a long time that the classical definition of probability suffices to establish the calculus of probability. From the point of view of the modern

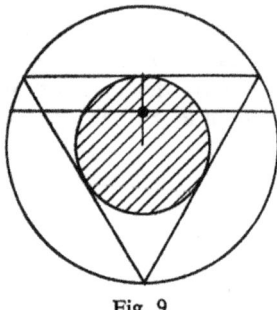

Fig. 9

theory, however, such a reduction is unnecessary (except in the case when a simplification of the calculations is brought about by it).

In what follows we continue to consider some further examples. Let us first deal with the so-called *Bertrand paradox*.

Example 3. Take a circle and choose at random a chord of it. What is the probability that this chord will be longer than the side of the regular triangle inscribed into the circle? The difficulty lies in the fact that it is not clear, what is meant by the expression that we choose a chord "at random". Each of the following interpretations seems to be more or less natural.

Interpretation 1. Since the length of a chord is uniquely determined by the position of its midpoint, we can accomplish the random choice of the chord by choosing at random a point in the interior of the circle and construct the chord whose midpoint is the chosen point. The probability that the point lies in a domain will be assumed to be proportional to the area of the domain. Clearly the chord will be longer than the side of the inscribed regular triangle, if the midpoint of the chord lies inside the circle drawn about the centre of the original circle with half of its radius (cf. Fig. 9); hence the answer is

$$\frac{\pi\left(\frac{r}{2}\right)^2}{\pi r^2} = \frac{1}{4}.$$

Interpretation 2. The length of the chord is uniquely determined by the distance of its midpoint from the centre of the circle. In view of the symmetry of the circle we may assume that the midpoint of the chord lies on a fixed radius of the circle and choose the midpoint of the chord so that the probability that it lies in a given segment of this fixed radius is assumed to be proportional to the length of this segment. The chord will be longer

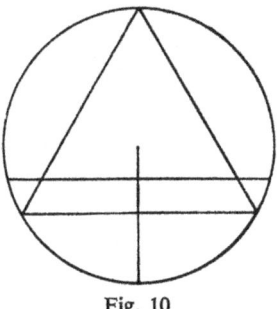

Fig. 10

than the side of the inscribed regular triangle, if its midpoint has a distance less than $\dfrac{r}{2}$ from the centre of the circle; the answer is thus $\dfrac{1}{2}$ (cf. Fig. 10).

Interpretation 3. Because of the symmetry of the circle one of the endpoints of the chord may be fixed, for instance in the point P_0; the other endpoint can be chosen on the circle at random. Let the probability that this other endpoint P lies on an arbitrary arc of the circle be proportional to the length of this arc. The regular triangle inscribed into the circle having for one of its vertices the fixed point P_0 divides the circumference into three equal parts. A chord drawn from the point P_0 will be longer than the side of the triangle, if its other endpoint lies on that one-third part of the circumference which is opposite to point P_0. Since the length of this latter is one third of the circumference, the answer is, according to this interpretation, equal to $\dfrac{1}{3}$.

From a well-known theorem of the elementary geometry concerning the central and peripheral angles it follows that the third interpretation is equivalent to the statement that the probability distribution of the intersection point of the chord and the semicircle of centre P_0 is uniform on this semicircle (Fig. 11).

Obviously, all interpretations discussed above can be realized in physical experiments. The example seemed once a paradox, because one did not pay attention to the fact that the three interpretations correspond to different experimental conditions concerning the random choice of the chord and these of course lead to different probability measures, defined on the same algebra of events. The obtained measure in the set of straight lines is, however, invariant with respect to motions of the plane only in the second interpretation,[1] in the other two interpretations congruent sets of lines do not necessarily have equal measure.

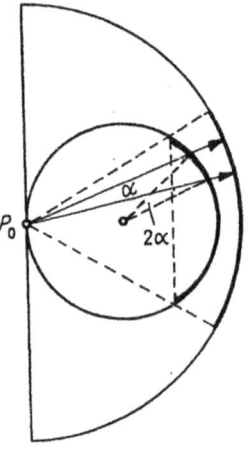

Fig. 11

Example 4. Decompose a unit segment into three subsegments by two points chosen at random. What is the probability that a triangle can be constructed from the three segments? Clearly we have to examine the probability that any one of the three segments is less than the sum of the remaining two. Compared with the above, the example contains something new, as here we have to choose two points at random. However, the problem can be reduced readily to one similar to those dealt with above. Let indeed be the segment in question the interval (0, 1) and the abscissas of the two points chosen at random be x and y. To these two points there corresponds a point of the plane with abscissa x and ordinate y. Thus there corresponds to any decomposition of the unit interval (0, 1) into three segments one point of the unit square of the plane and conversely. Now let the random choice of the two points on the interval (0, 1) be performed in such a manner that "the probability that the point representing the decomposition in question lies in a domain A of the unit square" be equal to the area of that domain (not only proportional, since the area of the unit square is 1). In this

[1] Cf. W. Blaschke [1].

case we only need to compute the area of the domain determined by the inequalities (Fig. 12)

$$0 < x < \frac{1}{2} < y < 1 \quad \text{and} \quad y - x < \frac{1}{2}$$

or

$$0 < y < \frac{1}{2} < x < 1 \quad \text{and} \quad x - y < \frac{1}{2}.$$

The area of this domain, i.e. the probability wanted, is equal to $\frac{1}{4}$.

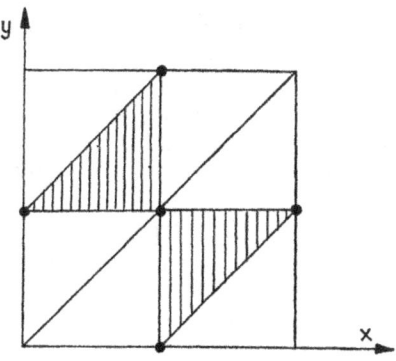

Fig. 12

The method just applied is often used, for instance in statistical physics. Here, to every state of the physical system a point of the "phase space" may be assigned, having for its coordinates the characterizing data of the state in question. Accordingly, the phase space has as many dimensions, as the state of the system has data to characterize it (the so-called degrees of freedom of the system). In our example we assigned a point of the phase space to a decomposition of the (0,1) interval by two points; the degree of freedom of the "system" is here equal to 2. The analogy can be made still more obvious by assigning to the decomposition of the (0,1) interval a physical system: two mass points moving in the interval (0,1).

Clearly the phase space may be chosen in many ways; by solving problems of probability in this way, however, one must not forget to verify in every given case separately the assumption that the probabilities belonging to the subdomains of the phase space are proportional to the area (volume).

Finally we shall discuss here a classical example, *Buffon's needle problem* (1777).

Example 5. A plane is partitioned with equidistant parallel lines of distance d into parallel strips of equal width. A needle of length l is thrown at random upon the plane. What is the probability that the needle intersects a line? For sake of brevity suppose $l < d$; in this case the needle can intersect no more than one of the parallel lines. The problem may then be solved as follows. The position of the needle on the plane may be characterized by three data: by the two coordinates of its midpoint and by the angle between the needle and the direction of the lines. Let the coordinate of the needle's midpoint perpendicular to the direction of the lines be denoted by x, that in the direction of the lines by y. Whether the needle does or does not intersect a line depends only on the coordinate x and the angle. So the coordinate y may be disregarded. Or, what comes to the same thing, we may draw a perpendicular to the parallels and assume that the midpoint of the needle lies always on it. If we take the origin of the coordinate system on one of the parallel lines, it may be assumed that the midpoint of the needle lies in the first parallel strip, since the strips are all of equal width $d < l$. Let φ denote the angle between the needle and the positive x-axis. The position of the needle is then characterized in the rectangular coordinate system (x, φ) (in the "phase space") by one point and it can be assumed that this point lies in the rectangle $(0 \leq x \leq d, \ 0 \leq \varphi \leq \pi)$. The probability that the point (x, φ) lies in an arbitrary domain of this rectangle is assumed to be proportional to the area of this domain. Loosely speaking, we assume that "all positions of the midpoint of the needle are equally probable and all directions of the needle are equally probable".

Fixing now the value of φ, the needle intersects the line $x = 0$ if $0 \leq x \leq \frac{l}{2} \sin \varphi$ and the line $x = d$ if $d - \frac{l}{2} \sin \varphi \leq x \leq d$. Thus the needle intersects the line $x = 0$, if and only if the point (x, φ) characterizing the position of the needle lies to the left of the sine curve drawn over the line $x = 0$ with an amplitude $\frac{l}{2}$ and the line $x = d$, if the characterizing point lies to the right of the sine curve drawn over the line $x = d$ with the same amplitude (Fig. 13).

Since the area under a half-wave of a sine curve of amplitude $\frac{l}{2}$ is equal to l, the area of the domain formed by the points which correspond to intersection will be $2l$. The area of the whole rectangle being πd, the sought probability is $\frac{2l}{\pi d}$. Thus in many repetitions of Buffon's experiment one will find intersection in approximately a fraction $\frac{2l}{\pi d}$ of the experiments. It was tried more than once to determine approximately the value of π by this

method. Since, however, the assumptions (especially the equiprobability of
the directions) are quite difficult to realize, even many thousands of experiments give the value of π only to a few digits. In principle, however, nothing
prevents us to carry out an experiment in which our assumptions about
position and direction of the needle are very nearly satisfied and thus to de-

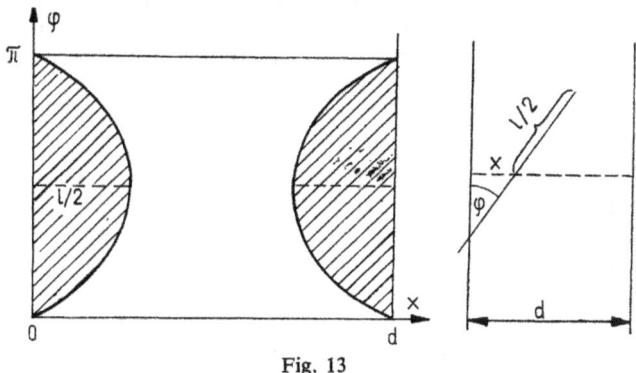

Fig. 13

termine the value of π with any prescribed precision. Of course this would
have no practical importance, since there are more straightforward and
reliable methods to compute the value of π. Still the question is of great
interest, since it shows that certain mathematical problems can be solved
approximately by performing experiments of a probabilistic nature. Nowadays difficult differential equations and other problems of numerical analysis are treated in this manner (this is the so-called Monte Carlo method).

   Questions dealt with in this paragraph are closely connected to integral
geometry.

§ 11. Conditional probability spaces

   The axiomatic foundation of probability theory given in 1933 by A. N.
Kolmogorov was of paramount importance and since then it furnishes the
very basis of this branch of mathematics. There are, however, problems either entirely outside of its range or leading in this theory to serious compli-
cations. In physics (e.g. in quantum mechanics) and in some parts of probability theory (especially in the theory of Markov chains and stochastic
processes) as well as in applications to number theory and integral geometry,
one often has to deal with so-called unbounded distributions, i.e. unbounded
measures. The use of unbounded measures, however, cannot be justified in
the theory of Kolmogorov. For instance one cannot speak (in the sense of

the preceding paragraph) about a uniform probability distribution in the whole Euclidean space. Similarly, it is nonsense to speak about the random choice of an integer such that every integer has the same probability to be chosen. At the first glance it might seem that this difficulty cannot be overcome, since the value of probability can never exceed 1. In spite of this, one can obtain by means of these unbounded, that is "nonsensical", distributions *conditional* probabilities which are in agreement with experience. Thus the necessity arose to generalize the theory of probability in a way justifying the use of such distributions.[1]

One can indeed give an axiomatic theory of probability which matches the above-mentioned requirements.[2]

This theory contains the theory of Kolmogorov as a special case. The fundamental concept of the theory is that of conditional probability; it contains cases where ordinary probabilities are not defined at all.

We start from the following definitions and axioms:

Let there be given a set Ω (called the space of elementary events) and let $\mathscr{A}$ denote a σ-algebra of subsets of Ω. The elements $A, B, \ldots$ etc. of $\mathscr{A}$ are called *events*. The set $\Omega - A$ will be denoted by $\bar{A}$. Let further $\mathscr{B}$ be a nonempty system of sets such that $\mathscr{B} \subseteq \mathscr{A}$. We assume that a set function $P(A \mid B)$ of two set variables is defined for $A \in \mathscr{A}$ and $B \in \mathscr{B}$. $P(A \mid B)$ will be called the *conditional probability of the event A with respect to the condition B*.

We postulate the following axioms:

a) $P(A \mid B) \geq 0$ and $P(B \mid B) = 1$ $(A \in \mathscr{A}, B \in \mathscr{B})$.

b) For any fixed $B \in \mathscr{B}$ $P(A \mid B)$, as a function of A, is a measure on $\mathscr{A}$, i.e. if $A_n \in \mathscr{A}$ and $A_n A_m = O$ for $n \neq m$, we have

$$P\left(\sum_{n=1}^{\infty} A_n \mid B \right) = \sum_{n=1}^{\infty} P(A_n \mid B).$$

c) If $A \in \mathscr{A}$, $B \in \mathscr{B}$, $C \in \mathscr{B}$, $B \subseteq C$ and $P(B \mid C) > 0$, then

$$P(A \mid B) = \frac{P(AB \mid C)}{P(B \mid C)}.$$

If the Axioms a), b), and c) are satisfied, we shall call the system $[\Omega, \mathscr{A}, \mathscr{B}, P(A \mid B)]$ a *conditional probability space*.

[1] The history of mathematics shows that on several occasions procedures, successful e.g. in physical applications but inexact in a mathematical sense of the word, were made exact later on by an extension of the mathematical notions involved.

[2] Rényi, A. [14], [15], [18]. The idea of such a theory is due to Kolmogorov himself; he, however, did not publish anything about it.

·If $P^*(A)$ is a measure defined on $\mathscr{A}$ and $P^*(\Omega) = 1$ (that is, if $[\Omega, \mathscr{A}, P^*]$ is a Kolmogorov probability field), further if $\mathscr{B}^*$ denotes the collection of all sets $B \in \mathscr{A}$ such that $P^*(B) > 0$, then — as it is easy to see — the system $[\Omega, \mathscr{A}, \mathscr{B}^*, P^*(A \mid B)]$ is a conditional probability space, provided $P^*(A \mid B)$ is defined by

$$P^*(A \mid B) = \frac{P^*(AB)}{P^*(B)} \quad (A \in \mathscr{A}, B \in \mathscr{B}^*).$$

$[\Omega, \mathscr{A}, \mathscr{B}^*, P^*(A \mid B)]$ will be called the *conditional probability space generated by the Kolmogorov probability space* $[\Omega, \mathscr{A}, P^*]$.

We shall prove some simple theorems which follow directly from our axioms.

THEOREM 1. *For $A \in \mathscr{A}$, and $B \in \mathscr{B}$ we have*

$$P(A \mid B) = P(AB \mid B).$$

PROOF. The statement follows from Axioms a) and c) by substitution of $C = B$.

THEOREM 2. *For $A \in \mathscr{A}$ and $B \in \mathscr{B}$ we have*

$$P(A \mid B) \leq 1.$$

PROOF. According to Theorem 1 and Axiom b) we have $P(A \mid B) \leq \leq P(B \mid B)$. Our statement follows by Axiom a).

THEOREM 3. *For $B \in \mathscr{B}$ we have $P(O \mid B) = 0$.*

PROOF. The statement is evident because of Axiom b).

THEOREM 4. *If $A \in \mathscr{A}$, $B \in \mathscr{B}$ and $AB = O$, then $P(A \mid B) = 0$.*

PROOF. The statement follows from Theorems 1 and 3.

THEOREM 5. *For $B \in \mathscr{B}$ we have $P(\Omega \mid B) = 1$.*

PROOF. The statement is obvious because of Axiom a) and Theorem 1.

THEOREM 6. *If for fixed $C \in \mathscr{B}$ we put $P_C^*(A) = P(A \mid C)$, the system $[\Omega, \mathscr{A}, P_C^*]$ is a Kolmogorov probability space. If B is an element of $\mathscr{A}$ such* *that* $BC \in \mathscr{B}$ *and* $P_C^*(B) > 0$, *further if* $P_C^*(A \mid B)$ *is, as usually, defined by*

$$P_C^* (A \mid B) = \frac{P_C^* (AB)}{P_C^* (B)},$$

we have

$$P_C^*(A \mid B) = P(A \mid BC).$$

PROOF. The first statement of the theorem is evident since P_C^* is a measure on $\mathscr{A}$, and $P_C^*(\Omega) = 1$. The second statement follows from Axiom c); indeed we have by Theorem 1

$$P_C^* (A \mid B) = \frac{P_C^* (AB)}{P^*(B)} = \frac{P(AB \mid C)}{P(B \mid C)} = \frac{P(ABC \mid C)}{P(BC \mid C)} = P(A \mid BC).$$

THEOREM 7. *Suppose* $\Omega \in \mathscr{B}$ *and put* $P^*(A) = P(A \mid \Omega)$. *Then* $[\Omega, \mathscr{A}, P^*]$ *is a Kolmogorov probability space. Further if* $P^*(B) > 0$ *we have*

$$P(A \mid B) = \frac{P^* (AB)}{P^* (B)}.$$

Remark. $\mathscr{B}$ may contain sets B such that $P^*(B) = 0$. On the other hand, sets B for which $P^*(B) > 0$ may not belong to $\mathscr{B}$. Hence $[\Omega, \mathscr{A}, \mathscr{B}, P(A \mid B)]$ is not necessarily identical to the Kolmogorov probability space $[\Omega, \mathscr{A}, P(A \mid \Omega)]$, not even in the case $\Omega \in \mathscr{B}$.

PROOF. Theorem 7 is a special case of Theorem 6.

From the theorems proved above one readily sees how the generalized theory of probability can be deduced from our axioms.

Let us mention here some further examples.

Example 1. Let Ω be the n-dimensional Euclidean space; let the points of Ω be denoted by $\omega = (\omega_1, \omega_2, \ldots, \omega_n)$. Let $\mathscr{A}$ denote the class of all measurable subsets of Ω, let further $f(\omega)$ be a nonnegative, measurable function defined on Ω and $\mathscr{B}$ the set of all measurable sets B such that $\int_B f(\omega)d\omega$ be finite and positive. Put

$$P(A \mid B) = \frac{\int\limits_{AB} f(\omega) \, d\omega}{\int\limits_{B} f(\omega) \, d\omega}.$$

$[\Omega, \mathscr{A}, \mathscr{B}, P(A \mid B)]$ is then a conditional probability space. If $\int\limits_{\Omega} f(\omega)d\omega < + \infty$, a conditional probability space generated by a Kolmogorov probability space is obtained; if, however, $\int_\Omega f(\omega)d\omega = + \infty$, this is not the case. Especially when $f(\omega) \equiv 1$, we obtain the uniform probability distribution in the whole n-dimensional space. In this case

$$P(A \mid B) = \frac{\mu_n(AB)}{\mu_n(B)},$$

where $\mu_n(C)$ denotes the n-dimensional Lebesgue measure of the set C.

Example 2. Let Ω be the set of the natural numbers, $\mathscr{A}$ the class of all subsets of Ω, further p_n ($n = 1, 2, \ldots$) a sequence of arbitrary nonnegative numbers not all equal to 0; let $\mathscr{B}$ denote the set of those subsets B of Ω for which $\sum_{n \in B} p_n$ is positive and finite. Let $\sum_{n \in A} p_n$ be denoted by $r(A)$ for $A \in \mathscr{A}$ and put

$$P(A \mid B) = \frac{r(AB)}{r(B)}.$$

Clearly $[\Omega, \mathscr{A}, \mathscr{B}, P(A \mid B)]$ is a conditional probability space. It is generated by a Kolmogorov probability space if and only if the series $\sum_{n=1}^{\infty} p_n$ is convergent.

Especially when $p_n = 1$ ($n = 1, 2, \ldots$),

$$P(A \mid B) = \frac{\sum_{n \in AB} 1}{\sum_{n \in B} 1}$$

is equal to the ratio of the number of elements of the set AB and the set B.[1]

Evidently the question arises how conditional probabilities are connected with relative frequencies, i.e. whether the generalized theory does have a frequency-interpretation too.

The answer is affirmative and even very simple. The conditional probability $P(A \mid B)$ can be interpreted in the generalized theory (as well as in the theory of Kolmogorov) as the number about which the relative frequency of A with respect to the condition B fluctuates. Thus the generalized theory has the same relation to the empirical world as Kolmogorov's theory.

---

[1] In both cases, $P(A \mid B)$ could have been represented as the ratio $\dfrac{\mu(A\,B)}{\mu(B)}$, where μ is an unbounded measure. (With respect to the conditions for the existence of such measures cf. Á. Császár [1], and A. Rényi [18]).

§ 12. Exercises

1. Let p_1, p_2, p_{12} be given real numbers. Prove that the validity of the four inequalities below is necessary and sufficient for the existence of two events A and B such that $P(A) = p_1$, $P(B) = p_2$, $P(AB) = p_{12}$.

$$1 - p_1 - p_2 + p_{12} \geq 0, \tag{1}$$

$$p_1 - p_{12} \geq 0, \tag{2}$$

$$p_2 - p_{12} \geq 0, \tag{3}$$

$$p_{12} \geq 0. \tag{4}$$

Hint. On the right hand side of the inequalities (1)–(4) we have the probabilities of $\bar{A}\bar{B}$, $A\bar{B}$, $\bar{A}B$, and AB. Of course they must be nonnegative, thus the conditions are necessary. Their sufficiency can be shown as follows: from (1)–(4) it is clear that

$$0 \leq p_{12} \leq p_1 \leq p_1 + p_2 - p_{12} \leq 1$$

and similarly

$$0 \leq p_{12} \leq p_2 \leq p_1 + p_2 - p_{12} \leq 1.$$

The numbers p_1, p_2, p_{12} are therefore nonnegative and do not exceed 1.

Consider the interval $I = (0, 1)$ and suppose that a random point P is uniformly distributed in this interval; i.e. let the probability that P lies in a subinterval of I be equal to the length of this subinterval. Let A denote the event that the point lies in the interval $0 < x < p_1$, and B that it lies in the interval $p_1 - p_{12} < x < p_1 + p_2 - p_{12}$. Then we have $P(A) = p_1$, $P(B) = p_2$, $P(AB) = p_{12}$.

2. Generalize the assertion of Exercise 1 to n events ($n = 3, 4, \ldots$).

3. Examine how the conditions of Exercise 2 can be simplified if we assume that $p_{i_1 i_2 \ldots i_k} = P(A_{i_1} A_{i_2} \ldots A_{i_k})$ $(1 \leq i_1 < i_2 < \ldots < i_k \leq n)$ depends *only* on k ($k = 1, 2, \ldots, n - 1$).

4. How can the conditions of Exercise 2 be simplified if we assume that for every $k = 2, 3, \ldots, n$

$$p_{i_1 i_2 \ldots i_k} = p_{i_1} p_{i_2} \cdots p_{i_k}.$$

5. Let $A_1, A_2, \ldots, A_n$ be any n events and suppose that the probabilities $P(A_{i_1} A_{i_2} \ldots A_{i_k})$ $(1 \leq k \leq n, 1 \leq i_1 < i_2 < \ldots < i_k \leq n)$ are known. Find the probability that at least k of the n events $A_1, A_2, \ldots, A_n$ will occur.

6. Prove the inequality

$$P(A \, \Delta \, C) \leq P(A \, \Delta \, B) + P(B \, \Delta \, C).$$

Remark. If we define the "distance" $d(A, B)$ of the events A and B as the probability $P(A \, \Delta \, B)$, then we have the "triangle inequality" $d(A, C) \leq d(A, B) + d(B, C)$.

7. If the distance of A and B is defined as

$$d^*(A, B) = \begin{cases} \dfrac{P(A \, \Delta \, B)}{P(A + B)} & \text{for } P(A + \text{B}) > 0, \\[2mm] 0 & \text{otherwise,} \end{cases}$$

then the triangle inequality is again valid.

8. What is the probability that in n throws of a die the sum of the obtained numbers is equal to k?

Hint. Determine the coefficient of x^k in the expansion of the generating function $(x + x^2 + x^3 + x^4 + x^5 + x^6)^n$.

9. What is the probability that the sum of the numbers thrown is larger than 10 in a throw with three dice?

Remark. This was the condition of gain in the "passe-dix" game which was current in the Seventeenth Century.

10. What is more probable: to get at least one six with four dice or at least one double six in 24 throws of two dice? (Chevalier de Méré's problem.)

11. In a party of n married couples everybody dances. Every gentleman dances with every one of the ladies with the same probability. What is the probability that nobody dances with his own wife? Find the limit of this probability for $n \to \infty$.

12. An urn contains n white and m red balls, $n \neq m$; balls are drawn from the urn at random without replacement. What is the probability that at some instant the numbers of white and red balls drawn are equal?

13. There is a queue of 100 men before the box-office of an exhibition. One ticket costs 1 shilling. 60 of the men in the queue have only 1 shilling coins, 40 only 2 shilling coins. The cash contains no money at the start. What is the probability that tickets can be sold without any trouble (i.e. that never comes a man, having only 2 shilling coins, before the cash desk at a moment when the latter contains no 1 shilling coin)?

14. A particle moves along the x-axis with unit velocity. If it reaches a point with integer abscissa it has one of two equiprobable possibilities: either it continues to proceed or it turns back. Suppose that at the moment $t = 0$ the particle was in the point $x = 0$. Find the probability that at a time t the particle will have a distance x from the origin (t is a positive integer, x an arbitrary integer).

15. Let the conditions of Exercise 14 be completed by the following: at the point with abscissa x_0 (a positive integer) there is an absorbing wall; if the particle arrives at the point of abscissa x_0 it will be absorbed and does not continue its movement. Answer the question of the preceding exercise for $x \leq x_0$.

16. A box contains M red and N white balls which are drawn one after the other without replacement. Let P_k denote the probability that the first red ball will be drawn at the k-th drawing. Since there are N white balls, clearly $k \leq N + 1$ and thus $P_1 + P_2 + \ldots + P_{N+1} = 1$. By substituting the explicit expression of P_k we obtain an identity. How can this identity be proved directly, without using probability theory?

17. Let us place eight rooks at random on a chessboard. What is the probability that no rook can take another?

Hint. One has to count the number of ways in which 8 rooks can be placed on a chessboard so that in every row and in every column there is exactly one rook.

18. Put

$$P_k(M, N) = \frac{\dbinom{M}{k}\dbinom{N - M}{n - k}}{\dbinom{N}{n}}$$

and

$$W_k = \binom{n}{k} p^k (1 - p)^{n-k} \qquad (k = 0, 1, \ldots, n),$$

where $p = \dfrac{M}{N}$. Prove that if M and N tend to infinity so that $\dfrac{M}{N} = p$ remains constant, then $P_k(M, N)$ tends to W_k.

19. Put

$$Q_k(M, N) = \frac{\binom{N-M}{k-1}(k-1)!\, M(N-k)!}{N!}$$

and

$$V_k = p(1 - p)^{k-1} \qquad (k = 0, 1, 2 \ldots),$$

where $p = \dfrac{M}{N}$. Show that if M and N tend to infinity so that $\dfrac{M}{N} = p$ remains constant, then $Q_k(M, N)$ tends to V_k (cf. § 5, Example 1b). Estimate the error $|Q_k(M, N) - V_k|$.

20. How many raisins are to be put into 20 ozs of dough in order that the probability is at least 0.99 that a cake of 1 oz contains at least one raisin?

21. The amount of water in a container may be determined as follows: a certain amount a of soluble stain is solved in 1 gallon of water taken from the container and the stained water is replaced. After perfect mixing 1 gallon water is taken again and its stain content determined. If the latter number is x and the mixing is supposed to be perfect, the vessel contains $\dfrac{100}{x}$ gallons of water. Similarly, the number of the fishes in a pond may be determined as follows: 100 fishes are caught, marked (e.g. by rings) and replaced into the pond. After an interval of some days 100 fishes are caught again and the marked ones are counted. If their number is $x > 0$ then the pond contains about $\dfrac{100}{x}$ fishes. If the pond contains 10 000 fishes, what is the probability that among the fishes of the second catch the number of marked fishes is 0, 1, 2, or 3?

22. A stick is broken at a random point and the longest part is again broken at random. What is the probability that a triangle can be formed from the three pieces so obtained? (Observe that the conditions of the breaking differ from those of Example 4 of § 10.)

23. Consider an undamped mathematical pendulum. Let the angle of the maximal elongation be 2°. What is the probability that at a randomly chosen instant the elongation will be greater than 1°?

24. Let Buffon's problem be modified by throwing upon the plane a disk instead of a needle. What is the probability that the disk will not cover any of the lines?

25. In a five storey building the first floor is 8 metres above the ground floor, while each subsequent storey is 6 meters high. Suppose that the elevator stops somewhere because of a short circuit. Let the height of the door of the elevator be 1.8 meter.

Compute the probability that at the time of the stopping only the wall of the elevator shaft can be seen from the elevator.

26. What conditions must the numbers p, q, r, s satisfy in order that there exist events A and B such that

$$P(A \mid B) = p, \; P(A \mid \bar{B}) = q, \; P(B \mid A) = r, \; P(B \mid \bar{A}) = s?$$

27. A box contains 1000 screws. These are tested at random so that the probability of a screw being tested is equal to $\frac{1}{10}$. Suppose that 2 per cent of the screws are defective; what is the probability that from the tested screws exactly two are defective?

28. If A and B are independent events and $A \subseteq B$, prove that either $P(A) = 0$ or $P(B) = 1$.

29. Show by an example that it is possible that the event A is independent of both BC and $B + C$, while B and C are also independent but A is not independent either of B or of C.

30. Prove that if A is independent of BC and of $B + C$, B of AC, and C of AB, further if $P(A)$, $P(B)$, and $P(C)$ are positive, then A, B, and C are completely independent.

31. We perform n independent experiments; suppose that the probability of the event A in the j-th experiment is p_j ($j = 1, 2, \ldots, n$). Let $\pi_{n,k}$ denote the probability that in the n experiments the event A occurs just k times. Prove that one has always $\pi_{n,k}^2 \geq \pi_{n,k-1} \cdot \pi_{n,k+1}$, regardless of the values of the probabilities $p_1, p_2, \ldots, p_n$.

Hint. Use the relation

$$\pi_{n+1,k} = \pi_{n,k-1} \, p_{n+1} + \pi_{n,k}(1 - p_{n+1})$$

and proceed by mathematical induction.

32. Let $A_1, A_2, \ldots, A_n$ be any distinct events. Let $P(A_k) = p_k$ ($k = 1, 2, \ldots, n$), further let U_r denote the probability that exactly r from the events A_k ($k = 1, 2, \ldots, n$) occur. Put

$$S_k = \sum_{1 \leq l_1 < l_2 < \ldots < l_k \leq n} P(A_{l_1} A_{l_2} \ldots A_{l_k}) \qquad (k = 1, 2, \ldots, n).$$

Then by Theorem 10 of § 3 we have the following relation:

$$U_r = S_r - \binom{r+1}{1} S_{r+1} + \binom{r+2}{2} S_{r+2} - \ldots + (-1)^{n-r} \binom{n}{n-r} S_n.$$

How will this expression be simplified if we assume that the events $A_1, A_2, \ldots, A_n$ are completely independent and equiprobable?

33. In the notations of the previous exercise prove that

$$S_r = U_r + \binom{r+1}{1} U_{r+1} + \ldots + \binom{n}{r} U_n.$$

34. Which simplification of the relation in the preceding exercise is possible if we assume that $A_1, A_2, \ldots, A_n$ are completely independent and equiprobable?

35. Let $A_1, A_2, \ldots, A_n$ be arbitrary events and B an event which is a function of the events $A_1, A_2, \ldots, A_n$. Prove that there exist numbers C_0 and $C_{i_1 i_2 \ldots i_r}$ ($1 \leq r \leq n$; $1 \leq i_1 < i_2 < \ldots < i_r \leq n$) independent of the choice of the events A_k such that

$$P(B) = C_0 + \sum_{r=1}^{n} \sum_{1 \leq i_1 < i_2 < \ldots < i_r \leq n} C_{i_1 i_2 \ldots i_r} P(A_{i_1} A_{i_2} \ldots A_{i_r}).$$

Hint. See the proof of Theorem 11 of § 3.

36. Prove Theorem 10 of § 3 using the results of Exercise 35, and by determining the coefficients $C_{i_1 i_2 \ldots i_r}$ in the particular case when the events A_k are independent. According to the statement of Exercise 35 the formula with these coefficients will be valid in the general case too.

37. As an application of Theorem 10 of § 3 solve the following problem: suppose that n persons throw their visit-cards into a hat and then everybody draws one visit-card from the hat. The probability that exactly r persons ($r = 0, 1, \ldots, n$) draw their own cards is:

$$W_r(n) = \frac{1}{r!} \left(1 - \frac{1}{1!} + \frac{1}{2!} - \frac{1}{3!} + \ldots + \frac{(-1)^{n-r}}{(n-r)!} \right)$$

$$\left(\text{e.g. for } n \to \infty \text{ we have } W_r(n) \to \frac{1}{e \cdot r!} \right).$$

38. The events $A_1, A_2, \ldots, A_n$ are said to be *exchangeable*[1], if the value of

$$P(A_{i_1} A_{i_2} \ldots A_{i_r}) \quad (1 \leq r \leq n; \ 1 \leq i_1 < i_2 < \ldots < i_r \leq n)$$

depends only on r and does not depend on the choice of the different indices $i_1, i_2, \ldots, i_r$ ($r = 1, 2, \ldots, n$). Thus if $A_1, A_2, \ldots, A_n$ are independent and equiprobable, they are also exchangeable. Show that from the exchangeability of the events $A_1, A_2, \ldots, A_n$ their independence does not follow.

39. a) Let an urn contain M red and $N - M$ white balls. n balls are drawn without replacement, $n \leq \min(M, N - M)$. Let A_k denote the event that the k-th drawing yields a red ball. Prove that the events $A_1, A_2, \ldots, A_n$ are exchangeable.

b) Prove that the events $A_1, A_2, \ldots, A_n$ defined in Exercise a) are even then exchangeable, if every replacement of a ball drawn from the urn is accompanied by throwing R balls of the same colour into the urn.

40. Each of N urns contains red and white balls. Let the number of the red balls in the r-th urn be a_r, that of the white balls b_r and let v_r be the probability of drawing a red ball from the r-th urn; that is, we put $v_r = \dfrac{a_r}{a_r + b_r}$. Perform the following experiment. Choose first one of the urns; suppose that the probability of choosing the r-th urn is $p_r > 0$ ($r = 1, 2, \ldots, N$). Draw from the chosen urn n balls with

----

[1] Also called "symmetrically dependent" or "equivalent" events.

replacement. Let A_k denote the event that the k-th drawing yields a red ball. Prove now the following statements:

   a) The events $A_1, A_2, \ldots, A_n$ are exchangeable.
   b) The events A_k are, generally, not even pairwise independent.
   c) Let W_k denote the probability that from the n drawings exactly k yield red balls. Compute the value of W_k.
   d) Let π_k denote the probability that the first red ball was drawn at the k-th drawing; compute the value of π_k.

   41. Let A_k denote the event that given the conditions of Exercise 37 the k-th person draws his own visiting card. Prove that the events A_k are exchangeable.

   42. Let N balls be distributed among n urns such that each ball can fall with the same probability into any one of the urns. Compute

   a) the probability $P_0 (n, N)$ that at least one ball falls into every urn;
   b) the probability $P_k (n, N)$ that exactly k $(k = 1, 2, \ldots, n - 1)$ of the urns remain empty.

   43. Let A_k denote in the preceding exercise the event that the k-th urn does not remain empty; show that the events A_k are exchangeable and

$$V_k = P(A_{i_1} A_{i_2} \ldots A_{i_k}) = \sum_{j=0}^{k} \binom{k}{j} (-1)^j \left(1 - \frac{j}{n} \right)^N .$$

Show that if $N = \lambda n$ and $n \to \infty$, then $\lim V_k = (1 - e^{-\lambda})^k$.

   (*Remark.* V_n is equal to the probability $P_0 (n, N)$ occurring in Exercise 42.)

   44. Banach was a passionate smoker and used to put one box of matches in both pockets in order to be never without matches. Every time he needed a match, he chose at random either the box in his right or that in his left pocket with the same probability $\frac{1}{2}$. One day he put into his pockets two full boxes, both containing n matches. Let P_k denote the probability that on first finding one of the boxes to be empty, the other box contained k matches. Calculate the value of P_k and find the value k which maximizes this probability.

   45. An urn contains M red and $N - M$ white balls, $\dfrac{M}{N} = p$. Let P_r denote the probability that in a sequence of drawings with replacement the r-th drawing of a red ball is preceded by an even number of drawings of white balls. Prove that we have $P_r > \dfrac{1}{2}$ for every value of p and r.

   46. Let an urn contain M red and $N - M$ white balls. Draw all balls from the urn in turn without replacement and note the serial numbers of the red drawings. Let these serial numbers be $k_1, k_2, \ldots, k_M$, and put $X = k_1 + k_2 + \ldots + k_M$. Let $P_n (M, N)$ denote the probability that $X = n \, (A \leq n \leq B)$, where

$$A = \frac{M (M + 1)}{2} \quad \text{and} \quad B = A + M(N - M).$$

Put

$$F(M, N, x) = \sum_{n=A}^{B} P_n (M, N) x^n.$$

Determine the polynomial $F(M, N, x)$ and thence the probabilities $P_n(M, N)$.[1]
Prove that

$$P_{B-n} (M, N) = P_{A+n} (M, N).$$

47. Prove by means of probability theory that if $\varphi(n)$ denotes the number of the positive integers less than n and relatively prime to n ($n = 1, 2, \ldots$), then[2]

$$\varphi(n) = n \prod_{p|n} \left(1 - \frac{1}{p} \right),$$

where the product is to be taken over all distinct prime factors p of n.

Hint. Choose at random one of the numbers $1, 2, \ldots, n$ such that each of these numbers is equally probable. Let A_p denote the event that the number chosen can be divided by the prime number p. Show that if $p_1, p_2, \ldots$ are the distinct prime factors of the number n, then the events $A_{p_1}, A_{p_2} \ldots$ are independent. The probability that the chosen number is relatively prime to n is, by definition, $\frac{\varphi(n)}{n}$. On the other hand we have $P(A_p) = \frac{1}{p}$, hence, because of the independence of the events A_p,

$$\frac{\varphi(n)}{n} = P(\prod_{p|n} \bar{A}_p) = \prod_{p|n} P(\bar{A}_p) = \prod_{p|n} \left(1 - \frac{1}{p} \right).$$

48. a) Let Ω be a countably infinite set, let its elements be $\omega_1, \omega_2, \ldots, \omega_n, \ldots$. Let $\mathcal{A}$ consist of all subsets of Ω and let the probability measure P be defined in the following manner: $P(\{\omega_n\}) = p_n$ where $p_n \geq p_{n+1} \geq 0$ ($n = 1, 2, \ldots$) and $\sum_{n=1}^{\infty} p_n = 1$. Prove that the set of those numbers x for which an $A \in \mathcal{A}$ can be found such that $P(A) = x$, is a perfect set.

b) Prove that, given the conditions of Exercise 48 a), the range of the set function $P(A)$ is identical to the interval [0, 1], if and only if

$$p_n \leq \sum_{k=n+1}^{\infty} p_k \qquad (n = 1, 2, \ldots).$$

c) Given the conditions of Exercise 48 a), prove that to every r-tuple of numbers $x_1, x_2, \ldots, x_r$ with

$$\sum_{i=1}^{r} x_i = 1, \quad x_i \geq 0, \quad (i = 1, 2, \ldots, r)$$

a complete system of events ·

$$A_1, A_2, \ldots, A_r \quad \text{with} \quad P(A_i) = x_i \quad (i = 1, 2, \ldots, r)$$

[1] This exercise is the basis of an important statistical method, called Wiicoxon's test.

[2] $\varphi(n)$ is called Euler's function.

can be found if and only if

$$p_n \leq \frac{1}{r} \sum_{k=n}^{\infty} p_k \qquad (n = 1, 2, \ldots).$$

Hint. a) A number x is said to be representable if there exists an event $A \in \mathcal{A}$ such that $P(A) = x$, i.e. if x can be represented in the form $x = \sum_{\omega_n \in A} p_n$. If x_n $(n = 1, 2, \ldots)$ is representable and $\lim_{n \to \infty} x_n = x \neq 0$, then it is readily seen that x is representable too. Indeed we can select from the sequence x_n an infinite subsequence x_{n_k} $(k = 1, 2, \ldots)$ such that in the representation of each x_{n_k} the greatest member is p_{i_1}. Take now from this sequence an infinite subsequence having in its representation for second greatest member p_{i_2}. By progressing in this manner we obtain a sequence p_{i_s} $(s = 1, 2, \ldots)$ and it is easy to verify that $\sum_{s=1}^{\infty} p_{i_s} = x$. The range of the function $P(A)$ is thus a closed set. Furthermore, if x is a number which can be represented as a sum of a finite number of the p_r-s, e.g. $x = \sum_{l=1}^{N} p_{ll}$, then

$$x = \lim_{n \to \infty} \left(\sum_{l=1}^{N} p_{ll} + p_n \right).$$

If $x = \sum_{l=1}^{\infty} p_{ll}$, then

$$x = \lim_{n \to \infty} \sum_{l=1}^{n} p_{ij}.$$

Thus the range of the function $P(A)$ is a perfect set.

c) It is easy to see that the condition is necessary. Its sufficiency can be shown in the following manner: suppose we have $x_1 \geq x_2 \geq \ldots \geq x_r$. Then $x_1 \geq \frac{1}{r}$, and on the other hand $p_1 \leq \frac{1}{r}$ hence p_1 can be used for the representation of x_1. Let now be $x_1' = \max(x_1 - p_1, x_2)$ then $x_1' \geq \frac{1}{r}(1 - p_1)$. Since $p_2 \leq \frac{1}{r}(1 - p_1)$, p_2 can therefore be used for the representation of x_1', that is for one of the x_r-s. Proceeding in this way we can see that every p_n can be used for the representation of an x_l. Since

$$\sum_{n=1}^{\infty} p_n = \sum_{j=1}^{r} x_j = 1,$$

we have obtained a decomposition of the series $\sum_{n=1}^{\infty} p_n$ into r disjoint subseries such that the sum of the j-th subseries is equal to x_j. If A_j consists of the elements ω_n for which p_n occurs in the representation of x_j, then the sets A_j have the required properties.

b) is the special case $r = 2$ of the statement c).

49. The Kolmogorov probability space $[\Omega, \mathcal{A}, P]$ is said to be *non-atomic*, if there exists to every event A of positive probability an event $B \subset A$ such that $0 < P(B) < < P(A)$. Prove that in the case of a non-atomic probability space $[\Omega, \mathcal{A}, P]$ the range of the function $P(A)$ is the whole interval $[0, 1]$.

Hint. Prove first that for any $\varepsilon > 0$ Ω can be decomposed into a finite number of disjoint subsets A_j $(A_j \in \mathcal{A}; j = 1, 2, \ldots, m)$ such that $P(A_j) \leq \varepsilon$. This can be seen as follows. If $A \in \mathcal{A}$, $P(A) > 0$, then A contains a subset $B \subseteq A$ such that $0 < P(B) \leq \varepsilon$. Indeed if $P(A) \leq \varepsilon$, we can choose $B = A$. If $P(A) > \varepsilon$, then (since P is non-atomic) a $B \subset A$ can be found such that $B \in \mathcal{A}$ and $0 < P(B) < P(A)$; here either $P(B)$ or $P(A - B)$ is not greater than $\dfrac{P(A)}{2}$. If $\dfrac{P(A)}{2} \leq \varepsilon$, we have completed the proof; if $\dfrac{P(A)}{2} > \varepsilon$, the procedure is continued. Since for large enough r we have $\dfrac{P(A)}{2^r} \leq \varepsilon$, there can be found in a finite number of steps a set B such that

$$B \subset A, \quad B \in \mathcal{A} \quad \text{and} \quad 0 < P(B) \leq \varepsilon.$$

Let us put now

$$\mu_\varepsilon(A) = \sup_{B \subseteq A, P(B) \leq \varepsilon} P(B) \quad \text{for} \quad A \in \mathcal{A}.$$

According to what was said above, $\mu_\varepsilon(A) > 0$ for $P(A) > 0$. Choose a set $A_1 \in \mathcal{A}$ for which $0 < P(A_1) \leq \varepsilon$, further a set $A_2 \subseteq \bar{A}_1$ for which

$$\varepsilon \geq P(A_2) \geq \frac{1}{2}\mu_\varepsilon(\bar{A}_1)$$

and then a set $A_3 \subseteq \overline{A_1 + A_2}$ for which $\varepsilon \geq P(A_3) \geq \dfrac{1}{2}\mu_\varepsilon(\overline{A_1 + A_2})$; generally, if the sets $A_1, A_2, \ldots, A_n$ are already chosen, we choose a set A_{n+1} such that the conditions

$$\overline{A_{n+1}} \subseteq \overline{A_1 + A_2 + \ldots + A_n}$$

and

$$\varepsilon \geq P(A_{n+1}) \geq \frac{1}{2}\mu_\varepsilon(\overline{A_1 + A_2 + \ldots + A_n})$$

are satisfied. Then $A_1, A_2, \ldots, A_n, \ldots$ are disjoint sets, hence $\sum\limits_{n=1}^{\infty} P(A_n) \leq 1$ and thus $\lim\limits_{n \to \infty} P(A_n) = 0$ and at the same time

$$\lim_{n \to \infty} \mu_\varepsilon(\overline{A_1 + A_2 + \ldots + A_n}) = 0.$$

Since $\mu_\varepsilon(A)$ is a monotonic set function, we get introducing the notation $\sum\limits_{n=2}^{\infty} A_n = B$ that $\mu_\varepsilon(\bar{B}) = 0$. But it follows that $P(\bar{B}) = 0$ and thus, introducing the notation $A_1' = A_1 + \bar{B}$, we obtain that

$$A_1' + \sum_{n=2}^{\infty} A_n = \Omega, \quad 0 < P(A_n) \leq \varepsilon \qquad (n = 2, 3, \ldots),$$

and $0 < P(A_1') \leq \varepsilon$.

Choose now N so large that $\sum_{n=N}^{\infty} P(A_n) \leq \varepsilon$. Then the sets $A_1', A_2, \ldots, A_{N-1}$ and

$A_N' = \sum_{n=1}^{\infty} A_n$ possess the required properties. Now we can construct for an arbitrary number x ($0 < x < 1$) an $A \in \mathscr{A}$ such that $P(A) = x$ in the following manner: Ω is decomposed first into a number N_1 of disjoint subsets A_{1j} such that

$$0 \leq P(A_{1j}) \leq \frac{x}{2} \quad (j = 1, 2, \ldots, N).$$

Let $x_{1,r} = P\left(\sum_{j=1}^{r} A_{1j}\right)$. Then x lies in one of the intervals $[x_{1,r}, x_{1,r+1})$, $r = 1, 2, \ldots,$ $N_1 - 1$, let it be e.g. the interval $[x_{1,r_1}, x_{1,r_1+1})$. If $x = x_{1,r_1}$, we have finished the construction. If $x_{1,r_1} < x < x_{1,r_1+1}$ we decompose A_{1,r_1+1} into subsets A_{2j} ($j = 1, 2, \ldots, N_2$) such that

$$0 \leq P(A_{2j}) \leq \frac{x - x_{1,r_1}}{2}.$$

Let

$$x_{2,r} = P\left(\sum_{j=1}^{r_1} A_{1j} + \sum_{j=1}^{r} A_{2j}\right) \quad (r = 1, 2, \ldots, N_2).$$

Then x lies in one of the intervals $[x_{2,r}, x_{2,r+1})$; e.g. $x \in [x_{2,r_2}, x_{2,r_2+1})$. By continuing this procedure we obtain a set

$$A = \sum_{j=1}^{r_1} A_{1j} + \sum_{j=1}^{r_2} A_{2j} + \ldots + \sum_{j=1}^{r_s} A_{sj} + \ldots$$

for which $P(A) = x$.

50. Prove for an arbitrary probability space that the range of $P(A)$ is a closed set.
Hint. A set $A \in \mathscr{A}$ will be called an *atom* (with respect to P), if $P(A) > 0$ and if $B \in \mathscr{A}$, $B \subseteq A$ imply either $P(B) = 0$ or $P(B) = P(A)$. Two atoms A and A' are, a set of zero measure excepted, either identical or disjoint. From this it follows that there can always be found either a finite or a countably infinite number of disjoint atoms A_n ($n = 1, 2, \ldots$) such that the set $\Omega - \sum_{n=1}^{\infty} A_n$ contains no further atoms. Put

$$\sum_{n=1}^{\infty} A_n = B, \quad \mu_1(A) = \mu(AB), \quad \mu_2(A) = \mu(A\bar{B}).$$

Then $\mu(A) = \mu_1(A) + \mu_2(A)$. Here $\mu_1(A)$ can be considered as a measure on the class of all subsets of the set Ω' having for its elements the sets A_n, and $\mu_2(A)$ is non-atomic. Hence the statement of Exercise 50 is reduced to the Exercises 48a) and 49.

CHAPTER III

DISCRETE RANDOM VARIABLES

§ 1. Complete systems of events and probability distributions

We have defined in § 2 of Chapter I the concept of a "complete system of events" with respect to finite probability algebras; this concept will now be extended to arbitrary Kolmogorov probability spaces. A finite or denumerably infinite system of events $\{A_n\}$ ($A_n \in \mathcal{A}$, $n = 1, 2, \ldots$) is said to be *complete* (in the wider sense), if for $i \neq j$ $A_i A_j = O$ and if the occurrence of an event A_n ($n = 1, 2, \ldots$) is "almost sure" (i.e. it has the probability 1):

$$P\left(\sum_n A_n\right) = \sum_n P(A_n) = 1. \tag{1}$$

Thus we do not require that $\sum_n A_n = \Omega$; only that $P(\overline{\Omega'}) = 0$ should hold, where

$$\Omega' = \sum_n A_n \subset \Omega.$$

The sequence of probabilities of a complete system of events will be called a *probability distribution* (or briefly *distribution*). From a purely mathematical point of view every sequence of nonnegative numbers $p_1, p_2, \ldots$ for which

$$\sum_n p_n = 1, \tag{2}$$

can be considered as a probability distribution.

The expression "probability distribution" hints at the interpretation that the probability 1 of the sure event is "distributed" among the events A_n ($n = 1, 2, \ldots$). There is a close analogy between probability distributions and mass-distributions in mechanical systems, since every sequence of nonnegative numbers $p_1, p_2, \ldots$ fulfilling (2) may be considered as a distribution of the unit mass among a finite or denumerably infinite number of points. Later on we shall often return to this analogy.

§ 2. The theorem of total probability and Bayes' theorem

Let $B_1, B_2, \ldots, B_n, \ldots$ be a complete system of events and let $P(B_i) > 0$ ($i = 1, 2, \ldots$). Then an arbitrary event $A \in \mathcal{A}$ can be decomposed according to the formula

$$A = \sum_{n=1}^{\infty} AB_n.$$

Since $B_i B_j = O$ holds for $i \neq j$, we obtain

$$P(A) = \sum_n P(AB_n). \tag{1}$$

According to the definition of conditional probabilities we have

$$P(AB_n) = P(A \mid B_n)(PB_n).$$

When substituted into (1) this gives

$$P(A) = \sum_n P(A \mid B_n) P(B_n). \tag{2}$$

This relation is called *the theorem of total probability*. Since $\sum_n P(B_n) = 1$, according to (2) the probability $P(A)$ is the weighted mean of the conditional probabilities $P(A \mid B_n)$ taken with the weights $P(B_n)$. From this it follows immediately that

$$\inf_n \; P(A \mid B_n) \leq P(A) \leq \sup_n \; P(A \mid B_n). \tag{3}$$

The theorem of total probability is closely connected to the following simple theorem of mechanics. The center of the gravity of a body can be obtained by decomposing the body into arbitrarily many parts and considering the mass of each part as concentrated in its center of gravity and then forming the center of gravity of the resulting point-system. Equation (2) is further analogous to the following chemical relation: Different solutions of the same salt are placed into N vessels, the total volume of the solutions being 1. Let $P(B_n)$ denote the volume of the n-th vessel and $P(A \mid B_n)$ the concentration of the solution of the n-th vessel. If we mix the contents of the vessels and denote by $P(A)$ the concentration of the resulting solution, Equation (2) will hold for this case too.

Example. Let an urn contain M red and $N - M$ white balls. Draw balls from the urn without replacement. Let A_k denote the event that we obtain a red ball at the k-th drawing. Clearly, $P(A_1) = \dfrac{M}{N}$. We shall show that

$P(A_k) = \dfrac{M}{N}$ $(k = 2, 3, \ldots, N)$. According to the theorem of total probability

$$P(A_2) = P(A_2 \mid A_1) P(A_1) + P(A_2 \mid \bar{A}_1) P(\bar{A}_1),$$

hence

$$P(A_2) = \frac{M-1}{N-1} \cdot \frac{M}{N} + \frac{M}{N-1} \cdot \frac{N-M}{N} = \frac{M}{N}.$$

Similarly we obtain that $P(A_k) = \dfrac{M}{N}$, if $k = 3, 4, \ldots, N$ (cf. Exercise 39a, § 12, Ch. II).

Let A and B be any two elements of an algebra of events with $P(A) > 0$ and $P(B) > 0$. From the values of $P(A)$, $P(B)$ and $P(A \mid B)$ one can obtain $P(B \mid A)$ as well; indeed (cf. (2), § 8, Ch. II)

$$P(B \mid A) = \frac{P(A \mid B) P(B)}{P(A)}. \tag{4}$$

If $\{B_n\}$ is a complete system of events and if in (4) B_k is substituted for B and Expression (2) for $P(A)$, we have

$$P(B_k \mid A) = \frac{P(A \mid B_k) P(B_k)}{\sum_n P(A \mid B_n) P(B_n)}. \tag{5}$$

This is *Bayes' theorem*.

There is hardly any other theorem of probability theory so much debated as this.

Bayes' theorem is well-proven, its validity cannot be doubted; only its practical applications are controversial. An often-used name of this theorem is for instance "theorem of the probability of causes". This name originates in the use of Bayes' theorem to infer the probabilities of the hypotheses (causes) B_k ($k = 1, 2, \ldots$) from the occurrence of an event A; i.e. if one wishes to examine how much the occurrence of an event A supports or refutes certain hypotheses. If the so-called *a priori probabilities* $P(B_k)$ are known, then Bayes' theorem can be applied and the *a posteriori probabilities* $P(B_k \mid A)$ can be computed. However, the probabilities $P(B_k)$ are often unknown. Then it is usual to give them arbitrary values, which is really a questionable procedure.

The name "theorem of the probability of causes" can lead to misunderstandings, hence we must discuss it a little further. Indeed, from (4) it follows that

$$\frac{P(A \mid B)}{P(A)} = \frac{P(B \mid A)}{P(B)}.$$

Thus if the occurrence of the event A increases (e.g. doubles) the probability of B, then the occurrence of the event B increases (doubles) the probability of the event A as well; hence it is entirely impossible to infer the direction of the causal relation from the value of the conditional probability only.

We mention finally the following chemical analogy of Bayes' theorem: Let N vessels contain solutions of different concentrations from the same salt. Let the total volume of the solutions be 1. Let $P(B_k)$ denote the volume of the solution in the k-th vessel and $P(A \mid B_k)$ the concentration of the salt in it, then formula (5) gives, what part of the total mass of salt is in the k-th vessel.

§ 3. Classical probability distributions

1. In the preceding Chapter we have already discussed independent repetitions of a simple alternative. Repeat n times an experiment with two possible outcomes A and $\bar{A}$, such that the repetitions are independent. If B_k ($k = 0, 1, \ldots, n$) denotes the event that A occurred at exactly k experiments, then the events B_k ($k = 0, 1, \ldots, n$) form a complete system of events and the corresponding probabilities are

$$W_k = P(B_k) = \binom{n}{k} p^k q^{n-k} \qquad (k = 0, 1, \ldots, n) \tag{1}$$

where $p = P(A)$ and $q = 1 - p = P(\bar{A})$. The sequence of numbers W_k is called the *binomial distribution of order n and parameter p*. The name hints at the fact that the numbers W_k are the terms of the expansion of $(p + q)^n$ according to the binomial theorem.

2. A natural extension of the binomial distribution is the *polynomial distribution*;[1] it is obtained by independent repetitions of an experiment having several possible outcomes. Let the possible outcomes of the experiment be $A_1, A_2, \ldots, A_r$ and let $P(A_j) = p_j$ ($j = 1, 2, \ldots, r$), then these probabilities fulfil the condition $p_1 + p_2 + \ldots + p_r = 1$. Let $B_{k_1, k_2, \ldots, k_r}$ denote the event that in n independent repetitions of the experiment the event A_1 occurs k_1 times, the event A_2 occurs k_2 times, $\ldots$, the event A_r occurs k_r times, where $k_1 + k_2 + \ldots + k_r = n$. Then we have

$$P(B_{k_1, k_2, \ldots, k_r}) = \frac{n!}{k_1! \, k_2! \ldots k_r!} \, p_1^{k_1} p_2^{k_2} \ldots p_r^{k_r}. \tag{2}$$

The name "polynomial distribution" comes from the fact that the terms $P(B_{k_1, k_2, \ldots, k_r})$ can be obtained by expanding $(p_1 + p_2 + \ldots + p_r)^n$ according to the polynomial theorem. If $r = 3$, we call the distribution the trinomial distribution.

---

[1] Called also "multinomial" distribution.

3. Let an experiment have only two possible outcomes, A and $\bar{A}$. Perform n independent repetitions of this experiment, but let the probability of A (and thus of $\bar{A}$) change from experiment to experiment. Let B_k denote the event that A occurred exactly k times $(k = 0, 1, \ldots, n)$, then

$$P(B_k) = \sum_i p_{i_1} p_{i_2} \cdots p_{i_k} (1 - p_{j_1})(1 - p_{j_2}) \cdots (1 - p_{j_{n-k}}) \qquad (3)$$

where p_i is the probability that A did occur at the i-th experiment. The summation is to be taken over all combinations $(i_1, i_2, \ldots, i_k)$ of the k-th order of the elements $(1, 2, \ldots, n)$ and $j_1, j_2, \ldots, j_{n-k}$ denote those numbers of the sequence $1, 2, \ldots, n$ which do not occur among $i_1, i_2, \ldots, i_k$. The numbers $P(B_k)$ form a probability distribution. If for instance all probabilities p_i are equal to each other, we obtain as a particular case the binomial distribution (1).

The distribution (3) occurs for instance in the following practical problem: In a factory there are n machines which do not work all the time. They are switched on and switched off independently from each other. Let p_i denote the probability that the i-th machine is working at a given moment, let $P(B_k)$ be the probability that at this instant exactly k machines are working, then $P(B_k)$ is given by the Formula (3). The fact that $\sum\limits_{k=0}^{n} P(B_k) = 1$ can be seen directly in the following manner: A simple calculation gives that

$$\sum_{k=0}^{n} P(B_k) x^k = \prod_{i=1}^{n} (1 - p_i + p_i x);$$

by substituting $x = 1$ we obtain $\sum\limits_{k=0}^{n} P(B_k) = 1$.

4. The following problem was discussed in the preceding Chapter. An urn contains M red and $N - M$ white balls $(M < N)$. Draw n times one ball from the urn without replacement $(n \leq N)$. What is the probability that there are k red balls among the n balls drawn? Denote this event by C_k. Then the events C_k [max $(0, n - (N - M)) \leq k \leq \min(n, M)$] form a complete system of events. The corresponding probabilities are, as we have already shown:

$$P(C_k) = \frac{\binom{M}{k}\binom{N - M}{n - k}}{\binom{N}{n}} \qquad (k = 0, 1, \ldots, n). \qquad (4)$$

This distribution is called the *hypergeometric distribution*.

5. This distribution can be generalized in the following manner: Suppose that the urn contains balls of r different colours, namely exactly N_i balls of the i-th colour $(i = 1, 2, \ldots, r)$. Let $N = \sum_{i=1}^{r} N_i$ be the total number of balls and let $C_{k_1, k_2, \ldots, k_r}$ denote the event that among n balls drawn without replacement the first colour occurs k_1 times, the second k_2 times, $\ldots$, the r-th colour k_r times $(k_1 + k_2 + \ldots + k_r = n)$. By a simple combinatorial consideration we obtain that

$$P(C_{k_1, k_2, \ldots, k_r}) = \frac{\binom{N_1}{k_1} \binom{N_2}{k_2} \cdots \binom{N_r}{k_r}}{\binom{N}{n}}. \tag{5}$$

Distribution (5) is called the *polyhypergeometric distribution*. It is, for example, applied in statistical quality control, when the commodities are classified into several categories. (Such categories are for instance: a) faultless; b) faulty but still serviceable; c) completely faulty.)

The events

$$C_{k_1, k_2, \ldots, k_r} \; (0 \le k_i \le \min(n, N_i); \quad \sum_{i=1}^{} k_i = n)$$

form a complete system of events, thus

$$\sum P(C_{k_1, k_2, \ldots, k_r}) = 1.$$

This can be seen directly, if we compare the coefficient of x^n on both sides of the identity

$$\prod_{i=1}^{r} (1 + x)^{N_i} = (1 + x)^N.$$

6. Let an urn contain M red and $N - M$ white balls. Let A_k $(k = 0, 1, \ldots, N - M)$ denote the event that at consecutive drawings without replacement we obtained the first red ball at the $(k + 1)$-st drawing. As was proved in § 5 of the preceding Chapter, we have

$$P(A_0) = \frac{M}{N},$$

$$P(A_k) = \frac{M}{N - k} \prod_{j=0}^{k-1} \left(1 - \frac{M}{N - j}\right) \; (k = 1, 2, \ldots, N - M). \tag{6}$$

Since the events A_k $(k = 0, 1, \ldots, N - M)$ form a complete system of events, we have the relation

$$\frac{M}{N} + \sum_{k=1}^{N-M} \frac{M}{N-k} \prod_{j=0}^{k-1} \left(1 - \frac{M}{N-j}\right) = 1.$$

This identity also has a direct proof, but it is not quite simple. It happens often that certain identities for which a mathematical proof may be rather elaborate, are readily obtained by means of probability calculus.

7. Let the preceding exercise be modified in the following manner: Let an urn again contain M red and $N - M$ white balls, but the drawn balls should now always be replaced. Let A_k denote the event that we obtain the first red ball at the $(k + 1)$-st drawing. The most marked difference between this problem and that of the drawing without replacement dealt with above is that the number k was there bounded ($k \le N - M$). Now, however, k can be arbitrarily large; in principle, it is even possible that we draw always a white ball. Hence it can be questioned whether the events A_k ($k = 0, 1, \ldots$) do form a complete system of events. Clearly, the events A_k mutually exclude each other, the only thing we have to examine is whether it is sure that one of the events does occur. By introducing the notation

$$\Omega' = \sum_{n=0}^{\infty} A_n$$

we have $\Omega' \ne \Omega$.

We shall prove, however, that the possibility to draw always a white ball in an infinite repetition of drawings has probability 0, thus in practice it does not count at all, i.e., the system of events $\{A_k\}$ is in a wider sense complete.

First of all we compute the probabilities $P(A_k)$. Put $\dfrac{M}{N} = p$, $1 - p = q$.

The probability that we obtain at the first k drawings white balls and at the $(k + 1)$-st drawing a red one is

$$P(A_k) = pq^k \qquad (k = 0, 1, \ldots). \tag{7}$$

From this

$$\sum_{k=0}^{\infty} P(A_k) = p \sum_{k=0}^{\infty} q^k = \frac{p}{1-q} = 1.$$

Hence the probability of Ω' is 1 and thus $P(\Omega - \Omega') = 0$. Though it is, in principle, possible that $\Omega - \Omega'$ occurs; this possibility can be neglected in practice. Hence the system $\{A_k\}$ of events is, in a wider sense of the word, complete.

The distribution pq^k ($k = 0, 1, \ldots$) is often called the *geometric distribution*, since the sum of the members pq^k is a geometric series. We shall see later on that this distribution belongs to a larger class of distributions, namely to the class of *negative binomial distributions*.

Examples 1–6 show finite probability algebras; the construction of these probability algebras causes scarcely any difficulty at all. As Example 7 deals with an infinite probability algebra, this deserves to be examined somewhat more thoroughly. This example deals with the infinite repetition of a simple alternative, the elementary events are thus infinite sequences consisting of the events A and $\bar{A}$. It is readily seen that the set of the sequences of this type has cardinal number of the continuum. Indeed, we associate to every sequence

$$E_1, E_2, \ldots, E_n, \ldots,$$

where the meaning of E_n can be only A or $\bar{A}$, the number x having the binary expansion $x = 0, \varepsilon_1 \varepsilon_2 \ldots \varepsilon_n \ldots$, where

$$\varepsilon_n = \begin{cases} 1 & \text{if } E_n = A, \\ 0 & \text{if } E_n = \bar{A}. \end{cases}$$

Thus the set of the elementary events of the sequence of experiments is mapped onto the interval $[0, 1]$. This mapping is one-to-one, with exception of the binary rational numbers $x = \dfrac{k}{2^l}$ (k and l are nonnegative integers).

Now we construct the probability space corresponding to this system. First of all let $\mathscr{A}_0$ denote the set of events obtained by prescribing the outcome of a finite number of experiments assuming nothing about the remaining experiments. Let $i_1 < i_2 < \ldots < i_k$ denote the indices of the experiments where the occurrence of A is prescribed and $j_1 < j_2 < \ldots < j_l$ similarly for the occurrence of $\bar{A}$. Let C denote the event so defined, then we have

$$P(C) = p^k q^l.$$

From this we can compute $P(C)$ for every $C \in \mathscr{A}_0$. Clearly $[\mathscr{A}_0, P]$ is a probability algebra, but $\mathscr{A}_0$ is not a σ-algebra. But if we consider the least σ-algebra $\mathscr{A}$ containing $\mathscr{A}_0$ and extend the set function $P(C)$ defined over $\mathscr{A}_0$ (readily seen to be a measure on $\mathscr{A}_0$) to $\mathscr{A}$, then we obtain the Kolmogorov probability space sought for (cf. Ch. II, § 6). In order to prove that $P(C)$ is a measure on $\mathscr{A}_0$, let us consider the above mapping of the sample space onto the interval $[0, 1]$; let the interval $[0, 1]$ be denoted by Ω^*. There corresponds to the algebra of sets $\mathscr{A}_0$ the class $\mathscr{A}_0^*$ of the subsets of Ω^* consisting of a finite number of pairwise disjoint intervals with binary rational endpoints. Just like in Chapter II, § 7 there can be given a function $F(x)$ so that the probability belonging to the interval $[a, b) = I$ be equal to $F(b) - F(a)$.

Indeed, if the interval $[a, b)$ is of the form $\left[\dfrac{m}{2^n}, \dfrac{m+1}{2^n}\right)$ (m being odd) and

$$\frac{m}{2^n} = \frac{1}{2^{i_1}} + \frac{1}{2^{i_2}} + \ldots + \frac{1}{2^{i_k}} \qquad (i_1 < i_2 < \ldots < i_k = n),$$

then we put

$$F(b) - F(a) = p^k q^{n-k}.$$

From this $F(x)$ can be determined at every binary rational point x. Thus for instance

$$F(0) = 0, \quad F(1) = 1, \quad F\left(\frac{1}{2}\right) = q, \quad F\left(\frac{1}{4}\right) = q^2,$$

$$F\left(\frac{3}{4}\right) = q + pq, \quad F\left(\frac{1}{8}\right) = q^3, \quad F\left(\frac{3}{8}\right) = q^2 + q^2 p,$$

$$F\left(\frac{5}{8}\right) = q + q^2 p, \quad F\left(\frac{7}{8}\right) = q + pq + p^2 q, \text{ etc.}$$

In general, if

$$x = \sum_{k=1}^{r} \frac{1}{2^{i_k}} \qquad (i_1 < i_2 < \ldots < i_r)$$

then

$$F(x) = \sum_{k=1}^{r} \left(\frac{p}{q}\right)^{k-1} q^{i_k}.$$

It is easy to see that $F(x)$ is an increasing continuous function and $F(0) = 0$, $F(1) = 1$. Hence the result of Chapter II, § 7 can be applied here. The extension $\mathscr{A}^*$ of $\mathscr{A}_0^*$ is in this case the collection of all Borel-measurable subsets of Ω^*. Especially, if $p = q = \dfrac{1}{2}$, then $F(x) = x$ and the measure P^* is the Lebesgue-measure. The fact that in an infinite sequence of experiments the probability of "obtaining except for a finite number of experiments always the same event" is zero, corresponds to the fact that in a binary expansion of almost every number both digits occur infinitely often. This is a special case of the well-known theorem of Borel, which will be discussed later on. The above construction of the probability space $[\Omega, \mathscr{A}, P]$ is a special case of a general theorem of Kolmogorov (the so-called *fundamental theorem of Kolmogorov*) which will be proved later.

8. The *negative binomial distribution* can be obtained as a generalization of the preceding problem. Consider an experiment having two possible outcomes A and $\bar{A}$ and let the probability of the event A be p, that of $\bar{A}$ be

$q = 1 - p$. Let $A_k^{(r)}$ denote the event that during independent repetitions of the experiment the event A occurred for the r-th time ($r \geq 1$) at the $(r + k)$-th experiment. We obtain by a simple combinatorial consideration that

$$P(A_k^{(r)}) = \binom{k + r - 1}{r - 1} p^r q^k \qquad (k = 0, 1, \ldots). \tag{8}$$

The events $A_k^{(r)}$ ($k = 0, 1, \ldots$) form a complete system of events in a wider sense. Since the events $A_k^{(r)}$ ($k = 0, 1, \ldots$) are pairwise disjoint, it is enough to show that $\sum_{k=0}^{\infty} P(A_k^{(r)}) = 1$. This follows from (8) in the following manner:

$$\sum_{k=0}^{\infty} \binom{k + r - 1}{r - 1} p^r q^k = p^r \sum_{k=0}^{\infty} \binom{-r}{k} (-q)^k = \left(\frac{p}{1 - q} \right)^r = 1.$$

The distribution (8) will be called the *negative binomial distribution of r-th order*, since the probabilities in question can be obtained as terms of the binomial series (for a negative exponent) of the expression $p^r(1 - q)^{-r}$. Since the events $A_k^{(r)}$ ($k = 0, 1, \ldots$) form a complete system of events, the probability, that the number of occurrences of an event in infinite repetitions of an alternative remains bounded, has the value zero.

Indeed, if C_n denotes the event that in the infinite sequence of experiments A occurred exactly n times; then, as proved above, $P(C_n) = 0$. If therefore C denotes the event that A occurs in the infinite sequence of events only a finite number of times, then $C = \sum_{n=0}^{\infty} C_n$, and

$$P(C) = \sum_{n=0}^{\infty} P(C_n) = 0.$$

Thus the event A occurs infinitely many times with the probability 1.[1]

9. Consider the following problem: let an urn contain M red and $N - M$ white balls. Draw a ball at random, replace the drawn ball and at the same time place into the urn R extra balls with the same colour as the one drawn.[2] Then we draw again a ball, and so on. What is the probability of the event that in n drawings we obtain exactly k times a red ball? Let this event be denoted by A_k. Of course we assume that at every drawing each ball of the

---

[1] Later on we shall prove more: let k_n denote the number of occurrences of A in the first n experiments, then not only $\lim_{n \to \infty} k_n = + \infty$ with probability 1, but more precisely $\lim_{n \to \infty} \dfrac{k_n}{n} = p$ with probability 1.

[2] R can be negative as well. In case of negative R we remove from the urn R balls of the same colour as the one drawn.

urn is selected with the same probability. We compute first the probability that we obtain at each of the first k drawings a red ball, and white balls at the remaining $n - k$ drawings. Clearly this probability is

$$\frac{\prod_{j=0}^{k-1} (M + jR) \prod_{h=0}^{n-k-1} (N - M + hR)}{\prod_{l=0}^{n-1} (N + lR)}. \tag{9}$$

From this it follows easily that

$$P(A_k) = \frac{\binom{n}{k} \prod_{j=0}^{k-1} (M + jR) \prod_{h=0}^{n-k-1} (N - M + hR)}{\prod_{l=0}^{n-1} (N + lR)}. \tag{10}$$

Distribution (10) is called the *Pólya distribution*.

If $R = 0$ and $\dfrac{M}{N} = p$, then we obtain from (10) the binomial distribution as a particular case. If $R = -1$, we get as a particular case from (10) the hypergeometric distribution.

We can also compute the probability that we obtain at the $(k + 1)$-th drawing the first red ball. Obviously, this probability is

$$\frac{M}{N + kR} \prod_{j=0}^{k-1} \left(\frac{N - M + jR}{N + jR} \right). \tag{11}$$

In the cases of $R = 0$ and $R = -1$ we have the particular cases already dealt with.

§ 4. The concept of a random variable

So far we have only considered whether a random event does or does not occur. Qualitative statements like this are often insufficient and quantitative investigations are necessary. In other words, for the description of random mass phenomena one needs numerical data. These numerical data are not constant, they show random fluctuations. Thus for instance the result of a throw in dicing is such a random number. Another example is the number of calls arriving at a telephone exchange during a given time-interval, or the number of disintegrating atoms of a radioactive substance during a given time-interval.

In order to characterize a random quantity we have to know its possible values and the probabilities of these values. Such random quantities are called *random variables*. In the present Chapter we shall discuss only random variables having a countable set of values; these are called *discrete random variables*. The random variables figuring in the above examples were all of the discrete type. The life time of a radioactive atom is for instance also a random variable but it is not a discrete one. General (not discrete) random variables will be dealt with in the following Chapters. In what follows random variables will be denoted by the letters $\xi, \eta, \zeta, \ldots$ of the Greek alphabet.

Let A be an arbitrary event. Let the random variable ξ_A be defined in the following way:

$$\xi_A = \begin{cases} 1 & \text{if } A \text{ occurs,} \\ 0 & \text{otherwise (i.e. if } \bar{A} \text{ occurs).} \end{cases}$$

Obviously, the value of ξ_A depends on chance, further we have

$$P(\xi_A = 1) = P(A),$$

and similarly

$$P(\xi_A = 0) = P(\bar{A}) = 1 - P(A).$$

A random variable ξ_A associated in this way to the event A is called the *indicator* of A. Conversely, we can assign to every random variable ξ assuming only two values, a and b, the event A that $\xi = a$ (where the event $\bar{A}$ means that $\xi = b$).

Starting from this trivial remark, we can make a further step forward. If $x_1, x_2, \ldots$ are the different possible values of the random variable ξ (i.e. the set of the possible values is finite or denumerable), a complete system of events $\{A_n\}$ can be associated to it. Indeed, let A_n denote the event that $\xi = x_n$, then clearly $A_n A_m = O$, if $n \neq m$ and $\sum_{n=1}^{\infty} A_n = \Omega$, hence $\sum_{n=1}^{\infty} P(A_n) = 1$. Conversely, there can be assigned (in several ways) to every complete system of events $\{A_n\}$ a random variable ξ such that in case of the occurrence of A_n the value of ξ should depend on the index n only. ξ can for instance be defined in the following manner:

$$\xi = n \qquad \text{if } A_n \text{ occurs} \qquad (n = 1, 2, \ldots).$$

The value n may be replaced by $f(n)$, where $f(n)$ is any function defined for the positive integers, for which $f(n) \neq f(m)$ if $n \neq m$. Thus we can see that a complete system of events can be assigned to every discrete random variable in a unique manner, while there can be assigned infinitely many different random variables to a complete system of events.

We shall deal in this Chapter with random variables assuming only real values. It must be said that probability theory deals also with random variables whose range does not consist of real numbers but for instance of

n-dimensional vectors. There are also random variables whose values are not vectors of a finite dimension but infinite sequences of numbers or functions, etc. Later on, we shall also examine such cases.

Now let us see, how the notion of a random variable is dealt with in the general theory of probability.

In Chapter II we were made familiar with Kolmogorov's foundation of probability theory. We started from a set Ω, the set of elementary events, and a σ-algebra $\mathscr{A}$ consisting of subsets of Ω. Here $\mathscr{A}$ consists of all events coming into our considerations. Further there was given a nonnegative σ-additive set function P defined on $\mathscr{A}$ such that $P(\Omega) = 1$. The value $P(A)$ of this function for the set A defines the probability of the event A. Naturally, we understand by a random variable a quantity depending on which one of the elementary events in question occurs. A random variable is therefore a function $\xi = \xi(\omega)$ assigning to every element ω of the set Ω (i.e., to every elementary event) a numerical value.

What kind of restrictions are to be prescribed for such a function? If we have a probability field where every subset of Ω corresponds to an event, no restriction is necessary at all. But if this is not the case, then the definition of a random variable calls for certain restrictions.

Since we consider in this Chapter discrete random variables only, we confine ourselves (for the present) to the following definition:

Let $[\Omega, \mathscr{A}, P]$ be a Kolmogorov probability space. A function $\xi = \xi(\omega)$ defined on Ω with a countable set of values is said to be a discrete random variable, if the set, for which $\xi(\omega)$ takes on a fixed value x belongs to $\mathscr{A}$ for every choice of this fixed value x.

Let $x_1, x_2, \ldots$ denote the different possible values of the random variable $\xi = \xi(\omega)$ and A_n the set of the elementary events $\omega \in \Omega$ for which $\xi(\omega) = x_n$, then A_n must belong to the algebra of sets $\mathscr{A}$ for every n. Only in this case the probability

$$P(\xi = x_n) = P(A_n)$$

is defined.

A complete system of events associated with a discrete random variable thus consists of those subsets of the space of events for which the random variable takes on the same value. Especially, if $\xi_A = \xi_A(\omega)$ is the indicator of the event A, then $\xi_A(\omega)$ is a random variable having the value 1 or 0 according as ω does or does not belong to the set A.

The sequence of probabilities of a complete system of events is said to be a probability distribution. Now that we have introduced the concept of random variable this probability distribution can be considered as the set of all probabilities corresponding to the different values taken on by a random variable. If for instance an experiment having the possible outcomes A and $\bar{A}$ is independently repeated n times, then the number ξ of the experiments showing the occurrence of the event A is a random variable with the binomial distribution, i.e.

$$P(\xi = k) = \binom{n}{k} p^k q^{n-k} \qquad (k = 0, 1, \ldots, n)$$

where $p = P(A)$ and $q = 1 - p$.

Let ξ be a random variable and $g(x)$ an arbitrary real valued function of the real variable x. Then $\eta = g(\xi)$ is also a random variable. Let further $\xi_1, \xi_2, \ldots, \xi_r$ be random variables and let $g(x_1, x_2, \ldots, x_r)$ be an arbitrary function of the r real variables $x_1, x_2, \ldots, x_r$, then $\eta = g(\xi_1, \xi_2, \ldots, \xi_r)$ is a random variable as well.

The distribution is the most important concept for the characterization of a random variable, it does not, however, characterize a random variable completely. If for instance we know the distributions of the random variables ξ and η, it is not, generally, possible to determine from this alone the distribution of the random variable $\zeta = g(\xi, \eta)$. In order to do this we have to know the "joint distribution of the random variables ξ, η, that is the probabilities $P(\xi = x_n, \eta = y_m)$. But if ξ and η are given as functions of the elementary event $\omega \in \Omega$, the joint distribution of ξ and η is herewith also given and the distribution of any function $\zeta = g(\xi, \eta)$ as well.

Let the possible values of the random variable ξ be $x_1, x_2, \ldots$ and let A be an arbitrary event of positive probability. We define the conditional distribution of the random variable ξ with respect to the condition A by the sequence of numbers

$$P(\xi = x_n \mid A) \qquad (n = 1, 2, \ldots).$$

We introduce further the notion of the distribution function of a random variable. If ξ is a random variable, then the function $F(x)$ defined by

$$F(x) = P(\xi < x)$$

for every real x is said to be the *distribution function*[1] of ξ. Here $P(\xi < x)$ stands for the probability of the event that the value of ξ is less than x; this event can be represented as the set A_x of the elements $\omega \in \Omega$ for which $\xi(\omega) < x$. If the discrete random variable ξ takes on the values x_n with the probabilities $p_n = P(\xi = x_n)$ $(n = 1, 2, \ldots)$ then we have, clearly,

$$F(x) = \sum_{x_k < x} p_k,$$

where the sum is to be extended over all values of k such that $x_k < x$.

---

[1] Called also cumulative distribution function.

The definition $F(x) = P(\xi \leq x)$ is also customary. This induces only minor modifications in its properties, e.g. this function is continuous from the right, while $P(\xi < x)$ is continuous from the left.

If, for instance, the distribution of ξ is a binomial distribution of order n and parameter p, then

$$F(x) = P(\xi < x) = \sum_{k < x} \binom{n}{k} p^k q^{n-k}. \tag{1}$$

Sometimes, an integral form of this distribution function is used. For this purpose we define the incomplete (and the complete) *beta integral* of Euler. If

$$\alpha > 0, \quad \beta > 0, \quad 0 \leq x \leq 1,$$

put

$$B(\alpha, \beta, x) = \int_0^x t^{\alpha-1} (1 - t)^{\beta-1} \, dt. \tag{2}$$

$B(\alpha, \beta, x)$ is called *Euler's incomplete beta integral of order* (α, β). It is well known that

$$B(\alpha, \beta) = B(\alpha, \beta, 1) = \frac{\Gamma(\alpha) \, \Gamma(\beta)}{\Gamma(\alpha + \beta)}, \tag{3}$$

where

$$\Gamma(\alpha) = \int_0^\infty x^{\alpha-1} e^{-x} \, dx \qquad (\alpha > 0)$$

is the so-called gamma function. $B(\alpha, \beta)$ is called *Euler's complete beta integral of order* (α, β). It is readily verified through integration by parts that

$$\sum_{k=0}^r \binom{n}{k} p^k q^{n-k} = (n - r) \binom{n}{r} \int_p^1 t^r (1 - t)^{n-r-1} \, dt =$$

$$= 1 - \frac{B(r + 1, n - r, p)}{B(r + 1, n - r)}, \tag{4}$$

hence

$$F(x) = P(\xi < x) = 1 - \frac{B(r + 1, n - r, p)}{B(r + 1, n - r)}$$

if

$$r < x \leq r + 1 \ (r = 0, 1, \ldots, n - 1). \tag{5}$$

The distribution function $F(x)$ of an arbitrary (not necessarily discrete) random variable $\xi(\omega)$ exists, iff the set A_x defined above belongs for every real x to $\mathscr{A}$. This will be always assumed in the following Chapters during the study of general random variables. In case of discrete random variables, however, this follows from the assumption that for every possible value n the set A_n of elements $\omega \in \Omega$ such that $\xi(\omega) = x_n$, belongs to $\mathscr{A}$. The distribution function $F(x)$ is always nondecreasing; further $\lim_{x \to -\infty} F(x) =$

$= 0$ and $\lim_{x \to +\infty} F(x) = 1$.

§ 5. The independence of random variables

It is obvious to call two random variables independent if the complete systems of events belonging to them are independent. This definition corresponds to the natural requirement that two random variables should be considered as independent, if the fact that one of them takes on a definite value has no influence on the random fluctuations of the other. Let ω denote any element of the space of events Ω; let $\xi = \xi(\omega)$ be the first, $\eta = \eta(\omega)$ the other random variable, let further A_n be the set of all ω-s for which $\xi(\omega) = x_n$ ($n = 1, 2, \ldots$) and B_n of those for which $\eta(\omega) = y_m$ ($m = 1, 2, \ldots$). ξ and η are said to be independent, if

$$P(A_n B_m) = P(A_n) P(B_m) \tag{1}$$

for every n and for every m; that is, if the complete systems of events $\{A_n\}$ and $\{B_m\}$ are independent. Or in a different notation: ξ and η are called independent, if

$$P(\xi = x_n, \eta = y_m) = P(\xi = x_n) P(\eta = y_m) \tag{1'}$$

for every n and m. Hence in case of two independent random variables the joint distribution of ξ and η is, according to (1'), determined by the distributions of ξ and η.

This definition can be generalized to the case of several random variables. The discrete random variables $\xi_1, \xi_2, \ldots, \xi_r$ are said to be (*completely*) *independent*, if for every system of values $x_{k_1}, x_{k_2}, \ldots, x_{k_r}$ the relation

$$P(\xi_1 = x_{k_1}, \xi_2 = x_{k_2}, \ldots, \xi_r = x_{k_r}) = \prod_{j=1}^{r} P(\xi_j = x_{k_j}) \tag{2}$$

holds. It is easy to see that any $s < r$ out of r independent random variables are also independent. The independence of the random variables $\xi_1, \xi_2, \ldots,$ ξ_s ($s < r$) can be verified by summing Formula (2) over all possible values of $x_{k_{s+1}}, \ldots, x_{k_r}$.

The converse of the statement is not true; from the pairwise independence of ξ_1, ξ_2, ξ_3 their complete independence does not follow. Let the random variables ξ_1, ξ_2, ξ_3 be the indicators of the events A_1, A_2, A_3, then relation (2) expresses just the complete independence of the events A_1, A_2, A_3. Since — as we know already — the complete independence of three events does not follow from their pairwise independence, the same holds for random variables, too.

A constant is, clearly, independent of any random variable. Indeed, if $\eta \equiv c$ ($c = $ constant) and ξ is an arbitrary random variable, then

$$P(\xi = x_k, y = c) = P(\xi = x_k) = P(\xi = x_k) P(\eta = c),$$

since the set defined by $\eta = c$ is the entire space Ω.

Next we prove the following theorem:

THEOREM 1. *Let* $\xi_1, \xi_2, \ldots, \xi_r$ *be independent discrete random variables and* $g_1(x), g_2(x), \ldots, g_r(x)$ *arbitrary functions of a real variable x. Then the random variables* $\eta_1 = g_1(\xi_1), \ldots, \eta_i = g_r(\xi_r)$ *are independent as well.*

PROOF. The proof will be given in detail only for $r = 2$, for $r > 2$ the procedure is essentially the same.

Let $\{x_{jk}\}$ be the sequence of the possible values of the random variable ξ_j ($j = 1, 2$) and $\{A_k\}$ the complete system of events belonging to the random variable ξ; A_{jk} is thus the set of those elementary events $\omega \in \Omega$ for which $\xi_j(\omega) = x_{jk}$.

If y_{jl} is one of the possible values of the random variable $\eta_j = g_j(\xi_j)$, then the set B_{jl} defined by $g_j(\xi_j) = y_{jl}$ can obviously be obtained as the union of finitely or denumerably many sets A_{jk}; B_{jl} is equal to the union of the sets A_{jk} whose indices satisfy the equation $g_j(x_{jk}) = y_{jl}$.

Since the complete systems of events $\{A_{1k}\}$ and $\{A_{2k}\}$ are independent, the sum of an arbitrary subsequence of the sets A_{1k} is independent of the sum of an arbitrary subsequence of the sets A_{2k}. From this our assertion follows.

We give a reformulation of the above theorem which we shall need later on. Let $\xi(\omega)$ be a discrete random variable with possible values $x_1, x_2, \ldots, x_n, \ldots$ and let A_n denote the set of those elementary events ω for which $\xi(\omega) = x_n$. Let further $\mathcal{A}_\xi$ be the least σ-algebra containing the sets A_n. $\mathcal{A}_\xi$ is called the *σ-algebra generated by* ξ. Clearly, $\mathcal{A}_\xi$ consists of the sets obtained as the union of finitely or denumberably many of the sets A_n. Obviously $\mathcal{A}_\xi \subseteq \mathcal{A}$. If $\xi_1, \xi_2, \ldots, \xi_r$ are independent random variables, $\mathcal{A}_{\xi_1}, \mathcal{A}_{\xi_2}, \ldots, \mathcal{A}_{\xi_r}$ are the σ-algebras generated by $\xi_1, \xi_2, \ldots, \xi_r$ and B_j is an arbitrary element of $\mathcal{A}_{\xi_j}$ ($j = 1, 2, \ldots, r$), then the events $B_1, B_2, \ldots, B_r$ are independent.

§ 6. Convolutions of discrete random variables

Let ξ and η be two random variables with possible values x_n and y_m ($n, m = 1, 2, \ldots$), respectively. Let the distributions of ξ and η be

$$P(\xi = x_n) = p_n \quad \text{and} \quad P(\eta = y_m) = q_m \quad (n, m = 1, 2, \ldots).$$

If $g(x, y)$ is any real valued function of two real variables, then — as mentioned above — $\zeta = g(\xi, \eta)$ is a random variable.

Let us determine the distribution of the random variable ζ. For every real number z

$$P(\zeta = z) = \sum_{g(x_n, y_m)=z} P(\xi = x_n, \eta = \eta_m). \tag{1}$$

The sum is here extended over those pairs (n, m) for which $g(x_n, y_m) = z$. If such pairs do not exist, the sum on the right hand side of (1) is zero.

In order to compute $P(\zeta = z)$ we have to know therefore, in general, the joint distribution of ξ and η. If ξ and η are independent, then $P(\xi = x_n, \eta = y_m) = P(\xi = x_n) P(\eta = y_m)$ and thus

$$P(\zeta = z) = \sum_{g(x_n, y_m)=z} p_n q_m. \tag{1'}$$

Let us consider now the important special case when ξ and η are independent and $g(x, y) = x + y$, hence $\zeta = \xi + \eta$. Then

$$P(\xi = z) = \sum_{x_n + y_m = z} p_n q_m. \tag{1''}$$

If ξ and η assume only integer values and $p_n = P(\xi = n)$, $q_m = P(\eta = m)$ $(n, m = 0, \pm 1, \pm 2, \ldots)$, then

$$P(\zeta = k) = \sum_{j=-\infty}^{+\infty} p_j q_{k-j} \qquad (k = 0, \pm 1, \pm 2, \ldots). \tag{2}$$

If ξ and η have only nonnegative integer values, then

$$P(\zeta = k) = \sum_{j=0}^{k} p_j q_{k-j}. \tag{3}$$

The distribution of $\zeta = \xi + \eta$ is called the *convolution* of the distributions ξ and η. In what follows we shall compute the convolution of some discrete distributions.

Let ξ and η be independent random variables having binomial distributions of order n_1 and n_2, respectively, with the same parameter p:

$$P(\xi = k) = \binom{n_1}{k} p^k q^{n_1-k} \qquad (k = 0, 1, \ldots, n_1),$$

$$P(\eta = l) = \binom{n_2}{l} p^l q^{n_2-l} \qquad (l = 0, 1, \ldots, n_2),$$

where $q = 1 - p$. If $\zeta = \xi + \eta$, then

$$P(\zeta = k) = \left[\sum_{j=0}^{k} \binom{n_1}{j} \binom{n_2}{k-j} \right] p^k q^{n_1+n_2-k}. \tag{4}$$

By the well-known identity

$$\sum_{j=0}^{k} \binom{n_1}{j} \binom{n_2}{k-j} = \binom{n_1 + n_2}{k}$$

it follows from (4) that

$$P(\zeta = k) = \binom{n}{k} p^k q^{n-k} \qquad (k = 0, 1, \ldots, n) \qquad (5)$$

where $n = n_1 + n_2$.

Hence the random variable ζ has a binomial distribution too. This result can also be obtained without any computation as follows: Consider an experiment with the possible outcomes A and $\bar{A}$; let $P(A) = p$. In the above example ξ, resp. η is equal to the number of occurrences of A in the course of n_1 resp. n_2 independent repetitions of the experiment. The assertion that ξ and η are independent means that we have two independent sequences of events. Perform a total number of $n = n_1 + n_2$ independent experiments, then $\zeta = \xi + \eta$ means the number of the occurrences of A in this sequence of experiments; hence ζ is a random variable having a binomial distribution of order n and parameter p; that is, Formula (5) is valid.

We encounter a practical application of this result when estimating the percentage of defective items. Consider a sampling with replacement from the population investigated. According to the above this can be done also by subdividing the whole population into two parts having the same percentage of defective items and selecting from one part a sample of n_1 elements and from the other one a sample of n_2 elements. This estimating procedure is equivalent to that which consists of the choice of a sample of $n = n_1 + n_2$ elements from the whole population.

It is to be noted here that the distribution of the sum of two independent random variables with hypergeometric distributions does not have a hypergeometric distribution. Hence the former assertion is not valid if the sampling is done without replacement. The difference is, however, negligible in practice, if the number of elements of the population is large with respect to that of the sample.

§ 7. Expectation of a discrete random variable

The random fluctuations of a random variable are described by its distribution function. In the practice, however, it is often necessary to characterize a distribution by a small number of data. The most important and simplest one of such data is the *expectation* defined below (first for discrete distributions only).

Let the possible values of the random variable ξ be $x_1, x_2, \ldots$ with corresponding probabilities $p_n = P(\xi = x_n)$ $(n = 1, 2, \ldots)$. Perform N independent observations of ξ; if N is a large number, then, according to the meaning of probability, at approximately Np_1 occasions we shall have $\xi = x_1$, at approximately Np_2 occasions $\xi = x_2$, and so on. Taking the arithmetic mean of the ξ-values obtained at the N observations, we obtain approximately the value

$$\frac{Np_1 \cdot x_1 + Np_2 \cdot x_2 + \ldots}{N} = \sum_k p_k x_k;$$

this is the value about which the arithmetic mean of the observed values of ξ fluctuates. Hence we define the *expectation* $E(\xi)$ of the discrete random variable ξ by the formula

$$E(\xi) = \sum_k p_k x_k. \tag{1}$$

Obviously, $E(\xi)$ is the weighted arithmetic mean of the values x_k with weights p_k.[1] In order that the definition should be meaningful we have to assume the absolute convergence of the series figuring on the right side of (1). Otherwise, namely, a rearrangement of the x_k values would give different values for the expectation.

If ξ can take on infinitely many values, then $E(\xi)$ does not always exist. E.g. if

$$P(\xi = 2^k) = \frac{1}{2^k} \qquad (k = 1, 2, \ldots),$$

then the series $\sum_k p_k x_k$ is divergent. Clearly, the expectation of discrete and bounded random variables always exists.

Sometimes, instead of "expectation", the expressions "mean value" or "average" are used. But they may lead to confusion with the average of the observed values. In order to discriminate the observed mean from the number about which the observed mean fluctuates we always call the latter "expectation".

Obviously, the expectation $E(\xi)$ depends only on the distribution of ξ; hence if ξ_1 and ξ_2 are two discrete random variables having the same distribution, then $E(\xi_1) = E(\xi_2)$. Therefore $E(\xi)$ can also be called the *expectation of the distribution* of ξ. The fluctuation about $E(\xi)$ of the averages

---

[1] Hence $E(\xi)$ lies always between the lower and the upper limit of the possible values of ξ.

formed from the observed values of ξ is described more precisely by the laws of large numbers, which we shall discuss later on. Here we mention only that the average of the observed values of ξ and the expectation $E(\xi)$ are essentially in the same relationship as the relative frequency and the probability of an event. This will be readily seen if we consider the indicator ξ_A of an event A having the probability p; indeed, $E(\xi_A) = p \cdot 1 + (1 - p) \cdot 0 = $ $= p$ and the average of the observed values of ξ_A is equal to the relative frequency of the event A.

Next we compute the expectations of some important distributions.

1. *The expectation of the binomial distribution.* The random variable ξ has a binomial distribution if it assumes the values $k = 0, 1, \ldots, n$ with probabilities

$$P(\xi = k) = \binom{n}{k} p^k q^{n-k},$$

where $q = 1 - p$ and $0 < p < 1$. Hence according to (1)

$$E(\xi) = \sum_{k=0}^{n} k \binom{n}{k} p^k q^{n-k} = np \sum_{r=0}^{n-1} \binom{n-1}{r} p^r q^{(n-1)-r} = np.$$

Example. The number of atoms disintegrating during a time interval t out of N atoms of a radioactive substance has a binomial distribution; indeed the probability that in the given time interval exactly k atoms will disintegrate is equal to $\binom{N}{k} p^k q^{N-k}$, where N means the number of atoms present at the beginning of the time interval and $p = 1 - e^{-\lambda t}$ (λ is the disintegration constant). Hence the expected value of the atoms disintegrating during the time interval t is given by $N(1 - e^{-\lambda t})$; thus the expected number of nondisintegrated atoms is $Ne^{-\lambda t}$. This exponential law of radioactivity does *not* state — as it is sometimes erroneously suggested —. that the number of the nondisintegrated atoms is an exponentially decreasing function of the time; on the contrary, it only states that the number of nondisintegrated atoms has an *expectation* which *is* an exponentially decreasing function of the time.

2. *The expectation of the negative binomial distribution.* The random variable ξ has a negative binomial distribution if its possible values are $r + k$ ($k = 0, 1, \ldots$) and if it takes on these values with probabilities

$$P(\xi = r + k) = \binom{k + r - 1}{r - 1} p^r q^k \qquad (k = 0, 1, \ldots),$$

where $0 < p < 1$, $q = 1 - p$. Because of (1) we have

$$E(\zeta) = \sum_{k=0}^{\infty} (r + k) \binom{k + r - 1}{r - 1} p^r q^k = \frac{r}{p} \sum_{k=0}^{\infty} \binom{k + r}{r} p^{r+1} q^k = \frac{r}{p}.$$

Example. In shooting at a target, suppose that every shot hits the target with the probability p and the outcomes of the shots are independent of each other. How many shots are necessary to hit the target r times?

The mathematical wording of the problem is as follows: Let the experiments of a sequence be independent of each other. Let the experiment have only two outcomes: A (the shot hits the target) and $\bar{A}$ (it does not). Let ζ denote the serial number of the experiment at which A occurred for the r-th time. As noted in § 3 of this Chapter, the probability that the event A occurs in the $(k + r)$-th experiment for the r-th time is

$$\binom{k + r - 1}{r - 1} p^r q^k;$$

hence ζ has a negative binomial distribution of order r. Thus in the average we need to fire $\dfrac{r}{p}$ shots in order to get r hits.

3. *The expectation of the hypergeometric distribution.* The random variable ζ has a hypergeometric distribution if it takes on the values $k = 0, 1, \ldots, n$ with probabilities

$$P(\zeta = k) = \frac{\binom{M}{k} \binom{N - M}{n - k}}{\binom{N}{n}}.$$

We obtain from (1) by a simple calculation that

$$E(\zeta) = n \frac{M}{N}.$$

Example. The hypergeometric distribution occurs for instance in sampling without replacement. Let $p = \dfrac{M}{N}$ denote the fraction of defective items in the lot examined. We want to estimate p from a sample of size n. The number of defective items has the same expectation np as in sampling with replacement.

§ 8. Some theorems on expectations

We shall now prove some basic theorems about expectations.

THEOREM 1. *If $E(\xi)$ and $E(\eta)$ exist, then $E(\xi + \eta)$ exists too and*

$$E(\xi + \eta) = E(\xi) + E(\eta).$$

The statement of this theorem is plausible because of the intuitive meaning of expectation. Indeed, if the observed values of ξ are $\xi_1, \xi_2, \ldots, \xi_n$ and those of η are $\eta_1, \eta_2, \ldots, \eta_n$, then $\dfrac{1}{n} \sum\limits_{k=1}^{n} \xi_k$ fluctuates about the number $E(\xi)$ and $\dfrac{1}{n} \sum\limits_{k=1}^{n} \eta_k$ about the number $E(\eta)$, hence $\dfrac{1}{n} \sum\limits_{k=1}^{n} (\xi_k + \eta_k)$ fluctuates about the number $E(\xi) + E(\eta)$; in consequence $E(\xi + \eta) = E(\xi) + E(\eta)$. Let us now give the proof of the theorem. Let the possible values of ξ be x_j ($j = 1, 2, \ldots$) and those of η y_k ($k = 1, 2, \ldots$), let further A_{jk} denote the event that $\xi = x_j$ and $\eta = y_k$. Clearly, the A_{jk} ($j, k = 1, 2, \ldots$) form a complete system of events. Further

$$\sum_j P(A_{jk}) = P(\eta = y_k)$$

and

$$\sum_k P(A_{jk}) = P(\xi = x_j).$$

On the other hand, the possible values of $\xi + \eta$ are the numbers z representable as $x_j + y_k$. It may happen that a number z can be represented in more than one way in the form $z = x_j + y_k$; in this case

$$z\,P(\xi + \eta = z) = z \sum_{x_j + y_k = z} P(A_{jk}) = \sum_{x_j + y_k = z} (x_j + y_k) P(A_{jk}).$$

Since the sum of two absolutely convergent series is itself absolutely convergent, we obtain that

$$E(\xi + \eta) = \sum_j \sum_k (x_j + y_k) P(A_{jk}) = E(\xi) + E(\eta)$$

and this is what we wished to prove.

The next theorem follows by mathematical induction from Theorem 1.

THEOREM 2. *If $E(\xi_i)$ ($i = 1, 2, \ldots, n$) exist, then $E(\xi_1 + \ldots + \xi_n)$ exists, too, and*

$$E(\xi_1 + \xi_2 + \ldots + \xi_n) = E(\xi_1) + E(\xi_2) + \ldots + E(\xi_n).$$

It is easy to prove the following theorem:

THEOREM 3. *Let $c_1, c_2, \ldots, c_n$ be constants and $\xi_1, \xi_2, \ldots, \xi_n$ random vari-*

ables the expectation of which exists, then

$$E\left(\sum_{k=1}^{n} c_k \xi_k\right) = \sum_{k=1}^{n} c_k E(\xi_k).$$

In other words, E is a linear operator.

It is further easy to show the following properties of the expectation: If $\xi \geq 0$, then $E(\xi) \geq 0$. If $|\xi| \leq |\eta|$ and $E(\eta)$ exists, then $E(\xi)$ exists as well.

Consider now some examples. We have proved already that a random variable with a binomial distribution of order n has the expectation $E(\xi) = np$. This can be deduced immediately from Theorem 2; indeed ξ can be written in the form $\xi = \sum_{j=1}^{n} \xi_j$, where ξ_j is the indicator of the event A at the j-th experiment. Since $E(\xi_j) = p$, it follows from Theorem 2 that $E(\xi) = np$.

Similarly a random variable having a negative binomial distribution of order r can be considered as the sum of r independent random variables each having a negative binomial distribution of the first order, with the same parameter p. Thus it follows from Theorem 2 that the negative binomial distribution of order r has the expectation $\dfrac{r}{p}$, as proved already.

Similarly, a random variable with a hypergeometric distribution can be represented as the sum of n indicator variables whose expectation is p (cf. the example after the theorem of complete probability). These indicator variables are not independent, but this does not affect the validity of Theorem 2.

THEOREM 4. *If $\eta = \xi - E(\xi)$, then $E(\eta) = 0$.*

PROOF. According to the additivity

$$E(\eta) = E(\xi) - E\big(E(\xi)\big).$$

Since the expectation of a constant is obviously the constant itself, we have $E(E(\xi)) = E(\xi)$, and our statement follows.

THEOREM 5. *If ξ and η are discrete random variables such that the expectations $E(\xi^2)$ and $E(\eta^2)$ exist, then $E(\xi\eta)$ exists as well and*

$$|E(\xi\eta)| \leq \sqrt{E(\xi^2)\,E(\eta^2)}. \tag{1}$$

Note. Essentially, the inequality (1) is Schwarz's inequality known from analysis.

PROOF. Consider the random variable

$$\zeta_\lambda = (\xi - \lambda\eta)^2,$$

where λ is a real parameter. Since $0 \leq \zeta_\lambda \leq 2\xi^2 + 2\lambda^2\eta^2$, $E(\zeta_\lambda)$ exists. Because of Theorem 3 we have

$$E(\zeta_\lambda) = E(\xi^2) - 2\lambda E(\xi\eta) + \lambda^2 E(\eta^2). \tag{2}$$

Since $\zeta_\lambda \geq 0$ we have $E(\zeta_\lambda) \geq 0$ for every real λ, therefore the polynomial (2) in λ of degree 2 is nonnegative. But as it is well known this is only possible if (1) holds, which is what we wished to prove.

Let ξ be a discrete random variable and A an event having positive probability. The *conditional expectation* of ξ with respect to the condition A is defined by the formula

$$E(\xi \mid A) = \sum_k P(\xi = x_k \mid A) x_k, \tag{3}$$

provided that the series on the right side is absolutely convergent (which is always fulfilled if $E(\xi)$ exists), where x_n $(n = 1, 2, \ldots)$ denote the possible values of ξ. $E(\xi \mid A)$ is therefore the expectation of the conditional distribution of ξ with respect to the condition A. If the events A_n $(n = 1, 2, \ldots)$ form a complete system of events, then in view of the theorem of total probability

$$E(\xi) = \sum_k P(\xi = x_k)x_k = \sum_k \sum_n P(\xi = x_k \mid A_n) P(A_n) x_k = \sum_n P(A_n) E(\xi \mid A_n).$$

Thus we proved the following theorem:

THEOREM 6. *If A_n $(n = 1, 2, \ldots)$ is a complete system of events and ξ is a discrete random variable, then*

$$E(\xi) = \sum_n P(A_n) E(\xi \mid A_n), \tag{4}$$

provided that $E(\xi)$ exists.

Particularly, if ξ_B is the indicator of the event B, then $E(\xi_B) = P(B)$, $E(\xi_B \mid A) = P(B \mid A)$ and we obtain the theorem of total probability as a special case of Theorem 6. Hence Theorem 6 is used to be called the *theorem of total expectation*.

Theorem 6 may also be interpreted in the following manner: The conditional expectation $E(\xi \mid A_n)$ can be considered as a random variable which takes on the value $E(\xi \mid A_n)$, if the event A_n $(n = 1, 2, \ldots)$ occurs. According to this interpretation the right side of (4) is the expectation of the discrete random variable $E(\xi \mid A_n)$. Let η be a random variable whose value depends on the event which actually occurs of the events A_n: e.g. put $\eta = n$ if A_n occurs $(n = 1, 2, \ldots)$. Since now $E(\xi \mid \eta)$ can be written instead of $E(\xi \mid A_n)$.

we have, according to the statement of Theorem 6,

$$E(E(\xi \mid \eta)) = E(\xi). \tag{5}$$

This relation will be used later on.

Example. Formula (5) can also be used to compute the expectation of the sum of a random number of random variables. Let $\xi_1, \xi_2, \ldots$ be independent random variables and let ν be a random variable independent of ξ_n ($n = 1, 2, \ldots$) and taking on the values $1, 2, \ldots$ with probabilities $q_1, q_2, \ldots$. Consider the random variable

$$\zeta = \xi_1 + \xi_2 + \ldots + \xi_\nu$$

which is the sum of a random number of random variables. It follows from (5) that

$$E(\zeta) = E\big(E(\zeta \mid \nu)\big).$$

If E_n is the expectation of ξ_n, then, in view of Theorem 2 and of the independence of the random variables ξ_n and ν, we obtain that

$$E(\zeta \mid \nu = n) = E_1 + E_2 + \ldots + E_n,$$

hence, according to (5),

$$E(\zeta) = \sum_{n=1}^{\infty} q_n (E_1 + E_2 + \ldots + E_n),$$

or, after rearrangement of the terms (which is admissible if the series

$$\sum_{n=1}^{\infty} q_n (|E_1| + |E_2| + \ldots + |E_n|)$$

converges), we have

$$E(\zeta) = \sum_{n=1}^{\infty} E_n \Big(\sum_{k=n}^{\infty} q_k \Big).$$

In the special case where the expectations of the random variables ξ_k are equal, i.e. $E_n = E$, then

$$E(\zeta) = E \sum_{n=1}^{\infty} n q_n = E \cdot E(\nu). \tag{6}$$

THEOREM 7. *If ξ and η are independent discrete random variables and if $E(\xi)$ and $E(\eta)$ exist, then $E(\xi \eta)$ exists as well and*

$$E(\xi \eta) = E(\xi) E(\eta). \tag{7}$$

PROOF. Let A_{jk} denote the event $\xi = x_j, \eta = y_k$ $(j, k = 1, 2, \ldots)$. Clearly, the possible values of $\xi\eta$ are the numbers which can be represented in the form $z = x_j y_k$. Further $zP(\xi\eta = z) = z \sum_{x_jy_k=z} P(A_{jk}) = \sum_{x_jy_k=z} x_jy_kP(A_{jk})$, hence

$$E(\xi\eta) = \sum_j \sum_k x_jy_k \, P(A_{jk}). \tag{8}$$

Because of the independence of ξ and η we have $P(A_{jk}) = P(\xi = x_j) \times \times P(\eta = y_k)$. Thus we obtain from (8) that

$$E(\xi\eta) = \left(\sum_j x_j P(\xi = x_j)\right) \left(\sum_k y_k P(\eta = y_k)\right) = E(\xi)\,E(\eta).$$

Since a series obtained as a sum of term-by-term products of two absolute convergent series is itself absolute convergent, Theorem 7 is herewith proved.

§ 9. The variance

The expectation of a random variable is the value about which the random variable fluctuates; but it does not give any information about the magnitude of this fluctuation. If we compute the expectation of the difference between a random variable and its expectation we obtain, as we have already seen, always zero. This is so because the positive and negative deviations from the expectation cancel each other. Thus it seems natural to consider the quantity

$$d(\xi) = E(\,|\,\xi - E(\xi)\,|\,) \tag{1}$$

as a measure of the fluctuations. Since, however, this expression is difficult to handle, it is the positive square root of the expectation of the random variable $(\xi - E(\xi))^2$ which is most frequently used as a measure of the magnitude of fluctuation. This quantity, called the *standard deviation* of ξ, is thus defined by the expression

$$D(\xi) = +\sqrt{E((\xi - E(\xi))^2)} \tag{2}$$

(provided that this value is finite) and $D^2(\xi)$ is called the *variance* of ξ.[1] The choice of $D(\xi)$ for measuring the fluctuations is advantageous from a mathematical point of view, as it makes computations easier. The real importance of the concept of variance is shown, however, by some basic theorems of probability theory discussed in the following Chapters, e.g. the central limit theorem.

---

[1] The letter D hints at the Latin word *dispersio*.

From the fact that E is a linear operator follows immediately

THEOREM 1. *If* $D(\xi)$ *exists, then*

$$D^2(\xi) = E(\xi^2) - [E(\xi)]^2.$$

This is the formula by which the standard deviation is most readily computed. If the discrete random variable ξ assumes the values x_n $(n = 1, 2, \ldots)$ with probabilities $p_n = P(\xi = x_n)$, then

$$D^2(\xi) = \sum_n p_n (x_n - E(\xi))^2 \tag{3}$$

and, according to Theorem 1,

$$D^2(\xi) = \sum_n p_n x_n^2 - (\sum_n p_n x_n)^2. \tag{4}$$

We obtain by a similar simple argument the somewhat more general

THEOREM 2. *For any real number* A *one has*

$$D^2(\xi) = E((\xi - A)^2) - [E(\xi) - A]^2.$$

From this we obtain immediately the following theorem:

THEOREM 3. *For any real number* A

$$E((\xi - A)^2) \geq D^2(\xi).$$

The equality holds if and only if $A = E(\xi)$.

Theorems 2 and 3 are similar (from a formal point of view even equal) to the well-known Steiner theorem in mechanics which states that the moment of inertia of a linear mass-distribution about an axis perpendicular to this line is equal to the sum of the moment of inertia about the axis through the center of gravity and the square of the distance of the axis from the center of gravity, provided that the total mass is unity; consequently, the moment of inertia has its minimal value if the axis passes through the center of gravity.

Theorem 3 exhibits an important relation between the expectation and the variance.

Theorem 2 is mostly used if the values of ξ lie near to a simple number A but the expectation has not exactly this value. For computational reasons it is then more convenient to calculate the value of $E(\xi - A)^2$.

Obviously, the standard deviation $D(\xi)$ is always nonnegative. If $D(\xi) = 0$, then ξ is equal to a constant with the probability 1. Indeed, because of $(\xi - E(\xi))^2 \geq 0$ the equality $D(\xi) = 0$ can only hold if $P(\xi = E(\xi)) = 1$, hence ξ is a constant with probability 1.

THEOREM 4. *For any random variable* ξ

$$d(\xi) \leq D(\xi).$$

PROOF. According to Theorem 5 of § 8

$$d^2(\xi) = E^2\left(\,|\,\xi - E(\xi)\,|\cdot 1\right) \leq D^2(\xi).$$

Equality can occur in other cases besides the trivial case when ξ is with probability 1 a constant, thus e.g. if ξ takes on the values $+1$ and -1 with the same probability $\dfrac{1}{2}$.

THEOREM 5. *If* $\eta = a\xi + b$ *(a and b are constant), then*

$$D(\eta) = |\,a\,|\cdot D(\xi).$$

PROOF. Since $E(\eta) = aE(\xi) + b$, we obtain that

$$D^2(\eta) = E\!\left(a^2(\xi - E(\xi))^2\right) = a^2 D^2(\xi).$$

Especially, we obtain that the standard deviation does not change if we add a constant to the random variable ξ, or multiply it by -1.

It is seen from (3) that the variance of a random variable depends on its distribution only. Hence we can speak about the variance of a distribution. We shall now compute the variances of certain discrete distributions and for sake of comparison we determine the values of $d(\xi)$ as well.

1. *The variance of the binomial distribution.* Let the distribution of the random variable ξ be a binomial distribution of order n:

$$P(\xi = k) = \binom{n}{k} p^k q^{n-k} \qquad (k = 0, 1, \ldots, n;\ \ 0 < p < 1;\ \ q = 1 - p).$$

In § 7 we have seen that the expectation of ξ is $E(\xi) = np$; similarly, we obtain here that

$$D^2(\xi) = \sum_{k=0}^{n} (k - np)^2 \binom{n}{k} p^k q^{n-k} = npq. \tag{5}$$

The value of $d(\xi)$, for sake of simplicity, will only be determined for a binomial distribution of an even order and parameter $\frac{1}{2}$. If $n = 2N$ and $p = \frac{1}{2}$, then $E(\xi) = N$ and thus

$$d(\xi) = \frac{1}{2^{2N}} \sum_{k=0}^{2N} \binom{2N}{k} |k - N| = \frac{N \binom{2N}{N}}{2^{2N}}.$$

By using Stirling's formula we obtain that for $N \to +\infty$

$$\frac{N \binom{2N}{N}}{2^{2N}} \approx \sqrt{\frac{N}{\pi}}.$$

Here $\approx$ is the sign of asymptotic equality. If a_N and b_N ($N = 1, 2, \ldots$) are two sequences of numbers ($b_N \neq 0$), we say that the two sequences are asymptotically equal ($a_N \approx b_N$) if

$$\lim_{N \to \infty} \frac{a_N}{b_N} = 1.$$

Since in the case of $n = 2N$, $p = \frac{1}{2}$ according to (5) we have

$$D(\xi) = \sqrt{\frac{N}{2}},$$

it follows that

$$d(\xi) \approx \sqrt{\frac{2}{\pi}} D(\xi).$$

Thus the quotient $\frac{d(\xi)}{D(\xi)}$ tends for $N \to \infty$ to the limit $\sqrt{\frac{2}{\pi}}$. We shall see later on that this holds for a whole class of distributions.

2. *The variance of a negative binomial distribution of the first order.* Let the distribution of the random variable ξ be a negative binomial distribution of the first order, i.e.

$$P(\xi = k + 1) = pq^k \qquad (k = 0, 1, \ldots),$$

where $0 < p < 1$, $q = 1 - p$. We have seen in § 7 that $E(\xi) = \dfrac{1}{p}$. Thus

$$D^2(\xi) = p \sum_{k=0}^{\infty} (k + 1)^2 q^k - \frac{1}{p^2},$$

and therefore

$$D^2(\xi) = \frac{q}{p^2}.$$

If p is small, then $D(\xi)$ is approximately equal to $E(\xi) = \dfrac{1}{p}$.

3. *The variance of the hypergeometric distribution.* Let ξ have a hypergeometric distribution, i.e.

$$P(\xi = k) = \frac{\binom{M}{k} \binom{N - M}{n - k}}{\binom{N}{n}} \qquad (k = 0, 1, \ldots, n).$$

As in the two preceding examples we obtain

$$D^2(\xi) = n \frac{M}{N} \left(1 - \frac{M}{N}\right) \left(1 - \frac{n - 1}{N - 1}\right).$$

Let us introduce the notations $\dfrac{M}{N} = p$, $q = 1 - p$, then

$$D(\xi) = \sqrt{npq \left(1 - \frac{n - 1}{N - 1}\right)},$$

hence the standard deviation of the hypergeometric distribution is somewhat less than that of the corresponding binomial distribution; the difference is, however, small if n is small with respect to N (cf. Example 18, § 12, Ch. II). Random fluctuations are for drawing from an urn without replacement less than for drawing with replacement. The quotient of the two standard deviations tends to 1 for $N \to +\infty$, if the value of $\dfrac{M}{N} = p$ remains fixed and n increases more slowly than N.

§ 10. Some theorems concerning the variance

In the present paragraph we shall prove several theorems concerning the variance, which will be used often later on.

THEOREM 1. *If $\xi_1, \xi_2, \ldots, \xi_n$ are pairwise independent, then*

$$D^2 \left(\sum_{k=1}^n \xi_k \right) = \sum_{k=1}^n D^2 (\xi_k).$$

PROOF. Let $E(\xi_k) = E_k$, then

$$D^2 \left(\sum_{k=1}^n \xi_k \right) = \sum_{k=1}^n D^2 (\xi_k) + 2 \sum_{j<k} E((\xi_j - E_j)(\xi_k - E_k)).$$

From the pairwise independence of ξ_k-s and from Theorem 7 of § 8 follows that

$$E((\xi_j - E_j)(\xi_k - E_k)) = 0 \qquad \text{if } j \neq k.$$

Thus we have proved our theorem.

It follows immediately, as a generalization of Theorem 1,

THEOREM 2. *If $\xi_1, \xi_2, \ldots, \xi_n$ are pairwise independent and $c_1, c_2, \ldots, c_n$ are real constants, then*

$$D^2 \left(\sum_{k=1}^n c_k \xi_k \right) = \sum_{k=1}^n c_k^2 D^2 (\xi_k).$$

Because of later applications, the following particular form of Theorem 1 deserves to be mentioned: If $\xi_1, \xi_2, \ldots, \xi_n$ are pairwise independent random variables having the same distribution and standard deviation D, their sum $\zeta_n = \xi_1 + \xi_2 + \ldots + \xi_n$ clearly has

$$D(\zeta_n) = D\sqrt{n}.$$

Let E denote the expectation of the distribution of ξ_k, then

$$E(\zeta_n) = nE.$$

Hence the ratio

$$\frac{D(\zeta_n)}{E(\zeta_n)} = \frac{D}{E\sqrt{n}}$$

tends to zero for $n \to \infty$, provided that E is distinct from zero. Consequences of this are dealt with in Chapter VII. If ξ is a positive random variable, the quotient $\dfrac{D(\xi)}{E(\xi)}$ is called the *coefficient of variation of* ξ.

As an interesting consequence of Theorem 2 we mention that if ξ and η are independent, then

$$D^2(\xi + \eta) = D^2(\xi - \eta).$$

Theorems 1 and 2 can be used to compute the variance of distributions.

1. *The variance of the binomial distribution.* If $\xi_1, \xi_2, \ldots, \xi_n$ are independent random variables assuming the value 1 with probability p and the value 0 with probability $q = 1 - p$, then their sum

$$\zeta_n = \xi_1 + \xi_2 + \ldots + \xi_n$$

is a random variable having a binomial distribution of order n. Since $D^2(\xi_k) = pq$, it follows from Theorem 1 that

$$D^2(\zeta_n) = npq.$$

Thus by applying Theorem 1 we can avoid the calculation used in § 9 for the determination of the variance of the binomial distribution.

2. *The variance of the negative binomial distribution.* In the former paragraph the variance of the negative binomial distribution of the first order was determined. If the independent random variables $\xi_1, \xi_2, \ldots, \xi_r$ have a negative binomial distribution of the first order, i.e. if

$$P(\xi_j = k + 1) = pq^k \qquad (k = 0, 1, \ldots; \quad j = 1, 2, \ldots r)$$

then, as we know already, $D^2(\xi_j) = \dfrac{q}{p^2}$. Applying Theorem 1 it follows for the negative binomially distributed random variable $\zeta_r = \xi_1 + \ldots + \xi_r$ of order r that

$$D^2(\zeta_r) = \frac{rq}{p^2}.$$

§ 11. The correlation coefficient

The correlation coefficient gives some information about the dependence of two random variables. If ξ and η are any two nonconstant discrete random variables, the value $R(\xi, \eta)$ defined by the formula

$$R(\xi, \eta) = \frac{E([\xi - E(\xi)][\eta - E(\eta)])}{D(\xi) D(\eta)} \tag{1}$$

is said to be the *correlation coefficient* of ξ and η. (If ξ or η is constant, we put $R(\xi, \eta) = 0$.)

From this definition follows immediately that $R(\eta, \xi) = R(\xi, \eta)$. If the possible values of ξ and η are x_m $(m = 1, 2, \ldots)$ and y_n $(n = 1, 2, \ldots)$, and $r_{mn} = P(\xi = x_m, \eta = y_n)$, then

$$R(\xi, \eta) = \frac{1}{D(\xi) D(\eta)} \sum_m \sum_n r_{mn} (x_m - E(\xi)) (y_n - E(\eta)).$$

If ξ is any nonconstant random variable, the random variable

$$\xi' = \frac{\xi - E(\xi)}{D(\xi)} \tag{2}$$

satisfies

$$E(\xi') = 0 \quad \text{and} \quad D(\xi') = 1.$$

The operation (2) which applied to ξ gives ξ' is called the *standardization* of the random variable ξ. It follows immediately from the definition of the correlation coefficient that

$$R(\xi, \eta) = E(\xi' \eta'). \tag{3}$$

Now we shall prove some theorems about the correlation coefficient.

THEOREM 1. *We have*

$$R(\xi, \eta) = \frac{E(\xi\eta) - E(\xi) E(\eta)}{D(\xi)D(\eta)}. \tag{4}$$

PROOF. It follows from the linearity of the operator E that

$$E([\xi - E(\xi)][\eta - E(\eta)]) = E(\xi\eta) - E(\xi) E(\eta).$$

THEOREM 2. *The value of $R(\xi, \eta)$ lies always between -1 and $+1$.*

PROOF. According to Theorem 5 of § 8

$$|E([\xi - E(\xi)][\eta - E(\eta)])| \le D(\xi) D(\eta).$$

Theorem 2 cannot be further sharpened, since

$$R(\xi, \xi) = +1$$

and

$$R(\xi, -\xi) = -1.$$

THEOREM 3. *If ξ and η are independent, then*

$$R(\xi, \eta) = 0.$$

PROOF. If ξ and η are independent, then, according to Theorem 7 of § 8

$$E(\xi\eta) = E(\xi)\,E(\eta).$$

Hence Theorem 3 follows from Theorem 1.

Remark. The converse of Theorem 3 does not hold. The independence of ξ and η does not follow, in general, from $R(\xi, \eta) = 0$. If $R(\xi, \eta) = 0$, then the random variables ξ and η are called *uncorrelated*. While uncorrelated random variables are not necessarily independent, nevertheless this is true for certain special cases (cf. e.g. Theorem 4).

THEOREM 4. *If ξ_A and ξ_B are indicators of the events A and B with positive probabilities, the condition $R(\xi_A, \xi_B) = 0$ is equivalent to the independence of ξ_A and ξ_B.*

PROOF. Since $E(\xi_A) = P(A)$, $E(\xi_B) = P(B)$ and $E(\xi_A\xi_B) = P(AB)$, it follows from the condition $R(\xi_A, \xi_B) = 0$ that

$$P(AB) = P(A)\,P(B),$$

which is equivalent to the independence of A and B.

The following is an example of two uncorrelated but not independent random variables. Let

$$P(\xi = 1, \eta = 1) = P(\xi = -1, \eta = 1) = P(\xi = 1, \eta = -1) =$$

$$= P(\xi = -1, \eta = -1) = \frac{p}{4},$$

$$P(\xi = 0, \eta = 1) = P(\xi = 0, \eta = -1) = P(\xi = 1, \eta = 0) =$$

$$= P(\xi = -1, \eta = 0) = \frac{1-p}{4}$$

where $0 < p < 1$. Then $E(\xi) = E(\eta) = E(\xi\eta) = 0$, hence ξ and η are uncorrelated. Since, however, $P(\xi = 0, \eta = 0) = 0 \neq P(\xi = 0)\,P(\eta = 0) = \dfrac{(1-p)^2}{4}$ they are not independent.

In what follows we shall study, what kind of consequences can be deduced from the knowledge of the value of the correlation coefficient. First we prove the following simple theorem:

THEOREM 5. $| R(\xi\eta) | = 1$ *holds, if and only if*

$$\eta = a\xi + b \tag{5}$$

with probability 1, *where a and b are real constants and* $a \neq 0$; *in this case* $R(\xi, \eta) = +1$ *or* -1 *according as* $a > 0$ *or* $a < 0$.

PROOF. Let $E(\xi) = m$. If the relation (5) holds between ξ and η, we have

$$R(\xi, \eta) = \frac{E(a(\xi - m)^2)}{|a| D^2(\xi)} = \text{sgn } a.^1$$

Suppose, for instance, $R(\xi, \eta) = +1$. (The case $R(\xi, \eta) = -1$ can be dealt with in the same manner.) Put

$$\xi' = \frac{\xi - E(\xi)}{D(\xi)}, \qquad \eta' = \frac{\eta - E(\eta)}{D(\eta)},$$

then by (3)

$$E(\xi' \eta') = 1,$$

hence

$$E((\xi' - \eta')^2) = 2 - 2 = 0.$$

From this it follows that

$$P(\xi' = \eta') = 1,$$

that is

$$\eta = E(\eta) + D(\eta) \frac{\xi - E(\xi)}{D(\xi)}$$

with the probability 1.

Thus, unless a linear relation of the form (5) holds between ξ and η, the absolute value of their correlation coefficient is less than 1.

---

[1] sgn x means the sign (signum) of x; it is defined by

$$\text{sgn } x = \begin{cases} 1 & \text{if } x > 0 \\ 0 & \text{if } x = 0 \\ -1 & \text{if } x < 0. \end{cases}$$

In the following we shall say that there is a *positive correlation* between ξ and η, if $R(\xi, \eta) > 0$ and a *negative correlation*, if $R(\xi, \eta) < 0$.

The following most instructive theorem is due to L. V. Kantorovich:

THEOREM 6. *Let ξ and η be discrete random variables assuming only a finite number of values. Let the possible different values of ξ be x_i ($i = 1, 2, \ldots, m$) and those of η y_j ($j = 1, 2, \ldots, n$). If ξ^h and η^k are for $h = 1, 2, \ldots, m - 1$ and $k = 1, 2, \ldots, n - 1$ uncorrelated, i.e.*

$$E(\xi^h \eta^k) = E(\xi^h) E(\eta^k) \qquad (h = 1, \ldots, m - 1; \ \ k = 1, \ldots, n - 1), \quad (6)$$

then ξ and η are independent.

PROOF. Let

$$P(\xi = x_i) = p_i, \ \ P(\eta = y_j) = q_j, \ \ P(\xi = x_i, \eta = y_j) = r_{ij},$$

then Equation (6) goes over into the equivalent form

$$\sum_{i=1}^{m} \sum_{j=1}^{n} r_{ij} x_i^h y_j^k = (\sum_{i=1}^{m} p_i x_i^h)(\sum_{j=1}^{n} q_j y_j^k);$$

the latter clearly holds also if $h = 0, k \leq n - 1$ and if $k = 0, h \leq m - 1$. By introducing the notation $\delta_{ij} = r_{ij} - p_i q_j$ we obtain for these new unknowns the following system of equations:

$$\sum_{i=1}^{m} \sum_{j=1}^{n} \delta_{ij} x_i^h y_j^k = 0 \qquad (h = 0, 1, \ldots, m - 1; \ \ k = 0, 1, \ldots, n - 1). \quad (7)$$

Introducing the notation

$$d_{ik} = \sum_{j=1}^{n} \delta_{ij} y_j^k, \qquad (8)$$

we have for the unknowns d_{ik} ($i = 1, 2, \ldots, m$) the system of linear equations

$$\sum_{i=1}^{m} d_{ik} x_i^h = 0 \qquad (h = 0, 1, \ldots, m - 1). \quad (9)$$

The determinant of this system is the so-called Vandermonde determinant. It is well known that its value is different from 0. Thus the system (9) of equations has no solution distinct from 0, i.e.:

$$d_{ik} = 0 \qquad (i = 1, 2, \ldots, m).$$

Since the above consideration holds for every $k = 0, 1, \ldots, n - 1$, we obtain

$$\sum_{j=1}^{n} \delta_{ij} y_j^k = 0 \qquad (k = 0, 1, \ldots, n - 1). \tag{10}$$

The determinant of Equations (10) is again a Vandermonde determinant, thus

$$\delta_{ij} = 0 \qquad (j = 1, 2, \ldots, n).$$

The same can be shown for every $i = 1, 2, \ldots, m$. From this follows

$$r_{ij} = p_i q_j,$$

thus ξ and η are independent.

Remark. The random variables ξ and η must fulfil $(m - 1)(n - 1)$ conditions in this theorem; as was seen in Chapter II, § 9, the same number of conditions is necessary to ensure the independence of two complete systems of events consisting of m and n events.

Finally, we give an example in which the correlation coefficients are effectively computed.

Let the r-dimensional distribution of the random variables

$$\xi_1, \xi_2, \ldots, \xi_r$$

be a polynomial distribution

$$P(\xi_1 = k_1, \xi_2 = k_2, \ldots, \xi_r = k_r) = \frac{n!}{k_1! \, k_2! \ldots k_r!} \, p_1^{k_1} p_2^{k_2} \ldots p_r^{k_r},$$

where $0 \le k_i \le n$ $(i = 1, 2, \ldots, r)$ and $k_1 + k_2 + \ldots + k_r = n$; furthermore $0 < p_i < 1$ and $\sum_{i=1}^{r} p_i = 1$. We compute the correlation coefficient $R(\xi_i, \xi_j)$. It follows from a simple calculation that

$$E(\xi_i \, \xi_j) = n(n - 1) p_i p_j.$$

It is easy to see that every component ξ_k of the polynomial distribution has a binomial distribution and thus

$$E(\xi_k) = np_k \qquad \text{and} \qquad D(\xi_k) = \sqrt{np_k(1 - p_k)},$$

i.e.

$$R(\xi_i, \xi_j) = -\sqrt{\frac{p_i p_j}{(1 - p_i)(1 - p_j)}} \qquad (i \ne j),$$

thus ξ_i and ξ_j are always negatively correlated.

§ 12. The Poisson distribution

Under certain conditions, the binomial distribution can be approximated by the so-called Poisson distribution. The Poisson distribution, dealt with in this and in the following paragraph, is one of the most important distributions in probability theory. Let us first consider a practical example.

The following problem occurs in the production of glass-bottles. In the melted glass, used for the production of the bottles, there remain little solid bodies briefly called "stones". If a stone gets into the mass of a bottle, the latter becomes defective. The stones are situated at random in the melted glass. But under constant circumstances of production, a given mass of glass contains in the average the same amount of stones. Suppose for instance that 100 kg of fluid glass contains an average number x of stones, let further the weight of a bottle be 1 kg. What per cent of the produced bottles will be defective, because of containing stones? At the first glance we could think that as the mass of 100 bottles contains in the average stones, approximately x per cent of the bottles will be defective. This consideration is, however, wrong, since it does not take into account that more than one of the stones can get into the mass of one bottle and thus the number of the defective items will usually be less.

The problem in question can be solved by means of probability theory. Let us first reduce the problem to a simplified model, nevertheless fulfilling the practical requirements. In practical applications of mathematics we generally work with such models. Whether such a model gives a true picture of the real situation depends on the adequate choice of the model.

We construct the following model for our problem. Suppose that every stone gets with the same probability into the mass of any of the bottles independently of what happens to the other stones. Thus the problem is reduced to an urn-problem: n balls are dropped at random into N urns, what is the probability that a randomly chosen urn contains exactly k balls? Since there are N equally probable possibilities for every one of the balls, the probability that an urn should contain just k balls is, according to the formula of the binomial distribution,

$$W_k = \binom{n}{k} \frac{1}{N^k} \left(1 - \frac{1}{N}\right)^{n-k} . \tag{1}$$

We ask for the percentage of defective items, if the production of N bottles requires M tons of liquid glass. In this case $N = 100\,M$ and $n = xM$. Since we are interested in the percentage of defective items in a long period of production, we may assume that M is very large. Let $\dfrac{x}{100} = \lambda$, then a simple calculation gives that

$$W_k = \frac{\lambda^k}{k!}\left(1 - \frac{\lambda}{n}\right)^{n-k} \cdot \prod_{j=1}^{k-1}\left(1 - \frac{j}{n}\right).$$ (2)

It is known that

$$\lim_{n\to\infty}\left(1 - \frac{\lambda}{n}\right)^n = e^{-\lambda}$$ (3)

hence from (2)

$$\lim_{n\to\infty} W_k = \frac{\lambda^k}{k!}e^{-\lambda} \qquad (k = 0, 1, \ldots).$$ (4)

Let

$$P_k = \frac{\lambda^k}{k!}e^{-\lambda} \qquad (k = 0, 1, \ldots).$$ (5)

From the power series of e^λ we have

$$\sum_{k=0}^{\infty} P_k = e^{-\lambda}\sum_{k=0}^{\infty}\frac{\lambda^k}{k!} = 1.$$ (6)

Thus the probabilities defined by (5) are the terms of a probability distribution, called the *Poisson distribution with parameter* λ: the meaning of λ in the above example is the average number of balls in one urn. It can be shown by direct calculation that λ is the expectation of the Poisson distribution (5). Namely from the relation

$$P(\xi = k) = \frac{\lambda^k}{k!}e^{-\lambda} \qquad (k = 0, 1, \ldots)$$

we have

$$E(\xi) = \sum_{k=0}^{\infty} k\,\frac{\lambda^k}{k!}e^{-\lambda} = \lambda\left(\sum_{k=1}^{\infty}\frac{\lambda^{k-1}}{(k-1)!}\right)e^{-\lambda} = \lambda e^\lambda e^{-\lambda} = \lambda.$$ (7)

Thus the expectation of the Poisson distribution (5) is λ; hence the distribution (5) can be called the *Poisson distribution with expectation* λ. The variance of the Poisson distribution can easily be calculated;

$$E(\xi^2) = \sum_{k=0}^{\infty} k^2\,\frac{\lambda^k}{k!}e^{-\lambda} = \sum_{k=2}^{\infty} k(k-1)\frac{\lambda^k}{k!}e^{-\lambda} + \lambda = \lambda^2 + \lambda,$$

hence

$$D^2(\xi) = \lambda^2 + \lambda - \lambda^2 = \lambda;$$

that is, the standard deviation of the Poisson distribution (5) is $D(\xi) = \sqrt{\lambda}$. Thus *the variance of a Poisson distribution is equal to the expectation.*

In the passage to the limit in (4) no use was made of the property that the probability for a ball to enter in a certain urn is $\dfrac{1}{N}$ with a natural number N. Therefore our result can also be stated in the following form: *The k-th term*

$$W_k = \binom{n}{k} p^k q^{n-k} \tag{8}$$

of the binomial distribution tends to the k-th term of the Poisson distribution, i.e. to the limit

$$P_k = \frac{\lambda^k}{k!} e^{-\lambda} \tag{9}$$

if $n \to \infty$ and $p \to 0$ in such a way that $np = \lambda$, where $\lambda > 0$ is a constant number. (Clearly, the condition $np = \lambda$ can be substituted by the condition $np \to \lambda$.)

The distribution function of the Poisson distribution can be expressed in integral form by means of *Euler's gamma incomplete function.* Let

$$\Gamma(z, x) = \int_0^x t^{z-1} e^{-t} \, dt \tag{10}$$

for $x > 0$, $z > 0$, denote the incomplete gamma function of Euler and

$$\Gamma(z) = \Gamma(z_1 + \infty) = \int_0^\infty t^{z-1} e^{-t} \, dt \tag{11}$$

the complete gamma function of Euler. Partial integration yields the formula

$$\sum_{k=0}^r \frac{\lambda^k e^{-\lambda}}{k!} = 1 - \frac{1}{r!} \int_0^\lambda t^r e^{-t} \, dt = 1 - \frac{\Gamma(r+1, \lambda)}{\Gamma(r+1)} . \tag{12}$$

Let us now return to our practical problem. Because of the relation between relative frequency and probability, the ratio of defective bottles and produced bottles is approximately equal to the probability of a bottle being defective, provided the number of manufactured bottles is sufficiently large. This probability, however, is $1 - W_0$ hence approximately $1 - e^{-\lambda}$. Since $\lambda = \dfrac{x}{100}$, the percentage of defective items is $100\left[1 - \exp\left(-\dfrac{x}{100}\right)\right]$. If x is very small, this is in fact nearly equal to x; in the case of large x, however, it is not. In the extreme case, when $x = 100$, the fraction of defective bottles is not 100 per cent as it would follow from the consideration mentioned at the beginning of this paragraph, but only $100(1 - e^{-1}) = 63.21\%$. Of course such a large fraction of defective items will not occur. If for instance $x = 30$, the fraction of defective items is $100(1 - e^{-0.3}) \approx 25.92\%$ instead of 30%. Clearly, if the number of stones is large, it is more economical to produce small bottles, provided of course that there is no way for clearing the liquid glass. Using 0.25 kg glass per bottle instead of 1 kg, the fraction of defective items decreases for $x = 30$ from 25.92% to 7.22%. As is seen from this example, probability theory can give useful hints for practical problems of production.

§ 13. Some applications of the Poisson distribution

In the previous paragraph the Poisson distribution was introduced as an approximation to the binomial distribution. Now we shall show that the Poisson distribution represents the exact solution of a problem of probability theory. This problem is of fundamental importance in physics, chemistry, biology, astronomy, and other fields.

First let us deal with the example of radioactive decay. The atoms of a radioactive element are randomly disintegrating. As experience shows, the probability for an atom (non-disintegrated until a certain moment) to disintegrate during the next time interval of length t depends only on the length t of this time interval. Let this probability be $F(t)$ and put $G(t) = 1 - F(t)$. As to the function $G(t)$ we know only that it is monotone decreasing and $G(0) = 1$. Let A_s denote the event that a certain atom does not disintegrate during the time interval $(0, s)$, then clearly $P(A_{s+t} \mid A_s) = G(t)$. It follows from the definition of conditional probability that

$$P(A_{s+t}) = P(A_{s+t} \mid A_s) P(A_s), \tag{1}$$

hence

$$G(s + t) = G(s)\, G(t). \tag{2}$$

Thus we obtained a functional equation for $G(t)$. If we assume further that $G(t)$ is differentiable at the point $t = 0$, $G(t)$ may be obtained in the following simple manner: Substitute in (2) Δt for s, then from (2) it follows

$$\frac{G(t + \Delta t) - G(t)}{\Delta t} = G(t)\frac{G(\Delta t) - 1}{\Delta t}. \tag{3}$$

Let Δt tend to 0. Because of $G(0) = 1$, we get

$$G'(t) = G'(0)\, G(t). \tag{4}$$

In the deduction of this equation the existence of the derivative of $G(t)$ was supposed only at the point $t = 0$; namely if the limit on the right side of (3) exists for $\Delta t \rightarrow 0$, then the same holds for the left side as well. $G'(0)$ is necessarily negative. It follows namely from the monotone decreasing property of $G(t)$ that $G'(0) \leq 0$. If we had $G'(0) = 0$, it would follow from (4) that $G(t) \equiv 1$, which means that no radioactive disintegration can occur. Thus putting $G'(0) = -\lambda$ we have $\lambda > 0$. The solution of (4) and $G(0) = 1$ is

$$G(t) = e^{-\lambda t}. \tag{5}$$

The same result can be obtained without the assumption of the existence of $G'(0)$; the assumption of $G(t)$ being monotone decreasing suffices. In fact, we have from (2)

$$G(2t) = G^2(t), \qquad G(3t) = G^3(t),$$

or, generally, for every positive integer

$$G(nt) = G^n(t). \tag{6}$$

Let $nt = s$, then

$$[G(s)]^{\frac{1}{n}} = G\left(\frac{s}{n}\right). \tag{7}$$

From (6) and (7) we obtain

$$G\left(\frac{m}{n}t\right) = [G(t)]^{\frac{m}{n}}$$

hence for every positive rational number r

$$G(r) = [G(1)]^r. \tag{8}$$

Since $G(1) < 1$, $G(1)$ can be written in the form $G(1) = e^{-\lambda}$. Thus we obtain from (8) that for every rational t

$$G(t) = e^{-\lambda t}. \tag{9}$$

However, because of the monotonicity of $G(t)$, (9) holds for every t. Therefore

$$F(t) = 1 - G(t) = 1 - e^{-\lambda t}. \tag{10}$$

Let us now examine the physical meaning of the constant λ. By expanding the function $F(\Delta t) = 1 - e^{-\lambda \Delta t}$ in powers of Δt we obtain the equality

$$F(\Delta t) = \lambda \Delta t + O((\Delta t)^2). \tag{11}$$

The left side of (11) is the probability that an atom, which did not disintegrate until the moment t, will disintegrate before the moment $t + \Delta t$. λ has thus the following physical meaning: the probability that an atom disintegrates during the time interval between t and $t + \Delta t$ is (up to higher powers of Δt) equal to $\lambda \Delta t$. The constant λ is called *constant of disintegration*; it characterizes the radioactive element in question and may serve for its identification. It is attractive to give another interpretation of the number λ, which enables us to measure it. The time during which approximately half of the mass of the radioactive substance disintegrates, is said to be the *half-life period*. More exactly, this is the time interval such that during it each of the atoms of the substance has probability $\dfrac{1}{2}$ of disintegrating. Consider a given mass of a radioactive element of disintegration constant λ. Since every atom disintegrates during the half-life period T with the probability $\dfrac{1}{2}$, we have $F(T) = \dfrac{1}{2}$. However, $G(T) = 1 - F(t) = e^{-\lambda T}$, thus $e^{-\lambda T} = \dfrac{1}{2}$ and

$$\lambda = \frac{\ln 2}{T} . \tag{12}$$

The disintegration constant is therefore inversely proportional to the half-life period. The obtained result may be expressed as follows: the life time of any atom of a radioactive element is a random variable ξ such that its distribution function $F(t) = P(\xi < t)$ has the form

$$F(t) = 1 - e^{-\lambda t} \qquad (t \geq 0),$$

where λ is a positive constant, the disintegration constant of the element in question. (For $t < 0$ clearly $F(t) = 0$ since the life time cannot be negative.) More concisely: the life time of a radioactive atom is an *exponentially distributed random variable*. Hence the custom to speak about the *exponential law of radioactive disintegration*.

Suppose that at time $t = 0$ there are N atoms. How many non-disintegrated atoms shall there be at time $t > 0$? The probability of disintegration during this time is for every atom $1 - e^{-\lambda t}$. So in view of the relation between relative frequency and probability, the number of disintegrations will be approximately $N(1 - e^{-\lambda t})$. Hence approximately $Ne^{-\lambda t}$ atoms remain non-disintegrated.

Let $P_k(t)$ be the probability that during the time interval $(0, t)$ exactly k atoms disintegrate. Suppose that the disintegration of each atom is an event independent of the disintegration of the others, then we have

$$P_k(t) = \binom{N}{k} (1 - e^{-\lambda t})^k \, e^{-(N-k)\lambda t}. \tag{13}$$

The number of disintegrations thus obeys the binomial law. If λt is small and k not too large, $P_k(t)$ may be approximated by a Poisson distribution; the probability $P_k(t)$ is approximately

$$P_k(t) = \frac{[N(1 - e^{-\lambda t})]^k \exp[-N(1 - e^{-\lambda t})]}{k!}. \tag{14}$$

As a further step we can replace for small λt values $(1 - e^{-\lambda t})$ simply by λt. Thus $P_k(t)$ is near to

$$P_k^*(t) = \frac{(N\lambda t)^k e^{-N\lambda t}}{k!}. \tag{15}$$

The half-life period of radium is 1580 years. Taking a year for unit we obtain $\lambda = 0.000439$. If t is less than a minute, λt is of the order 10^{-9}. For 1 g uranium mineral, containing approximately 10^{15} radium atoms, the relative errors committed in replacing $P_k(t)$ by $P_k^*(t)$ are of the order 10^{-3}. If we restrict ourselves to the case where t is small with respect to the half-life period, we can choose the model so that the Poisson distribution represents the exact distribution of the number of radioactive disintegrations. Consider a certain mass of radioactive substance and assume

1. If $t_1 < t_2 < t_3$ and $A_k(t_1, t_2)$ denote the event that "during the time interval (t_1, t_2) k disintegrations occur", then the events $A_k(t_1, t_2)$ and $A_l(t_2, t_3)$ are independent for all nonnegative integer values of k and l.

2. The events $A_k(t_1, t_2)$, $k = 0, 1, \ldots$ form a complete system. If k is given, $P[A_k(t_1, t_2)]$ depends only on the difference $t_2 - t_1$. In other words, the process of radioactive disintegration is homogeneous with respect to time. Let $W_k(t)$ denote the probability of k disintegrations during a time interval of length t $(t_2 - t_1 = t)$.

3. If t is small enough, the probability that during a time interval t there occurs more than one disintegration is negligibly small compared to the probability that there occurs exactly one. That is

$$\lim_{t \to 0} \frac{1 - W_0(t) - W_1(t)}{W_1(t)} = 0, \tag{16}$$

or equivalently

$$\lim_{t \to 0} \frac{1 - W_0(t)}{W_1(t)} = 1. \tag{17}$$

In words: the probability that there occurs at least one disintegration is, in the limit, equal to the probability that there occurs exactly one.

Clearly, $W_0(0) = 1$ and $W_k(0) = 0$ for $k \geq 1$. Further $W_0(t)$ is a monotone decreasing function of t. From this and from conditions 1 and 2 it follows that

$$W_0(t + s) = W_0(t) \, W_0(s);$$

hence we have·

$$W_0(t) = e^{-\mu t} \qquad \text{where} \qquad \mu > 0. \tag{18}$$

In order to determine the functions $W_k(t)$ we show first that

$$\lim_{\Delta t \to 0} \frac{W_k(\Delta t)}{\Delta t} = 0 \quad \text{if} \quad k = 2, 3, \ldots . \tag{19}$$

Obviously, this is a consequence of (16) and of the relation

$$\sum_{k=2}^{\infty} W_k(\Delta t) = 1 - W_0(\Delta t) - W_1(\Delta t). \tag{20}$$

Since for $k \geq 1$ $\quad W_k(0) = 0$, (19) can be written in the form

$$W_k'(0) = 0 \qquad (k = 2, 3, \ldots,). \tag{21}$$

It is to be noted here that the existence of $W_k'(0)$ was not assumed, but proved.

The event that k disintegrations occur during the time interval $(0, t + \Delta t)$, can happen in three ways:

a) $k - 1$ disintegrations occur between 0 and t and one between t and $t + \Delta t$;

b) k disintegrations occur between 0 and t and 0 between t and $t + \Delta t$;

c) at most $k - 2$ disintegrations occur between 0 and t and at least 2 between t and $t + \Delta t$.

Thus, because of conditions 1 and 2, we get

$$W_k(t + \Delta t) = W_k(t) \, W_0(\Delta t) + W_{k-1}(t) \, W_1(\Delta t) + R, \tag{22}$$

where $R = o(\Delta t)$, according to condition 3 and relation (19). In view of (17) and (18) we obtain from (22) for $\Delta t \to 0$

$$W_k'(t) = \mu\big(W_{k-1}(t) - W_k(t)\big) \qquad (k = 1, 2, \ldots). \tag{23}$$

Thus we obtained for the $W_k(t)$ a readily solvable system of differential equations. Put

$$V_k(t) = W_k(t) \, e^{\mu t}, \tag{24}$$

then, from (23)

$$V_k'(t) = \mu V_{k-1}(t) \qquad (k = 1, 2, \ldots). \qquad (25)$$

From $W_0(t) = e^{-\mu t}$ follows $V_0(t) = 1$ and we obtain

$$V_1(t) = \mu t,$$

$$V_2(t) = \frac{\mu^2 t^2}{2},$$

and, in general,

$$V_k(t) = \frac{(\mu t)^k}{k!}.$$

Hence

$$W_k(t) = \frac{(\mu t)^k e^{-\mu t}}{k!} \qquad (k = 0, 1, \ldots).$$

Thus we have proved that the number of disintegrations during a time interval t, given conditions 1–3, has a Poisson distribution with expectation proportional to t.

The Poisson distribution can also be used in studying the number of telephone calls during a given time interval. Let $A_k(t_1, t_2)$ be the event: "between the moments t_1 and t_2 a telephone exchange receives exactly k calls"; the assumptions introduced for the radioactive disintegration are here approximately valid (at least during the "rush hours"). The number of the calls has thus a Poisson distribution. The situation is analogous for the number of electrons emitted by the glowing cathode of an electron tube during a time interval t; also for the number of shooting stars observed during a time interval t as well as for other phenomena exhibiting random fluctuations.

As an application of the Poisson distribution in astronomy let us consider now the mean density λ of the stars in some region of the Milky Way. This density càn be considered to be constant. We understand by this that in a volume V there are in the average $V\lambda$ stars.

In the same manner as in the case of radioactive disintegration (reformulating of course conditions 1–3 adequately), it can be shown that the probability, that a region of volume V of the space contains exactly k stars, is equal to

$$\frac{(\lambda V)^k e^{-\lambda V}}{k!} \qquad (k = 0, 1, \ldots). \qquad (26)$$

The distribution of the stars thus follows the same law as the radioactive disintegration; the only difference is that here the volume plays the role of time. The same reasoning holds for particular kinds of stars as well, e.g. for double stars. In the same manner the distribution of red and white cells in the blood can be determined. Let A_k denote the event that there are exactly k cells to be seen in the visual field of the microscope, then we have

$$P(A_k) = \frac{(\lambda T)^k e^{-\lambda T}}{k!} \qquad (k = 0, 1, \ldots), \qquad (27)$$

where T is the area of the visual field and λ is the average number of cells per unit area.

§ 14. The algebra of probability distributions

In the present paragraph we shall summarize systematically the relations between probability distributions, which we encountered in the previous paragraphs. In particular, we shall deal with relations which permit to construct other distributions from a given one. We shall consider probability distributions belonging to discrete random variables ξ taking on positive values only; such distributions will be denoted by $\{p_0, p_1, \ldots, p_k, \ldots\}$ where $p_k = P(\xi = k)$ $(k = 0, 1, \ldots)$. For the sake of brevity the notation $\mathscr{P} = \{p_0, p_1, \ldots, p_k, \ldots\}$ will be used as well.

A fundamental operation is the *mixing* of probability distributions. Let $\{\alpha_n\}$ $(n = 0, 1, \ldots)$ be nonnegative numbers with sum equal to 1 and let $\mathscr{P}_n = \{p_{nk}\}$ be for each value of n $(n = 0, 1, \ldots)$ a probability distribution. Let us form the expression

$$\pi_k = \sum_{n=0}^{\infty} \alpha_n p_{nk}. \qquad (1)$$

Obviously, the numbers π_k $(k = 0, 1, \ldots)$ form again a probability distribution; indeed $\pi_k \geq 0$ and

$$\sum_{k=0}^{\infty} \pi_k = \sum_{n=0}^{\infty} \alpha_n \sum_{k=0}^{\infty} p_{nk} = \sum_{n=0}^{\infty} \alpha_n = 1. \qquad (2)$$

Let the probability distribution $\Pi = \{\pi_k\}$ be defined by

$$\Pi = \sum_{n=0}^{\infty} \alpha_n \mathscr{P}_n;$$

Π will be called the *mixture of the probability distributions* $\mathscr{P}_n$ *taken with the weights* α_n.

For instance, the mixture of the binomial distributions

$$\mathscr{B}_n(p) = \left\{ \binom{n}{k} p^k q^{n-k} \right\}$$

taken with the weight $\alpha_n = \dfrac{\lambda^n e^{-\lambda}}{n!}$ is a Poisson distribution. In fact

$$\sum_{n=k}^{\infty} \frac{\lambda^n e^{-\lambda}}{n!} \binom{n}{k} p^k q^{n-k} = \frac{(\lambda p)^k e^{-\lambda p}}{k!} . \tag{3}$$

Another example is the mixture of hypergeometric distributions

$$\mathscr{H}_n(M,N) = \left\{ \frac{\binom{M}{k} \binom{N-M}{n-k}}{\binom{N}{n}} \right\}$$

with weights $\alpha_n = \dbinom{N}{n} p^n q^{N-n}$. This leads to the binomial distribution $\mathscr{B}_M(p)$, as is seen from the relation

$$\sum_{n=k}^{N-(M-k)} \binom{N}{n} p^n q^{N-n} \frac{\binom{M}{k} \binom{N-M}{n-k}}{\binom{N}{n}} = \binom{M}{k} p^k q^{M-k}. \tag{4}$$

Geometrically, mixtures of distributions can be represented in the following way: Two distributions $\mathscr{P}_1 = \{p_{1k}\}$ and $\mathscr{P}_2 = \{p_{2k}\}$ can be considered as two points in an infinite dimensional space having the coordinates p_{1k} and p_{2k} respectively. The mixture

$$\alpha \mathscr{P}_1 + \beta \mathscr{P}_2 = \{\alpha p_{1k} + \beta p_{2k}\} \qquad (0 < \alpha < 1, \ \beta = 1 - \alpha)$$

subdivides the "segment" $\mathscr{P}_1 \mathscr{P}_2$ in proportion $\alpha : \beta$. All distributions of probability $\mathscr{P} = \{p_n\}$ are on the "hyperplane" of this space with equation $\sum_{n=0}^{\infty} p_n = 1$, namely in that part of this hyperplane for which $p_n \geq 0$. These points constitute thus a "simplex" S. Since

$$\alpha \mathscr{P}_1 + \beta \mathscr{P}_2 \qquad (0 < \alpha < 1, \ \beta = 1 - \alpha)$$

is a probability distribution as well, it follows that S contains with two points the segment joining them. S is thus convex.

Another often-used operation is the *convolution* of probability distributions. The convolution of the distributions $\mathscr{P} = \{p_k\}$ and $Q = \{q_k\}$ is the distribution $\mathscr{R} = \{r_k\}$, where

$$r_k = \sum_{j=0}^{k} p_j\, q_{k-j}. \tag{5}$$

As it was seen in § 6 $\mathscr{R}$ is the distribution of the sum $\xi + \eta$ of two independent random variables ξ and η having the distributions $\mathscr{P}$ and Q respectively. Even without the knowledge of this result, it is readily shown that $\mathscr{R}$ is a probability distribution. In fact $r_k \geq 0$ and

$$\sum_{k=0}^{\infty} r_k = \sum_{j=0}^{\infty} p_j \sum_{h=0}^{\infty} q_h = 1. \tag{6}$$

The convolution of $\mathscr{P}$ and Q is denoted by $\mathscr{P}Q$. Since

$$\sum_{=0}^{k} p_j\, q_{k-j} = \sum_{j=0}^{k} q_j\, p_{k-j},$$

we have

$$\mathscr{P}Q = Q\mathscr{P}. \tag{7}$$

The convolution is thus a commutative operation. It is associative as well:

$$\mathscr{P}_1(\mathscr{P}_2\mathscr{P}_3) = (\mathscr{P}_1\mathscr{P}_2)\mathscr{P}_3 = \mathscr{P}_1\mathscr{P}_2\mathscr{P}_3. \tag{8}$$

In fact, if $\mathscr{P}_j = \{p_{jk}\}$ $(j = 1, 2, 3)$, the k-th term of the distribution $\mathscr{P}_1(\mathscr{P}_2\mathscr{P}_3)$ as well as of the distribution $(\mathscr{P}_1\mathscr{P}_2)\mathscr{P}_3$ is equal to

$$\sum_{i+j+h=k} p_{1i}\, p_{2j}\, p_{3h}.$$

In this manner multiple convolutions and convolution-powers of a distribution may be defined. By the *n*-th *convolution-power* of a distribution $\mathscr{P}$ we understand the *n*-fold convolution of the distribution $\mathscr{P}$ with itself, in symbols $\mathscr{P}^n$. Thus for instance if $p_0 = q = 1 - p$ and $p_1 = p$, we obtain as convolution of power *n* of the binomial distribution $\mathscr{B}_1(p) = \{q, p\}$ of order 1 the binomial distribution

$$\mathscr{B}_n(p) = (\mathscr{B}_1(p))^n. \tag{9}$$

In fact

$$\sum_{j=0}^{k} \binom{m}{j} p^j q^{m-j} \binom{n}{k-j} p^{k-j} p^{n-k+j} = \binom{m+n}{k} p^k q^{m+n-k},$$

hence

$$\mathscr{B}_m(p)\,\mathscr{B}_n(p) = \mathscr{B}_{m+n}(p). \tag{10}$$

Relation (9) can be obtained from (10) by mathematical induction.

Similarly, it can be shown that for the negative binomial distribution $\mathscr{C}_r(p) = [\mathscr{C}_1(p)]^r$, where $\mathscr{C}_r(p) = \{p_k^{(r)}\}$ with $p_k^{(r)} = 0$ for $k < r$ and $p_k^{(r)} =$
$= \begin{pmatrix} k-1 \\ r-1 \end{pmatrix} p^r q^{k-r}$ for $k \geq r$.

It can be shown finally that the convolution of two Poisson distributions is again a Poisson distribution. If $\mathscr{P}(\lambda) = \left\{ \dfrac{\lambda^k e^{-\lambda}}{k!} \right\}$, then

$$\mathscr{P}(\lambda) \cdot \mathscr{P}(\mu) = \mathscr{P}(\lambda + \mu) \tag{11}$$

since

$$\sum_{j=0}^{k} \frac{\lambda^j e^{-\lambda}}{j!}\, \frac{\mu^{k-j} e^{-\mu}}{(k-j)!} = \frac{(\lambda + \mu)^k e^{-(\lambda+\mu)}}{k!}$$

i.e. the distribution obtained as the convolution "product" of two Poisson distributions has for its parameter the sum of the parameters of the two "factors".

Let us now introduce the *degenerate distribution* $\mathscr{E}_0$. It is defined by

$$\mathscr{E}_0 = \{1, 0, 0, \ldots, 0, \ldots\}.$$

Obviously, for any distribution $\mathscr{P}$ one has

$$\mathscr{P}\mathscr{E}_0 = \mathscr{P}. \tag{12}$$

Thus the distribution $\mathscr{E}_0$ plays the role of the unit element with respect to the convolution operation.[1] The distributions $\mathscr{E}_n$, defined by $p_n = 1$, $p_m = 0$ for $m \neq n$, are also degenerate distributions. It is easy to show that

$$\mathscr{E}_r \mathscr{E}_s = \mathscr{E}_{r+s}, \qquad \mathscr{E}_r = \mathscr{E}_1^r. \tag{13}$$

It is readily seen that the operations mixture and convolution commute:

$$(\sum_{n=0}^{\infty} \alpha_n \mathscr{P}_n)Q = \sum_{n=0}^{\infty} \alpha_n (\mathscr{P}_n Q). \tag{14}$$

By means of the operations mixture and convolution functions of probabil-

---

[1] The probability distributions form a commutative semi-group with respect to convolution with unit element $\mathscr{E}_0$.

ity distributions can be defined in the following manner: Let $g(z) = \sum_{n=0}^{\infty} W_n z^n$ be a power series with nonnegative coefficients such that $g(1) = \sum_{n=0}^{\infty} W_n = 1$. If $\mathscr{S}$ is an arbitrary probability distribution, let $g(\mathscr{S})$ be defined by

$$g(\mathscr{S}) = \sum_{n=0}^{\infty} W_n \mathscr{S}^n \qquad (\mathscr{S}^0 = \mathscr{E}_0). \tag{15}$$

If for instance $\mathscr{E}_1$ is the degenerate distribution defined above and if $g(z) = (pz + q)^n$ $(0 < p < 1)$, then, because of (13), we have

$$\mathscr{B}_n(p) = (p\mathscr{E}_1 + q)^n. \tag{16}$$

Similarly, if $g(z) = e^{\lambda(z-1)}$ $(\lambda > 0)$, then

$$\mathscr{P}(\lambda) = \exp[\lambda(\mathscr{E}_1 - 1)], \tag{17}$$

where $\mathscr{P}(\lambda)$ is a Poisson distribution of parameter λ. In fact

$$\exp[\lambda(\mathscr{E}_1 - 1)] = e^{-\lambda} \sum_{k=0}^{\infty} \frac{\lambda^k \mathscr{E}_1^k}{k!} = e^{-\lambda} \sum_{k=0}^{\infty} \frac{\lambda^k}{k!} \mathscr{E}_k = \left\{ \frac{\lambda^k e^{-\lambda}}{k!} \right\}.$$

§ 15. Generating functions

In the present paragraph, we shall again deal with random variables taking on nonnegative integer values only. Let ξ be such a random variable and put $P(\xi = k) = p_k$ $(k = 0, 1, \ldots)$. The *generating function*[1] $G_\xi(z)$ of the random variable ξ is defined by the power series

$$G_\xi(z) = \sum_{k=0}^{\infty} p_k z^k, \tag{1}$$

where z is a complex variable. The power series (1) is certainly convergent for $|z| \leq 1$, since

$$\sum_{k=0}^{\infty} p_k = 1 \tag{2}$$

and represents an analytic function which is regular in the open unit disk. The introduction of the generating function makes it possible to treat some problems of probability theory by the methods of the theory of functions of a complex variable.

---

[1] Called sometimes probability generating function.

Since the generating function is uniquely determined by the distribution of a random variable, we may speak about the *generating function of a probability distribution* on the set of the nonnegative integers.

It follows immediately from the definition of the generating function that the distribution of a random variable is uniquely determined by its generating function; in fact

$$p_0 = G_\xi(0), \qquad p_k = \frac{G_\xi^{(k)}(0)}{k!} \qquad (k = 1, 2, \ldots) \qquad (3)$$

where $G_\xi^{(k)}(z)$ is the k-th derivative of $G_\xi(z)$. The series (1) may converge in a circle larger than $|z| \leq 1$, or even in the entire plane.

Examples

1. *Generating function of the binomial distribution.* Let ξ be a random variable having a binomial distribution of order n, then

$$G_\xi(z) = (1 + p(z-1))^n = (pz + q)^n.$$

2. *Generating function of the Poisson distribution.* Let ξ be a random variable having a Poisson distribution with expectation λ, then

$$G_\xi(z) = e^{\lambda(z-1)}.$$

(Compare these with the corresponding Formulas (16) and (17) of the preceding paragraph.)

3. *Generating function of the negative binomial distribution.* Let ξ be a random variable of negative binomial distribution with expectation $\dfrac{r}{p}$, then we have

$$G_\xi(z) = \left(\frac{pz}{1 - (1-p)z} \right)^r.$$

From the generating function of a distribution one can obviously get all characteristics (expectation, variance, etc.) of the distribution. We shall now show that these quantities can all be expressed indeed by the derivatives of the generating function at the point $z = 1$. Since the generating function is, in general, defined only for $|z| \leq 1$, we understand by the "derivative at the point $z = 1$" always the left side derivative (provided it exists).

If the derivatives $G_\xi^{(r)}(z)$ of $G_\xi(z)$ exist at $z = 1$, we have the following relations:

$$G_\xi'(1) = \sum_{k=1}^\infty k p_k,$$

$$G_\xi''(1) = \sum_{k=2}^\infty k(k-1) p_k$$

and, in general

$$G_{\xi}^{(r)}(1) = \sum_{k=r}^{\infty} k(k-1)\ldots(k-r+1)p_k \qquad (r = 1, 2, \ldots), \qquad (4)$$

where the series on the right is convergent. Conversely, it is easy to show that if the series in (4) converges, the derivative $G_{\xi}^{(r)}(1)$ exists and Formula (4) is valid. The number

$$M_s = E(\xi^s) = \sum_{k=1}^{\infty} k^s p_k \qquad (s = 1, 2, \ldots) \qquad (5)$$

is called the *moment of order s of* ξ (hence M_1 is the expectation). Thus we have

$$G_{\xi}(1) = M_0 = 1,$$

$$G'_{\xi}(1) = M_1,$$

$$G''_{\xi}(1) = M_2 - M_1,$$

and, in general,

$$G_{\xi}^{(r)}(1) = \sum_{j=1}^{r} S_r^{(j)} M_j \qquad (r = 1, 2, \ldots) \qquad (6)$$

where the $S_r^{(j)}$ are Stirling numbers of the first kind defined by the relation

$$x(x-1)\ldots(x-r+1) = \sum_{j=1}^{r} S_r^{(j)} x^j.$$

Equations (6), if solved with respect to M_j, give

$$M_1 = G'_{\xi}(1),$$

$$M_2 = G'_{\xi}(1) + G''_{\xi}(1)$$

and, in general,

$$M_s = \sum_{j=1}^{s} \sigma_s^{(j)} G_{\xi}^{(j)}(1), \qquad (7)$$

where $\sigma_s^{(j)}$ are Stirling numbers of the second kind, defined by

$$x^s = \sum_{j=1}^{s} \sigma_s^{(j)} x(x-1)\ldots(x-j+1).$$

Equations (7) allow the calculation of the *central moments* of ξ, i.e. the moments of $\xi - E(\xi)$:

$$m_s = E([\xi - E(\xi)]^s) \qquad (s = 2, 3, \ldots). \qquad (8)$$

In fact

$$m_s = \sum_{r=0}^{s} \binom{s}{r} (-1)^r M_{s-r} M_1^r.$$ (9)

For $s = 2$ we obtain the often used formula

$$D^2(\xi) = m_2 = M_2 - M_1^2 = G''_\xi(1) + G'_\xi(1) - [G'_\xi(1)]^2.$$ (10)

A convenient procedure to calculate moments (central or not) of higher orders by means of the generating function is the following: Substitute $z = e^w$ into $G_\xi(z)$ and expand the function $G_\xi(e^w)$ in powers of w:

$$G_\xi(e^w) = \sum_{k=0}^{\infty} p_k \sum_{s=0}^{\infty} \frac{(kw)^s}{s!} = \sum_{s=0}^{\infty} \frac{w^s}{s!} \sum_{k=0}^{\infty} p_k k^s$$ (11)

or

$$H_\xi(w) = G_\xi(e^w) = \sum_{s=0}^{\infty} \frac{M_s}{s!} w^s.$$ (12)

The function $H_\xi(w)$ is called the *moment generating function* of the random variable ξ. In order to calculate central moments we put

$$I_\xi(w) = e^{-wM_1} H_\xi(w).$$ (13)

A simple computation furnishes

$$I_\xi(w) = 1 + \sum_{s=2}^{\infty} \frac{m_s w^s}{s!}.$$ (14)

$I_\xi(w)$ is called the *central moment generating function* of ξ. $H_\xi(w)$ and $I_\xi(w)$ exist only if $G_\xi(w)$ is regular at $z = 1$. The necessary and sufficient condition for this is the existence of all moments of ξ and the finiteness of the expression

$$\varlimsup_{s \to +\infty} \sqrt[s]{\frac{M_s}{s!}}.$$

This condition is always fulfilled for bounded random variables and also in case of certain unbounded distributions, e.g. the Poisson distribution and the negative binomial distribution.

If $H_\xi(w)$ exists, then $I_\xi(0) = 1$, since $G_\xi(1) = 1$. But then there can be found a circle $|w| < r$ in which $I_\xi(w) \neq 0$, hence $\ln I_\xi(w)$ is regular. Put $K_\xi(w) = \ln I_\xi(w)$. Since $K_\xi(0) = 0$ and $K'_\xi(0) = I'_\xi(0) = 0$, we have for $|w| < r$

$$K_\xi(w) = \sum_{l=2}^{\infty} \frac{k_l w^l}{l!}.$$ (15)

The coefficients $k_l = k_l(\xi)$ $(l = 2, 3, \ldots)$ are called *cumulants* or *semi-invariants* of the random variable ξ. If $\eta = \xi + C$ (C being a fixed positive integer), then $k_l(\eta)$ and $k_l(\xi)$ are identical (since $G_\eta(z) = z^c G_\xi(z)$, thus $I_\eta(w) = I_\xi(w)$), hence the name semi-invariant. The meaning of the name "cumulants" will be explained later.

Between the first cumulants and the first central moments we have the following simple relations

$$k_2 = m_2 = D^2(\xi),$$

$$k_3 = m_3, \tag{16}$$

$$k_4 = m_4 - 3m_2^2.$$

These can be established by differentiating the equality $K_\xi(w) = \ln I_\xi(w)$. The function $K_\xi(w)$ is called the *cumulant generating function* of ξ.

Example. The cumulants of the Poisson distribution. Let ξ be a random variable having a Poisson distribution with expectation λ. We have

$$G_\xi(z) = e^{\lambda(z-1)},$$

$$H_\xi(w) = e^{\lambda(e^w - 1)},$$

$$I_\xi(w) = e^{\lambda(e^w - 1 - w)},$$

hence

$$K_\xi(w) = \lambda(e^w - 1 - w) = \lambda \sum_{l=2}^{\infty} \frac{w^l}{l!}. \tag{17}$$

In consequence, all cumulants $k_i(\xi)$ are equal to λ. In particular, not only the variance of ξ, but also its third central moment are equal to λ. This can also be seen by direct calculation.

In what follows we shall prove some properties of generating functions, properties which make the application of these functions a very fruitful device in probability theory.

THEOREM 1. *If ξ and η are two independent random variables, we have*

$$G_{\xi+\eta}(z) = G_\xi(z) \cdot G_\eta(z) \tag{18}$$

and, consequently,

$$K_{\xi+\eta}(w) = K_\xi(w) + K_\eta(w). \tag{19}$$

Relation (19) states that in adding independent random variables, their cumulant generating function as well as their cumulants themselves are added (or "cumulated"); since (19) implies

$$k_l(\xi + \eta) = k_l(\xi) + k_l(\eta) \qquad (l = 2, 3, \ldots). \tag{20}$$

Remark. For $l=2$ relation (20) is already well known to us: the variance of the sum of independent random variables is equal to the sum of the variances. For $l = 3$, relation (20) shows that this holds for the third central moments, too.

PROOF. Equality (18) is proved by direct calculation; (19) follows immediately from (18) and from $E(\xi + \eta) = E(\xi) + E(\eta)$.

THEOREM 2. *If the distribution of the random variable η is the mixture with weights α_n ($\alpha_n \geq 0$, $\sum\limits_{n=0}^{\infty} \alpha_n = 1$), of the random variables ξ_n ($n = 0, 1, \ldots$), then*

$$G_\eta(z) = \sum_{n=0}^{\infty} \alpha_n G_{\xi_n}(z). \tag{21}$$

PROOF. The probability that the quantity η is equal to the random variable ξ_n is, by assumption, equal to α_n. Thus, if $q_k = P(\eta = k)$ and $p_{nk} = P(\xi_n = k)$, we have

$$q_k = \sum_{n=0}^{\infty} \alpha_n p_{nk}. \tag{22}$$

Consequently

$$G_\eta(z) = \sum_{k=0}^{\infty} q_k z^k = \sum_{n=0}^{\infty} \alpha_n \sum_{k=0}^{\infty} p_{nk} z^k, \tag{23}$$

where the order of the summations may be interchanged because of the absolute convergence of the double series. Relation (21) is herewith proved.

THEOREM 3. *Assume that the random variables $\xi_1, \xi_2, \ldots, \xi_n, \ldots$ have the same distribution; let $G(z)$ be their common generating function. Let further be ν a random variable taking on positive integer values only, which is independent from the ξ_n-s. The generating function of the sum*

$$\eta = \xi_1 + \xi_2 + \ldots + \xi_\nu \tag{24}$$

of a random number of random variables is equal to $G_\nu[G(z)]$.

PROOF. Theorem 3 is a consequence of Theorems 1 and 2. In fact, the distribution of η is the mixture of the distributions $\mathscr{P}^n$ ($n = 1, 2, \ldots$) with weights $\alpha_n = P(\nu = n)$, where $\mathscr{P}$ stands for the distribution of ξ_n. According to

Theorem 1 the generating function of $\mathscr{S}^n$ is $[G(z)]^n$, hence by Theorem 2:

$$G_\eta(z) = \sum_{n=1}^\infty \alpha_n [G(z)]^n.$$

But, by definition, $G_\nu(z) = \sum_{n=1}^\infty \alpha_n z^n$. Hence

$$G_\eta(z) = G_\nu\big(G(z)\big), \qquad (25)$$

which finishes the proof of our theorem.

The generating function of the joint distribution of several nonnegative integer valued random variables can be defined analogously. Let for instance ξ and η be two random variables assuming nonnegative integer values only (independence is here not supposed); the joint distribution of the random variables ξ and η is defined by the probabilities

$$r_{hk} = P(\xi = h, \eta = k) \qquad (h, k = 0, 1, \ldots). \qquad (26)$$

The generating function of the joint distribution of the random variables ξ and η is defined by the series

$$G(x, y) = \sum_{h=0}^\infty \sum_{k=0}^\infty r_{hk}\, x^h y^k, \qquad (27)$$

where x and y are complex variables satisfying the conditions $|x| \le 1$, $|y| \le 1$. Obviously, $G(x, 1)$ and $G(1, y)$ are the respective generating functions of ξ and η. The probabilities r_{hk} are uniquely determined by $G(x, y)$, namely

$$r_{hk} = \frac{1}{h!\, k!} \frac{\partial^{h+k} G(x, y)}{\partial x^h \partial y^k} \bigg|_{\substack{x=0 \\ y=0}}. \qquad (28)$$

If ξ and η are independent, $r_{hk} = p_h q_k$ with $p_h = P(\xi = h)$ and $q_k = P(\eta = k)$. From this it follows that $G(x, y) = G_\xi(x)\, G_\eta(y)$. Conversely, from the latter relation follows the independence of ξ and η.

Example. The binomial distribution. Let the possible outcomes A, B, C of an experiment mutually exclude each other and have the respective probabilities p, q, r $(p + q + r = 1)$. Let us perform n independent trials. Let ξ denote the number of trials leading to the outcome A, η the number of those leading to the outcome B. The random variables ξ and η are not independent. The joint distribution of the random variables ξ and η can be given by the probabilities

$$r_{hk} = P(\xi = h, \eta = k) = \frac{n!}{h!\, k!\, (n - h - k)!}\, p^h q^k r^{n-h-k}. \qquad (29)$$

The generating function $G_n(x, y)$ is given by

$$G_n(x, y) = (px + qy + r)^n. \tag{30}$$

If the number n of the trials is a random variable having a Poisson distribution with expectation N, ξ and η become independent from each other. In fact, according to Theorem 2 (which can immediately be generalized to the case of two dimensions) we obtain that the generating function of the mixture of the trinomial distributions (30) with weights $\dfrac{N^n e^{-N}}{n!}$, $n = 0, 1, \ldots$, is

$$G(x, y) = \sum_{n=0}^{\infty} \frac{N^n e^{-N}}{n!} G_n(x, y) = e^{N(px + qy + r - 1)} \tag{31}$$

and since $p + q + r = 1$, we have

$$G(x, y) = e^{Np(x-1)} e^{Nq(y-1)}; \tag{32}$$

therefore ξ and η are independent random variables with Poisson distribution and with expectations Np and Nq, respectively.

Conversely, ξ and η are only independent if the number of trials has a Poisson distribution. In fact, if $\alpha_n = P(v = n)$, further if ξ and η are independent, then

$$G(x, y) = G(x, 1) G(1, y) = \sum_{n=0}^{\infty} \alpha_n G_n(x, y). \tag{33}$$

Let $A(z)$ be the generating function of v, then according to (30) and (33) we have

$$G(x, y) = A(p(x - 1) + q(y - 1) + 1).$$

Hence from (33)

$$A(p(x - 1) + q(y - 1) + 1) = A(p(x - 1) + 1) A(q(y - 1) + 1). \tag{34}$$

If we put $g(z) = A(z + 1)$, $g(z)$ satisfies the functional equation

$$g(a + b) = g(a) g(b). \tag{35}$$

But from this follows because of the regularity of $g(z)$ that $g(z) = e^{Nz}$. Hence $A(z) = e^{N(z-1)}$; that is v has a Poisson distribution.

Now we shall prove the following theorem:

THEOREM 4. *If the sequence of distributions of random variables* $\xi_1, \xi_2, \ldots$ *(assuming nonnegative integer values only) converges to a probability distri-*

bution, i.e. if

$$\lim_{n \to \infty} p_{nk} = p_k \qquad (k = 0, 1, \ldots) \tag{36}$$

and

$$\sum_{k=0}^{\infty} p_k = 1 \tag{37}$$

are valid for

$$p_{nk} = P(\xi_n = k) \qquad (k = 0, 1, \ldots), \tag{38}$$

then the generating functions of the ξ_n converge, in the closed unit circle, to the generating function of the distribution $\{p_k\}$. Hence we have

$$\lim_{n \to \infty} G_n(z) = G(z) \ for \ |z| \leq 1 \tag{39}$$

where

$$G_n(z) = \sum_{k=0}^{\infty} p_{nk} z^k \tag{40}$$

and

$$G(z) = \sum_{k=0}^{\infty} p_k z^k. \tag{41}$$

Conversely, if the sequence $G_n(z)$ tends to a limit $G(z)$ for every z with $|z| \leq \leq 1$, then (36) and (37) are valid, i.e. $G(z)$ is the generating function of a distribution $\{p_k\}$ and the distributions $\{p_{nk}\}$ converge to this distribution $\{p_k\}$.

Remark. If (36) does hold while (37) does not, then (39) is valid only in the interior of the unit circle. This can be seen from the following example. Let $\xi_n = n$, hence

$$p_{nk} = \begin{cases} 1 & \text{for } k = n, \\ 0 & \text{otherwise;} \end{cases}$$

consequently,

$$\lim_{n \to \infty} p_{nk} = 0 \qquad (k = 0, 1, \ldots),$$

but

$$\lim_{n \to \infty} G_n(z) = \lim_{n \to \infty} z^n = \begin{cases} 0 & \text{for } |z| < 1, \\ 1 & \text{for } z = 1, \end{cases}$$

while for $z = e^{i\vartheta}$ with $0 < \vartheta < 2\pi$ there exists no limit.

It can be seen from the same example that if we assume (39) to hold for $|z| < 1$ only, $G(z)$ will not necessarily be a generating function.

PROOF OF THEOREM 4. First we show that (39) follows from (36) and (37).

Let $\varepsilon > 0$ be an arbitrary number; choose N such that

$$\sum_{k=N}^{\infty} p_k < \frac{\varepsilon}{4}, \tag{42}$$

where p_k has the sense given in (37); this will be always possible because of (37). Choose next a number n so large that

$$|p_{nk} - p_k| < \frac{\varepsilon}{4N} \qquad (k = 0, 1, \ldots, N-1) \tag{43}$$

holds, which is possible because of (36). Since $\sum_{k=0}^{\infty} p_{nk} = 1$, it follows from (42) and (43) that for n large enough

$$\sum_{k=N}^{\infty} p_{nk} < \frac{\varepsilon}{2}. \tag{44}$$

In fact

$$\sum_{k=N}^{\infty} p_{nk} = 1 - \sum_{k=0}^{N-1} p_{nk} \leq 1 - \sum_{k=0}^{N-1} p_k + \frac{\varepsilon}{4} = \sum_{k=N}^{\infty} p_k + \frac{\varepsilon}{4} < \frac{\varepsilon}{2}.$$

It follows from relations (42), (43) and (44) that for $|z| \leq 1$ and for sufficiently large n

$$|G(z) - G_n(z)| \leq \sum_{k=0}^{N-1} |p_k - p_{nk}| + \sum_{k=N}^{\infty} p_k + \sum_{k=N}^{\infty} p_{nk} < \varepsilon,$$

which was to be proved.

Now we shall prove that (39) implies (36) and (37). From the assumption

$$\lim_{n \to \infty} G_n(z) = G(z) \quad \text{for} \quad |z| \leq 1$$

and from

$$|G_n(z)| \leq G_n(1) = 1 \quad \text{for} \quad |z| \leq 1, \; n = 1, 2, \ldots$$

it follows according to the known theorem of Vitali that $G(z)$ is regular for $|z| < 1$ and that $G_n(z)$ converges uniformly to $G(z)$ in the entire circle $|z| \leq r < 1$. Putting

$$G(z) = \sum_{k=0}^{\infty} p_k z^k$$

and denoting by C_r the circle $|z| = r < 1$, we obtain that

$$\lim_{n \to \infty} p_{nk} = \lim_{n \to \infty} \frac{1}{2\pi i} \oint_{C_r} \frac{G_n(z)}{z^{k+1}} \, dz = \frac{1}{2\pi i} \oint_{C_r} \frac{G(z)}{z^{k+1}} \, dz = p_k.$$

From this (36) follows. Since $G(1) = \lim_{n \to \infty} G_n(1) = 1$, we get (37).

Example. By means of Theorem 4 another proof can be given of the fact that the binomial distribution converges to the Poisson distribution. Let $G_n(z)$ be the generating function of the binomial distribution $\mathscr{B}_n\left(\dfrac{\lambda}{n}\right)$, then

$$G_n(z) = \left(1 + \frac{\lambda(z-1)}{n}\right)^n.$$

Clearly

$$\lim_{n \to \infty} G_n(z) = e^{\lambda(z-1)},$$

and since $e^{\lambda(z-1)}$ is the generating function of the Poisson distribution $\mathscr{P}(\lambda)$, our statement follows from the second part of Theorem 4.

It can be proved in the same manner that the negative binomial distribution $\mathscr{C}_r(p)$ converges to the Poisson distribution $\mathscr{P}(\lambda)$ for $r \to \infty$, if $(1-p)r = \lambda$ is constant. In other words, if

$$P(\xi_r = k) = \binom{r+k-1}{k} p^r q^k \qquad (k = 0, 1, \ldots),$$

where $p = 1 - \dfrac{\lambda}{r}$ and $q = 1 - p = \dfrac{\lambda}{r}$, then the distribution of ξ_r converges to the Poisson distribution $\mathscr{P}(\lambda)$. Since the generating function $G_n(z)$ of the distribution $\mathscr{C}_r\left(1 - \dfrac{\lambda}{r}\right)$ is given by

$$G_n(z) = \left(\frac{1 - \dfrac{\lambda}{r}}{1 - \dfrac{\lambda z}{r}}\right)^r$$

and

$$\lim_{n \to \infty} \left(\frac{1 - \dfrac{\lambda}{r}}{1 - \dfrac{\lambda z}{r}}\right)^r = e^{\lambda(z-1)},$$

our statement follows from Theorem 4.

The reader may have noticed that the present and the preceding paragraph deal substantially with the same problems. The only difference is that instead of the algebraic point of view the analytical viewpoint is favored here. Obviously, it means the same to say that the distribution $\mathscr{P}$ can be exhibited in the form $\mathcal{P} = G(\mathcal{E}_1)$, where $G(z)$ is a power series of nonnegative coefficients such that $G(1) = 1$ and $\mathcal{E}_1$ denotes the distribution $\{0, 1, 0, \ldots, 0, \ldots\}$, or to say that the distribution $\mathcal{P}$ has the generating function $G(z)$. In dealing with algebraic relations between distributions, the first point of view is entirely sufficient and the analytic point of view is superfluous. If, however, theorems of convergence are considered, the analytic point of view is preferable.

As an example of the application of generating functions, let us consider now a problem taken from the theory of chain reactions. Consider the chain reaction occurring in an electron multiplier. This instrument consists of so-called "screens". If an electron hits a screen, secondary electrons are generated, whose number is a random variable. These electrons hit a second screen, making free new electrons from it, whose number is again a random variable, etc. Suppose that the distribution of the secondary electrons produced by one primary electron is the same for each screen. Calculate the probability that exactly k electrons are produced from the n-th screen. Let ξ_{nr} $(r = 1, 2, \ldots)$ be the number of secondary electrons produced from the n-th screen by the r-th electron; assume that $\xi_{n1}, \xi_{n2}, \ldots$ are independent random variables with the same distribution which take on nonnegative integer values only. Let p_k denote the probability $p_k = P\,(\xi_{nr} = k)$ $(k = 0, 1, \ldots)$. Let further η_n denote the number of electrons issued from the n-th screen. We have then

$$\eta_n = \xi_{n1} + \xi_{n2} + \ldots + \xi_{n\eta_{n-1}}, \tag{45}$$

in fact, the number of electrons emerging from the n-th screen is the sum of the electrons liberated by those emerging from the $(n - 1)$-th screen. Thus the random variable η_n is exhibited as the sum of independent random variables, the number of terms of the sum being equal to the random variable η_{n-1}. Put

$$G(z) = \sum_{k=0}^{\infty} p_k z^k \tag{46}$$

and let $G_n(z)$ be the generating function of η_n. We have $G_1(z) = G(z)$ and it follows from Theorem 3 that

$$G_n(z) = G_{n-1}\big(G(z)\big) \qquad (n = 2, 3, \ldots), \tag{47}$$

hence

$$G_2(z) = G\big(G(z)\big), \qquad G_3(z) = G\big(G(G(z))\big) \quad \text{etc.}$$

The generating function $G_n(z)$ is thus the n-th iterate of $G(z)$. Sometimes it is convenient to employ the recursive formula

$$G_n(z) = G\big(G_{n-1}(z)\big) \qquad (n = 2, 3, \ldots). \tag{48}$$

In general, we have

$$G_{n+m}(z) = G_n(G_m(z)). \tag{49}$$

Let us compute from the generating function $G_n(z)$ the expectation M_n of η_n. Put

$$M = \sum_{k=1}^{\infty} kp_k = G'(1). \tag{50}$$

It is here to be mentioned that an electron multiplication, in the true sense of the word, takes place only if $M > 1$; in fact, only then can it be expected to observe an increase of the number of electrons (cf. the calculations below). In order to calculate M_n differentiate (47), put $z = 1$, then we have

$$M_n = G_n'(1) = G_{n-1}'(1) G'(1) = M_{n-1} M. \tag{51}$$

Consequently

$$M_n = M^n \qquad (n = 1, 2, \ldots). \tag{52}$$

The expectation of the number of electrons emitted from the n-th screen is thus the n-th power of the expectation of the number of electrons emitted from the first screen. For $M > 1$ this expectation increases beyond every bound for $n \to \infty$; for $M < 1$ it tends to 0. In the latter case the process stops sooner or later. Let us see now, what is the probability of this. Let $P_{n,k}$ be the probability that k electrons are emitted from the n-th screen; particularly, we have

$$P_{n,0} = G_n(0). \tag{53}$$

It can be supposed that $G(0) = p_0$ is positive, since if $G(0) = 0$, obviously $P_{n,0} = 0$ for $n = 1, 2, \ldots$.

The sequence $P_{n,0}$ $(n = 1, 2, \ldots)$ is monotone increasing. This can be seen immediately: in fact if no electron is emitted from the n-th screen, the same will hold for the $(n + 1)$-st screen too; the converse, however, is not true. According to (53) we have

$$P_{n+1,0} = G_n(G(0)) > G_n(0) = P_{n,0}, \tag{54}$$

the sequence $P_{n,0}$ is thus monotone increasing. Since for every n $P_{n,0} \leq 1$, the limit

$$\lim_{n \to \infty} P_{n,0} = P \tag{55}$$

exists. It follows from (53) that

$$P_{n,0} = G(P_{n-1,0}); \tag{56}$$

P is therefore a root of the equation

$$P = G(P). \tag{57}$$

Since $G(1) = 1$, 1 is also a root of this equation. We shall show that for $M \leq 1$ there exist no other real roots. In this case therefore the probability

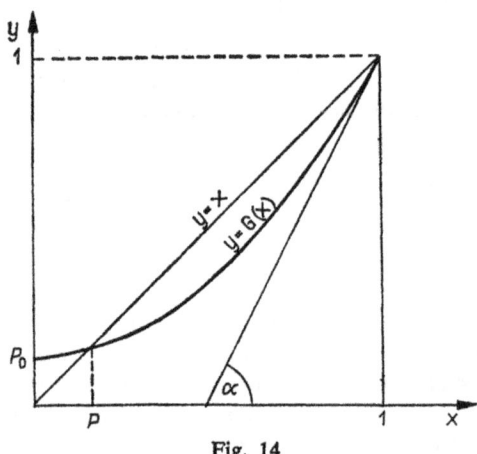

Fig. 14

that no electrons are emitted from the n-th screen, tends to 1 if $n \to \infty$. To prove this draw the curve $y = G(x)$. Since $G(x)$ is a power series with non-negative coefficients, the same holds for all its derivatives, $G(x)$ is therefore monotone increasing in the interval $0 \leq x \leq 1$ and is also convex. The equation $P = G(P)$ means that P is the abscissa of the intersection of the curve $y = G(x)$ and the line $y = x$. Since $G(0) > 0$, $G(x) - x$ is positive for $x = 0$. Now if $G'(1) = M > 1$, $G(x) - x$ is, because of $G(1) = 1$, negative in an appropriate left hand side neighbourhood of the point $x = 1$ (see Fig. 14). As $G(x)$ is continuous, there exists a value $P(0 < P < 1)$ satisfying (57). Because of the convexity of $G(x)$ there can exist no further points of intersection.

It can be proved in the same manner that for $M \leq 1$ Equation (57) has no real roots other than $P = 1$. (There can of course exist complex roots of (57).)

It is yet to be shown that for $M > 1$ the sequence $P_{n,0}$ $(n = 1, 2, \ldots)$ converges to the smaller of the two roots of Equation (57). This can be seen immediately from Fig. 15 by relation (47) which gives in case of $M > 1$ for every z $(0 < z < 1)$ the relation

$$\lim_{n \to \infty} G_n(z) = P, \tag{58}$$

hence

$$\lim_{n \to \infty} P_{n,k} = 0 \qquad (k = 1, 2, \ldots). \tag{59}$$

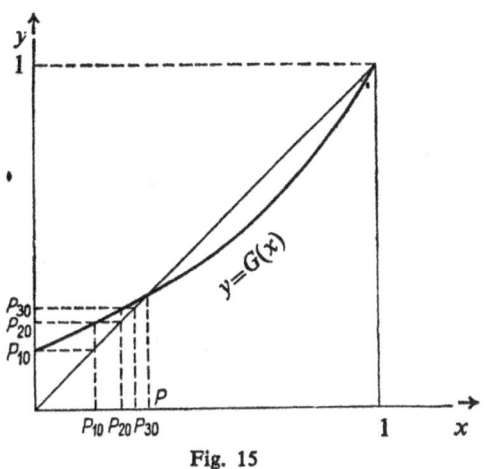

Fig. 15

Thus the probability that from the n-th screen there are exactly $k \geq 1$ electrons issued tends to 0 for $n \to \infty$ for each fixed value of k. From

$$\lim_{n \to \infty} P_{n,0} = P < 1 \quad \text{and} \quad \sum_{k=0}^{\infty} P_{n,k} = 1$$

it follows that for large enough n the number of the emitted electrons (provided that the process did not stop) will be arbitrarily large with a (conditional) probability near to 1. This is in accordance with experience.

§ 16. Approximation of the binomial distribution by the normal distribution

In probability theory *Stirling's formula* is often employed in the following form:

$$n! = \left(\frac{n}{e}\right)^n \sqrt{2\pi n} \, \exp\left(\frac{\theta_n}{12n}\right) \qquad (0 < \theta_n < 1). \tag{1}$$

This can be proved by means of Euler's summation formula and Wallis' formula. We employ Euler's summation formula in the following form:

Let $f(x)$ be a continuously differentiable function in the closed interval $[a, b]$; let further be

$$\varrho(x) = x - [x] - \frac{1}{2}, \tag{2}$$

where $[x]$ denotes as usually the integral part of x; i.e. $[x] = k$ for $k \leq$ $\leq x < k + 1$ $(k = 0, 1, \ldots)$. Then we have

$$\sum_{a < k \leq b} f(k) = \int_a^b f(x)\, dx - [\varrho(b)f(b) - \varrho(a)f(a)] + \int_a^b \varrho(x)f'(x)\, dx. \tag{3}$$

Remark. If $a = A - \frac{1}{2}$, $b = B + \frac{1}{2}$, A and B integers, we have $\varrho(a) = \varrho(b) = 0$ and instead of (3) we may simply write[1]

$$\sum_{k=A}^{B} f(k) = \int_{A-\frac{1}{2}}^{B+\frac{1}{2}} f(x)\, dx + \int_{A-\frac{1}{2}}^{B+\frac{1}{2}} \varrho(x)f'(x)\, dx. \tag{4}$$

In the present paragraph an approximation will be given for the terms

$$W_k = \binom{n}{k} p^k q^{n-k} \qquad (k = 0, 1, \ldots, n; \; 0 < p < 1; \; q = 1 - p) \tag{5}$$

of the binomial distribution.

Put $z = k - np$, hence

$$k = np + z \quad \text{and} \quad n - k = nq - z. \tag{6}$$

Evaluating asymptotically the binomial coefficient $\binom{n}{k}$ figuring in (5) by Stirling's formula, a simple calculation gives

$$W_k = \sqrt{\frac{n}{2\pi\,(np + z)\,(nq - z)}} \left(1 - \frac{z}{np + z}\right)^{np+z} \left(1 + \frac{z}{nq - z}\right)^{nq-z} e^\delta \tag{7}$$

with

$$\delta = \frac{\theta_n}{12n} - \frac{\theta_k}{12k} - \frac{\theta_{n-k}}{12(n - k)}, \tag{8}$$

where θ_n is defined by (1). We assume that the quantity

$$x = \frac{z}{\sqrt{npq}} \tag{9}$$

---

[1] The proof of this formula can be found e.g. in K. KNOPP [1].

remains bounded:

$$|x| \leq A \qquad (A = \text{constant}). \tag{10}$$

For the different factors on the right hand side of (7) we obtain

$$\sqrt{\frac{n}{2\pi(np+z)(nq-z)}} = \frac{1}{\sqrt{2\pi npq}}\left[1 - \frac{x(q-p)}{2\sqrt{npq}} + O\left(\frac{1}{n}\right)\right] \tag{11}$$

and

$$\left(1 - \frac{z}{np+z}\right)^{np+z}\left(1 + \frac{z}{nq-z}\right)^{nq-z} =$$

$$= e^{-\frac{x^2}{2}}\left[1 + \frac{(q-p)x^3}{6\sqrt{npq}} + O\left(\frac{1}{n}\right)\right]. \tag{12}$$

According to assumption (10) we have[1]

$$\delta = O\left(\frac{1}{n}\right), \tag{13}$$

and the constant figuring in the residual term $O\left(\frac{1}{n}\right)$ in Equations (11), (12) and (13) depends on A only; thus we obtain from the relations (7), (11), (12), and (13) the following theorem:

THEOREM 1. *If* $0 < p < 1$, $q = 1 - p$, *and*

$$W_k = \binom{n}{k} p^k q^{n-k} \qquad (k = 0, 1, \ldots, n), \tag{14}$$

further if

$$|x| = \left|\frac{k-np}{\sqrt{npq}}\right| \leq A, \tag{15}$$

then

$$W_k = \frac{e^{-\frac{x^2}{2}}}{\sqrt{2\pi npq}}\left[1 + \frac{(x^3 - 3x)(q-p)}{6\sqrt{npq}} + O\left(\frac{1}{n}\right)\right], \tag{16}$$

[1] Here, as well as in what follows, the notation $a_N = O(b_N)$ is employed. If a_N and b_N ($N = 1, 2, \ldots$) are sequences of numbers such that $b_N \neq 0$ and there exists a constant $C > 0$ for which $|a_N| \leq C |b_N|$, this fact will be denoted by $a_N = O(b_N)$. (Read: "a_N is of order not exceeding that of b_N".) If, however, $\lim_{N \to \infty} \frac{a_N}{b_N} = 0$, this will be denoted by $a_N = o(b_N)$. (Read: "a_N is of smaller order than b_N".)

where the constant intervening in $O\left(\dfrac{1}{n}\right)$ depends on A only.

In practice, usually the following weaker form

$$W_k = \frac{\exp\left[-\dfrac{(k-np)^2}{2npq}\right]}{\sqrt{2\pi npq}}\left[1 + O\left(\frac{1}{\sqrt{n}}\right)\right] \qquad (17)$$

of Theorem 1 suffices.

Thus the probabilities $\binom{n}{k}p^k q^{n-k}$ are approximated by values of the function

$$f(x) = \frac{1}{\sqrt{2\pi}\sigma}\exp\left[-(x-m)^2/2\sigma^2\right] \qquad (18)$$

at the point $x = k$, where the constants m and σ have the values $m = np$ and $\sigma = \sqrt{npq}$. This function is represented graphically by a bell-shaped curve (Fig. 16) called Gauss' curve (or Laplace curve, or "normal" curve). Function (18) plays a central role in probability theory.

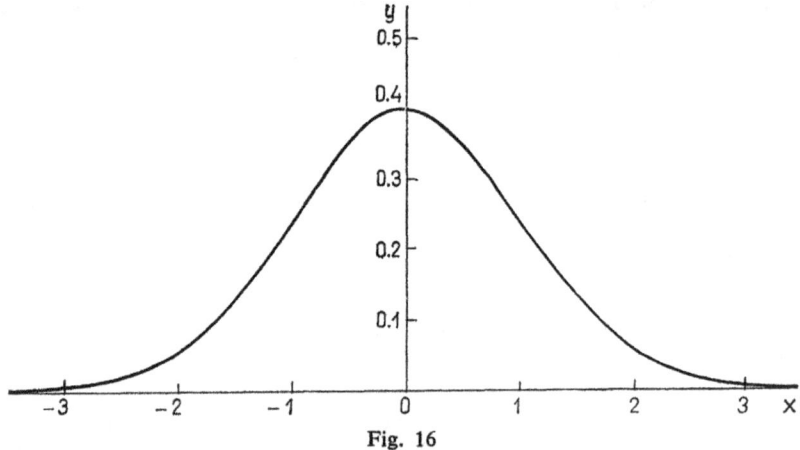

Fig. 16

Theorem 1 can be "verified" experimentally for $p = \dfrac{1}{2}$ by means of Galton's desk.

Galton's desk (Fig. 17) is a triangular inclined plane provided with nails arranged regularly in n horizontal lines, the k-th line containing k nails. A ball, launched from the vertex of the triangle, will be diverted at every line either to the left or to the right, with the same probability $\frac{1}{2}$. Under the last line of nails there follows a line of $n + 1$ boxes in which the balls are accumulated. In order to fall into the k-th box (numbered from the left, $k = 0, 1, \ldots, n$) a ball has to be diverted k times to the right and $n - k$

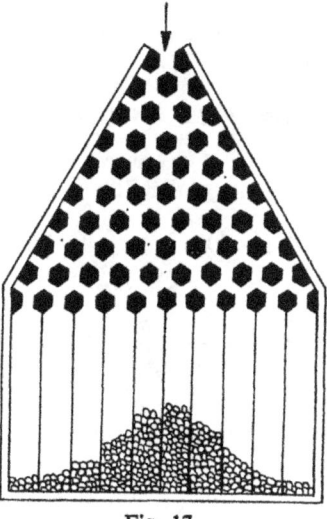

Fig. 17

times to the left. If the directions taken at each of the lines are independent, the probability of this event will be $\binom{n}{k} 2^{-n}$. By letting a large enough number of balls roll down Galton's desk, their distribution in the boxes exhibits quite neatly a curve similar to the Laplace–Gauss curve. Theorem 1 states that the limit relation

$$\lim_{n \to \infty} \frac{\binom{n}{k} p^k q^{n-k}}{\frac{1}{\sqrt{2\pi npq}} \exp\left(-\left[\frac{(k - np)^2}{2npq}\right]\right)} = 1 \tag{19}$$

holds, if with n also k tends to infinity so that

$$\frac{|k - np|}{\sqrt{npq}}$$

remains bounded; with these conditions the convergence is even uniform. Formula (19) is the so-called de Moivre–Laplace theorem.

This result can be expressed in a more concise and practical, though weaker form. Let W_k be the probability that during n repetitions of an alternative the event A occurs exactly k times. If n is very large, it is more reasonable to ask for the probability that the number k of occurrences of A lies between two given limits, than for the probability that it assumes a fixed value.

Our problem may conveniently be phrased as follows: what is the probability that the inequality

$$np + a\sqrt{npq} \leq k < np + b\sqrt{npq} \qquad (20)$$

should hold, where a and b ($a < b$) are two given real numbers. It follows from Formula (16) that this probability is

$$W^{(n)}(a, b) = \sum_{a \leq \frac{k-np}{\sqrt{npq}} < b} W_k =$$

$$= \frac{1}{\sqrt{2\pi npq}} \sum_{a \leq x_k < b} e^{-\frac{x_k^2}{2}} \left(1 + \frac{q-p}{6\sqrt{npq}}(x_k^3 - 3x_k) + O\left[\frac{1}{n}\right]\right), \qquad (21)$$

where $x_k = \dfrac{k - np}{\sqrt{npq}}$ was substituted. It will be seen that there exists a limit

$$\lim_{n \to \infty} W^{(n)}(a, b) = W(a, b) \qquad (22)$$

which can be calculated and the residual term estimated.

Choose the numbers a and b such that

$$A = \frac{1}{2} + np + a\sqrt{npq} \quad \text{and} \quad B = -\frac{1}{2} + np + b\sqrt{npq}$$

are integers; this can always be done without changing the value of $W^{(n)}(a, b)$. It follows then from (4) that

$$W^{(n)}(a, b) = \frac{1}{\sqrt{2\pi}} \int_a^b e^{-\frac{x^2}{2}} \left(1 + \frac{(q-p)(x^3 - 3x)}{6\sqrt{npq}}\right) dx + O\left(\frac{1}{n}\right). \qquad (23)$$

Since $\int e^{-\frac{x^2}{2}}(x^3 - 3x)dx$ can be given explicitly, we have

THEOREM 2. *If*

$$np + a\sqrt{npq} + \frac{1}{2} = A \qquad and \qquad np + b\sqrt{npq} - \frac{1}{2} = B$$

are integers $(A < B)$, then

$$\sum_{A \leq k \leq B} \binom{n}{k} p^k q^{n-k} = \frac{1}{\sqrt{2\pi}} \int_a^b e^{-\frac{x^2}{2}} dx + R \qquad (24a)$$

where

$$R = \frac{q-p}{6\sqrt{npq}} \left((1-b^2) e^{-\frac{b^2}{2}} - (1-a^2) e^{-\frac{a^2}{2}} \right) + O\left(\frac{1}{n}\right). \qquad (24b)$$

From (24a) and (24b) follows the limit relation

$$\lim_{n \to \infty} \sum_{\alpha \leq \frac{k-np}{\sqrt{npq}} \leq \beta} \binom{n}{k} p^k q^{n-k} = \frac{1}{\sqrt{2\pi}} \int_\alpha^\beta e^{-\frac{x^2}{2}} dx \qquad (25)$$

for each given pair (α, β) of real numbers $(\alpha < \beta)$; it suffices in fact to replace α by a_n, β by b_n, where a_n is the least number such that

$$A = np + a_n \sqrt{npq} + \frac{1}{2} \geq np + \alpha \sqrt{npq}$$

(A integer) and b_n is the largest number such that (B integer)

$$B = np + b_n \sqrt{npq} - \frac{1}{2} \leq np + \beta \sqrt{npq}.$$

Obviously,

$$a_n - \alpha = O\left(\frac{1}{\sqrt{n}}\right) \qquad \text{and} \qquad b_n - \beta = O\left(\frac{1}{\sqrt{n}}\right),$$

hence

$$\lim_{n \to \infty} \int_{a_n}^{b_n} e^{-\frac{x^2}{2}} dx = \int_\alpha^\beta e^{-\frac{x^2}{2}} dx.$$

Thus the right hand side of (25) gives an approximate value for the probability that the number of occurrences of an event A (having the probability $P(A) = p$) in an experiment consisting of n independent trials, lies between the limits $np + \alpha \sqrt{npq}$ and $np + \beta \sqrt{npq}$. To use this result we must have the values of the integral

$$\frac{1}{\sqrt{2\pi}} \int_\alpha^\beta e^{-\frac{x^2}{2}} dx$$

for every pair (α, β). The integral $\int e^{-\frac{x^2}{2}} dx$ cannot be expressed by elementary functions; however, the function

$$\Phi(y) = \frac{1}{\sqrt{2\pi}} \int_{-\infty}^{y} e^{-\frac{x^2}{2}} \, dx \tag{26}$$

is tabulated with a great precision and a table of its values is given at the end of this volume (cf. Table 6). The curve $y = \Phi(x)$ is shown in Fig. 18.

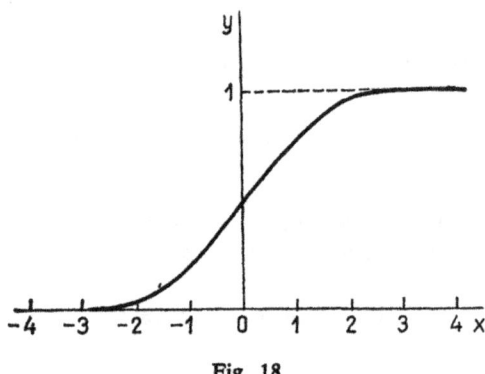

Fig. 18

It is easy to see that

$$\Phi(+\infty) = \frac{1}{\sqrt{2\pi}} \int_{-\infty}^{+\infty} e^{-\frac{x^2}{2}} \, dx = 1. \tag{27}$$

In fact, when introducing polar coordinates we find

$$\Phi^2(+\infty) = \frac{1}{2\pi} \int_{-\infty}^{+\infty} \int_{-\infty}^{+\infty} e^{-\frac{x^2+y^2}{2}} \, dx \, dy = \int_{0}^{\infty} re^{-\frac{r^2}{2}} \, dr = 1.$$

From Theorem 2 follows immediately

THEOREM 3. *For every real y*

$$\lim_{n \to \infty} \sum_{\frac{k-np}{\sqrt{npq}} \le y} \binom{n}{k} p^k q^{n-k} = \Phi(y). \tag{28}$$

In fact, it follows from (25) that for every sufficiently large N

$$\lim_{n\to\infty} \sum_{\frac{k-np}{\sqrt{npq}} \leq y} \binom{n}{k} p^k q^{n-k} \geq \Phi(y) - \Phi(-N) \tag{29}$$

and

$$\overline{\lim_{n\to\infty}} \sum_{\frac{k-np}{\sqrt{npq}} \leq y} \binom{n}{k} p^k q^{n-k} \leq 1 - \Phi(N) + \Phi(y). \tag{30}$$

Because of (27), (28) follows.

The function $\Phi(x)$ is one of the most important distribution functions. A random variable ξ having for its distribution function $\Phi(x)$ or more generally $\Phi\left(\dfrac{x-m}{\sigma}\right)$ (with $\sigma > 0$ and m an arbitrary real number) is said to be *normally distributed* or a *Gaussian random variable*. $\Phi(x)$ itself is called the standard *normal* or *Gaussian* distribution function, or distribution function of Laplace–Gauss.

Theorem 2 may be expressed also by saying that for large n the binomial distribution is approximated by the normal distribution.

§ 17. Bernoulli's law of large numbers

The results obtained in the preceding paragraph allow us to prove a very important theorem, Bernoulli's "law of large numbers".

THEOREM 1. *Let the event A be one of the possible outcomes of an experiment, with probability*

$$P(A) = p \qquad (0 < p < 1)$$

and let $f_A(n)$ be the relative frequency of the event A in a sequence of n independent repetitions of the experiment. Given two arbitrarily small positive numbers ε and δ, there exists a number N depending on ε and δ only such that for $n \geq N$

$$P\big(|f_A(n) - P(A)| < \varepsilon\big) \geq 1 - \delta. \tag{1}$$

PROOF. We have

$$P\big(|f_A(n) - P(A)| < \varepsilon\big) = \sum_{|k-np| < n\varepsilon} \binom{n}{k} p^k q^{n-k}.$$

Choose a number Y such that

$$\Phi(Y) - \Phi(-Y) \geq 1 - \frac{\delta}{2}. \tag{2}$$

For $n \geq \dfrac{Y^2 pq}{\varepsilon^2} = N_1$ we have $Y\sqrt{npq} < n\varepsilon$, and it follows

$$P(\,|f_A(n) - P(A)\,| < \varepsilon) \geq \sum_{|k-np| < Y\sqrt{npq}} \binom{n}{k} p^k q^{n-k}. \tag{3}$$

According to Formula (25) of § 16 N_2 can be chosen such that for $n \geq N_2$

$$\sum_{|k-np| \leq Y\sqrt{npq}} \binom{n}{k} p^k q^{n-k} \geq \Phi(Y) - \Phi(-Y) - \frac{\delta}{2}\,; \tag{4}$$

from (2), (3) and (4) it follows that (1) is verified for $n \geq N = \max\,(N_1, N_2)$.

Bernoulli's law of large numbers can also be proved directly, without the use of the de Moivre–Laplace theorem.

Formula (1) is equivalent to

$$\sum_{|k-np| < \varepsilon n} \binom{n}{k} p^k q^{n-k} \geq 1 - \delta \tag{5}$$

for every $n \geq N$.

The identity

$$\sum_{k=0}^{n} (k - np)^2 \binom{n}{k} p^k q^{n-k} = npq \tag{6}$$

(given as relation (5) in § 9) states that the variance of the binomial distribution is equal to npq. Thus we have

$$npq \geq \sum_{|k-np| \geq \varepsilon n} \binom{n}{k} p^k q^{n-k} (k - np)^2 \geq \varepsilon^2 n^2 \sum_{|k-np| \geq \varepsilon n} \binom{n}{k} p^k q^{n-k}$$

and, consequently,

$$\sum_{|k-np| < \varepsilon n} \binom{n}{k} p^k q^{n-k} = 1 - \sum_{|k-np| \geq \varepsilon n} \binom{n}{k} p^k q^{n-k} \geq 1 - \frac{pq}{\varepsilon^2 n}\,.$$

Thus for $n \geq \dfrac{pq}{\varepsilon^2 \delta} = N$ relation (1) is verified. Since $pq = p(1 - p) \leq$ $\leq \dfrac{1}{4}$, it suffices to take for N the value $N = \dfrac{1}{4\varepsilon^2 \delta}$. We shall see in Chapter VI that one can take for N a much smaller value as well.

The method of proof employed above is often used in probability theory. Later on (in Ch. VII) it will be formulated in a more general form as the inequality of Tchebychev.

Finally, some remarks should be added here concerning Bernoulli's law of large numbers.

In introducing the concept of *probability*, a number called probability was assigned to events whose relative frequency possessed a certain stability in the course of a long sequence of experiments. This stability of relative frequency is now proved mathematically. It is quite remarkable that the theory leads to a precise formulation of this stability; it is undoubtedly a proof of its power.

At the same time it can be understood, why the stability of relative frequency could not be defined precisely at the introduction of the concept of probability. Indeed, in its formulation occurs the concept of probability: the law of large numbers states just that after a long sequence of experiments a large deviation between relative frequency and probability becomes very improbable.

It may seem that there lurks some vicious circle here: probability was indeed defined by means of the stability of relative frequency, and yet in the definition of the stability of relative frequency the concept of probability is hidden. In reality there is no logical fault. The "definition" of the probability stating that the probability is the numerical value around which the relative frequency is fluctuating at random is not a mathematical definition: it is an intuitive description of the realistic background of the concept of probability. Bernoulli's law of large numbers, on the other hand, is a theorem deduced from the mathematical concept of probability; there is thus no vicious circle.

The theorem dealt with above is a particular case of more general theorems which will be discussed in Chapter VI. Similarly, the approximation of the binomial distribution by the normal distribution is a particular case of the general limit theorems, to be dealt with in Chapter VII of the present book.

§ 18. Exercises

1. Suppose a calculator is so good that he does not make more than three errors in the average in doing 1000 additions. Suppose he checks his additions by testing the addition modulo 9 and corrects the errors thus discovered. There can, however, still remain undetected errors: in fact, it may occur that the erroneous result differs from the exact sum by a multiple of 9. How many errors remain in the average among his additions?

Hint. It can be assumed that, if the sum is erroneous, the error lies with an equal probability $\frac{1}{9}$ in any of the residue classes 0, 1, 2, 3, 4, 5, 6, 7, 8 mod 9. Let A denote the event "the sum is erroneous", B the event "the error could be detected by testing the sum modulo 9". The probability sought is the conditional probability $P(A \mid \bar{B})$; according to Bayes' rule it has the value $\frac{1}{2992}$.

2. A missing letter is to be found with the probability P in one of the eight drawers of a secretary. Suppose that seven drawers were already tried in vain. What is the probability to find the letter in the last drawer?

3. Determine the maximal term of the binomial, multinomial, hypergeometric and polyhypergeometric distributions.

4. Consider the terms

$$P_{k_1, k_2, \ldots, k_r} = \frac{\binom{N_1}{k_1} \binom{N_2}{k_2} \cdots \binom{N_r}{k_r}}{\binom{N}{n}}$$

of the polyhypergeometric distribution, where

$$k_1 + k_2 + \ldots + k_r = n, \quad N_1 + N_2 + \ldots + N_r = N.$$

Prove that if the numbers N_j $(j = 1, 2, \ldots, r)$ tend to infinity so that

$$\lim_{N \to \infty} \frac{N_j}{N} = p_j > 0 \quad (j = 1, 2, \ldots, r),$$

we have for fixed values of $k_1, k_2, \ldots, k_r$

$$\lim P_{k_1, k_2, \ldots, k_r} = \frac{n!}{k_1! \, k_2! \ldots k_r!} \, p_1^{k_1} p_2^{k_2} \cdots p_r^{k_r}.$$

Thus under the above conditions the terms of the polyhypergeometric distribution converge to the corresponding terms of the multinomial distribution.

5. If

$$\lim_{n \to +\infty} n p_j = \lambda_j > 0 \quad (j = 1, 2, \ldots, r - 1),$$

the multinomial distribution

$$\left\{ \frac{n!}{k_1! \, k_2! \ldots k_r!} \, p_1^{k_1} \cdots p_r^{k_r} \right\}$$

tends to an $(r - 1)$-dimensional Poisson distribution. That is, for fixed $k_1, k_2, \ldots, k_{r-1}$ $(k_r = n - k_1 - k_2 - \ldots - k_{r-1})$ we have for $n \to +\infty$

$$\lim_{n \to +\infty} \frac{n!}{k_1! \, k_2! \ldots k_r!} \, p_1^{k_1} p_2^{k_2} \cdots p_r^{k_r} = \frac{\lambda_1^{k_1} \cdots \lambda_{r-1}^{k_{r-1}}}{k_1! \ldots k_{r-1}!} e^{-(\lambda_1 + \ldots + \lambda_{r-1})}.$$

6. Deduce Formula (4) of § 4 from Formula (12) of § 12, using the convergence of the binomial distribution to the Poisson distribution.

7. Determine the maximal term of the Poisson distribution $\dfrac{\lambda^k e^{-\lambda}}{k!}$ $(k = 0, 1, \ldots;$ $\lambda > 0)$.

8. If λ is constant and $N = n \ln n + \lambda n$, there exists a limit of the probabilities $P_k(n, N)$ (cf. Ch. II, § 12, Exercise 42.b) for $n \to \infty$: we have for any fixed real value of λ and any fixed nonnegative integer k

$$\lim_{n \to \infty} P_k(n, n \ln n + \lambda n) = \frac{(e^{-\lambda})^k \exp(-e^{-\lambda})}{k!} \quad (k = 0, 1, \ldots).$$

Thus, if we distribute $N = n$ In $n + \lambda n$ balls into n urns, the number of the urns which remain empty will be, for large n, approximately distributed according to a Poisson distribution with expectation $e^{-\lambda}$.

9. If $\dfrac{M}{N} = p$, $\dfrac{R}{N} = r$, and $n \to \infty$ so that

$$\lim_{n \to +\infty} np = \lambda > 0 \quad \text{and} \quad \lim_{n \to \infty} nr = \mu > 0$$

then

$$\lim_{n \to \infty} \binom{n}{k} \cdot \frac{\prod\limits_{j=0}^{k-1}(M+jR) \prod\limits_{j=0}^{n-k-1}(N-M+jR)}{\prod\limits_{j=0}^{n-1}(N+jR)} =$$

$$= \binom{\frac{\lambda}{\mu}+k-1}{k}\left(\frac{1}{1+\mu}\right)^{\frac{\lambda}{\mu}}\left(\frac{\mu}{1+\mu}\right)^{k}.$$

Thus under the above conditions, the Pólya distribution tends to a negative binomial distribution. If $\mu = 0$, the above limit becomes $\dfrac{\lambda^k e^{-\lambda}}{k!}$; the limit distribution is then a Poisson distribution.

10. A roll of a certain fabric contains in the average five faults per 100 yards . The cloth will be cut into pieces of 3 yards. How many faultless pieces does one expect to find?

Hint. It can be supposed that the number of faults has a Poisson distribution. The probability of finding k faults in an x yards long piece is therefore equal to

$$\frac{\left(\dfrac{x}{20}\right)^k e^{-\frac{x}{20}}}{k!} \qquad (k = 0, 1, \ldots).$$

11. In a forest there are on the average ten trees per 100 m². For sake of simplicity suppose that all trees have a circular section with a diameter of 20 cm. A gun is fired in a direction in which the edge of the forest is 100 m away. What is the probability that the shot will hit a trunk?

Hint. It can be assumed that the trees have a Poisson distribution; the probability that on a surface area T m² there are k trees is equal to

$$\frac{\left(\dfrac{T}{10}\right)^k e^{-\frac{T}{10}}}{k!} \qquad (k = 0, 1, \ldots).$$

Each tree can be considered and represented by its centre.

12. In a summer evening there can be observed on the average one shooting star in every ten minutes. What is the probability to observe two during a quarter of an hour?

13. At a certain post office 1017 letters without address were posted during one year. Estimate the number of days on which more than two letters without address were posted.

14. Let $\mathcal{B}_1(p) = \{p, 1 - p\}$ be a binomial distribution of order 1; let $g(z) =$
$$= \frac{1 - \alpha}{1 - \alpha z}.$$ Determine the distribution $g[\mathcal{B}_1(p)]$.

15. Let $\mathcal{B}_1(p)$ be the same as in the preceding exercise. Show that

$$\exp\left[\lambda(\mathcal{B}_1(p) - 1)\right] = \mathcal{P}(\lambda p)$$

is the Poisson distribution with expectation λp.

16. Let p be the probability of an event A. Perform n independent trials and denote by f the relative frequency of A in this sequence of trials. With the aid of the approximation of the binomial distribution by the normal distribution answer the following questions:

a) If $p = 0.4$ and $n = 1500$, what is the probability for f to lie between 0.40 and 0.44?

b) If $p = 0.375$, how many independent trials have to be performed in order that the probability of $|f - p| \leq 0.01$ is at least 0.995?

c) Let $p = \dfrac{2}{3}$, $n = 1200$. How should ε be chosen in order that the probability of $|f - p| < \varepsilon$ be at least 0.985?

d) Suppose $n = 14\,400$. For which values of p will the probability of $|f - p| < 0.01$ be at least 0.99?

17. Put

$$\Phi(x) = \frac{1}{\sqrt{2\pi}} \int_{-\infty}^{x} e^{-\frac{u^2}{2}}\, du.$$

Prove that the expansion

$$\Phi(x) = \frac{1}{2} + \frac{1}{\sqrt{2\pi}}\, e^{-\frac{x^2}{2}} \left(\frac{x}{1} + \frac{x^3}{1 \cdot 3} + \frac{x^5}{1 \cdot 3 \cdot 5} + \cdots \right)$$

holds.

18. Prove that for $x > 1$,

$$\Phi(x) = 1 - \frac{e^{-\frac{x^2}{2}}}{\sqrt{2\pi}\, x \left(1 + \dfrac{\theta}{x^2} \right)} \qquad \text{where} \quad 0 < \theta < 1.$$

Hint. Use the identity

$$\int_{x}^{\infty} \left(1 + \frac{1}{u^2} \right) e^{-\frac{u^2}{2}}\, du = \frac{1}{x}\, e^{-\frac{x^2}{2}}.$$

19. Suppose $P(A) = \dfrac{1}{2}$. Perform N independent experiments; let $f(n)$ be the number of the occurrences of A among the first n experiments $(n = 1, 2, \ldots, N)$; put $f(0) = 0$. Verify the formula

$$P_N(M) = P\Big(\max_{1 \le n \le N} (2f(n) - n) < M \Big) = 1 - \frac{1}{2M} \sum_{k=0}^{[\frac{1}{2}(N-M)]} \binom{M + 2k}{k} \frac{M}{M + 2k} \cdot \frac{1}{4^k}$$

for $M = 1, 2, \ldots$.

20. Applying the result of Exercise 19 prove that

$$\lim_{N \to +\infty} P_N(x \sqrt{N}) = 2\Phi(x) - 1 \qquad (x > 0).$$

21. The function of two variables $\Psi(x, y) = \Phi\left(\dfrac{x}{\sqrt{y}}\right)$ fulfils the partial differential equation of heat-conduction

$$\frac{\partial \Psi}{\partial y} = \frac{1}{2} \frac{\partial^2 \Psi}{\partial x^2} ;$$

the function

$$U(x, n) = \sum_{k < \frac{n+x}{2}} \binom{n}{k} \frac{1}{2^n}$$

fulfils the difference equation

$$\Delta_n U = \frac{1}{2} \Delta_x^2 U$$

where

$$\Delta_n U = U(x, n + 1) - U(x, n),$$
$$\Delta_x^2 U = U(x + 1, n) - 2U(x, n) + U(x - 1, n).$$

22. Prove the asymptotic relation

$$\binom{n}{k} p^k q^{n-k} \approx \frac{1}{\sqrt{2\pi\, npq}} \exp\left[-\frac{(k - np)^2}{2npq} \right]$$

under the following conditions: p is a constant, n and k tend to infinity so that

$$\lim_{n \to +\infty} \frac{(k - np)^3}{n^2} = 0 .$$

23. Prove the asymptotic relation

$$\frac{\dbinom{M}{k} \dbinom{N - M}{n - k}}{\dbinom{N}{n}} \approx \frac{1}{\sqrt{2\pi\, npq}} \exp\left[-\frac{(k - np)^2}{2npq} \right] .$$

where

$$N \to \infty, \quad M = pN, \quad 0 < p < 1, \quad q = 1 - p, \quad n \to \infty, \quad n = o(\sqrt{N})$$

and
$$|k - np| = o(n^{\frac{2}{3}}).$$

Thus also the hypergeometric distribution can be approximated by the normal distribution.

24. Establish the following power series expansion:

$$\Phi(x) =$$

$$= \frac{1}{2} + \frac{1}{\sqrt{2\pi}}\left[x - \frac{x^3}{1!\,2\cdot 3} + \frac{x^5}{2!\,4\cdot 5} - \frac{x^7}{3!\,8\cdot 7} + \cdots + \frac{(-1)^k x^{2k+1}}{k!\,2^k(2k+1)} + \cdots\right].$$

How many terms are to be taken to calculate $\Phi(2)$ with an accuracy of four decimal digits?

25. If x is positive, the difference

$$1 - \Phi(x) - \frac{1}{\sqrt{2\pi}}\, e^{-\frac{x^2}{2}} \times$$

$$\times \left[\frac{1}{x} - \frac{1}{x^3} + \frac{1\cdot 3}{x^5} - \frac{1\cdot 3\cdot 5}{x^7} + \cdots + \frac{(-1)^k\, 1\cdot 3\ldots(2k-1)}{x^{2k+1}}\right]$$

is positive or negative according as k is odd or even. The value of the function $1 - \Phi(x)$ is thus always contained between the k-th and $(k+1)$-st partial sums of the divergent series

$$\frac{1}{\sqrt{2\pi}}\, e^{-\frac{x^2}{2}}\left[\frac{1}{x} - \frac{1}{x^3} + \frac{1\cdot 3}{x^5} - \frac{1\cdot 3\cdot 5}{x^7} + \cdots\right].$$

How many terms are there to be taken to calculate $\Phi(4)$ with an accuracy of 10^{-8}?

26. Show that

$$\frac{1 + \left(1 - e^{-\frac{x^2}{2}}\right)^{\frac{1}{2}}}{2} < \Phi(x) < \frac{1 + (1 - e^{-x^2})^{\frac{1}{2}}}{2} \qquad \text{for} \quad x > 0.$$

27. (Method of Laplace.) Let $f(x)$ be a complex valued function continuous in a neighbourhood of $x = a$ such that

$$f(a) = 1, \quad |f(x)| < 1 \quad \text{for } x \neq a, \quad f''(a) = -b < 0.$$

Let further $\{g_n(t)\}$ be a sequence of complex-valued functions such that

$$\lim_{n \to \infty} g_n\!\left(a + \frac{t}{\sqrt{nb)}}\right) = A$$

be uniformly fulfilled for t in every finite interval. Show that for every $x > 0$, $y > 0$ we have

$$\lim_{n \to \infty} \sqrt{\frac{nb}{2\pi}} \int_{a-\frac{y}{\sqrt{nb}}}^{a+\frac{x}{\sqrt{nb}}} g_n(t)\, f^n(t)\, dt = \frac{A}{\sqrt{2\pi}} \int_{-y}^{x} e^{-\frac{u^2}{2}}\, du.$$

28. Show that for every real value of x

$$\lim_{\lambda \to \infty} \sum_{k < \lambda + x\sqrt{\lambda}} \frac{\lambda^k e^{-\lambda}}{k!} = \Phi(x).$$

Hint. Use the result of the preceding Exercise.

29. Show with the aid of the result of Exercise 27 and the relation

$$\sum_{j=0}^{k} \binom{n}{j} p^j q^{n-j} = (n-k) \binom{n}{k} \int_{p}^{1} t^k (1-t)^{n-k-1} \, dt$$

$(0 < p < 1; q = 1 - p)$ directly the validity of the limit relation

$$\lim_{n \to \infty} \sum_{\frac{k-np}{\sqrt{npq}} < x} \binom{n}{k} p^k q^{n-k} = \Phi(x).$$

30. Prove the following strong form of Stirling's formula

$$n! = \left(\frac{n}{e}\right)^n \sqrt{2\pi n} \, \exp\left(-\frac{1}{12n} - \frac{\gamma_n}{360n^3}\right), \, 0 < \gamma_n < 1.$$

31. In a factory at any instant t each of n machines is at work with probability p and is under repair with probability $1 - p$. The machines are operated independently of each other. What is the probability that at a given instant at least m machines are working? Calculate an approximate value of this probability for $n = 200$; $p = 0.95$; $m = 180$.

32. Prove the well-known approximation-theorem of Weierstrass in the following manner: Show with the aid of the inequality

$$\sum_{|k-np| > n\varepsilon} \binom{n}{k} p^k q^{n-k} < \frac{p(1-p)}{n\varepsilon^2}$$

deduced in course of the proof of Bernoulli's law of large numbers, that for any function $f(x)$, continuous in the closed interval $[0, 1]$, the so-called *Bernstein poly-nomials* of $f(x)$

$$B_n(x) = \sum_{k=0}^{n} \binom{n}{k} f\left(\frac{k}{n}\right) x^k (1-x)^{n-k}$$

converge uniformly to $f(x)$ in the interval $[0, 1]$ if $n \to \infty$.

33. Find the limit

$$B(\alpha) = \lim_{n \to \infty} \frac{n^{\frac{\alpha-1}{2}}}{2^{n\alpha}} \sum_{k=0}^{n} \binom{n}{k}^{\alpha} \quad \text{for} \quad \alpha > 0.$$

34. The following problem occurs in statistical mechanics: Let $E_1, E_2, \ldots, E_n$ be the possible values of the energy of a particle belonging to a system of N particles. If a particle has the energy E_k, it is said to be in the k-th state. The state of the whole system can be characterized by giving the "occupation numbers" $N_1, N_2, \ldots, N_n$;

N_k being the number of particles having the energy E_k ($k = 1, 2, \ldots, n$). Let W_k be the probability of a particle being in the state k, $W(N_1, N_2, \ldots, N_n)$ the probability that the system is in the state characterized by the occupation numbers $N_1, N_2, \ldots, N_n$. By assuming that the states of the particles are independent from each other, we have, obviously

$$W(N_1, N_2, \ldots, N_n) = \frac{N!}{N_1! \, N_2! \ldots N_n!} \, W_1^{N_1} \, W_2^{N_2} \ldots W_n^{N_n}, \tag{1}$$

with

$$N = N_1 + N_2 + \ldots + N_n. \tag{2}$$

The probabilities of the possible states have therefore a multinomial distribution. If, however, the total energy E of the system is given, not all these states can actually occur: besides condition (2) the following one must be fulfilled, too:

$$\sum_{k=1}^{n} N_k \, E_k = E. \tag{3}$$

According to the definition of the conditional probability, probabilities (1) fulfilling (3) are simply multiplied by a constant factor. Find the values of $N_1, N_2, \ldots, N_n$ fulfilling (2) and (3) for which the expression (1) takes on its maximal value.

Hint. Consider the numbers N_k ($k = 1, 2, \ldots, n$) as continuous variables and replace in (1) the factorials $N_k!$ by $\Gamma(N_k + 1)$, then apply the well-known identity

$$\ln \Gamma(N + 1) = \left(N + \frac{1}{2}\right) \ln N - N + \ln \sqrt{2\pi} - \int_0^\infty \frac{\varrho(x)dx}{x + N},$$

where $\varrho(x) = x - [x] - \frac{1}{2}$. By differentiation[1] we obtain

$$\frac{\Gamma'(N + 1)}{\Gamma(N + 1)} = \ln N + \frac{1}{2N} + \int_0^\infty \frac{\varrho(x)dx}{(x + N)^2}.$$

By Lagrange's method of multipliers, the conditional extremum of (1) under the conditions (2) and (3) can be found. Thus it follows that

$$N_k \approx N \frac{W_k \, e^{-\beta E_k}}{\sum_{j=1}^{n} W_j \, e^{-\beta E_j}},$$

where the constant β must be chosen so that (3) is satisfied. This is *Boltzmann's energy distribution.*

35. Let $0 < p < \frac{1}{2}$, $q = 1 - p$, n an integer such that $np = m$ is also an integer. Let A be an event with probability p. Show that during the course of n independent

---

[1] Often this formula is deduced by differentiating Stirling's formula; this of course is inadmissible. Our procedure is correct, since we differentiate an identity and not an asymptotic relation.

repetitions of an experiment, having possible outcomes A and $\bar{A}$, the probability that A occurs less than m times is greater than the probability that A occurs more than m times (*Simmons' Theorem*).

Hint. The inequality to be proven is

$$\sum_{k=0}^{m-1} \binom{n}{k} p^k q^{n-k} > \sum_{k=m+1}^{n} \binom{n}{k} p^k q^{n-k}.$$

By putting   (1)

$$B_r = \binom{n}{m-r} p^{m-r} q^{n-m+r} \qquad (r = 0, 1, \ldots, m),$$

$$C_r = \binom{n}{m+r} p^{m+r} q^{n-m-r} \qquad (r = 0, 1, \ldots, n-m),$$

(1) may be written in the form

$$\sum_{r=0}^{m} B_r < \sum_{r=0}^{n-m} C_r .$$ (1')

Put $\dfrac{B_r}{C_r} = D_r$, then we have

$$\frac{D_{r+1}}{D_r} - 1 = \frac{(p-q)(r^2 + r - npq)}{(n-m-r)(n-m+r+1)p^2} ,$$

thus $\dfrac{D_{r+1}}{D_r} - 1$ is positive for small values of r. It decreases as r increases and is negative for $r \geq s$, where s is the least integer for which $s(s+1) > npq$. As $D_0 = 1$, $D_1 = \dfrac{B_1}{C_1} = \dfrac{npq + q}{npq + p} > 1$ there exists an integer $k > 1$ such that $\dfrac{B_r}{C_r} \geq 1$ for $r = 0, 1, \ldots, k-1, \dfrac{B_r}{C_r} < 1$ for $k \leq r \leq n - m$. For this value of k we get

$$\sum_{r=0}^{k-1} (k-r-1) B_r > \sum_{r=0}^{k-1} (k-r-1) C_r$$ (2)

and

$$\sum_{r=k}^{m} (r-k+1) B_r > - \sum_{r=k}^{n-m} (r-k+1) C_r .$$ (3)

From the identity

$$\sum_{k=0}^{n} (k-m) \binom{n}{k} p^k q^{n-k} = 0$$

it follows

$$\sum_{r=0}^{m} r B_r = \sum_{r=0}^{n-m} r C_r .$$ (4)

From (2), (3) and (4) it follows

$$(k-1) \sum_{r=0}^{m} B_r > (k-1) \sum_{r=0}^{n-m} C_r ,$$

which was to be proved. This proof is due to E. Feldheim.

36. Prove the following asymptotic relation for the terms of the multinomial distribution. For

$$n \to +\infty, \quad \sum_{j=1}^{r} k_j = n, \quad |k_j - np_j| = O(\sqrt{n})$$

we have

$$\frac{n!}{k_1! \, k_2! \dots k_r!} \, p_1^{k_1} p_2^{k_2} \dots p_r^{k_r} \approx$$

$$\approx \frac{1}{(\sqrt{2\pi n})^{r-1} \sqrt{p_1 p_2 \dots p_r}} \exp\left[-\frac{1}{2n} \sum_{j=1}^{r} \frac{(k_j - np_j)^2}{p_j} \right].$$

Hint. Use Stirling's formula.

37. An urn contains N cards labelled from 1 to N. Perform n drawings without replacement. Let ξ denote the least number drawn. Find the distribution of the random variable ξ.

38. Let ξ and η be nonnegative integer valued random variables such that if the value of η is fixed ξ has a Poisson distribution and conversely. Show that

$$R_{jk} = P(\xi = j, \eta = k) = C \frac{\lambda^j \, \mu^k \, \nu^{jk}}{j! \, k!} \qquad (j, k = 0, 1, \dots),$$

where λ, μ, ν are positive constants, and

$$\frac{1}{C} = \sum_{j=0}^{\infty} \sum_{k=0}^{\infty} \frac{\lambda^j \, \mu^k \, \nu^{jk}}{j! \, k!} \, .$$

For the independence of ξ and η it is necessary and sufficient that $\nu = 1$ should hold. The distribution R_{jk} is therefore a generalization of the Poisson distribution for two dimensions. (Distribution of N. G. Obreskov.)

39. Let ξ and η be two independent random variables both having a Poisson distribution with expectation λ. Determine the distribution of $\xi - \eta$.

40. Each of two urns contains $N - 1$ white balls and one red ball. Draw from both urns n balls ($n < N$) without replacement. Put now all $2N$ balls into one and the same urn and draw $2n$ balls without replacement. In which one of the two cases is it more probable to obtain at least one red ball?

41. Let λ be the disintegration constant of a radioactive material. Let the probability of observing the disintegration of any one of the atoms be denoted by c (c is proportional to the solide angle under which the counter is seen from the point from where the radiation starts). Let N denote the number of the atoms at the time $t = 0$, ξ_t the number of disintegrations observed in the time interval $(0, t)$. Prove by applying the theorem of total probability that. ξ_t has a binomial distribution.

Hint. The probability that exactly n atoms disintegrate during the interval $(0, t)$ is

$$\binom{N}{n} (1 - e^{-\lambda t})^n \, e^{-\lambda t(N-n)};$$

the probability that among them k disintegrations are observed, is $\binom{n}{k} c^k (1 - c)^{n-k}$.

The theorem of total probability gives

$$P(\xi_t = k) = \sum_{n=k}^{N} \binom{N}{n}(1 - e^{-\lambda t})^n \, e^{-\lambda t(N-n)} \binom{n}{k} c^k (1 - c)^{n-k} =$$

$$= \binom{N}{k} \left(c(1 - e^{-\lambda t})\right)^k \left(1 - c(1 - e^{-\lambda t})\right)^{N-k}.$$

Note that because of $c(1 - e^{-\lambda t}) < 1 - e^{-c\lambda t}$ somewhat fewer disintegrations are observed than when the value of the disintegration constant would be λc and all disintegrations would be visible. But this difference is only important for large values of t.

42. Let $\xi_1, \xi_2, \ldots, \xi_r$ be independent random variables with the same negative binomial distribution of order 1:

$$P(\xi_k = n) = (1 - p)p^{n-1} \quad (n = 1, 2, \ldots; \ k = 1, 2, \ldots, r; \ 0 < p < 1).$$

Let ν be a random variable, independent from ξ_j, with a Poisson distribution and expectation λ. Determine the distribution of

$$\zeta = \xi_1 + \xi_2 + \ldots + \xi_{\nu+1}.$$

Hint. By using the notations of § 14, the distribution of $\xi_1 + \xi_2 + \ldots + \xi_{k+1}$ can be written as

$$\mathscr{E}_1^{k+1} \left(\frac{1-p}{1-p\mathscr{E}_1}\right)^{k+1},$$

that of ζ is given therefore by

$$\mathscr{P} = \sum_{k=0}^{\infty} \frac{\lambda^k e^{-\lambda}}{k!} \mathscr{E}_1^{k+1} \left(\frac{1-p}{1-p\mathscr{E}_1}\right)^{k+1} = \frac{\mathscr{E}_1 (1-p) e^{-\lambda}}{1 - p\mathscr{E}_1} \, e^{-\frac{p\mathscr{E}_1 x}{1-p\mathscr{E}_1}},$$

where $x = -\dfrac{(1-p)\lambda}{p}$. It is known that

$$\frac{1}{1-z} \, e^{-\frac{xz}{1-z}} = \sum_{k=0}^{\infty} L_k(x) \, z^k,$$

where the

$$L_k(x) = \frac{e^x}{x^k} \frac{d^k}{dx^k} \, (x^k e^{-x})$$

are the Laguerre-polynomials.[1] It follows

$$P(\zeta = n) = (1 - p)e^{-\lambda} L_{n-1}\left(-\frac{(1-p)}{p}\lambda\right)p^{n-1} \qquad (n = 1, 2, \ldots).$$

43. Calculate the expectation of the number of marked fishes at the second capture (cf. Ch. II, § 12, Exercise 21), if there are 10 000 fishes in the lake and if at the first capture 100 fishes are marked.

44. Calculate the expectation of the number of matches in one of the boxes in Banach's pockets at the moment when he found the other box empty for the first time (cf. Ch. II, § 12, Exercise 14).

[1]Cf. e.g. G. Pólya and G. Szegő [1].

45. Calculate the expectation of the sum defined in Chapter II, § 12, Exercise 46:

$$X = k_1 + k_2 + \ldots + k_M.$$

Hint. Let $\mathscr{F}_k$ be the distribution of a random variable which assumes the values $0, 1, 2, \ldots, k - 1$ with the probabilities $\dfrac{1}{k}$:

$$\mathscr{F}_k = \left[\frac{1}{k}, \frac{1}{k}, \ldots, \frac{1}{k}, 0, 0, \ldots \right].$$

Show that the distribution of $X - M(M + 1)/2$ can be written in the form

$$\frac{\mathscr{F}_N \mathscr{F}_{N-1} \ldots \mathscr{F}_{N-M+1}}{\mathscr{F}_1 \mathscr{F}_2 \ldots \mathscr{F}_M}.$$

From this it follows that $E(X) = \dfrac{M(N + 1)}{2}$.

46. Suppose that a player gambles according to the following strategy at a play of coin tossing: he bets always on "tail"; if "head" occurs, he doubles his stake in the next tossing. He plays until tail occurs for the first time. What is the expectation of his gain?

Hint. If the tail occurs at the n-th toss for the first time (the probability of this event is $\dfrac{1}{2^n}$), the gain of the player, if his bet at the first toss was 1 shilling, will be 1 shilling, since

$$2^n - \sum_{k=0}^{n-1} 2^k = 1.$$

The expectation of the gain is thus

$$\sum_{n=1}^{\infty} \frac{1}{2^n} = 1.$$

It seems that with this strategy the player could ensure for himself a gain. This, however, would be true only if he would dispose over an infinite sum of money. His fortune being limited, it is easy to show by a simple calculation that the expectation of his gain is 0 even if he doubles the stake always when a head appears.

47. Calculate the expectation of the Pólya distribution.

48. The chevalier de Méré asked Pascal the following. Two gamblers play a game where the chances are equal. They deposit at the beginning of the game the same amount of money. They agree that he who is the first to have won N games gets the whole deposit. They are, however, obliged to interrupt the game at a moment when the one player gained $N - n$ times and the other $N - m$ times ($1 \leq n \leq N$; $1 \leq m \leq N$). How is the deposited money to be distributed? Calculate this proportion for $n = 2$ and $m = 3$.

Hint. The distribution of the deposited money is said to be "fair" if the money is distributed in the proportion $p_n : p_m$, p_n denoting the probability that the first gambler would win and p_m the probability that the second. Thus each gambler receives a sum equal to this expectation. The problem is thus to calculate the probability that the first (or the second) wins, under the condition that he already won $N - n$ (i. e. $N - m$) games.

49. In playing bridge, 52 cards are distributed among four players. The values of the cards distributed are measured by the number of "tricks" in the following manner: If a player has the ace and the king of the same suit, this amounts to 2 tricks; ace and queen of the same suit without the king to $1 \frac{1}{2}$; king and queen without ace to 1; ace alone to 1; king alone to $\frac{1}{2}$ trick. What is the expectation of the total number of tricks in the hand of a player?

Hint. Obviously, the expectation of the number of tricks is the same for all players and in each of the suits. Hence the expectation of the total number of tricks for a player in all four suits is equal to the expectation of sum of tricks for the four players in one suit. Thus it suffices to consider one suit only, e.g. spades. The expectation of the tricks in the hand of one player is equal to the sum of the expectations of all tricks present in the spades. However, this sum is equal to 2, except in the case when the ace, the king, and the queen of spades are in the hands of different players; in this case the sum of tricks is $\frac{3}{2}$. Hence the expectation looked for is 1.801.

50. a) There are M red and $N - M$ white balls in an urn. We put $\frac{M}{N} = p$. Draw n balls without replacement from the urn and let the random variables ξ_k ($k = 1, 2, ..., n$) be defined as follows:

$$\xi_k = \begin{cases} 1 & \text{if at the } k\text{-th drawing a red ball is drawn,} \\ 0 & \text{otherwise.} \end{cases}$$

Calculate $R(\xi_j, \xi_k)$ $(1 \leq j < k \leq n)$.

b) If

$$\zeta_n = \xi_1 + \xi_2 + \ldots + \xi_n$$

prove that

$$D^2(\zeta_n) = np(1 - p)\left(1 - \frac{n-1}{N-1}\right).$$

CHAPTER IV

GENERAL THEORY OF RANDOM VARIABLES

§ 1. The general concept of a random variable

We have already introduced in Chapter III the general concept of a random variable. Let $[\Omega, \mathscr{A}, P]$ be a Kolmogorov probability space. We understand by a random variable a function of a real variable $\xi = \xi(\omega)$, defined for each $\omega \in \Omega$, such that every level-set of ξ belongs to $\mathscr{A}$. The level-sets of $\xi = \xi(\omega)$ are the sets A_x defined by $\xi(\omega) < x$, where x is an arbitrary real number. The function $F(x) = P(A_x) = P(\xi < x)$ is called (see Ch. III) the distribution function of the random variable ξ.

THEOREM 1. *If ξ is a random variable and $g(x)$ a Borel-measurable[1] function of the real variable x, then $\eta = g(\xi)$ is also a random variable.*

PROOF. Let $\xi^{-1}(A)$ be the set of those elementary events $\omega \in \Omega$ for which $\xi(\omega) \in A$. We have clearly

$$\xi^{-1}\Big(\sum_n A_n\Big) = \sum_n \xi^{-1}(A_n) \tag{1a}$$

and

$$\xi^{-1}(A - B) = \xi^{-1}(A) - \xi^{-1}(B). \tag{1b}$$

Let I_x denote the interval $(-\infty, x)$ and $I_{a,b}$ the half-open interval $[a, b)$. By assumption, $\xi^{-1}(I_x) = A_x \in \mathscr{A}$. Hence, according to (1b), $\xi^{-1}(I_{a,b}) \in \mathscr{A}$ for every pair of real numbers (a, b), $a < b$. Since $\mathscr{A}$ is a σ-algebra, it follows from (1a) and (1b) that $\xi^{-1}(A) \in \mathscr{A}$ for every Borel-set A of the real line. Theorem 1 follows immediately.

THEOREM 2. *If the distribution function $F(x)$ of a random variable ξ is given for every real x, then $P[\xi^{-1}(A)]$ is uniquely determined for every Borel subset A of the set of real numbers.*

---

[1] A function $g(x)$ is said to be Borel-measurable if the level-set defined by $g(x) < c$ is a Borel-set for every real c. In particular, every continuous function is Borel-measurable.

PROOF. This theorem follows immediately from Theorem 2 of Chapter II, § 7.

§ 2. Distribution functions and density functions

Let $F(x) = P(\xi < x)$ be the distribution function of the random variable ξ. If the random variables ξ and η are almost surely equal (i.e. if $P(\xi \neq \eta) = 0$), then their distribution functions are obviously identical. In what follows we shall establish some properties of distribution functions.

1. *A distribution function $F(x)$ is a nondecreasing function.*
According to the definition of level-sets, we have $A_x \subset A_y$ for $x < y$; hence

$$F(x) = P(A_x) \leq P(A_y) = F(y).$$

A distribution function $F(x)$ is not necessarily continuous. It follows however from the monotonicity of $F(x)$ that at any discontinuity point $F(x)$ possesses both a left-hand side and a right hand side limit.

2. *For any distribution function $F(x)$ we have*

$$\lim_{h \to +0} F(x - h) = F(x).$$

Hence a distribution function is continuous from the left at every discontinuity point. In fact, $F(x) - F(x - h)$ is the probability that $x - h \leq \xi < x$, i.e. $F(x) - F(x - h) = P(B_h)$, where $B_h = A_x \bar{A}_{x-h}$. Obviously, the sample space does not contain any element which belongs to B_h *for every* $h > 0$. If an element ω of Ω belongs to B_h, we have always $\xi(\omega) < x$. Choose now a small enough $h' > 0$ such that $h' < x - \xi(\omega)$, then ω will not belong any more to B_h. Let $\{h_n\}$ ($n = 1, 2, \ldots$) be an arbitrary monotonic sequence of positive numbers tending to zero. To prove the left-continuity of $F(x)$ it is sufficient to prove that

$$\lim_{n \to +\infty} P(B_{h_n}) = 0 \cdot$$

But this is a particular case of Theorem 3, Chapter II, § 7.

3. *For every distribution function*

$$\lim_{x \to -\infty} F(x) = 0 \quad and \quad \lim_{x \to +\infty} F(x) = 1$$

and thus we may write

$$F(-\infty) = 0 \quad and \quad F(+\infty) = 1.$$

In fact, if $\{x_n\}$ $(n = 1, 2, \ldots)$ is any sequence of real numbers such that $x_n < x_{n+1}$ and $\lim_{n \to \infty} x_n = + \infty$, then A_{x_n} is a subset of $A_{x_{n+1}}$ hence the sets $A_{x_{n+1}} - A_{x_n}$ $(n = 1, 2, \ldots)$ and A_{x_1} are disjoint and

$$F(x_n) = P(A_{x_1}) + \sum_{k=1}^{n-1} P(A_{x_{k+1}} - A_{x_k}).$$

Since

$$P(A_{x_1}) + \sum_{k=1}^{\infty} P(A_{x_{k+1}} - A x_1) = P(\sum_{k=1}^{\infty} A_{x_k}) = P(\Omega) = 1,$$

it follows that

$$\lim_{n \to +\infty} F(x_n) = 1.$$

Similarly it can be shown that

$$\lim_{x \to -\infty} F(x) = 0.$$

All these results may be combined in the following theorem:

THEOREM 1. *The distribution function of an arbitrary random variable is a nondecreasing left-continuous function, which has for* $-\infty$ *and for* $+\infty$ *the limits* 0 *and* 1, *respectively.*

The converse of this theorem is also true: Every function $F(x)$ having these properties can be considered as a distribution function. The proof runs as follows: Let $x = G(y)$ be the inverse function of $y = F(x)$. (The definition of $G(y)$ is unique, if the following conventions are adopted: if $F(x)$ has a jump at x_0, i.e. if $F(x_0) = a$ and $F(x_0 + 0) = b > a$, we put $G(y) = x_0$ for $a < y \leq b$; if $F(x)$ is constant and equal to y_0 in the interval $c < x \leq d$ but $F(x) < y_0$ for $x < c$, we put $G(y_0) = c$.) If Ω is the interval $(0, 1)$, $\mathscr{A}$ the system of all Borel-measurable subsets of Ω and $P(A)$ is for $A \in \mathscr{A}$ the Lebesgue measure of A, then the function $\eta(y) = G(y)$ defined for all $y \in \Omega$ is a random variable on the probability space $[\Omega, \mathscr{A}, P]$ and the distribution function of $\eta(y)$ is

$$P(\eta < x) = P(G(y) < x) = P(y < F(x)) = F(x).$$

Hence η is a random variable with distribution function $F(x)$.

If ξ is a bounded random variable, i.e. if there exist two constants c and C such that for every element ω of the sample space Ω the inequality $c \leq \xi(\omega) < C$ holds, then clearly $F(x) = 0$ for $x \leq c$ and $F(x) = 1$ for $x \geq C$.

If the random variable is "almost surely" constant, i.e. if there exists a set $A \in \mathscr{A}$ such that $P(\bar{A}) = 0$ and $\xi(\omega) = c$ for every $\omega \in A$, then we obtain for the distribution function of ξ

$$F(x) = \begin{cases} 0 \text{ for } x \leq c \\ 1 \text{ otherwise.} \end{cases}$$

The distribution function of a constant is said to be a *degenerate* distribution function. Eveiy nondegenerate distribution function has at least two points of increase, i.e. at least two points where $F(x + h) - F(x) > 0$ for every $h > 0$. If $F(x)$ is the distribution function of a random variable ξ which assumes only a finite number of values, x is a point of increase of $F(x)$ if and only if ξ takes on the value x with positive probability. The set of jumps of a monotonic function, and thus in particular of a distribution function, is necessarily finite or denumerably infinite. In fact, if the jumps of the distribution function are projected upon the y-axis, a system of disjoint intervals is obtained because of the monotonicity of the function. Our statemert can be deduced from the fact that every interval contains a rational number and the set of all rational numbers is denumerable.

Distribution functions which are not only continuous but absolutely continuous, deserve particular attention. A distribution function is said to be *absolutely continuous* if for any given positive number ε there exists a $\delta > 0$ such that for every system of disjoint intervals

$$(a_k, b_k) \ (k = 1, 2, \ldots, n; \ a_k < b_k)$$

the inequality

$$\sum_{k=1}^{n} (b_k - a_k) < \delta$$

implies

$$\sum_{k=1}^{n} |F(b_k) - F(a_k)| < \varepsilon.$$

Every absolutely continuous function, as is known, is almost everywhere differentiable and is equal to the indefinite integral of its derivative. This is a necessary and sufficient condition for the absolute continuity of a function.

If the distribution function $F(x)$ is absolutely continuous we put $f(x) = F'(x)$. If $F(x)$ is at a point non-differentiable, $f(x)$ is not defined there, but it can be defined arbitrarily. But such points are known to form a set of measure zero. The function $f(x)$ is called the *density function* of the probability distribution given by $F(x)$. If $F(x)$ is the distribution function of the random variable ξ, $f(x)$ is called also the density function of ξ.

Example. We have already seen (Ch. III, § 13) that the function defined by

$$F(x) = \begin{cases} 1 - e^{-\lambda x} & \text{for } \geq 0 \\ 0 & \text{otherwise,} \end{cases}$$

with $\lambda > 0$, is the distribution function of the life-time of a radioactive atom. This function is absolutely continuous and we have

$$f(x) = F'(x) = \begin{cases} \lambda e^{-\lambda x} & \text{for } x > 0 \\ 0 & \text{for } x < 0. \end{cases}$$

($f(0)$ is not defined, as $F'(0)$ does not exist.)

Thus the density function of the life-time of radioactive atoms exists and is equal to $\lambda e^{-\lambda x}$ for $x > 0$.

Let ξ be any random variable; let A_y be the event $\xi < y$ and $A_{a,b}$ the event $a \leq \xi < b$; then we have $A_{a,b} = A_b \bar{A}_a$. Now, if $F(x)$ is absolutely continuous and $F'(x) = f(x)$, we have

$$P(A_{a,b}) = F(b) - F(a) = \int_a^b f(x)\, dx. \tag{1}$$

From this follows immediately

$$\int_{-\infty}^{+\infty} f(x)\, dx = F(+\infty) - F(-\infty) = 1. \tag{2}$$

Conversely, every nonnegative measurable function $f(x)$ fulfilling (2) can be considered as a probability density. Indeed, the function

$$F(x) = \int_{-\infty}^{x} f(t)\, dt \tag{3}$$

obviously has every property of a distribution function and $F'(x) = f(x)$ holds almost everywhere.

Consider now the case when $b - a = \Delta a$ is small and $F'(a)$ exists. Since, by definition,

$$\lim_{\Delta a \to 0} \frac{F(a + \Delta a) - F(a)}{\Delta a} = f(a),$$

it follows from this that

$$P(a \leq \xi < a + \Delta a) = f(a)\Delta a + o(\Delta a), \tag{4}$$

where, as usually, $o(\Delta a)$ represents a quantity which, divided by Δa, tends to zero for $\Delta a \to 0$.

Example. In Chapter III, § 16 we encountered the standard normal distribution having the distribution function

$$F(x) = \frac{1}{\sqrt{2\pi}} \int_{-\infty}^{x} e^{-\frac{t^2}{2}} \, dt.$$

The density function of the standard normal distribution is therefore

$$f(x) = \frac{1}{\sqrt{2\pi}} e^{-\frac{x^2}{2}}.$$

We need now the following well-known theorem: Every nondecreasing monotonic function may be represented as a sum of three nondecreasing monotonic functions, the first of which is a step function, the second is absolutely continuous and the third is a continuous "singular" function (i.e. a continuous nondecreasing function whose derivative is almost everywhere equal to zero). From this it follows readily that every distribution function can be written in the form

$$F(x) = p_1 F_1(x) + p_2 F_2(x) + p_3 F_3(x),$$

where p_1, p_2, p_3 are nonnegative numbers having sum 1 and $F_i(x)$ ($i = 1$, 2, 3) are the three distribution functions such that $F_1(x)$ is the distribution function of a discrete random variable, $F_2(x)$ is an absolutely continuous distribution function and $F_3(x)$ is a singular distribution function. This decomposition is evidently unique.

§ 3. Probability distributions in several dimensions

By a random vector of n dimensions we understand an n-dimensional vector $\zeta = (\xi_1, \ldots, \xi_n)$ whose components ξ_i are random variables on the same probability space. The distribution function of the random vector ζ is defined as the function of n variables

$$F(x_1, x_2, \ldots, x_n) = P(\xi_1 < x_1, \xi_2 < x_2, \ldots, \xi_n < x_n). \tag{1}$$

The probability figuring on the right side of (1) is always defined; in fact, let $A_x^{(k)}$ denote the level set of all $\omega \in \Omega$ such that $\xi_k(\omega) < x$ ($k = 1, 2, \ldots, n$), then $A_x^{(k)} \in \mathscr{A}$ and

$$\prod_{k=1}^{n} A_{x_k}^{(k)} \in \mathscr{A}.$$

The right hand side of Equation (1) is exactly

$$P(\prod_{k=1}^{n} A_{x_k}^{(k)}).$$

If a problem of probability theory involves n random variables, these can always be considered as components of an n-dimensional random vector. In general, the function defined by (1) will be called the *joint distribution function* of the random variables $\xi_1, \xi_2, \ldots, \xi_n$.

For example the value $F(x_1, y_1) = P(\xi < x_1, \eta < y_1)$ of the distribution function of a 2-dimensional random vector represents the probability that the endpoint of a random vector $\zeta = (\xi, \eta)$ beginning in $(0, 0)$ lies in the quadrant of the (x, y) plane defined by $x < x_1, y < y_1$.

Let us consider now some general properties of multidimensional distribution functions.

1. $F(x_1, \ldots, x_n)$ *is a nondecreasing function of every one of its variables.*

2. $F(x_1, \ldots, x_n)$ *is left-continuous in every variable.*

3. $F(x_1, \ldots, x_n) = 0$, *if at least one of its variables is equal to* $-\infty$.

4. $F(x_1, \ldots, x_n) = 1$ *if all variables are equal to* $+\infty$.

Besides these trivial properties, every n-dimensional distribution has another characteristic property which for $n = 1$ follows from the above property 1. For $n > 1$, however, it does not follow from it. The probability $P(a_k \leq \xi_k < b_k; k = 1, 2, \ldots)$ may be written in the form

$$P(a_k \leq \xi_k < b_k; \ k = 1, 2, \ldots, n) = \sum (-1)^{\sum_{k=1}^{n} \varepsilon_k} F(c_1, c_2, \ldots, c_n) \quad (2)$$

where $c_k = \varepsilon_k a_k + (1 - \varepsilon_k)b_k$. The numbers ε_k assume independently of each other the values 0 and 1. The sum on the right hand side of (2) has thus 2^n terms. Thus, for instance, for $n = 2$ we obtain

$$P(a_1 \leq \xi_1 < b_1, a_2 \leq \xi_2 < b_2) = F(b_1, b_2) - F(a_1, b_2) - F(b_1, a_2) +$$
$$+ F(a_1, a_2).$$

Formula (2) is a direct consequence of Theorem 9, Chapter II, § 3. In fact, let A_k be the event $\xi_k < a_k$, B_k the event $\xi_k < b_k$ $(k = 1, 2, \ldots, n)$. If we put

$$A = \sum_{k=1}^{n} A_k, \qquad B = \prod_{k=1}^{n} B_k,$$

we find that

$$P(a_k \le \xi_k < b_k; \quad k = 1, 2, \ldots, n) = P(\bar{A}B) = P(B) - P(\sum_{k=1}^{n} A_k B).$$

If we now put $C_k = A_k B$ $(k = 1, 2, \ldots, n)$ we obtain (2) by applying the above-mentioned theorem to the events C_k.

As the left hand side of (2) is a probability, it is certainly nonnegative. Thus we have established the sought property:

5. *We have*

$$\sum (-1)^{\sum_{k=1}^{n} \varepsilon_k} F(\varepsilon_1 a_1 + (1 - \varepsilon_1) b_1, \ldots, \varepsilon_n a_n + (1 - \varepsilon_n) b_n) \ge 0,$$

where $\varepsilon_1, \varepsilon_2, \ldots, \varepsilon_n$ assume the values 0 and 1 independently of each other and $a_k < b_k$ $(k = 1, 2, \ldots, n)$ are arbitrary real numbers.

Property 5 does not follow from properties 1–4. If for instance $n = 2$ and

$$F(x_1, x_2) = \begin{cases} 1 \text{ if } x_1 + x_2 > 0, \\ 0 \text{ otherwise}, \end{cases}$$

properties 1–4 are fulfilled but property 5 is not, since for instance

$$F(2, 2) - F(-1, 2) - F(2, -1) + F(-1, -1) = -1 < 0.$$

By introducing the notation

$$\Delta_h^{(k)} F(x_1, \ldots, x_n) = F(x_1, \ldots, x_{k-1}, x_k + h, x_{k+1}, \ldots, x_n) - F(x_1, \ldots, x_n)$$

we may write condition 5 in the following form:

5'. *We have*

$$\Delta_{h_1}^{(1)} \Delta_{h_2}^{(2)} \ldots \Delta_{h_n}^{(n)} F(x_1, x_2, \ldots, x_n) \ge 0$$

for $h_k \ge 0$ and for any real numbers x_k $(k = 1, 2, \ldots, n)$. Here the "product" of the (commutative) operations $\Delta_{h_k}^{(k)}$ means that they are to be performed one after the other. It is easy to prove that if condition 5' holds for $h_1 = h_2 = \ldots = h_n = h > 0$ it is valid in general.

Conversely, it can be shown that every function $F(x_1, x_2, \ldots, x_n)$ fulfilling conditions 1–5 may be considered as a distribution function. This follows from § 7 of Chapter II.

If the distribution function of the random vector $\zeta = (\xi_1, \xi_2, \ldots, \xi_n)$ is $F(x_1, x_2, \ldots, x_n)$ and B is a Borel-set of the n-dimensional space, then

$$P(\zeta \in B) = \int \ldots \int_B d \, F(x_1, \ldots, x_n),$$

where on the right-hand side figures a Stieltjes-integral; in other words $P(\zeta \in B)$ is equal to the value on the set B of the measure defined by the function F in the n-dimensional space.

If the distribution function of an n-dimensional random vector is absolutely continuous, then the density function

$$f(x_1, x_2, \ldots, x_n) = \frac{\partial^n F(x_1, x_2, \ldots, x_n)}{\partial x_1 \partial x_2 \ldots \partial x_n} \tag{3}$$

exists almost everywhere, and we have always

$$f(x_1, x_2, \ldots, x_n) \geq 0$$

because of property 5. This follows from

$$f(x_1, x_2, \ldots, x_n) = \lim_{h \to 0} \frac{\Delta_h^{(1)} \Delta_h^{(2)} \ldots \Delta_h^{(n)} F(x_1, x_2, \ldots, x_n)}{h^n} .$$

Further we have

$$F(x_1, \ldots, x_n) = \int_{-\infty}^{x_1} \ldots \int_{-\infty}^{x_n} f(t_1, \ldots, t_n) \, dt_1 \ldots dt_n ; \tag{4}$$

hence in particular

$$\int_{-\infty}^{+\infty} \ldots \int_{-\infty}^{+\infty} f(x_1, \ldots, x_n) \, dx_1 \ldots dx_n = 1. \tag{5}$$

Further

$$P(a_k \leq \xi_k < b_k; \ k = 1, 2, \ldots, n) = \int_{a_1}^{b_1} \ldots \int_{a_n}^{b_n} f(x_1, \ldots, x_n) \, dx_1 \ldots dx_n, \tag{6}$$

or, more generally, if B is a Borel subset of the n-dimensional space, then

$$P(\zeta \in B) = \int \ldots \int_B f(x_1, \ldots, x_n) \, dx_1 \ldots dx_n. \tag{7}$$

In other words: the probability that the endpoint of the random vector ζ lies in a Borel-set B of the n-dimensional space is equal to the integral on B of $f(x_1, \ldots, x_n)$.

THEOREM 1. *If $\varphi(x_1, \ldots, x_n)$ is a Borel-measurable function of n variables and if $\xi_1, \ldots, \xi_n$ are random variables, then $\eta = \varphi(\xi_1, \ldots, \xi_n)$ is also a random variable.*

PROOF. Let $\zeta^{-1}(B)$ be the set of those points $\omega \in \Omega$ for which $\zeta(\omega) \in B$, where ζ is an n-dimensional random vector and B is a Borel-set of the n-dimensional space; clearly $\zeta^{-1}(B) \in \mathscr{A}$. From this Theorem 1 follows in the same manner as Theorem 1 of § 1 was proved.

Let us remark that to every 3-dimensional probability distribution there can be assigned a 3-dimensional distribution of the unit mass such that any domain D contains the mass $P(D)$. If $f(x, y, z)$ is the density function of the probability distribution in question, this same function will represent the density of the corresponding mass distribution.

§ 4. Conditional distributions and conditional density functions

Let ξ be any random variable, B an event of positive probability. Of course B is assumed to belong to the probability algebra on which ξ is defined. The *conditional distribution function of ξ with respect to the condition B* is defined as the function

$$F(x \mid B) = P(\xi < x \mid B) = P(A_x \mid B),$$

where A_x has the same meaning as in § 1. If the conditional distribution function thus defined is absolutely continuous, its derivative $f(x \mid B) = F'(x \mid B)$ will be called the *conditional density function of ξ with respect to B*. Evidently, if $P(B) = 1$, the ordinary distribution function and density function are obtained.

Take for instance the conditional distribution function of the life-time of a radioactive atom with respect to the condition that it did not disintegrate until the moment t_0. As is already known, this is equal to the ordinary distribution function if t is replaced by $t - t_0$. Let B_0 be the event: the atom did not disintegrate until the moment t_0. Then we have

$$F(t \mid B_0) = \begin{cases} 1 - e^{-\lambda(t-t_0)} & \text{for } t > t_0, \\ 0 & \text{otherwise} \end{cases}$$

and

$$f(t \mid B_0) = \begin{cases} \lambda e^{-\lambda(t-t_0)} & \text{for } t > t_0, \\ 0 & \text{for } t < t_0. \end{cases}$$

If $\{B_n\}$ $(n = 1, 2, \ldots)$ is a complete system of events with $P(B_n) > 0$, we have

$$F(x) = \sum_n P(B_n) F(x \mid B_n) \tag{1}$$

and

$$f(x) = \sum_n P(B_n) f(x \mid B_n). \tag{2}$$

For the generalization of the concept of the conditional distribution function and conditional density function see Chapter V, § 2.

§ 5. Independent random variables

Two random variables ξ and η are said to be (stochastically) independent, if for every real x and y

$$P(\xi < x,\ \eta < y) = P(\xi < x)\,P(\eta < y), \tag{1}$$

i.e., if the two-dimensional distribution function of (ξ, η) is equal to the product of the distribution functions of ξ and η. From (1) is readily deduced that

$$P(a \le \xi < b,\ c \le \eta < d) = P(a \le \xi < b)\ P(c \le \eta < d) \tag{2}$$

and, more generally, for any two Borel-sets A and B (cf. Theorem 2 below):

$$P(\xi \in A,\ \eta \in B) = P(\xi \in A)\,P(\eta \in B). \tag{2'}$$

For discrete random variables, this definition of independence coincides with that given in Chapter III, § 5.

The independence of several random variables may be defined in a similar manner. The random variables $\xi_1, \xi_2, \ldots, \xi_n$ are said to be independent, if for every system of real numbers $x_1, x_2, \ldots, x_n$ the relation

$$P(\xi_1 < x_1, \ldots, \xi_n < x_n) = \prod_{k=1}^{n} P(\xi_k < x_k) \tag{3}$$

holds. If the random variables $\xi_1, \xi_2, \ldots, \xi_n$ are independent, any k $(k < n)$ chosen arbitrarily from them are independent as well. To see this, it suffices to substitute $x_j = +\infty$, where the j-s are the indices of the random variables which do not figure among the chosen k.

The converse of this relation does not hold. For example the fact that ξ, η, ζ are pairwise independent does not imply their independence. We have already seen this in the preceding Chapter.

If $\xi_1, \xi_2, \ldots, \xi_n$ are discrete random variables, the above definition of independence is equivalent to the definition given in the preceding Chapter.

We shall prove now some simple theorems about independent random variables.

THEOREM 1. *A constant is independent of every random variable.*

PROOF. If $\eta = c$ (c = constant), we have

$$P(\xi < x, \eta < y) = \begin{cases} P(\xi < x) \text{ for } c < y, \\ 0 \text{ otherwise,} \end{cases}$$

hence (1) is valid.

THEOREM 2. *Let* $\xi_1, \xi_2, \ldots, \xi_n$ *be independent random variables and let* $g_k(x)$ ($k = 1, 2, \ldots, n$) *be Borel-measurable functions; then the random variables* $\eta_k = g_k(\xi_k)$ *are independent.*

PROOF. If $B_1, \ldots, B_n$ are Borel subsets of the real axis, it follows from (3) that

$$P(\xi_1 \in B_1, \ldots, \xi_n \in B_n) = \prod_{k=1}^{n} P(\xi_k \in B_k). \qquad (4)$$

In fact, if $B_1, \ldots, B_n$ are unions of finitely many intervals, (4) follows from (3). Let now $B_2, B_3, \ldots, B_n$ be fixed and let B_1 alone be considered as variable: thus both sides of (4) represent a measure. The theorem about the unique extension of a measure (Ch. II, § 7, Theorem 2) can be applied here and it follows that (4) is true for any Borel-set B_1. Let now be B_1 an arbitrary, fixed Borel-set and let $B_3, \ldots, B_n$ be fixed sets, each of them being the union of finitely many intervals. By repeating the preceding reasoning it can be seen that (4) remains valid, if B_2 too is an arbitrary Borel-set. By progressing in this manner (4) can be proved. Theorem 2 follows immediately from (4).

In particular it follows from Theorem 2 that the random variables

$$\eta_k = a_k \xi_k + b_k \quad (k = 1, 2, \ldots, n)$$

where a_k and b_k are arbitrary constants, are independent if $\xi_1, \ldots, \xi_n$ are independent.

Furthermore, it follows from Theorem 2 that for independent random variables $\xi_1, \ldots, \xi_n$ Formula (3) remains valid if for one or several values of k on both sides one of the expressions $\leq x_k$, $> x_k$, or $\geq x_k$ will be written instead of $< x_k$.

THEOREM 3. *If* $\xi_1, \xi_2, \ldots, \xi_n$ *are independent random variables with density functions* $f_1(x), f_2(x), \ldots, f_n(x)$, *respectively, then the joint distribution of the random variables* $\xi_1, \ldots, \xi_n$ *is absolutely continuous with density function*

$$f(x_1, \ldots, x_n) = \prod_{k=1}^{n} f_k(x_k). \qquad (5)$$

Conversely, (5) implies the independence of the random variables $\xi_1, \ldots, \xi_n$.

PROOF. (5) follows from (3) because of Formula (3) of § 3. Conversely, (3) is obtained by integrating (5).

THEOREM 4. *Let $\xi_1, \xi_2, \ldots, \xi_n$ be independent random variables and let $h(x_1, \ldots, x_k)$ be a Borel-measurable function of k variables $(k < n)$. Then the random variables*

$$h(\xi_1, \ldots, \xi_k), \ \xi_{k+1}, \ldots, \xi_n$$

are independent.

The proof is similar to that of Theorem 2.

The independence of two random vectors, $\xi = (\xi_1, \ldots, \xi_n)$ and $\eta = (\eta_1, \ldots, \eta_m)$ can be defined as follows: ξ and η are said to be independent, if the equality

$$P(\xi_1 < x_1, \ldots, \xi_n < x_n; \eta_1 < y_1, \ldots, \eta_m < y_m) =$$

$$= P(\xi_1 < x_1, \ldots, \xi_n < x_n) \, P(\eta_1 < y_1, \ldots, \eta_m < y_m) \tag{6}$$

is identically fulfilled in the variables x_j and y_k.

§ 6. The uniform distribution

The random variable ξ is said to be *uniformly distributed* on the interval (a, b) $(a < b)$ if its density function is

$$f(x) = \begin{cases} 0 \text{ for } x < a \text{ and for } b < x, \\ \dfrac{1}{b-a} \text{ for } a < x < b. \end{cases} \tag{1}$$

At the points $x = a$ and $x = b$ $f(x)$ can be defined arbitrarily[1]. The corresponding distribution function is

$$F(x) = \begin{cases} 0 & \text{for } x \le a, \\ \dfrac{x-a}{b-a} & \text{for } a \le x \le b, \\ 1 & \text{for } x > b. \end{cases} \tag{2}$$

The uniform distribution of a random vector can be defined in a similar manner. An *n*-dimensional random vector $\zeta = (\xi_1, \ldots, \xi_n)$ is said to be uniformly distributed on a nonempty open set G of the *n*-dimensional space

---

[1] Sometimes $f(x)$ is also called "rectangular" density function.

with finite n-dimensional Lebesgue-measure, if the density function of the random vector is given by

$$f(x_1, \ldots, x_n) = \begin{cases} \dfrac{1}{\mu_n(G)} & \text{for } (x_1, \ldots, x_n) \in G \\ 0 & \text{otherwise,} \end{cases} \tag{3}$$

where $\mu_n(G)$ is the "volume" (the n-dimensional Lebesgue-measure) of G. We already encountered uniformly distributed random variables in Chapter II, § 10 in connection with the geometric probabilities. The geometrical determination of probabilities is nothing else than the reduction of the problem to certain uniformly distributed random variables. In fact when one deals with geometric probabilities, it is always assumed that the probability of a point to lie in an interval of the real axis (in a domain of the plane, space, or more generally, of the n-dimensional space) is proportional to the length of the interval (to the area or volume of the domain in question). But this means that the random variable considered (or the random vector) is uniformly distributed. If ξ is uniformly distributed on the interval (a, b), then, according to (1) and Formula (1) of § 2, the probability that ξ lies in a subinterval (c, d) $(a \leq c < d \leq b)$ of (a, b) is given by

$$\int_c^d f(x) \, dx = \frac{d - c}{b - a}, \tag{4}$$

thus it is indeed proportional to the length of the interval (c, d).

A similar statement holds also in the multidimensional case. If ζ is a random vector uniformly distributed on an n-dimensional domain G, the probability that the endpoint of ζ lies in a domain G_1 which is a subset of G is equal to

$$\int \ldots \int_{G_1} f(x_1, \ldots, x_n) \, dx_1 \ldots dx_n = \frac{\mu_n(G_1)}{\mu_n(G)}. \tag{5}$$

The case when G is a parallelepiped with its edges parallel to the axes deserves particular consideration: we have

$$f(x_1, \ldots, x_n) = \begin{cases} \dfrac{1}{\mu_n(G)} & \text{for } a_k < x_k < b_k \;\; (k = 1, 2, \ldots n), \\ 0 & \text{otherwise,} \end{cases}$$

where

$$\mu_n(G) = \prod_{k=1}^n (b_k - a_k).$$

Hence

$$f(x_1, \ldots, x_n) = \prod_{k=1}^{n} f_k(x_k), \tag{6}$$

where $f_k(x_k)$ $(k = 1, \ldots, n)$ is the density function of a random variable uniformly distributed on the interval (a_k, b_k); consequently $\xi_1, \ldots, \xi_n$ are independent. Conversely: if ξ_k is uniformly distributed on (a_k, b_k) and if the ξ_k are independent, the vector $\zeta = (\xi_1, \ldots, \xi_n)$ is uniformly distributed in the parallelepiped $a_k < x_k < b_k$ $(k = 1, 2, \ldots, n)$.

For an infinite interval (or for a domain of infinite volume) the uniform distribution can be defined by means of the theory of conditional probability spaces. We shall return to this in Chapter V.

§ 7. The normal distribution

We already encountered the normal distribution. It was introduced as the limit-distribution of the binomial distribution. It has a paramount role in probability theory. Many random variables dealt with in practice have a nearly normal distribution. Often, the normal distribution is called the "law of errors", since random errors in the result of measurements are often normally distributed. In Chapter VIII it will be proved that the distribution of the sum of a large number of independent random variables has approximately a normal distribution under quite general conditions.

First of all let two general notions be defined: that of the similarity of distributions and that of a family of distributions. Two distribution functions $F_1(x)$ and $F_2(x)$ are said to be *similar*, if there exist two numbers $\sigma \neq 0$ and m such that if $F_1(x)$ is the distribution function of a random variable ξ, then $F_2(x)$ is the distribution function of $\eta = \sigma\xi + m$. As the inequality $\sigma\xi + m < x$ is for $\sigma > 0$ equivalent to $\xi < \dfrac{x - m}{\sigma}$ and for $\sigma < 0$ to $\xi > \dfrac{x - m}{\sigma}$, we have either

$$F_2(x) = F_1\left(\frac{x - m}{\sigma}\right) \tag{1a}$$

(for $\sigma > 0$), or

$$F_2(x) = 1 - F_1\left(\frac{x - m}{\sigma} + 0\right) \tag{1b}$$

(for $\sigma < 0$).

If $F_1(x)$ is absolutely continuous, $F_2(x)$ is absolutely continuous as well. In this case, we obtain for the density functions $f_i(x) = F_i'(x)$ $(i = 1, 2)$

$$f_2(x) = \frac{1}{|\sigma|} f_1\left(\frac{x - m}{\sigma}\right). \tag{2}$$

Clearly the relation of similarity is symmetric, reflexive, and transitive. Thus it permits the classification of the distributions into types called *families*. Every family of distributions is a set depending on two parameters (m and σ). All uniform distributions (on the line) are thus similar to the distribution uniform on the interval $(0, 1)$. In fact, the uniform distribution on (a, b) has the density function

$$\frac{1}{b-a} f\left(\frac{x-a}{b-a}\right)$$

where $f(x)$ is the density function of the uniform distribution on $(0, 1)$, that is

$$f(x) = \begin{cases} 1 \text{ for } 0 < x < 1 \\ 0 \text{ for } x < 0 \text{ and } 1 < x. \end{cases}$$

One can also define families of multidimensional distributions. The distribution functions $F_1(x_1, \ldots, x_n)$ and $F_2(x_1, \ldots, x_n)$ are said to be *similar* if there exists a linear transformation

$$\xi_k' = a_{k0} + \sum_{i=1}^{n} a_{ki} \xi_i \quad (k = 1, 2, \ldots, n) \tag{3}$$

with a non-zero determinant $D = |a_{ik}|$ such that the random vector $\xi = (\xi_1, \ldots, \xi_n)$ with the distribution function $F_1(x_1, \ldots, x_n)$ is transformed by (3) into the vector $\xi' = (\xi_1', \ldots, \xi_n')$ with distribution function $F_2(x_1, \ldots, x_n)$.

If the functions $F_1(x_1, \ldots, x_n)$ and $F_2(x_1, \ldots, x_n)$ are absolutely continuous and have the density functions $f_1(x_1, \ldots, x_n)$ and $f_2(x_1, \ldots, x_n)$, then by a well-known property of linear transformations it follows that

$$f_2(x_1', \ldots, x_n') = \frac{1}{|D|} f_1(x_1, \ldots, x_n), \tag{4}$$

where

$$x_k' = a_{k0} + \sum_{i=1}^{n} a_{ki} x_i \quad (k = 1, \ldots, n).$$

For $n = 1$, (4) reduces to (2).

Let us now return to the normal distribution. We shall call every distribution normal which is similar to that obtained as the limit of the binomial distribution, i.e. to the distribution with density function $\dfrac{e^{-\frac{x^2}{2}}}{\sqrt{2\pi}}$. Thus the density function of a normal distribution has the form

$$f(x) = \frac{1}{\sigma} \varphi\left(\frac{x-m}{\sigma}\right) = \frac{1}{\sqrt{2\pi}\,\sigma} \exp\left(-\frac{(x-m)^2}{2\sigma^2}\right), \tag{5a}$$

where

$$\varphi(x) = \frac{1}{\sqrt{2\pi}}\, e^{-\frac{x^2}{2}}. \tag{5b}$$

(We have taken $\sigma > 0$; this restriction to positive values of σ is permissible since $\varphi(x)$ is an even function.) In other words: a normal distribution function has the form

$$F(x) = \Phi\left(\frac{x-m}{\sigma}\right), \tag{6a}$$

where

$$\Phi(x) = \frac{1}{\sqrt{2\pi}} \int\limits_{-\infty}^{x} e^{-\frac{t^2}{2}} dt. \tag{6b}$$

If the distribution function of a random variable ξ is given by (6a), we shall call ξ for the sake of brevity $N(m, \sigma)$ distributed. Let us now consider the multidimensional normal distributions. For the sake of simplicity, let us first restrict ourselves to the case of two dimensions.

If ξ and η are independent normally distributed random variables with density functions

$$\frac{1}{\sigma_1} \varphi\left(\frac{x-m_1}{\sigma_1}\right) \quad \text{and} \quad \frac{1}{\sigma_2} \varphi\left(\frac{x-m_2}{\sigma_2}\right),$$

then the density function of the random vector $\zeta = (\xi, \eta)$ is equal to the product of the density functions of ξ and η; i.e. to

$$h(x, y) = \frac{1}{2\pi\sigma_1\sigma_2} \exp\left\{-\frac{1}{2}\left[\frac{(x-m_1)^2}{\sigma_1^2} + \frac{(y-m_2)^2}{\sigma_2^2}\right]\right\}. \tag{7a}$$

A random vector having a density function of the form (7a) or one similar to it is said to be *normally distributed* (or *Gaussian*). Since all distributions having density functions of type (7a) are similar to each other, the two-dimensional normal distributions form a family. The density function (7a) (with $m_1 = m_2 = 0$) is represented on Fig. 19.

A simple calculation shows that the most general form of the two-dimensional normal density function is given by

$$\frac{\sqrt{AC-B^2}}{2\pi}\exp\left[-\frac{1}{2}\left(A(x-m_1)^2 + 2B(x-m_1)(y-m_2) + C(y-m_2)^2\right)\right],\text{(7b)}$$

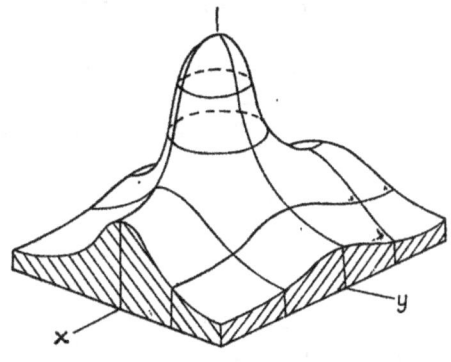

Fig. 19

where A and C are positive, B is a real number such that $B^2 < AC$, m_1 and m_2 are arbitrary real numbers. If $B \neq 0$, ζ and η are not independent. In fact, in this case the density function cannot be decomposed into two factors, one depending only on x and the other only on y.

We introduce now the concept of the *projection* of a probability distribution. Let $\zeta = (\xi_1, \ldots, \xi_n)$ be an n-dimensional random vector. The projection of the distribution of ζ upon the line g having the cosines of direction

$$g_k \ (k = 1, 2, \ldots, n; \ \sum_{k=1}^{n} g_k^2 = 1),$$

is defined as the distribution of the real random variable

$$\zeta_g = \sum_{k=1}^{n} g_k \xi_k.$$

If the distribution of ζ is known, all its projections are known as well. In particular, the distribution of $\xi_k \ (k = 1, \ldots, n)$ is thus the projection of the distribution of ζ upon the x_k-axis. Let $F(x_1, \ldots, x_n)$ be the distribution function and $f(x_1, \ldots, x_n)$ the density function of ζ, $F_k(x)$ and $f_k(x)$ those of ξ_k. We have

$$F_k(x_k) = F(+\infty, \ldots, +\infty, x_k, +\infty, \ldots, +\infty) \tag{8}$$

and, similarly (for almost every x_k)

$$f_k(x_k) = \int_{-\infty}^{+\infty} \ldots \int_{-\infty}^{+\infty} f(x_1, \ldots, x_n) \, dx_1 \ldots dx_{k-1} \, dx_{k+1} \ldots dx_n. \qquad (9)$$

To understand the notion of a "projection" it is useful to consider the analogous notion for a mass-distribution. For instance let a distribution of the unit mass over the plane be determined by the density function $h(x, y)$ and let us "project" it upon the x-axis, in the sense that we assign to the interval (a, b) the total mass contained in the strip $a \leq x < b$, $-\infty < < y < +\infty$. This mass is equal to

$$\int_a^b \int_{-\infty}^{+\infty} h(x, y) \, dy dx.$$

Consider now the projection of an arbitrary two-dimensional normal distribution (7b) upon the x-axis (and upon the y-axis). For the density function $f(x)$ (and $g(y)$) of these projections, a simple calculation gives, as we have

$$\frac{1}{\sqrt{2\pi}\sigma} \int_{-\infty}^{+\infty} \exp\left(-\frac{(x-m)^2}{2\sigma^2}\right) dx = 1,$$

the results

$$f(x) = \frac{1}{\sigma_1} \varphi\left(\frac{x - m_1}{\sigma_1}\right) \quad \text{and} \quad g(y) = \frac{1}{\sigma_2} \varphi\left(\frac{y - m_2}{\sigma_2}\right), \qquad (10)$$

where

$$\sigma_1 = \sqrt{\frac{C}{AC - B^2}} \quad \text{and} \quad \sigma_2 = \sqrt{\frac{A}{AC - B^2}}. \qquad (11)$$

Thus the projections upon the axes of a two-dimensional normal distribution with density function (7b) are one-dimensional normal distributions.[1] The projection on an arbitrary line may be calculated in the same manner and the result is always a one-dimensional normal distribution. Suppose that the components $\xi_1, \xi_2, \ldots, \xi_n$ of an n-dimensional random vector ζ are independent and ξ_k is normally distributed with the density function $\frac{1}{\sigma_k} \varphi\left(\frac{x}{\sigma_k}\right)$ $(k = 1, \ldots, n)$. The density function of the random vector ζ is

---

[1] The projections of a distribution in n-space on the coordinate axes are also called its *marginal distributions*.

$$f(x_1, \ldots, x_n) = \frac{1}{(2\pi)^{\frac{n}{2}} \prod\limits_{k=1}^{n} \sigma_k} \exp\left[-\frac{1}{2}\sum_{k=1}^{n}\frac{x_k^2}{\sigma_k^2}\right]. \tag{12}$$

If the density function of a random vector has the form (12), it is said to be *normally distributed* or *Gaussian*. Every distribution similar to this is said to be an *n-dimensional normal (or Gaussian) distribution.* In order to obtain the general form of the density of an n-dimensional normal distribution put

$$\xi_j' = \sum_{k=1}^{n} c_{jk}\,\xi_k + m_j, \tag{13}$$

where (c_{jk}) is an orthogonal matrix, i.e.

$$\sum_{k=1}^{n} c_{ik}\,c_{jk} = \delta_{ij} = \begin{cases} 1 & \text{for} \quad i = j, \\ 0 & \text{otherwise,} \end{cases} \tag{14}$$

and where the m_j $(j = 1, \ldots, n)$ are real numbers.[1]

Consider now the random variables $\xi_1', \ldots, \xi_n'$ as coordinates of a vector ζ'. Determine the density function $g(x_1', \ldots, x_n')$ of ζ'. By (13) and (14) we have

$$\xi_k = \sum_{j=1}^{n} c_{jk}\,(\xi_j' - m_j) \tag{15}$$

and in consequence of (4) we obtain (as in the two-dimensional case),

$$g(x_1', \ldots, x_n') = \frac{1}{(2\pi)^{\frac{n}{2}} \prod\limits_{k=1}^{n} \sigma_k} \exp\left[-\frac{1}{2}\sum_{k=1}^{n}\frac{1}{\sigma_k^2}\left(\sum_{j=1}^{n} c_{jk}\,(x_j' - m_j)\right)^2\right], \tag{16}$$

or, by putting

$$b_{ij} = \sum_{k=1}^{n} \frac{c_{ik}\,c_{jk}}{\sigma_k^2},$$

$$g(x_1', \ldots, x_n') = \frac{1}{(2\pi)^{\frac{n}{2}} \prod\limits_{k=1}^{n} \sigma_k} \times$$

$$\times \exp\left[-\frac{1}{2}\sum_{i=1}^{n}\sum_{j=1}^{n} b_{ij}\,(x_i' - m_i)\,(x_j' - m_j)\right], \tag{17}$$

---

[1] We can restrict ourselves here to orthogonal transformations, since the most general nondegenerate linear transformation may be decomposed into an orthogonal transformation and a transformation of the form $x_k'' = \lambda_k x_k$ $(k = 1, \ldots, n)$.

where (b_{ij}) is a symmetrical matrix such that the quadratic form

$$\sum_{i=1}^{n} \sum_{j=1}^{n} b_{ij} z_i z_j$$

is positive definite. It is known that a positive definite quadratic form can be transformed into a sum of squares. Thus if the density function of ζ' has the form (17), there exists an orthogonal transformation with matrix $C = (c_{ij})$ such that the n-dimensional density function of the random variables

$$\xi_k = \sum_{j=1}^{n} c_{jk} (\zeta'_j - m_j)$$

has the form (12). Note that the factor $1/\prod_{k=1}^{n} \sigma_k$ is equal to the positive square root of the determinant $|b_{ij}|$. The matrix $B = (b_{ij})$ can be written as CSC^*, where C^* is the transpose of C and S is the diagonal matrix

$$S = \begin{bmatrix} \dfrac{1}{\sigma_1^2} & 0 & \cdots & 0 \\[2mm] 0 & \dfrac{1}{\sigma_2^2} & \cdots & 0 \\[1mm] \cdots & \cdots & \cdots & \cdots \\[1mm] 0 & 0 & \cdots & \dfrac{1}{\sigma_n^2} \end{bmatrix}.$$

For the determinants, since $|C| = |C^*| = \pm 1$, we have

$$|B| = |S| = \prod_{k=1}^{n} \frac{1}{\sigma_k^2}.$$

Consequently, the density function (17) can be written as

$$g(x_1, \ldots, x_n) = \sqrt{\frac{|B|}{(2\pi)^n}} \exp\left[-\frac{1}{2} \sum_{i=1}^{n} \sum_{j=1}^{n} b_{ij} (x_i - m_i)(x_j - m_j) \right], \quad (18)$$

where the quadratic form $\Sigma b_{ij} z_i z_j$ is positive definite, $m_1, \ldots, m_n$ are arbitrary real numbers, and $|B|$ is the determinant of the matrix $B = (b_{ij})$. *Every density function of the form* (18) *is the density function of an n-dimensional normal distribution;* a suitable orthogonal transformation leads from (18) to a density function of the form (12). Since evidently all distributions with a density function of the form (12) are similar to each other, the n-dimensional normal distributions form a family. It has some interest to study the case of an m-dimensional vector $\zeta = (\xi_1, \ldots, \xi_n, 0, \ldots, 0)$ where $m > n$, and where the n-dimensional vector $(\xi_1, \ldots, \xi_n)$ has a density function of the form (12). By applying the orthogonal transformation

$$\xi'_j = \sum_{k=1}^{n} c_{jk} \xi_k + m_j \quad (j = 1, 2, \ldots, m), \tag{19}$$

we obtain an m-dimensional vector which, however, is not really m-dimensional; indeed (19) implies

$$\xi_k = \sum_{j=1}^{m} c_{jk} (\xi'_j - m_j) \quad \text{for} \quad k = 1, 2, \ldots, n \tag{20}$$

and

$$0 = \sum_{i=1}^{m} c_{jk} (\xi'_j - m_j) \quad \text{for} \quad k = n+1, \ldots, m. \tag{21}$$

Formula (21) expresses that the point $(\xi'_1, \ldots, \xi'_n)$ lies in an n-dimensional subspace of the m-dimensional space. A distribution of this kind is said to be a *degenerate* m-dimensional normal distribution.

§ 8. Distribution of a function of a random variable

Let ξ be a random variable with known distribution and let $y = \psi(x)$ be a Borel-measurable function. It is then easy to determine the distribution of the random variable $\eta = \psi(\xi)$. Let $\psi^{-1}(E)$ denote the set of real numbers x for which $\psi(x)$ belongs to the Borel-set E; let further I_y denote the interval $(-\infty, y)$ and let $F(x)$ be the distribution function of ξ and $G(y)$ that of η. It follows that

$$G(y) = P(\eta < y) = P(\xi \in \psi^{-1}(I_y)) = \int_{\psi^{-1}(I_y)} dF(x).$$

Let us first consider some particular cases. If ξ is a discrete random variable, the calculation of the distribution of η is almost trivial. In fact, let x_k $(k = 1, 2, \ldots)$ denote the possible values of ξ, then

$$P(\eta = y) = \sum_{\psi(x_k) = y} P(\xi = x_k),$$

where the summation extends over those values of k for which $\psi(x_k) = y$.

Let us now consider the case of an absolutely continuous distribution function. Let $f(x)$ be the density function of ξ. Assume $\psi(x)$ to be monotonic and differentiable and suppose $\psi'(x) \neq 0$ for every x. If $g(y)$ is the density function of $\eta = \psi(\xi)$, one easily finds that

$$g(y) = \begin{cases} \dfrac{f(\psi^{-1}(y))}{|\psi'(\psi^{-1}(y))|} & \text{for } \inf \psi(x) < y < \sup \psi(x), \\ 0 & \text{otherwise,} \end{cases} \tag{1}$$

where $x = \psi^{-1}(y)$ is the inverse function of $y = \psi(x)$.

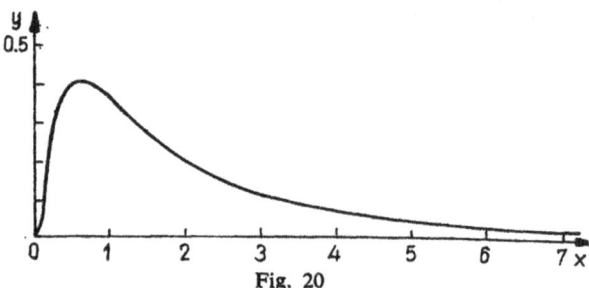

Fig. 20

If for instance ξ is a normally distributed random variable and $\eta = e^\xi$, we have by (1) putting $M = e^m$,

$$g(y) = \begin{cases} \dfrac{1}{\sqrt{2\pi}\,\sigma y} \exp\left[-\dfrac{\left(\ln \dfrac{y}{M}\right)^2}{2\sigma^2} \right] & \text{for } y > 0, \\ 0 & \text{for } y \le 0. \end{cases} \tag{2}$$

A random variable having the density function (2) is said to be *lognormal*. The lognormal distribution is of great importance in the theory of crushing of materials. The distribution of the grains of a granular material (stone, metal or crystal powder, etc.), in particular of a product produced by a breaking-process, is lognormal under rather general conditions. This density function is represented by the curve seen on Fig. 20.

Take now another example. Let the random vector ζ be uniformly distributed on the circumference of the unit circle; what is the density of the distribution of the projection ξ of ζ on the x-axis? We obtain from (1)

$$g(y) = \begin{cases} \dfrac{1}{\pi\sqrt{1 - y^2}} & \text{for } -1 < y < +1, \\ 0 & \text{otherwise.} \end{cases} \tag{3}$$

§ 9. The convolution of distributions

Let two independent random variables ξ and η be given having the distribution functions $F(x)$ and $G(y)$ respectively. Consider the sum $\zeta = \xi + \eta$; let $H(z)$ be its distribution function. We have clearly

$$H(z) = \iint_{x+y<z} dF(x)\, dG(y) = \tag{1}$$

$$= \int_{-\infty}^{+\infty} F(z-y)\, dG(y) = \int_{-\infty}^{+\infty} G(z-x)\, dF(x).$$

The distribution function $H(z)$ is called the *convolution of the distribution functions $F(x)$ and $G(y)$*. The convolution operation is denoted by $H = F * G$ Clearly it is commutative and associative; in fact, if ξ_1, ξ_2, ξ_3 are independent random variables, we have

$$\xi_1 + \xi_2 = \xi_2 + \xi_1 \text{ and } (\xi_1 + \xi_2) + \xi_3 = \xi_1 + (\xi_2 + \xi_3).$$

From this follows for the distribution functions that

$$F_1 * F_2 = F_2 * F_1 \text{ and } (F_1 * F_2) * F_3 = F_1 * (F_2 * F_3).$$

Suppose that ξ and η are independent random variables having absolutely continuous distribution functions; let $f(x)$ and $g(y)$ be their density functions. It will be shown that $H = F * G$ is also absolutely continuous and that the density function of $\zeta = \xi + \eta$ is

$$h(z) = \int_{-\infty}^{+\infty} f(x)\, g(z-x)\, dx = \int_{-\infty}^{+\infty} f(z-y)\, g(y)\, dy, \tag{2}$$

(the equality of the integrals in (2) can be shown e.g. by a transformation of the variable).

Formula (2) can be proved as follows: (1) is equivalent to

$$H(z) = \int_{-\infty}^{+\infty} \int_{-\infty}^{z} f(x-y)\, dx\, dG(y) =$$

$$= \int_{-\infty}^{z} \int_{-\infty}^{+\infty} f(x-y)\, dG(y)\, dx. \tag{3}$$

By differentiating (3) we obtain

$$h(z) = \int_{-\infty}^{+\infty} f(z-y)\, dG(y). \tag{4}$$

From (4) follows immediately (2). Further it can be seen that the distribution of $\zeta = \xi + \eta$ is absolutely continuous, provided that one of ξ and η has such a distribution, regardless of the other distribution.

The function $h(x)$ defined by (2) is called the *convolution* of the density functions $f(x)$ and $g(x)$ and is denoted by $h = f * g$. It is easy to show that $h(x)$ is a density function; as a matter of fact (2) implies $h(x) \geq 0$ and

$$\int_{-\infty}^{+\infty} h(x)\,dx = \int_{-\infty}^{+\infty}\int_{-\infty}^{+\infty} f(x-y)g(y)\,dy\,dx = \int_{-\infty}^{+\infty} f(x)\,dx \int_{-\infty}^{+\infty} g(y)\,dy = 1.$$

In what follows, we shall give some examples of the convolution of absolutely continuous distributions (the convolution of discrete distributions was already dealt with in Chapter III, § 6).

1. *Convolution of uniform distributions.*
Suppose that

$$f(x) = \begin{cases} \dfrac{1}{b-a} & \text{for } a < x < b, \\ 0 & \text{otherwise} \end{cases} \tag{5}$$

and

$$g(x) = \begin{cases} \dfrac{1}{d-c} & \text{for } c < x < d, \\ 0 & \text{otherwise.} \end{cases} \tag{6}$$

Assume $d - c \geq b - a$. The convolution of the density functions of two independent random variables ξ and η with the respective density functions (5) and (6) is equal to

$$h(x) = \begin{cases} 0 & \text{for } x \leq a+c \ \text{ or } \ b+d \leq x, \\[2mm] \dfrac{x-(a+c)}{(b-a)(d-c)} & \text{for } a+c \leq x \leq b+c, \\[2mm] \dfrac{1}{d-c} & \text{for } b+c \leq x \leq a+d, \\[2mm] \dfrac{(b+d)-x}{(b-a)(d-c)} & \text{for } a+d \leq x \leq b+d. \end{cases} \tag{7}$$

The graph of the function $y = h(x)$ is an isosceles trapezoid with its base on the x-axis (Fig. 21 represents the case $a = -1, b = 0, c = -1, d = +1$).

Note that $h(x)$ is everywhere continuous, though $f(x)$ and $g(x)$ have jumps. (The convolution in general smoothes out discontinuities.)

In particular, if ξ and η have the same uniform distribution, the graph $h(x)$ is an isosceles triangle; this is the so-called *Simpson distribution*.

By repeated application of (2) one can determine the density function of the sum of several independent random variables with absolutely continuous

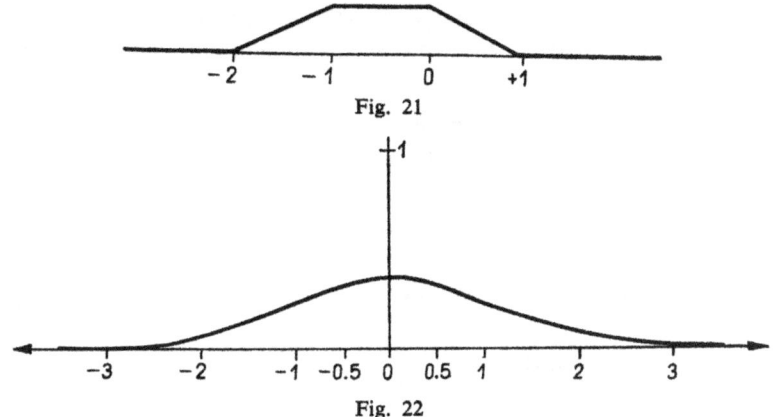

Fig. 21

Fig. 22

distribution. Thus for instance the density function of the sum of three independent random variables uniformly distributed on $(-1, +1)$ is given by

$$
h(x) = \begin{cases} 0 & \text{for} \quad |x| \geq 3, \\[2mm] \dfrac{(3 - |x|)^2}{16} & \text{for} \quad 1 \leq |x| \leq 3, \\[2mm] \dfrac{3 - x^2}{8} & \text{for} \quad 0 \leq |x| \leq 1. \end{cases} \tag{8}
$$

The function $h(x)$ (cf. Fig. 22) is not only continuous but also everywhere differentiable. The curve has already a bell-shaped form as the Gaussian curve; by adding more and more independent random variables with uniform distribution on $(-1, +1)$, this similarity becomes still closer: we have here a particular case of the central limit theorem to be dealt with later. The density function of the sum of n mutually independent random variables with uniform distribution on $(-1, +1)$ is

$$
f_n(x) = \begin{cases} \dfrac{1}{2^n (n-1)!} \displaystyle\sum_{k=0}^{\left[\frac{n+x}{2}\right]} (-1)^k \binom{n}{k}(n + x - 2k)^{n-1} & \text{for} \ |x| < n, \\[3mm] 0 & \text{otherwise} \end{cases} \tag{9}
$$

as it is readily proved by mathematical induction. The graph of the function $f_n(x)$ consists of arcs of functions of degree $n - 1$; it is $(n - 2)$-times differentiable, i.e. the first $n - 2$ derivatives of these functions are equal at the endpoints of these arcs.

This distribution was first studied by N. I. Lobatchewski. He wanted to use it to evaluate the error of astronomical measurements, in order to decide whether the Euclidean or the non-Euclidean geometry is valid in the Universe.

2. The convolution of normal distributions.

Let ξ and η be two independent random variables with density functions

$$f(x) = \frac{1}{\sigma_1} \varphi\left(\frac{x - m_1}{\sigma_1}\right) \quad \text{and} \quad g(x) = \frac{1}{\sigma_2} \varphi\left(\frac{x - m_2}{\sigma_2}\right);$$

it follows from (2) by an easy calculation that putting $h = f * g$ one has

$$h(x) = \frac{1}{\sqrt{\sigma_1^2 + \sigma_2^2}} \varphi\left(\frac{x - (m_1 + m_2)}{\sqrt{\sigma_1^2 + \sigma_2^2}}\right). \tag{10}$$

The sum of ξ and η is thus also a normally distributed random variable; the parameters of the distribution are $m = m_1 + m_2$ and $\sigma = (\sigma_1^2 + \sigma_2^2)^{\frac{1}{2}}$. It follows that the sum of any number of independent and normally distributed random variables is again a normally distributed random variable.

3. Pearson's χ^2-distribution.[1]

The distribution of the sum of the squares of n independent random variables $\xi_1, \ldots, \xi_n$ with the same normal distribution, plays an important role in mathematical statistics. We shall determine the density function of this sum for any n. Let $\varphi(x)$ be the density function of the random variables ξ_k $(k = 1, 2, \ldots, n)$. Let the sum of the squares of the ξ_k be denoted by

$$\chi_n^2 = \sum_{k=1}^{n} \xi_k^2. \tag{11}$$

Let $h_n(x)$ be the density function of χ_n^2. The statement

$$h_n(x) = \begin{cases} \dfrac{x^{\frac{n}{2}-1} e^{-\frac{x}{2}}}{2^{\frac{n}{2}} \Gamma\left(\dfrac{n}{2}\right)} & \text{for} \quad x > 0, \\[4mm] 0 & \text{for} \quad x < 0 \end{cases} \tag{12}$$

[1] This distribution was already used by Helmert, before Pearson.

can be proven by mathematical induction. For we have by Formula (1) of § 8

$$h_1(x) = \begin{cases} \dfrac{1}{\sqrt{2\pi x}}\, e^{-\frac{x}{2}} & \text{for } x > 0, \\ 0 & \text{for } x < 0, \end{cases} \tag{13}$$

which shows that (12) is valid for $n = 1$. Suppose that (12) is valid for a certain value of n. Given (2) and the induction assumption we have

$$h_{n+1}(x) = \frac{e^{-\frac{x}{2}}\, x^{\frac{n+1}{2}-1}}{\sqrt{2\pi}\, 2^{\frac{n}{2}}\, \Gamma\left(\dfrac{n}{2}\right)} \int_0^1 \frac{y^{\frac{n}{2}-1}}{\sqrt{1-y}}\, dy. \tag{14}$$

As for Euler's beta function $B(a, b)$ the formula

$$B(a, b) = \int_0^1 t^{a-1}(1-t)^{b-1}\, dt \quad (a > 0,\ b > 0) \tag{15}$$

is valid, we have[1]

$$B(a, b) = \frac{\Gamma(a)\, \Gamma(b)}{\Gamma(a+b)}. \tag{16}$$

Since $\Gamma\left(\dfrac{1}{2}\right) = \sqrt{\pi}$, from (14) follows

$$h_{n+1}(x) = \frac{x^{\frac{n+1}{2}-1}\, e^{-\frac{x}{2}}}{2^{\frac{n+1}{2}}\, \Gamma\left(\dfrac{n+1}{2}\right)} \quad \text{for } x > 0.$$

Thus (12) holds with $n + 1$ instead of n; thus it holds for every n.

From (12) we obtain that the density function $g_n(x)$ of $\chi_n = +\sqrt{\xi_1^2 + \ldots + \xi_n^2}$ is

$$g_n(x) = 2\, \frac{x^{n-1}\, e^{-\frac{x^2}{2}}}{2^{\frac{n}{2}}\, \Gamma\left(\dfrac{n}{2}\right)} \quad \text{for } x > 0. \tag{17}$$

The distribution with density function (12) is called *Pearson's χ^2-distribution with n degrees of freedom.* The distribution with density function (17) is called the *χ-distribution with n degrees of freedom.*

[1] For the proof of this formula cf. e.g. F. Lösch and F. Schoblik [1] or V. I. Smirnov [1].

For $n = 3$ Equation (17) gives the density function of *Maxwell's velocity distribution*, which is of great importance in the kinetic theory of gases.

Consider a gas contained in a vessel. The velocity of a molecule has for its components in the directions x, y, and z the random variables ξ, η, and ζ,

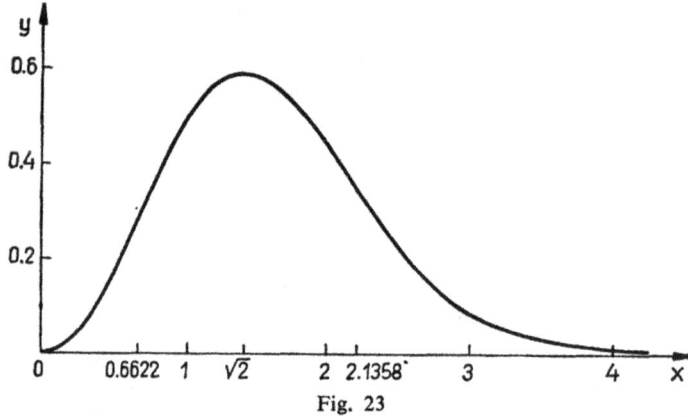

Fig. 23

respectively. It is shown in the kinetic theory of gases that these three random variables are independent, normally distributed, and have the same density function:

$$\frac{1}{\sigma} \varphi \left(\frac{x}{\sigma} \right) .$$

The physical meaning of ξ, η and ζ having identical distributions is that the pressure of the gas has the same value in every direction; $m = 0$ means that the gas does not move as a whole, only its molecules move at random. We wish to determine the density function of the absolute value of the velocity

$$v = \sqrt{\xi^2 + \eta^2 + \zeta^2} \; . \tag{18}$$

Clearly, $\dfrac{\xi}{\sigma}$, $\dfrac{\eta}{\sigma}$, $\dfrac{\zeta}{\sigma}$ have the density function $\varphi(x)$; hence, by (17), the density

function of $\dfrac{v}{\sigma}$ is $g_3(x)$. According to Formula (1) of § 8 the density $v(x)$ of v

is $\dfrac{1}{\sigma} g_3 \left(\dfrac{x}{\sigma} \right)$. Since $\Gamma \left(\dfrac{3}{2} \right) = \dfrac{1}{2} \Gamma \left(\dfrac{1}{2} \right) = \dfrac{1}{2} \sqrt{\pi}$, we have

$$v(x) = \frac{1}{\sigma^2} \sqrt{\frac{2}{\pi}} x^2 e^{-\frac{x^2}{2\sigma^2}} \qquad (x \geq 0). \tag{19}$$

(The curve representing $y = v(x)$ is drawn on Fig. 23 for $\sigma = 1$.) Note that σ has the physical meaning

$$\sigma = \sqrt{\frac{kT}{M}}, \tag{20}$$

where T is the absolute temperature, M the mass of the molecules, and k is Boltzmann's constant.

Let further be noted that $h_2(x) = \frac{1}{2} e^{-\frac{x}{2}}$: the χ^2-distribution with 2 degrees of freedom is an exponential distribution.

4. Convolution of exponential distributions.

The exponential distribution was introduced in the previous section in connection with radioactive disintegration; but it occurs also in many other problems of physics and technology. In what follows, we give an example from the textile industry; namely the problem of the tearing of the yarn on the loom. At a given moment, the yarn is or is not torn, according as the section of the yarn, submitted at this moment to a certain stress, does or does not yield to the latter. Evidently this does not depend on the time during which the loom worked uninterruptedly. Let ξ be the random variable representing this time-interval, i.e. the time between the start of the work and the first rupture of the yarn; let $F(x)$ denote the distribution function and $f(x)$ the density function of ξ; for $F(x)$ one obtains, as in the case of the radioactive disintegration, the functional equation

$$\frac{1 - F(t + s)}{1 - F(s)} = 1 - F(t), \tag{21}$$

from which it follows that

$$F(t) = 1 - e^{-\lambda t} \quad \text{for} \quad t \geq 0 \tag{22}$$

and

$$f(t) = \lambda e^{-\lambda t} \quad (t > 0) \tag{23}$$

(where λ is a positive constant). Hence the random variable ξ has an exponential distribution.

Consider now the functioning of the loom during a sufficiently long time interval. Let ζ_n denote the time interval until the n-th rupture of the yarn. For the sake of simplicity assume the time wasted between the rupture and the tieing of the yarn to be so small that it can be neglected. Then we have

$$\zeta_n = \xi_1 + \xi_2 + \ldots + \xi_n,$$

where $\xi_1, \ldots, \xi_n$ are independent and every one of them has distribution (22). Let $F_n(t)$ be the distribution function and $f_n(t)$ the density function of ζ_n.

It can be shown by induction that

$$f_n(t) = \frac{\lambda^n \, t^{n-1} \, e^{-\lambda t}}{(n-1)!} \quad \text{for} \quad t > 0; \; n = 1, 2, \dots \dots \tag{24}$$

By (23), Formula (24) holds for $n = 1$. Assume its validity for a certain value of n. Since $\zeta_{n+1} = \zeta_n + \xi_{n+1}$ and further ζ_n is independent of ξ_{n+1}, Formula (2) can be applied here. Thus we obtain

$$f_{n+1}(t) = \int_0^t f_n(u) f_1(t-u) \, du = \frac{\lambda^{n+1} \, t^n \, e^{-\lambda t}}{n!}$$

and (24) is hereby proved. It follows

$$F_n(t) = \Gamma_n(\lambda t), \qquad (t \geq 0) \tag{25}$$

where

$$\Gamma_n(x) = \frac{1}{(n-1)!} \int_0^x u^{n-1} e^{-u} \, du \quad (x \geq 0) \tag{26}$$

is the incomplete Γ-function.

The distribution with the distribution function (25) is called the Γ-*distribution of order n and parameter λ*. For $\lambda = \dfrac{1}{2}$, $f_n(t)$ is equal to the function $h_{2n}(t)$ defined by (12); thus the χ^2-distribution with $2n$ degrees of freedom is the same as the Γ-distribution of order n.

This result permits us to calculate the probability that the yarn is torn exactly n times during a time interval $(0, T)$. Let v_T be the number of breakings of the yarn in the time interval $(0, T)$; clearly v_T can only assume non-negative integer values. The event $v_T = n$ means that $\zeta_n < T$, but $\zeta_{n+1} \geq T$. Let A_n denote the event $\zeta_n < T$; then, because of $A_{n+1} \subset A_n$, we have

$$P(v_T = n) = F_n(T) - F_{n+1}(T). \tag{27}$$

Substituting here for $F_n(T)$ and $F_{n+1}(T)$ and integrating by parts, we find

$$P(v_T = n) = \frac{(\lambda T)^n \, e^{-\lambda T}}{n!}. \tag{28}$$

Thus the random variable v_T has a Poisson distribution of parameter λT. Here we encountered an important further property of the Poisson distribution: *if a sequence of events has the property that time intervals between consecutive events do not depend on each other and have distribution function $1 - e^{-\lambda t}$ $(t \geq 0)$, then the number of events occurring in a fixed interval $(0, T)$ has a Poisson distribution with parameter λT.*

The above reasoning can be applied to a large number of technical problems (e.g. the breaking of machine parts).

One can also determine the distribution of a sum of independent random variables having exponential distributions with different parameters. Let $\xi_1, \ldots, \xi_n$ be independent random variables with exponential distributions, let $\lambda_k e^{-\lambda_k t}$ be the density function of ξ_k for $t > 0$ where the numbers $\lambda_1, \ldots, \lambda_n$ are all different. It can be shown by induction that the density function $g_n(t)$ of $\eta_n = \xi_1 + \ldots + \xi_n$ is given by

$$g_n(t) = (-1)^{n-1} \lambda_1 \lambda_2 \ldots \lambda_n \sum_{k=1}^{n} \frac{e^{-\lambda_k t}}{\prod_{i \neq k} (\lambda_k - \lambda_i)} \quad \text{for} \quad t > 0. \quad (29)$$

Formula (29) has the following physical application: Let A_1 be a radioactive substance with the disintegration constant λ_1. The disintegration of an A_1 atom means its transformation into some other kind of atom A_2; suppose that the A_2 atoms are radioactive as well and have the disintegration constant λ_2. Similarly, let A_k ($k = 3, 4, \ldots, n$) be the result of the disintegration of an A_{k-1} atom, the disintegration constant of A_k being λ_n for $k \leq n$. Assume that the substance A_{n+1} is not radioactive. Denoting by η the time necessary for the transformation of an atom A_1 into an atom A_{n+1}, η_n clearly has the density function (29). For instance if A_1 is uranium, A_{n+1} is lead and η_n is the time necessary for an uranium atom to change into lead.

§ 10. Distribution of a function of several random variables

Let $\xi_1, \ldots, \xi_n$ be arbitrary random variables with the joint (n-dimensional) distribution function $F(x_1, \ldots, x_n)$ and let $g(x_1, \ldots, x_n)$ be a Borel-measurable function. Evidently, the distribution function of $\eta = g(\xi_1, \ldots, \xi_n)$ is

$$P(\eta < y) = \int \ldots \int_{g(x_1, \ldots, x_n) < y} dF(x_1, \ldots, x_n).$$

Let us consider some important particular cases. Let ξ and η be independent random variables with absolutely continuous distribution functions; let us consider the random variables $\zeta_1 = \xi\eta$ and $\zeta_2 = \dfrac{\xi}{\eta}$. Let the density functions of ξ and η be $f(x)$ and $g(y)$; we have

$$P(\xi\eta < z) = \iint_{xy < z} f(x) g(y) \, dx \, dy \quad (1a)$$

and

$$P\left(\frac{\xi}{\eta} < z\right) = \iint_{\frac{x}{y} < z} f(x) g(y) \, dx \, dy. \quad (1b)$$

By differentiating we obtain the corresponding density functions $p(z)$ and $q(z)$ of ζ_1 and ζ_2:

$$p(z) = \int_{-\infty}^{+\infty} g(y) f\left(\frac{z}{y}\right) \frac{dy}{|y|}, \tag{2}$$

$$q(z) = \int_{-\infty}^{+\infty} |y| g(y) f(zy) \, dy. \tag{3}$$

Let us give some examples.

1. *Student's distribution.*

We shall determine the distribution of the random variable

$$\zeta = \frac{\xi_0}{\sqrt{\xi_1^2 + \ldots + \xi_n^2}}, \tag{4}$$

where $\xi_0, \xi_1, \ldots, \xi_n$ are independent random variables having the same normal distribution with density function

$$\varphi(x) = \frac{1}{\sqrt{2\pi}} e^{-\frac{x^2}{2}}.$$

Let $q_n(z)$ be the density function of ζ. We know already the density function of the denominator of (4) (cf. Formula (17) of § 9), hence we obtain from (3)

$$q_n(z) = \frac{1}{\sqrt{\pi}} \frac{\Gamma\left(\frac{n+1}{2}\right)}{\Gamma\left(\frac{n}{2}\right)} \frac{1}{(1+z^2)^{\frac{n+1}{2}}}. \tag{5}$$

The distribution with density function (5) is called *Student's distribution with n degrees of freedom*. It plays an important role in mathematical statistics, since "Student's *t*-test" is based on it. The particular case $n = 1$ gives the *Cauchy distribution*, with the density function

$$q_1(z) = \frac{1}{\pi(1+z^2)}. \tag{6}$$

2. *Distribution of the ratio of two independent random variables having χ^2-distributions.*

In mathematical statistics one is often interested in the density function $h(z)$ of the ratio of two independent random variables ξ and η having χ^2-distributions with n and m degrees of freedom, respectively.

It follows easily from Formula (12) of § 9 and from (3) of the present section that

$$h(z) = \frac{\Gamma\left(\dfrac{n+m}{2}\right)}{\Gamma\left(\dfrac{n}{2}\right)\Gamma\left(\dfrac{m}{2}\right)} \frac{z^{\frac{n}{2}-1}}{(1+z)^{\frac{n+m}{2}}} \quad \text{for} \quad z > 0. \tag{7}$$

3. The beta distribution.

If ζ is the ratio considered in the previous example, let τ denote the random variable $\tau = \dfrac{\zeta}{1+\zeta}$ and $k(x)$ the density function of τ. By (1) of § 8 we obtain

$$k(x) = \frac{\Gamma\left(\dfrac{n+m}{2}\right)}{\Gamma\left(\dfrac{n}{2}\right)\Gamma\left(\dfrac{m}{2}\right)} x^{\frac{n}{2}-1}(1-x)^{\frac{m}{2}-1} \quad \text{for} \quad 0 < x < 1. \tag{8}$$

The distribution function $K(x) = \int_0^x k(t)\,dt$ is thus

$$K(x) = \frac{\Gamma\left(\dfrac{n+m}{2}\right)}{\Gamma\left(\dfrac{n}{2}\right)\Gamma\left(\dfrac{m}{2}\right)} \int_0^x t^{\frac{n}{2}-1}(1-t)^{\frac{m}{2}-1}\,dt = B_{\frac{n}{2},\frac{m}{2}}(x) \text{ for } 0 \le x \le 1, \tag{9}$$

where

$$B_{a,b}(x) = \frac{\Gamma(a+b)}{\Gamma(a)\Gamma(b)} \int_0^x t^{a-1}(1-t)^{b-1}\,dt \tag{10}$$

is, up to a numerical factor, Euler's incomplete beta integral. The distribution $B(a, b)$ $(a > 0, b > 0)$ having (10) for its distribution function is called the *beta-distribution of order* (a, b).

4. Order statistics.

In nonparametric statistics the following problem is of importance: Let $\xi_1, \xi_2, \ldots, \xi_n$ be independent random variables with the same continuous distribution: let $F(x)$ be the distribution function of ξ_k. Arrange the values of $\xi_1, \ldots, \xi_n$ in increasing order,[1] and denote by ξ_k^* the k-th of these ordered values; hence, in particular

$$\xi_1^* = \min_{1 \le k \le n} \xi_k, \quad \xi_n^* = \max_{1 \le k \le n} \xi_k. \tag{11}$$

ξ_k^* is called the k-th order statistics of the sample $(\xi_1, \ldots, \xi_n)$.

[1] The probability that equal values occur is 0.

Determine now the distribution function $F_k(x)$ of ξ_k^* $(k = 1, 2, \ldots, n)$; clearly $\xi_k^* < x$ means that among the values taken on by $\xi_1, \ldots, \xi_n$ there are at least k which are less than x. The probability that r given variables among the ξ_k are less than x and the other $n - r$ greater than or equal to x, is given by $[F(x)]^r [1 - F(x)]^{n-r}$; since the first r can be chosen in $\binom{n}{r}$ different ways, we have

$$F_k(x) = \sum_{r=k}^{n} \binom{n}{r} [F(x)]^r [1 - F(x)]^{n-r}. \tag{12}$$

This expression can be simplified by taking into account the identity

$$\sum_{r=k}^{n} \binom{n}{r} p^r (1 - p)^{n-r} = \frac{n!}{(k-1)! \, (n-k)!} \int_0^p x^{k-1} (1 - x)^{n-k} \, dx, \tag{13}$$

which gives

$$F_k(x) = B_{k, n+1-k}\left(F(x)\right), \tag{14}$$

where $B_{k, n+1-k}(x)$ is the incomplete beta function of order $(k, n + 1 - k)$ (cf. (10)). In the case when $F(x) = x$ for $0 \le x \le 1$, i.e. if the ξ_k-s are uniformly distributed on the interval $(0, 1)$, ξ_k^* has a beta distribution of order $(k, n + 1 - k)$, and, in particular, for $0 \le x \le 1$,

$$F_1(x) = P\left(\min_{1 \le k \le n} \xi_k < x\right) = 1 - (1 - x)^n \tag{15}$$

and

$$F_n(x) = P\left(\max_{1 \le k \le n} \xi_k < x\right) = x^n. \tag{16}$$

If $\xi_1, \ldots, \xi_n$ are independent and have the same continuous, monotone, and strictly increasing distribution function $F(x)$, then the random variables $\eta_k = F(\xi_k)$ $(k = 1, \ldots, n)$ are independent and uniformly distributed on the interval $(0, 1)$. In fact, if $x = F^{-1}(y)$ is the inverse function of $y = F(x)$, we have

$$P\left(\eta_k < x\right) = P\left(\xi_k < F^{-1}(x)\right) = F\left(F^{-1}(x)\right) = x \tag{17}$$

for $0 \le x \le 1$.

If now η_k^* is the k-th among the random variables $\eta_1, \ldots, \eta_n$ ranked according to increasing order, it is clear that $\eta_k^* = F(\xi_k^*)$ and we have

$$P\left(\eta_k^* < x\right) = B_{k, n+1-k}(x). \tag{18}$$

5. *Mixtures.*

Let $F_k(x)$ $(k = 1, 2, \ldots)$ be arbitrary distribution functions and $\{p_k\}$

a discrete probability distribution. Then

$$F(x) = \sum_{k=1}^{\infty} p_k F_k(x) \tag{19}$$

is also a distribution function. It is called the *mixture of the distribution functions $F_k(x)$ ($k = 1, 2, \ldots$) taken with the weights p_k.* This concept was already defined in the foregoing Chapter for the particular case where the functions $F_k(x)$ are discrete distribution functions.

Consider the following example: a physical quantity is measured by two different procedures, the errors of the measurements being in both cases normally distributed with density functions $\dfrac{1}{\sigma_1} \varphi\left(\dfrac{x}{\sigma_1}\right)$ and $\dfrac{1}{\sigma_2} \varphi\left(\dfrac{x}{\sigma_2}\right)$. N_1 measurements were performed by the first, and N_2 measurements by the second method without registering, which of the results was furnished by the first and which by the second of the methods (the measurements were mixed). What will be the distribution function of the error of a measurement chosen at random from these $N = N_1 + N_2$ measurements? If

$$\Phi(x) = \frac{1}{\sqrt{2\pi}} \int_{-\infty}^{x} e^{-\frac{t^2}{2}} \, dt,$$

it follows from the theorem of total probability that this distribution function $F(x)$ is given by

$$F(x) = \frac{N_1}{N} \Phi\left(\frac{x}{\sigma_1}\right) + \frac{N_2}{N} \Phi\left(\frac{x}{\sigma_2}\right),$$

i.e. $F(x)$ is the mixture of the distribution functions of the errors of the two methods, taken with the weights $\dfrac{N_1}{N}$ and $\dfrac{N_2}{N}$.

It is easy to extend the notion of the mixture to a nondenumerable set of distribution functions. If $F(t, x)$ for each value of the parameter t is a distribution function and for each fixed value of x $F(t, x)$ is a measurable function of t and if $G(t)$ is an arbitrary distribution function, the Stieltjes integral

$$H(x) = \int_{-\infty}^{+\infty} F(t, x) \, dG(t) \tag{20}$$

defines a distribution function called the mixture of the distribution functions $F(t, x)$ mixed with the distribution function $G(t)$. If $G(t)$ is a discrete distribution function, (20) reduces to (19). It is easy to see that the function $H(x)$ defined by (20) is in fact a distribution function.

Let us consider an important application. One has often to determine the distribution function of the sum

$$\eta = \xi_1 + \xi_2 + \ldots + \xi_\nu \qquad (21)$$

such that the number ν of the terms is a random variable. Assume that the ξ_k are mutually independent and ν is independent of the ξ_k. Let $F_k(x)$ denote the distribution function of ξ_k, $G_n(x)$ the distribution function of $\zeta_n = \xi_1 + \ldots + \xi_n$ and $H(x)$ the distribution function of the random variable η defined by (21); let further be $P(\nu = n) = p_n$ $(n = 1, 2, \ldots)$. Then, by the theorem of total probability,

$$H(x) = \sum_{n=1}^{\infty} p_n G_n(x), \qquad (22)$$

i.e. $H(x)$ is a mixture of the distribution functions $G_n(x)$.

Example. If $p_n = \binom{n-1}{k-1} p^k q^{n-k}$ $(n = k, k+1, \ldots)$ and $F_k(x) = 1 - e^{-\lambda x}$ for $x \geq 0$, further if the random variables $\nu, \xi_1, \xi_2, \ldots$ are independent, then

$$G_n(x) = \int_0^x \frac{\lambda^n t^{n-1} e^{-\lambda t}}{(n-1)!} \, dt,$$

and, by (22)

$$H(x) = \int_0^x \frac{(p\lambda)^k t^{k-1} e^{-p\lambda t}}{(k-1)!} \, dt, \qquad (23)$$

hence η has a Γ-distribution of order k a:

§ 11. The general notion of expectation

We shall now extend the notion of expectation to an arbitrary random variable ξ. In order to do this assume that a great number of independent observations were made on the value of ξ. Arrange the observed values into classes such that the k-th class should contain the values between kh (included) and $(k+1)h$ (excluded) $(h > 0; k = 0, \pm 1, \pm 2, \ldots)$. According to the law of large numbers the arithmetic mean of the observed values will be near to

$$\sum_{k=-\infty}^{+\infty} kh P \left(kh \leq \xi < (k+1)h \right), \qquad (1)$$

provided of course that the series is convergent; the approximation will be the closer the smaller is the value of h. Hence it is natural to define the *expec-*

tation of ξ by

$$E(\xi) = \lim_{h \to 0} \sum_{k=-\infty}^{+\infty} khP(kh \leq \xi < (k+1)h),\qquad(2)$$

if this limit exists. If ξ is a discrete random variable, this definition coincides with that given in the preceding Chapter.

Obviously, if the limit (2) exists, it represents the Lebesgue integral of the function $\xi = \xi(\omega)$ with respect to the probability measure P, i.e.

$$E(\xi) = \int_{\Omega} \xi(\omega) \, dP .\qquad(3)$$

(2) can be interpreted in a different manner too. Let $\xi_h = h\left[\dfrac{\xi}{h}\right]$, where

$[x]$ denotes the entire part of the real number x; ξ_h is a discrete random variable and

$$\sum_{k=-\infty}^{+\infty} khP(kh \leq \xi < (k+1)h) = \sum_{k=-\infty}^{+\infty} khP(\xi_h = kh) = E(\xi_h)$$

is the expectation of ξ_h. (2) can be written in the form

$$E(\xi) = \lim_{h \to 0} E(\xi_h);\qquad(2')$$

ξ_h is the greatest multiple of h not exceeding ξ. For $h = 10^{-r}$, ξ_h is nothing else than the value of ξ rounded off to r decimal places.

In what follows, the knowledge of the Lebesgue integral will be taken for granted; we shall give without proof the properties of $E(\xi)$ which follow directly from the properties of the Lebesgue integral. Theorems, which in the general case can be proved in the same manner as in the case of discrete distributions and which were proved for the latter in § 7 of Chapter III, will be formulated here without proof. But the reader may profit from carrying through these proofs for the general case too.

Evidently, the expectation $E(\xi)$ depends only on the distribution function of ξ; hence one may call $E(\xi)$ the *expectation of the distribution of* ξ.

If ξ is a random variable with distribution function $F(x)$, then

$$E(\xi) = \int_{-\infty}^{+\infty} x dF(x) .\qquad(4)$$

If ξ is bounded with probability 1, then $E(\xi)$ exists. If $P(A \leq \xi \leq B) = 1$, then $A \leq E(\xi) \leq B$; in particular, if $P(\xi \geq 0) = 1$, we have $E(\xi) \geq 0$, the equality being valid if and only if $P(\xi = 0) = 1$. If the distribution function of ξ is absolutely continuous and if $f(x)$ is the density function of ξ, then

$$E(\xi) = \int_{-\infty}^{+\infty} xf(x)dx.$$ (5)

E.g. for the Cauchy distribution with the density function

$$f(x) = \frac{1}{\pi(1 + x^2)}$$

the expectation does not exist, since in this case the integral (5) does not converge.

Let us now consider some examples.

1. *Expectation of the uniform distribution.*
If ξ is a random variable uniformly distributed on the interval (a, b), it follows from (5) that

$$E(\xi) = \frac{a + b}{2},$$

which is also evident because of the symmetry of the uniform distribution.

2. *Expectation of the normal distribution*
If ξ is a normally distributed random variable, its density function has the form

$$f(x) = \frac{1}{\sqrt{2\pi}\,\sigma} \exp\left(-\frac{(x - m)^2}{2\sigma^2}\right) \quad (m \text{ real}, \ \sigma > 0).$$

By applying (5) we obtain easily

$$E(\xi) = m.$$

Thus we have found the probabilistic meaning of one of the parameters of the normal distribution.[1]

3. *Expectation of the gamma distribution.*
If the random variable ξ has a Γ distribution of order k, its density function is of the form

$$f(x) = \frac{\lambda^k x^{k-1} e^{-\lambda x}}{(k - 1)!} \quad (x > 0),$$

---

[1] Later on we shall see that σ is the standard deviation of ξ.

where λ is a positive constant; from this it follows by (5) that

$$E(\zeta) = \frac{k}{\lambda}.$$

In particular, the ordinary exponential distribution with the distribution function $1 - e^{-\lambda x}$ for $x \geq 0$ has expectation $\frac{1}{\lambda}$. Thus we found another probabilistic meaning of the parameter λ of the exponential distribution. The disintegration constant of a radioactive substance is thus the inverse of the mean life-time of a radioactive atom. As we have seen, the relation $\frac{1}{\lambda} = \frac{h}{\ln 2}$ holds between the constant and the half-period h; from this it follows that the mean life is equal to the product of the half-period and $\frac{1}{\ln 2}$ (i.e. it is 1.34 times the half-period).

4. *Expectation of the χ^2- and χ-distributions.*
According to Formula (12) of § 9 the density function of χ_n^2 is

$$h_n(x) = \frac{x^{\frac{n}{2}-1} e^{-\frac{x}{2}}}{2^{\frac{n}{2}} \Gamma\left(\frac{n}{2}\right)}.$$

A simple calculation gives

$$E(\chi_n^2) = n.$$

Similarly, for the expectation of χ_n

$$E(\chi_n) = \sqrt{2}\, \frac{\Gamma\left(\frac{n+1}{2}\right)}{\Gamma\left(\frac{n}{2}\right)}.$$

By applying Stirling's formula, we find for $n \to \infty$

$$E(\chi_n) \approx \sqrt{n},$$

hence

$$E(\chi_n^2) \approx [E(\chi_n)]^2 \quad \text{if} \quad n \to \infty.$$

5. *Expectation of the beta distribution.*
Let ζ be a random variable with a beta distribution; its density function is

$$B'_{a,b}(x) = \frac{\Gamma(a+b)}{\Gamma(a)\Gamma(b)}\, x^{a-1}(1-x)^{b-1} \quad (0 < x < 1).$$

From this, by (5),

$$E(\xi) = \frac{a}{a+b}.$$

6. Order statistics.

Let $\xi_1, \ldots, \xi_n$ be independent random variables each uniformly distributed on the interval $(0, 1)$. Let ξ_k^* be the random variable which assumes the k-th of the values $\xi_1, \ldots, \xi_n$ ranked according to increasing magnitude; by Formula (14) of § 10

$$F_k(x) = B_{k,\,n+1-k}(x).$$

Hence $E(\xi_k^*) = \dfrac{k}{n+1}$; the expectations of the ξ_k^*-s subdivide the interval $(0, 1)$ into $n + 1$ equal intervals, as could also be guessed by a symmetry argument.

We hinted already at the analogy between probability distributions and distributions of masses. Consider now the distribution of the unit mass on a line, such that between the abscissas a and $b > a$ there should lie a mass $F(b) - F(a)$, where $F(x)$ is a given distribution function. If x_0 is the center of gravity of this distribution, we know that

$$x_0 = \int_{-\infty}^{+\infty} x\,dF(x),$$

hence x_0 is equal to the expectation of the probability distribution which has for its distribution function $F(x)$.

Let ξ be an arbitrary random variable and A an event of positive probability. We define the *conditional expectation* $E(\xi \mid A)$ *of* ξ *with respect to the condition* A as a limit

$$E(\xi \mid A) = \lim_{h \to 0} \sum_{k=-\infty}^{+\infty} khP(kh \le \xi < (k+1)h \mid A).$$

Since $P(B \mid A) \le \dfrac{P(B)}{P(A)}$, the existence of $E(\xi)$ implies the existence of the conditional expectation $E(\xi \mid A)$ for any event A such that $P(A) > 0$.

If $F(x \mid A)$ is the conditional distribution function of ξ with respect to the condition A, then

$$E(\xi \mid A) = \int_{-\infty}^{+\infty} x\,dF(x \mid A). \tag{6}$$

Clearly, since

$$E(\xi \mid A) = \frac{\int_A \xi(\omega)\, dP}{P(A)} = \int_A \xi(\omega)\, dQ = \int_\Omega \xi(\omega)\, dQ$$

where $Q(B) = P(B \mid A)$, and $Q(B)$ is a probability measure, all results valid for ordinary expectations are also valid for conditional expectations.

We shall now give some often used theorems.

THEOREM 1. *The relation*

$$E(\sum_{k=1}^n c_k \xi_k) = \sum_{k=1}^n c_k E(\xi_k)$$

holds for any random variables ξ_k with finite expectation and for any constants c_k. Thus the functional E is linear.

This theorem is a direct consequence of (3) and of the corresponding properties of the integral.

Let ξ and η be two normally distributed independent random variables with density functions

$$\frac{1}{\sigma_1} \varphi\left(\frac{x - m_1}{\sigma_1}\right) \quad \text{and} \quad \frac{1}{\sigma_2} \varphi\left(\frac{x - m_2}{\sigma_2}\right).$$

The density function of the random variable $\xi + \eta$ is, as we have seen already, $\frac{1}{\sigma} \varphi\left(\frac{x - m}{\sigma}\right)$, where $m = m_1 + m_2$ and $\sigma = \sqrt{\sigma_1^2 + \sigma_2^2}$. It was proved above that the parameter m figuring in the density function is the expectation of the distribution. Hence the relation $m = m_1 + m_2$ is a consequence of Theorem 1.

Similarly, because of Theorem 1, the expectation of the gamma distribution of order n is $\frac{n}{\lambda}$, since the gamma distribution is the distribution of the sum of n independent random variables with the same exponential distribution of parameter λ (i.e. having the same expectation $\frac{1}{\lambda}$). The sum figuring in this example was one of independent random variables; one should, however, realize that Theorem 1 holds for any random variables, without any assumption about their independence.

THEOREM 2. *Let A_n $(n = 1, 2, \ldots)$ $(P(A_n) > 0)$ be a complete system of events and ξ a random variable such that its expectation $E(\xi)$ exists, then*

$$E(\xi) = \sum_{n=1}^\infty E(\xi \mid A_n) P(A_n). \tag{7}$$

This follows immediately from the theorem of total probability.

The statement of Theorem 2 may be expressed in the following manner: Consider $\eta = E(\xi \mid A_\nu)$ as a random variable with its values depending on which one of the events A_k $(k = 1, 2, \ldots)$ occurred; i.e. $\eta = E(\xi \mid A_\nu)$, if event A_ν occurred. Then the right hand side of (7) is just the expectation of this discrete random variable η, hence

$$E(\xi) = E\big(E(\xi \mid A_\nu)\big). \tag{8}$$

THEOREM 3. *If ξ and η are independent random variables such that $E(\xi)$ and $E(\eta)$ exist, then the expectation of $\xi\eta$ exists as well and*

$$E(\xi\eta) = E(\xi)\,E(\eta). \tag{9}$$

PROOF. Assume first $\xi \geq 0$. Let A_k be the event $kh \leq \eta < (k + 1)h$; evidently, the events A_k $(k = 0, \pm1, \pm2, \ldots)$ form a complete system of events. Hence, by Theorem 2,

$$E(\xi\eta) = \sum_{h=-\infty}^{+\infty} P(A_k)\,E(\xi\eta \mid A_k). \tag{10}$$

The conditional expectations $E(\xi\eta \mid A_k)$ exist, since η is bounded, under condition A_k.

Since, however, ξ and η are independent, we have

$$E(\xi)\,kh \leq E(\xi\eta \mid A_k) \leq E(\xi)\,(k + 1)\,h. \tag{11}$$

If we put this into (10), the series on the right side can be seen to converge, thus $E(\xi\eta)$ exists; further (9) holds since the sums

$$\sum_{k=-\infty}^{+\infty} khP(A_k) \quad \text{and} \quad \sum_{k=-\infty}^{+\infty} (k + 1)\,hP(A_k)$$

tend to $E(\eta)$, if $h \to 0$. Thus (9) is proved for $\xi \geq 0$. The restriction $\xi \geq 0$ can be eliminated as follows: Put

$$\xi_1 = \frac{|\xi| + \xi}{2}, \quad \xi_2 = \frac{|\xi| - \xi}{2}; \tag{12}$$

then $\xi_1 \geq 0$, $\xi_2 \geq 0$ and $\xi = \xi_1 - \xi_2$. Since η is independent of ξ_1 and ξ_2, we have

$$E(\xi\eta) = E(\xi_1\,\eta) - E(\xi_2\,\eta) = [E(\xi_1) - E(\xi_2)]\,E(\eta) = E(\xi)\,E(\eta) \tag{13}$$

and herewith Theorem 3 is proved.

THEOREM 4. *If $F(x)$ is the distribution function of ξ and if $E(\xi)$ exists, the following limit relations are valid:*

$$\lim_{x \to +\infty} x(1 - F(x)) = 0, \tag{14}$$

$$\lim_{x \to -\infty} xF(x) = 0. \tag{15}$$

PROOF. Since $E(\xi)$ exists, the integral

$$\int_{-\infty}^{+\infty} |x| \, dF(x)$$

exists. Hence

$$0 \leq \lim_{x \to +\infty} x(1 - F(x)) \leq \lim_{x \to +\infty} \int_{x}^{+\infty} y \, dF(y) = 0.$$

The proof of (15) is similar.

THEOREM 5. *If $E(\xi)$ exists, it can be expressed by ordinary integrals:*

$$E(\xi) = \int_{0}^{\infty} (1 - F(y)) \, dy - \int_{-\infty}^{0} F(y) \, dy. \tag{16}$$

Conversely, the existence of the integrals on the right-hand side of (16) implies the existence of the expectation $E(\xi)$.

PROOF. An integration by parts gives

$$\int_{0}^{x} y \, dF(y) = -x(1 - F(x)) + \int_{0}^{x} (1 - F(y)) \, dy \tag{17}$$

and

$$\int_{-x}^{0} y \, dF(y) = xF(-x) - \int_{-x}^{0} F(y) \, dy. \tag{18}$$

If we add term by term Equations (17) and (18) and let x tend to infinity we obtain, by (14) and (15), Formula (16).

Conversely, the existence of the integrals on the right-hand side of (16) implies the existence of the expectation $E(\xi)$. In fact, the convergence of the integrals implies for $x > 0$

$$x(1 - F(x)) \leq 2 \int_{\frac{x}{2}}^{x} (1 - F(y)) \, dy \quad \text{and} \quad xF(-x) \leq 2 \int_{-x}^{-\frac{x}{2}} F(y) \, dy,$$

hence (14) and (15) are valid. Because of (17) and (18), the second part of Theorem 5 follows.

Theorem 5 has the following graphical interpretation: Draw the curve representing $F(x)$ and the line $y = 1$. The expectation is equal to the differ-

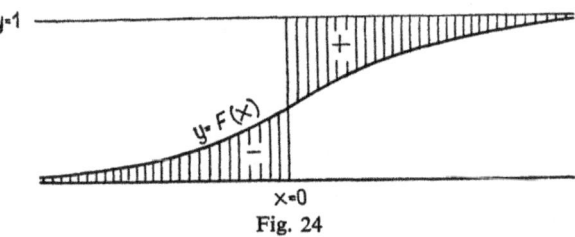

Fig. 24

ence of the areas of the domains marked by $+$ and $-$ on Fig. 24. The (evident) fact follows that a distribution symmetric with respect to $x = a$ has expectation a if this expectation exists. A distribution is said to be *symmetric* with respect to a if

$$F(a - x) = 1 - F(a + x + 0).$$

THEOREM 6. *If $H(x)$ is a continuous function, which is on every finite interval of bounded variation[1] and ξ is a random variable with the distribution function $F(x)$, then*

$$E(H(\xi)) = \int_{-\infty}^{+\infty} H(x)\, dF(x) \tag{19}$$

whenever $E[H(x)]$ exists.

PROOF. Since every function of bounded variation is the difference of two monotone functions, it suffices to prove the theorem for monotone $H(x)$. Let $x = H^{-1}(y)$ be the inverse function of $y = H(x)$. If $H(x)$ is monotone increasing, $P[H(\xi) < x] = P[\xi < H^{-1}(x)]$, hence

$$E(H(\xi)) = \int_{-\infty}^{+\infty} y\, dF(H^{-1}(y)). \tag{20}$$

Relation (19) results from (20) by a transformation $x = H^{-1}(y)$ of the variable of integration.

Examples. 1. The expectations $E(\xi^n)$, if they exist, are expressed by

$$E(\xi^n) = \int_{-\infty}^{+\infty} x^n\, dF(x) \tag{21}$$

---

[1] Relation (19) holds for every Borel function $H(x)$ provided that its expectation $E[H(x)]$ exists; cf. § 17, Exercise 47.

and are called *moments of order* n ($n = 1, 2, \ldots$) of the random variable ξ.

2. $$\varphi_\xi(t) = E(e^{it\xi}) = E(\cos t\xi) + iE(\sin t\xi) = \int_{-\infty}^{+\infty} e^{itx}\, dF(x)$$

is the *characteristic function* of the random variable ξ.

Characteristic functions play an important role in the study of distribution functions; Chapter VI will deal with them.

Theorems 4 and 5 of Chapter III, § 8 are also valid in the general case. Their proof is almost the same as for discrete random variables.

§ 12. Expectation vectors of higher dimensional probability distributions

If the distribution of an n-dimensional random vector

$$\zeta = (\xi_1, \ldots, \xi_n)$$

is known, then so are the components $E(\xi_k)$ ($k = 1, \ldots, n$) of its expectation. They can be considered as the components of an n-dimensional vector

$$E(\zeta) = \big(E(\xi_1), \ldots, E(\xi_n)\big),$$

called the *expectation vector* of the random vector ζ. In the three-dimensional case, the expectation vector specifies the center of gravity of the corresponding mass-distribution.

Let us calculate for example the expectation vector of a normally distributed n-dimensional random vector $\zeta = (\eta_1, \ldots, \eta_n)$, where the density function of ζ is given by Formula (18) of § 7. By the definition of the n-dimensional normal distribution, the components η_k can be exhibited in the form

$$\eta_k = m_k + \sum_{j=1}^{n} c_{kj} \xi_j,$$

where the ξ_j are normally distributed independent random variables with density function $\dfrac{1}{\sigma_j} \varphi\left(\dfrac{x}{\sigma_j}\right)$ and thus expectation $E(\xi_j) = 0$; hence $E(\eta_k) = m_k$. Thus we have found the probabilistic meaning of the parameters m_k figuring in Formula (18) of § 7.

§ 13. The median and the quantiles

The notion of the median is related to that of the expectation. Let ξ be a random variable with continuous distribution function $F(x)$, strictly increasing for every x such that $0 < F(x) < 1$. The *median* of ξ is the (unique) number x for which $F(x) = \dfrac{1}{2}$. If the distribution is symmetric with respect to a certain point, then the median always coincides with the expectation if the latter exists. There are certain distributions for which the expectation does not exist, but the median always exists. Consider for instance the Cauchy distribution, with density function $f(x) = \dfrac{1}{\pi} (1 + x^2)^{-1}$. Here the expectation does not exist, but the median does and is evidently equal to zero.

We introduce the somewhat more general notion of a quantile. The q-quantile $(0 < q < 1)$ denoted by $Q(q)$, of a random variable ξ (or, more precisely, of the corresponding distribution function $F(x)$, continuous and strictly increasing for $0 < F(x) < 1$, by assumption) is defined as that value of x for which $F(x) = q$. In this notation the median is equal to $Q\left(\dfrac{1}{2}\right)$.

In particular, $Q\left(\dfrac{1}{4}\right)$ is called the lower *quartile*, $Q\left(\dfrac{3}{4}\right)$ the upper *quartile*.

For the normal distribution with distribution function $\Phi\left(\dfrac{x - m}{\sigma}\right)$, where

$$\Phi(x) = \frac{1}{\sqrt{2\pi}} \int_{-\infty}^{x} e^{-\frac{t^2}{2}} \, dt,$$

the lower and upper quartiles are

$$Q\left(\frac{1}{4}\right) = m - 0.6745 \, \sigma \quad \text{and} \quad Q\left(\frac{3}{4}\right) = m + 0.6745 \, \sigma,$$

as follows from the tables of the normal distribution function. Since $F[Q(q)] = q$, the function $x = Q(q)$ is the inverse of the distribution function $q = F(x)$.

Now we shall prove a simple but important inequality.

THEOREM 1. (*Markov-inequality*). *Let ξ be a positive random variable with finite expectation $E(\xi)$. Then for every $\lambda > 1$ we have*

$$P(\xi \geq \lambda E(\xi)) \leq \frac{1}{\lambda}. \tag{1}$$

(The inequality also holds for $0 < \lambda \leq 1$, but in this case it is trivial, since every probability is at most equal to 1.)

PROOF. From

$$m = E(\xi) = \int_{0}^{\infty} x \, dF(x)$$

follows

$$m \geq \int\limits_{\lambda m}^{\infty} x dF(x) \geq \lambda m \int\limits_{\lambda m}^{\infty} dF(x) = \lambda m (1 - F(\lambda m)),$$

which proves (1).

If $F(x)$ is continuous and strictly increasing and if $x = Q(y)$ is the y-quantile, i.e. the inverse function of $y = F(x)$, then (1) can be written in the form

$$Q \left(1 - \frac{1}{\lambda} \right) \leq \lambda m.$$

In particular (for $\xi \geq 0$), the upper quartile can never exceed the fourfold of the expectation.

§ 14. The general notions of standard deviation and variance

As in the discrete case, the quantity

$$D(\xi) = + \sqrt{E([\xi - E(\xi)]^2)} \tag{1}$$

is used as a measure of the magnitude of fluctuations of the random variable ξ around its expectation. $D^2(\xi)$ is called the *variance*, $D(\xi)$ the *standard deviation* of ξ. $D(\xi)$ is a nonnegative number which is zero if and only if $P(\xi = c) = 1$ for some constant c. According to Theorem 6 of § 11, (1) can be written in the form

$$D^2(\xi) = \int\limits_{-\infty}^{+\infty} (x - E(\xi))^2 dF(x) = \int\limits_{-\infty}^{+\infty} x^2 dF(x) - \left(\int\limits_{-\infty}^{+\infty} x dF(x) \right)^2 \tag{2}$$

where $F(x)$ is the distribution function of the random variable ξ. If this distribution function is absolutely continuous and if we put $F'(x) = f(x)$, then we have

$$D^2(\xi) = \int\limits_{-\infty}^{+\infty} (x - E(\xi))^2 f(x) \, dx = \int\limits_{-\infty}^{+\infty} x^2 f(x) \, dx - \left(\int\limits_{-\infty}^{+\infty} x f(x) \, dx \right)^2. \tag{3}$$

Theorems 1–5 of § 9 and 1–2 of § 10 of Chapter III about the variance are also valid in the general case. The proofs are essentially the same, since they rest upon the corresponding theorems concerning the expectation. We shall now calculate the variances of some particular distributions.

1. *Uniform distribution.*

If ξ is a random variable uniformly distributed on (a, b), then by (3)

$$D(\xi) = \frac{b - a}{2\sqrt{3}}.$$

2. Normal distribution.

Let ξ be a random variable with density function

$$\frac{1}{\sigma} \varphi \left(\frac{x - m}{\sigma} \right) = \frac{1}{\sqrt{2\pi} \, \sigma} \exp \left(- \frac{(x - m)^2}{2\sigma^2} \right).$$

We know that $E(\xi) = m$. By a transformation of the variable $\dfrac{x - m}{\sigma} = u$ we obtain

$$D^2(\xi) = \frac{1}{\sqrt{2\pi} \, \sigma} \int\limits_{-\infty}^{+\infty} (x - m)^2 \exp \left(- \frac{(x - m)^2}{2\sigma^2} \right) dx =$$

$$= \frac{\sigma^2}{\sqrt{2\pi}} \int\limits_{-\infty}^{+\infty} u^2 e^{-\frac{u^2}{2}} \, du.$$

From here follows by a simple calculation

$$D^2(\xi) = \sigma^2.$$

Thus we have found the probabilistic meaning of the parameter σ of the normal distribution.

3. Exponential distribution.

If the density function of the random variable ξ is given by $\lambda e^{-\lambda x}$ for $x > 0$, then we have seen that $E(\xi) = \dfrac{1}{\lambda}$. Hence

$$D^2(\xi) = \lambda \int\limits_{0}^{\infty} \left(x - \frac{1}{\lambda} \right)^2 e^{-\lambda x} \, dx = \frac{1}{\lambda^2}$$

and

$$D(\xi) = \frac{1}{\lambda}.$$

The standard deviation of the exponential distribution is numerically equal to its expectation.

4. Student's distribution.

Let ξ be a random variable having Student's distribution with n degrees of freedom; its density function is given by Formula (5) of § 10. Since $f(x)$ is an even function, $E(\xi) = 0$ for $n \geq 2$. [For $n = 1$ (i.e. in the case of the

Cauchy distribution) the expectation does not exist.] By applying (3) we obtain

$$D^2(\zeta) = \frac{1}{\sqrt{\pi}} \frac{\Gamma\left(\dfrac{n+1}{2}\right)}{\Gamma\left(\dfrac{n}{2}\right)} \int_{-\infty}^{+\infty} \frac{x^2}{(1+x^2)^{\frac{n+1}{2}}}\, dx.$$

Take for new variable of integration $y = \dfrac{x^2}{1+x^2}$; then

$$D^2(\zeta) = \frac{1}{n-2} \quad \text{for} \quad n \geq 3;$$

for $n = 2$ the variance is infinite.

5. Beta distribution.

If ζ has a beta distribution $B(a, b)$, then $E(\zeta) = \dfrac{a}{a+b}$ as we have seen. From this, by (3)

$$D^2(\zeta) = \frac{\Gamma(a+b)}{\Gamma(a)\,\Gamma(b)} \int_0^1 x^{a+1}(1-x)^{b-1}\, dx - \left(\frac{a}{a+b}\right)^2 =$$

$$= \frac{ab}{(a+b)^2(a+b+1)}.$$

6. Convolution of normal distributions.

Let ζ and η be independent normally distributed random variables with densities

$$\frac{1}{\sigma_1}\varphi\left(\frac{x-m_1}{\sigma_1}\right) \quad \text{and} \quad \frac{1}{\sigma_2}\varphi\left(\frac{x-m_2}{\sigma_2}\right).$$

The density function of $\zeta + \eta$ is $\dfrac{1}{\sigma}\varphi\left(\dfrac{x-m}{\sigma}\right)$ with $m = m_1 + m_2$ and $\sigma = (\sigma_1^2 + \sigma_2^2)^{\frac{1}{2}}$. The relation $m = m_1 + m_2$ is valid since the expectation of the sum of two random variables is equal to the sum of the expectations of the terms. The relation $\sigma = (\sigma_1^2 + \sigma_2^2)^{\frac{1}{2}}$ follows from Theorem 1 of § 10 in Chapter III, since we have seen that the parameter σ represents the standard deviation of the normal distribution.

7. Variance of the gamma distribution of order n.

Let $\zeta_1, \ldots, \zeta_n$ be independent random variables with the same density

function $\lambda e^{-\lambda x}$ for $x > 0$; their sum $\zeta_n = \xi_1 + \ldots + \xi_n$ has a gamma distribution of order n. Since (cf. No. 3) $D^2(\xi_k) = \dfrac{1}{\lambda^2}$, Theorem 1 of § 10,
Chapter III implies $D^2(\zeta_n) = \dfrac{n}{\lambda^2}$. Of course a direct proof is also possible.

8. *Variance of Pearson's χ^2-distribution.*

The χ^2-distribution with n degrees of freedom was defined as the sum of
the squares of n independent random variables having the same normal
distribution. The variance of the square of a normally distributed random
variable with density function

$$\varphi(x) = \frac{1}{\sqrt{2\pi}} \, e^{-\frac{x^2}{2}}$$

is according to Theorem 6 of § 11 and Formula (3), equal to

$$D^2(\xi^2) = \frac{1}{\sqrt{2\pi}} \int_{-\infty}^{+\infty} x^4 e^{-\frac{x^2}{2}} \, dx - [E(\xi^2)]^2 = 2.$$

Consequently, according to Chapter III, § 10, Theorem 1, the standard
deviation of the χ^2-distribution with n degrees of freedom is $\sqrt{2n}$.

§ 15. On some other measures of fluctuation

The difference of the quartiles characterizes to some extend the fluctua-
tions of a random variable. The quantity $\dfrac{1}{2}\left[Q\left(\dfrac{3}{4}\right) - Q\left(\dfrac{1}{4}\right) \right]$ is called the
quartile deviation and is denoted by $q(\xi)$. If ξ is uniformly distributed on
$(0, 1)$, then $q(\xi) = \dfrac{1}{4} = \dfrac{\sqrt{3}\, D(\xi)}{2}$; if ξ is normally distributed, the tables
for the normal distribution give $q(\xi) \approx 0.6745 \, \sigma$. It is to be noted that in
some (chiefly older) books the density function of the normal distribution
is not given in the form

$$\Phi(x) = \frac{1}{\sqrt{2\pi}} \int_{-\infty}^{x} e^{-\frac{t^2}{2}} \, dt,$$

but by

$$\Psi(x) = \Phi(\varrho \sqrt{2}\, x) = \frac{\varrho}{\sqrt{\pi}} \int_{-\infty}^{x} e^{-\varrho^2 t^2} \, dt,$$

where $\varrho \approx 0.477 \approx \dfrac{0.6745}{2}\sqrt{2}$. Anyone of these two forms can be taken as the "standard form"; it is a question of convention which is chosen. $\Phi(x)$ has the advantage that for a normal distribution of the form $\Phi\left(\dfrac{x-m}{\sigma}\right)$ the expectation m and standard deviation σ can be obtained immediately, without calculation; if a normal distribution is brought to the form $\Psi\left(\dfrac{x-m}{q}\right)$ expectation m and quartile deviation q can be read off without any further computation.

If the distribution function $F(x)$ of the random variable ξ is continuous and strictly increasing for $0 < F(x) < 1$, then the value of ξ lies with probability $\dfrac{1}{2}$ in the interval $\left[Q\left(\dfrac{1}{4}\right), Q\left(\dfrac{3}{4}\right)\right]$. Clearly, every interval $\left[Q(\delta), Q\left(\delta+\dfrac{1}{2}\right)\right]$ $\left(0 \le \delta \le \dfrac{1}{2}\right)$ possesses the same property. If the distribution is symmetric with respect to the origin and if its density function is monotone decreasing for $x > 0$, then $\left[Q\left(\dfrac{1}{4}\right), Q\left(\dfrac{3}{4}\right)\right]$ is the smallest interval possessing this property. In this case

$$Q\left(\frac{1}{4}\right) = -Q\left(\frac{3}{4}\right), \quad \text{hence} \quad q(\xi) = Q\left(\frac{3}{4}\right).$$

THEOREM 1. *For every random variable ξ symmetrically distributed about the origin with a continuous distribution function $F(x)$ that is strictly increasing for $0 < F(x) < 1$, the inequality*

$$q(\xi) \le \sqrt{2}\,D(\xi) \tag{1}$$

is valid.

PROOF. Let $F(x)$ be the distribution function of ξ. As ξ is symmetric with respect to the origin, $D^2(\xi) = E(\xi^2)$. Put $\lambda = \dfrac{q^2(\xi)}{D^2(\xi)}$ and apply the Markov inequality (§ 13, Theorem 1) to the random variable ξ^2. Then we obtain

$$P\left(|\xi| \ge Q\left(\frac{3}{4}\right)\right) = P(\xi^2 \ge q^2(\xi)) \le \frac{D^2(\xi)}{q^2(\xi)}. \tag{2}$$

On the other hand, because of the symmetry of the distribution,

$$P\left(|\xi| \ge Q\left(\frac{3}{4}\right)\right) = \frac{1}{2}. \tag{3}$$

From (2) and (3) it follows that $\frac{1}{2} \leq \frac{D^2(\xi)}{q^2(\xi)}$ which proves (1).

The inequality (1) is sharp. This is shown by the following example: Let the distribution of the random variable ξ be the mixture, with weights $\frac{1}{4}, \frac{1}{2}, \frac{1}{4}$ of three normal distributions with the same standard deviation $\varepsilon(> 0)$ and expectations $-1, 0, +1$. Since ε can be chosen arbitrarily small, it follows from the example that $\sqrt{2}$ figuring in (1) cannot be replaced by a smaller number.

The quartile deviation $q(\xi)$ is mostly used when the standard deviation of ξ is infinite, e.g. in the case of the Cauchy distribution.

The standard deviation of a random variable that is uniformly distributed on the interval $(m - a, m + a)$ is given by $\frac{a}{\sqrt{3}}$. If ξ is an arbitrary random variable with $E(\xi) = m$ and $D(\xi) = \sigma$, the interval

$$(m - \sigma\sqrt{3}, \ m + \sigma\sqrt{3}) \tag{4}$$

may be characterized by the fact that a new variable uniformly distributed on this interval has the same expectation and the same standard deviation as ξ. The interval (4) is called the *interval of concentration* of ξ, the inverse of its length is called *concentration* of ξ and is denoted by $k(\xi)$.

Sometimes the absolute mean deviation

$$d(\xi) = E(\,|\,\xi - E(\xi)\,|\,)$$

is also used as a measure of fluctuations. By Theorem 6 of § 11

$$d(\xi) = \int\limits_{-\infty}^{+\infty} |x - E(\xi)|\,dF(x).$$

For the normal distribution

$$d(\xi) = \sqrt{\frac{2}{\pi}}\,D(\xi),$$

for the uniform distribution on an interval

$$d(\xi) = \frac{\sqrt{3}}{2}\,D(\xi),$$

and for the exponential distribution

$$d(\xi) = \frac{2}{e}\,D(\xi).$$

Of course Theorem 4 of § 9 is also valid in the general case.

§ 16. Variance in the higher dimensional case

Let $\zeta = (\xi_1, \ldots, \xi_n)$ be an n-dimensional random vector with distribution function $F(x_1, \ldots, x_n)$. Clearly the fluctuation of ζ cannot be characterized by a single number. The standard deviations $D(\xi_k)$ furnish certain information. But this is insufficient, since these standard deviations are only concerned with the fluctuations of the projections on the coordinate axes and the choice of the axes is arbitrary. More information about the fluctuations of ζ is obtained by considering its projections on all possible straight lines. Put $m_k = E(\xi_k)$ and let P_0 be the point $(m_1, \ldots, m_n)$. Let g be a line passing through P_0 with direction cosines $\alpha_1, \ldots, \alpha_n$ ($\alpha_1, \ldots, \alpha_n$ are real numbers for which $\sum_{k=1}^{n} \alpha_k^2 = 1$). Put

$$\zeta_g = \sum_{k=1}^{n} \alpha_k (\xi_k - m_k) \tag{1}$$

and calculate $D^2(\zeta_g)$. (1) implies $E(\zeta_g) = 0$, hence

$$D^2(\zeta_g) = E(\zeta_g^2).$$

If we put

$$D_{ij} = E\big((\xi_i - m_i)(\xi_j - m_j)\big), \tag{2}$$

then

$$D^2(\zeta_g) = \sum_{i=1}^{n} \sum_{j=1}^{n} D_{ij}\, \alpha_i\, \alpha_j. \tag{3}$$

Let D denote the matrix of coefficients D_{ij}:

$$D = \begin{pmatrix} D_{11} \ldots D_{1n} \\ \cdots\cdots\cdots \\ D_{n1} \ldots D_{nn} \end{pmatrix}. \tag{4}$$

Because of (3), the determinant $|D|$ is always nonnegative. If $|D| = 0$, we have a so-called degenerate distribution. In what follows, it will be always assumed that $|D| > 0$. From the coefficients D_{ij} the standard deviation of the projection of ζ on an arbitrary line can be calculated. Thus the fluctuation of ζ can be characterized by the matrix (4), called the *dispersion matrix*[1] of ζ. The n-dimensional ellipsoid with equation

$$\sum_{i=1}^{n} \sum_{j=1}^{n} D_{ij}\, x_i\, x_j = c^2$$

[1] Since D_{ii} is called the *covariance* of ξ_i and ξ_j, the dispersion matrix is also called the *covariance matrix*.

is called the *dispersion ellipsoid* of the distribution. It is easy to see that the dispersion matrix is invariant under a shift of the coordinate system. Under the rotation of the coordinate system D is transformed as a matrix of a tensor. Let in fact $C = (c_{ij})$ be an orthogonal matrix and

$$\xi'_k = \sum_{j=1}^{n} c_{kj} (\xi_j - m_i),$$

then $E(\xi'_k) = 0$ and

$$D'_{ij} = E(\xi'_i \xi'_j) = \sum_{k=1}^{n} c_{ik} \sum_{h=1}^{n} c_{jh} D_{kh}.$$

If the matrix (D'_{ij}) is denoted by D', we have

$$D' = CDC^*,$$

where C^* is the transpose of the orthogonal matrix C. Hence we may speak about the *dispersion tensor*, which does not depend on the choice of the coordinate system. Again, the similarity to the moments of inertia should be noticed: in case of several dimensions the moment of inertia is also characterized by a tensor and by ellipsoids of inertia.

We have now to deal with the notion of *ellipsoid of concentration*. For the sake of simplicity let us restrict ourselves to the case of two dimensions. Consider the ellipse

$$E(x, y) = Ax^2 + 2Bxy + Cy^2 = 4 \qquad (AC - B^2 > 0) \qquad (5)$$

and suppose that the random vector $\vartheta = (\eta_1, \eta_2)$ is uniformly distributed inside this ellipse. The elements of the dispersion matrix of ϑ, i.e. the numbers $d_{ij} = E(\eta_i \eta_j)$ $(i,j = 1, 2)$ are defined by

$$d_{11} = \frac{1}{F} \iint_E x^2 dx dy, \qquad d_{12} = d_{21} = \frac{1}{F} \iint_E xy dx dy,$$

$$d_{22} = \frac{1}{F} \iint_E y^2 dx dy, \qquad (6)$$

where E denotes the interior of the ellipse with Equation (5) and F its area. Calculation of the integrals in (6) gives

$$d_{11} = \frac{C}{AC - B^2}, \qquad d_{12} = d_{21} = -\frac{B}{AC - B^2}, \qquad d_{22} = \frac{A}{AC - B^2}. \qquad (7)$$

Let $\zeta = (\xi_1, \xi_2)$ be any random vector. Choose the numbers A, B, C such that the dispersion matrix of a random vector uniformly distributed in the ellipse (5) coincides with that of ζ. We put, therefore

$$A = \frac{D_{22}}{\Delta}, \quad B = -\frac{D_{12}}{\Delta}, \quad C = \frac{D_{11}}{\Delta}, \tag{8}$$

where $\Delta = D_{11}D_{22} - D_{12}^2$. Hence the ellipse

$$D_{22}\, x^2 - 2D_{12}\, xy + D_{11}\, y^2 = 4\Delta \tag{9}$$

has the property that a random vector uniformly distributed in it possesses the same dispersion matrix as the given random vector ζ. The ellipse (9) is called the *ellipse of concentration* of the random vector $\zeta = (\xi_1, \xi_2)$ and the number

$$k(\zeta) = \frac{1}{4\pi\sqrt{\Delta}}, \tag{10}$$

i.e. the reciprocal of the area of the ellipse (9), is called the *concentration* of ζ.

If A, B, C are chosen according to (8), the matrix $\begin{pmatrix} A\ B \\ B\ C \end{pmatrix}$ is the inverse of

$$\begin{pmatrix} D_{11}\ D_{12} \\ D_{21}\ D_{22} \end{pmatrix}.$$

The case of higher dimensions turns out to be quite similar. The equation of the ellipsoid of concentration is here

$$\sum_{i=1}^{n} \sum_{j=1}^{n} \frac{\Delta_{ij}\, x_i x_j}{\Delta} = n + 2, \tag{11}$$

where Δ is the value of the determinant $|D_{ij}|$ and Δ_{ij} the value of the co-factor of the element in the i-th row and j-th column. The concentration, that is to say the reciprocal of the volume of the ellipsoid (11), is equal to

$$k(\zeta) = \frac{\Gamma\left(\dfrac{n}{2} + 1\right)}{(n+2)^{\frac{n}{2}}\, \pi^{\frac{n}{2}}\, \sqrt{\Delta}}. \tag{12}$$

Of course, this holds only for $\Delta > 0$. If $\Delta = 0$, the point $(\xi_1, \ldots, \xi_n)$ lies, with probability 1, on a hyperplane of at most $n - 1$ dimensions;

hence the distribution effectively is not n-dimensional. Indeed, $|D| = 0$ implies the existence of numbers $x_1, \ldots, x_n$ which do not all vanish and satisfy

$$\sum_{j=1}^{n} D_{ij} x_j = 0 \qquad (i = 1, \ldots, n).$$

But then

$$E([\sum_{j=1}^{n} x_j (\xi_j - m_j)]^2) = 0;$$

consequently the random vector $(\xi_1, \ldots, \xi_n)$ lies with probability 1 on the hyperplane

$$\sum_{j=1}^{n} x_j (\xi_j - m_j) = 0.$$

Consider now in some detail the two-dimensional normal distribution. Let

$$f(x, y) = \frac{\sqrt{AC - B^2}}{2\pi} \exp\left(-\frac{1}{2}(Ax^2 + 2Bxy + Cy^2)\right) \qquad (13)$$

be the density function of the two-dimensional random vector $\zeta = (\xi, \eta)$. The expectations of ξ and η are equal to zero and the elements of the dispersion matrix are

$$D_{11} = \frac{C}{AC - B^2}, \qquad D_{12} = -\frac{B}{AC - B^2}, \qquad D_{22} = \frac{A}{AC - B^2}.$$

It follows that

$$A = \frac{D_{22}}{|D|}, \qquad B = -\frac{D_{12}}{|D|}, \qquad C = \frac{D_{11}}{|D|},$$

where $|D| = D_{11}D_{22} - D_{12}^2$. If we put

$$\varrho = \frac{D_{12}}{\sqrt{D_{11} D_{22}}}, \qquad \sigma_1 = \sqrt{D_{11}}, \qquad \sigma_2 = \sqrt{D_{22}}, \qquad (14)$$

we find

$$f(x, y) = \frac{1}{2\pi\sigma_1\sigma_2\sqrt{1 - \varrho^2}} \times$$

$$\times \exp\left[-\frac{1}{2(1 - \varrho^2)}\left(\frac{x^2}{\sigma_1^2} - \frac{2\varrho xy}{\sigma_1\sigma_2} + \frac{y^2}{\sigma_2^2}\right)\right]. \qquad (15)$$

The number ϱ is the *correlation coefficient* $R(\xi, \eta)$ of the random variables ξ and η. We have already introduced this quantity for discrete distributions. It is similarly defined in the general case and its properties are the same. Thus

$$R(\xi, \eta) = \frac{E\big([\xi - E(\xi)]\,[\eta - E(\eta)]\big)}{D(\xi)\,D(\eta)} = \frac{E(\xi\eta) - E(\xi)\,E(\eta)}{D(\xi)\,D(\eta)}\,. \tag{16}$$

Theorems 1, 2, 3 and 5 of Chapter III, § 11 are valid and can be proved in nearly the same way.

THEOREM 1. *If the random vector $\zeta = (\xi, \eta)$ is normally distributed in the plane and if $R(\xi, \eta) = 0$, then ξ and η are independent.*

PROOF. If $\varrho = 0$, we see from (15) that the density function is given by

$$f(x, y) = \frac{1}{\sigma_1}\,\varphi\left(\frac{x}{\sigma_1}\right) \cdot \frac{1}{\sigma_2}\,\varphi\left(\frac{y}{\sigma_2}\right); \tag{17}$$

hence ξ and η are independent. This theorem is easily generalized to any number of dimensions.

THEOREM 2. *If the random vector $\zeta = (\eta_1, \ldots, \eta_n)$ is normally distributed in the n-dimensional space and $R(\eta_i, \eta_j) = 0$ if $i \neq j$, then the random variables $\eta_1, \ldots, \eta_n$ are independent.*

PROOF. By § 7 we can write

$$\eta_j = \sum_{k=1}^{n} c_{jk}\,\xi_k + m_j \qquad (j = 1, 2, \ldots, n). \tag{18}$$

Now the random variables ξ_k ($k = 1, 2, \ldots, n$) are pairwise independent, each of them is normally distributed, the expectation of ξ_k is zero and its standard deviation σ_k; hence $C = (c_{jk})$ is an orthogonal matrix. Since by assumption $R(\eta_i, \eta_j) = 0$ for $i \neq j$, we have

$$\sum_{k=1}^{n} c_{ik}\,c_{jk}\,\sigma_k^2 = 0 \quad \text{for} \quad j \neq i. \tag{19}$$

The meaning of this is, however, that the vector

$$(c_{i1}\,\sigma_1^2, \ldots, c_{in}\,\sigma_n^2) \tag{20}$$

is orthogonal to every one of the vectors $(c_{j1}, \ldots, c_{jn})$, $j \neq i$. This means that the vector (20) must be parallel to the vector $(c_{i1}, \ldots, c_{in})$. Hence there exists a constant $\lambda_i \neq 0$ such that

$$c_{ik}\sigma_k^2 = \lambda_i\, c_{ik} \qquad (k = 1, 2, \ldots, n). \tag{21}$$

On the other hand, as the inverse of C is its transpose, we can write, according to (18),

$$\xi_k = \sum_{j=1}^{n} c_{jk}(\eta_j - m_j).$$

Thus we obtain for the density function of the random vector $(\eta_1, \ldots, \eta_n)$

$$g(y_1, \ldots, y_n) = \frac{1}{(2\pi)^{\frac{n}{2}} \prod_{k=1}^{n} \sigma_k} \exp\left[-\frac{1}{2}\sum_{k=1}^{n}\frac{1}{\sigma_k^2}\left(\sum_{j=1}^{n} c_{jk}(y_j - m_j)\right)^2\right]. \tag{22}$$

But it follows from (21), that

$$\sum_{k=1}^{n}\frac{1}{\sigma_k^2}\left(\sum_{j=1}^{n} c_{jk}(y_j - m_j)\right)^2 = \sum_{i=1}^{n}\frac{(y_i - m_i)^2}{\lambda_i}; \tag{23}$$

and consequently

$$g(y_1, \ldots, y_n) = \frac{1}{(2\pi)^{\frac{n}{2}} \prod_{k=1}^{n} \sigma_k} \exp\left[-\frac{1}{2}\sum_{i=1}^{n}\frac{(y_i - m_i)^2}{\lambda_i}\right],$$

which proves the independence of the random variables $\eta_1, \ldots, \eta_n$.

Remark. If, instead of assuming that the random vector $(\eta_1, \ldots, \eta_n)$ has an n-dimensional normal distribution, the weaker condition is assumed that the components $\eta_1, \ldots, \eta_n$ are each normally distributed, then the assertion of Theorem 2 is false. This can be seen from the following example: Let the density function of the random vector (ξ, η) be

$$h(x, y) = \frac{1}{2\pi}\left[\left(\sqrt{2}\, e^{-\frac{x^2}{2}} - e^{-x^2}\right)e^{-y^2} + \left(\sqrt{2}\, e^{-\frac{y^2}{2}} - e^{-y^2}\right)e^{-x^2}\right].$$

(It is readily verified that $h(x, y)$ is in fact a density function.) The density functions $f(x)$ and $g(x)$ of ξ and η are

$$f(x) = g(x) = \frac{1}{\sqrt{2\pi}}\, e^{-\frac{x^2}{2}},$$

i.e. ξ and η are normally distributed with expectation 0 and standard deviation 1. Since $h(x, y)$ is an even function of both x and y, it follows that $R(\xi, \eta) = 0$. The random variables ξ and η, however, are not independent, since evidently $h(x, y) \not\equiv f(x)\, g(y)$; thus ξ and η are each normally distributed and are uncorrelated, but they still are dependent.

From Theorem 2 follows

THEOREM 3. *Let* $\xi_1, \ldots, \xi_n$ *be mutually independent random variables with the same normal distribution with* $E(\xi_k) = 0$, $D(\xi_k) = 1$, $(k = 1, \ldots, n)$. *If* $a_1, \ldots, a_n$ *and* $b_1, \ldots, b_n$ *are real numbers, not all equal to zero, then the random variables*

$$\eta_1 = \sum_{k=1}^{n} a_k \xi_k \quad and \quad \eta_2 = \sum_{k=1}^{n} b_k \xi_k$$

are independent if and only if $\sum_{k=1}^{n} a_k b_k = 0.$

PROOF. Since

$$R(\eta_1, \eta_2) = \frac{\sum_{k=1}^{n} a_k b_k}{\sqrt{\sum_{k=1}^{n} a_k^2 \sum_{k=1}^{n} b_k^2}},$$

the necessary condition that η_1 and η_2 should be uncorrelated is that

$$\sum_{k=1}^{n} a_k b_k = 0.$$

We shall now show that the random vector (η_1, η_2) is normally distributed. There can be found an orthogonal matrix (c_{kl}) such that

$$c_{1k} = \lambda a_k, \quad c_{2k} = \mu b_k \quad (k = 1, \ldots, n).$$

The n-dimensional distribution of the random variables

$$\eta_j' = \sum_{k=1}^{n} c_{jk} \xi_k \qquad (j = 1, 2, \ldots, n)$$

is thus a normal distribution and the same holds for the two-dimensional distribution of $\eta_1 = \dfrac{\eta_1'}{\lambda}$ and $\eta_2 \doteq \dfrac{\eta_2'}{\mu}$. Since

$$\frac{a_k}{\sqrt{\sum_{k=1}^{n} a_k^2}} \quad and \quad \frac{b_k}{\sqrt{\sum_{k=1}^{n} b_k^2}}$$

are the direction cosines of two directions,

$$\sum_{k=1}^{n} a_k b_k = 0$$

means that these two directions are orthogonal and η_1, η_2 are (up to a numerical factor) the projections of the random vector $(\xi_1, \ldots, \xi_n)$ on these directions. Our result may thus be formulated as follows: If $\xi_1, \ldots, \xi_n$ are mutually independent random variables with the same normal distribution, then the projections of the random vector $\zeta = (\xi_1, \ldots, \xi_n)$ on two lines d_1, d_2 are independent iff d_1 and d_2 are orthogonal.

§ 17. Exercises

1. Let the distribution function $F(x)$ of the random variable ξ be continuous and strictly increasing for $-\infty < x < +\infty$. Determine the distribution function of the following random variables:

$$\text{a) } \eta_1 = F(\xi), \quad \text{b) } \eta_2 = \ln \frac{1}{F(\xi)}, \quad \text{c) } \eta_3 = \Psi\big(F(\xi)\big)$$

where $x = \Psi(y)$ is the inverse function of the normal distribution function

$$y = \Phi(x) = \frac{1}{\sqrt{2\pi}} \int_{-\infty}^{x} e^{-\frac{t^2}{2}} \, dt \, .$$

2. Draw the curve of the density function

$$y = f(x) = \frac{1}{\sqrt{2\pi}\,\sigma} \exp\left(-\frac{(x-m)^2}{2\sigma^2}\right)$$

and determine its points of inflexion. Let A and B be their abscissas. Calculate the probability that the value of a random variable with density function $f(x)$ lies between A and B.

3. Draw the curve of the density function

$$y = f(x) = \begin{cases} \dfrac{1}{\sqrt{2\pi}\,\sigma x} \exp\left[-\dfrac{(\ln x - m)^2}{2\sigma^2}\right] & \text{for } x > 0, \\ 0 & \text{for } x < 0 \end{cases}$$

of the lognormal distribution and calculate its extrema and points of inflexion. Calculate the expectation and standard deviation of the lognormal distribution.

4. a) Show that if the random variable ξ has a lognormal distribution, the same holds for $\eta = c\xi^\alpha$ ($c > 0$; $\alpha \neq 0$).

b) Suppose that the diameters of the particles of a certain kind of sand possess a lognormal distribution; let $f(x)$ be the density function of this distribution (cf. Exercise 3), with $m = -0.5$, $\sigma = 0.3$; x is measured in millimeters. The sand particles are supposed to have spherical form. Find the total weight of the sand particles which have diameters less than 0.5 mm, if the total weight of a certain amount of sand is given.

5. Let the random variable η have a lognormal distribution with density function

$$f(x) = \frac{1}{\sqrt{2\pi}\,\sigma x} \exp\left[-\frac{(\ln x - m)^2}{2\sigma^2}\right] \quad \text{for } x > 0.$$

If the curve of $y = f(x)$ is drawn on a paper where the horizontal axis has a logarithmic subdivision, then (apart from a numerical factor) one obtains a normal curve. It does not coincide with the density function of $\ln \eta$, but is shifted to the left over a distance σ^2.

6. Let the random point (ξ, η) have a normal distribution on the plane, with density function

$$f(x, y) = \frac{1}{2\pi\sigma^2} \exp\left[-\frac{x^2 + y^2}{2\sigma^2}\right].$$

Find the density function of $\zeta = \max(|\xi|, |\eta|)$.

7. a) Let the random point (ξ, η) have the same distribution as in Exercise 6. Show that the angle θ between the vector $\zeta = (\xi, \eta)$ and the x-axis is uniformly distributed on the interval $(0, 2\pi)$.

b) Determine the density function of θ, if the point (ξ, η) has density

$$\frac{1}{2\pi\sigma_1\sigma_2} \exp\left[-\frac{1}{2}\left(\frac{x_1^2}{\sigma_1^2} + \frac{x_2^2}{\sigma_2^2}\right)\right].$$

8. Let the density function of the probability distribution of the life-time of the tubes of a radio receiver with 6 tubes be $\lambda^2 t\, e^{-\lambda t}$ for $t > 0$, where $\lambda = 0.25$ if the unit of time is a year. Find the probability that during 6 years no one of the tubes has to be replaced. (The life-times of the individual tubes are supposed to be independent of each other.)

9. A distribution with density function $y = f(x)$ satisfying the differential equation

$$\frac{y'}{y} = \frac{x + \alpha}{\beta + \gamma x + \delta x^2} \qquad (\alpha, \beta, \gamma, \delta \text{ are constants})$$

is called a *Pearson distribution*. Show that the following are Pearson distributions:

a) the normal distribution
b) the "exponential" distribution
c) the gamma-distribution
d) the beta-distribution
e) Student's distribution
f) the χ^2 distribution
g) Cauchy's distribution

h) the distribution with density function $f(x) = \dfrac{c^{m-1}}{(m-2)!} x^{-m} e^{-\frac{c}{x}}$ for $x > 0$; $(c > 0; m = 2, 3, \ldots)$.

10. a) Let the point (ξ, η) be uniformly distributed in the interior of the unit circle. We put

$$\varrho = \sqrt{\xi^2 + \eta^2}, \quad \varphi = \text{arc tan } \frac{\eta}{\xi}.$$

Show that ϱ and φ are independent.

b) Let the point (ξ, η, ζ) be uniformly distributed on the surface of the unit sphere. Introduce spherical coordinates θ and φ (geographical longitude and latitude) and show that θ and φ are independent.

c) Let the point (ξ, η, ζ) be uniformly distributed in the cylinder $\xi^2 + \eta^2 \leq 1$, $0 \leq \zeta \leq 1$. Show that $\varphi = \text{arc tan } \dfrac{\eta}{\xi}$, $\varrho = \sqrt{\xi^2 + \eta^2}$, and ζ are independent.

d) Find the general theorem of which a), b) and c) are particular cases.

Hint. The independence of the new coordinates results, in the three cases, from the fact that the functional determinant of the transformation can be decomposed into factors each containing only one of the new variables.

11. Let ξ and η be independent random variables with the same density function

$$f(x) = \frac{1}{2} e^{-|x|} \qquad (-\infty < x < +\infty).$$

Find the distribution of $\zeta = \xi + \eta$.

12. Show that a two-dimensional normal distribution is uniquely determined by its projection on three non-parallel lines.

13. Let ξ and η be independent random variables with the same density function

$$f(x) = \frac{2}{\pi} \frac{1}{e^x + e^{-x}}.$$

Find the distribution of $\zeta = \xi + \eta$.

14. Let ξ be a random variable with density function

$$f(x) = \frac{1}{2\sqrt{2\pi}} \left[\exp\left(-\frac{1}{2} (x - m)^2 \right) + \exp\left(-\frac{1}{2} (x + m)^2 \right) \right].$$

Find the values of m for which $f(x)$ has two maxima.

15. Let $\xi_1, \xi_2, \ldots, \xi_n$ be independent random variables having a Cauchy distribution with density function

$$f(x) = \frac{1}{\pi(1 + x^2)}.$$

Find the density function of

$$\zeta = \frac{1}{n} \sum_{k=1}^{n} \xi_k.$$

16. Let the random variables $\xi_1, \xi_2, \ldots, \xi_n$ be independent and uniformly distributed on the interval $(0, 1)$. Determine the density function of $\zeta = \sum_{k=1}^{n} \xi_k^2$.

17. Let the random variables $\xi_1, \xi_2, \ldots .. \xi_n$ be independent and uniformly distributed on the interval $(0, 1)$. Let $\xi_k^* = R_k(\xi_1, \xi_2, \ldots, \xi_n)$ $k = 1, 2, \ldots, n$ be the k-th among the values $\xi_1, \ldots, \xi_n$ arranged in increasing order. (ξ_k^* is called the k-th *order statistic* of the sample $(\xi_1, \ldots, \xi_n)$.)

a) Find the distribution function of the random variable $\xi_{k-h}^* - \xi_k^*$ $(1 \leq k < k + h \leq n)$ and show that it is independent of k.

b) Find the distribution function of the ratio $\dfrac{\xi_k^*}{\xi_{k+h}^*}$ $(1 \leq k < k + h \leq n)$.

c) Show that $\dfrac{\xi_1^*}{\xi_2^*}, \dfrac{\xi_2^*}{\xi_3^*}, \ldots, \dfrac{\xi_{n-1}^*}{\xi_n^*}, \xi_n^*$ are independent and that their n-dimensional density function is

$$f(x_1, x_2, \ldots, x_n) = n! \, x_2 x_3^2 \ldots x_n^{n-1} \qquad (0 \leq x_k \leq 1; k = 1, 2, \ldots, n).$$

d) Show that the random variables $\left(\dfrac{\xi_k^*}{\xi_{k+1}^*} \right)^k$ are uniformly distributed in the interval $(0, 1)$.

18. The random variables $\xi_1, \xi_2, \ldots, \xi_n$ are called *exchangeable,* if their n-dimensional distribution function $F(x_1, x_2, \ldots, x_n)$ is a symmetric function of its variables. (Exchangeable random variables have thus the same distribution and consequently the same expectation.)

a) Choose at random and independently, with a constant probability density, n points in the interval $(0, 1)$. Let their abscissas be $\xi_1, \xi_2, \ldots, \xi_n$. The interval $(0, 1)$ is subdivided by these points into $n + 1$ subintervals of the respective lengths $\eta_1, \eta_2, \ldots, \eta_{n+1}$. Show that

$$E(\eta_k) = \frac{1}{n + 1} .$$

Hint. The $\eta_1, \eta_2, \ldots, \eta_{n+1}$ are exchangeable random variables and we have

$$\sum_{k=1}^{n+1} \eta_k = 1.$$

b) Calculate the standard deviation of the random variables η_k.

c) Let ξ_k^* be the k-th order statistic of the sample $(\xi_1, \xi_2, \ldots, \xi_n)$ (see Exercise 17). Show that

$$E(\xi_k^*) = \frac{k}{n + 1} .$$

Hint. $\xi_k^* = \eta_1 + \eta_2 + \ldots + \eta_k$.

d) Which is larger:

$$D^2(\xi_k^*) = D^2(\sum_{i=1}^{k} \eta_i) \quad \text{or} \quad \sum_{i=1}^{k} D^2(\eta_i) ?$$

e) Calculate the correlation coefficient

$$R(\xi_i^*, \xi_j^*) \quad 1 \leq i < j \leq n \,.$$

19. The mixture with equal weights of the distributions with distribution function $B_{k,n+1-k}\,(x)\,(k = 1, 2, \ldots, n)$ is the uniform distribution in the interval $(0, 1)$. How could this be shown without calculation?

20. If the probability that a car runs at least x miles without a puncture is $e^{-\lambda x}$ with $\lambda = 0.0001$, is it worth while to carry three spare tires on a trip of 12 000 miles?

21. Let the random variables ξ and η be independent, let ξ have an exponential distribution with density function $\lambda e^{-\lambda x}$ $(x > 0)$, and let η be uniformly distributed on $(0, 2\pi)$. Put $\zeta_1 = \sqrt{\xi} \cdot \cos \eta$, $\zeta_2 = \sqrt{\xi} \cdot \sin \eta$. Show that ζ_1 and ζ_2 are independent and have the same density function

$$\sqrt{\frac{\lambda}{\pi}}\, e^{-\lambda x^2} \,.$$

22. Let $\xi_1, \xi_2, \ldots, \xi_n$ be independent random variables, let the density function of $\xi_k\,(k = 1, 2, \ldots, n)$ be

$$\lambda\,(k + h - 1)\,e^{-\lambda(k+h-1)x} \text{ for } x > 0\,,$$

where $\lambda > 0$ and h is a real number. Find the distribution function of the sum $\eta = \sum_{k=1}^{n} \xi_k$ and show that $\zeta = \exp\,(-\lambda \eta)$ has a beta distribution.

23. Let $h_n(x)$ be the density function of Student's distribution with n degrees of freedom. Show that

$$\lim_{n \to \infty} \frac{1}{\sqrt{n}}\, h_n\left(\frac{x}{\sqrt{n}}\right) = \frac{1}{\sqrt{2\pi}}\, e^{-\frac{x^2}{2}} \,.$$

24. The substances $A_1, A_2, \ldots, A_{n+1}$ form a radioactive chain, i.e. if an A_1 atom disintegrates it is transformed into an A_2 atom, similarly the A_2 atoms into A_3 atoms, and so on. The A_{n+1} atoms are not radioactive. Suppose that at the instant $t = 0$ the number of A_1 atoms is N_1, the number of A_2 atoms is $N_2, \ldots$, while there are N_n atoms of A_n. Find the density function of the time interval needed for an atom chosen at random to change into an A_{n+1} atom.

25. Let λ be the disintegration constant of a radioactive atom. Let there be N atoms present at the time 0.

a) Calculate the standard deviation of the number of atoms disintegrated up to the time t.

b) Calculate the expectation and the standard deviation of the half-period (i.e. of the random time interval till the $\frac{N}{2}$-th disintegration, if N is even).

26. a) Let $\eta_k\,(k = 1, 2, \ldots)$ be the time required for the transformation of a radioactive atom A_1 into an A_{k+1} atom, through the intermediary states $A_2, \ldots, A_k$, i.e. the duration of the process

$$A_1 \to A_2 \to \ldots \to A_{k+1} \,.$$

Let further λ_k be the disintegration constant of the A_k atoms, $g_k(t)$ the density function of η_k and $\xi_k(t)$ the number of A_k atoms which are present at the time t. It is assumed that at the moment 0 there are only A_1 atoms present and their number is equal to N. Find the distribution function of η_k and of $\xi_k(t)$ $(k = 1, 2, \ldots)$.

Hint. Let $P_k(t)$ be the probability that at the time t an atom is in the state A_k. These probabilities can be calculated in the following way: The probability that an atom A_k changes into an atom A_{k+1} during a time interval $(t, t + \Delta t)$ is, by definition of $g_k(t)$, equal to $g_k(t) \Delta t + o(\Delta t)$. On the other hand, the probability of this event is as well expressed by $P_k(t)\lambda_k \Delta t + o(\Delta t)$; the possibility that during the time interval $(t, t + \Delta t)$ an atom passes through several successive disintegrations can be neglected. Hence we have

$$P_k(t) = \frac{1}{\lambda_k}\, g_k(t)\,. \tag{1}$$

Since the disintegrations of the individual atoms are independent, we obtain

$$P\big(\xi_k(t) = r\big) = \left(\frac{g_k(t)}{\lambda_k}\right)^r \left(1 - \frac{g_k(t)}{\lambda_k}\right)^{N-r} \qquad (r = 0, 1, \ldots, N). \tag{2}$$

The expectation and the standard deviation of the number of A_k atoms at the moment t can now be calculated, since we know that

$$g_k(t) = (-1)^{k-1} \lambda_1 \lambda_2 \ldots \lambda_k \sum_{\substack{j=1 \\ i \neq j}}^{k} \frac{e^{-\lambda_j t}}{\prod (\lambda_i - \lambda_j)} \qquad (t > 0) \tag{3}$$

(cf. Ch. IV, § 9, (29)).

b) Put $M_k(t) = E(\xi_k(t))$. Show that the functions $M_k(t)$ satisfy Bateman's system of differential equations

$$M_k'(t) = \lambda_{k-1} M_{k-1}(t) - \lambda_k M_k(t) \qquad (M_0(t) \equiv 0; \ k = 1, 2, \ldots). \tag{4}$$

Hint. By (2) we have

$$M_k(t) = \frac{N g_k(t)}{\lambda_k}\,. \tag{5}$$

If we differentiate the identity

$$g_k(t) = \int_0^t \lambda_k\, e^{-\lambda_k(t-u)}\, g_{k-1}(u)\, du, \tag{6}$$

we obtain

$$g_k'(t) = \lambda_k\, (g_{k-1}(t) - g_k(t)). \tag{7}$$

Because of (5), (4) follows from (7).

Remark. If the number $M_k(t)$ is very large, the fluctuation of the number $\xi_k(t)$ of the atoms A_k about $M_k(t)$ is small with respect to $M_k(t)$, since by (2)

$$D(\xi_k(t)) = \sqrt{M_k(t) \left(1 - \frac{M_k(t)}{N}\right)} \le \sqrt{M_k(t)}.$$

Hence as a fiirst approach $M_k(t)$ may be considered as the námber of A_k atoms existing at the time t. However, one should not forget that this number is in reality a random variable with expectation $M_k(t)$.

c) Show shat the graph of the function $y = M_k(t)$ has for any $\lambda_1, \lambda_2, \ldots, \lambda_k$ only one maximum. Show further that $0 = m_1 < m_2 < \ldots < m_n$, where m_k denotes the abscissa of the maximum of the function $M_k(t)$.

Remark. The atoms A_{n+1} not being radioactive, $M_{n+1}(t)$ is evidently an increasing function of time, hence $m_{n+1} = +\infty$.

d) Show that $t = 0$ is a zero of order $k - 1$ of the function $M_k(t)$.

27. Let ξ, η, ζ be the components of the velocity of a molecule of a gas in a container. Let the random variables ξ, η, ζ be independent and uniformly distributed on the interval $(-A, +A)$. Calculate the density function $f_A(x)$ of the energy of this molecule. Determine further the limit

$$\lim_{A \to +\infty} A^3 f_A(x) = w(x).$$

Hint. Let the mass of the molecule be denoted by m and its energy by E, then

$$E = \frac{m}{2} (\xi^2 + \eta^2 + \zeta^2),$$

hence

$$P(E < t) = \frac{1}{8A^3} \iiint\limits_{\frac{m}{2}(x^2+y^2+z^2)<t} dx \, dy \, dz = \frac{\pi}{6A^3} \left(\frac{2t}{m}\right)^{\frac{3}{2}} \text{ for } \sqrt{\frac{2t}{m}} < A,$$

since the integral is equal to the volume of a sphere with radius $\sqrt{\dfrac{2t}{m}}$. Thus

$$f_A(t) = \left(\frac{2}{m}\right)^{\frac{3}{2}} \frac{\pi}{4A^3} \sqrt{t} \text{ for } \sqrt{\frac{2t}{m}} < A,$$

hence

$$w(t) = c \sqrt{t} \quad (c = \text{constant}).$$

28. In Exercise 34 of Chapter III, § 18 we studied the most probable energy distribution of a gas consisting of N particles, when the total energy E of the gas was given. The probability p_k of the energy E_k was found to be given by

$$p_k = \frac{w_k \, e^{-\beta E_k}}{\sum\limits_{j=1}^{n} w_j \, e^{-\beta E_j}}.$$

This result was obtained under the assumption that E can only take on the discrete values E_k. Let now the energy be considered as a continuous random variable. For the density function of the energy we obtain in a similar way the expression

$$p(t) = \frac{w(t) \, e^{-\beta t}}{\int\limits_0^\infty w(u) \, e^{-\beta u} \, du},$$

where β can be determined from

$$N \int_0^\infty t w(t) \, e^{-\beta t} \, dt = E \int_0^\infty w(t) \, e^{-\beta t} \, dt.$$

Let $w(t)$ be chosen such that

$$w(t) = c \sqrt{t} \quad \text{for } 0 \le t < c',$$

where c' is a positive constant and $c = \dfrac{3}{2c'^{\frac{3}{2}}}$. Calculate under these conditions, for the limiting case $c' \to +\infty$, the value of β, the function $p(t)$, and the distribution of the velocity of the molecule.

Hint. With the above notations we have for $c' \to +\infty$

$$\frac{E}{N} = \frac{3}{2\beta} .$$

It is known from statistical mechanics that $\dfrac{E}{N} = \dfrac{3kT}{2}$, where k is Boltzmann's constant and T the absolute temperature. So $\beta = \dfrac{1}{kT}$ and

$$p(t) = \frac{2\sqrt{t}\, \exp\left[-\dfrac{t}{kT}\right]}{\sqrt{\pi}\,(kT)^{\frac{3}{2}}} .$$

Let the velocity of a molecule be denoted by v and its kinetic energy by E_{kin}, then the density function of v will be given by

$$f(v) = p(E_{\text{kin}}) \frac{dE_{\text{kin}}}{dv} = \left(\frac{m}{kT}\right)^{\frac{3}{2}} \sqrt{\frac{2}{\pi}}\, v^2 \exp\left[-\frac{v^2}{2}\left(\frac{m}{kT}\right)\right] .$$

This derivation of the Maxwell distribution coincides essentially with one usually given in textbooks of statistical mechanics. (We return to this question in Chapter V, § 3.)

29. a) Calculate from the Maxwell distribution the mean velocity of the molecules of a gas having the absolute temperature T and consisting of molecules of mass m.

b) Show that the average kinetic energy at the absolute temperature T of the molecules of a gas is equal to $\dfrac{3}{2}\, kT$ (k is Boltzmann's constant).

c) Compare the mean kinetic energy of a molecule with the kinetic energy of a molecule moving with mean velocity. Which of the two is larger?

30. a) Consider a gas containing in 1 cm³ N molecules and calculate the mean free path of a molecule.

Hint. The molecules are considered as spheres of radius r and are supposed to be distributed in the space according to a Poisson distribution, i.e. the probability that a volume V contains no molecules is expressed by e^{-NV}. The probability that the volume ΔV contains just one molecule is given by $N\Delta V + o(\Delta V)$. The meaning of the statement that "a molecule covers a distance s without collision and then collides on a segment of length Δs with another molecule" is just the following: a cylinder of radius $2r$ and height s does not contain the center of any of the molecules and another cylinder of radius $2r$ and height Δs contains the center of at least one of the molecules. Thus the probability in question is

$$4r^2\pi N e^{-4\pi N r^2 s}\, \Delta s + o(\Delta s),$$

i. e. the distribution of the free path is an exponential distribution with density function

$$4\pi N r^2 e^{-4\pi N r^2 s} .$$

Hence the length of the mean free path is $\dfrac{1}{4\pi N r^2}$.

b) Calculate the mean time interval between two consecutive collisions of a molecule.

Hint. Let the length of the free path be denoted by s, the velocity of the molecule by v, then $\tau = \dfrac{s}{v}$, where τ denotes the time interval studied. s and v can be assumed to be independent, thus $E(\tau) = E(s)E\left(\dfrac{1}{v}\right)$; the first of these two factors is known from Exercise 30 a), the second can be computed from the Maxwell distribution.

31. Calculate the standard deviation of the velocity and kinetic energy of a gas molecule, if the absolute temperature of the gas is T and the mass of its molecules m.

32. Let the endpoint of a three-dimensional random vector ζ possess a uniform distribution on the surface of the unit sphere. Let θ be the angle between the vector ζ and the positive x-axis. Show that the density function of θ is given by $\dfrac{\sin t}{2}$ $(0 \leq t < \pi)$.

33. Choose at random a chord in the unit circle and determine the expectation of its length under the three suppositions considered in the discussion of Bertrand's paradox (Ch. II, § 10).

34. Let $\xi_1, \ldots, \xi_{n+m}$ be independent normally distributed random variables with density function $\dfrac{1}{\sqrt{2\pi}} e^{-\frac{x^2}{2}}$. Calculate the expectation and the standard deviation of

$$\zeta = \frac{\xi_1^2 + \ldots + \xi_n^2}{\xi_{n+1}^2 + \ldots + \xi_{n+m}^2} .$$

35. Let m_k be the median of the gamma distribution of parameter λ and order k. Show that $\lim\limits_{k \to +\infty} \dfrac{m_k}{k} = \dfrac{1}{\lambda}$.

36. Let the distribution of the random variable ζ be the mixture of the distribution of the random variables $\xi_1, \ldots, \xi_n$ with weights p_k $(k = 1, 2, \ldots, n)$. Show that $E(\xi) = \sum\limits_{k=1}^{n} p_k E(\xi_k)$.

37. Under the same assumptions as in Exercise 36 put

$$M_k = E(\xi_k) \qquad (k = 1, 2, \ldots, n) .$$

Show that

$$D^2(\xi) = \sum_{k=1}^{n} p_k D^2(\xi_k) + D^2(\mu),$$

where μ is a random variable assuming the values $M_1, M_2, \ldots, M_n$ with probabilities

$p_1, p_2, \ldots, p_n$. From this follows

$$D^2(\xi) \geq \sum_{k=1}^{n} p_k\, D^2(\xi_k);$$

equality holds iff $M_1 = M_2 = \ldots = M_n$.

38. a) Let ξ be a normally distributed random variable with density function

$$f(x) = \frac{1}{\sqrt{2\pi}\sigma}\, \exp\left[-\frac{(x-m)^2}{2\sigma^2}\right].$$

Deduce $E(\xi) = m$ from the fact that the function $y = f(x)$ satisfies the differential equation $\sigma^2 y' = -(x-m)\,y$.

b) Let the density function of the random variable ξ be given by

$$f(x) = \frac{\lambda^{m-1}}{(m-2)!}\, x^{-m}\, e^{-\frac{\lambda}{x}} \qquad (x > 0),$$

where $m \geq 3$ is a positive integer and $\lambda > 0$. Calculate $E(\xi)$ from the fact that the function $y = f(x)$ satisfies the differential equation

$$y' = \left(\frac{\lambda}{x^2} - \frac{m}{x}\right) y.$$

c) Apply the same method in general to Pearson's distributions (cf. Exercise 9).

39. Suppose that there are 9 barbers working in a hairdressing-saloon. One shaving takes 10 minutes. Someone coming in sees that all barbers are working and 3 customers are waiting for service. What waiting time can he expect till he is served?

Hint. Assume that the moments of the finishing of the individual shavings are independent and are uniformly distributed on the time interval $(0, 10')$.

40. Let $\xi_1, \ldots, \xi_n$ be independent random variables having the same distribution. Prove that

$$E\left(\frac{\xi_1 + \ldots + \xi_k}{\xi_1 + \ldots + \xi_n}\right) = \frac{k}{n} \qquad (1 \leq k \leq n).$$

41. Prove that if the standard deviation of the random variable ξ with the distribution function $F(x)$ exists, then

$$\lim_{x \to +\infty} x^2\,(1 - F(x) + F(-x)) = 0$$

and

$$E(\xi^2) = 2 \int_0^{\infty} x\,(1 - F(x) + F(-x))\ dx.$$

42. Calculate the dispersion matrix of a nondegenerate n-dimensional normal distribution.

Hint. Let the n-dimensional density function of the random variables $\eta_1, \ldots, \eta_n$ be

$$g(y_1, \ldots, y_n) = \left(\frac{|B|}{(2\pi)^n}\right)^{\frac{1}{2}} \exp\left[-\frac{1}{2}\left(\sum_{i=1}^{n} \sum_{j=1}^{n} b_{ij}\, y_i\, y_j\right)\right], \tag{1}$$

where $|B|$ is the determinant of the matrix $B = (b_{ij})$. There can be given independent normally distributed random variables ξ_k such that $E(\xi_k) = 0$ and

$$\eta_j = \sum_{k=1}^{n} c_{jk} \xi_k \qquad (j = 1, 2, \ldots, n), \tag{2}$$

where $C = (c_{jk})$ is an orthogonal matrix. Let $\sigma_k = D(\xi_k)$ and let S be the diagonal matrix having for its elements the numbers $\dfrac{1}{\sigma_k^2}$. Then B, S, and C are connected by the relation $B = CSC^*$, where C^* is the transpose of the matrix C. If we put $D_{ij} = E(\eta_i \eta_i)$, then we have by (2)

$$D_{ij} = \sum_{k=1}^{n} c_{ik} c_{jk} \sigma_k^2. \tag{3}$$

Hence the matrix $D = (D_{ij})$ can be written in the form $D = CS^{-1}C^*$, where S^{-1} denotes the inverse of the matrix S and thus $BD = CSC^*CS^{-1}C^* = E$, where E is the unit matrix of order n. Thus the dispersion matrix D of the normal distribution is the inverse of the matrix B of the quadratic form figuring in the exponent of (1).

43. a) Using the result of the preceding exercise, find a new proof for Theorem 2 of § 16.

b) Let $\xi_1, \ldots, \xi_n$ be independent normally distributed random variables with $E(\xi_k) = 0$, $D(\xi_k) = \sigma$; show that if the matrix $C = (c_{ij})$ is orthogonal, then the random variables

$$\eta_j = \sum_{k=1}^{n} c_{jk} \xi_k$$

are independent.

c) Determine the ellipsoid of concentration of the n-dimensional normal distribution and prove Formula (12) of § 16.

d) What is the geometric meaning of Exercise b)?

Hint. The components $\xi_1, \xi_2, \ldots, \xi_n$ of an n-dimensional normally distributed random vector are independent, iff the axes of the ellipsoid of concentration are parallel to the coordinate axes. If the random variables $\xi_1, \xi_2, \ldots, \xi_n$ have the same normal distribution, then the ellipsoid of concentration is an n-dimensional sphere; thus the condition required is fulfilled for every choice of the coordinate system.

44. a) When considering errors of measurements the following rule is often used: If the random variables $\xi_1, \ldots, \xi_n$ are independent, further if the first partial derivatives of the function $g(x_1, \ldots, x_n)$ are continuous and if $\eta = g(x_1, \ldots, x_n)$, then

$$D^2(\eta) \sim \sum_{k=1}^{n} \left(\frac{\partial g}{\partial x_k} \right)^2 D^2(\xi_k),$$

where the partial derivatives are to be taken at the points $x_k = E(\xi_k)$ $(k = 1, \ldots, n)$. Discuss the validity of this rule.

b) Let ξ and η be independent random variables. Prove that

$$D^2(\xi\eta) = D^2(\xi)D^2(\eta) + E^2(\xi) D^2(\eta) + E^2(\eta)D^2(\xi).$$

45. Counters used in the study of cosmic rays, radioactivity and other physical phenomena do not register all particles hitting the apparatus; in fact, the latter remains in a passive state for some time interval $h > 0$ after a hit by a particle, and does not

register any particle arriving before the end of this time interval. The number of particles counted is thus smaller than the number of the particles actually coming in. The average number of particles registered during unit time is said to be the "virtual density of events" and is denoted by P; the average number of the particles actually arriving during the unit time is said to be the "actual density of events" and is denoted by p. (Every arriving particle renders the apparatus insensitive for a time interval h, regardless whether the particle was registered by the apparatus.) As to the arrival of the particles, the usual assumption made in the study of radioactive radiations is introduced, namely that the probability of the arrival of n particles during a time interval t is given by $\dfrac{(pt)^n\, e^{-pt}}{n!}$ $(n = 0, 1, \ldots)$.

a) Determine the virtual density of events P.

b) Determine that value of the actual density of events which makes the virtual density maximal.

Hint. The probability that a particle arrives during a time interval Δt and is registered is equal to the probability that the particle arrived during the time interval considered and no other particle did arrive during the preceding time interval of length h. This probability is approximately $pe^{-ph}\Delta t$, hence $P = pe^{-ph}$. If the passive period h is known and P was experimentally determined, the above transcendental equation is obtained for p. By differentiating we find that P has its maximal value, if $p = \dfrac{1}{h}$; then $P = \dfrac{1}{eh}$.

c) Calculate the distribution, expectation and standard deviation of two consecutive registered particle-arrivals.

Hint. Suppose that an arrival was registered at the time $t = 0$ and let $W(t)$ be the probability that at least a time interval t will pass till the following registered arrival. It is easy to see that $W(t)$ satisfies the following (retarded) difference-differential equation

$$W'(t) = P\big(1 - W(t - h)\big) \qquad (t > h) \tag{1}$$

and fulfils the initial condition $W(t) = 0$ for $0 \le t \le h$. The solution of (1) is given by

$$W(t) = \sum_{k=1}^{n} \frac{(-1)^{k-1} P^k (t - kh)^k}{k!} \text{ for } nh \le t < (n + 1) h \qquad (n = 1, 2, \ldots). \tag{2}$$

Integrate (1) from h to $+\infty$, then

$$1 = P \int_0^\infty t W'(t)\, dt.$$

Hence the expectation M of the time spent between two consecutive registered arrivals is given by $\dfrac{1}{P}$. The standard deviation D of this time interval can be determined in a similar manner and one obtains

$$D = \frac{\sqrt{1 - 2hP}}{P} . \tag{3}$$

Observe that for $h = 0$ we have $D = \dfrac{1}{P} = M$. For $h > 0$, we have $\dfrac{D}{M} < 1$. Hence the fact that the apparatus has a passive period diminishes the relative standard deviation of the distribution.

d) If the radiation has a too high intensity, a "scaler" is commonly used in order to make the observations more easy. This apparatus registers only every k-th particle. (In practice k is a power of 2.) Calculate the virtual density of events for this case too.

Hint. First calculate the probability that during the interval $(t, t + \Delta t)$ there arrives a "k-th particle", i.e. a particle having in the list of arriving particles a serial number which is divisible by k. Clearly, the probability of this event is

$$\left(\sum_{n=1}^{\infty} \frac{p^{nk} t^{nk-1} e^{-pt}}{(nk-1)!} \right) \Delta t + o(\Delta t).$$

As the factor of Δt depends also on t, the process is not stationary. But this dependence on t is very weak when t is large; in fact, it can be shown that

$$\lim_{t \to +\infty} \sum_{n=1}^{\infty} \frac{p^{nk} t^{nk-1} e^{-pt}}{(nk-1)!} = \frac{p}{k}. \tag{1}$$

Relation (1) follows from

$$\sum_{n=1}^{\infty} \frac{p^{nk} t^{nk-1} e^{-pt}}{(nk-1)!} = \frac{p}{k} \sum_{r=1}^{k-1} \omega_r \, e^{pt(\omega_r - 1)} \tag{2}$$

where $\omega_r = \exp\left(\dfrac{2\pi i r}{k} \right)$ $(r = 0, 1, \ldots, k-1)$, since the real parts of $\omega_r - 1$ $(r = 1, 2, \ldots, k-1)$ are negative and $\omega_0 - 1 = 0$. Hence the probability that a particle arriving between t and $t + \Delta t$ is a "k-th particle", is given, for a sufficiently large t, approximately by $\dfrac{p}{k} \Delta t$. Thus the registered density of events is $\dfrac{p}{k}$. Here we neglected the passivity of the apparatus following the arriving of a particle.

46. Suppose that the expectation of the random variable ξ exists and let a be a real number. Prove that $E(|\xi - a|)$ takes on its minimum if a is the median of ξ.

47. Let ξ be a random variable with distribution function $F(x)$. Show that

$$E(H(\xi)) = \int_{-\infty}^{+\infty} H(x) dF(x)$$

holds without restriction for every Borel-measurable function $H(x)$, such that the expectation $E(H(\xi))$ exists.

Hint. The value of $E(H(\xi))$ only depends on the distribution of $H(\xi)$, hence on the distribution of ξ, since for every Borel set B $P(H(\xi) \in B) = P(\xi \in H^{-1}(B))$, where $H^{-1}(B)$ denotes the set of the real numbers x for which $H(x) \in B$. Hence $E(H(\xi))$ does not depend on the fact on what probability space $[\Omega, \mathcal{A}, P]$ the random variable ξ is defined; thus let Ω be the real axis, $\mathcal{A}$ the set of all Borel subsets of Ω and P the Lebesgue-Stieltjes measure defined on Ω by $P(I_{ab}) = F(b) - F(a)$, where I_{ab} is an arbitrary interval $a \le x < b$. Under these conditions $\xi(x) = x(-\infty < x < +\infty)$ has distribution function $F(x)$, hence

$$E(H(\xi)) = \int_{\Omega} H(\xi) dP = \int_{-\infty}^{+\infty} H(x) \, dF(x).$$

CHAPTER V

MORE ABOUT RANDOM VARIABLES

§ 1. Random variables on conditional probability spaces

Let $\mathscr{F} = [\Omega, \mathscr{A}, \mathscr{B}, P]$ be a conditional probability space (cf. Ch. II, § 11). A real valued function $\xi = \xi(\omega)$ defined for $\omega \in \Omega$ is said to be a *random variable on* $\mathscr{F}$, if the level sets A_x of ξ (A_x is the set of all $\omega \in \Omega$ such that $\xi(\omega) < x$) belong to $\mathscr{A}$ for every real x. A vector valued function $\zeta = = (\xi_1, \ldots, \xi_r)$ on Ω is said to be *a random vector on* $\mathscr{F}$, if all of its components $\xi_1, \ldots, \xi_r$ are random variables on $\mathscr{F}$. Since, by assumption, $\mathscr{A}$ is a σ-algebra of subsets of Ω it follows that for every Borel set B of the r-dimensional Euclidean space the set $\zeta^{-1}(B)$ of all $\omega \in \Omega$ for which $\zeta(\omega) \in B$ belongs to the σ-algebra $\mathscr{A}$.

If C is any fixed element of $\mathscr{B}$, $\mathscr{F}_C = [\Omega, \mathscr{A}, P(A \mid C)]$ is a Kolmogorov probability space (cf. Theorem 6, § 11, Ch. II). Since every random variable ξ on $\mathscr{F}$ is an ordinary random variable on $\mathscr{F}_C$, the usual notions can be applied to the random variables on $\mathscr{F}_C$. Thus there can be defined for every random variables ξ on $\mathscr{F}$ its conditional distribution function, its conditional expectation, etc. with respect to the condition $C \in \mathscr{B}$. All theorems proved for ordinary random variables are valid for the random quantities defined on $\mathscr{F}$ with respect to a conditional probability space $\mathscr{F}_C$ for every given C. New problems, however, arise if we let C vary.

Let ξ be a (real) random variable on $\mathscr{F}$, I_{ab} the interval $a \le x < b$ and $\xi^{-1}(I_{ab})$ the set of all $\omega \in \Omega$ with $\xi(\omega) \in I_{ab}$. Clearly, for every I_{ab} the set $\xi^{-1}(I_{ab})$ belongs to $\mathscr{A}$ but it does not necessarily belong to $\mathscr{B}$. Let $\mathscr{M}$ be the set of all intervals $I_{ab} \subseteq I_0$ with $\xi^{-1}(I_{ab}) \in \mathscr{B}$, where I_0 is a given, (possibly infinite), interval. The following two conditions are assumed to hold for $\mathscr{M}$:

Condition N_1. The set $\mathscr{M}$ is not empty; for $I_1 \in \mathscr{M}$ and $I_2 \in \mathscr{M}$ there exists an $I_3 \in \mathscr{M}$ such that $I_1 + I_2 \subseteq I_3$.

Condition N_2. For $I_1 \in \mathscr{M}$, $I_2 \in \mathscr{M}$, and $I_1 \subset I_2$ we have

$$P\big(\xi^{-1}(I_1) \mid \xi^{-1}(I_2)\big) > 0.$$

Conditions N_1 and N_2 are evidently fulfilled if $\mathscr{M}$ consists of a single element only. Let J be the union of all intervals $I \in \mathscr{M}$. The set $J \subseteq I_0$ is an open or half open interval with endpoints α and β (α may be equal to $-\infty$ and β to $+\infty$). Let c_0 be any point in the interior of J, i.e. $\alpha < c_0 < \beta$. Take a sequence of intervals $I_{a_n b_n} \in \mathcal{M}$ ($n = 1, 2, \ldots$) with

$$a_{n+1} \leq a_n < c_0 < b_n \leq b_{n+1}, \lim_{n \to +\infty} a_n = \alpha, \lim_{n \to +\infty} b_n = \beta.$$

There can always be found such a sequence when the condition N_1 is fulfilled. Put for $x \in I_{a_n b_n}$

$$F_n(x) = \frac{P(\xi^{-1}(I_{c_0 x}) \mid \xi^{-1}(I_{a_n b_n}))}{P(\xi^{-1}(I_{a_1 b_1}) \mid \xi^{-1}(I_{a_n b_n}))} \qquad \text{for} \quad c_0 \leq x \leq b_n$$

and

$$F_n(x) = -\frac{P(\xi^{-1}(I_{x c_0}) \mid \xi^{-1}(I_{a_n b_n}))}{P(\xi^{-1}(I_{a_1 b_1}) \mid \xi^{-1}(I_{a_n b_n}))} \qquad \text{for} \quad a_n \leq x \leq c_0.$$

From Axioms A, B, and C (Ch. II, § 11) of conditional probability spaces follows then that for $a_n \leq c \leq a \leq b \leq d \leq b_n$ and $I_{cd} \in \mathcal{M}$

$$P(\xi^{-1}(I_{ab}) \mid \xi^{-1}(I_{cd})) = \frac{F_n(b) - F_n(a)}{F_n(d) - F_n(c)}$$

($c < d$ and $F_n(d) - F_n(c) > 0$ follow from our assumptions). Furthermore, for $a_n \leq x \leq b_n$ and for $N > n$

$$F_n(x) = F_N(x).$$

Therefore the value of $F_n(x)$ does not depend on n and we can omit the index n by writing simply

$$F(x) = F_n(x) \quad \text{for} \quad a_n \leq x \leq b_n \quad (n = 1, 2, \ldots). \tag{1}$$

The function $F(x)$ is defined everywhere on the interval (α, β), it is non-decreasing and leftcontinuous; for $I_{cd} \in \mathcal{M}$ we have $F(d) - F(c) > 0$ and for $c \leq a \leq b \leq d$ the relation

$$P(\xi^{-1}(I_{ab}) \mid \xi^{-1}(I_{cd})) = \frac{F(b) - F(a)}{F(d) - F(c)} \tag{2}$$

is valid. Thus the following theorem can be stated:

THEOREM 1. *Let ξ be a random variable on a conditional probability space $\mathcal{F} = [\Omega, \mathcal{A}, \mathcal{B}, P]$ and let $\mathcal{M}$ be the set of the half open intervals I_{ab} contained in an interval I_0 such that $\xi^{-1}(I_{ab}) \in \mathcal{B}$. Let $\mathcal{M}$ fulfil the conditions N_1 and N_2. Let further J be the union of all $I \in \mathcal{M}$; J is an interval contained* in I_0; let α and β denote the endpoints of J. Then there exists a nondecreasing, leftcontinuous function $F(x)$ defined on (α, β), such that for $I_{cd} \in \mathcal{M}$ we have $F(d) - F(c) > 0$ and for $I_{ab} \subseteq I_{cd}$ the relation

$$P\big(\xi^{-1}(I_{ab}) \mid \xi^{-1}(I_{cd})\big) = \frac{F(b) - F(a)}{F(d) - F(c)} \tag{3}$$

holds.

A function $F(x)$ having the above properties will be called a *distribution function of* ξ *on* (α, β). Under the assumptions of Theorem 1, the random variable ξ thus possesses a distribution function on (α, β). The distribution function $F(x)$ of ξ is evidently not uniquely determined since for $\lambda > 0$ and for arbitrary μ, the function $G(x) = \lambda F(x) + \mu$ is also a distribution function of ξ on (α, β). Conversely, if $F(x)$ and $G(x)$ are distribution functions of ξ on (α, β) and if the conditions of Theorem 1 are fulfilled, then for any two subintervals I_{cd} and $I_{\gamma\delta}$ of (α, β) with $I_{cd} \in \mathcal{M}$ and $I_{\gamma\delta} \in \mathcal{M}$ there can be found an interval $I_{ef} \in \mathcal{M}$ such that $I_{cd} \subseteq I_{ef}$ and $I_{\gamma\delta} \subseteq I_{ef}$. Thus we have

$$\frac{F(d) - F(c)}{G(d) - G(c)} = \frac{F(f) - F(e)}{G(f) - G(e)} = \frac{F(\delta) - F(\gamma)}{G(\delta) - G(\gamma)} \,.$$

Hence

$$\lambda = \frac{F(d) - F(c)}{G(d) - G(c)} \quad \text{for} \quad I_{cd} \in \mathcal{M} \tag{4}$$

is a constant. And since for every $I_{ab} \subseteq I_{ef}$

$$\frac{F(b) - F(a)}{F(f) - F(e)} = \frac{G(b) - G(a)}{G(f) - G(e)} \,, \tag{5}$$

it follows that

$$F(b) - \lambda G(b) = F(a) - \lambda G(a) = \mu \tag{6}$$

is also a constant; thus

$$G(x) = \lambda F(x) + \mu, \tag{7}$$

where λ and μ are the constants defined by (4) and (6).

Thus the distribution function of ξ on (α, β) is uniquely defined up to a linear transformation.

When the distribution function $F(x)$ of ξ on (α, β) is absolutely continuous on every closed subinterval of (α, β), then

$$f(x) = F'(x) \tag{8}$$

is called a *density function of* ξ. According to what was said above, the density function of ξ is uniquely determined up to a positive constant; $f(x)$ is nonnegative, measurable and integrable on every closed subinterval $[a, b]$ of the interval (α, β).

Example 1. Let Ω be the set of all real numbers and $\mathcal{A}$ the set of all Borel subsets of Ω, let further be $g(x)$ a function which is nonnegative, measurable and integrable on every finite interval of the real number axis. Let $\mathcal{B}$ be the set of all intervals I_{ab} such that

$$0 < \int_a^b g(x)\,dx < +\infty;$$

assume that $\mathcal{B}$ is not empty. Define conditional probabilities by

$$P(A \mid B) = \frac{\int_{AB} g(x)\,dx}{\int_A g(x)\,dx} \quad \text{for} \quad A \in \mathcal{A} \text{ and } B \in \mathcal{B}. \tag{9}$$

Put $\xi(\omega) = \omega\,(-\infty < \omega < +\infty)$. Then ξ is a random variable on the conditional probability space $\mathcal{F} = [\Omega, \mathcal{A}, \mathcal{B}, P]$. If $I_0 = (-\infty, +\infty)$, $\mathcal{M}$ is identical to $\mathcal{B}$ and all conditions of Theorem 1 are fulfilled; hence ξ has a distribution function $F(x)$ and indeed

$$F(x) = \begin{cases} \lambda \int_0^x g(t)\,dt + \mu & \text{for} \quad x \geq 0, \\[2mm] -\lambda \int_x^0 g(t)\,dt + \mu & \text{for} \quad x \leq 0 \end{cases} \tag{10}$$

is a distribution function of ξ for any choice of the constants $\lambda > 0$ and μ. Furthermore $\lambda g(x)$ is a density function of ξ for any $\lambda > 0$. In particular, for $g(x) \equiv 1$

$$F(x) = \lambda x + \mu \qquad (-\infty < x < +\infty)$$

is a distribution function of ξ and the density function of ξ is an (arbitrary) positive constant λ; in this case we say that ξ *is uniformly distributed on the whole real axis.* It should be remarked that the distribution function of a random variable on a conditional probability space may assume negative values and is not necessarily bounded. It is easy to see that the following theorem holds:

THEOREM 2. *Let $F(x)$ be a distribution function and $f(x)$ a density function of a random variable ξ defined on a conditional probability space $\mathcal{F}$. Let*

$y = h(x)$ be a monotone function and $x = h^{-1}(y)$ its inverse. Then

$$F(h^{-1}(y)) = G(y)$$

is a distribution function and, if $y = h(x)$ is absolutely continuous,

$$\frac{f(h^{-1}(y))}{|h'(h^{-1}(y))|} = g(y)$$

is a density function of $\eta = h(\xi)$.

Example 2. Suppose that ξ is uniformly distributed on the whole real axis, then the same holds for $\eta = a\xi + b$ (for any constants $a \neq 0$ and b).

Example 3. Let ξ be uniformly distributed on the whole real axis. Then $\eta = e^{\xi}$ has on $(0, +\infty)$ the density function

$$f(x) = \frac{1}{x} \qquad (x > 0).$$

The distribution of η is said to be *logarithmically uniform* on the half line $(0, +\infty)$.

Example 4. Let ξ possess a logarithmically uniform distribution on the half line $x > 0$. Then $\eta = a\xi^b$ has the same distribution as ξ for any $a > 0$ and $b \neq 0$.

Example 5. If ξ has on the interval $(0, +\infty)$ the density function

$$f(x) = x^{\alpha},$$

then $\eta = c\xi$ $(c > 0)$ has the same density function.

THEOREM 3. *Let ξ be a random variable defined on a conditional probability space $\mathscr{F}$ and let $F(x)$ be a distribution function and $f(x)$ a density function of ξ on the interval (α, β). If $I_{ab} \subseteq I_{\alpha\beta}$ and $\xi^{-1}(I_{ab}) \in \mathscr{B}$, then the conditional expectation of ξ with respect to the condition $a \leq \xi < b$ is given by*

$$E(\xi \mid a \leq \xi < b) = \frac{\int\limits_a^b x\, dF(x)}{F(b) - F(a)} = \frac{\int\limits_a^b x f(x)\, dx}{\int\limits_a^b f(x)\, dx}. \tag{11}$$

(Clearly, the value of $E(\xi \mid a \leq \xi < b)$ does not depend on the choice of $F(x)$ or $f(x)$.)

Example 6. If ξ is uniformly distributed on the whole real axis, then

$$E(\xi \mid a \leq \xi < b) = \frac{a+b}{2}$$

for all $a < b$.

Example 7. If ξ is logarithmically uniformly distributed on the positive semi-axis, then

$$E(\xi \mid a \leq \xi < b) = \frac{b-a}{\ln \dfrac{b}{a}} \quad \text{for} \quad 0 < a < b.$$

Example 8. If ξ is uniformly distributed on the whole real axis, $|\xi|$ is uniformly distributed on the positive semi-axis. The distribution function of ξ^2 is thus $\sqrt{x}$ and its density function is $\dfrac{1}{\sqrt{x}}$ for $x > 0$. Hence for $0 \leq a < b$

$$E(\xi^2 \mid a \leq \xi < b) = E(\xi^2 \mid a^2 \leq \xi^2 < b^2) = \frac{\displaystyle\int_{a^2}^{b^2} \sqrt{x}\, dx}{\displaystyle\int_{a^2}^{b^2} \frac{dx}{\sqrt{x}}} = \frac{a^2 + ab + b^2}{3}$$

and consequently

$$D^2(\xi \mid a \leq \xi < b) = \frac{a^2 + ab + b^2}{3} - \left(\frac{a+b}{2}\right)^2 = \frac{(b-a)^2}{12},$$

in accordance with the fact that under the condition $a \leq \xi < b$, ξ is uniformly distributed on the interval (a, b) and the standard deviation of such a distribution is $\dfrac{b-a}{2\sqrt{3}}$. (Cf. Ch. IV, § 14.)

Distribution functions and density functions of an r-dimensional random vector on a conditional probability space can be defined in a similar way. Let I be an "interval" of the r-dimensional space, i.e. the set of the points $x = (x_1, \ldots, x_r)$ whose coordinates satisfy $a_k \leq x_k < b_k$ $(k = 1, 2, \ldots, r)$ and let $F(x_1, \ldots, x_r)$ be a function of r variables. Like in Chapter IV, § 3, we introduce the notation

$$\Delta_I F = \Delta_{h_1}^{(1)} \Delta_{h_2}^{(2)} \ldots \Delta_{h_r}^{(r)} F(a_1, \ldots, a_r),$$

where $h_k = b_k - a_k$ $(k = 1, 2, \ldots, r)$. We have the following theorem:

THEOREM 4. *Let ζ be an r-dimensional $(r = 2, 3, \ldots)$ random vector on a conditional probability space $\mathscr{F} = [\Omega, \mathscr{A}, \mathscr{B}, P]$ and let $\zeta^{-1}(I)$ denote the*

set of those $\omega \in \Omega$ for which $\zeta(\omega) \in I$, where I is an interval of the r-dimensional space E^r. Let I_0 denote a fixed interval of E^r and $\mathcal{M}$ the set of those
intervals $I \subseteq I_0$ for which $\zeta^{-1}(I) \in \mathcal{B}$. Assume that the conditions N_1 and N_2
given above are fulfilled. Then if J is the union of all intervals $I \in \mathcal{M}$ (J is
also an interval of E^r), there exists a function F on J such that $\Delta_I F \geq 0$ for
every $I \subseteq J$ and for $I_2 \in \mathcal{M}$ and $I_1 \subseteq I_2$ the relation

$$P(\zeta^{-1}(I_1) \mid \zeta^{-1}(I_2)) = \frac{\Delta_{I_1} F}{\Delta_{I_2} F} \tag{12}$$

is valid.

PROOF. If $\mathcal{M}$ consists of just one interval, the statement of the theorem
is trivial. Otherwise, let $I_1 \in \mathcal{M}, I_2 \in \mathcal{M}$ with $I_1 \subseteq I_2$, and let

$$(x_1^{(0)}, x_2^{(0)}, \ldots, x_r^{(0)}) \in I_1 .$$

For $x = (x_1, x_2, \ldots, x_r) \in I_2$ put

$$F(x_1, \ldots, x_r) = (-1)^k \frac{P(\zeta^{-1}(I_x) \mid \zeta^{-1}(I_2))}{P(\zeta^{-1}(I_1) \mid \zeta^{-1}(I_2))} , \tag{13}$$

where I_x is the interval $a_i \leq t_i < b_i$ $(i = 1, \ldots, r)$ with

$$a_i = \min(x_i^{(0)}, x_i), \quad b_i = \max(x_i^{(0)}, x_i)$$

and k is the number of the values of i for which $x_i < x_i^{(0)}$.

Like in the proof of Theorem 1, we see that $F(x_1, \ldots, x_r)$ does not
depend on the choice of I_2. Clearly, F is nondecreasing with respect to
each of its variables, $\Delta_I F \geq 0$ and (12) is true. Theorem 4 is thus proved.

Every function F fulfilling (12) is said to be a *distribution function of* ζ
on J. The distribution function is not uniquely determined; if F is a distribution function of ζ and μ is any nondecreasing function of $r - 1$ of the
variables $x_1, \ldots, x_r$ then for every $\lambda > 0$

$$G(x_1, \ldots, x_r) = \lambda F(x_1, \ldots, x_r) + \mu \tag{14}$$

is also a distribution function of ζ.

If F is absolutely continuous on every $I \in \mathcal{M}$ we call

$$f(x_1, \ldots, x_r) = \frac{\partial^r F}{\partial x_1 \ldots \partial x_r} \tag{15}$$

the *density function of* ζ *on* J. It is determined up to a positive constant
factor.

Let $\xi_1, \ldots, \xi_r$ be random variables on the conditional probability
space $\mathcal{F} = [\Omega, \mathcal{A}, \mathcal{B}, P(A \mid B)]$ and put $\zeta = (\xi_1, \ldots, \xi_r)$. We shall say

that the random variables $\xi_1, \ldots, \xi_r$ are *mutually independent* if

$$\Delta_I F = \prod_{i=1}^{r} (F_i(b_i) - F_i(a_i)), \tag{16}$$

where F is a distribution function of ζ, I is any interval $I = (a_k \leq x_k < b_k;$ $k = 1, \ldots, r)$ with $I \subseteq J$ and the F_i are nondecreasing functions. If F is absolutely continuous and the random variables $\xi_1, \ldots, \xi_r$ are independent, the density function f of ζ is

$$f = \prod_{i=1}^{r} f_i(x_i), \tag{17}$$

where the nonnegative function $f_i(x)$ is equal to $F_i'(x)$. Conversely, from (17) follows (16) and thus the independence of the random variables $\xi_1, \ldots, \xi_r$.

Example 9. Let $\Omega = E^r$ be the r-dimensional Euclidean space; let $g(x)$, where $x = (x_1, x_2, \ldots, x_r)$, be a function which is nonnegative, measurable and integrable on every finite interval I of E^r; let $\mathscr{A}$ be the set of the Borel subsets of E^r, let $\mathscr{B}$ be the set of all nonempty $B \in \mathscr{A}$ for which $0 < < \int_B g(x)\,dx < +\infty$ and put, for $A \in \mathscr{A}$, $B \in \mathscr{B}$

$$P(A \mid B) = \frac{\int_{AB} g(x)\,dx}{\int_B g(x)\,dx}.$$

Put $\zeta(x) = x$. Then $\mathscr{F} = [\Omega, \mathscr{A}, \mathscr{B}, P]$ is a conditional probability space and ζ is a random vector on $\mathscr{F}$. If I_x denotes the interval

$$\min(0, x_i) \leq t_i < \max(0, x_i) \qquad (i = 1, 2, \ldots, r),$$

then the distribution function of ζ is given by

$$F(x_1, \ldots, x_r) = (-1)^k \int_{I_x} g(x)\,dx,$$

where k is the number of the values of i for which $x_i < 0$ and $g(x)$ is the density function of ζ.

In the case $g(x) \equiv 1$, ζ is uniformly distributed on the whole space E^r. In this particular case we can put

$$F(x_1, \ldots, x_r) = x_1 x_2 \ldots x_r.$$

Let ζ be an r-dimensional random vector, I an interval and $B \subseteq I$ a Borel subset of E^r, furthermore let $\zeta^{-1}(I) \in \mathscr{B}$. Let F be a distribution function of ζ. Then we have

$$P(\zeta^{-1}(B) \mid \zeta^{-1}(I)) = \frac{\int \ldots \int_B dF}{\Delta_I F}.$$

If $\zeta^{-1}(B)$ belongs also to $\mathscr{B}$ and if $C \subset B$ is another Borel subset, it follows that

$$P(\zeta^{-1}(C) \mid \zeta^{-1}(B)) = \frac{P(\zeta^{-1}(C) \mid \zeta^{-1}(I))}{P(\xi^{-1}(B) \mid \zeta^{-1}(I))} = \frac{\int \ldots \int_C dF}{\int \ldots \int_B dF}.$$

Thus we have proved the following theorem:

THEOREM 5. *If F is a distribution function of the r-dimensional random vector ζ on a conditional probability space $\mathscr{F} = [\Omega, \mathscr{A}, \mathscr{B}, P]$ and if B and $C \subseteq B$ are Borel subsets and I is an interval of E^r, further if $B \subseteq I$, $\zeta^{-1}(B) \in \mathscr{B}$, $\zeta^{-1}(I) \in \mathscr{B}$, then*

$$P(\zeta^{-1}(C) \mid \zeta^{-1}(B)) = \frac{\int \ldots \int_C dF}{\int \ldots \int_B dF}.$$

From Theorem 5 we can easily deduce

THEOREM 6. *Let ξ and η be independent nonnegative random variables on the conditional probability space $\mathscr{F}$. Let (ξ, η) have distribution function $F(x) G(y) (0 < x < +\infty; 0 < y < +\infty)$ and let $\lim_{x \to +0} F(x) = F(0)$ be finite. Then the sum $\zeta = \xi + \eta$ has distribution function*

$$H(x) = \int_0^x (F(x - y) - F(0)) \, dG(y).$$

Remark. If we put $F(y) = F(0)$ for $y < 0$, we can also write

$$H(x) = \int_0^\infty (F(x - y) - F(0)) \, dG(y).$$

If we assume further that $F(0) = 0$ (which does not restrict generality), then we can simply write

$$H(x) = \int_0^\infty F(x - y) \, dG(y).$$

A similar theorem holds for more than two nonnegative independent random variables.

PROOF OF THEOREM 6. By Theorem 5, if $\zeta^{-1}(I_{cd}) \in \mathscr{B}$ and $I_{ab} \subseteq I_{cd}$,

$$P(a \leq \xi + \eta < b \mid c \leq \xi + \eta < d) = \frac{\iint\limits_{a \leq x+y<b} dF(x)\, dG(y)}{\iint\limits_{c \leq x+y<d} dF(x)\, dG(y)} =$$

$$= \frac{H(b) - H(a)}{H(d) - H(c)},$$

hence Theorem 6 follows.

If F is absolutely continuous,

$$h(x) = \int_0^x f(x - y)\, dG(y)$$

is a density function of $\zeta = \xi + \eta$. Finally, if $G(y)$ is absolutely continuous and $g(y) = G'(y)$,

$$h(x) = \int_0^x f(x - y)\, g(y)\, dy.$$

Example 10. Let the random vector $(\xi_1, \ldots, \xi_n)$ be uniformly distributed on the n-dimensional space, and put

$$\chi_n^2 = \xi_1^2 + \ldots + \xi_n^2.$$

The random vector $(\xi_1^2, \ldots, \xi_n^2)$ has density function

$$f(x_1, \ldots, x_n) = \frac{1}{\sqrt{x_1 \ldots x_n}} \quad \text{for} \quad x_k > 0 \quad (k = 1, 2, \ldots, n).$$

It follows by Theorem 6 that the density function of χ_n^2 is given by

$$h_n(x) = \int_0^x h_{n-1}(x - y)\, \frac{dy}{\sqrt{y}}.$$

We obtain by induction

$$h_n(x) = x^{\frac{n}{2} - 1} \quad \text{for} \quad x > 0.$$

In particular, $\xi_1^2 + \xi_2^2$ is thus uniformly distributed on the positive semi-axis.

§ 2. Generalization of the notion of conditional probability on Kolmogorov probability spaces

Let ξ be a discrete random variable defined on a probability space $[\Omega, \mathcal{A}, P]$, let x_k $(k = 1, 2, \ldots)$ be the values taken on by ξ with positive probability. Let A_k denote the event $\xi = x_k$ and B an arbitrary event. Let the random variable η be defined by

$$\eta = P(B \mid A_k) \quad \text{for} \quad \xi = x_k \tag{1}$$

($\eta = P(B \mid A_k)$ for every $\omega \in \Omega$ such that $\xi(\omega) = x_k$). Instead of (1), the notation $\eta = P_\xi(B)$ will be also used.

Let U denote any Borel set of real numbers and $\xi^{-1}(U)$ the set of all $\omega \in \Omega$ such that $\xi(\omega) \in U$. Let further $\mathcal{A}_\xi$ be the family of the sets $\xi^{-1}(U)$. We have thus $\mathcal{A}_\xi \subseteq \mathcal{A}$. The family $\mathcal{A}_\xi$ is a σ-algebra since

$$\xi^{-1}(\Sigma U_k) = \Sigma \xi^{-1}(U_k) \quad \text{and} \quad \xi^{-1}(U - V) = \xi^{-1}(U) - \xi^{-1}(V),$$

it is the minimal σ-algebra with respect to which ξ is measurable.

It is easy to see that

$$P(AB) = \int_A P_\xi(B) dP \qquad (A \in \mathcal{A}_\xi, B \in \mathcal{A}). \tag{2}$$

In fact, since $A \in \mathcal{A}_\xi$, for every k such that $A_k A \neq O$, we have $A_k \subseteq A$, hence

$$\int_A P_\xi(B) \, dP = \sum_{k=1}^{\infty} P(B \mid A_k) P(AA_k) = \sum_{k=1}^{\infty} P(ABA_k) = P(AB).$$

Obviously, $P_\xi(B)$ can be interpreted as the *conditional probability of the event B for a given value of* ξ. Of course, the question arises whether this definition may be extended to any random variable so that formula (2) should remain valid. We shall show that this extension is possible. The difficulty of the problem is seen from the fact that for instance a random variable with absolutely continuous distribution function assumes each of its values with probability zero and up to now we defined conditional probabilities only for conditions having a positive probability.

Let ξ be an arbitrary random variable. Let us fix a $B \in \mathcal{A}$ with $P(B) > 0$ and consider the measures $P(A)$ and $P(AB)$ on the σ-algebra $\mathcal{A}_\xi$. Clearly, $0 \leq P(AB) \leq P(A)$. Hence $P(A) = 0$ implies $P(AB) = 0$: the measure $P(AB)$ is thus absolutely continuous with respect to $P(A)$.

In what follows, we shall need the *Radon–Nikodym theorem:*

Let $\mathcal{A}$ be a σ-algebra of the subsets of a set Ω, let $\mu(A)$ be a measure and $\nu(A)$ a σ-additive real set function on $\mathcal{A}$. The measure μ is assumed *to be σ-finite, i.e. Ω is assumed to be decomposable into denumerably many subsets Ω_k with $\Omega_k \in \mathcal{A}$, $\mu(\Omega_k) < +\infty$. Let further $v(A)$ be absolutely continuous with respect to $\mu(A)$, i.e. let $\mu(A) = 0$ imply $v(B) = 0$ for every $B \in \mathcal{A}$, $B \subset A$. Under these conditions there exists a function $f(\omega)$, measurable with respect to the σ-algebra $\mathcal{A}$, such that for every $A \in \mathcal{A}$ the relation*

$$v(A) = \int_A f(\omega)\, d\mu \tag{3}$$

holds. If v is nonnegative (i.e. if it is a measure), then $f(\omega) \geq 0$. The function $f(\omega)$ is determined in an essentially unique manner in the sense that whenever $g(\omega)$ is another function fulfilling the conditions of the theorem, then $f(\omega) = g(\omega)$ holds almost everywhere (with respect to μ). That is to say, if D denotes the set of the points ω at which $f(\omega) \neq g(\omega)$, then $\mu(D) = 0$.

The function $f(\omega)$ figuring in (3) is denoted by $\dfrac{dv}{d\mu}$ and is called *the (Radon–Nikodym) derivative of the set function v with respect to μ.*

Consider now on the σ-algebra $\mathcal{A}$ the measures $\mu(A) = P(A)$ and $v(A) = P(AB)$ $(A \in \mathcal{A}_\xi;\ B \in \mathcal{A}$ is fixed) and apply the Radon–Nikodym theorem. Since $P(\Omega) = 1$, $P(A)$ is not only σ-finite but even finite. Further we have seen that $P(AB)$ is absolutely continuous with respect to $P(A)$. Hence there exists a function $f(\omega)$ which is measurable with respect to $\mathcal{A}_\xi$ such that

$$P(AB) = \int_A f(\omega)\, dP \qquad (A \in \mathcal{A}_\xi). \tag{4}$$

According to the Radon–Nikodym theorem, $f(\omega)$ is determined up to a set of measure zero; $f(\omega)$ is a nonnegative measurable random variable with respect to $\mathcal{A}_\xi$. Obviously, we have *almost everywhere* $f(\omega) \leq 1$. Indeed, (4) implies

$$P(A\bar{B}) = \int_A \left(1 - f(\omega)\right)\, dP. \tag{5}$$

Thus if we would have the inequality $1 - f(\omega) < 0$ on a set C with positive measure, this would imply $P(C\bar{B}) < 0$, which is impossible.

We shall call the random variable $f(\omega)$ the *conditional probability of the event B for a given value of ξ* and shall denote it by $P_\xi(B)$. Thus we can write

$$P(AB) = \int_A P_\xi(B)\, dP \qquad (A \in \mathcal{A}_\xi), \tag{6}$$

which is a generalization of (2). If we want to emphasize that $P_\xi(B)$ depends on ω, we shall write $P_\xi(B;\ \omega)$ instead of $P_\xi(B)$.

In particular, if we put $A = \Omega$ in (6) we find

$$P(B) = \int_\Omega P_\xi(B)\, dP = E\big(P_\xi(B)\big), \tag{7}$$

i.e. *the expectation of the random variable* $P_\xi(B)$ *is equal to* $P(B)$. If we put now $B = \Omega$ in (6), we have

$$P(A) = \int_A P_\xi(\Omega)\, dP \quad \text{for every } A \in \mathcal{A}_\xi.$$

On the other hand, however,

$$P(A) = \int_A 1 \cdot dP,$$

hence, with probability 1 we have

$$P_\xi(\Omega) = 1. \tag{8}$$

One can prove in a similar manner that with probability 1

$$P_\xi(B_1) \leq P_\xi(B_2) \quad \text{for} \quad B_1 \subset B_2.$$

In particular, when ξ is a discrete random variable, $P_\xi(B)$ coincides almost everywhere with the random variable defined in (1) at the beginning of this section, since (2) determines the value of $P_\xi(B)$ for almost every ω.

It is seen from (1) that the value of $P_\xi(B)$, for a discrete random variable ξ, only depends on the value of ξ. But this holds for the general case as well and is expressed by the fact that $P_\xi(B)$ is measurable with respect to $\mathcal{A}_\xi$, which may be rephrased by saying that $P_\xi(B) = h(\xi)$, where $y = h(x)$ is a Borel-measurable function. This can be seen from the following, somewhat modified, definition of $P_\xi(B)$.

Apply the Radon–Nikodym theorem to the measure defined on the σ-algebra of the Borel subsets U of the real numbers by $P(\xi^{-1}(U)) = \mu(U)$ and $P(B\xi^{-1}(U)) = \nu(U)$. The Radon–Nikodym theorem states the existence of a Borel-measurable function $g(x)$, defined for the real numbers x, such that

$$P(B\xi^{-1}(U)) = \int_U g(x)\, dF(x) \tag{9}$$

holds for every Borel set of U of the real axis. Here $F(x)$ denotes the distribution function of ξ .

Obviously, the relation $g(\xi(\omega)) = f(\omega)$ holds for almost every $\omega \in \Omega$.

If the random variable $P(B \mid \xi = x)$ is defined by the function $g(x)$ of formula (9), then by definition it only depends on x; further for almost every $\omega \in \Omega$ $P(B \mid \xi = x) = P_\xi(B; \omega)$ where $x = \xi(\omega)$.

If A is a fixed set, $A \in \mathcal{A}, P(A) > 0$, then $P(B \mid A)$, considered as a function of the set $B \in \mathcal{A}$, is a probability measure. We shall now discuss, how far this remains valid for $P_\xi(B)$. Suppose $B_k \in \mathcal{A}$ $(k = 1, 2, \ldots)$, $B_j B_k = 0$ for $j \neq k$, and $\sum_{k=1}^{\infty} B_k = B$. Consider an arbitrary random variable

ξ and define the random variables $P_\xi(B_k)$ $(k = 1, 2, \ldots)$ and $P_\xi(B)$ as above. Then, for $A \in \mathscr{A}_\xi$

$$P(AB_k) = \int_A P_\xi(B_k)\,dP \tag{10}$$

and

$$P(AB) = \int_A P_\xi(B)\,dP.$$

But from (10) and from

$$\sum_{k=1}^\infty P(AB_k) = P(AB)$$

it follows that

$$P(AB) = \int_A \left(\sum_{k=1}^\infty P_\xi(B_k)\right)\,dP,$$

hence $\sum_{k=1}^\infty P_\xi(B_k)$ fulfils relation (6) which defines $P_\xi(B)$. Thus with probability 1

$$P_\xi(B) = \sum_{k=1}^\infty P_\xi(B_k). \tag{11}$$

The elements ω for which the relation (11) does not hold form thus a set C of measure zero, i.e. $P(C) = 0$. Since $P_\xi(B)$ is determined only almost everywhere, one cannot expect to prove more than this. The exceptional set C depends on the sets B_k and the union of the exceptional sets corresponding to the individual sequences $\{B_k\}$ is not necessarily a set of measure zero since the set of all sequences $\{B_k\}$ is nondenumerable if $\mathscr{A}$ has infinitely many elements. Thus we cannot state that for a fixed ξ, $P_\xi(B)$ as a function of B is a measure; in general this is not true.

In practice, however, this fact causes scarcely any difficulty at all. In most cases, the conditional probability $P_\xi(B) = P(B \mid \xi = x)$ is studied simultaneously for nondenumerably infinitely many B only when the conditional distribution of a random variable η is to be determined with respect to the condition $\xi = x$; i.e. if the probabilities

$$P(\eta < y \mid \xi = x)$$

are to be considered for every real value of y. If these conditional probabilities can be defined in such a manner that $P(\eta < y \mid \xi = x)$ is a distribution function with probability 1, then this function is said to be the *conditional distribution function of η with respect to the condition $\xi = x$* and is denoted by $F(y \mid x)$:

$$F(y \mid x) = P(\eta < y \mid \xi = x).$$

If $F(y \mid x)$ is an absolutely continuous function of y and if

$$F(y \mid x) = \int_{-\infty}^{y} f(t \mid x) \, dt$$

is valid, then $f(y \mid x)$ is said to be the *conditional density function of η with respect to the condition* $\xi = x$.

Since conditional probabilities are determined only with probability 1, it can always be achieved that for almost every x the random variable $P(\eta < y \mid \xi = x)$ as a function of y should be a distribution function. The proof of this statement will just be sketched.

The conditional probabilities $P(\eta < y \mid \xi = x)$ are first defined, by means of the Radon–Nikodym theorem, for *rational* numbers $y \in \mathscr{R}$ only. Then there exists a set V with $P(\xi^{-1}(V)) = 0$ such that for $x \notin V$ the function $P(\eta < y \mid \xi = x)$ as a function of y (y rational) is nondecreasing, left-continuous, and fulfils the conditions

$$\lim_{\substack{y \to -\infty \\ y \in \mathscr{R}}} P(\eta < y \mid \xi = x) = 0 \quad \text{and} \quad \lim_{\substack{y \to +\infty \\ y \in \mathscr{R}}} P(\eta < y \mid \xi = x) = 1.$$

Extend now the definition of $P(\eta < y \mid \xi = x)$ to irrational values of y in the following manner:

$$P(\eta < y \mid \xi = x) = \sup_{y' < y, y' \in \mathscr{R}} P(\eta < y' \mid \xi = x).$$

Then $P(\eta < y \mid \xi = x)$ as a function of y is a distribution function and we have

$$P(\eta < y, \xi \in U) = \int_{U} P(\eta < y \mid \xi = x) \, dF(x).$$

In fact, this relation is valid for every rational y and hence for every real y as well. Herewith our statement is proved.

Thus we have defined the conditional probabilities $P(B \mid A)$ even for $P(A) = 0$; but let it be emphasized that in the latter case the conditional probability $P(B \mid A)$ is only defined, if A can be considered as a level set of a random variable ξ, i.e., if there exists an x such that A is the set of the elements ω of Ω for which $\xi(\omega) = x$. Then $P(B \mid A)$ is defined by $P(B \mid A) = P(B \mid \xi = x)$. However, a set of probability zero can be obtained as a level set of different random variables, thus e.g. A may be defined by any one of the conditions $\xi_1 = x_1$ and $\xi_2 = x_2$. Thus it is possible that

$$P(B \mid \xi_1 = x_1) \neq P(B \mid \xi_2 = x_2),$$

though the conditions $\xi_1 = x_1$ and $\xi_2 = x_2$ define the same set A. A conditional probability with respect to a condition of probability zero is therefore defined only if this condition is an element of the decomposition of the sample space into pairwise disjoint subsets and is considered as an element of this decomposition. The corresponding conditional probability $P(B \mid A)$ depends thus on the decomposition in which A was imbedded.

With the Radon–Nikodym theorem, we proved so far the *existence* of the conditional probability $P_\xi(B)$ only. Let us now see, how $P_\xi(B) = P(B \mid \xi = x)$ can be *effectively determined*. In order to do this let us remark that relation (9), in the case of $P(B) > 0$, may be brought to the form

$$P(B)\,(F_B(b) - F_B(a)) = \int_a^b P(B \mid \xi = x)\,dF(x) \tag{12}$$

where $F_B(x)$ denotes the distribution function of ξ with respect to the condition B and where we have chosen for U the interval $[a, b]$. It follows by a well-known theorem of Lebesgue that (if $F(x)$ is the distribution function of ξ)

$$\frac{P(B \mid \xi = x)}{P(B)} = \lim_{h \to 0} \frac{F_B(x + h) - F_B(x)}{F(x + h) - F(x)} \tag{13}$$

for almost every x (i.e. for every $x \notin C$, with $P(\xi^{-1}(C)) = 0$).

In particular, if $F(x)$ and $F_B(x)$ are absolutely continuous and if $F'(x) = f(x)$, $F'_B(x) = f_B(x)$, then for almost every x

$$P(B \mid \xi = x) = P(B) \frac{f_B(x)}{f(x)} \tag{14}$$

whenever $f(x) > 0$.

Examples.

1. Let (ξ, η) be a random vector with absolutely continuous distribution and with density function $h(x, y)$. Let

$$f(x) = \int_{-\infty}^{+\infty} h(x, y)\,dy$$

be the density function of ξ. Let $\xi^{-1}(U)$ and $\xi^{-1}(V)$ denote the events $\xi \in U$ and $\eta \in V$ respectively, where U and V are Borel sets on the real axis. Assume that the function $f(x)$ is positive for $x \in U$. Then

$$P(\eta \in V, \xi \in U) = \iint_{\substack{x \in U \\ y \in V}} h(x, y)\,dx\,dy = \int_{x \in U} \left(\int_{y \in V} \frac{h(x, y)}{f(x)}\,dy \right) f(x)\,dx,$$

hence

$$P(\eta = V \mid \xi = x) = \int\limits_{y \in V} \frac{h(x, y)}{f(x)} \, dy;$$

thus the conditional density function $g(y \mid x)$ of η with respect to the condition $\xi = x$ is given for the x values which fulfill $f(x) > 0$ by

$$g(y \mid x) = \frac{h(x, y)}{f(x)} . \tag{15}$$

$g(y \mid x)$ is not defined for those x values for which $f(x) = 0$.

Similarly, if $g(y)$ is the density function of η and $f(x \mid y)$ is the conditional density function of ξ with respect to a given value y of η (i.e. with respect to the condition $\eta = y$), we find for $g(y) > 0$ that

$$f(x \mid y) = \frac{h(x, y)}{g(y)} . \tag{16}$$

2. Let ξ and η be independent random variables and $F(x)$ the distribution function of ξ. Then

$$P(\eta \in V, \xi \in U) = P(\eta \in V)P(\xi \in U) = \int\limits_{x \in U} P(\eta \in V) \, dF(x),$$

where U and V are arbitrary Borel sets. Hence

$$P(\eta \in V \mid \xi = x) = P(\eta \in V). \tag{17}$$

Consequently, if the random variables ξ and η are independent, then the conditional distribution function of η with respect to the condition $\xi = x$ is identical with the ordinary (unconditional) distribution function of η. Conversely, if (17) is valid for every Borel set V and for every $x \in U$ with $P(\xi \in U) = 1$, then ξ and η are independent.

3. Let (ξ, η) be a normally distributed random vector with the density function

$$\frac{1}{2\pi} \exp\left[-\frac{1}{2} (x^2 + y^2) \right].$$

Let ϱ and ϑ $(0 \le \vartheta < 2\pi)$ be the polar coordinates of the point (ξ, η). Find the conditional distribution of ϑ with respect to the condition $\varrho = r > 0$. We have

$$P(0 \le \vartheta < \varphi, \xi^2 + \eta^2 \le R^2) = \frac{1}{2\pi} \iint\limits_{\substack{0 \le \vartheta < \varphi \\ x^2 + y^2 \le R^2}} \exp\left[-\frac{1}{2} (x^2 + y^2) \right] dx \, dy,$$

or, by introducing polar coordinates,

$$P(0 < \vartheta < \varphi, \xi^2 + \eta^2 \le R^2) = \frac{\varphi}{2\pi} \int_0^R re^{-\frac{r^2}{2}} dr.$$

Since the density function of ϱ is given by $re^{-\frac{r^2}{2}}$, we obtain

$$P(0 \le \vartheta < \varphi \mid \varrho = r) = \frac{\varphi}{2\pi} \qquad (0 \le \varphi < 2\pi).$$

Hence ϑ is uniformly distributed in the interval $[0, 2\pi)$ under the condition $\varrho = r$ for every $r > 0$ and thus ϑ and ϱ are independent.

4. Let ξ and η be independent random variables. We shall determine the conditional distribution of $H(\xi, \eta)$ with respect to the condition $\xi = x$, here $H(x, y)$ is assumed to be a Borel measurable function.

If U is an arbitrary Borel-measurable set and if $F(x)$ and $G(y)$ are the distribution functions of ξ and η, then

$$P(H(\xi, \eta) < Z, \xi \in U) = \int_{\substack{H(x,y)<Z \\ x \in U}} dF(x)\, dG(y) = \int_{x \in U} \left(\int_{H(x,y)<Z} dG(y) \right) dF(x);$$

the conditional distribution function in question is thus the distribution function of $H(x, \eta)$.

5. Let U be a Borel set and $B = \xi^{-1}(U)$. Then, with probability 1

$$P_\xi(B) = \begin{cases} 1 & \text{for } \omega \in B, \\ 0 & \text{otherwise.} \end{cases}$$

In fact, if $A = \mathscr{A}_\xi$,

$$P(AB) = \int_{AB} dP = \int_A \chi_B(\omega)\, dP,$$

where

$$\chi_B(\omega) = \begin{cases} 1 & \text{for } \omega \in B, \\ 0 & \text{otherwise.} \end{cases}$$

Since χ_B is measurable with respect to $\mathscr{A}_\xi$, we have, with probability 1,

$$P_\xi(B) = \chi_B(\omega).$$

6. (Particular case of 5.) Let Ω be the interval $[0, 1]$, let $\mathscr{A}$ be the set of Borel subsets of Ω and P the Lebesgue measure. Put

$$\xi(\omega) = \omega \qquad (0 \le \omega < 1).$$

Then, for $B \in \mathscr{A}$ (with probability 1),

$$P_\xi(B) = \begin{cases} 1 & \text{for } \omega \in B, \\ 0 & \text{otherwise.} \end{cases}$$

7. Let Ω be the unit square of the plane (x, y), $\mathscr{A}$ the class of the Borel subsets of Ω and P the two-dimensional Lebesgue measure. Put $\xi(x, y) = x$.

Since, for every $B \in \mathscr{A}$ and for any Borel set U of the real axis (according to the theorem of Fubini),

$$P(B\xi^{-1}(U)) = \int_U (\int_{(x,y)\in B} dy)\, dx,$$

we find

$$P(B \mid \xi = x_0) = \int_{(x_0,y)\in B} dy = \mu(B_{x_0}),$$

where B_{x_0} represents the intersection of B by the line $x = x_0$ and μ the one-dimensional Lebesgue measure. In this case $P(B \mid \xi = x_0)$ is thus, as a function of B, a measure on the σ-algebra $\mathscr{A}$, for every x_0.

§ 3. Generalization of the notion of conditional probability on conditional probability spaces

Let $\mathscr{F} = [\Omega, \mathscr{A}, \mathscr{B}, P(A \mid B)]$ be a conditional probability space and ξ a random variable on $\mathscr{F}$. Let $B \in \mathscr{A}$ and $C \in \mathscr{B}$ be given sets, with $P(B \mid C) > 0$ and let $\mathscr{A}_\xi$ be the least σ-algebra with respect to which ξ is measurable. Consider the measures $\mu_C(A) = P(A \mid C)$ and $v_C(A) = P(AB \mid C)$ on $\mathscr{A}_\xi$. $v_C(A)$ is absolutely continuous with respect to $\mu_C(A)$; there exists thus, by the Radon–Nikodym theorem, a function measurable with respect to $\mathscr{A}_\xi$, $f(\omega) = P_\xi(B \mid C)$ such that

$$P(AB \mid C) = \int_A P_\xi(B \mid C)\, d\mu_C \quad \text{for} \quad A \in \mathscr{A}_\xi. \tag{1}$$

The random variable $P_\xi(B \mid C)$ will be called the *conditional probability of the event B with respect to the condition C and for a given value of ξ*; this of course depends on ξ, but also on C; but the dependence is quite obvious in the most important particular cases.

If C is fixed, $P_\xi(B \mid C)$ can be considered as the conditional probability of the event B on the ordinary Kolmogorov probability space $\mathscr{F}_C = [\Omega, \mathscr{A}, P(A \mid C)]$ with respect to the condition that ξ assumes a given value. The random variable $P_\xi(B \mid C)$ has thus, for fixed C, all the properties proved in § 2 for $P_\xi(B)$.

Let us point out the following circumstance. If $A(x)$ is the set of all $\omega \in \Omega$ for which $\xi(\omega) = x$, it may happen that the sets $CA(x)$ belong to the family $\mathscr{B}$ for some values of x or even for every one of its values and thus $P(B \mid CA(x))$ is defined. But *a priori* it is not at all certain that $P(B \mid CA(x))$ coincides with $P_\xi(B \mid C)$, i.e. that

$$P(AB \mid C) = \int_A P(B \mid CA(\xi)) \, d\mu_C \quad \text{for} \quad A \in \mathscr{A}_\xi.$$

This regularity property does not follow from the axioms and if necessary it must be postulated as an additional axiom.

Consider now the following important particular case:

Let Ω be an arbitrary set and $\mathscr{A}$ a σ-algebra of subsets of Ω. Let further μ be a σ-finite measure on $\mathscr{A}$ and let $\mathscr{B}$ be the family of sets

$$B \in \mathscr{A} \quad \text{with} \quad 0 < \mu(B) < +\infty .$$

We define

$$P(A \mid B) = \frac{\mu(AB)}{\mu(B)} ,$$

when $A \in \mathscr{A}$ and $B \in \mathscr{B}$.

Let ξ be a random variable on the conditional probability algebra

$$F = [\Omega, \mathscr{A}, \mathscr{B}, P(A \mid B)]$$

and let $\mathscr{A}_\xi$ be the least σ-algebra with respect to which ξ is measurable.

Let $B (B \in \mathscr{A})$ be fixed; since the measure $\nu(A) = \mu(AB)$ is absolutely continuous on $\mathscr{A}$ with respect to $\mu(A)$, there exists a function $f(\omega, B)$ which is measurable with respect to $\mathscr{A}_\xi$ and has the property

$$\mu(AB) = \int_A f(\omega, B) \, d\mu \quad \text{for every} \quad A \in \mathscr{A}_\xi. \tag{2}$$

If $C \in \mathscr{B}$ and $B \subseteq C$, it follows from (2) by putting $\mu_C(A) = \dfrac{\mu(AC)}{\mu(C)}$ that

$$P(AB \mid C) = \frac{\mu(AB)}{\mu(C)} = \int_A f(\omega, B) \, d\mu_C. \tag{3}$$

The function $f(\omega, B)$ obviously does not depend on C. Hence introducing the notation $P_\xi(B) = f(\omega, B)$ we have

$$P(AB \mid C) = \int_A P_\xi(B) \, d\mu_C \tag{4}$$

for $A \in \mathscr{A}_\xi$, $B \in \mathscr{A}$ and $B \subseteq C \in \mathscr{B}$.

Clearly $P_\xi(B)$ is with respect to μ almost everywhere uniquely defined, further almost everywhere with respect to μ holds

$$0 \le P_\xi(B) \le 1. \tag{5}$$

With exception of a set of μ-measure zero we also have

$$P_\xi(B) = \sum_{k=1}^{\infty} P_\xi(B_k) \quad \text{if} \quad \sum_{k=1}^{\infty} B_k = B \text{ and } B_j B_k = 0 \text{ for } j \ne k.$$

If η is another random variable on $\mathscr{F}$, it can be shown as in § 2 that the values of $P_\xi(\eta^{-1}(V))$ can be chosen such that for every $\omega \notin D$, with $\mu(D) = 0$, $P_\xi(\eta^{-1}(V))$ is a measure on the Borel subsets of the real number axis. This measure will be called the *conditional distribution of η for given ξ*. If

$$P_\xi\left(\eta^{-1}(V)\right) = \int_V g(y \mid x)\,dy \quad \text{for} \quad \xi(\omega) = x, \tag{6}$$

then $g(y|x)$ will be called the *conditional density function of η with respect to the condition $\xi = x$*.

Let ξ and η be random variables defined on $\mathscr{F}$ with the two-dimensional density function $h(x, y)$; assume that the integral

$$f(x) = \int_{-\infty}^{+\infty} h(x, y)\,dy \tag{7}$$

exists for every x. Then $f(x)$ is a density function of ξ. In fact, if U and V are two intervals, $U \subseteq V$, we have

$$P(\xi \in U \mid \xi \in V) = \frac{\int_U f(x)\,dx}{\int_V f(x)\,dx}, \tag{8}$$

if $\xi^{-1}(V) \in \mathscr{B}$.

In this case the conditional density function of η with respect to the condition $\xi = x$ is equal, for $f(x) > 0$, to

$$g(y \mid x) = \frac{h(x.\,y)}{f(x)}. \tag{9}$$

In fact, for $U \subseteq V$ and $\xi^{-1}(V) \in \mathscr{B}$ we have

$$P(\eta \in W, \xi \in U \mid \xi \in V) = \frac{\displaystyle\int_{\substack{x \in U \\ y \in W}} \int h(x,y)\,dx\,dy}{\displaystyle\int_V f(x)\,dx} = \frac{\displaystyle\int_U \left(\int_W g(y \mid x)\,dy\right) f(x)\,dx}{\displaystyle\int_V f(x)\,dx}. \tag{10}$$

Finally, let the relation

$$\int\limits_{-\infty}^{+\infty} g(y \mid x)\,dy = \frac{f(x)}{f(x)} = 1 \quad \text{for} \quad f(x) > 0 \tag{11}$$

be mentioned, expressing the fact that the conditional distribution of η for given ξ is an ordinary distribution.

Let us consider now some examples.

1. Let the point (ξ, η) be uniformly distributed in the domain of the plane defined by $|x^2 - y^2| \le 1$. The density function $h(x, y)$ of (ξ, η) is

$$h(x, y) = \begin{cases} 1 & \text{for } |x^2 - y^2| < 1, \\ 0 & \text{otherwise.} \end{cases}$$

The density function $f(x)$ of ξ is

$$f(x) = \int_{-\infty}^{+\infty} h(x, y)\,dy,$$

hence

$$f(x) = \begin{cases} 2(\sqrt{x^2 + 1} - \sqrt{x^2 - 1}) & \text{for } |x| > 1, \\ 2\sqrt{x^2 + 1} & \text{otherwise.} \end{cases}$$

Similarly, if $g(y)$ is the density function of η, we have

$$g(y) = \begin{cases} 2(\sqrt{y^2 + 1} - \sqrt{y^2 - 1}) & \text{for } |y| > 1, \\ 2\sqrt{y^2 + 1} & \text{otherwise.} \end{cases}$$

It follows that

$$f(x \mid y) = \begin{cases} \dfrac{1}{2(\sqrt{y^2 + 1} - \sqrt{y^2 - 1})} & \text{for } |y| > 1, \sqrt{y^2 - 1} < |x| < \sqrt{y^2 + 1}, \\ \dfrac{1}{2\sqrt{y^2 + 1}} & \text{for } |y| \le 1, 0 \le |x| < \sqrt{y^2 + 1}, \\ 0 & \text{otherwise.} \end{cases}$$

Hence ξ is, for $\eta = y$ ($|y| < 1$), uniformly distributed on the interval $(-\sqrt{y^2 + 1}, +\sqrt{y^2 + 1})$.

2. Let ξ and η be two independent random variables with absolutely continuous distributions. The density function of their joint distribution

is thus

$$h(x, y) = f(x) g(y),$$

where $g(y)$ is an ordinary density function. Hence the conditional density function $g(y \mid x)$ of η with respect to the condition $\xi = x$ is

$$g(y \mid x) = g(y).$$

The conditional density function $g(y \mid x)$ does not depend on the value of x.

3. Let $(\xi_1, \xi_2, \ldots, \xi_n)$ be a random vector uniformly distributed in the whole n-dimensional space and let $\eta_n = \xi_1^2 + \xi_2^2 + \ldots + \xi_n^2$. Determine the conditional density function of ξ_k with respect to the condition $\eta_n = y (y > 0)$. We know already that the density function of η_n is $y^{\frac{n}{2}-1}$ for $y > 0$ (§ 1, Example 10). It follows that the two-dimensional density function of ξ_k and η_k is

$$h_n(x, y) = \begin{cases} (y - x^2)^{\frac{n-3}{2}} & \text{for } y > x^2, \\ 0 & \text{otherwise.} \end{cases}$$

For the conditional density function of ξ_k with respect to the condition $\eta_n = y$ we find thus

$$f_n(x \mid y) = \frac{C_n}{\sqrt{y}} \left(1 - \frac{x^2}{y}\right)^{\frac{n-3}{2}} \quad \text{for} \quad |x| < \sqrt{y}, \tag{12}$$

where the constant C_n will be determined by

$$\int_{-\sqrt{y}}^{+\sqrt{y}} f_n(x \mid y) \, dx = 1. \tag{13}$$

From (12) and (13) it follows

$$C_n = \frac{1}{\sqrt{\pi}} \frac{\Gamma\left(\dfrac{n}{2}\right)}{\Gamma\left(\dfrac{n-1}{2}\right)}; \tag{14}$$

and finally, we obtain thus

$$f_n(x \mid y) = \frac{1}{\sqrt{\pi y}} \frac{\Gamma\left(\dfrac{n}{2}\right)}{\Gamma\left(\dfrac{n-1}{2}\right)} \cdot \left(1 - \frac{x^2}{y}\right)^{\frac{n-3}{2}} \quad \text{for } -\sqrt{y} < x < +\sqrt{y}. \tag{15}$$

From (15) follows

$$\lim_{n \to +\infty} f_n(x \mid n\sigma^2) = \frac{1}{\sqrt{2\pi} \, \sigma} \, e^{-\frac{x^2}{2\sigma^2}} \qquad (-\infty < x < +\infty), \qquad (16)$$

hence every ξ_k (k fixed) has in the limit a normal (conditional) distribution, if the condition imposed is $\eta_n = n\sigma^2$ and n tends to infinity.

4. We deduce now the Maxwell distribution from the preceding example.[1] Let ξ_k, η_k, ζ_k ($k = 1, 2, \ldots, n$) be the components of the velocities of n atoms of a certain amount of gas. We assume that the (a priori) distribution of the point $(\xi_1, \eta_1, \zeta_1, \ldots, \xi_n, \eta_n, \zeta_n)$ is uniform on the whole $3n$-dimensional phase space. Consider the conditional distribution of the velocity components with respect to the condition that the total kinetic energy of the gas be constant. This kinetic energy is given by

$$E = \frac{m}{2} \sum_{k=1}^{n} (\xi_k^2 + \eta_k^2 + \zeta_k^2),$$

where m represents the mass of a particle of the gas. The conditional density function of the distribution studied is, by the above example,

$$f_{3n}\left(x \,\middle|\, \frac{2E}{m}\right) = \sqrt{\frac{m}{2\pi E}} \, \frac{\Gamma\left(\dfrac{3n}{2}\right)}{\Gamma\left(\dfrac{3n-1}{2}\right)} \left(1 - \frac{x^2 m}{2E}\right)^{\frac{3n-3}{2}}$$

$$\text{for } |x| < \sqrt{\frac{2E}{m}}. \qquad (17)$$

By taking into account that $E = \dfrac{3}{2} kTn$ (k Boltzmann's constant, T the absolute temperature of the gas) we find for the conditional density function $h_n(x \mid T)$ of the velocity components ξ_k, η_k, ζ_k at constant temperature T

$$h_n(x \mid T) = \frac{1}{\sqrt{\dfrac{3\pi kTn}{m}}} \, \frac{\Gamma\left(\dfrac{3n}{2}\right)}{\Gamma\left(\dfrac{3n-1}{2}\right)} \left(1 - \frac{x^2 m}{3kTn}\right)^{\frac{3n-3}{2}} \qquad (18)$$

$$\text{for } |x| < \sqrt{\frac{3kTn}{m}}$$

[1] Cf. A. Rényi [19].

hence

$$\lim_{n \to +\infty} h_n(x \mid T) = \frac{1}{\sqrt{\dfrac{2\pi kT}{m}}} \; e^{-\frac{x^2 m}{2kT}}. \tag{19}$$

Thus the distribution of each component of the velocity tends to a normal distribution with the expectation 0 and standard deviation $\sqrt{\dfrac{kT}{m}}$.

Since for large n under the condition $E = $ constant, ξ_k, η_k, and ζ_k tend to be independent, it follows already that the distribution of the random variables

$$v_k = \sqrt{\xi_k^2 + \eta_k^2 + \zeta_k^2},$$

i.e. of the velocities of the particles, tends to the Maxwell distribution. But it is profitable to perform exactly the above calculations, i.e. to calculate the conditional distribution of v_k for every finite n. It is quite natural to call this distribution the *Maxwell distribution of order n*; for this distribution tends to the ordinary Maxwell distribution if $n \to +\infty$.

The calculations are entirely similar to those in the preceding example. Put

$$v_k = \sqrt{\xi_k^2 + \eta_k^2 + \zeta_k^2};$$

if we put

$$\eta_{3n} = \sum_{k=1}^{n} \xi_k^2 + \eta_k^2 + \zeta_k^2$$

and if $h_n(v, y)$ is the two-dimensional density function of v_k and η_{3n}, we have

$$h_n(v, y) = v^2 (y - v^2)^{\frac{3n-5}{2}} \quad \text{for} \quad 0 < v < \sqrt{y}. \tag{20}$$

Thus if $V_n(v \mid y)$ denotes the conditional density function of v_k with respect to the condition $\eta_{3n} = y$ we obtain

$$V_n(v \mid y) = D_n \frac{v^2}{y^{\frac{3}{2}}} \left(1 - \frac{v^2}{y}\right)^{\frac{3n-5}{2}} \quad \text{for} \quad 0 < v < \sqrt{y}, \tag{21}$$

the constant D_n being determined by

$$\int_0^{\sqrt{y}} V_n(v \mid y) \, dv = 1. \tag{22}$$

Hence

$$D_n = \frac{4}{\sqrt{\pi}} \frac{\Gamma\left(\dfrac{3n}{2}\right)}{\Gamma\left(\dfrac{3n-3}{2}\right)} . \tag{23}$$

If $W_n(v \mid T)$ denotes the conditional density function of the velocity of the particles at a given absolute temperature T, we have

$$W_n(v \mid T) = V_n\left(v \left| \frac{3nkT}{m}\right.\right) = \frac{4v^2}{\sqrt{\pi}} \left(\frac{m}{3nkT}\right)^{\frac{3}{2}} \frac{\Gamma\left(\dfrac{3n}{2}\right)}{\Gamma\left(\dfrac{3n-3}{2}\right)} \times$$

$$\times \left(1 - \frac{mv^2}{3nkT}\right)^{\frac{3n-5}{2}} \quad \text{for} \quad 0 \le v < \sqrt{\frac{3nkT}{m}}. \tag{24}$$

The distribution with density function (24) is called the *Maxwell distribution of order n*. As we have already seen, it tends for $n \to \infty$ to the ordinary Maxwell distribution, i.e.

$$\lim_{n \to \infty} W_n(v \mid T) = \sqrt{\frac{2}{\pi}} \left(\frac{m}{kT}\right)^{\frac{3}{2}} v^2 \, e^{-\frac{v^2 m}{2kT}} \quad (0 < v < +\infty). \tag{25}$$

§ 4. Generalization of the notion of conditional mathematical expectation in Kolmogorov probability spaces

In § 2 we have defined the conditional probability of an event with respect to the condition that a random variable assumes a given value. Similarly, we can define the conditional expectation of a random variable η with respect to the condition that the random variable ξ assumes a given value.

Let ξ be a random variable and $\mathscr{A}_\xi$ the least σ-algebra with respect to which ξ is measurable; let η be any other random variable with finite expectation. If ξ is a *discrete* random variable assuming the values x_k $(k = 1, 2, \ldots)$ with positive probabilities and if A_k is the event $\xi = x_k$, then let $E(\eta \mid \xi)$ denote the random variable such that $E(\eta \mid \xi) = E(\eta \mid A_k)$ for $\xi = x_k$ (i.e. for every $\omega \in \Omega$ with $\xi(\omega) = x_k$); we have thus, for $A \in \mathscr{A}_\xi$

$$\int_A E(\eta \mid \xi)\,dP = \sum_{k=1}^{+\infty} E(\eta \mid A_k) P(AA_k),$$

hence

$$\int_A E(\eta \mid \xi)\, dP = \int_A \eta\, dP, \tag{1}$$

provided that $A \in \mathcal{A}_\xi$ (this means in case of a discrete random variable ξ that $A = \sum_r A_{j_r}$, $(j_1 < j_2 < \ldots)$ is valid).

In the *general case*, we want to define the random variable $E(\eta \mid \xi)$ so that it is measurable with respect to $\mathcal{A}_\xi$ and the relation (1) is valid. Put

$$v(A) = \int_A \eta\, dP \qquad (A \in \mathcal{A}_\xi);$$

because of the known properties of the integral, $v(A)$ is σ-additive on $\mathcal{A}_\xi$ and absolutely continuous with respect to $P(A)$. Hence, by the Radon–Nikodym theorem, there exists a function $f(\omega) = \dfrac{dv}{dP}$ which is measurable with respect to $\mathcal{A}_\xi$ and fulfils the relation

$$v(A) = \int_A f(\omega)\, dP$$

whenever $A \in \mathcal{A}_\xi$. Therefore if $E(\eta \mid \xi)$ is defined by $E(\eta \mid \xi) = f(\omega)$, (1) is satisfied. It follows from the definition that for $A = \Omega$

$$E\big(E(\eta \mid \xi)\big) = E(\eta). \tag{2}$$

In particular, if $\eta = \eta_B$, where η_B is the indicator of the set B, i.e.

$$\eta_B(\omega) = \begin{cases} 1 & \text{for } \omega \in B, \\ 0 & \text{otherwise,} \end{cases}$$

then

$$v(A) = \int_A \eta_B\, dP = P(AB),$$

and

$$E(\eta_B \mid \xi) = P_\xi(B).$$

The conditional probability $P_\xi(B)$ of B for a given value of ξ may thus also be considered as a conditional expectation.

Of course one may ask whether $E(\eta \mid \xi)$ is with probability 1 equal to the expectation of the conditional distribution of η for a given value of ξ (i.e. to the expectation of the distribution $P_\xi(\eta^{-1}(V))$). The response is affirmative, provided that $P_\xi(\eta^{-1}(V))$ is with probability 1 a probability distribution. This can always be achieved, as we have already seen. In this case

$$E(\eta \mid \xi) = \int_\Omega \eta\, dP_\xi, \tag{3}$$

with probability 1. In order to prove this it suffices to show that for every $A \in \mathcal{A}_\xi$ the relation

$$\int_A \left(\int_\Omega \eta \, dP_\xi \right) dP = \int_A \eta \, dP \qquad (4)$$

holds. Obviously, this relation is fulfilled for $\eta = \eta_B$, where η_B is the indicator of the set B; indeed in this case

$$\int_\Omega \eta_B \, dP_\xi = P_\xi(B), \qquad \int_A \eta_B \, dP = P(AB),$$

and (4) will be reduced to the relation

$$\int_A P_\xi(B) \, dP = P(AB)$$

defining $P_\xi(B)$. Hence (4) holds when η takes on a denumerable set of values. From this, because of the known properties of the Lebesgue integral, it can be shown that (4) is generally valid.

If ξ and η are independent, it follows from (3) that we have with probability 1

$$E(\eta \mid \xi) = E(\eta). \qquad (5)$$

Furthermore the following theorem can be stated for arbitrary random variables ξ and η: If $f(x)$ is a Borel-measurable function such that $E(f(\xi)\eta)$ exists, then we have, with probability 1,

$$E(f(\xi)\eta \mid \xi) = f(\xi) E(\eta \mid \xi). \qquad (6)$$

To prove this it suffices to show that

$$\int_A f(\xi) E(\eta \mid \xi) \, dP = \int_A f(\xi) \eta \, dP \quad \text{for} \quad A \in \mathcal{A}_\xi. \qquad (7)$$

It follows from (3) that

$$\int_A f(\xi) E(\eta \mid \xi) \, dP = \int_A f(\xi) \left(\int_\Omega \eta \, dP_\xi \right) dP = \int_A E(f(\xi)\eta \mid \xi) \, dP = \int_A f(\xi) \eta \, dP.$$

Relation (6) furnishes a new proof of the fact that, for independent ξ and η, $E(\xi\eta) = E(\xi) \cdot E(\eta)$ (cf. Ch. IV). In fact, it follows from (2) and (6) that

$$E(\xi\eta) = E\big(E(\xi\eta \mid \xi)\big) = E\big(\xi E(\eta \mid \xi)\big). \qquad (8)$$

Thus if ξ and η are independent, $E(\eta \mid \xi) = E(\eta)$ with probability 1 and from this follows the desired result.

Consider now another important property of the conditional expectation. Let ξ and η be two random variables and $g(x)$ a Borel-measurable function.

We have then with probability 1

$$E(E(\eta \mid \xi) \mid g(\xi)) = E(\eta \mid g(\xi)). \tag{9}$$

In order to see this it suffices to prove the relation

$$\int_A E(E(\eta \mid \xi) \mid g(\xi)) dP = \int_A \eta \, dP \tag{10}$$

for every $A \in \mathscr{A}_{g(\xi)}$. By applying twice the definition of conditional probabilities and by taking into account that $\mathscr{A}_{g(\xi)} \subset \mathscr{A}_\xi$, we obtain

$$\int_A E(E(\eta \mid \xi) \mid g(\xi)) dP = \int_A E(\eta \mid \xi) \, dP = \int_A \eta \, dP,$$

which proves (10) and hence (9) too.

The ordinary expectation is known to be a linear functional. How far does this hold for the conditional expectation? If c_1 and c_2 are two constants, we have with probability 1,

$$E(c_1 \eta_1 + c_2 \eta_2 \mid \xi) = c_1 E(\eta_1 \mid \xi) + c_2 E(\eta_2 \mid \xi). \tag{11}$$

Indeed we have by (1) for every $A \in \mathscr{A}_\xi$

$$\int_A \left(c_1 E(\eta_1 \mid \xi) + c_2 E(\eta_2 \mid \xi) \right) dP = c_1 \int_A E(\eta_1 \mid \xi) \, dP + c_2 \int_A E(\eta_2 \mid \xi) \, dP =$$

$$= \int_A (c_1 \eta_1 + c_2 \eta_2) dP.$$

Nevertheless we cannot state that $E(\eta \mid \xi)$ is a linear functional with probability 1, since (11) holds with probability 1 only and the exceptional sets corresponding to every pair (η_1, η_2) may together even cover the whole space Ω.

§ 5. Generalization of Bayes' theorem

Let ξ and η be two random variables with absolutely continuous distributions and a two-dimensional density function $h(x, y)$. Put further

$$f(x) = \int_{-\infty}^{+\infty} h(x, y) \, dy, \tag{1}$$

$$g(y) = \int_{-\infty}^{+\infty} h(x, y) \, dx, \tag{2}$$

$$f(x \mid y) = \frac{h(x, y)}{g(y)} \quad \text{for} \quad g(y) > 0, \quad \text{otherwise arbitrary,} \tag{3}$$

$$g(y \mid x) = \frac{h(x, y)}{f(x)} \quad \text{for} \quad f(x) > 0, \quad \text{otherwise arbitrary.} \tag{4}$$

Clearly we have

$$f(x) = \int\limits_{-\infty}^{+\infty} f(x \mid y) \, g(y) dy \tag{5}$$

and

$$g(y) = \int\limits_{-\infty}^{+\infty} g(y \mid x) f(x) dx. \tag{6}$$

It follows from (3) and (4) that

$$g(y \mid x) = \frac{f(x \mid y) \, g(y)}{f(x)} \quad \text{for} \quad f(x) > 0, \tag{7}$$

hence by (5)

$$g(y \mid x) = \frac{f(x \mid y) \, g(y)}{\int\limits_{-\infty}^{+\infty} f(x \mid t) \, g(t) dt}. \tag{8}$$

Formula (8) may be considered as a *generalization of Bayes' theorem* for the case of absolutely continuous distributions. With this formula one can express the conditional density function of η for a given value of ξ by means of the conditional density function of ξ for a given value of η and the unconditional density function of η. It follows from (8) that

$$P(a \le \eta < b \mid \xi = x) = \int\limits_{a}^{b} g(y \mid x) \, dy =$$

$$= \frac{\int\limits_{a}^{b} f(x \mid y) \, g(y) dy}{\int\limits_{-\infty}^{+\infty} f(x \mid t) \, g(t) dt} \tag{9}$$

or

$$P(a \le \eta < b \mid \xi = x) = \frac{\int\limits_{a}^{b} f(x \mid y) \, dG(y)}{\int\limits_{-\infty}^{+\infty} f(x \mid t) \, dG(t)}, \tag{10}$$

where $G(y)$ is the ordinary distribution function of η.

Formula (10) is therefore valid even if η does not have an absolutely continuous distribution.

Relation (8) is in certain cases also valid for ξ and η defined on a conditional probability space. This holds if the two-dimensional density function (in the sense explained in § 1) exists. For the functions $f(x)$ and $g(y)$ defined in (1) and in (2) — provided that they exist, i.e. the integrals (1) and (2) are finite — we have, in general,

$$\int_{-\infty}^{+\infty} f(x)dx = \int_{-\infty}^{+\infty} g(y)\,dy = +\infty.$$

Let it be mentioned that $h(x, y), f(x), g(y)$ are only defined up to a constant factor. If $f(x \mid y)$ and $g(y \mid x)$ are computed by (3) and (4) or (8), this factor disappears. The obtained density functions $f(x \mid y)$ and $g(y \mid x)$ are already so normed that their integral from $-\infty$ to $+\infty$ has the value 1.

§ 6. The correlation ratio

Let ξ and η be two random variables on a Kolmogorov probability space; suppose that $E(\eta)$ and $D^2(\eta)$ exist, let further $E(\eta \mid \xi)$ denote the conditional probability of η for a given value of ξ. We know that

$$E(E(\eta \mid \xi)) = E(\eta). \tag{1}$$

For the variance of $E(\eta \mid \xi)$ we have

$$D_\xi^2(\eta) = D^2(E(\eta \mid \xi)) = E(E^2(\eta \mid \xi)) - E^2(\eta). \tag{2}$$

THEOREM 1. *If $E(\eta)$ and $D^2(\eta)$ exist, we have*

$$D^2(\eta) = D_\xi^2(\eta) + E([E(\eta \mid \xi) - \eta]^2). \tag{3}$$

PROOF. We have

$$\eta - E(\eta) = [\eta - E(\eta \mid \xi)] + [E(\eta \mid \xi) - E(\eta)],$$

therefore

$$D^2(\eta) = E([E(\eta \mid \xi) - \eta]^2) + D_\xi^2(\eta) + 2E([\eta - E(\eta \mid \xi)][E(\eta \mid \xi) - E(\eta)]). \tag{4}$$

By (2) and (6) of § 4

$$E([\eta - E(\eta \mid \xi)][E(\eta \mid \xi) - E(\eta)]) = E([\eta - E(\eta \mid \xi)]E(\eta \mid \xi)) =$$
$$= E(E([\eta - E(\eta \mid \xi)]E(\eta \mid \xi) \mid \xi)) = E(E(\eta \mid \xi)E(\eta - E(\eta \mid \xi) \mid \xi)) = 0.$$

Thus (4) implies (3).

Remarks.

1. It was implicitly shown in proving (3) that the random variables $\eta - E(\eta \mid \xi)$ and $E(\eta \mid \xi)$ are uncorrelated.

2. The assertion of Theorem 1 may be written in the form

$$D^2(\eta) = D^2\big(E(\eta \mid \xi)\big) + E\big(D^2(\eta \mid \xi)\big) \tag{5}$$

where $D^2(\eta \mid \xi)$ is the *conditional variance* of η for a given value of ξ, defined by

$$D^2(\eta \mid \xi) = E\big([\eta - E(\eta \mid \xi)]^2 \mid \xi\big).$$

According to Formula (2) of § 4 we have thus

$$E\big(D^2(\eta \mid \xi)\big) = E\big(E([\eta - E(\eta \mid \xi)]^2 \mid \xi)\big) = E\big([\eta - E(\eta \mid \xi)]^2\big).$$

Assuming $D(\eta) > 0$, put

$$K_\xi(\eta) = \frac{D_\xi(\eta)}{D(\eta)}. \tag{6}$$

Then by Theorem 1

$$0 \le K_\xi(\eta) \le 1. \tag{7}$$

$K_\xi(\eta)$ will be called *correlation ratio of η with respect to ξ*; it is defined only if $D(\eta) > 0$. This notion was introduced by K. Pearson in a somewhat less general form, and in full generality by A. N. Kolmogorov. It gives a certain information about the mutual dependence of ξ and η. This is shown by the following two theorems.

THEOREM 2. *If ξ and η are independent, $K_\xi(\eta) = 0$. The converse, however, does not hold: $K_\xi(\eta) = 0$ does not imply the independence of ξ and η, though it implies the vanishing of the correlation coefficient $R(\xi, \eta)$.*

THEOREM 3. *The relation $K_\xi(\eta) = 1$ is valid iff $\eta = g(\xi)$, where $g(\xi)$ is a Borel-measurable function.*

PROOF OF THEOREM 2. If ξ and η are independent, $D_\xi(\eta) = 0$, hence $K_\xi(\eta) = 0$. If $K_\xi(\eta) = 0$, $E(\eta \mid \xi)$ is equal to $E(\eta)$ with probability 1, therefore by relation (8) of § 4

$$E(\xi\eta) = E\big(\xi E(\eta \mid \xi)\big) = E(\xi)E(\eta),$$

thus $R(\xi, \eta) = 0$. The following example shows that $K_\xi(\eta) = 0$ does not imply the independence of ξ and η: Let the point (ξ, η) be uniformly distributed in the circle $x^2 + y^2 < 1$; let $g(y \mid x)$ be the conditional density function of η with respect to the condition $\xi = x$. We have

$$g(y \mid x) = \frac{1}{2\sqrt{1-x^2}} \quad \text{for} \quad |y| < \sqrt{1-x^2}; \quad -1 < x < +1,$$

hence $E(\eta \mid \xi) \equiv 0$ and, consequently, $K_\xi(\eta) = 0$ though ξ and η are evidently not independent.

PROOF OF THEOREM 3. If $K_\xi(\eta) = 1$, (3) shows that

$$E([\eta - E(\eta \mid \xi)]^2) = 0,$$

hence, with the probability 1,

$$\eta = E(\eta \mid \xi); \tag{8}$$

η is thus measurable with respect to $\mathscr{A}_\xi$ and therefore it can be written in the form $\eta = g(\xi)$. Conversely, if $\eta = g(\xi)$, then η is measurable with respect to $\mathscr{A}_\xi$, thus (8) is valid with probability 1, therefore it follows that $K_\xi(\eta) = 1$.

Unlike the correlation coefficient, the correlation ratio is not symmetrical. To characterize the dependence between ξ and η both quantities $K_\xi(\eta)$ and $K_\eta(\xi)$ can be used, provided that the variances of both ξ and η exist and are positive. The conditional expectation $E(\eta \mid \xi)$ can be characterized by the following property:

THEOREM 4. *If ξ and η are any two random variables and $D^2(\eta)$ is finite and if $g(x)$ is a Borel-measurable function, then the expression*

$$E([\eta - g(\xi)]^2)$$

takes on its minimum for $g(\xi) = E(\eta \mid \xi)$.

PROOF. By Formula (2) of § 4

$$E([\eta - g(\xi)]^2) = E(E([\eta - g(\xi)]^2 \mid \xi)). \tag{9}$$

It follows by a basic property of the expectation (see Theorem 2 of § 9, Ch. III) that

$$E([\eta - g(\xi)]^2 \mid \xi) = \int (\eta - g(\xi))^2 \, dP_\xi \geq \int (\eta - E(\eta \mid \xi))^2 \, dP_\xi. \tag{10}$$

It follows from (9) and (10) that

$$E([\eta - g(\xi)]^2) \geq E([\eta - E(\eta \mid \xi)]^2), \tag{11}$$

q.e.d. Equality in (11) can occur, if and only if the relation

$$g(\xi) = E(\eta \mid \xi)$$

is valid with probability 1.

Remark. The curve $y = E(\eta \mid \xi = x)$ is called the regression curve of η with respect to ξ.

In particular, it follows from Theorem 4 that for any two real numbers a and b

$$E([\eta - (a\xi + b)]^2) \geq E([\eta - E(\eta \mid \xi)]^2).\quad\quad (12)$$

The left-hand side is minimal for

$$a = \frac{D(\eta)}{D(\xi)} R(\xi, \eta), \quad b = E(\eta) - aE(\xi). \quad\quad (13)$$

The line

$$y - E(\eta) = \frac{D(\eta)}{D(\xi)} R(\xi, \eta) [x - E(\xi)] \quad\quad (14)$$

is called the *regression curve of η with respect to ξ.* If a and b are given by (13), we have

$$E([\eta - (a\xi + b)]^2) = D^2(\eta) [1 - R^2(\xi, \eta)]. \quad\quad (15)$$

On the other hand, because of (3) and (6),

$$E([\eta - E(\eta \mid \xi)]^2) = D^2(\eta) [1 - K_\xi^2(\eta)]. \quad\quad (16)$$

From (12), (15) and (16) it follows that

$$R^2(\xi, \eta) \leq K_\xi^2(\eta). \quad\quad (17)$$

This permits to restate the proposition of Theorem 2: If $K_\xi(\eta) = 0$ then

$R(\xi, \eta) = 0$. Inequality (17) may be sharpened as follows:

THEOREM 5. *If ξ is an arbitrary random variable and η a random variable with finite expectation and variance, then*

$$K_\xi^2(\eta) = \sup_g R^2(g(\xi), \eta), \quad\quad (18)$$

where $y = g(x)$ *runs through the set of all Borel-measurable functions for which the expectation and variance of* $g(\xi)$ *exist. The relation*

$$K_\xi^2(\eta) = R^2\left(g(\xi), \eta\right) \tag{19}$$

holds, iff

$$g(\xi) = aE(\eta \mid \xi) + b, \tag{20}$$

where $a \neq 0$ *and* b *are constants.*

PROOF. One can assume without restriction of generality that $E(\eta) = E(g(\xi)) = 0$ and $D(\eta) = D(g(\xi)) = 1$. By (2) and (6) of § 4 and by the Schwarz inequality,

$$R^2\left(g(\xi), \eta\right) = E^2\left(\eta g(\xi)\right) = E^2\left(E(\eta g(\xi) \mid \xi)\right) = E^2\left(g(\xi) E(\eta \mid \xi)\right) \leq$$

$$\leq E\left(E^2(\eta \mid \xi)\right) = K_\xi^2(\eta).$$

The condition for equality is here easily verified.

Theorem 5 permits to give a new definition of the correlation ratio $K_\xi(\eta)$ and even a new definition of the conditional expectation $E(\eta \mid \xi)$. Certainly, Formula (19) defines $E(\eta \mid \xi) = g(\xi)$ up to a linear transformation only. But it is easy to obtain a unique definition. In effect, $E(\eta \mid \xi)$ may be characterized as the function $g_0(\xi)$ fulfilling the relation

$$R^2\left(\eta, g_0(\xi)\right) = \sup_g R^2\left(\eta, g(\xi)\right) = K_\xi^2(\eta) \tag{21}$$

and the relations

$$\left.\begin{array}{c} E\left(g_0(\xi)\right) = E(\eta), \\ D^2\left(g_0(\xi)\right) = D^2(\eta) K_\xi^2(\eta), \\ E\left(\eta\, g_0(\xi)\right) > 0. \end{array}\right\} \tag{22}$$

§ 7. On some other measures of the dependence of two random variables

Another measure of the dependence of two random variables is given by the *contingency*. This notion was introduced for discrete distributions by K. Pearson (mean square contingency).[1]

Let ξ and η be discrete random variables assuming the values x_k ($k = 1, 2, \ldots$) and y_j ($j = 1, 2, \ldots$), and only these with positive probabilities. Let A_k and B_j denote the events $\xi = x_k$ and $\eta = y_j$ respectively. The contingency $\varphi(\xi, \eta)$ is defined by

$$\varphi(\xi, \eta) = \left(\sum_j \sum_k \frac{[P(A_k B_j) - P(A_k) P(B_j)]^2}{P(A_k) P(B_j)}\right)^{\frac{1}{2}}, \tag{1}$$

[1] For the general case see A. Rényi [28].

or with an obvious transformation,

$$\varphi^2(\xi, \eta) = \sum_j \sum_k \frac{P^2(A_k B_j)}{P(A_k) P(B_j)} - 1. \tag{2}$$

It is clear that $\varphi(\xi, \eta)$ is zero iff ξ and η are independent. If the number of the values x_k is n and that of the y_j-s is m with $m \geq n$, then

$$\varphi^2(\xi, \eta) \leq n - 1, \tag{3}$$

as because of $P(A_k B_j) \leq P(B_j)$ it follows from (2) that

$$1 + \varphi^2(\xi, \eta) \leq \sum_k \sum_j \frac{P(A_k B_j)}{P(A_k)} = n. \tag{4}$$

It can be seen from (4) that in (3) the sign of equality holds iff for every k and for every j either $P(A_k B_j) = P(B_j)$ or $P(A_k B_j) = 0$. Since, however

$$\sum_{k=1}^{n} P(A_k B_j) = P(B_j) \qquad (P(B_j) > 0),$$

this cannot occur unless for one k_j the relation $P(A_{k_j} B_j) = P(B_j)$ and for the other $k \neq k_j$ the relation $P(A_k B_j) = 0$ is valid. But then $\xi = x_{k_j}$ for $\eta = y_j$ and consequently $\xi = f(\eta)$.

Conversely, if $\xi = f(\eta)$, $\varphi^2(\xi, \eta) = n - 1$. If both ξ and η assume infinitely many values, the series on the right of (1) may be divergent; in this case $\varphi(\xi, \eta) = +\infty$.

Before defining the contingency for arbitrary random variables, the notion of *regular dependence* will be introduced. Let ξ and η be any two random variables. If C is an arbitrary two-dimensional Borel set, we put

$$P(C) = P((\xi, \eta) \in C). \tag{5}$$

Let A and B be Borel sets on the x-axis and the y-axis respectively; put

$$P_1(A) = P(\xi^{-1}(A)) \tag{6a}$$

and

$$P_2(B) = P(\eta^{-1}(B)). \tag{6b}$$

Let $A \times B$ denote the set of the points of the (x, y)-plane for which $x \in A$ and $y \in B$. Define the measure $Q(C)$ for the two-dimensional Borel sets of the form $C = A \times B$ by

$$Q(A \times B) = P_1(A) P_2(B). \tag{7}$$

This measure can be extended to all two-dimensional Borel sets of the plane in a unique manner, since the values of its extension are uniquely determined by the values on the "parallelograms" $A \times B$.

If P is absolutely continuous with respect to Q, the dependence between ξ and η is said to be *regular*. This is evidently the case if ξ and η are independent, since then $P \equiv Q$. It is easy to see that the dependence between two discrete random variables is always regular.

If the dependence between ξ and η is regular, there exists according to the Radon–Nikodym theorem, a Borel-measurable function $k(x, y) = \dfrac{dP}{dQ}$ such that for every two-dimensional Borel set C the relation

$$P(C) = \int_C k(x, y)\, dQ \qquad (8)$$

holds. If $F(x)$ and $G(y)$ are the distribution functions of ξ and η, respectively, and if A and B are any two Borel subsets of the real axis, then the function $k(x, y)$ satisfies the relation

$$P(\xi \in A, \eta \in B) = \int_{x \in A} \int_{y \in B} k(x, y)\, dF(x)\, dG(y). \qquad (9)$$

In particular, if ξ and η are discrete random variables,

$$k(x, y) = \begin{cases} \dfrac{P(A_k B_j)}{P(A_k)\, P(B_j)} & \text{for} \quad x = x_k, \; y = y_j, \\[2mm] 0 & \text{otherwise.} \end{cases} \qquad (10)$$

If the joint distribution of ξ and η is absolutely continuous with the density function $h(x, y)$ and if $f(x)$ and $g(y)$ are the density functions of ξ and η respectively, then we evidently have

$$k(x, y) = \frac{h(x, y)}{f(x)\, g(y)} . \qquad (11)$$

We can now define the contingency for arbitrary regularly dependent random variables ξ and η by

$$\varphi(\xi, \eta) = \left(\int_{-\infty}^{+\infty} \int_{-\infty}^{+\infty} [k(x, y) - 1]^2\, dF(x)\, dG(y) \right)^{\frac{1}{2}} \qquad (12)$$

or equivalently by

$$\varphi^2(\xi, \eta) = \int_{-\infty}^{+\infty} \int_{-\infty}^{+\infty} k^2(x, y)\, dF(x)\, dG(y) - 1. \qquad (13a)$$

In particular, if ξ and η are discrete random variables, relation (1) is obtained from (12), because of (10). If the joint distribution of ξ and η is absolutely continuous, we obtain, because of (11),

$$\varphi^2(\xi, \eta) = \int_{-\infty}^{+\infty} \int_{-\infty}^{+\infty} \frac{h^2(x, y)}{f(x) g(y)} \, dxdy - 1. \tag{13b}$$

Obviously $\varphi(\xi, \eta) = 0$ holds iff $P \equiv Q$, i.e. if ξ and η are independent.

Now we prove a theorem establishing a relation between the correlation coefficient and the contingency.

THEOREM 1. *Let ξ and η be regularly dependent random variables, $u(x)$ and $v(y)$ Borel-measurable functions such that the variances $D^2(u(\xi))$ and $D^2(v(\eta))$ exist and are positive. Then we have*

$$R^2\left(u(\xi), v(\eta)\right) \leq \varphi^2(\xi, \eta). \tag{14}$$

PROOF. We may assume without restricting the generality that $E(u(\xi)) = E(v(\eta)) = 0$ and $D(u(\xi)) = D(v(\eta)) = 1$. Then by the definition of the correlation coefficient,

$$R(u(\xi), v(\eta)) = \int_{-\infty}^{+\infty} \int_{-\infty}^{+\infty} u(x) \, v(y) \, k(x, y) \, dF(x) \, dG(y). \tag{15}$$

But by assumption

$$\int_{-\infty}^{+\infty} \int_{-\infty}^{+\infty} u(x) \, v(y) dF(x) \, dG(y) = 0. \tag{16}$$

From (15) and (16) follows

$$R(u(\xi), v(\eta)) = \int_{-\infty}^{+\infty} \int_{-\infty}^{+\infty} u(x) \, v(y) \left[k(x, y) - 1 \right] dF(x) \, dG(y). \tag{17}$$

By applying the Schwarz inequality and (12), we obtain

$$R^2\left(u(\xi), v(\eta)\right) \leq \varphi^2(\xi, \eta),$$

which proves the theorem.

The quantity

$$\psi(\xi, \eta) = \sup_{u,v} | R(u(\xi), v(\eta)) |, \tag{18}$$

where $u(x)$ and $v(y)$ assume all Borel-measurable functions for which expectations and variances of $u(\xi)$ and $v(\eta)$ exist, can also be considered as a measure of the dependence between ξ and η. This quantity is called *maximal correlation* of ξ and η and was introduced first, for discrete random variables, by Hirschfeld, for absolutely continuous distributions by Gebelein. Its most simple properties are contained in the following theorem:

THEOREM 2. *If* $\psi(\xi, \eta)$ *is the maximal correlation of* ξ *and* η, *defined by* (18), *we have always*

a) $\psi(\xi, \eta) = \psi(\eta, \xi)$;

b) $0 \le \psi(\xi, \eta) \le 1$;

c) *if* $y = a(x)$ *and* $y = b(x)$ *are strictly monotonic functions, then*

$$\psi(a(\xi), b(\eta)) = \psi(\xi, \eta);$$

d) $\psi(\xi, \eta) = 0$ *iff* ξ *and* η *are independent;*

e) *if there exists between* ξ *and* η *a relation of the form* $U(\xi) = V(\eta)$, *where* $U(x)$ *and* $V(y)$ *are Borel-measurable functions with* $D(U(\xi)) > 0$, *then* $\psi(\xi, \eta) = 1$;

f) *we have*

$$| R(\xi, \eta) | \le \min (K_\xi(\eta), K_\eta(\xi)) \le \max (K_\xi(\eta), K_\eta(\xi)) \le \psi(\xi, \eta) \le \varphi(\xi, \eta).$$

PROOF. Properties a), b), and c) are direct consequences of the definition. If ξ and η are independent, clearly $\psi(\xi, \eta) = 0$. Conversely, if

$$\psi(\xi, \eta) = 0, \text{ then } R(u(\xi), v(\eta)) = 0$$

for every u and v. If we choose

$$u_a(x) = \begin{cases} 1 & \text{for } x < a, \\ 0 & \text{for } x \ge a, \end{cases} \qquad v_b(x) = \begin{cases} 1 & \text{for } x < b, \\ 0 & \text{for } x \ge b, \end{cases}$$

it follows from $R(u_a(\xi), v_b(\eta)) = 0$ that

$$P(\xi < a, \eta < b) = P(\xi < a) P(\eta < b).$$

As a and b are arbitrary this means that ξ and η are independent, hence d) is proved. If $U(\xi) = V(\eta)$ with $D(U(\xi)) > 0$ we know that

$$R(U(\xi), V(\eta)) = 1,$$

hence $\psi(\xi, \eta) = 1$. Property f) can be deduced by comparing the definition of maximal correlation to Theorem 1 of this § and to Theorem 5 of § 6.

A further notion which we want to study here is that of the *modulus of (pairwise) dependence* of a sequence of random variables. Let $\xi_1, \xi_2, \ldots, \xi_n, \ldots$ be a (finite or infinite) sequence of arbitrary random variables. We define the *modulus of dependence* of the sequence $\{\xi_n\}$ as the smallest positive real number A satisfying for all sequences $\{x_n\}$ with $\sum x_n^2 < +\infty$ the inequality

$$|\sum_n \sum_m \psi(\xi_n, \xi_m) x_n x_m| \leq A \sum_n x_n^2, \tag{19}$$

i.e. the least upper bound of the quadratic form

$$\sum_n \sum_m \psi(\xi_n, \xi_m) x_n x_m$$

under condition $\sum x_h^2 = 1$. (If it is unbounded, the modulus of dependence is infinite.)

In particular, if the sequence $\{\xi_n\}$ contains only two elements, its modulus of dependence will be $1 + \psi(\xi_1, \xi_2)$. If the sequence $\{\xi_n\}$ is finite, A is finite as well; if the sequence contains infinitely many elements, A is not necessarily finite. If the elements of $\{\xi_n\}$ are pairwise independent, $A = 1$, otherwise $A > 1$.

The following theorem furnishes an inequality between the correlation ratio and the modulus of dependence of a sequence of random variables.

THEOREM 3. *Let $\{\xi_n\}$ be a sequence of random variables with a finite modulus of dependence A. If η is a random variable with a finite variance, we have*

$$\sum_n K_{\xi_n}^2 (\eta) \leq A. \tag{20}$$

For the proof we need a lemma which is a generalization of the Bessel inequality, well known from the theory of orthogonal series, to the case of *quasiorthogonal* functions.

LEMMA. *Let $\{\zeta_n\}$ be a finite or infinite sequence of random variables such that $E(\zeta_n^2)$ exists $(n = 1, 2, \ldots)$. Suppose that the quadratic form*

$$\sum_n \sum_m E(\zeta_n \zeta_m) x_n x_m$$

is bounded and that we have

$$|\sum_n \sum_m E(\zeta_n \zeta_m) x_n x_m| \leq B \sum_n x_n^2; \tag{21}$$

then for every random variable η for which $E(\eta^2)$ exists we have

$$\sum_n E^2 (\eta \zeta_n) \leq BE(\eta^2). \tag{22}$$

Note. If $E(\zeta_n \zeta_m) = 0$ for $n \neq m$ and if $E(\zeta_n^2) = 1$, i.e. if the sequence $\{\zeta_n\}$ is orthonormal, (21) is valid with $B = 1$ and (22) reduces to Bessel's inequality

$$\sum_n E^2(\eta \zeta_n) \leq E(\eta^2). \tag{23}$$

PROOF OF THE LEMMA. Put

$$a_n = E(\eta \zeta_n). \tag{24}$$

Obviously,

$$E\left(\left[\eta - \frac{1}{B} \sum_n a_n \zeta_n\right]^2\right) \geq 0. \tag{25}$$

By carrying out the calculations, we find

$$E(\eta^2) - \frac{2}{B} \sum_n a_n^2 + \frac{1}{B^2} \sum_n \sum_m a_n a_m E(\zeta_n \zeta_m) \geq 0. \tag{26}$$

Because of (21) we have

$$\frac{1}{B^2} \left| \sum_n \sum_m a_n a_m E(\zeta_n \zeta_m) \right| \leq \frac{1}{B} \sum_n a_n^2, \tag{27}$$

hence it follows from (26) that

$$\frac{1}{B} \sum_n a_n^2 \leq E(\eta^2), \tag{28}$$

which is, by (24), equivalent to (22).

PROOF OF THEOREM 3. Let $f_n(x)$ be a Borel-measurable function such that $E(f_n(\xi_n)) = 0$ and $D(f_n(\xi_n)) = 1$. Put

$$\zeta_n = f_n(\xi_n). \tag{29}$$

Then by definition of the maximal correlation

$$|E(\zeta_n \zeta_m)| = |R(f_n(\xi_n), f_m(\xi_m))| \leq \psi(\xi_n, \xi_m). \tag{30}$$

Hence according to (19)

$$\left| \sum_n \sum_m E(\zeta_n \zeta_m) x_n x_m \right| \leq \sum_n \sum_m \psi(\xi_n, \xi_m) |x_n| \cdot |x_m| \leq A \sum_n x_n^2. \tag{31}$$

Thus the lemma can be applied to the sequence $\{\zeta_n\}$ with $B = A$, provided that $E(\eta^2)$ exists. Then

$$\sum E^2(\zeta_n [\eta - E(\eta)]) \leq AD^2(\eta) \cdot \tag{32}$$

But

$$E(\zeta_n [\eta - E(\eta)]) = D(\eta) R(f_n(\xi_n), \eta). \tag{33}$$

Let $f_n(x)$ be chosen such that

$$f_n(\xi_n) = \frac{E(\eta \mid \xi_n) - E(\eta)}{D_{\xi_n}(\eta)}, \tag{34}$$

where $D_{\xi_n}(\eta) = D(E(\eta \mid \xi_n))$ is the standard deviation of $E(\eta \mid \xi_n)$. Then according to Theorem 5 of § 6

$$R^2(f_n(\xi_n), \eta) = K_{\xi_n}^2(\eta). \tag{35}$$

Hence by (32) and (33)

$$D^2(\eta) \sum_n K_{\xi_n}^2(\eta) \leq AD^2(\eta). \tag{36}$$

After division by $D^2(\eta)$ we obtain (20) and thus Theorem 3 is proved.

Theorem 3 is a probabilistic generalization of the *large sieve* of Yu. V. Linnik, which has important applications in number theory.[1]

§ 8. The fundamental theorem of Kolmogorov

In what follows, we shall often prove theorems concerning an infinite sequence of random variables. The conditions of these theorems involve the simultaneous distributions of a finite number of the random variables considered. We shall not prove for each particular theorem the existence on some probability space of a sequence of random variables fulfilling the assumptions of the theorem; the solution of this existence problem is furnished by a general theorem due to Kolmogorov. Kolmogorov proved this fundamental theorem for an arbitrary (not necessarily denumerable) set of random variables; we restrict ourselves to the case of denumerable sets.

THEOREM 1. (*Kolmogorov's fundamental theorem*). *For any integer n let* $F_n(x_1, x_2, \ldots, x_n)$ *be an n-dimensional distribution function, fulfilling the*

---

[1] With the help of this generalization Rényi succeeded to prove that every positive integer n can be written in the form $n = p + P$, where p is a prime and P is the product of at most K prime factors; K denotes here a universal constant. Cf. A. Rényi [2].

following conditions of compatibility:

$$F_{n+m}(x_1, x_2, \ldots, x_n, +\infty, \ldots, +\infty) = F_n(x_1, x_2, \ldots, x_n)$$

$$(n, m = 1, 2, \ldots). \qquad (1)$$

Then there exists a Kolmogorov probability space on which the random variables ξ_n ($n = 1, 2, \ldots$) can be so defined that for every n the n-dimensional distribution function of the random variables $\xi_1, \xi_2, \ldots, \xi_n$ is equal to $F_n(x_1, x_2, \ldots, x_n)$.

PROOF. Let Ω be the set of all infinite sequences

$$\omega = (\omega_1, \omega_2, \ldots, \omega_n, \ldots)$$

of real numbers, Π_n the function, defined on Ω, projecting Ω upon the subspace Ω_n on the first n coordinates of ω; i.e., for $\omega = (\omega_1, \omega_2, \ldots, \omega_n, \ldots)$ we put

$$\Pi_n \omega = (\omega_1, \omega_2, \ldots, \omega_n). \qquad (2)$$

For $A \subset \Omega$, let $\Pi_n A$ denote the set of all elements of Ω_n which can be brought to the form $y = \Pi_n \omega$ with $\omega \in A$.

Let now $A \subset \Omega_n$ be any subset of Ω_n. We shall call the set of elements $\omega = (\omega_1, \omega_2, \ldots, \omega_n, \ldots)$ such that $\Pi_n \omega = (\omega_1, \ldots, \omega_n) \in A$ an *n-dimensional cylinder with base A*; we shall denote this set by $\Pi_n^{-1}(A)$.

If A is Borel-measurable, the corresponding cylinder set is said to be a *Borel cylinder set*. Let $\mathscr{A}$ be the set of all Borel cylinder sets; $\mathscr{A}$ is an algebra of sets. To see this let us remark that an n-dimensional cylinder set is at the same time an $(n + m)$-dimensional cylinder set as well. In fact

$$\Pi_n^{-1}(A) = \Pi_{n+m}^{-1}\left(\Pi_{n+m}\Pi_n^{-1}(A)\right). \qquad (3)$$

Hence if A is an n-dimensional Borel set, $\Pi_{n+m}(\Pi_n^{-1}(A))$ is an $(n + m)$-dimensional Borel set. Thus for a finite number of cylinders it can always be assumed that their bases have the same number of dimensions, e.g. N. If $A = \Pi_N^{-1}(A')$, $B = \Pi_N^{-1}(B')$, where A' and B' are Borel sets of the N-dimensional space, then

$$A + B = \Pi_N^{-1}(A' + B'),$$

$$A - B = \Pi_N^{-1}(A' - B');$$

$A + B$ and $A - B$ are thus Borel cylinders again. Finally, since $\Omega = \Pi_N^{-1}(\Omega_N)$, the set Ω itself is a Borel cylinder as well.

Let $\mathscr{A}^*$ be the least σ-algebra of subsets of Ω which contains all Borel cylinders of Ω. The probability measure P on $\mathscr{A}^*$ is defined in the following manner: First we define P on the Boolean algebra $\mathscr{A}$ and then we extend the definition as was done in Chapter II.

Let A be a Borel cylinder of Ω and N an integer with $A = \Pi_N^{-1}(A_N)$, A_N being a Borel set of Ω_N. $F_N(x_1, \ldots, x_N)$ generates on Ω_N a probability measure which we denote by P_N. We put $P(A) = P_N(A_N)$.

The definition is unique, since from

$$\Pi_N^{-1}(A_N) = \Pi_{N+M}^{-1}(A_{N+M})$$

follows, because of (1),

$$P_{N+M}(A_{N+M}) = P_N(A_N).$$

Consequently, the definition of $P(A)$ does not depend on the base figuring in the construction of A.

Clearly, the set function $P(A)$ is nonnegative; it is easy to show that it is (finitely) additive. If $A \in \mathscr{A}$, $B \in \mathscr{A}$, $AB = 0$, then, because of

$$A = \Pi_N^{-1}(A_N), \quad B = \Pi_N^{-1}(B_N),$$

we have $A_N B_N = 0$. Hence

$$P(A + B) = P_N(A_N + B_N) = P_N(A_N) + P_N(B_N) = P(A) + P(B).$$

(We made use of the fact that the value of $P(A)$ does not depend on the dimension of the chosen base of A.) It is further clear that $P(\Omega) = P_N(\Omega_N) = 1$. It remains to prove that $P(A)$ is not only additive but also σ-additive on $\mathscr{A}$. By Theorem 3, § 7 of Chapter II it suffices to show that P has the following property:

Property K. *If* $A_n \in \mathscr{A}, A_{n+1} \subseteq A_n$ $(n = 1, 2, \ldots)$ *and* $\prod_{n=1}^{\infty} A_n = 0$, *then*
$$\lim_{n \to +\infty} P(A_n) = 0.$$

We shall show by an indirect proof that property K is fulfilled. The inequality $P(A_n) \geq P(A_{n+1})$, $n = 1, 2, \ldots$ is obviously true. Hence $\lim_{n \to +\infty} P(A_n) = p$ exists. Assume $p > 0$. We show that then $D = \prod_{n=1}^{\infty} A_n$ cannot be empty.

It can be assumed without restriction of generality that A_n is an n-dimensional cylinder; in fact if d_n denotes the exact (minimal) number of dimensions of A_n, we have $d_n \leq d_{n+1}$. Further $\lim_{n \to +\infty} d_n = +\infty$ can be assumed, since in the case of $d_n \leq d$ our assertion would follow from $P_d(A)$ being a σ-additive measure. If $d_n \to +\infty$, we can replace the sequence $\{A_n\}$ by another sequence $\{A'_n\}$, where A'_n is an n-dimensional cylinder and $\{A'_n\}$ contains the sequence $\{A_n\}$.

Put $A_n = \Pi_n^{-1}(B_n)$, where B_n is an n-dimensional Borel set. Since $P(A_n) = P_n(B_n) \geq p > 0$, we can find in Ω_n a compact set Z_n with $Z_n \subseteq B_n$ such that

$$P_n(Z_n) \geq P_n(B_n) - \frac{p}{2^{n+1}} \qquad (n = 1, 2, \ldots).$$

Put $C_n = \Pi_n^{-1}(Z_n)$. C_n is also a Borel cylinder, and

$$P(C_n) = P_n(Z_n) \geq P(A_n) - \frac{p}{2^{n+1}}.$$

Let now $D_n = C_1 C_2 \ldots C_n$. We have

$$P(A_n - D_n) = P\left(A_n(\bar{C}_1 + \ldots + \bar{C}_n)\right) \leq \sum_{k=1}^{n} P(A_k - C_k) \leq$$

$$\leq \sum_{k=1}^{n} \frac{p}{2^{k+1}} \leq \frac{p}{2},$$

hence

$$P(D_n) = P(A_n) - P(A_n - D_n) \geq \frac{p}{2} > 0.$$

Thus the set D_n cannot be empty for any value of n. Choose now in D_n a point $\omega^{(n)} = (\omega_1^{(n)}, \omega_2^{(n)}, \ldots, \omega_n^{(n)})$. Then a sequence $\{n_j\}$ can be given with

$$\lim_{j \to +\infty} \omega_k^{(n_j)} = \omega_k \qquad (k = 1, 2, \ldots)$$

(G. Cantor's "diagonal method"). Since all Z_n are closed, for every n

$$(\omega_1, \ldots, \omega_n) \in Z_n;$$

hence $\omega = (\omega_1, \omega_2, \ldots, \omega_n, \ldots)$ belongs to D_n and $\omega \in \prod_{n=1}^{\infty} D_n$; therefore $\prod_{n=1}^{\infty} D_n$ is not empty. Similarly, $\prod_{n=1}^{\infty} A_n$ cannot be empty either, and thus our assumption leads to a contradiction. Hence we must have $p = \lim_{n \to +\infty} P(A_n) = 0$ and P is σ-additive on $\mathscr{A}$; it follows that the extension of P is σ-additive on the σ-algebra $\mathscr{A}^*$. We have proved that $[\Omega, \mathscr{A}^*, P]$ is a Kolmogorov probability space. Put therefore

$$\xi_k(\omega) = \omega_k \quad (k = 1, 2, \ldots) \quad \text{for} \quad \omega = (\omega_1, \ldots, \omega_k, \ldots). \qquad (4)$$

Then $\xi_k = \xi_k(\omega)$ is a random variable on $[\Omega, \mathscr{A}^*, P]$, since if U is a Borel set on the real axis, $\xi_k^{-1}(U)$ is a k-dimensional Borel cylinder which belongs thus to $\mathscr{A}$ and, consequently, to $\mathscr{A}^*$. On the other hand, obviously

$$P(\xi_1 < x_1, \xi_2 < x_2, \ldots, \xi_n < x_n) = F(x_1, x_2, \ldots, x_n); \tag{5}$$

the n-dimensional distribution function of the random variables $\xi_1, \xi_2, \ldots, \xi_n$ is thus identical with the function $F_n (x_1, x_2, \ldots, x_n)$. Herewith Theorem 1 is proved.

Example. Let $\{G_n(x)\}$, $n = 1, 2, \ldots$ be any sequence of distribution functions. There can be constructed a probability space and on it a sequence of random variables $\xi_n (n = 1, 2, \ldots)$ in such a manner that the ξ_n are mutually independent and the distribution function of ξ_n is $G_n(x)$. To see this it suffices to note that the functions

$$F_n(x_1, \ldots, x_n) = \prod_{k=1}^{n} G_k(x_k)$$

fulfil all conditions of Theorem 1.

§ 9. Exercises

1. Let there be given in the plane a circle C_1 of radius R with its center in the origin, and a circle C_2 concentrical with C_1 having a radius $r < R$. Let us draw a line d at random which intersects C_1, so that if the equation of d is written in the form

$$x \cos \varphi + y \sin \varphi = \varrho,$$

φ and ϱ are independent random variables, φ being uniformly distributed in $(0, \pi)$ and ϱ in $(-R, +R)$. Let ξ denote the length of the chord of d inside C_2. Determine the distribution function, expectation, and standard deviation of ξ.

Hint. Let first ξ be fixed. Then

$$P(\xi < x \mid \varphi) = \begin{cases} 0 & \text{for } x \leq 0, \\ 1 - \dfrac{1}{R} \sqrt{r^2 - \dfrac{x^2}{4}} & \text{for } 0 < x \leq 2r, \\ 1 & \text{for } 2r \leq x. \end{cases}$$

$\left(\text{At the point } x = 0 \text{ the distribution function has thus a jump of the value } 1 - \dfrac{r}{R}.\right)$

This expression being independent of φ, the conditional density function of ξ under the condition $\xi > 0$ is

$$f(x) = \begin{cases} 0 & \text{for } x < 0 \text{ and } x > 2r, \\ \dfrac{x}{4R\sqrt{r^2 - \dfrac{x^2}{4}}} & \text{for } 0 < x < 2r. \end{cases}$$

This leads to

$$E(\xi) = \frac{r^2\pi}{2R} \quad \text{and} \quad D(\xi) = \frac{r^2}{2R}\sqrt{\frac{32R}{3r} - \pi^2}.$$

2. Let d be a line chosen at random as in Exercise 1. Let B be a convex domain in the circle C_1. Let ξ denote the length of the chord of d inside B. Calculate the expectation of ξ.

Hint. We have $E(\xi) = E\big(E(\xi \mid \varphi)\big)$, where φ has the same meaning as in Exercise 1. $E(\xi \mid \varphi)$ is equal to the integral along the chords of the domain B lying in a given direction, divided by $2R$; for fixation of φ means restriction to the chords which form an angle $\varphi + \dfrac{\pi}{2}$ with the x-axis. Hence $E(\xi \mid \varphi) = \dfrac{|B|}{2R}$, $|B|$ being the area of B. We see that $E(\xi) = \dfrac{|B|}{2R}$. It is not necessary to require the convexity of B neither that it be simply connected.

3. Let there be given in the plane a curve L consisting of a finite number of convex arcs and contained in a circle C of radius R. Choose at random (in the sense explained in Exercise 1) a line d intersecting C. What is the expectation of the number of the points of intersection of this line with L?

Hint. Consider first the particular case when L is a segment of length l of a straight line. In this case the number v of points of intersection is 0 or 1. If φ is the angle between the normal to d and the segment L, the expectation under the condition of fixed φ is $E(v \mid \varphi) = \dfrac{l \cos \varphi}{2R}$. This leads to

$$E(v) = \frac{l}{\pi R}\int_0^{\frac{\pi}{2}} \cos \varphi d\varphi = \frac{l}{\pi R}.$$

From this it follows for polygons, and by a limit procedure for all piecewise convex (or concave) curves L, that $E(v) = \dfrac{|L|}{\pi R}$, where $|L|$ is the length of the curve L.

4. Calculate $E(\xi^n)$ for $n = 2, 3, \ldots$ the conditions of Exercise 1.

Hint. We have

$$E(\xi^n) = \frac{2^n \, r^{n+1}}{R}\int_0^{\frac{\pi}{2}} \sin^{n+1} \vartheta \, d\vartheta.$$

Note. Exercises 1 to 4 present well-known results of integral geometry[1] from a probabilistic point of view.

5. Establish the law $PV = RT$ for ideal gases on the basis of the kinetic theory of gases. V denotes here the molar volume, P the pressure, T the absolute temperature of the gas, further $R = Nk$ where N is Avogadro's number and k Boltzmann's constant.

[1] Cf. W. Blaschke [1].

Hint. The pressure of the gas is equal to the expectation of the quantity of motion imparted by the molecules of the gas during unit time to a unit surface of the vessel wall. We assume that the shocks are perfectly elastic. If a molecule of mass m and velocity v strikes the wall in a direction which forms an angle ϑ with the normal vector of the wall, then the quantity of motion imparted by the molecule will be $2\,mv\cos\vartheta$. In order to strike a unit surface K of the wall during a time interval $(t, t + 1)$, the molecule of velocity v moving in a direction which makes an angle with the normal vector to the wall has to be included at the time t in an oblique cylinder of (unit) base K and height $v\cos\vartheta$. Under the assumption that the molecules are uniformly distributed in the recipient, the probability of the shock in question is $\dfrac{v\cos\vartheta}{W}$, where W denotes the volume of the vessel. Hence the expectation of the quantity of motion imparted to the wall by the considered molecule will be $\dfrac{2\,v^2 m\cos^2\vartheta}{W} = \dfrac{4\,e\cos^2\vartheta}{W}$, where e is the kinetic energy of the molecule. The quantity $\dfrac{4e\cos^2\vartheta}{W}$ is a random variable. Hence we have to calculate its expectation. (Here the relation $E\big(E(\xi \mid \eta)\big) = E(\xi)$ is to be applied.) If the velocity components are supposed to be independent and to have normal distributions with the density function $\dfrac{1}{\sigma\sqrt{2\pi}}\exp\left(-\dfrac{x^2}{2\sigma^2}\right)$ where $\sigma = \sqrt{\dfrac{kT}{m}}$, then ϑ and e are independent and the distribution of the direction of the velocity vector is uniform. Hence

$$E\left(\frac{4e}{W}\cos^2\vartheta\right) = \frac{4}{W}\,E(e)\,E(\cos^2\vartheta).$$

We know already (Ch. IV, § 17, Exercise 29b) that $E(e) = \dfrac{3}{2}\,kT$. Since $E(\cos^2\vartheta) = \dfrac{1}{6}$, we find

$$E\left(\frac{4e}{W}\cos^2\vartheta\right) = \frac{kT}{W}$$

for the expectation of the "pressure" exerted upon the wall by one molecule. Since there are N molecules in a gram molecule of gas, we find for n gram molecules, because of the additivity of the expectation, the value

$$P = \frac{nNkT}{W} = \frac{NkT}{V} = \frac{RT}{V},$$

where $V = \dfrac{W}{n}$ is the molar volume and $R = Nk$ the ideal gas constant.

6. Let $\xi_1, \xi_2, \ldots, \xi_n$ be independent random variables uniformly distributed in the interval $(0, 1)$. Let them be arranged into an increasing sequence and let the k-th element of this sequence be denoted by ξ_k^*.

a) Show that the conditional density function of $\xi_1^*, \xi_2^*, \ldots, \xi_k^*$ with respect to the condition $\xi_{k+1}^* = c$ is given by

$$f_k(x_1, x_2, \ldots, x_k \mid \xi_{k+1}^* = c) = \begin{cases} \dfrac{k!}{c^k} & \text{for } 0 < x_1 < x_2 < \ldots < x_k < c, \\[2mm] 0 & \text{otherwise.} \end{cases}$$

b) Show that under the condition $\xi^*_{k+1} = c$ the random vectors $(\xi^*_1, \ldots, \xi^*_k)$ and $(\xi^*_{k+2}, \ldots, \xi^*_n)$ are independent.

7. Let $\xi_1, \xi_2, \ldots, \xi_n, \ldots$ be independent random variables. Consider the sums $\zeta_n = \xi_1 + \xi_2 + \ldots + \xi_n$. Show that under the condition $\zeta_n = x$ the random variables ζ_k and ζ_l are independent for $k < n < l$.

8. Let the random vector (ξ, η) have the normal density function

$$h(x, y) = \frac{\sqrt{AC - B^2}}{2\pi} \exp\left[-\frac{1}{2}(Ax^2 + 2Bxy + Cy^2)\right].$$

Prove the following relations:

$$R(\xi, \eta) = -\frac{B}{\sqrt{AC}},$$

$$E(\eta \mid \xi) = -\frac{B}{C}\,\xi,$$

$$E(\xi \mid \eta) = -\frac{B}{A}\,\eta,$$

$$K_\eta(\xi) = K_\xi(\eta) = |R(\xi, \eta)| = \frac{|B|}{\sqrt{AC}}.$$

9. If the random vector (ξ, η) has a nondegenerate normal distribution, show that

$$\varphi(\xi, \eta) = \frac{|r|}{\sqrt{1 - r^2}} \quad \text{and} \quad \psi(\xi, \eta) = |r|,$$

where $r = R(\xi, \eta)$ is the correlation coefficient, $\varphi(\xi, \eta)$ the contingency, and $\psi(\xi, \eta)$ the maximal correlation of the random variables ξ and η.

10. If the functions $a(x)$ and $b(x)$ are strictly monotone, then

$$\varphi\big(a(\xi), b(\eta)\big) = \varphi(\xi, \eta).$$

11. If (ξ, η) is uniformly distributed in a circle, $\psi(\xi, \eta) = \frac{1}{3}$.

12. If ξ and η are the indicators of the events A and B, i.e. if

$$\xi(\omega) = \begin{cases} 1 & \text{for } \omega \in A, \\ 0 & \text{otherwise,} \end{cases}$$

$$\eta(\omega) = \begin{cases} 1 & \text{for } \omega \in B, \\ 0 & \text{otherwise,} \end{cases}$$

then

$$\psi^2(\xi, \eta) = \varphi^2(\xi, \eta) = K_\xi^2(\eta) = K_\eta^2(\xi) = R^2(\xi, \eta) =$$

$$= \frac{[P(AB) - P(A)\,P(B)]^2}{P(A)\,[1 - P(A)]\,P(B)\,[1 - P(B)]},$$

provided that $0 < P(A) < 1$ and $0 < P(B) < 1$.

13. Prove the following variant of Bayes' theorem: Let ζ be a random variable with an absolutely continuous distribution with the density function $f(x)$ and let η be a discrete random variable. Let y_k $(k = 1, 2, \ldots)$ denote the possible values of η and $p_k(x)$ the conditional probability $P(\eta = y_k \mid \zeta = x)$. Let $f_k(x)$ be the conditional density function of ζ given $\eta = y_k$. We have

$$f_k(x) = \frac{p_k(x) f(x)}{\displaystyle\int_{-\infty}^{+\infty} p_k(t) f(t)\, dt}.$$

Hint. By definition

$$\int_A p_k(x) f(x)\, dx = P(\xi \in A, \eta = y_k),$$

hence

$$P(\xi < x \mid \eta = y_k) = \frac{P(\xi < x, \eta = y_k)}{P(\eta = y_k)} = \frac{\displaystyle\int_{-\infty}^{x} p_k(t) f(t)\, dt}{\displaystyle\int_{-\infty}^{+\infty} p_k(t) f(t)\, dt}.$$

14. Suppose that the probability of an event A is a random variable ζ with density function $p(x)$ $(p(x) = 0$ for $x < 0$ and $1 < x)$. Perform n independent experiments for which the value $P(A) = \zeta$ is constant and denote by η_n the number of the experiments in which A occurred. Let $p_{nk}(x)$ be the conditional (*a posteriori*) density function of ζ with respect to the condition $\eta_n = k$ $(k = 0, 1, 2, \ldots n)$; according to the preceding exercise

$$p_{nk}(x) = \frac{x^k (1 - x)^{n-k} p(x)}{\displaystyle\int_0^1 t^k (1 - t)^{n-k} p(t)\, dt}.$$

a) Show that if ζ has a beta distribution of order (r, s), then ζ has under condition $\eta_n = k$ a beta distribution of order $(k + r, n - k + s)$.

b) If $p(x)$ is continuous and positive on $(0, 1)$ and if f is a constant $(0 < f < 1)$, then

$$\lim_{n \to +\infty} \sqrt{\frac{f(1 - f)}{n}} \cdot p_{n,[fn]}\left(f + y\sqrt{\frac{f(1 - f)}{n}}\right) = \frac{1}{\sqrt{2\pi}}\, e^{-\frac{y^2}{2}}.$$

15. Let ζ be a random variable and let $\xi_1, \xi_2, \ldots, \xi_n$ be random variables which are for every fixed value of ζ independent and have a normal distribution with expectation ζ and standard deviation σ $(\sigma > 0$ is a constant). Let $p(x)$ be the density function of ζ. Study the conditional density function $p_n(x \mid y)$ of ζ under the condition

$$\frac{\xi_1 + \xi_2 + \ldots + \xi_n}{n} = y$$

and show that if $p(x)$ is positive and continuous, we have, for fixed x and y,

$$\lim_{n \to +\infty} \frac{p_n\left(y + \dfrac{x}{\sqrt{n}} \,\middle|\, y\right)}{\sqrt{n}} = \frac{1}{\sqrt{2\pi}\sigma}\, e^{-\frac{x^2}{2\sigma^2}}.$$

16. Let ζ be a random variable with an exponential distribution. For every given value of ζ, let $\xi_1, \xi_2, \ldots, \xi_n$ be independent normally distributed random variables with expectation ζ and standard deviation $\sigma > 0$. Determine the conditional distribution of ζ with respect to the condition

$$\frac{\xi_1 + \xi_2 + \ldots + \xi_n}{n} = y.$$

17. Let μ be a random variable having the density function $p(t)$. Let for every given value of μ the random variables $\xi_1, \ldots, \xi_n$ be independent, normally distributed, with expectation μ and standard deviation $\sigma > 0$. Show that $\xi_1, \ldots, \xi_n$ are exchangeable (cf. Ch. IV, § 17, Exercise 18).

18. Let $\xi_1, \ldots, \xi_N$ be independent random variables having the same distribution and finite variance. Put $\eta_n = \xi_1 + \xi_2 + \ldots + \xi_n$ $(n = 1, 2, \ldots, N)$. Calculate the correlation ratio $K_{\eta_n}(\eta_N)$ $(n < N)$.

19. Let $\xi_1, \ldots, \xi_N$ be independent random variables with the same distribution. Put $\eta_n = \xi_1 + \ldots + \xi_n$ $(n = 1, 2, \ldots, N)$. Calculate the contingency

$$\varphi(\eta_n, \eta_m) \qquad (n < m \leq N).$$

20. Let the random variables $\xi_1, \xi_2, \ldots, \xi_n$ be independent and uniformly distributed in the interval $(0, 1)$. Let ξ_k^* denote the k-th order statistic of the sample $\xi_1, \ldots, \xi_n$. (See Exercise 17 of Ch. IV.) Compute $K_{\xi_k^*}(\xi_l^*)$ and $\varphi(\xi_k^*, \xi_l^*)$ for $k < l \leq n$.

21. Suppose that the probability p of an event A is a random variable on a conditional probability space. Let $g(t) = \dfrac{1}{t(1 - t)}$ $(0 < t < 1)$ be its density function. Let p be constant during the course of n independent experiments and let η_n denote the number of those experiments in which the event A occurred. Calculate the a posteriori density function and the conditional expectation of the random variable p with respect to the condition $\eta_n = k$ $(0 < k < n)$.

Hint. According to Bayes' theorem the a posteriori density function of p with respect to the condition $\eta_n = k$ is

$$g_k(p) = \frac{p^{k-1}(1 - p)^{n-k-1}}{\int_0^1 t^{k-1}(1 - t)^{n-k-1}\, dt};$$

the "a posteriori distribution" of p is thus a beta distribution of order $(k, n - k)$ and the conditional expectation is $\dfrac{k}{n}$.

22. Let ζ be a random variable with Poisson distribution and expectation λ, where λ is a random variable with a logarithmically uniform distribution on $(0, +\infty)$. Calculate the a posteriori density function and the conditional expectation of λ with respect to the condition $\zeta = n \geq 1$.

Hint. Bayes' theorem gives for the conditional density function of λ with respect to the condition $\zeta = n$:

$$g_n(\lambda) = \frac{\lambda^{n-1} e^{-\lambda}}{(n - 1)!};$$

the a posteriori distribution of λ is thus a gamma distribution of order n.

23. Let the random variable ξ have a normal distribution $N(\mu, \sigma)$, where μ is a random variable uniformly distributed on the whole real axis. Determine the a posteriori density function and the expectation of μ with respect to the condition $\xi = a$.

Hint. According to Bayes' theorem

$$g(\mu \mid \xi = a) = \frac{1}{\sqrt{2\pi}\sigma} \exp\left[-\frac{(\mu - a)^2}{2\sigma^2}\right].$$

The a posteriori distribution of μ is thus a normal distribution with expectation a and standard deviation σ.

24. Let $\xi_1, \xi_2, \ldots, \xi_n$ be independent random variables having the same normal distribution $N(m, \sigma)$. Put $\zeta_n = \dfrac{\xi_1 + \xi_2 + \ldots + \xi_n}{n}$. Determine the conditional distribution of $(\xi_1, \xi_2, \ldots, \xi_n)$ for a given value of ζ_n.

Hint. By assumption, the n-dimensional density function of the $\xi_k (k = 1, 2, \ldots, n)$ is

$$f(x_1, \ldots, x_n) = \frac{1}{\left(\sigma\sqrt{2\pi}\right)^n} \exp\left[-\frac{1}{2\sigma^2}\sum_{k=1}^{n}(x_k - m)^2\right].$$

Put

$$\bar{x} = \frac{1}{n}\sum_{k=1}^{n} x_k.$$

We have

$$\sum_{k=1}^{n}(x_k - m)^2 = \sum_{k=1}^{n}(x_k - \bar{x})^2 + n(\bar{x} - m)^2,$$

hence

$$f(x_1, \ldots, x_n) = \frac{1}{\sigma}\sqrt{\frac{n}{2\pi}}\exp\left[-\frac{n(\bar{x} - m)^2}{2\sigma^2}\right] \cdot \frac{1}{\left(\sigma\sqrt{2\pi}\right)^{n-1}\sqrt{n}} \times$$

$$\times \exp\left[-\frac{1}{2\sigma^2}\sum_{k=1}^{n}(x_k - \bar{x})^2\right].$$

The density function of ζ_n is

$$g(\bar{x}) = \frac{1}{\sigma}\sqrt{\frac{n}{2\pi}}\exp\left[-\frac{n(\bar{x} - m)^2}{2\sigma^2}\right];$$

hence the conditional density function of the random vector $(\xi_1, \ldots, \xi_n)$ for $\zeta_n = \bar{x}$ is

$$\frac{1}{\left(\sigma\sqrt{2\pi}\right)^{n-1}\sqrt{n}}\exp\left[-\frac{1}{2\sigma^2}\sum_{k=1}^{n}(x_k - \bar{x})^2\right].$$

This function does not depend on m; a property which is expressed by saying that ζ_n is a sufficient statistic for the parameter m.

25. a) Let there be given n independent random variables $\xi_1, \ldots, \xi_n$ with the same normal distribution $N(0, \sigma)$. Put

$$\zeta = \frac{1}{n}\sum_{k=1}^{n}\xi_k \quad \text{and} \quad \tau = \sum_{k=1}^{n}(\xi_k - \zeta)^2.$$

Show that ζ and τ are independent.

Hint. Let (c_{jk}) $(j, k = 1, 2, \ldots, n)$ be an orthogonal matrix with $c_{1k} = \dfrac{1}{\sqrt{n}}$
$(k = 1, 2, \ldots, n)$. Put

$$\eta_j = \sum_{k=1}^{n} c_{jk}\, \xi_k \qquad (j = 1, 2, \ldots, n).$$

Then $\eta_1 = \sqrt{n}\,\zeta$ and

$$\sum_{j=1}^{n} \eta_j^2 = \sum_{k=1}^{n} \xi_k^2 = \eta_1^2 + \sum_{k=1}^{n} (\xi_k - \zeta)^2,$$

hence

$$\tau = \sum_{j=2}^{n} \eta_j^2.$$

We know (cf. Ch. IV, § 17, Exercise 43) that $\eta_1, \ldots, \eta_n$ are independent normally distributed random variables with expectation 0 and standard deviation σ; hence

$$\zeta = \frac{\eta_1}{\sqrt{n}} \quad \text{and} \quad \tau = \sum_{j=2}^{n} \eta_j^2$$

are independent; τ has a χ^2-distribution with $(n - 1)$ degrees of freedom.

b) Let $\xi_1, \ldots, \xi_n$ be independent random variables with the same normal distribution. Let the expectation μ and the standard deviation σ of the ξ_k be independent random variables on a conditional probability space, μ being uniformly distributed on the whole real axis and σ logarithmically uniformly distributed in the interval $(0, +\infty)$. Put

$$\zeta = \frac{\xi_1 + \xi_2 + \cdots + \xi_n}{n} \quad \text{and} \quad \tau = \sum_{k=1}^{n} (\xi_k - \zeta)^2.$$

Determine the *a posteriori* distribution of μ and σ^2 under condition $\zeta = x$ and $\tau = z$. Show that given these conditions σ and $\dfrac{x - \mu}{\sigma}$ are independent.

Hint. The density function of the vector (μ, σ^2) with respect to the condition $\zeta = x$, $\tau = z$ is, according to Bayes' theorem and the result of Exercise 25 a)

$$\sqrt{\frac{n}{2\pi}}\; \frac{z^{\frac{n-1}{2}}\, \exp\left[-\dfrac{z}{2\sigma^2}\right] \exp\left[-\dfrac{n(x - \mu)^2}{2\sigma^2}\right]}{2^{\frac{n-3}{2}}\, \sigma^{n+1}\, \Gamma\!\left(\dfrac{n-1}{2}\right)},$$

thus σ and $\dfrac{x - \mu}{\sigma}$ are independent.

26. Let there be given a sequence of pairwise independent events A_n $(n = 1, 2, \ldots)$ with $P(A_n) \geq a > 0$ $(n = 1, 2, \ldots)$ and an arbitrary random variable η of finite variance. Show that

$$\lim_{n \to +\infty} E(\eta \mid A_n) = E(\eta). \tag{1}$$

If η is the indicator of the event B $\left(0 < P(B) < 1\right)$, it follows from (1) that

$$\lim_{n \to +\infty} P(B \mid A_n) = P(B). \tag{2}$$

Hint. Let ξ_n be the indicator of the event A. We have

$$K_{\xi_n}^2(\eta) = \frac{P(A_n)}{1 - P(A_n)} \cdot \frac{[E(\eta \mid A_n) - E(\eta)]^2}{D^2(\eta)}$$

hence, by Theorem 3 of § 7,

$$\sum_{n=1}^{\infty} \frac{P(A_n)}{1 - P(A_n)} \, [E(\eta \mid A_n) - E(\eta)]^2 \leq D^2(\eta).$$

Thus

$$\lim_{n \to +\infty} \frac{P(A_n)}{1 - P(A_n)} \, [E(\eta \mid A_n) - E(\eta)]^2 = 0,$$

which proves (1).

Remark. Cf. Ch. VII, § 10, Theorem 1.

27. Let a sequence of pairwise independent events A_n $(n = 1, 2, \ldots)$ be given and assume

$$\sum_{n=1}^{\infty} P(A_n) = +\infty.$$

Let

$$B = \limsup_{n \to +\infty} A_n = \prod_{n=1}^{\infty} \left(\sum_{k=n}^{\infty} A_k \right)$$

denote the event that infinitely many of the events A_n occur simultaneously. Show that $P(B) = 1$.

Hint. Let C be any event with $0 < P(C) < 1$. Like in Exercise 26, it follows from Theorem 3, § 7 that

$$\sum_{n=1}^{\infty} \frac{P(A_n)}{1 - P(A_n)} \, [P(C \mid A_n) - P(C)]^2 \leq P(C) [1 - P(C)].$$

Since $\sum_{n=1}^{\infty} P(A_n)$ diverges, clearly

$$\liminf_{n \to +\infty} [P(C \mid A_n) - P(C)]^2 = 0. \tag{1}$$

Apply (1) to $C = C_k = \sum_{n=k}^{\infty} A_n$. Obviously, $P(C_k) > 0$. It follows from (1), in view of $P(C_k \mid A_n) = 1$ for $n \geq k$, that $P(C_k) = 1$; hence $P(\bar{C}_k) = 0$. Since $B = \prod_{k=1}^{\infty} C_k$, we have $\bar{B} = \sum_{k=1}^{\infty} \bar{C}_k$ and hence $P(\bar{B}) = 0$, which is equivalent to $P(B) = 1$.

Remark. The assertion of Exercise 27 is a sharper form of the Borel-Cantelli lemma (cf. Ch. VII, § 5).

28. Let ξ and η be arbitrary random variables, $f(x)$ and $g(x)$ Borel-measurable functions such that

$$E\big(f(\xi)\big) = E\big(g(\eta)\big) = 0, \quad D\big(f(\xi)\big) = D\big(g(\eta)\big) = 1$$

and

$$R\big(f(\xi), g(\eta)\big) = E\big(f(\xi)\, g(\eta)\big) = \psi(\xi, \eta)\,,$$

or to put it otherwise, suppose that $R\big(u(\xi), v(\eta)\big)$ assumes its maximal value for $u = f$ and $v = g$. Then the following equations hold with probability 1, where $\lambda = \psi\,(\xi, \eta)$:

$$E\big(f(\xi) \mid \eta\big) = \lambda g(\eta) \tag{1}$$

and

$$E\big(g(\eta) \mid \xi\big) = \lambda f(\xi)\,, \tag{2}$$

hence also

$$E\big(E(f(\xi) \mid \eta) \mid \xi\big) = \lambda^2 f(\xi) \tag{3}$$

and

$$E\big(E(g(\eta) \mid \xi) \mid \eta\big) = \lambda^2 g(\eta)\,. \tag{4}$$

Hint. We have

$$\psi(\xi, \eta) = E\big((f(\xi)g(\eta)\big) = E\big(E(f(\xi)g(\eta) \mid \xi)\big) = E\big(f(\xi)E(g(\eta)|\xi)\big)\,,$$

hence according to Schwarz' inequality

$$\psi^2(\xi, \eta) \le E\big(E^2(g(\eta)|\xi)\big) = D^2.$$

On the other hand, if $f^*(\xi) = \dfrac{E\big(g(\eta)|\xi\big)}{D}$ fulfils $E\big(f^*(\xi)\big) = 0$ and $D\big(f^*(\xi)\big) = 1$, then

$$R\big(f^*(\xi), g(\eta)\big) = E\big(f^*(\xi)g(\eta)\big) \le \psi(\xi, \eta)\,.$$

But as

$$E\big(f^*(\xi)g(\eta)\big) = \frac{E\big(g(\eta)E(g(\eta)|\xi)\big)}{D} = \frac{E\big(E^2(g(\eta)|\xi)\big)}{D} = D\,,$$

we conclude that $D^2 \le \psi^2(\xi, \eta)$. Hence $D^2 = \psi^2(\xi, \eta)$. Since in Schwarz' inequality equality holds only in the case of proportionality, we must have $E\big(g(\eta) \mid \xi\big) = \lambda f(\xi)$ which proves (2). But

$$E\big(f(\xi)E(g(\eta)|\xi)\big) = \psi(\xi, \eta)\,.$$

On the other hand, by (2)

$$E\big(f(\xi)E(g(\eta) \mid \xi)\big) = \lambda E\,\big(f^2(\xi)\big) = \lambda\,,$$

hence $\lambda = \psi(\xi, \eta)$. Equation (1) is proved in a similar way.

29. With the notations of Exercise 28 we have

$$E\big(f(\xi) \mid g(\eta)\big) = \lambda g(\eta)$$

and

$$E\big(g(\eta) \mid f(\xi)\big) = \lambda f(\xi).$$

Hence the regression curves of $\xi^* = f(\xi)$ with respect to $\eta^* = g(\eta)$ as well as that of η^* with respect to ξ^* are straight lines (or, as it is expressed, the regression of ξ^* and η^* is *linear*).

Hint. The proof is similar to that of Exercise 28.

30. Let L_ξ^2 be the set of all random variables $f(\xi)$ such that $f(x)$ is a Borel-measurable function with $E\big(f(\xi)\big) = 0$ and $E\big(f^2(\xi)\big)$ is finite. If we put

$$\big(f_1(\xi),\ f_2(\xi)\big) = E\big(f_1(\xi)f_2(\xi)\big),$$

L_ξ^2 is a Hilbert space. Further we define $Af(\xi)$, for $f(\xi) \in L_\xi^2$, by

$$Af(\xi) = E\big(E(f(\xi)|\eta)|\xi\big).$$

Show that $Af(\xi) = f_1(\xi)$ belongs also to L_ξ^2 and the linear transformation $Af(\xi)$ of the space L_ξ^2 is positive and symmetric, i.e. it fulfils the relations

$$\big(Af(\xi), f(\xi)\big) \geq 0 \quad \text{and} \quad \big(Af(\xi), g(\xi)\big) = \big(f(\xi), Ag(\xi)\big).$$

CHAPTER VI

CHARACTERISTIC FUNCTIONS

§ 1. Random variables with complex values

Characteristic functions are useful analytic tools of probability theory, especially for proving limit theorems. This Chapter presents the definition and the properties of characteristic functions; the following two chapters will deal with the limit theorems themselves.

The characteristic function of a random variable ξ is defined as the expectation of the complex valued random variable $e^{i\xi t}$. Thus we have to study complex random variables first; we shall see how theorems on real random variables can be extended to complex random variables. If ξ and η are real random variables, we say that the quantity $\zeta = \xi + i\eta$ is a complex random variable. The distribution of ζ can be characterized by the joint distribution of ξ and η.

We define the expectation of $\zeta = \xi + i\eta$ by

$$E(\zeta) = \int_\Omega \zeta\, dP, \tag{1}$$

which implies

$$E(\zeta) = E(\xi) + iE(\eta). \tag{2}$$

The random variables $\zeta_1 = \xi_1 + i\eta_1$ and $\zeta_2 = \xi_2 + i\eta_2$ are said to be independent if the two-dimensional random vectors (ξ_1, η_1) and (ξ_2, η_2) are independent. The independence of several complex random variables is defined in a similar way.

If $\zeta_1, \zeta_2, \ldots, \zeta_n$ are independent complex random variables and if the expectations $E(\zeta_k)$ $(k = 1, 2, \ldots, n)$ exist, one can see at once that

$$E(\prod_{k=1}^n \zeta_k) = \prod_{k=1}^n E(\zeta_k). \tag{3}$$

If $A(x) = a(x) + ib(x)$ is a complex valued Borel function of the real variable x and ξ is a real random variable, further if the expectation of $\zeta = A(\xi)$ exists, then the latter can be calculated by

$$E(\zeta) = \int_{-\infty}^{+\infty} A(x)\, dF(x), \tag{4}$$

where $F(x)$ is the distribution function of ζ. In fact, according to Exercise 47, § 17 of Chapter IV,

$$E(\zeta) = \int_{-\infty}^{+\infty} a(x)\, dF(x) + i \int_{-\infty}^{+\infty} b(x)\, dF(x).$$

It is easy to prove that for every random variable with complex values

$$|E(\zeta)| \le E(|\zeta|). \tag{5}$$

§ 2. Characteristic functions and their basic properties

We define the *characteristic function* of a random variable ξ as the expectation of $e^{i\xi t}$; thus it is a function of the real variable t. It is denoted by $\varphi_\xi(t)$. Thus by definition

$$\varphi_\xi(t) = E(e^{i\xi t}). \tag{1}$$

According to Formula (4) of § 1

$$\varphi_\xi(t) = \int_{-\infty}^{+\infty} e^{ixt}\, dF(x) \tag{2}$$

where $F(x)$ is the distribution function of ξ; hence $\varphi_\xi(t)$ is the *Fourier–Stieltjes transform of $F(x)$*. If the distribution of ξ is discrete and ξ assumes the values x_k ($k = 1, 2, \ldots$) with probabilities p_k ($k = 1, 2, \ldots$), then $\varphi_\xi(t)$ can be written in the form

$$\varphi_\xi(t) = \sum_{k=1}^{\infty} p_k\, e^{itx_k}. \tag{3}$$

If the distribution function of ξ is absolutely continuous with the density function $f(x) = F'(x)$, we have

$$\varphi_\xi(t) = \int_{-\infty}^{+\infty} e^{itx} f(x)\, dx. \tag{4}$$

Hence $\varphi_\xi(t)$ is the *Fourier transform of $f(x)$*.

Thus we see that the characteristic function of an arbitrary random variable depends only on its distribution; characteristic functions of random variables with the same distribution are identical. The function defined by (2) can thus be called the characteristic function of $F(x)$ (as well as the characteristic function of a random variable with distribution function $F(x)$).

First of all, let it be noted that every distribution function has a characteristic function since the Stieltjes integral (2) exists always, in view of $|e^{ixt}| = 1$.

If ξ assumes positive integer values only, with

$$P(\xi = k) = p_k \qquad (k = 0, 1, \ldots),$$

we see that

$$\varphi_\xi(t) = \sum_{k=0}^{n} p_k e^{ikt} = G_\xi(e^{it}),$$

where

$$G_\xi(z) = \sum_{k=0}^{n} p_k z^k \qquad (|z| \leq 1)$$

is the generating function of ξ, discussed already in Chapter III, § 15. In this case the characteristic function is therefore equal to the generating function on the boundary of the unit circle. In the general case when ξ may take on other than positive integral values, the generating function is not defined; the characteristic function, however, exists for every random variable.

We shall now prove some elementary theorems concerning the characteristic functions of probability distributions.

THEOREM 1. *We have always* $|\varphi_\xi(t)| \leq 1$; *equality holds for* $t = 0$.

PROOF. Since $|e^{i\xi t}| = 1$ and because of Formula (5) of § 1 we have

$$|\varphi_\xi(t)| \leq E(|e^{i\xi t}|) = 1.$$

Further $\varphi_\xi(0) = E(e^0) = 1$.

THEOREM 2. *The function* $\varphi_\xi(t)$ *is uniformly continuous on the whole real axis* $-\infty < t < +\infty$.

PROOF. Let $\varepsilon > 0$ be given. Choose a $\lambda > 0$ such that

$$P(|\xi| > \lambda) < \frac{\varepsilon}{3}.$$

If we denote by A_λ the event $|\xi| > \lambda$, we evidently have

$$\varphi_\xi(t) = E(e^{i\xi t} \mid A_\lambda) P(A_\lambda) + E(e^{i\xi t} \mid \bar{A}_\lambda) P(\bar{A}_\lambda). \tag{5}$$

Since $|E(e^{i\xi t} \mid A_\lambda)| \leq 1$, we conclude that

$$|\varphi_\xi(t) - E(e^{i\xi t} \mid \bar{A}_\lambda) P(\bar{A}_\lambda)| \leq P(A_\lambda) < \frac{\varepsilon}{3}. \tag{6}$$

Consequently,

$$| \varphi_\xi(t_2) - \varphi_\xi(t_1) | \leq E(| e^{i\xi t_2} - e^{i\xi t_1} | \, | \bar{A}_\lambda) + \frac{2\varepsilon}{3}. \tag{7}$$

From

$$| e^{ib} - e^{ia} | = | i \int_a^b e^{iz} \, dz | \leq b - a \quad \text{for} \quad a < b \tag{8}$$

follows

$$| e^{i\xi t_2} - e^{i\xi t_1} | < \frac{\varepsilon}{3} \quad \text{for} \quad | \xi | \leq \lambda \quad \text{and} \quad | t_2 - t_1 | < \frac{\varepsilon}{3\lambda} = \delta,$$

hence

$$E(| e^{i\xi t_2} - e^{i\xi t_1} | \, \bar{A}_\lambda) < \frac{\varepsilon}{3} \quad \text{for} \quad | t_2 - t_1 | < \delta. \tag{9}$$

From (7) and (9) we conclude that

$$| \varphi_\xi(t_2) - \varphi_\xi(t_1) | < \varepsilon \quad \text{for} \quad | t_2 - t_1 | < \delta,$$

where $\delta > 0$ depends only on ε. This proves Theorem 2.

THEOREM 3. *If a and b are constants and if $\eta = a\xi + b$, then*

$$\varphi_\eta(t) = e^{ibt} \, \varphi_\xi(at).$$

PROOF.

$$\varphi_\eta(t) = E(e^{i(a\xi + b)t}) = e^{ibt} \, E(e^{i\xi at}).$$

THEOREM 4. *If $t_1, t_2, \ldots, t_n$ are arbitrary real numbers and $z_1, z_2, \ldots, z_n$ arbitrary complex numbers, further if $\varphi_\xi(t)$ is the characteristic function of a random variable ξ and if $\bar{z} = x - iy$ is the conjugate of the complex number $z = x + iy$, then we have*

$$\sum_{h=1}^n \sum_{k=1}^n \varphi_\xi(t_h - t_k) z_h \bar{z}_k \geq 0. \tag{10}$$

Remark. Functions satisfying (10) are said to be *positive definite*. A remarkable theorem of Bochner says that every positive definite function $\varphi(t)$ for which $\varphi(0) = 1$ is the characteristic function of a probability distribution. We shall not give the proof of this theorem.

PROOF OF THEOREM 4. We have

$$\sum_{h=1}^n \sum_{k=1}^n \varphi_\xi(t_h - t_k) z_h \bar{z}_k = E(| \sum_{k=1}^n e^{i t_k \xi} z_k |^2).$$

THEOREM 5. *For every real t,* $\varphi_\xi(-t) = \overline{\varphi_\xi(t)}$. *In particular, if the distribution function of ξ is symmetric with respect to the origin, $\varphi_\xi(t)$ is a real even function of t.*

PROOF. Let ζ be a random variable with complex values; then $E(\overline{\zeta}) = \overline{E(\zeta)}$. This leads to

$$\varphi_\xi(-t) = E(e^{-i\xi t}) = \overline{E(e^{i\xi t})}.$$

If the distribution function of ξ is symmetric, i.e. if ζ and $-\zeta$ have the same distribution, their characteristic functions are identical; we have thus

$$\varphi_\xi(t) = \varphi_{-\xi}(t) = \varphi_\xi(-t) = \overline{\varphi_\xi(t)}.$$

Consequently, $\varphi_\xi(t)$ is real and, since $\varphi_\xi(t) = \varphi_\xi(-t)$, $\varphi_\xi(t)$ is an even function.

THEOREM 6. *If $\xi_1, \xi_2, \ldots, \xi_n$ are mutually independent random variables, the characteristic function of their sum is equal to the product of the characteristic functions of the individual terms:*

$$\varphi_{\xi_1 + \xi_2 + \ldots + \xi_n}(t) = \prod_{k=1}^{n} \varphi_{\xi_k}(t).$$

PROOF. This follows from Formula (3) of § 1.

Remarks:

1. Theorem 6 expresses a property of the characteristic functions which exhibits their successful applicability to probability theory. Indeed, the distribution of a sum of independent random variables is the convolution of the distribution functions of the individual terms; the calculation of this convolution is in most cases rather complicated. On the contrary, Theorem 6 allows a very simple calculation of the characteristic function of a sum of independent random variables from the characteristic functions of its terms, as it is just their product. Further, as we shall see in § 4, from the properties of the characteristic function the properties of the corresponding distribution function can be deduced.

2. The converse of Theorem 6 does not hold. From

$$\varphi_{\xi_1 + \xi_2}(t) = \varphi_{\xi_1}(t)\,\varphi_{\xi_2}(t)$$

the independence of ξ_1 and ξ_2 does not follow. Let for instance be $\xi_1 = \xi_2 = \xi$, where ξ has a Cauchy distribution: $\varphi_\xi(t) = e^{-|t|}$.

(cf. Example 4 of § 3). According to Theorem 3 we have thus

$$\varphi_{\xi_1 + \xi_1}(t) = \varphi_{2\xi}(t) = e^{-2|t|} = \varphi_{\xi_1}(t) \, \varphi_{\xi_1}(t),$$

though ξ is obviously not independent of itself.

THEOREM 7. *If the first n moments* $E(\xi^k) = M_k$ $(k = 1, 2, \ldots, n)$ *exist for the random variable* ξ, *then the characteristic function* $\varphi_\xi(t)$ *is n times differentiable and*

$$\varphi_\xi^{(k)}(0) = i^k M_k \qquad (k = 1, 2, \ldots, n). \tag{11}$$

PROOF. Let $F(x)$ be the distribution function of ξ. If

$$\int_{-\infty}^{+\infty} |x| \, dF(x)$$

exists, the integral

$$\int_{-\infty}^{+\infty} x e^{ixt} \, dF(x)$$

converges uniformly in t. Hence

$$\varphi_\xi'(t) = \int_{-\infty}^{+\infty} ixe^{ixt} \, dF(x);$$

in particular

$$\varphi_\xi'(0) = iM_1. \tag{12}$$

By iterating the operation we obtain

$$\varphi_\xi^{(k)}(t) = i^k \int_{-\infty}^{+\infty} x^k e^{ixt} \, dF(x) \qquad (k = 1, 2, \ldots, n); \tag{13}$$

from here (11) follows by putting $t = 0$.

THEOREM 8. *Let the distribution function of the random variable* ξ *be absolutely continuous. If the density function* $f(x)$ *of* ξ *is k times differentiable* $(k = 0, 1, \ldots)$ *and if*

$$C_j = \int_{-\infty}^{+\infty} |f^{(j)}(x)| \, dx$$

exists for $j = 1, 2, \ldots, k$, *we have*[1]

$$\lim_{|t| \to +\infty} |t|^k |\varphi_\xi(t)| = 0. \tag{14}$$

---

[1] It suffices to assume the finiteness of C_k; this implies the finiteness of $C_1, \ldots, C_{k-1}$. Cf. S. Bochner and K. Chandrasekharan [1], p. 29.

PROOF. If we perform k times an integration by parts on

$$\varphi_\xi(t) = \int_{-\infty}^{+\infty} f(x) e^{ixt} \, dx \tag{15}$$

and consider that by our assumption $\lim_{|x| \to \infty} f^{(j)}(x) = 0$ for $j = 1, 2, \ldots, k-1$, we obtain

$$\varphi_\xi(t) = \left(\frac{i}{t}\right)^k \int_{-\infty}^{+\infty} f^{(k)}(x) e^{ixt} \, dx. \tag{16}$$

From (16) it follows that

$$|\varphi_\xi(t)| \le \frac{C_k}{|t|^k}. \tag{17}$$

Since by assumption $|f^{(k)}(x)|$ is integrable on $(-\infty, +\infty)$, (14) follows from (16) by Riemann's lemma concerning the Fourier integral.[1]

Inequality (17) is obviously of interest for the study of the behaviour of $\varphi_\xi(t)$ for large values of $|t|$.

Remark. According to Theorem 7, the "smoothness" (differentiability) of $\varphi_\xi(t)$ is determined by the behaviour of $f(x)$ for $|x| \to +\infty$; by Theorem 8 the "smoothness" of $f(x)$ determines the behaviour of $\varphi_\xi(t)$ for $|t| \to \infty$. The two theorems are therefore in a certain sense dual.

THEOREM 9. *If the first n moments of ξ, $M_k = E(\xi^k)$ $(k = 1, 2, \ldots, n)$ exist, we have (with $M_0 = 1$), for $t \to 0$ that*

$$\varphi_\xi(t) = \sum_{k=0}^{n} \frac{M_k (it)^k}{k!} + o(t^n). \tag{18}$$

PROOF. This follows immediately from Theorem 7.

THEOREM 10. *If all the moments $M_k = E(\xi^k)$ $(k = 1, 2, \ldots, n)$ of the random variable ξ exist and if*

$$\limsup_{n \to +\infty} \sqrt[n]{\frac{|M_n|}{n!}} = \frac{1}{R} \tag{19}$$

is finite, then the domain of definition of $\varphi_\xi(t)$ can be extended to complex t-values. We have, for $|t| < R$,

$$\varphi_\xi(t) = \sum_{n=0}^{\infty} \frac{M_n (it)^n}{n!}; \tag{20}$$

---

[1] Cf. G. H. Hardy and W. W. Rogosinski [1], p. 23.

$\varphi_{\xi}(t)$ *is even a holomorphic function in the whole band* $|v| < R$ *of the complex plane* $t = u + iv$.

PROOF. If the assumptions of the theorem are fulfilled, $\varphi_{\xi}(t)$ is, because of (11), arbitrarily often differentiable at the point $t = 0$ and we have $\varphi_{\xi}^{(n)}(0) = i^n M_n$. From this (20) follows immediately.

Because of (13) for every real t_0 and every n

$$|\varphi_{\xi}^{(2n)}(t_0)| \le M_{2n}. \tag{21}$$

Hence, according to Schwarz's inequality,

$$|\varphi_{\xi}^{(2n+1)}(t_0)| \le \int_{-\infty}^{+\infty} |x|^{2n+1}\, dF(x) \le \sqrt{M_{2n} M_{2n+2}} \le \frac{M_{2n} + M_{2n+2}}{2}. \tag{22}$$

We obtain from (19), (21) and (22) for every real t_0

$$\limsup_{n \to +\infty} \sqrt[n]{\frac{|\varphi^{(n)}(t_0)|}{n!}} \le \frac{1}{R}. \tag{23}$$

Hence $\varphi_{\xi}(t)$ is regular in every circle $|t - t_0| < R$ which leads already to our assertion.

Remark. It follows from Theorem 10 that the function $\varphi_{\xi}(t)$ is uniquely determined by the sequence M_n ($n = 1, 2, \ldots$), whenever (19) holds. In fact, by (20) $\varphi_{\xi}(t)$ is determined by the sequence $\{M_n\}$ in the circle $|t| < R$, hence the value of $\varphi_{\xi}(t)$ for every real t can be determined by analytic continuation. We shall see in § 4 that a distribution function is uniquely determined by its characteristic function. Hence if (19) is fulfilled, the distribution of ξ is uniquely determined by the sequence of moments

$$M_n = E(\xi^n) \qquad (n = 1, 2, \ldots).$$

The question, whether the moments $M_n = E(\xi^n)$ do or do not determine uniquely the distribution function $F(x)$ of ξ, is called the *Stieltjes moment problem*. In general, $F(x)$ is not uniquely determined by the sequence of the moments.

DEFINITION. The random variable ξ has a *lattice distribution* with span d, if it takes on only values of the form $dk + r$ ($k = 0, \pm 1, \pm 2, \ldots$), where $d > 0$ and r are real constants.

THEOREM 11. *If ξ has a lattice distribution with span d, then*

$$\left| \varphi_\xi \left(\frac{2\pi n}{d} \right) \right| = 1 \quad \text{for} \quad n = 0, \pm 1, \pm 2, \ldots;$$

if ξ does not have a lattice distribution, we have $|\varphi_\xi(t)| < 1$ for every $t \neq 0$.

PROOF. If all values of ξ are of the form $dk + r$ and if $P(\xi = dk + r) = p_k$ $(k = 0, \pm 1, \pm 2, \ldots)$ we have, for any integer n,

$$\left| \varphi_\xi \left(\frac{2\pi n}{d} \right) \right| = \sum_{k=-\infty}^{+\infty} p_k = 1 .$$

Conversely, if for a $t_0 \neq 0$ we have $\varphi_\xi(t_0) = e^{i\alpha}$ with real α, we conclude

$$\int_{-\infty}^{+\infty} e^{i(t_0 x - \alpha)} \, dF(x) = 1 = \int_{-\infty}^{+\infty} dF(x),$$

hence

$$\int_{-\infty}^{+\infty} [1 - \cos(t_0 x - \alpha)] \, dF(x) = 0 .$$

Since $1 - \cos(t_0 x - \alpha)$ is positive except for $x = \dfrac{2k\pi}{t_0} + \dfrac{\alpha}{t_0}$ $(k = 0,$ $\pm 1, \ldots)$ (for which values it is equal to 0), all jumps of $F(x)$ must therefore belong to the arithmetic progression $dk + r$ with $d = \dfrac{2\pi}{t_0}$ and $r = \dfrac{\alpha}{t_0}$.

THEOREM 12. *If the distribution of ξ is the mixture of the distributions of the random variables ξ_k with weights p_k $(k = 1, 2, \ldots)$, then*

$$\varphi_\xi(t) = \sum_k p_k \varphi_{\xi_k}(t).$$

PROOF. Let $F(x)$ be the distribution function of ξ, $F_k(x)$ that of ξ_k. We know that

$$F(x) = \sum_k p_k F_k(x).$$

From this Theorem 12 follows immediately.

Remark. The characteristic function may be considered as an operator which assigns to the distribution function $F(x)$ the function $\varphi(t)$. Then Theorem 12 expresses the fact that this operator is linear.

§ 3. Characteristic functions of some important distributions

We determine now explicitly the characteristic functions of some distributions.

Example 1. The characteristic function of the normal distribution.
Let ξ be a normally distributed random variable with $E(\xi) = 0$, $D(\xi) = 1$. Then

$$\varphi_\xi(t) = \frac{1}{\sqrt{2\pi}} \int_{-\infty}^{+\infty} e^{ixt - \frac{x^2}{2}} dx = \frac{1}{\sqrt{2\pi}} e^{-\frac{t^2}{2}} \int_L e^{-\frac{z^2}{2}} dz,$$

where L is the horizontal line $z = x - it$ $(-\infty < x < +\infty)$ of the complex plane; $e^{-\frac{z^2}{2}}$ is an entire function, its integral is thus zero along any closed curve and in particular along the quadrangle R_x with the vertices $-x - it$, $x - it$, x, $-x$. The relation

$$\left| \int_{x-it}^{x} e^{-\frac{z^2}{2}} dz \right| \le e^{-\frac{x^2}{2}} \int_0^{|t|} e^{\frac{u^2}{2}} du \tag{1}$$

implies

$$\lim_{|x| \to \infty} \left| \int_{x-it}^{x} e^{-\frac{z^2}{2}} dz \right| = 0. \tag{2}$$

Hence

$$\frac{1}{\sqrt{2\pi}} \int_L e^{-\frac{z^2}{2}} dz = \frac{1}{\sqrt{2\pi}} \int_{-\infty}^{+\infty} e^{-\frac{x^2}{2}} dx \tag{3}$$

and consequently

$$\varphi_\xi(t) = e^{-\frac{t^2}{2}}. \tag{4}$$

If the random variable ξ is $N(m, \sigma)$ the random variable $\xi' = \dfrac{\xi - m}{\sigma}$ is $N(0, 1)$ and $\xi = \sigma\xi' + m$. From $\varphi_{\xi'}(t) = e^{-\frac{t^2}{2}}$ and from Theorem 3 in § 2 follows

$$\varphi_\xi(t) = e^{imt - \frac{\sigma^2 t^2}{2}}. \tag{5}$$

Example 2. The characteristic function of the exponential distribution.
Let ξ be a random variable with an exponential distribution and of

expectation $\frac{1}{\lambda}$. The density function is thus $\lambda e^{-\lambda x}$ for $x > 0$ and

$$\varphi_\xi(t) = \lambda \int_0^\infty e^{-x(\lambda - it)}\, dx = \frac{1}{1 - \dfrac{it}{\lambda}}. \tag{6}$$

From this it follows immediately by Theorem 6 of § 2 that a random variable ξ_k having a Γ-distribution of order k and expectation $\frac{k}{\lambda}$ has characteristic function

$$\varphi_{\xi_k}(t) = \frac{1}{\left(1 - \dfrac{it}{\lambda}\right)^k}.$$

Example 3. The characteristic function of the uniform distribution.
Let ξ be a random variable uniformly distributed on the interval $(-A, +A)$; then

$$\varphi_\xi(t) = \frac{1}{2A} \int_{-A}^{+A} e^{ixt}\, dx = \frac{\sin At}{At}.$$

Example 4. The characteristic function of the Cauchy distribution.
Let ξ be a random variable with a Cauchy distribution. Then

$$\varphi_\xi(t) = \frac{1}{\pi} \int_{-\infty}^{+\infty} \frac{e^{ixt}}{1 + x^2}\, dx = e^{-|t|}. \tag{7}$$

(The integral can be evaluated by the method of residues.)
It should be noted that $\varphi_\xi(t)$ is not differentiable at the point $t = 0$. According to Theorem 7 of § 2 this is linked with the fact that ξ does not have an expectation.

Example 5. The characteristic function of Pearson's χ^2-distribution.
If χ_n^2 is a random variable having a χ^2-distribution with n degrees of freedom, we can write

$$\chi_n^2 = \sum_{k=1}^n \xi_k^2,$$

where $\xi_1, \xi_2, \ldots, \xi_n$ are normally distributed independent random variables for which $E(\xi_k) = 0$ and $D(\xi_k) = 1$. It is easy to see that

$$\varphi_{\xi_k^2}(t) = \sqrt{\frac{2}{\pi}} \int_0^\infty e^{-\frac{x^2}{2}(1-2it)} \, dx = \frac{1}{\sqrt{1-2it}}.$$

(One has to take that branch of the square root which is equal to 1 for $t = 0$.) Hence Theorem 6 of § 2 leads to

$$\varphi_{\chi_n^2}(t) = \frac{1}{(1-2it)^{\frac{n}{2}}}. \tag{8}$$

Example 6. The characteristic function of the binomial distribution.
Let ξ be a random variable having a binomial distribution of order n with parameter p; according to § 15 of Chapter III

$$\varphi_\xi(t) = G_\xi(e^{it}) = [1 + p(e^{it} - 1)]^n.$$

§ 4. Some fundamental theorems on characteristic functions

In this paragraph properties of characteristic functions will be discussed which are essential for the proof of limit distribution theorems of probability theory.

THEOREM 1a. *If $\varphi(t)$ is the characteristic function of the distribution function $F(x)$ and if a and b are continuity points of $F(x)$ $(a < b)$, then*

$$F(b) - F(a) = \frac{1}{2\pi} \int_{-\infty}^{+\infty} \left(\varphi(t) \, \frac{e^{-ita} - e^{-itb}}{2it} - \varphi(-t) \, \frac{e^{ita} - e^{itb}}{2it} \right) dt. \tag{1}$$

THEOREM 1b. *Every distribution function is uniquely determined by its characteristic function.*

Theorem 1b follows immediately from Theorem 1a; in fact if $\varphi(t)$ is known, (1) gives the increment of $F(x)$ on every interval the endpoints of which are points of continuity of $F(x)$. The set of discontinuity points of $F(x)$ being denumerable, a may tend to $-\infty$ through a sequence consisting only of continuity points of $F(x)$; hence (1) gives the value of $F(b)$ at every point of continuity b. As $F(x)$ is by definition leftcontinuous, the values of $F(x)$ at a point of discontinuity can be obtained by letting b tend from the left to such a point.

Since the unicity Theorem 1b follows from the inversion Formula (1), it suffices to prove the latter. Before beginning the proof we have to make first some remarks. It was pointed out in § 2 that $\varphi(-t) = \overline{\varphi(t)}$. Thus if $\mathrm{Re}\{z\}$ denotes the real part of the complex number z, (1) can be rewritten in the form

$$F(b) - F(a) = \frac{1}{2\pi} \int_{-\infty}^{+\infty} \mathrm{Re}\left\{\varphi(t) \,\frac{e^{-ita} - e^{-itb}}{it}\right\} dt. \tag{2}$$

The real parts of $\varphi(t)$ and of $\dfrac{e^{-ita} - e^{-itb}}{it}$ are even functions, while their imaginary parts are odd functions. Therefore the same holds for

$$\Psi(t) = \varphi(t)\,\frac{e^{-ita} - e^{-itb}}{it} \tag{3}$$

as well. Consequently, $\overline{\Psi(-t)} = \Psi(t)$. If $\mathrm{Im}\{z\}$ denotes the imaginary part of z, we have

$$\frac{1}{2\pi} \int_{-T}^{+T} \mathrm{Im}\{\Psi(t)\}\, dt = 0 \quad \text{for every } T > 0, \tag{4}$$

hence by (2)

$$F(b) - F(a) = \lim_{T \to \infty} \frac{1}{2\pi} \int_{-T}^{+T} \varphi(t)\,\frac{e^{-ita} - e^{-itb}}{it}\, dt. \tag{5}$$

In many textbooks the inversion formula is given in the form (5). Nevertheless, while the improper integral (1) always exists, the same cannot be stated regarding the integral

$$\frac{1}{2\pi} \int_{-\infty}^{+\infty} \varphi(t)\,\frac{e^{-ita} - e^{-itb}}{it}\, dt. \tag{6}$$

But if this integral exists, its value is by Formula (5) equal to $F(b) - F(a)$. For the proof of Formula (2), we need two simple lemmas.

LEMMA 1. *Put*

$$S(\alpha, T) = \frac{2}{\pi} \int^{T} \frac{\sin \alpha t}{t}\, dt. \tag{7}$$

For every real α and for every positive T we have

$$|S(\alpha, T)| \le 2. \tag{8}$$

Furthermore

$$\lim_{T \to +\infty} S(\alpha, T) = \frac{2}{\pi} \int_0^\infty \frac{\sin \alpha t}{t}\, dt = \begin{cases} +1 & \text{for } \alpha > 0, \\ 0 & \text{for } \alpha = 0, \\ -1 & \text{for } \alpha < 0; \end{cases} \tag{9}$$

and the convergence is uniform for $|\alpha| \ge \delta > 0$, where δ is an arbitrarily small positive number.

PROOF. If we put

$$S(x) = \frac{2}{\pi} \int_0^x \frac{\sin u}{u}\, du,$$

we have

$$S(\alpha, T) = S(\alpha T). \tag{10}$$

Put

$$c_n = \frac{2}{\pi} \int_{n\pi}^{(n+1)\pi} \frac{\sin u}{u}\, du,$$

then we have

$$c_n = (-1)^n \frac{2}{\pi} \int_0^\pi \frac{\sin u}{n\pi + u}\, du \qquad (n = 0, 1, 2, \ldots); \tag{11}$$

the numbers c_n have alternating signs, their absolute value decreases, hence the series $\sum c_n$ is convergent. From

$$S(x) = \sum_{k=0}^{n-1} c_k + \frac{2}{\pi} \int_{n\pi}^x \frac{\sin u}{u}\, du \quad \text{for } n\pi \le x \le (n+1)\pi \tag{12}$$

it follows that for even values of n

$$\sum_{k=0}^{n-1} c_k \le S(x) \le \sum_{k=0}^n c_k \quad \text{for } n\pi \le x \le (n+1)\,\pi, \tag{13}$$

and for odd values of n

$$\sum_{k=0}^n c_k \le S(x) \le \sum_{k=0}^{n-1} c_k \quad \text{for } n\pi \le x \le (n+1)\,\pi. \tag{14}$$

Hence in every case

$$0 \leq S(x) \leq c_0 \leq 2 \quad \text{for} \quad x \geq 0. \tag{15}$$

Since $S(-x) = -S(x)$, we have for every real x

$$|S(x)| \leq 2. \tag{16}$$

Thus (8) is proved. (9) follows from the well-known formula

$$S(\infty) = \frac{2}{\pi} \int\limits_0^\infty \frac{\sin u}{u}\, du = 1. \tag{17}$$

The uniform convergence follows from (10).

LEMMA 2. *Put*

$$D(T, z, a, b) = \frac{1}{2\pi} \int\limits_{-T}^{+T} \frac{\sin t(z-a) - \sin t(z-b)}{t}\, dt \tag{18}$$

and

$$D(z, a, b) = D(+\infty, z, a, b) = \frac{1}{2\pi} \int\limits_{-\infty}^{+\infty} \frac{\sin t(z-a) - \sin t(z-b)}{t}\, dt. \tag{19}$$

For every real z, a, b and for every positive T

$$|D(T, z, a, b)| \leq 2; \tag{20}$$

further if $a < b$, then

$$\lim_{T \to \infty} D(T, z, a, b) = D(z, a, b) = \begin{cases} 1 & \text{for } a < z < b, \\ \dfrac{1}{2} & \text{for } z = a \text{ or } z = b, \\ 0 & \text{for } z < a \text{ or } b < z. \end{cases} \tag{21}$$

The convergence is uniform for $|z-a| \geq \delta$, $|z-b| \geq \delta$ ($\delta > 0$ arbitrary).

PROOF. Since

$$D(T, z, a, b) = \frac{1}{2}[S(z-a, T) - S(z-b, T)],$$

Lemma 2 is an immediate consequence of Lemma 1.

Now we turn to the proof of Theorem 1a. We have

$$\frac{1}{2\pi} \int_{-\infty}^{+\infty} \text{Re}\left\{ \varphi(t)\, \frac{e^{-ita} - e^{-itb}}{it} \right\} dt =$$

$$= \frac{1}{2\pi} \int_{-\infty}^{+\infty} \int_{-\infty}^{+\infty} \left(\int \frac{\sin t(z-a) - \sin t(z-b)}{t}\, dF(z) \right) dt. \qquad (22)$$

On the other hand, since a and b are points of continuity of $F(x)$, we have by Lemma 2

$$F(b) - F(a) = \int_{-\infty}^{+\infty} D(z, a, b)\, dF(z). \qquad (23)$$

In order to prove (2) it suffices thus to prove that the order of integration may be reversed in the right hand side of Formula (22). The difficulty is that the integral (19) representing $D(z, a, b)$ is not absolutely convergent. But by Lemma 2 we know that $D(T, z, a, b) - D(z, a, b)$ tends uniformly to zero on the whole real axis, except for the intervals $a - \delta < z < a + \delta$ and $b - \delta < z < b + \delta$, where δ is a small positive number. Furthermore on these intervals $|D(T, z, a, b)| \leq 2$. Since a and b are continuity points of $F(x)$, we have

$$\lim_{T \to \infty} \int_{-\infty}^{+\infty} D(T, z, a, b)\, dF(z) = \int_{-\infty}^{+\infty} D(z, a, b)\, dF(z) = F(b) - F(a). \qquad (24)$$

On the other hand

$$\int_{-\infty}^{+\infty} D(T, z, a, b)\, dF(z) =$$

$$= \frac{1}{2\pi} \int_{-T}^{+T} \left(\int_{-\infty}^{+\infty} \frac{\sin t(z-a) - \sin t(z-b)}{t}\, dF(z) \right) dt. \qquad (25)$$

Here the order of the integrations can evidently be interchanged, because of the absolute integrability of the integrand in the domain $-\infty < z < +\infty$, $|t| \leq T$. If we let T tend to $+\infty$, then (25) leads to Theorem 1a because of (22), (23) and (24). If a and b are points of discontinuity of $F(x)$ we find, by a slight modification of the proof,

$$\frac{F(b+0) + F(b)}{2} - \frac{F(a+0) + F(a)}{2} =$$

$$= \frac{1}{2\pi} \int_{-\infty}^{+\infty} \text{Re}\left\{ \varphi(t)\, \frac{e^{-ita} - e^{-itb}}{it} \right\} dt. \qquad (26)$$

Of course the density function $f(x) = F'(x)$ of an absolutely continuous $F(x)$ may also be expressed in terms of $\varphi(t)$. We restrict ourselves to the case where the integral

$$\int_{-\infty}^{+\infty} |\varphi(t)|\, dt \tag{27}$$

exists. Then

$$f(x) = \lim_{h \to 0} \frac{F(x+h) - F(x-h)}{2h} =$$

$$= \lim_{h \to 0} \frac{1}{4\pi} \int_{-\infty}^{+\infty} \frac{\sin th}{th} [\varphi(t)\, e^{-itx} + \varphi(-t)\, e^{itx}]\, dt. \tag{28}$$

Since (27) exists and because of

$$\left| \frac{\sin th}{th} [\varphi(t)\, e^{-itx} + \varphi(-t)\, e^{itx}] \right| \leq 2\,|\varphi(t)|,$$

the limit and the integration can be interchanged according to the theorem of Lebesgue, hence

$$f(x) = \frac{1}{2\pi} \int_{-\infty}^{+\infty} \varphi(t)\, e^{-itx}\, dt. \tag{29}$$

It is easy to show that the integral figuring on the right hand side of (29) is a uniformly continuous and bounded function of x. This leads to

THEOREM 2. *If $\varphi(t)$ is the characteristic function of the random variable ζ and if the integral (27) exists, then ζ has a uniformly continuous and bounded density function given by*

$$f(x) = \frac{1}{2\pi} \int_{-\infty}^{+\infty} \varphi(t)\, e^{-itx}\, dt. \tag{30}$$

We shall prove now

THEOREM 3. *The distribution functions $F_n(x)$ $(n = 1, 2, \ldots)$ tend to a distribution function $F(x)$ at every point of continuity of $F(x)$, iff the characteristic functions $\varphi_n(t)$ of $F_n(x)$ tend for $n \to \infty$ to a function $\varphi(t)$ continuous for $t = 0$. In this case $\varphi(t)$ is the characteristic function of $F(x)$ and the functions $\varphi_n(t)$ converge uniformly to $q(t)$ on every finite interval.*

PROOF. We show first that the condition is necessary, i.e. we have to show that if

$$\lim_{n\to\infty} F_n(x) = F(x) \tag{31}$$

at every point of continuity of the distribution function $F(x)$, then

$$\lim_{n\to\infty} \varphi_n(t) = \varphi(t) \tag{32}$$

holds, where

$$\varphi(t) = \int_{-\infty}^{+\infty} e^{itx}\, dF(x), \tag{33}$$

and the convergence in (32) is uniform in every finite t-interval. Let $\varepsilon > 0$ be given. Choose a number $\lambda > 0$ such that $+\lambda$ and $-\lambda$ are continuity points of $F(x)$ and

$$F(-\lambda) < \frac{\varepsilon}{8}, \quad F(+\lambda) > 1 - \frac{\varepsilon}{8},$$

in this case

$$\left| \int_{|x|>\lambda} e^{ixt}\, dF(x) \right| < \frac{\varepsilon}{4}. \tag{34}$$

For $n \geq n_1$, where n_1 depends only on ε, the inequalities

$$F_n(-\lambda) < \frac{\varepsilon}{8}, \quad F_n(+\lambda) > 1 - \frac{\varepsilon}{8}$$

hold, hence

$$\left| \int_{|x|>\lambda} e^{ixt}\, dF_n(x) \right| < \frac{\varepsilon}{4} \quad \text{for} \quad n \geq n_1. \tag{35}$$

Consequently, for $n \geq n_1$,

$$|\varphi_n(t) - \varphi(t)| \leq \left| \int_{-\lambda}^{+\lambda} e^{ixt}\, d[F_n(x) - F(x)] \right| + \frac{\varepsilon}{2}. \tag{36}$$

Integrating by parts we obtain for $|t| \leq T$

$$\left| \int_{-\lambda}^{+\lambda} e^{ixt}\, d[F_n(x) - F(x)] \right| \leq$$

$$\leq |F_n(\lambda) - F(\lambda)| + |F_n(-\lambda) - F(-\lambda)| + T\int_{-\lambda}^{+\lambda} |F_n(x) - F(x)|\, dx. \tag{37}$$

Now $|F_n(x) - F(x)| \leq 2$ and according to the theorem of Lebesgue limit and integration can be interchanged, hence the right hand side of (37), and by (36) $\varphi_n(t) - \varphi(t)$ too, tend for $n \to \infty$ uniformly to zero if $|t| \leq T$. Thus we proved that the condition of Theorem 3 is necessary.

We show now that it is sufficient as well, i.e. that from (32), with $\varphi(t)$ continuous for $t = 0$, follows (31). According to a well-known theorem of Helly every sequence $\{F_n(x)\}$ possesses a subsequence $\{F_{n_k}(x)\}$ that converges to a monotone nondecreasing function $F(x)$ at all continuity points of the latter.

We show first that this function $F(x)$ is necessarily a distribution function. It suffices to show that $F(+\infty) = 1$, $F(-\infty) = 0$, and that $F(x)$ is left-continuous. This latter condition can always be realized by a suitable modification of $F(x)$ at its points of discontinuity. Since $F(x)$ is a limit of distribution functions, we have always $0 \leq F(x) \leq 1$. Hence it suffices to prove that $F(+\infty) - F(-\infty) = 1$. First we prove the following formula:

$$\int_0^x [F_n(y) - F_n(-y)]\, dy = \frac{1}{\pi} \int_{-\infty}^{+\infty} \frac{1 - \cos xt}{t^2}\, \varphi_n(t)\, dt \quad \text{if } x > 0. \tag{38}$$

In fact

$$I_n(x) = \frac{1}{\pi} \int_{-\infty}^{+\infty} \frac{1 - \cos xt}{t^2}\, \varphi_n(t)\, dt =$$

$$= \frac{1}{\pi} \int_{-\infty}^{+\infty} \left[\int_{-\infty}^{+\infty} \frac{1 - \cos xt}{t^2} \cos yt\, dt \right] dF_n(y) \tag{39}$$

(the order of integrations can be interchanged because of the integrability of $\dfrac{1 - \cos xt}{t^2}$ and because of $|\varphi_n(t)| \leq 1$). It is known that

$$\frac{1}{\pi} \int_{-\infty}^{+\infty} \frac{1 - \cos xt}{t^2}\, dt = |x|. \tag{40}$$

From (40) it follows that for $x > 0$

$$\frac{1}{\pi} \int_{-\infty}^{+\infty} \frac{1 - \cos xt}{t^2} \cos yt\, dt = \begin{cases} 0 & \text{for } y \leq -x, \\ x + y & \text{for } -x \leq y \leq 0, \\ x - y & \text{for } 0 \leq y \leq x, \\ 0 & \text{for } x \leq y. \end{cases} \tag{41}$$

Hence by (39)

$$I_n(x) = \int_{-x}^{x} (x - |y|) \, dF_n(y). \tag{42}$$

An integration by parts in (42) leads to (38).

Since $F_n(y) - F_n(-y)$ is a nondecreasing function of y, we obtain from (38)

$$F_n(x) - F_n(-x) \geq \frac{1}{\pi} \int_{-\infty}^{+\infty} \frac{1 - \cos xt}{xt^2} \, \varphi_n(t) \, dt, \tag{43}$$

or

$$F_n(x) - F_n(-x) \geq \frac{1}{\pi} \int_{-\infty}^{+\infty} \frac{1 - \cos u}{u^2} \, \varphi_n\left(\frac{u}{x}\right) du. \tag{44}$$

Suppose that x and $-x$ are both continuity points of $F(x)$ and that n runs through the sequence $\{n_k\}$. Then from the theorem of Lebesgue concerning the interchangeability of the limit and the integral it follows that

$$F(x) - F(-x) \geq \frac{1}{\pi} \int_{-\infty}^{+\infty} \frac{1 - \cos u}{u^2} \, \varphi\left(\frac{u}{x}\right) du. \tag{45}$$

$\varphi(t)$ is continuous for $t = 0$ and because of $\varphi_n(0) = 1$ we have $\varphi(0) = 1$ as well; hence we obtain, by applying Lebesgue's theorem again and by taking (40) into account

$$F(+\infty) - F(-\infty) \geq 1. \tag{46}$$

Consequently, $F(+\infty) = 1$ and $F(-\infty) = 0$; $F(x)$ is therefore a distribution function.

It remains still to prove that 1) $\varphi(t)$ is the characteristic function of $F(x)$ and 2) that the whole sequence $\{F_n(x)\}$ converges to $F(x)$; for the latter, according to the theorem of Helly, it suffices to show that the sequence $\{F_n(x)\}$ possesses no subsequence converging to a function other than $F(x)$. Both of these statements follow immediately from Theorem 1b and from the already proved first part of the present theorem.

Hence from the sequence of distribution functions $(F_n(x))$ $(n = 1, 2, \ldots)$ there cannot be selected any subsequence which does not converge to $F(x)$; this means that $\lim_{n \to \infty} F_n(x) = F(x)$. The uniformity of the convergence in the relation

$$\lim_{n \to \infty} \varphi_n(t) = \varphi(t) \quad \text{for } |t| \leq T$$

$(T > 0$ fixed arbitrarily) follows from the already proved necessity of the condition of Theorem 3. Herewith our theorem is completely proved.
Let us add some remarks.

1. We have seen that if the sequence of distribution functions $\{F_n(x)\}$ $(n = 1, 2, \ldots)$ converges to a distribution function $F(x)$, then the sequence $\varphi_n(t)$ of characteristic functions of the distribution functions $F_n(x)$ converges for every t, when $n \to \infty$, to the characteristic function $\varphi(t)$ of $F(x)$. If the condition that $F(x)$ be a distribution function is omitted, the sequence of the characteristic functions $\varphi_n(t)$ does not necessarily converge; let e.g. be

$$F_n(x) = \begin{cases} 1 & \text{for } x > n, \\ 0 & \text{for } x \le n. \end{cases}$$

For every finite x, $\lim_{n \to \infty} F_n(x) = 0$, nevertheless $\varphi_n(t) = e^{int}$ does not tend to a limit (except for $t = 2k\pi$ $(k = 0, \pm1, \pm2, \ldots)$).

2. We have proved that if the characteristic functions $\varphi_n(t)$ of the functions $F_n(x)$ converge to a function $\varphi(t)$ continuous at $t = 0$, then the functions $F_n(x)$ converge to a distribution function $F(x)$ with characteristic function $\varphi(t)$. If we omit the condition that $\varphi(t)$ is continuous at the origin, our proposition is no longer valid. Thus for instance let $F_n(x)$ be the distribution function of the uniform distribution on the interval $(-n, +n)$, that is

$$F_n(x) = \frac{1}{2} + \frac{x}{2n} \quad \text{for } |x| \le n;$$

then $\varphi_n(t) = \dfrac{\sin nt}{nt}$, and thus the limit

$$\lim_{n \to \infty} \varphi_n(t) = \varphi(t)$$

exists for every real t and is given by

$$\varphi(t) = \begin{cases} 1 & \text{for } t = 0, \\ 0 & \text{otherwise,} \end{cases}$$

thus $\varphi(t)$ is not continuous for $t = 0$. The sequence $F_n(x)$ converges when $n \to \infty$ for every x to $\dfrac{1}{2}$. $F(x)$ is therefore identically equal to the constant $\dfrac{1}{2}$, thus it is not a distribution function.

We show finally that the characteristic functions of two different distributions may coincide on a finite interval.

Consider the random variable ξ which assumes the values $\pm (2k + 1)$ $(k = 0, 1, \ldots)$ with probabilities

$$P(\xi = 2k + 1) = P(\xi = -(2k + 1)) = \frac{4}{\pi^2 (2k + 1)^2} \qquad (k = 0, 1, \ldots).$$

We know that

$$\sum_{n=0}^{\infty} \frac{1}{(2n + 1)^2} = \frac{\pi^2}{8}, \tag{47}$$

hence

$$\sum_{n=-\infty}^{+\infty} P(\xi = 2n + 1) = 1,$$

and we find

$$\varphi_\xi (t) = \frac{8}{\pi^2} \sum_{n=0}^{\infty} \frac{\cos (2n + 1) t}{(2n + 1)^2} = 1 - \frac{2|t|}{\pi} \quad \text{for } |t| \leq \pi, \tag{48}$$

further $\varphi_\xi(t)$ is periodic with period 2π.

Let now η be a random variable assuming the values $0, \pm (4k + 2)$ $(k = 0, 1, \ldots)$ with the probabilities

$$P(\eta = 0) = \frac{1}{2} \quad \text{and} \quad P(\eta = \pm (4k + 2)) = \frac{2}{\pi^2 (2k + 1)^2} \qquad (k = 0, 1, \ldots).$$

Clearly the condition

$$P(\eta = 0) + \sum_{n=-\infty}^{+\infty} P(\eta = 4n + 2) = 1$$

is fulfilled because of (47) and we obtain

$$\varphi_\eta (t) = \frac{1}{2} + \frac{4}{\pi^2} \sum_{k=0}^{\infty} \frac{\cos (4k + 2) t}{(2k + 1)^2} = 1 - \frac{2|t|}{\pi} \quad \text{for } |t| \leq \frac{\pi}{2}. \tag{49}$$

According to (48) and (49) we have thus

$$\varphi_\xi (t) = \varphi_\eta (t) \quad \text{for } |t| \leq \frac{\pi}{2}. \tag{50}$$

The function $\varphi_\eta(t)$ is periodic with period π. Let the real axis be partitioned into subintervals

$$\frac{2k - 1}{2} \pi \leq t \leq \frac{2k + 1}{2} \pi \qquad (k = 0, \pm 1, \pm 2, \ldots),$$

then we see that the functions $\varphi_\xi(t)$ and $\varphi_\eta(t)$ are identical on intervals with an even index k and are of the opposite sign on intervals with an odd index k.

§ 5. Characteristic properties of the normal distribution

Let ξ and η be independent random variables with the same normal distribution; it is easy to see that $\xi + \eta$ and $\xi - \eta$ are independent. It is quite remarkable that this property is characteristic for the normal distribution. In fact, Bernstein has proved the following

THEOREM 1. *If ξ and η are independent random variables with the same distribution and finite variance, further if $\xi + \eta$ and $\xi - \eta$ are independent, then ξ and η are normally distributed.*

PROOF. We may assume without restricting generality that $E(\xi) = E(\eta) = 0$ and $D(\xi) = D(\eta) = 1$. If $\varphi(t)$ is the characteristic function of the common distribution of ξ and η, the characteristic function of $\xi + \eta$ is $\varphi^2(t)$ and that of $\xi - \eta$ is $\varphi(t) \cdot \varphi(-t)$. Since $\xi + \eta$ and $\xi - \eta$ are independent, the characteristic function of their sum is equal to the product of their characteristic functions. The characteristic function of $(\xi + \eta) + (\xi - \eta) = = 2\xi$ is, by Theorem 3 of § 2, equal to $\varphi(2t)$. Hence

$$\varphi(2t) = \varphi^3(t)\,\varphi(-t). \tag{1}$$

Now $\varphi(t)$ can never be zero. In fact, if for a value t_0 we would have $\varphi(t_0) = 0$, then by (1) we would have $\varphi\left(\dfrac{t_0}{2}\right) = 0$ and thus also $\varphi\left(\dfrac{t_0}{2^n}\right) = 0$ $(n = 1, 2, \ldots)$. As $\varphi(t)$ is continuous, we would have $\varphi(0) = 0$; this is impossible as $\varphi(0) = 1$ (cf. Theorem 1 of § 2). Put

$$\psi(t) = \ln \varphi(t) \tag{2}$$

then, by (1)

$$\psi(2t) = 3\psi(t) + \psi(-t). \tag{3}$$

Put

$$\delta(t) = \psi(t) - \psi(-t). \tag{4}$$

If in (3) t is replaced by $-t$ and the equality so obtained is subtracted from (3) we find

$$\delta(2t) = 2\delta(t). \tag{5}$$

By assumption $\varphi(t)$ is twice differentiable and $\varphi'(0) = 0$, $\varphi''(0) = -1$ (cf. § 2, Theorem 7). Since $\varphi(t) \neq 0$, $\psi(t)$ and $\delta(t)$ are twice differentiable as well and we have $\delta(0) = 0$, $\delta'(0) = 0$.

We obtain from (5)

$$\frac{\delta(t)}{t} = \frac{\delta\left(\dfrac{t}{2^n}\right)}{\dfrac{t}{2^n}} \qquad (n = 1, 2, \ldots). \tag{6}$$

The right side of (6) tends for $n \to \infty$ to $\delta'(0)$, i.e. to zero. Hence

$$\delta(t) \equiv 0. \tag{7}$$

It follows that $\psi(t) \equiv \psi(-t)$ and by (3) that

$$\psi(2t) = 4\psi(t). \tag{8}$$

This leads to

$$\frac{\psi(t)}{t^2} = \frac{\psi\left(\dfrac{t}{2^n}\right)}{\left(\dfrac{t}{2^n}\right)^2}. \tag{9}$$

The right hand side of (9) tends to $-\dfrac{1}{2}$ since $\psi(0) = \psi'(0) = 0$, $\psi''(0) = -1$. Hence

$$\psi(t) = -\frac{t^2}{2} \quad \text{and} \quad \varphi(t) = e^{-\frac{t^2}{2}};$$

ξ and η are thus normally distributed random variables.

Similarly, one can prove

THEOREM 2. *Let ξ and η be independent random variables having the same distribution with zero expectation and finite variance. If $\dfrac{\xi + \eta}{\sqrt{2}}$ has the same distribution as ξ and η, then ξ and η are normally distributed.*

PROOF. Assume $D(\xi) = D(\eta) = 1$. If $\varphi(t)$ denotes the characteristic function of ξ and η, we have

$$\varphi(0) = 1, \quad \varphi'(0) = 0, \quad \varphi''(0) = -1.$$

By assumption, the characteristic function of $\dfrac{\xi + \eta}{\sqrt{2}}$ is also equal to $\varphi(t)$. By Theorems 3 and 6 of § 2, however, the characteristic function of $\dfrac{\xi + \eta}{\sqrt{2}}$ is $\varphi^2\left(\dfrac{t}{\sqrt{2}}\right)$, hence

$$\varphi^2\left(\frac{t}{\sqrt{2}}\right) = \varphi(t). \tag{10}$$

From this follows, as in the proof of Theorem 1, that $\varphi(t) \neq 0$ for every t. If we put again $\ln \varphi(t) = \psi(t)$, then $\psi(t)$ is twice differentiable,

$$\psi(0) = \psi'(0) = 0 \text{ and } \psi''(0) = -1.$$

From (10)

$$\psi(t) = 2\psi\left(\frac{t}{\sqrt{2}}\right), \tag{11}$$

hence for every positive n

$$\frac{\psi(t)}{t^2} = \frac{\psi\left(\dfrac{t}{2^{\frac{n}{2}}}\right)}{\left(\dfrac{t}{2^{\frac{n}{2}}}\right)^2}, \tag{12}$$

hence $\psi(t) = -\dfrac{t^2}{2}$ and $\varphi(t) = \exp\left(-\dfrac{t^2}{2}\right)$ which proves our theorem.

Theorem 2 can be rephrased, by using the notion of families of distributions, as follows:

THEOREM 2'. *Let $F(x)$ be a distribution function such that*

$$\int\limits_{-\infty}^{+\infty} x\, dF(x) = 0, \quad \int\limits_{-\infty}^{+\infty} x^2\, dF(x) = 1.$$

If the family of distributions $\left\{F\left(\dfrac{x-m}{\sigma}\right)\right\}$, $\sigma > 0$, *is closed with respect to the operation of convolution, i.e. if for any real numbers m_1, m_2 and for any positive numbers σ_1, σ_2 there can be found constants m and σ (m real, σ positive) such that*

$$F\left(\frac{x-m_1}{\sigma_1}\right) * F\left(\frac{x-m_2}{\sigma_2}\right) = F\left(\frac{x-m}{\sigma}\right), \tag{13}$$

then

$$F(x) = \frac{1}{\sqrt{2\pi}} \int\limits_{-\infty}^{x} e^{-\frac{u^2}{2}} \, du; \qquad (14)$$

$\left\{ F\left(\dfrac{x-m}{\sigma} \right) \right\}$ *is thus the family of the normal distributions.*

PROOF. Obviously, $m = m_1 + m_2$, $\sigma = \sqrt{\sigma_1^2 + \sigma_2^2}$. If we put

$$\varphi(t) = \int\limits_{-\infty}^{+\infty} e^{ixt} \, dF(x),$$

then

$$\varphi(\sigma_1 t)\, \varphi(\sigma_2 t) = \varphi(\sqrt{\sigma_1^2 + \sigma_2^2}\, t). \qquad (15)$$

For $\sigma_1 = \sigma_2 = \dfrac{1}{\sqrt{2}}$ (15) reduces to (10); hence Theorem 2′ follows from Theorem 2.

Theorem 2′ explains to some extent the fact that errors of measurements are usually normally distributed. In effect, the condition, that the sum of two independent errors of measurement belongs to the same family of distributions as the two errors themselves, cannot be fulfilled, in the case of a finite variance, by other than normal distributions. The condition that $F(x)$ should have finite variance is necessary for the validity of Theorem 2′. Thus, for instance, for the distribution function

$$F(x) = \frac{1}{2} + \frac{1}{\pi} \, \text{arc tan } x$$

of the Cauchy distribution we have the relation

$$F\left(\frac{x - m_1}{\sigma_1} \right) * F\left(\frac{x - m_2}{\sigma_2} \right) = F\left(\frac{x - (m_1 + m_2)}{\sigma_1 + \sigma_2} \right). \qquad (16)$$

(16) follows easily by taking into account that the characteristic function of $F(x)$ is equal to $e^{-|t|}$.

If the family of distrubtions $\left\{ F\left(\dfrac{x - m}{\sigma} \right) \right\}$ is closed under the operation of convolution, $F(x)$ is said to be *stable*. According to Theorem 2′, the normal distribution is the only stable distribution having finite variance, as pointed out above. There exist, however, other stable distributions, e.g. the Cauchy distribution. Stable distributions will be dealt with in § 8.

We deal now with some further remarkable properties of normal distributions. If ξ and η are independent normally distributed random variables, their sum $\xi + \eta$ is, as we know already, normally distributed too. We shall now prove that the converse of this statement is also true: this result is due to H. Cramér.

THEOREM 3. *If ξ and η are independent random variables and if $\xi + \eta$ is normally distributed, then ξ and η are normally distributed themselves.*

PROOF. We may suppose $E(\xi + \eta) = 0$, and $D(\xi + \eta) = 1$; the characteristic function of $\xi + \eta$ is then $\exp\left(-\dfrac{t^2}{2}\right)$. Let $\varphi_\xi(t)$ and $\varphi_\eta(t)$ be the characteristic functions of ξ and η, respectively. We have thus

$$\varphi_\xi(t)\,\varphi_\eta(t) = e^{-\frac{t^2}{2}}. \tag{17}$$

If $F(x)$ and $G(x)$ denote the distribution functions of ξ and η, respectively, we have

$$\varphi_\xi(t) = \int_{-\infty}^{+\infty} e^{ixt}\, dF(x), \quad \varphi_\eta(t) = \int_{-\infty}^{+\infty} e^{ixt}\, dG(x). \tag{18}$$

We show now that the definition of $\varphi_\xi(t)$ and $\varphi_\eta(t)$ can be extended to all complex values of t, so that $\varphi_\xi(t)$ and $\varphi_\eta(t)$ are entire functions of the complex variable t. Let us first suppose $t = iv$ (v real) and let A and B be any two positive numbers. We have

$$\int_{-A}^{+A} e^{-vx}\, dF(x) \cdot \int_{-B}^{+B} e^{-vy}\, dG(y) \le \int_{-\infty}^{+\infty}\int_{-\infty}^{+\infty} e^{-v(x+y)}\, dF(x)\, dG(y) =$$

$$= \varphi_{\xi+\eta}(iv) = e^{\frac{v^2}{2}}. \tag{19}$$

Since $e^{-vx} > 0$, the following integrals exist:

$$\varphi_\xi(iv) = \int_{-\infty}^{+\infty} e^{-vx}\, dF(x) \tag{20a}$$

and

$$\varphi_\eta(iv) = \int_{-\infty}^{+\infty} e^{-vy}\, dG(y). \tag{20b}$$

If now $t = u + iv$, we have

$$|\varphi_\xi(t)| = \left| \int_{-\infty}^{+\infty} e^{itx}\, dF(x) \right| \le \int_{-\infty}^{+\infty} e^{-vx}\, dF(x) = \varphi_\xi(iv). \tag{21}$$

The definition of $\varphi_\xi(t)$ and $\varphi_\eta(t)$ can thus be extended to every complex t. It is easy to see that $\varphi_\xi(t)$ and $\varphi_\eta(t)$ are holomorphic on the whole complex plane, hence they are entire functions of t.

Because of (17), $\varphi_\xi(t) \neq 0$, $\varphi_\eta(t) \neq 0$ for every t. Hence $\ln \varphi_\xi(t)$ and $\ln \varphi_\eta(t)$ are entire functions too, where that branch of the logarithmic function is to be taken, for which $\ln 1 = 0$. If $a > 0$ and $b > 0$ are such that $F(a) - F(-a) > \dfrac{1}{2}$ and $G(b) - G(-b) > \dfrac{1}{2}$, then

$$\varphi_\xi(i v) = \int_{-\infty}^{+\infty} e^{-xv} \, dF(x) \geq \frac{e^{-a|v|}}{2} \tag{22}$$

and

$$\varphi_\eta(i v) = \int_{-\infty}^{+\infty} e^{-xv} \, dG(x) \geq \frac{e^{-b|v|}}{2}. \tag{23}$$

Hence, for $t = u + i v$,

$$|\varphi_\xi(t)| \leq \varphi_\xi(iv) = \frac{e^{\frac{v^2}{2}}}{\varphi_\eta(iv)} \leq 2e^{\frac{v^2}{2}+b|v|} \leq 2e^{\frac{|t|^2}{2}+b|t|}. \tag{24}$$

Similarly, we obtain

$$|\varphi_\eta(t)| \leq 2e^{\frac{|t|^2}{2}+a|t|}. \tag{25}$$

If the real part of z is denoted by $\mathrm{Re}(z)$, we have by (24)

$$\left| \mathrm{Re}\left(\ln \varphi_\xi(t) \right) \right| = \left| \ln \frac{1}{|\varphi_\xi(t)|} \right| \leq |t^2| + \max(a,b)|t| + \ln 2. \tag{26}$$

We have $\varphi_\xi(0) = \varphi_\eta(0) = 1$; furthermore we may suppose without restricting generality, $\varphi_\xi'(0) = \varphi_\eta'(0) = 0$. Indeed if we would have $\varphi_\xi'(0) = \alpha$, and consequently, $\varphi_\eta'(0) = -\alpha$, we could always consider instead of ξ and η the random variables $\xi - \alpha$ and $\eta + \alpha$ whose characteristic functions satisfy the above conditions. From this we conclude that the functions $\dfrac{\ln \varphi_\xi(t)}{t^2}$ and $\dfrac{\ln \varphi_\eta(t)}{t^2}$ are everywhere holomorphic, furthermore

$$\frac{\left| \mathrm{Re}\left(\ln \varphi_\xi(t) \right) \right|}{|t|^2} \quad \text{and} \quad \frac{\left| \mathrm{Re}\left(\ln \varphi_\eta(t) \right) \right|}{|t|^2}$$

are, because of (25) and (26), bounded on the whole t-plane. According to a well-known theorem of H. A. Schwarz the relation

$$f(z) = \operatorname{Im}\left(f(0)\right) + \frac{1}{2\pi} \int\limits_0^{2\pi} \operatorname{Re}\left(f(Re^{i\theta})\right) \frac{Re^{i\theta} + z}{Re^{i\theta} - z}\, d\theta \qquad (27)$$

holds for $|z| < R$ for every function $f(z)$ holomorphic on $|z| \le R$. It follows from (27) that $\dfrac{|\ln \varphi_\xi(t)|}{|t|^2}$ and $\dfrac{|\ln \varphi_\eta(t)|}{|t|^2}$ are bounded on the whole plane; they are thus, according to Liouville's theorem, constant. Hence $\varphi_\xi(t) = \exp(ct^2)$ and $\varphi_\eta(t) = \exp(dt^2)$. Because of

$$\varphi_\xi(-t) = \overline{\varphi_\xi(t)}, \quad \varphi_\eta(-t) = \overline{\varphi_\eta(t)}, \quad |\varphi_\xi(t)| \le 1, \quad |\varphi_\eta(t)| \le 1,$$

there follows

$$\varphi_\xi(t) = \exp\left[-\frac{t^2}{2\sigma_1^2}\right] \qquad (28)$$

and

$$\varphi_\eta(t) = \exp\left[-\frac{t^2}{2\sigma_2^2}\right], \qquad (29)$$

and herewith Theorem 3 is proved.

If $f_1(t), \ldots, f_r(t)$ are characteristic functions and $\alpha_1, \ldots, \alpha_r$ are positive rational numbers, further if

$$\prod_{k=1}^r \left(f_k(t)\right)^{\alpha_k} = e^{-\frac{t^2}{2}}, \qquad (30)$$

then it follows from Cramér's theorem that the functions $f_k(t)$ $(k = 1, 2, \ldots, r)$ are characteristic functions of normal distributions. In fact, if N denotes the common denominator of the numbers $\alpha_1, \ldots, \alpha_r$, we have

$$\prod_{k=1}^r \left(f_k(t)\right)^{N\alpha_k} = e^{-\frac{Nt^2}{2}}, \qquad (31)$$

where $N\alpha_k$ $(k = 1, 2, \ldots, r)$ are integers. Hence

$$f_k(t) g_k(t) = e^{-\frac{Nt^2}{2}}, \qquad (k = 1, \ldots, r), \qquad (32)$$

where we have put

$$g_k(t) = (f_k(t))^{N\alpha_k - 1} \prod_{j \ne k} (f_j(t))^{N\alpha_j}.$$

$g_k(t)$ is also a characteristic function, hence by Cramér's theorem $f_k(t)$

is the characteristic function of a normal distribution. If not all α_k are rational, Cramér's theorem does not guarantee the validity of the proposition; however, Yu. V. Linnik and A. A. Singer have proved that it holds in this more general case too; i.e. they proved the following

THEOREM 4. *If the functions $f_k(t)$ $(k = 1, \ldots, r)$ are characteristic functions and if we have in some interval $|t| \leq \delta$ $(\delta > 0)$ identically*

$$\prod_{k=1}^{r} (f_k(t))^{\alpha_k} = e^{imt - \frac{\sigma^2 t^2}{2}}, \qquad (33)$$

where m is a real number and σ, α_1, α_2, $\ldots$, α_r are positive numbers, then the functions $f_k(t)$ $(k = 1, 2, \ldots, r)$ are characteristic functions of normal distributions.

PROOF. The following proof is due to Yu. V. Linnik and A. A. Singer [1]. It consists of five steps.

Step 1. Put

$$g_k(t) = f_k\left(\frac{t}{\sigma\sqrt{2}}\right) f_k\left(-\frac{t}{\sigma\sqrt{2}}\right), \qquad (k = 1, 2, \ldots, r).$$

Clearly $g_k(t)$ is a characteristic function too; in fact if ξ and η are independent random variables possessing the same distribution with characteristic function $f_k(t)$, then $g_k(t)$ is the characteristic function of $\dfrac{\xi - \eta}{\sigma\sqrt{2}}$. Thus the identity

$$\prod_{k=1}^{r} (g_k(t))^{\alpha_k} = e^{-\frac{t^2}{2}} \qquad (34)$$

holds if $|t| \leq \delta$. Furthermore $g_k(t)$ is a real and even function. If we prove from (34) that $g_k(t)$ is the characteristic function of a normal distribution, then the theorem of Cramér implies the same conclusion for $f_k(t)$. It follows from (34) that $g_k(t) \neq 0$ for $|t| \leq \delta$, hence we may take the logarithm of the two sides of Equation (34):

$$\sum_{k=1}^{r} \alpha_k \ln \frac{1}{g_k(t)} = \frac{t^2}{2} . \qquad (35)$$

Let $G_k(x)$ be the distribution function corresponding to the characteristic function $g_k(t)$. It follows from the assumptions concerning $g_k(t)$ that $G_k(x)$ is symmetric with respect to the origin; hence we have for any $a > 0$

$$g_k(t) = \int_{-\infty}^{+\infty} \cos tx \cdot dG_k(x) \leq 1 - \int_{-a}^{+a} (1 - \cos tx)\, dG_k(x).$$

Since for $|t| < \dfrac{\pi}{2a}$ the relation

$$\int\limits_{-a}^{+a} (1 - \cos tx)\, dG_k(x) < 1$$

holds and since for $0 < x < 1$ we have $x \le \ln\left(\dfrac{1}{1-x}\right)$, it follows from (35) for $a > \dfrac{\pi}{2\delta}$ that

$$\sum_{k=1}^{r} \alpha_k \int\limits_{-a}^{+a} (1 - \cos tx)\, dG_k(x) \le \frac{t^2}{2} \quad \text{for} \quad |t| < \frac{\pi}{2a}. \tag{36}$$

If we divide both sides of (36) by t^2 and let t tend to zero, we obtain

$$\sum_{k=1}^{r} \alpha_k \int\limits_{-a}^{+a} x^2\, dG_k(x) \le 1. \tag{37}$$

Since (37) holds for every $a > \dfrac{\pi}{2\delta}$, the integrals

$$\int\limits_{-\infty}^{+\infty} x^2\, dG_k(x) \qquad (k = 1, \ldots, r)$$

exist; by Theorem 7 of § 2 $g_k(t)$ is thus twice differentiable and, $g_k(t)$ being an even function,

$$g_k'(0) = 0 \qquad (k = 1, \ldots, r). \tag{38}$$

From (38) and (35) we conclude that

$$-\sum_{k=1}^{r} \alpha_k g_k''(0) = \sum_{k=1}^{r} \alpha_k \int\limits_{-\infty}^{+\infty} x^2\, dG_k(x) = 1. \tag{39}$$

Step 2. We show now that $g_k(t)$ possesses derivatives of every order. For this we need the classical formula of Faà di Bruno concerning the successive derivatives of composite functions.[1] This formula states: if

$$z = H(y), \quad y = h(t)$$

and if we put

$$h_0(t) = h(t), \quad h_v(t) = \frac{1}{v!}\frac{d^v h(t)}{dt^v} \qquad (v = 1, 2, \ldots),$$

then we have for every integer p

$$\frac{d^p z}{dt^p} = \sum \frac{d^l H(y)}{dy^l} \cdot \frac{p!}{i_1!\, i_2! \ldots i_s!}\, h_{i_1}^{i_1}(t) \ldots h_{i_s}^{i_s}(t), \tag{40}$$

[1] Cf. e.g. E. Lukács [2].

where the summation is extended over all nonnegative integers $i_1, \ldots, i_s$ and $l_1, \ldots, l_s$ satisfying the conditions

$$\sum_{j=1}^{s} i_j = l \quad \text{and} \quad \sum_{j=1}^{s} l_j i_j = p,$$

where l assumes the values $l = 1, 2, \ldots, p$.

In particular, if $H(y) = \ln y$ and $p = 2q$, we obtain

$$\frac{d^{2q} \ln h(t)}{dt^{2q}} = \sum \frac{(-1)^{l-1} (2q)! (l-1)!}{i_1! \, i_2! \ldots i_s!} \prod_{j=1}^{s} \left(\frac{h^{(l_j)}(t)}{l_j! \, h(t)} \right)^{i_j}, \tag{41}$$

where the summation is to be extended over $l = 1, 2, \ldots, 2q$ and over i_j, l_j such that

$$\sum_{j=1}^{s} i_j = l, \quad \sum_{j=1}^{s} i_j l_j = 2q.$$

Now we show by induction that the integrals

$$\int_{-\infty}^{+\infty} x^{2q} \, dG_k(x) \qquad (q = 1, 2, \ldots; \ k = 1, 2, \ldots, r)$$

exist. Suppose that this holds for a given integer q; from this it follows that $g_k(t)$ $(k = 1, \ldots, r)$ is exactly $2q$ times differentiable and that

$$g_k^{(2l-1)}(0) = 0 \qquad (l = 1, \ldots, q; \ k = 1, \ldots, r).$$

According to (35) we have

$$\frac{d^{2q} \left(\frac{t^2}{2} \right)}{dt^{2q}} = (2q)! \sum_{k=1}^{r} \alpha_k \sum_{l=1}^{2q} (-1)^l (l-1)! \sum \prod_{j=1}^{s} \frac{\left(\frac{g_k^{(l_j)}(t)}{g_k(t) l_j!} \right)^{i_j}}{i_j!} \tag{42}$$

where the summation is as in (41).

Put $t = 0$ in (42) and subtract the relation thus obtained from (42). Separate the terms with the indices $l = 1$, $i_1 = 1$, $l_1 = 2q$ and consider that the left side of (42) is either 1 or 0 for $q = 1$ or $q \geq 2$, respectively. Thus we obtain

$$\sum_{k=1}^{r} \alpha_k \left[\frac{g_k^{(2q)}(t)}{g_k(t)} - g_k^{(2q)}(0) \right] =$$

$$= (2q)! \sum_{k=1}^{r} \alpha_k \sum_{l=2}^{2q} (-1)^{l-1} (l-1)! [S_{kl}(t) - S_{kl}(0)], \tag{43}$$

with

$$S_{kl}(t) = \sum \prod_{j=1}^{s} \frac{\left(\frac{g_k^{(l_j)}(t)}{g_k(t) \cdot l_j!} \right)^{i_j}}{i_j!},$$

the summation being extended over the i_j, l_j such that

$$\sum_{j=1}^{s} i_j = l, \quad \sum_{j=1}^{s} i_j l_j = 2q.$$

We show now that the right hand side of (43) has the order of magnitude $O(t^2)$ when $t \to 0$. In fact if $v < 2q$ is an odd number, then by the induction hypothesis $g_k^{(v)}(t) = O(|t|)$. Hence it suffices to consider terms for which all the l_j are even. If v is even, $v \leq 2q - 2$, then we have

$$\frac{g_k^{(v)}(t)}{g_k(t)} - g_k^{(v)}(0) = O(t^2),$$

from which our statement follows. But

$$\frac{1}{g_k(t)} - 1 = O(t^2)$$

is also valid; hence it follows from (43) that

$$\sum_{k=1}^{r} \alpha_k [g_k^{(2q)}(t) - g_k^{(2q)}(0)] = O(t^2). \tag{44}$$

In consequence, the expression

$$\sum_{k=1}^{r} \alpha_k \int_{-\infty}^{+\infty} \frac{1 - \cos tx}{t^2} x^{2q} dG_k(x)$$

is bounded. If we let t tend to zero, we see that the integrals

$$\int_{-\infty}^{+\infty} x^{2q+2} dG_k(x) \qquad (k = 1, \ldots, r)$$

exist and this means that $g_k(t)$ $(k = 1, \ldots, r)$ is at least $(2q + 2)$ times differentiable. As we know already that the integrals

$$\int_{-\infty}^{\infty} x^2 dG_k(x) \qquad (k = 1, \ldots, r)$$

exist, the proof is finished; $g_k(t)$ $(k = 1, \ldots, r)$ is thus infinitely often differentiable.

Step 3. We shall show now that the $g_k(t)$ are holomorphic in a circle $|t| \leq R$ $(R > 0)$. In order to show this we have to evaluate the order of magnitude of the derivatives $g_k^{(2q)}(0)$. We can restrict ourselves to the case when $\alpha_k \geq 1$ $(k = 1, \ldots, r)$; otherwise there exists an integer N_0 such that $N_0\alpha_k \geq 1$ $(k = 1, \ldots, r)$. Since (34) leads to the equality

$$\prod_{k=1}^{r}\left[g_k\left(\frac{t}{\sqrt{N_0}}\right)\right]^{N_0\alpha_k} = e^{-\frac{t^2}{2}},$$

(34) is satisfied by the functions

$$g_k^*(t) = g_k\left(\frac{t}{\sqrt{N_0}}\right)$$

for $\alpha_k^* = N_0\alpha_k$. Without restriction of generality we may thus assume $\alpha_k \geq 1$ $(k = 1, \ldots, r)$. Now raise the two sides of (34) to the power $2q$, differentiate $2q$ times and put $t = 0$. By introducing the notation $\gamma_k(t) = [g_k(t)]^{2\alpha_k q}$ we obtain thus

$$\sum_{l_1+\ldots+l_r=2q}\frac{2q!}{l_1!\ldots l_r!}\gamma_1^{(l_1)}(0)\ldots\gamma_r^{(l_r)}(0) = \frac{d^{2q}}{dt^{2q}}e^{-qt^2}\Bigg|_{t=0}. \tag{45}$$

The quantities $\gamma_k^{(l)}(t)$ can be evaluated by means of the formula of Faà pi Bruno:

$$\gamma_k^{(l)}(t) = \sum_{v=1}^{l}2q\alpha_k(2q\alpha_k-1)\ldots(2q\alpha_k-v+1)[g_k(t)]^{2q\alpha_k-v} \times$$

$$\times \sum\frac{l!}{i_1!\ldots i_s!}\prod_{j=1}^{s}\left(\frac{g_k^{(l_j)}(t)}{l_j!}\right)^{i_j}, \tag{46}$$

where in the inner sum the summation is to be taken over the i_j-s and l_j-s such that $\sum i_j = v$, $\sum i_j l_j = l$. Because of

$$g_k^{(2l-1)}(0) = 0, \quad \text{sgn } g_k^{(2l)}(0) = (-1)^l$$

and

$$2q\alpha_k(2q\alpha_k-1)\ldots(2q\alpha_k-v+1) \geq 0 \quad \text{for } v \leq 2q,$$

it follows that all nonzero terms on the left hand side of (45) have the sign of $(-1)^q$. The right hand side is

$$\frac{d^{2q}e^{-t^2q}}{dt^{2q}}\Bigg|_{t=0} =: (2q)^q(2q)!\,H_{2q}(0), \tag{47}$$

where $H_{2q}(x)$ is the *Hermite polynomial of order* $2q$:

$$H_{2q}(x) = \frac{1}{(2q)!}e^{\frac{x^2}{2}}\cdot\frac{d^{2q}}{dx^{2q}}e^{-\frac{x^2}{2}}. \tag{48}$$

Thus

$$H_{2q}(0) = \frac{(-1)^q}{q! \, 2^q}.$$

Since on the left hand side of Equation (45) there occur the terms $2q\alpha_k g_k^{(2q)}(0)$ too, the relation

$$g_k^{(2q)}(0) \leq \frac{(2q)^q \, (2q)!}{q! \, 2^q}$$

must hold, wherefrom

$$\limsup_{q \to \infty} \left(\frac{g_k^{(2q)}(0)}{(2q)!} \right)^{\frac{1}{2q}} \leq \sqrt{e}, \qquad (49)$$

thus $g_k(t)$ is holomorphic in the circle $|t| < \dfrac{1}{\sqrt{e}}$ $(k = 1, \ldots, r)$.

Step 4. We show that the functions $g_k(t)$ are also entire functions. Put $h_k(t) = g_k \, i \, \sqrt{t}$. Since $g_k(t)$ is an even function, $h_k(t)$ is holomorphic in the circle $|t| < \dfrac{1}{e}$. Suppose that not all $g_k(t)$ are entire functions; then the same holds for the functions $h_k(t)$. Let $h_{k_0}(t)$ be the function $h_k(t)$ which has the smallest radius of convergence, which radius we denote by R. Take $0 < r < R$; put $\mathscr{H}_k(t) = h_k(r + t)$. Then

$$\prod_{k=1}^{r} \mathscr{H}_k^{\alpha_k}(t) = e^{-\frac{r+t}{2}} \qquad (50)$$

and

$$\prod_{k=1}^{r} \left(\frac{\mathscr{H}_k(t)}{\mathscr{H}_k(0)} \right)^{\alpha_k} = e^{\frac{t}{2}}. \qquad (51)$$

Since $\mathscr{H}_k(t)$ $(k = 1, \ldots, r)$ too can be represented by a power series with positive coefficients, we obtain, by raising (51) to the n-th power and differentiating n-times,

$$n\alpha_{k_0} \mathscr{H}_{k_0}^{(n)}(0) \leq \left(\frac{n}{2} \right)^n \mathscr{H}_{k_0}(0),$$

hence

$$\limsup_{n \to \infty} \left(\frac{\mathscr{H}_{k_0}^{(n)}(0)}{n!} \right)^{\frac{1}{n}} \leq \frac{e}{2}. \qquad (52)$$

$\mathscr{H}_{k_0}(t)$ is thus holomorphic in the circle $|t| < \dfrac{2}{e}$, i.e. $h_{k_0}(t)$ is holomorphic in the circle $|t - r| < \dfrac{2}{e}$. From this it follows for an r sufficiently near to R that $h_{k_0}(t)$ is regular at the point $t = R$. This, however, contradicts the known theorem according to which the sum of a power series with positive coefficients having a radius of convergence equal to R, is singular at the point $+R$.[1]

Step 5. The proof of Theorem 4 can now be finished like that of Cramér's theorem. If we choose the numbers $a_k > 0$ $(k = 1, 2, \ldots, r)$ such that

$$\int\limits_{-a_k}^{a_k} dG_k(x) \geq \frac{1}{2} \qquad (k = 1, \ldots, r),$$

then

$$\ln |g_k(t)| \leq \frac{|t|^2}{2\alpha_k} + C|t|,$$

where C is a positive constant. Because of (34) and step 4, the function $\ln g_k(t)$ is an entire function, hence by Liouville's theorem $\ln g_k(t)$ is a polynomial of at most the second degree, which leads to the statement of Theorem 4.

Theorem 4 enables us to generalize Theorem 1. In fact, as Darmois and Skitovitch[2] have shown:

THEOREM 5. *If $\xi_1, \xi_2, \ldots, \xi_r$ are independent random variables and $a_1, \ldots, a_r, b_1, \ldots, b_r$ are real numbers different from 0, further if the random variables*

$$\eta_1 = \sum_{k=1}^r a_k \xi_k \quad and \quad \eta_2 = \sum_{k=1}^r b_k \xi_k$$

are independent, then the random variables $\xi_1, \xi_2, \ldots, \xi_r$ are normally distributed.

PROOF. Linnik has shown that the proof of this theorem can be reduced to that of Theorem 4 as follows: Take $a_1 = a_2 = \ldots = a_r = 1$, which does not restrict the generality. η_1 and η_2 are by assumption independent, hence we have

$$E(e^{i(u\eta_1 + v\eta_2)}) = E(e^{iu\eta_1}) E(e^{iv\eta_2}). \tag{53}$$

[1] Cf. e.g. E. C. Titchmarsh [1], p. 214.
[2] G. Darmois [1], V. R. Skitovitch [1].

If $\varphi_k(t)$ is the characteristic function of ξ_k, we have by (53) because of the independence of the ξ_k,

$$\prod_{k=1}^{r} \varphi_k(u + b_k v) = \prod_{k=1}^{r} \varphi_k(u)\, \varphi_k(b_k v). \tag{54}$$

In a neighbourhood of the origin the $\varphi_k(t)$ are not all zero. Put therefore $\psi_k(t) = \ln \varphi_k(t)$, then

$$\sum_{k=1}^{r} \psi_k(u + b_k v) = \sum_{k=1}^{r} \psi_k(u) = \sum_{k=1}^{r} \psi_k(b_k v). \tag{55a}$$

Replace in this equality u by $-u$, v by $-v$ and add the equality so obtained to the former one; it follows, by putting $\psi_k^*(t) = \psi_k(t) + \psi_k(-t)$, that

$$\sum_{k=1}^{r} \psi_k^*(u + b_k v) = \sum_{k=1}^{r} \psi_k^*(u) + \sum_{k=1}^{r} \psi_k^*(b_k v). \tag{55b}$$

We prove now that $\psi_k^*(t) = -c_k t^2$. This means that $\varphi_k(t)\varphi_k(-t)$ is the characteristic function of a normal distribution and the proof of Theorem 5 is finished by Cramér's theorem.

Multiply the two sides of (55b) by $x - u$ and integrate from 0 to x, thus

$$\sum_{k=1}^{r} \int_{0}^{x} (x - u)\, \psi_k^*(u + b_k v)\, du = \frac{x^2}{2} \left(\sum_{k=1}^{r} \psi_k^*(b_k v) \right) +$$

$$+ \int_{0}^{x} (x - u) \sum_{k=1}^{r} \psi_k^*(u)\, du.$$

Then, by variable-transformation and integration by parts, we get

$$\int_{0}^{x} (x - u)\, \psi_k^*(u + b_k v)\, du = - x \int_{0}^{b_k v} \psi_k^*(\tau)\, d\tau + \int_{b_k v}^{x + b_k v} \left(\int_{0}^{t} \psi_k^*(\tau) d\tau \right) dt.$$

If we put

$$B(v) = \sum_{k=1}^{r} \psi_k^*(b_k v), \tag{56}$$

we obtain

$$\sum_{k=1}^{r} \int_{b_k v}^{x + b_k v} \left(\int_{0}^{t} \psi_k^*(\tau)\, d\tau \right) dt - x \int_{0}^{b_k v} \psi_k^*(\tau) d\tau = \tag{57}$$

$$= \frac{x^2}{2} B(v) + \int_{0}^{x} (x - u) \sum_{k=1}^{r} \psi_k^*(u)\, du.$$

The left hand side of (57) is obviously a differentiable function of v. Hence

$$\sum_{k=1}^{r} b_k \int_{b_k v}^{x+b_k v} \psi_k^*(\tau)\, d\tau = \frac{x^2}{2} B'(v) + A(v)\, x, \qquad (58)$$

where

$$A(v) = \sum_{k=1} b_k \psi_k^*(b_k v).$$

Replacing in (58) x by $-x$ and adding the equation so obtained to (58), we get

$$\sum_{k=1}^{r} b_k \left(\int_{b_k v}^{x+b_k v} \psi_k^*(\tau)\, d\tau + \int_{b_k v}^{-x+b_k v} \psi_k^*(\tau)\, d\tau \right) = x^2 B'(v). \qquad (59)$$

Clearly, this equation can be differentiated with respect to v; doing this and putting $v = 0$ we find, because of $\psi_k^*(-x) = \psi_k^*(x)$ and $\psi_k^*(0) = 0$

$$\sum_{k=1}^{r} b_k^2 \psi_k^*(x) = \frac{x^2}{2} B''(0). \qquad (60)$$

From this follows

$$\prod_{k=1}^{r} (\varphi_k^*(x))^{b_k^2} = e^{B''(0)\frac{x^2}{2}},$$

where

$$\varphi_k^*(x) = e^{\psi_k^*(x)} = \varphi_k(x)\, \varphi_k(-x).$$

Since $|\varphi_k^*(x)| \leq 1$, the relation $B''(0) \leq 0$ must hold. Equality cannot hold here, since then the ξ_k would be constants with probability 1. Hence we can put $B''(0) = -\sigma^2 < 0$ and we have thus

$$\prod_{k=1}^{r} (\varphi_k^*(x))^{b_k^2} = e^{-\sigma^2 \frac{x^2}{2}} \qquad (\sigma^2 > 0) \qquad (61)$$

in a neighbourhood of the origin. By Theorem 4 the functions $\varphi_k^*(x)$, and consequently the functions $\varphi_k(x)$, are characteristic functions of normal distributions. Theorem 5 is herewith proved.

Finally, we shall prove one further characteristic property of the normal distribution: the following theorem of E. Lukács:

THEOREM 6. *If $\xi_1, \xi_2, \ldots, \xi_n$ are independent random variables having the same distribution of finite expectation and variance, then this distribution* *is a normal one iff*

$$\xi = \sum_{k=1}^{n} \xi_k \quad and \quad \eta = \sum_{k=1}^{n} \left(\xi_k - \frac{\xi}{n} \right)^2$$

are independent.

Remark. R. A. Fisher proved the independence of ξ and η for normally distributed ξ_k. The converse theorem was proved by E. Lukács in the case where the ξ_k have a finite variance. It was already proved before by R. C. Geary under the stronger condition that all moments of the ξ_k exist. Later on it was proved by J. Kawata and H. Sakamoto[1] as well as by A. A. Singer[2] that even the existence of the variance is unnecessary; however, we shall not deal with this more general case.

PROOF OF THEOREM 6. *The condition is necessary.* The ξ_k are normally distributed; we may assume $E(\xi_k) = 0$, $D(\xi_k) = 1$. If (c_{ij}) is an orthogonal matrix of n rows and n columns with $c_{1j} = \dfrac{1}{\sqrt{n}}$ $(j = 1, 2, \ldots, n)$ we know (cf. Ch. IV, § 17, Exercise 43.b) that the random variables

$$\zeta_j^* = \sum_{k=1}^{n} c_{jk}\xi_k \qquad (j = 1, 2, \ldots, n)$$

are mutually independent, too.

We have thus $\xi = \sqrt{n}\,\zeta_1^*$ and

$$\eta = \sum_{k=1}^{n} \xi_k^2 - \frac{\xi^2}{n} = \sum_{j=1}^{n} \zeta_j^{*2} - \zeta_1^{*2} = \sum_{j=2}^{n} \zeta_j^{*2},$$

which shows the independence of ξ and η.

2. *The condition is sufficient.* We may assume $E(\xi_k) = 0$. By the assumption of the theorem

$$E(e^{i(u\xi+v\eta)}) = E(e^{iu\xi})\,E(e^{iv\eta}). \tag{62}$$

If we differentiate on both sides of (62) with respect to v (which is allowed because of Theorem 7 of § 2), and substitute $v = 0$ afterwards, we obtain

$$E(\eta e^{iu\xi}) = E(e^{iu\xi})\,E(\eta) = \big(\varphi(u)\big)^n E(\eta), \tag{63}$$

where

$$\varphi(u) = E(e^{iu\xi_k})$$

is the characteristic function of the random variables ξ_k. From

$$\eta = \left(1 - \frac{1}{n}\right) \sum_{k=1}^{n} \xi_k^2 - \frac{2}{n} \sum_{j=1}^{n-1} \sum_{j<k} \xi_j \xi_k \tag{64}$$

[1] J. Kawata and H. Sakamoto [1].
[2] A. A. Singer [1].

follows $E(\eta) = (n - 1) \sigma^2$, by putting $\sigma^2 = E(\xi_k^2)$. Since

$$E(\xi_k e^{iu\xi_k}) = - i\varphi'(u) \tag{65}$$

and

$$E(\xi_k^2 e^{iu\xi_k}) = - \varphi''(u), \tag{66}$$

(63) can be written in the form

$$- (n - 1) \varphi''(u) (\varphi(u))^{n-1} + \frac{2}{n} \binom{n}{2} (\varphi'(u))^2 (\varphi(u))^{n-2} =$$

$$= (n - 1) \sigma^2 (\varphi(u))^n. \tag{67}$$

If we divide by $(n - 1) (\varphi(u))^n$, we find

$$\frac{\varphi''(u)}{\varphi(u)} - \left(\frac{\varphi'(u)}{\varphi(u)} \right)^2 = - \sigma^2. \tag{68}$$

The left hand side of (68) is the second derivative of $\ln \varphi(u)$. If we integrate twice and consider that $\varphi'(0) = iE(\xi_k) = 0$, we find

$$\ln\varphi(u) = - \frac{\sigma^2 u^2}{2} \tag{69}$$

which proves the theorem of Lukács.

§ 6. Characteristic functions of multidimensional distributions

Let $\xi = (\xi_1, \xi_2, \ldots, \xi_n)$ be an n-dimensional random vector with distribution function $F(x_1, x_2, \ldots, x_n)$.

For any two vectors $t = (t_1, t_2, \ldots, t_n)$ and $x = (x_1, x_2, \ldots, x_n)$ we put

$$(x, t) = \sum_{k=1}^{n} x_k t_k. \tag{1}$$

For sake of brevity we write $F(x) = F(x_1, x_2, \ldots, x_n)$. We define the characteristic function of ξ by

$$\varphi_\xi(t) = E(e^{i(\xi,t)}) = \int_{-\infty}^{+\infty} \ldots \int_{-\infty}^{+\infty} e^{i(x,t)} dF(x). \tag{2}$$

The characteristic function of an n-dimensional distribution function is thus a function of n real variables. As is readily seen, it has the following properties:

1. For $0 = (0, 0, \ldots, 0)$ we have

$$\varphi_\xi(0) = 1.$$

2. For every t

$$|\varphi_\xi(t)| \le 1.$$

3. $\varphi_\xi(t)$ is a uniformly continuous function of t.

4. If

$$\eta = (\eta_1, \ldots, \eta_n), \quad \eta_j = \sum_{k=1}^n c_{jk}\xi_k + b_j \qquad (j = 1, \ldots, n)$$

and

$$u = (u_1, \ldots, u_n), \qquad u_k = \sum_{j=1}^n c_{jk}t_j \qquad (k = 1, \ldots, n),$$

then

$$\varphi_\eta(t) = e^{i(b,t)}\,\varphi_\xi(u).$$

5. $\dfrac{\partial \varphi_\xi(t)}{\partial t_k}\bigg|_{t=0} = iE(\xi_k)$ when $E(\xi_k)$ exists.

6. $\dfrac{\partial^2 \varphi_\xi(t)}{\partial t_j \partial t_k}\bigg|_{t=0} = -E(\xi_j\xi_k)$ when $E(\xi_j\xi_k)$ exists.

7. If the density function $f(x)$ of ξ exists, then

$$\varphi_\xi(t) = \int_{-\infty}^{+\infty} \cdots \int_{-\infty}^{+\infty} e^{i(t,x)}f(x)\,dx,$$

where dx is an abbreviation for $dx_1 dx_2 \ldots dx_n$.

Example 1. The characteristic function of the multinomial distribution.
If

$$p_j \ge 0 \quad (j = 1, 2, \ldots, n), \quad \sum_{j=1}^n p_j = 1$$

and

$$P(\xi_1 = k_1, \ldots, \xi_n = k_n) = \frac{N!}{k_1! \ldots k_n!}\, p_1^{k_1}, \ldots, p_n^{k_n},$$

where $\sum_{j=1}^n k_j = N$ ($k_j \ge 0$ being an integer), we obtain

$$\varphi_\xi(t) = \left(\sum_{k=1}^n p_k e^{it_k}\right)^N.$$

Example 2. The characteristic function of the n-dimensional normal distribution.

If the vector $\eta = (\eta_1, \ldots, \eta_n)$ is by definition normally distributed, there exist normally distributed independent random variables $\xi_1, \ldots, \xi_n$ with $E(\xi_k) = 0$, $D^2(\xi_k) = \sigma_k^2$ and an orthogonal matrix (c_{jk}) such that

$$\eta_j = \sum_{k=1}^{n} c_{jk}\xi_k + m_j \qquad (j = 1, 2, \ldots, n). \tag{3}$$

Then by property 4 of the characteristic function

$$\varphi_\eta(t) = e^{i(m,t)} \varphi_\xi(u), \tag{4}$$

where

$$u_k = \sum_{j=1}^{n} c_{jk}t_j, \quad u = (u_1, \ldots, u_n), \quad m = (m_1, \ldots, m_n). \tag{5}$$

It suffices therefore to determine the characteristic function of ξ. Since, by assumption, the random variables ξ_k are independent, we have

$$\varphi_\xi(u) = E(e^{i(u,\xi)}) = \prod_{k=1}^{n} E(e^{iu_k\xi_k}) = \exp\left[-\frac{1}{2}\sum_{k=1}^{n}\sigma_k^2 u_k^2\right]. \tag{6}$$

(4), (5) and (6) lead to

$$\varphi_\eta(t) = \exp\left[i(m, t) - \frac{1}{2}\sum_{h=1}^{n}\sum_{j=1}^{n} b_{hj}t_h t_j\right], \tag{7}$$

where

$$b_{hj} = \sum_{k=1}^{n}\sigma_k^2 c_{hk}c_{jk}. \tag{8}$$

It is easy to see that the quadratic form

$$\sum_{h=1}^{n}\sum_{j=1}^{n} b_{hj}t_h t_j$$

is positive definite and the matrix $B = (b_{hj})$ is the inverse of the matrix $A = (a_{hj})$, the elements of which are the coefficients in the expression of the density function of η. The matrix B is thus the dispersion matrix of η. In fact, a simple calculation shows that the matrix B can be written in the form $B = CSC^{-1}$, where S is the diagonal matrix with elements σ_k^2. On the other hand, we have proved (cf. Ch. IV, § 17, Exercise 42) that the density function $g(y)$ of η with $y = (y_1, \ldots, y_n)$ is given by

$$g(y) = \frac{\sqrt{|A|}}{(2\pi)^{\frac{n}{2}}} \exp\left[-\frac{1}{2}\sum_{h=1}^{n}\sum_{j=1}^{n} a_{hj}(y_h - m_h)(y_j - m_j)\right], \tag{9}$$

where the matrix $A = (a_{hj})$ can be written as $A = CS^{-1}C^{-1}$. Hence we have obtained

THEOREM 1. *If η is an n-dimensional normally distributed random variable with density function (9) such that the quadratic form*

$$\sum_{h=1}^{n} \sum_{j=1}^{n} a_{hj} z_h z_j$$

is positive definite and its determinant is denoted by $|A|$, then the characteristic function of η is given by (7) where $m = (m_1, \ldots, m_n)$ and where $(b_{hj}) = B = = A^{-1}$ is the dispersion matrix of η.

There exists an inversion formula for n-dimensional characteristic functions too. It is given by

THEOREM 2. *If $\varphi_\xi(t)$ is the characteristic function of the n-dimensional vector ξ, we have*

$$P(\xi \in I) = \lim_{T \to \infty} \frac{1}{(2\pi)^n} \int_{-T}^{+T} \cdots \int_{-T}^{+T} \prod_{k=1}^{n} \frac{e^{-it_k a_k} - e^{-it_k b_k}}{it_k} \cdot \varphi_\xi(t)\, dt, \qquad (10)$$

whenever the distribution functions of all ξ_k are continuous at the points a_k and b_k $(k = 1, \ldots, n)$; here I is the n-dimensional interval $a_k \leq x_k < b_k$ $(k = 1, \ldots, n)$.

Like in the one-dimensional case it follows from this theorem that an n-dimensional distribution function is uniquely determined by its characteristic function. The proof is analogous to that of the uniqueness theorem for $n = 1$; we leave it to the reader.

By means of the uniqueness theorem we prove

THEOREM 3. *The distribution of the vector $\xi = (\xi_1, \ldots, \xi_n)$ is uniquely determined by the distribution functions of the projections of ξ upon all lines passing through the origin.*

PROOF. Let d_α be an arbitrary line passing through the origin having the direction-cosines $\alpha_1, \ldots, \alpha_n$; put $\alpha = (\alpha_1, \ldots, \alpha_n)$. The projection of ξ on d_α is thus

$$\xi_\alpha = (\alpha, \xi). \qquad (11)$$

Hence the characteristic function of the random variable ξ_α is

$$\varphi_{\xi_\alpha}(t) = E(e^{i(\alpha, \xi)t}) = \varphi_\xi(\alpha t), \qquad (12)$$

where αt is the vector with components $\alpha_k t$ $(k = 1, \ldots, n)$. Since every system $t_1, \ldots, t_n$ of real numbers can be written in the form $t_k = t\alpha_k$ where t is real and

$$\sum_{k=1}^{n} \alpha_k^2 = 1,$$

the theorem follows from the uniqueness theorem.

THEOREM 4. *If* $\xi = (\xi_1, \ldots, \xi_n)$ *and* $\eta = (\eta_1, \ldots, \eta_n)$ *are n-dimensional independent random vectors and if* $\zeta = \xi + \eta$, *we have*

$$\varphi_\zeta(t) = \varphi_\xi(t)\, \varphi_\eta(t).$$

PROOF. Let d_α be any line passing through the origin with direction cosines α_k $(k = 1, \ldots, n)$, and put $\alpha = (\alpha_1, \ldots, \alpha_n)$. If $\xi_\alpha, \eta_\alpha, \zeta_\alpha$ are the projections of ξ, η, ζ upon d_α, we have $\zeta_\alpha = \xi_\alpha + \eta_\alpha$. Hence because of the independence of ξ and η:

$$\varphi_{\zeta_\alpha}(t) = \varphi_{\xi_\alpha}(t)\, \varphi_{\eta_\alpha}(t). \tag{13}$$

It follows from (12) that

$$\varphi_\zeta(t\alpha) = \varphi_\xi(t\alpha)\, \varphi_\eta(t\alpha), \tag{14}$$

and the proof can be finished as that of Theorem 3.

THEOREM 5. *If* $\xi_1, \ldots, \xi_n$ *are independent,* $\xi = (\xi_1, \ldots, \xi_n)$, $t = (t_1, \ldots, t_n)$, *(where* $t_1, \ldots, t_n$ *are real numbers), then*

$$\varphi_\xi(t) = \prod_{k=1}^{n} \varphi_{\xi_k}(t_k). \tag{15}$$

Conversely, (15) *implies the independence of* $\xi_1, \ldots, \xi_n$.

PROOF. It follows from the assumption that the random variables $e^{it_k \xi_k}$ are also independent, hence

$$\varphi_\xi(t) = E\left(\prod_{k=1}^{n} e^{it_k \xi_k}\right) = \prod_{k=1}^{n} E(e^{it_k \xi_k}) = \prod_{k=1}^{n} \varphi_{\xi_k}(t_k). \tag{16}$$

If on the other hand $x = (x_1, \ldots, x_n)$ and if $F(x)$ is the distribution function of ξ, $F_k(x)$ the distribution function of ξ_k $(k = 1, \ldots, n)$, further if

$$G(x) = \prod_{k=1}^{n} F_k(x_k),$$

then by (15) it follows that the characteristic functions of $F(x)$ and $G(x)$ are identical. Hence, because of the uniqueness, $F(x) \equiv G(x)$ and the random variables ξ_k are independent.

As an application of the preceding theorem we prove now the following theorem due to M. Kac (cf. Ch. III, § 11, Theorem 6).

THEOREM 6. *If ξ and η are two bounded random variables fulfilling for any integers k and l the relation*

$$E(\xi^k \eta^l) = E(\xi^k) E(\eta^l), \tag{17}$$

then ξ and η are independent.

PROOF. Our assumption implies the absolute convergence of the series

$$\varphi_\xi(t_1) = \sum_{k=0}^{\infty} \frac{E(\xi^k)(it_1)^k}{k!} \quad \text{and} \quad \varphi_\eta(t_2) = \sum_{l=0}^{\infty} \frac{E(\eta^l)(it_2)^l}{l!}$$

for every complex value of t_1 and t_2. If we put $\zeta = (\xi, \eta)$ and $t = (t_1, t_2)$, it follows from (17) that

$$\varphi_\zeta(t) = \sum_{k=0}^{\infty} \sum_{l=0}^{\infty} \frac{E(\xi^k \eta^l)(it_1)^k (it_2)^l}{k! \, l!} = \varphi_\xi(t_1) \varphi_\eta(t_2).$$

Hence the theorem is proved.

Theorem 3 of § 4 may also be extended to the case of higher dimensions:

THEOREM 7. *Let ξ_N $(N = 1, 2, \ldots)$ be a sequence of n-dimensional random vectors; let $F_N(x)$ $(x = (x_1, \ldots, x_n))$ denote the distribution function of ξ_N and $\varphi_N(t)$ $(t = (t_1, \ldots, t_n))$ its characteristic function. If at every point of continuity x of $F(x)$ the relation*

$$\lim_{N \to +\infty} F_N(x) = F(x) \tag{18}$$

is valid and if $F(x)$ is a distribution function as well, then

$$\lim_{N \to \infty} \varphi_N(t) = \varphi(t), \tag{19}$$

where $\varphi(t)$ is the characteristic function of $F(x)$. The convergence is uniform on every n-dimensional interval.

Conversely, if (19) holds for every system of values $(t_1, \ldots, t_n) = t$ and if $\varphi(t)$ is continuous for $t = 0$, then (18) holds as well, the function $F(x)$ figuring in it is a distribution function and φ is its characteristic function;

*furthermore, for any $A > 0$ the convergence in (19) is uniform for $|t| \leq A$,
where $|t| = \sqrt{t_1^2 + \ldots + t_n^2}$.*

The proof of this theorem is omitted here, as it is essentially the same
as that of Theorem 3 of § 4.

As an application of Theorem 7 we show that the multinomial distribution tends to the normal distribution. Suppose that the random vector
$\xi_N = (\xi_{N1}, \ldots, \xi_{Nn})$ $(N = 1, 2, \ldots)$ has a multinomial distribution

$$P(\xi_{N1} = k_1, \ldots, \xi_{Nn} = k_n) = \frac{N!}{k_1! \ldots k_n!} p_1^{k_1} \ldots p_n^{k_n}, \tag{20}$$

where $k_1, \ldots, k_n$ are integers and

$$\sum_{j=1}^{n} k_j = N, \quad p_j \geq 0, \quad \sum_{j=1}^{n} p_j = 1.$$

Let η_N denote the random vector $(\eta_{N1}, \ldots, \eta_{Nn})$ with

$$\eta_{Nj} = \frac{\xi_{Nj} - Np_j}{\sqrt{Np_j}}. \tag{21}$$

We obtain for the characteristic function of η_N

$$\varphi_{\eta_N}(t) = \exp\left[-i\sqrt{N}\left(\sum_{j=1}^{n} t_j\sqrt{p_j}\right)\right] \cdot \left[\sum_{j=1}^{n} p_j \exp\left(\frac{it_j}{\sqrt{Np}}\right)\right]^N,$$

hence

$$\ln \varphi_{\eta_N}(t) = -i\sqrt{N}\left(\sum_{j=1}^{n} t_j\sqrt{p_j}\right) +$$

$$+ N\ln\left[1 + \sum_{j=1}^{n} p_j\left(\exp\left(\frac{it_j}{\sqrt{Np_j}}\right) - 1\right)\right]. \tag{22}$$

By substituting the expansions of e^x and $\ln(1 + x)$ we have

$$\lim_{N \to \infty} \ln \varphi_{\eta_N}(t) = -\frac{1}{2}\left[\sum_{j=1}^{n} t_j^2 - \left(\sum_{j=1}^{n} t_j\sqrt{p_j}\right)^2\right]. \tag{23}$$

The limit distribution has the characteristic function

$$\varphi(t) = \exp\left[-\frac{1}{2}\left(\sum_{j=1}^{n} t_j^2 - \left(\sum_{j=1}^{n} t_j\sqrt{p_j}\right)^2\right)\right]. \tag{24}$$

This is the characteristic function of a degenerate normal distribution;

in effect the random variables η_{Nk} are connected by the linear relation

$$\sum_{k=1}^{n} \eta_{Nk} \sqrt{p_k} = 0;$$

the limit distribution too is thus concentrated on the hyperplane defined by the equation

$$\sum_{k=1}^{n} x_k \sqrt{p_k} = 0.$$

§ 7. Infinitely divisible distributions

In this section we shall deal with certain types of distributions which can be described most conveniently by means of their characteristic functions.

The probability distribution of a random variable ξ is said to be *infinitely divisible* if for every $n = 2, 3, \ldots$ there exists a probability distribution the n-fold convolution of which is equal to the distribution of ξ. Or to put it otherwise: the distribution of ξ is infinitely divisible if, for every n, ξ can be represented in the form $\xi = \xi_1 + \xi_2 + \ldots + \xi_n$, where $\xi_1, \ldots, \xi_n$ are mutually independent random variables with the same distribution.[1]

Let $\varphi(t)$ be the characteristic function of ξ. Obviously, to say that the distribution of ξ is infinitely divisible means the same as to say that for every integer n the function $[\varphi(t)]^{\frac{1}{n}}$ is again a characteristic function.

The infinitely divisible distributions can be characterized by the following property:

THEOREM 1. *The function $\varphi(t)$ is the characteristic function of an infinitely divisible distribution iff $\ln \varphi(t)$ is of the form*

$$\ln \varphi(t) = i\gamma t - \frac{\sigma^2 t^2}{2} + \int_{-\infty}^{+\infty} \left(e^{iut} - 1 - \frac{iut}{1+u^2} \right) \frac{1+u^2}{u^2} \, dG(u), \qquad (1)$$

where γ and $\sigma > 0$ are real constants and $G(u)$ is a nondecreasing bounded function. (Formula of Lévy and Khinchin.)

If the distribution has a finite variance, (1) may be written in the form

$$\ln \varphi(t) = imt + \sigma^2 \int_{-\infty}^{+\infty} (e^{iut} - 1 - iut) \frac{dK(u)}{u^2}, \qquad (2)$$

---

[1] This second definition is not completely exact, since it is not certain that the ξ_k-s can in fact be realized on the probability space in question. However, if the words: "for a suitable choice of the probability space" are added, then the second formulation becomes correct and equivalent to the first one.

where m is a real number, σ > 0, and K(u) is a distribution function. (Formula of Kolmogorov.)

In particular, if

$$K(u) = \begin{cases} 0 & \text{for } u \le 0, \\ 1 & \text{for } u > 0, \end{cases}$$

then

$$\ln \varphi(t) = imt - \frac{\sigma^2 t^2}{2};$$

hence in this case the distribution in question is a normal one. It can be seen from (1) as well as from (2) that if $\varphi(t)$ is the characteristic function of an infinitely divisible distribution, not only $[\varphi(t)]^{\frac{1}{n}}$ but also $[\varphi(t)]^{\alpha}$ is a characteristic function for every $\alpha > 0$; furthermore, (1) shows that $\varphi(t)$ differs from zero for real t.

If we put in (2) $m = \lambda h$, $\sigma^2 = \lambda h^2$ with $\lambda > 0$, and

$$K(u) = \begin{cases} 0 & \text{for } u \le h, \\ 1 & \text{for } u > h, \end{cases}$$

then

$$\varphi(t) = \exp\{\lambda(e^{ith} - 1)\}.$$

The distribution is thus in this case a (generalized) Poisson distribution in the following sense: A random variable so distributed takes on the values kh ($k = 0, 1, 2, \ldots$) with probabilities:

$$P(\xi = kh) = \frac{\lambda^k e^{-\lambda}}{k!} \qquad (k = 0, 1, \ldots).$$

It follows immediately from the definition that the convolution of infinitely divisible distributions is itself an infinitely divisible distribution. Thus a distribution which is the convolution of a normal distribution and of a finite number of generalized Poisson distributions (of the above type) is infinitely divisible.

It follows from (1) that every infinitely divisible distribution can be obtained as the limit of convolutions of a normal distribution and of a finite number of generalized Poisson distributions. It can be shown that the limit of a convergent sequence of infinitely divisible distributions is itself an infinitely divisible distribution.

The proof of Theorem 1, however, will not be dealt with here.[1]

[1] Cf. e.g. P. Lévy [4] or B. V. Gnedenko and A. N. Kolmogorov [1].

§ 8. Stable distributions

A distribution function $F(x)$ is said to be *stable* if for any two given real numbers m_1, m_2 and for any two positive numbers σ_1, σ_2, there exist a real number m and a positive number σ such that

$$F\left(\frac{x - m_1}{\sigma_1}\right) * F\left(\frac{x - m_2}{\sigma_2}\right) = F\left(\frac{x - m}{\sigma}\right). \tag{1}$$

Here the sign $*$ denotes the operation of convolution. As we have seen the normal distribution is a stable distribution, in fact the function

$$F(x) = \frac{1}{\sqrt{2\pi}} \int_{-\infty}^{x} e^{-\frac{u^2}{2}} \, du,$$

satisfies (1) with

$$m = m_1 + m_2 \quad \text{and} \quad \sigma = \sqrt{\sigma_1^2 + \sigma_2^2}. \tag{2}$$

Theorem 2′ of § 5 implies that the only stable distribution of finite variance is the normal distribution. There exist, however, stable distributions of infinite variance; thus for instance the Cauchy distribution with the distribution function

$$F(x) = \frac{1}{2} + \frac{1}{\pi} \arctan x$$

fulfils (1) with $m = m_1 + m_2$ and $\sigma = \sigma_1 + \sigma_2$.

We state now without proof the following theorem:

THEOREM 1. *A distribution with the characteristic function $\varphi(t)$ is stable iff $\varphi(t)$ can be written in the form*

$$\ln \varphi(t) = i\gamma t - c \, |t|^\alpha \left[1 + i\beta \, \frac{t}{|t|} \, \omega(t, \alpha)\right], \tag{3}$$

where the constants α, β, c fulfil the inequalities

$$-1 \leq \beta \leq 1, \quad 0 < \alpha \leq 2, \quad c \geq 0, \tag{4}$$

γ is a real number and

$$\omega(t, \alpha) = \begin{cases} \tan \dfrac{\pi\alpha}{2} & \text{for } \alpha \neq 1, \\[2mm] \dfrac{2}{\pi} \ln |t| & \text{for } \alpha = 1. \end{cases} \tag{5}$$

The number α is called the *characteristic exponent* of the stable distribution defined by (3). For the normal distribution $\alpha = 2$, $c > 0$; in the case of $\alpha = 1$, $\beta = 0$ we obtain the Cauchy distribution.

It can be proved without any difficulty that for a stable distribution with characteristic exponent $\alpha < 2$, the moments of order $\delta < \alpha$ exist but those of order $\delta \geq \alpha$ do not exist.

It follows from Theorem 1 that every stable distribution is infinitely divisible. In fact, if $\varphi(t)$ fulfils (3), the same holds for $[\varphi(t)]^{\frac{1}{n}}$ with the same α and β and with $\dfrac{\gamma}{n}$ or $\dfrac{c}{n}$ instead of γ or c, respectively. This can be seen directly as well, since if the distribution with the characteristic function $\varphi(t)$ is stable, we have

$$\sqrt[n]{\varphi(t)} = \varphi(q_n t)\, e^{i\gamma_n t}$$

with $q_n > 0$; hence $[\varphi(t)]^{\frac{1}{n}}$ $(n = 1, 2, \ldots)$ is again a characteristic function.

For a detailed study of stable distributions we refer to the books cited in the footnote of the preceding paragraph. Lévy calls only those nondegenerate distribution functions $F(x)$ stable, for which to any two positive numbers c_1 and c_2 there exists a positive number c such that $F(c_1 x) * F(c_2 x) = F(cx)$. Distributions, which we called stable above, are called *quasi-stable* by Lévy. It can be shown that a distribution with characteristic function $\varphi(t)$ is stable in the sense of P. Lévy, iff $\ln \varphi(t)$ may be written in the form (3) with $\gamma = 0$ for $\alpha \neq 1$ and $\beta = 0$ for $\alpha = 1$. Thus the following result is valid:

THEOREM 2. *If a distribution with the characteristic function $\varphi(t)$ is stable in the sense of P. Lévy, $\ln \varphi(t)$ can be written in the form*

$$\ln \varphi(t) = -\left(c_0 + ic_1 \frac{t}{|t|}\right)|t|^\alpha \quad \text{with } 0 < \alpha \leq 2, \tag{6a}$$

where c_0 is positive.

It can be shown further that the inequality

$$\frac{|c_1|}{c_0} \leq \left|\tan \frac{\pi\alpha}{2}\right| \tag{6b}$$

holds; however, this will not be proved here.

PROOF OF THEOREM 2. If $\varphi(t)$ is the characteristic function of a distribution which is stable in the sense of Lévy, there exists for every pair

$c_1 > 0$, $c_2 > 0$ a number $c > 0$ with

$$\varphi(c_1 t)\, \varphi(c_2 t) = \varphi(ct). \tag{7}$$

In particular, if $c_1 = c_2 = 1$, we have

$$\varphi^2(t) = \varphi(qt). \tag{8}$$

It may be supposed $q \neq 1$, since $q = 1$ would imply, because of $\varphi(0) = 1$ and the continuity of $\varphi(t)$, the relation $\varphi(t) \equiv 1$ and $\varphi(t)$ would then be the characteristic function of the constant zero. Furthermore (8) implies $\varphi(t) \neq 0$ for every real t: since if $\varphi(t_0) = 0$ could hold, (8) would lead to

$$\varphi\left(\frac{t_0}{q^n}\right) = 0 \qquad (n = 1, 2, \ldots) \tag{9a}$$

and

$$\varphi(q^n t_0) = 0 \qquad (n = 1, 2, \ldots) \tag{9b}$$

which is impossible both for $q > 1$ (9a) and for $q < 1$ (9b) in view of $\varphi(0) = 1$ and the fact that $\varphi(t)$ is continuous at $t = 0$.

Thus $\psi(t) = \ln \varphi(t)$ is continuous too, and $\psi(0) = 0$. Let $c_1, c_2, \ldots, c_n$ be any n positive numbers; according to (7) there exists a $c > 0$ with

$$\sum_{k=1}^{n} \psi(c_k t) = \psi(ct).$$

Hence, for $c_1 = c_2 = \ldots = c_n = 1$ there exists a $c(n)$ with

$$n\psi(t) = \psi(c(n)\, t). \tag{10}$$

Thus

$$n\psi(t) = \psi(c(n)\, t) = m\psi\left(\frac{c(n)}{c(m)}\; t\right).$$

If we put $c\left(\dfrac{n}{m}\right) = \dfrac{c(n)}{c(m)}$, we obtain

$$\frac{n}{m}\, \psi(t) = \psi\left(c\left(\frac{n}{m}\right) t\right). \tag{11}$$

Consequently, there corresponds to every rational r a number $c(r)$ such that

$$r\psi(t) = \psi(c(r)\, t). \tag{12}$$

We show now that $c(r)$ is uniquely determined and the relation

$$c(rs) = c(r)\, c(s) \tag{13}$$

holds for any two positive rational numbers r and s.

From $\psi(at) \equiv \psi(bt)$ and $a < b$ follows $\psi(t) = \psi\left(\left(\dfrac{a}{b}\right)^n t\right)$, and because of the continuity of $\psi(t)$, we have $\psi(t) = \psi(0) = 0$, which is impossible since the distribution was supposed to be nondegenerate. Hence necessarily $a = b$ and $c(r)$ is thus unique. Since further

$$rs\psi(t) = \psi\big(c(rs)\,t\big) = \psi\big(c(r)\,c(s)\,t\big)$$

for every t, (13) holds for any two positive rational numbers r, s. Let now q be a rational number $q > 1$ and t a real number such that $\psi(t) \neq 0$. Then

$$\psi\big([c(q)]^n\, t\big) = q^n\, \psi(t). \tag{14}$$

We have necessarily $c(q) > 1$, since $c(q) \leq 1$ would imply

$$\max_{|u| \leq |t|} |\psi(u)| \geq q^n |\psi(t)|$$

for every n, which contradicts the continuity of $\psi(t)$ on every finite interval. Since $\dfrac{c(r)}{c(s)} = c\left(\dfrac{r}{s}\right)$, $r > s$ implies $c(r) > c(s)$. The function $c(r)$ is thus increasing. For every irrational $\lambda > 0$ we define $c(\lambda)$ by

$$c(\lambda) = \lim_{r \to \lambda} c(r),$$

where r tends to λ through rational values. Because of the continuity of $\psi(t)$ it follows from (12) that the equalities

$$\lambda\psi(t) = \psi\big(c(\lambda)\,t\big) \tag{15}$$

and

$$c(\lambda\mu) = c(\lambda)\,c(\mu) \tag{16}$$

are valid for every positive value of λ and μ and that $c(\lambda)$, as a function of the real variable $\lambda > 0$ is increasing. Put

$$g(x) = \ln c(e^x) \qquad (-\infty < x < +\infty), \tag{17}$$

then it follows that $g(x)$ is increasing and

$$g(x + y) = g(x) + g(y) \tag{18}$$

is valid. Hence (cf. Ch. III, § 13)

$$g(x) = \frac{x}{\alpha} \quad \text{for} \quad \alpha > 0 \tag{19}$$

and

$$c(x) = x^{\frac{1}{\alpha}}, \quad \text{for} \quad x > 0. \tag{20}$$

This leads to

$$\lambda^\alpha \psi(t) = \psi(\lambda t) \tag{21}$$

for every real t and positive λ. Thus for $t > 0$ we have

$$\psi(\lambda t) = \lambda^\alpha \psi(t) = t^\alpha \psi(\lambda), \tag{22}$$

and therefore for $\lambda = 1$:

$$\psi(t) = |t|^\alpha \psi(1), \quad \text{for} \quad t > 0. \tag{23}$$

Put $\psi(1) = -(c_0 + ic_1)$. Since $\varphi(-t) = \overline{\varphi(t)}$, and thus also $\psi(-t) = \overline{\psi(t)}$, we obtain for every real t

$$\psi(t) = - \left(c_0 + i \frac{t}{|t|} c_1 \right) |t|^\alpha. \tag{24}$$

Because of $|\varphi(t)| \le 1$, $c_0 > 0$ holds. It remains to show that $0 < \alpha \le 2$. But $\alpha > 2$ would imply $\varphi''(0) = 0$ and hence $D^2(\xi) = 0$, and thus ξ would be a constant. Herewith (6a) is proved.

§ 9. Characteristic functions of conditional probability distributions

In the present paragraph we need a generalization of the concept of "function" due to L. Schwartz. In order to construct the theory of *generalized functions* (called *distributions* by L. Schwartz) we follow the way suggested by J. Mikusinski and worked out by G. Temple and M. J. Lighthill.[1]

Let C denote the set of infinitely often differentiable complex-valued functions, for which

$$f^{(k)}(x) = O\left(\frac{1}{|x|^N} \right) \quad \text{for} \quad |x| \to +\infty \tag{1}$$

for any nonnegative integers k and N. If $f(x) \in C$ and if a, b are real, and λ is a complex number, then $\lambda f(x) \in C$, $f(ax + b) \in C$, and $f^{(k)}(x) \in C$ $(k = 1, 2, \ldots)$; furthermore $f(x) \in C$, $g(x) \in C$ imply $f(x) + g(x) \in C$. Let K denote the set of all infinitely often differentiable functions $g(x)$ such that

$$g^{(k)}(x) = O(|x|^{N_k}) \quad \text{for} \quad |x| \to \infty \text{ and } k = 1, 2, \ldots,$$

[1] Cf. M. J. Lighthill [1].

where the numbers N_k are integers. It is easy to see that $f(x) \in C$ and $g(x) \in K$ imply $f(x) \cdot g(x) \in C$.

A sequence of functions $\{f_n(x)\}$ $(f_n(x) \in C; n = 1, 2, \ldots)$ is said to be *regular*, if for every $h(x) \in C$ the limit

$$\lim_{n \to +\infty} \int_{-\infty}^{+\infty} f_n(x) h(x) \, dx \tag{2}$$

exists. Two regular sequences $\{f_n(x)\}$ and $\{g_n(x)\}$ are said to be *equivalent* if *for every* $h(x) \in C$

$$\lim_{n \to +\infty} \int_{-\infty}^{+\infty} f_n(x) h(x) \, dx = \lim_{n \to +\infty} \int_{-\infty}^{+\infty} g_n(x) h(x) \, dx. \tag{3}$$

This equivalence relation defines a partition of the set of regular sequences of functions into classes. A *generalized function* is an equivalence class of regular sequences of functions.

In the present paragraph generalized functions will be denoted by capitals $(F(x), G(x), \ldots)$ and the ordinary functions by lower case letters $(f(x), g(x), \ldots)$. If the regular sequence of functions $\{f_n(x)\}$ defines a generalized function $F(x)$ we express this fact by a sign $\sim$: $F(x) \sim \{f_n(x)\}$. If $F(x) \sim \{f_n(x)\}$ and if $h(x) \in C$, we define the "integral"

$$\int_{-\infty}^{+\infty} F(x) h(x) \, dx$$

by

$$\int_{-\infty}^{+\infty} F(x) h(x) \, dx = \lim_{n \to +\infty} \int_{-\infty}^{+\infty} f_n(x) h(x) \, dx, \tag{4}$$

where on the right hand side the limit exists by assumption and remains the same when $\{f_n(x)\}$ is replaced by another sequence equivalent to it.

If $F(x) \sim \{f_n(x)\}$ is regular, the sequence $\{\lambda f_n(ax + b)\}$ is evidently regular for any two real numbers a, b and for any complex number λ. We put therefore $\lambda F(ax + b) \sim \{\lambda f_n(ax + b)\}$. If $F(x) \sim \{f_n(x)\}$ and $G(x) \sim \{g_n(x)\}$, the sequence $\{f_n(x) + g_n(x)\}$ is again regular; we put $F(x) + G(x) \sim \{f_n(x) + g_n(x)\}$. Finally if $\{f_n(x)\}$ is regular and if $g(x) \in K$, the sequence $\{f_n(x) g(x)\}$ is regular. For $F(x) \sim \{f_n(x)\}$ we put $F(x)g(x) \sim \{f_n(x) \cdot g(x)\}$.

If $\{f_n(x)\}$ is regular, $\{f_n'(x)\}$ is also regular because if $h(x) \in C$ we have

$$\int_{-\infty}^{+\infty} f_n'(x) h(x) \, dx = -\int_{-\infty}^{+\infty} f_n(x) h'(x) \, dx, \tag{5}$$

from which the existence of the limit

$$\lim_{n\to\infty} \int_{-\infty}^{+\infty} f_n'(x)\, h(x)\, dx = - \int_{-\infty}^{+\infty} F(x)\, h'(x)\, dx$$

follows. The derivative of a generalized function $F(x) \sim \{f_n(x)\}$ is defined by

$$F'(x) \sim \{f_n'(x)\}.$$

Thus we have

$$\int_{-\infty}^{+\infty} F'(x)\, h(x)\, dx = - \int_{-\infty}^{+\infty} F(x)\, h'(x)\, dx. \qquad (6)$$

A generalized function is thus infinitely often differentiable. It is easy to prove the following rules of calculation:

$$\lambda(F(x) + G(x)) = \lambda F(x) + \lambda G(x),$$

$$(cF(x))' = cF'(x),$$

$$(F(x) + G(x))' = F'(x) + G'(x),$$

$$(F(ax))' = aF'(ax).$$

Example. Let us put

$$f_n(x) = \sqrt{\frac{n}{2\pi}}\; e^{-\frac{nx^2}{2}}.$$

If $h(x) \in C$, it is clear that

$$\lim_{n\to\infty} \int_{-\infty}^{+\infty} f_n(x)\, h(x)\, dx = h(0).$$

The generalized function $\{f_n(x)\}$, which we denote by $\delta(x)$, is called *Dirac's delta function.* For every $h(x) \in C$ we have thus

$$\int_{-\infty}^{+\infty} \delta(x)\, h(x)\, dx = h(0).$$

We prove now some theorems.

THEOREM 1. *If* $f(x) \in C$ *and if*

$$\varphi(t) = \int_{-\infty}^{+\infty} f(x)\, e^{itx}\, dx \qquad (-\infty < t < +\infty), \qquad (7)$$

$\varphi(t)$ *belongs then to* C *as well.*

PROOF. $f(x) = O(|x|^{-k})$ for every integer k, hence the integral (7) exists for every real number t; further

$$\int_{-\infty}^{+\infty} |f(x)| \cdot |x|^k \, dx$$

exists too. The function $\varphi(t)$ is thus infinitely often differentiable and

$$\varphi^{(k)}(t) = \int_{-\infty}^{+\infty} (ix)^k f(x) e^{itx} \, dx \qquad (k = 1, 2, \ldots). \qquad (8)$$

If we integrate (7) N times by parts we obtain

$$\varphi(t) = \left(\frac{i}{t}\right)^N \int_{-\infty}^{+\infty} f^{(N)}(x) e^{itx} \, dx \qquad (N = 1, 2, \ldots). \qquad (9)$$

The integral on the right hand side exists, since $f(x) \in C$, hence $\varphi(t) =$ $= O(|t|^{-N})$ for $|t| \to +\infty$ and for every integer N. By (8) the same holds for $\varphi^{(k)}(t)$ since $(ix)^k f(x) \in C$, hence Theorem 1 is proved.

The function $\varphi(t)$ is the *Fourier transform* of $f(x)$.

THEOREM 2. *If* $f(x) \in C$ *and if*

$$\varphi(t) = \int_{-\infty}^{+\infty} f(x) e^{itx} \, dx,$$

we have the following inversion formula:

$$f(x) = \frac{1}{2\pi} \int_{-\infty}^{+\infty} \varphi(t) e^{-itx} \, dt. \qquad (10)$$

PROOF. The preceding theorem guarantees the existence of $\int_{-\infty}^{+\infty} |\varphi(t)| dt$ and (10) follows from Theorem 2 of § 4.

THEOREM 3 (Parseval's theorem). *If* $f(x) \in C, g(x) \in C$, *further if*

$$\varphi(t) = \int_{-\infty}^{+\infty} f(x) e^{itx} \, dx,$$

$$\gamma(t) = \int_{-\infty}^{\infty} g(x) e^{itx} \, dx,$$

then we have

$$\int\limits_{-\infty}^{+\infty} f(x)\,g(x)\,dx = \frac{1}{2\pi} \int\limits_{-\infty}^{+\infty} \varphi(-t)\,\gamma(t)\,dt. \tag{11}$$

PROOF. By Theorem 2

$$g(x) = \frac{1}{2\pi} \int\limits_{-\infty}^{+\infty} \gamma(t)\,e^{-itx}\,dt.$$

Hence

$$\int\limits_{-\infty}^{+\infty} f(x)\,g(x)\,dx = \frac{1}{2\pi} \int\limits_{-\infty}^{+\infty} \int\limits_{-\infty}^{+\infty} f(x)\,\gamma(x)\,e^{-itx}\,dt\,dx =$$

$$= \frac{1}{2\pi} \int\limits_{-\infty}^{+\infty} \gamma(t)\,\varphi(-t)\,dt$$

since the order of integration can be interchanged.

THEOREM 4. *If* $\{f_n(x)\} \sim F(x)$ *is a regular sequence of functions and if* $\varphi_n(t)$ *is the Fourier transform of* $f_n(x)$, *then* $\{\varphi_n(t)\} \sim \Phi(t)$ *is again a regular sequence of functions; the generalized function* $\Phi(t)$ *remains invariant when* $\{f_n(x)\}$ *is replaced by an equivalent sequence. For this relation between the two generalized functions* $F(x)$ *and* $\Phi(t)$ *we can write*

$$\Phi(t) = \int\limits_{-\infty}^{+\infty} F(x)\,e^{itx}\,dx. \tag{12}$$

If $h(x)$ *is a function of the class* C *and* $\chi(t)$ *is its Fourier transform, we have*

$$\int\limits_{-\infty}^{+\infty} \Phi(t)\chi(t)\,dt = 2\pi \int\limits_{-\infty}^{+\infty} F(x)\,h(-x)\,dx. \tag{13}$$

We say that the generalized function $\Phi(t)$ is the *Fourier transform* of $F(x)$.

PROOF. Let $\chi(t) \in C$ and

$$h(x) = \frac{1}{2\pi} \int\limits_{-\infty}^{+\infty} \chi(t)\,e^{-itx}\,dt.$$

Then, according to Theorem 3

$$\int_{-\infty}^{+\infty} \varphi_n(t)\chi(t)\,dt = 2\pi \int_{-\infty}^{+\infty} f_n(x)h(-x)\,dx.$$

Thus if $F(x) \sim \{f_n(x)\}$ we have

$$\lim_{n\to\infty} \int_{-\infty}^{+\infty} \varphi_n(t)\chi(t)\,dt = 2\pi \int_{-\infty}^{+\infty} F(x)h(-x)\,dx, \qquad (14)$$

hence $\{\varphi_n(t)\}$ is indeed regular.

It can be seen from (14) that if $\{f_n(x)\}$ is replaced by an equivalent sequence, $\{\varphi_n(t)\}$ will be replaced by an equivalent sequence as well. (13) follows immediately from (14).

Let D denote the class of the ordinary measurable functions $f(x)$ such that $|f(x)| < A(1 + |x|^N)$, $-\infty < x < +\infty$ for any integer N and $A > 0$. From $f(x) \in D$, $h(x) \in C$ follows immediately the existence of the integral

$$\int_{-\infty}^{+\infty} f(x)h(x)\,dx.$$

THEOREM 5. *If $f(x) \in D$, there exists a generalized function $F(x)$ such that for every $h(x) \in C$*

$$\int_{-\infty}^{+\infty} f(x)h(x)\,dx = \int_{-\infty}^{+\infty} F(x)h(x)\,dx. \qquad (15)$$

PROOF. Put

$$f_n(x) = \sqrt{\frac{n}{2\pi}} \int_{-\infty}^{+\infty} f(y)\,e^{\frac{-y^2}{n}}\,e^{\frac{-n(x-y)^2}{2}}\,dy.$$

A simple calculation shows that $f_n(x) \in C$; furthermore from the fact that a Lebesgue integrable function is almost everywhere equal to the derivative of its indefinite integral we find that $\lim_{n\to\infty} f_n(x) = f(x)$ for almost every x. According to the convergence theorem of Lebesgue,

$$\lim_{n\to\infty} \int_{-\infty}^{+\infty} f_n(x)h(x)\,dx = \int_{-\infty}^{+\infty} f(x)h(x)\,dx,$$

q.e.d.

Let $f(x) \in D$ and let $F(x) \sim \{f_n(x)\}$ be the generalized function corresponding to it (in a unique manner) according to Theorem 5. Write $f(x) \sim F(x)$.

Let now ξ be a random variable on a conditional probability space with density function $f(x)$ and assume that $f(x) \in D$. The characteristic function $\Phi_\xi(t)$ of ξ is defined by

$$\Phi_\xi(t) = \int_{-\infty}^{+\infty} F(x) e^{itx} \, dx \tag{16}$$

with $F(x) \sim f(x)$ in the sense of (12); $\Phi_\xi(t)$ is thus a generalized function. If

$$\int_{-\infty}^{+\infty} f(x) \, dx = 1$$

i.e. if $f(x)$ is an ordinary density function, and if we put as usually

$$\varphi_\xi(t) = \int_{-\infty}^{+\infty} f(x) e^{ixt} \, dx,$$

then $\varphi_\xi(t) \sim \Phi_\xi(t)$; the definition of $\Phi_\xi(t)$ is thus consistent with the definition of ordinary characteristic functions $\varphi_\xi(t)$. In fact, in this case $\varphi_\xi(t)$ is continuous and $|\varphi_\xi(t)| \leq 1$, hence $\varphi_\xi(t) \in D$. It suffices thus to prove that for every $\chi(t) \in C$ the relation

$$\int_{-\infty}^{+\infty} \varphi_\xi(t) \chi(t) \, dt = \int_{-\infty}^{+\infty} \Phi_\xi(t) \chi(t) \, dt \tag{17}$$

holds, where the integral on the left is an ordinary integral. If

$$h(x) = \frac{1}{2\pi} \int_{-\infty}^{+\infty} \chi(t) e^{-itx} \, dt,$$

we obtain, by proceeding as in the proof of Theorem 3,

$$\int_{-\infty}^{+\infty} \varphi_\xi(t) \chi(t) \, dt = 2\pi \int_{-\infty}^{+\infty} h(-x) f(x) \, dx. \tag{18}$$

Furthermore, by Theorem 1,

$$\int_{-\infty}^{+\infty} \Phi_\xi(t) \chi(t) \, dt = 2\pi \int_{-\infty}^{+\infty} F(x) h(-x) \, dx. \tag{19}$$

By the definition of $F(x)$, the right hand sides of (18) and (19) coincide; we get thus (17).

If $f(x)$ is the density function of a conditional distribution, $f(x)$ and consequently $\Phi_\xi(t)$ are only determined up to a constant factor.

Example 1. If ξ is uniformly distributed on $(-\infty, +\infty)$ then $f(x) \equiv 1$. If $h(x) \in C$, we can write

$$\int_{-\infty}^{+\infty} f(x)\, h(x)\, dx = \int_{-\infty}^{+\infty} h(x)\, dx = \lim_{n \to \infty} \int_{-\infty}^{+\infty} h(x)\, e^{-\frac{x^2}{2n}}\, dx; \qquad (20)$$

hence for the generalized function $F(x)$ corresponding to $f(x)$, the relation $F(x) \sim \left\{\exp\left(-\dfrac{x^2}{2n}\right)\right\}$ is valid. Since

$$\int_{-\infty}^{+\infty} e^{-\frac{x^2}{2n}} e^{itx}\, dx = \sqrt{2\pi n}\; e^{-\frac{nt^2}{2}}, \qquad (21)$$

we obtain

$$\Phi_\xi(t) = \int_{-\infty}^{+\infty} F(x)\, e^{ixt}\, dx \sim 2\pi \left\{\sqrt{\frac{n}{2\pi}}\, e^{-\frac{nt^2}{2}}\right\} \sim 2\pi\delta(t). \qquad (22)$$

The characteristic function of ξ is thus the Dirac delta function.

Example 2. Suppose $f_k(x) = e^{ikx}$. Find the Fourier transform of $f_k(x)$. If $h(x) \in C$ we have

$$\int_{-\infty}^{+\infty} f_k(x)\, h(x)\, dx = \lim_{n \to \infty} \int_{-\infty}^{+\infty} h(x)\, e^{ikx - \frac{x^2}{2n}}\, dx,$$

hence $f_k(x) \sim \left\{\exp\left(ikx - \dfrac{x^2}{2n}\right)\right\}$. And since

$$\int_{-\infty}^{+\infty} e^{ikx - \frac{x^2}{2n}} e^{ixt}\, dx = \sqrt{2\pi n}\; e^{-\frac{n(k+t)^2}{2}},$$

we find for the Fourier transform $\Phi_k(t)$ of $f_k(t)$

$$\Phi_k(t) \sim 2\pi \left\{\sqrt{\frac{n}{2\pi}}\, e^{-\frac{n(k+t)^2}{2}}\right\} = 2\pi\delta(k+t).$$

We introduce now the concept of *convergence of generalized functions.* Let $F_k(x)$ $(k = 1, 2, \ldots)$ be a sequence of generalized functions. We say that $F_k(x)$ tends to the generalized function $F(x)$ (in signs $F_k(x) \to F(x)$), if for every $h(x) \in C$

$$\lim_{k \to \infty} \int_{-\infty}^{+\infty} F_k(x)\, h(x)\, dx = \int_{-\infty}^{+\infty} F(x)\, h(x)\, dx. \qquad (23)$$

THEOREM 6. *Let* $\{F_k(x)\}$ $(k = 1, 2, \ldots)$ *be a sequence of generalized functions and put*

$$\Phi_k(t) = \int\limits_{-\infty}^{+\infty} F_k(x)\, e^{itx}\, dx \qquad (k = 1, 2, \ldots). \tag{24}$$

Then $F_k(x) \rightharpoonup F(x)$ *iff* $\Phi_k(t) \rightharpoonup \Phi(t)$. *Thus*

$$\Phi(t) = \int\limits_{-\infty}^{+\infty} F(x)\, e^{itx}\, dx.$$

The proof follows immediately from Theorem 3.

Theorem 6 permits the use of characteristic functions for the establishment of the convergence of distribution functions also in case of conditional distributions. In this case the characteristic functions are usually not ordinary, but generalized functions. As an example we prove a limit distribution theorem.

THEOREM 7. *Let* $\xi_1, \xi_2, \ldots, \xi_n, \ldots$ *be independent random variables with the same distribution. Assume further that their distribution function is absolutely continuous with finite variance and bounded density function* $f(x)$. *Suppose* $E(\xi_k) = 0$. *Put*

$$\zeta_n = \xi_1 + \xi_2 + \ldots + \xi_n. \tag{25}$$

Then the distribution of ζ_n *tends to the conditional distribution uniform on the whole real axis, that is for any four real numbers* $c \leq a < b \leq d$ *we have*

$$\lim_{n \to \infty} P(a \leq \zeta_n < b \mid c \leq \zeta_n < d) = \frac{b-a}{d-c}. \tag{26}$$

PROOF. Let $f_n(x)$ denote the density function of ζ_n and σ the standard deviation of the ξ_k. $f(x) \leq M$ implies $f_n(x) \in D$. We show that for every $h(x) \in C$

$$\lim_{n \to \infty} \sigma \sqrt{\frac{n}{2\pi}} \int\limits_{-\infty}^{+\infty} f_n(x)\, h(x)\, dx = \frac{1}{2\pi} \int\limits_{-\infty}^{+\infty} h(x)\, dx. \tag{27}$$

Relation (27) proves the theorem. Indeed if it holds for every $h(x) \in C$, let then be $h(a, b, \varepsilon, x)$ a function of the class C such that

$$0 \leq h(a, b, \varepsilon, x) \leq 1 \tag{28}$$

and

$$h(a, b, \varepsilon, x) = \begin{cases} 0 & \text{for } x \leq a - \varepsilon, \\ 1 & \text{for } a \leq x \leq b, \\ 0 & \text{for } b + \varepsilon \leq x.^{1} \end{cases} \tag{29}$$

We have then

$$\frac{\int_{-\infty}^{+\infty} f_n(x) h(a + \varepsilon, b - \varepsilon, \varepsilon, x) dx}{\int_{-\infty}^{+\infty} f_n(x) h(c, d, \varepsilon, x) dx} \leq \frac{\int_a^b f_n(x) dx}{\int_c^d f_n(x) dx} \leq$$

$$\leq \frac{\int_{-\infty}^{+\infty} f_n(x) h(a, b, \varepsilon, x) dx}{\int_{-\infty}^{+\infty} f_n(x) h(c + \varepsilon, d - \varepsilon, \varepsilon, x) dx}. \tag{30}$$

Since

$$P(a \leq \zeta_n < b \mid c \leq \zeta_n < d) = \frac{\int_a^b f_n(x) dx}{\int_c^d f_n(x) dx}, \tag{31}$$

(27) and (30) enable us to write

$$\frac{\int_{-\infty}^{+\infty} h(a + \varepsilon, b - \varepsilon, \varepsilon, x) dx}{\int_{-\infty}^{+\infty} h(c, d, \varepsilon, x) dx} \leq l \leq L \leq \frac{\int_{-\infty}^{+\infty} h(a, b, \varepsilon, x) dx}{\int_{-\infty}^{+\infty} h(c + \varepsilon, d - \varepsilon, \varepsilon, x) dx}, \tag{32}$$

[1] These conditions are e.g. fulfilled by the function

$$h(a, b, \varepsilon, x) = k\left(\frac{x - a + \varepsilon}{\varepsilon}\right) - k\left(\frac{x - b}{\varepsilon}\right),$$

where we put

$$k(x) = \begin{cases} 0 & \text{for } x < 0, \\ \dfrac{\int_0^x \exp\left[\dfrac{1}{-t(1-t)}\right] dt}{\int_0^1 \exp\left[\dfrac{1}{-t(1-t)}\right] dt} & \text{for } 0 \leq x \leq 1, \\ 1 & \text{for } x > 1. \end{cases}$$

where

$$l = \liminf_{n \to \infty} P(a \le \zeta_n < b \,|\, c \le \zeta_n < d)$$

and

$$L = \limsup_{n \to \infty} P(a \le \zeta_n < b \,|\, c \le \zeta_n < d).$$

When $\varepsilon \to 0$, the first and the last member of the threefold inequality (32) tend to $\dfrac{b-a}{d-c}$, hence (27) implies (26).

Let now be $F_n(x) \sim f_n(x)$ and let $\varphi_n(t)$ be the Fourier transform of $\sigma \sqrt{2\pi n}\, f_n(x)$. By Theorem 6 it suffices to prove that $\Phi_n(t) \to \delta(t)$, where $\Phi_n(t)$ is the generalized function corresponding to $\varphi_n(t)$ $(n = 1, 2, \ldots)$ and $\delta(t)$ is Dirac's delta.

Put

$$\varphi(t) = \int\limits_{-\infty}^{+\infty} f(x)\, e^{itx}\, dx.$$

We see that $\varphi_n(t) = \sigma \sqrt{2\pi n}\, \varphi^n(t)$. We have to show that for every $\chi(t) \in C$ one has

$$\lim_{n \to \infty} \sigma \sqrt{\frac{n}{2\pi}} \int\limits_{-\infty}^{+\infty} \varphi^n(t)\, \chi(t)\, dt = \chi(0). \tag{33}$$

The proof can be carried out by means of the method of Laplace (cf. Ch. III, § 18, Exercise 27).

By Theorem 11 of § 2 we have $|\varphi(t)| < 1$ for $t \ne 0$; furthermore by Theorem 8 of § 2

$$\lim_{|t| \to +\infty} \varphi(t) = 0.$$

Hence there can be assigned to every $\varepsilon > 0$ a $q = q(\varepsilon)$ with $0 < q(\varepsilon) < 1$ such that $|\varphi(t)| < q(\varepsilon)$ for $|t| > \varepsilon$. But then we have

$$\sigma \sqrt{\frac{n}{2n}} \left| \int\limits_{|t| > \varepsilon} \varphi^n(t)\, \chi(t)\, dt \right| < \sigma \sqrt{n}\, [q(\varepsilon)]^n \int\limits_{-\infty}^{+\infty} |\chi(t)|\, dt. \tag{34}$$

On the other hand, for every $|t| \le \varepsilon$

$$\ln \varphi(t) = -\frac{\sigma^2 t^2}{2}\, (1 + \eta(t)) \quad \text{with} \quad \lim_{t \to \infty} \eta(t) = 0.$$

By introducing a new variable $u = t\sigma \sqrt{n}$, we obtain

$$\sigma \sqrt{\frac{n}{2n}} \int\limits_{-\varepsilon}^{+\varepsilon} \varphi^n(t) \chi(t)\, dt =$$

$$= \frac{1}{\sqrt{2\pi}} \int\limits_{-\varepsilon\sigma\sqrt{n}}^{+\varepsilon\sigma\sqrt{n}} \exp\left[-\frac{u^2}{2}\left(1 + \eta\left(\frac{u}{\sigma\sqrt{n}}\right)\right)\right] \chi\left(\frac{u}{\sigma\sqrt{n}}\right) du.$$

Since $\chi(t)$ is continuous for $t = 0$ and is bounded, it follows from Lebesgue's theorem that

$$\lim_{n\to\infty} \sigma \sqrt{\frac{n}{2\pi}} \int\limits_{-\varepsilon}^{+\varepsilon} \varphi^n(t) \chi(t)\, dt = \chi(0). \qquad (35)$$

(35) and (34) lead to (33).

Let us remark that the assumptions of Theorem 7 can be considerably weakened.

The product of two generalized functions is generally not defined. The way which would seem quite natural to follow leads astray: the regularity of $\{f_n(x)\}$ and $\{g_n(x)\}$ *does not*, in general, imply the regularity of $\{f_n(x)g_n(x)\}$. Just take as an example

$$f_n(x) = g_n(x) = \sqrt{\frac{n}{2\pi}}\, e^{-\frac{nx^2}{2}}$$

Here for every $h(x) \in C$ with $h(0) \neq 0$ we have

$$\lim_{n\to\infty} \frac{n}{2\pi} \left| \int\limits_{-\infty}^{+\infty} h(x)\, e^{-nx^2}\, dx \right| = +\infty.$$

Consequently, $\delta^2(x)$ has no sense at all.

So far we did not define the characteristic function of a random variable defined on a conditional probability space unless the random variable had a density function. Now we have to deal with the general case. Let $\mathscr{F}(x)$ be the distribution function of ξ. Suppose that there exists a generalized function such that

$$\int\limits_{-\infty}^{+\infty} F(x)\, h(x)\, dx = \int\limits_{-\infty}^{+\infty} h(x)\, d\mathscr{F}(x) \qquad (36)$$

for every $h(x) \in C$, where on the right hand side figures an ordinary Stieltjes integral. Then $\Phi(t)$, the Fourier transform of $F(x)$, will be considered as the *characteristic function* of the random variable ζ.

Example. Suppose that ζ is uniformly distributed on the set of the integers, i.e. the distribution function of ζ is given by $[x]$ ($[x]$ represents the integer part of x, i.e. the largest integer smaller than or equal to x). In this case

$$\int_{-\infty}^{+\infty} h(x)\,d\mathscr{F}(x) = \sum_{k=-\infty}^{+\infty} h(k); \tag{37}$$

there exists a generalized function fulfilling (36), namely

$$F(x) = \sum_{k=-\infty}^{+\infty} \delta(x - k). \tag{38}$$

it is easy to show that

$$\Phi(t) = 2\pi \sum_{k=-\infty}^{+\infty} \delta(t - 2k\pi). \tag{39}$$

If we apply (13) to any function $h(x) \in C$ and to an $F(x)$ defined by (38), we find

$$\sum_{k=-\infty}^{+\infty} h(k) = \sum_{k=-\infty}^{+\infty} \chi(2k\pi), \tag{40}$$

where $\chi(t)$ is the Fourier transform of $h(x)$. (40) is *Poisson's* well-known *summation formula*. In particular, if $h(x) = \exp(-x^2\lambda^2)$, then

$$\chi(t) = \frac{\sqrt{\pi}}{\lambda} \exp\left[-\frac{t^2}{4\lambda^2}\right];$$

and it follows from (40) that

$$\sum_{k=-\infty}^{+\infty} e^{-k^2\lambda^2} = \frac{\sqrt{\pi}}{\lambda} \sum_{k=-\infty}^{+\infty} e^{-\frac{k^2\pi^2}{\lambda^2}} \tag{41}$$

This is a formula known from the theory of θ-functions. We shall need it later on.

Now follows a theorem similar to Theorem 7.[1]

THEOREM 8. *Let $\zeta_1, \zeta_2, \ldots, \zeta_n$ be independent integer valued random variables having the same distribution. Suppose that their expectation is zero*

---

[1] It was proved by K. L. Chung and P. Erdős [1] under weaker conditions.

and their variance finite. Suppose further that the greatest common divisor of the values assumed by $\xi_1 - \xi_2$ with positive probabilities is equal to 1. Put $\zeta_n = \xi_1 + \xi_2 + \ldots + \xi_n$ ($n = 1, 2, \ldots$), then for any two integers k and l

$$\lim_{n \to \infty} \frac{P(\zeta_n = k)}{P(\zeta_n = l)} = +1. \tag{42}$$

Hence when $n \to \infty$, the distribution of ζ_n tends to the uniform distribution on the set of integers.

PROOF. Let $D(\xi_k) = \sigma$. If we show that

$$\lim_{n \to \infty} \sigma \sqrt{2\pi n}\, P(\zeta_n = k) = 1 \qquad (k = 1, 2, \ldots), \tag{43}$$

the theorem is proved. Let

$$\varphi(t) = \sum_{k=-\infty}^{+\infty} P(\xi_1 = k)\, e^{ikt} \tag{44}$$

be the characteristic function of the random variables ξ_n. We have

$$P(\zeta_n = k) = \frac{1}{2\pi} \int_{-\pi}^{+\pi} \varphi^n(t)\, e^{-ikt}\, dt. \tag{45}$$

Since by assumption $\varphi(0) = 1$, $\varphi'(0) = 0$, $\varphi''(0) = -\sigma^2$ and $|\varphi(t)| < 1$ for $0 < |t| \leq \pi$, the method of Laplace (cf. Ch. III, § 18, Exercise 27) leads immediately to the result.

This result can be rewritten in the following manner:

$$\lim_{n \to \infty} \sigma \sqrt{2\pi n} \int_{-\infty}^{+\infty} \varphi^n(t)\, h(t)\, dt = 2\pi \sum_{k=-\infty}^{+\infty} h(2k\pi) \tag{46}$$

for every $h(x) \in C$; hence $\sigma \sqrt{2\pi n}\, \varphi^n(t)$ tends for $n \to \infty$ to the generalized function (39). Thus if $F_n(x)$ is a generalized function such that

$$\int_{-\infty}^{+\infty} F_n(x)\, h(x)\, dx = \sigma \sqrt{2\pi n} \sum_{k=-\infty}^{+\infty} P(\zeta_n = k)\, h(k),$$

$F_n(x)$ tends for $n \to \infty$ to the generalized function (38).

§ 10. Exercises

1. Prove the following

THEOREM. *If ζ is a discrete-valued random variable taking on the values x_k ($x_j \neq x_k$ for $j \neq k$) with probabilities p_k ($k = 1, 2, \ldots$) and if $\varphi_\zeta(t)$ is the characteristic function of ζ, we have*

$$p_n = \lim_{T \to \infty} \frac{1}{2T} \int_{-T}^{T} \varphi_\zeta(t)\, e^{-ix_n t}\, dt \qquad (n = 1, 2, \ldots).$$

Hint. The series

$$\varphi_\zeta(t) = \sum_{n=1}^{\infty} p_n\, e^{ix_n t}$$

is absolutely and uniformly convergent, hence it can be integrated term by term. Furthermore, since for every nonzero real number x

$$\lim_{T \to \infty} \int_{-T}^{T} e^{ixt}\, dt = 0,$$

the theorem follows immediately.

2. Let ζ be an integer-valued random variable and let $\varphi_\zeta(t)$ be its characteristic function. Prove that

$$P(\xi = k) = \frac{1}{2\pi} \int_{-\pi}^{\pi} \varphi_\zeta(t)\, e^{-ikt}\, dt \qquad (k = 0, \pm 1, \pm 2, \ldots).$$

3. Prove the theorem of Moivre and Laplace by means of the result of the preceding exercise.

Hint. By Exercise 2

$$\binom{n}{k} p^k q^{n-k} = \frac{1}{2\pi} \int_{-\pi}^{\pi} (pe^{it} + q)^n\, e^{-ikt}\, dt.$$

For $|k - np| = O\!\left(n^{\frac{2}{3}}\right)$ the method of Laplace leads after some calculations to

$$\frac{1}{2\pi} \int_{-\pi}^{\pi} (pe^{it} + q)^n\, e^{-ikt}\, dt = \frac{1}{\sqrt{2\pi npq}} \exp\left[-\frac{(k - np)^2}{2npq}\right] + O\!\left(\frac{1}{n}\right).$$

4. Prove the following characteristic property of the normal distribution: Let $F(x)$ be an absolutely continuous distribution function, let $F'(x) = f(x)$ and

$$\int_{-\infty}^{+\infty} x^2 f(x)\, dx = 1.$$

If we put

$$H\big(f(x)\big) = -\int_{-\infty}^{+\infty} f(x)\, \ln f(x)\, dx,$$

we have

$$H\big(f(x)\big) \le \ln \sqrt{2\pi e},$$

where equality holds only for $f(x) = (2\pi)^{-\frac{1}{2}} \exp\left(-\dfrac{x^2}{2}\right)$. Hence $H\big(f(x)\big)$ assumes its largest value in the case of the normal distribution. (In information theory the number $H(f(x))$ is called the *entropy* of the distribution with the density function $f(x)$; cf. Appendix.)

5. If $E(\xi)$ exists, then we know that $\varphi_\xi(t)$ is differentiable at $t = 0$ and $\varphi'_\xi(0) = iE(\xi)$. Show that the differentiability of $\varphi_\xi(t)$ does not necessarily imply the existence of $E(\xi)$.

Hint. Put

$$P(\xi = n) = P(\xi = -n) = \frac{c}{n^2 \ln n} \qquad (n \ge 2),$$

with

$$c = \left(2 \sum_{n=2}^{\infty} \frac{1}{n^2 \ln n}\right)^{-1}$$

We find

$$\varphi_\xi(t) = 2c \sum_{n=2}^{\infty} \frac{\cos nt}{n^2 \ln n}, \qquad \varphi'_\xi(t) = -2c \sum_{n=2}^{\infty} \frac{\sin nt}{n \ln n}.$$

The trigonometric series $\varphi'_\xi(t)$ is uniformly convergent[1] and $\varphi'_\xi(0) = 0$. Nevertheless. $E(\xi)$ does not exist.

6. Let ξ be a random variable and $M_\alpha = E(|\xi|^\alpha)$, $\alpha > 0$. Suppose that M_α is finite. Show that if $0 < \beta < \alpha$,

$$(M_\beta)^{\frac{1}{\beta}} \le (M_\alpha)^{\frac{1}{\alpha}}.$$

Hint. For positive a and b $p > 1$, $q = \dfrac{p}{p-1}$ we have[2] $ab \le \dfrac{a^p}{p} + \dfrac{b^q}{q}$. Apply this inequality with

$$a = \frac{|\xi|^\beta}{(M_\alpha)^{\frac{\beta}{\alpha}}}, \qquad b = 1, \qquad p = \frac{\alpha}{\beta}.$$

7. Study the limit distribution of the multinomial distribution

$$p_{k_1 k_2 \ldots k_r} = \frac{N!}{k_1! \, k_2! \ldots k_r!} \, p_{N_1}^{k_1} \ldots p_{N_r}^{k_r} \qquad \left(k_j \ge 0, \sum_{j=1}^{r} k_j = N\right)$$

when $N \to \infty$, with $\sum_{j=1}^{r} p_{N_j} = 1$ and $\lim_{N \to \infty} N p_{N_j} = \lambda_j$ $(j = 1, 2, \ldots, r-1)$.

Hint. The $(r-1)$-dimensional characteristic function of the multinomial distribution is

$$\left(1 + \sum_{j=1}^{r-1} p_{N_j}(e^{it_j} - 1)\right)^N,$$

[1] Cf. e.g. A. Zygmund [1], p. 108.
[2] Cf. e.g. G. H. Hardy, J. E. Littlewood and G. Pólya [1], p. 111.

hence from $\lim_{N \to \infty} Np_{N_j} = \lambda_j$ follows

$$\lim_{N \to \infty} \left(1 + \sum_{j=1}^{r-1} p_{N_j} (e^{it_j} - 1)\right)^N = \prod_{j=1}^{r-1} \exp\left[\lambda_j (e^{it_j} - 1)\right].$$

8. a) Let ξ be a random variable of zero expectation and put

$$d(\xi) = \int_{-\infty}^{+\infty} |x| \, dF(x).$$

If $\varphi(t)$ is the characteristic function of ξ, we have

$$d(\xi) = \frac{1}{\pi} \int_{-\infty}^{\infty} \frac{1 - \text{Re}\big(\varphi(t)\big)}{t^2} \, dt. \tag{1}$$

(Here as usual $\text{Re}(z)$ denotes the real part of z.)

Hint. From

$$1 - \text{Re}\big(\varphi(t)\big) = \int_{-\infty}^{\infty} (1 - \cos xt) dF(x)$$

follows

$$\frac{1}{\pi} \int_{-\infty}^{\infty} \frac{1 - \text{Re}\big(\varphi(t)\big)}{t^2} \, dt = \frac{1}{\pi} \int_{-\infty}^{\infty} \left(\int_{-\infty}^{\infty} \frac{1 - \cos xt}{t^2} \, dt \right) dF(x).$$

Hence the relation (1) follows according to Formula (40) of § 4.

Remark. It can be shown that the necessary and sufficient condition for the existence of $d(\xi)$ is the existence of the integral

$$\int_{-\infty}^{\alpha} \frac{1 - \text{Re}\big(\varphi(t)\big)}{t^2} \, dt.$$

b) If we add to the assumption a) the other one that the variance of ξ exists, we have further

$$d(\xi) = -\frac{1}{\pi} \int_{-\infty}^{+\infty} \frac{\text{Re}\big(\varphi'(t)\big)}{t} \, dt.$$

Hint. This can be obtained from a) using integration by parts.

9. Let $\xi_1, \xi_2, \ldots, \xi_n$ be independent random variables having the same distribution which is symmetric with respect to the origin and has variance 1. Consider the sums $\zeta_n = \xi_1 + \xi_2 + \ldots + \xi_n$. Show that

$$\lim_{N \to \infty} \frac{d(\zeta_n)}{D(\zeta_n)} = \sqrt{\frac{2}{\pi}}.$$

Hint. If $\varphi(t)$ is the characteristic function of the random variables ζ_n, we have

$$\varphi_{\zeta_n / \sqrt{n}}(t) = \left[\varphi\left(\frac{t}{\sqrt{n}}\right)\right]^n.$$

Since $\varphi(t)$ is real for every real t, we obtain, by taking into account Exercise 8,

$$\frac{d(\zeta_n)}{D(\zeta_n)} = -\frac{\sqrt{n}}{\pi} \int_{-\infty}^{\infty} \varphi^{n-1}(u) \, \frac{\varphi'(u)}{u} \, du.$$

From this we obtain the required result by the method of Laplace.

10. If $\Phi(t)$ is the Fourier transform of the generalized function $F(x)$, the Fourier transform of $F(ax + b)$ is

$$\frac{1}{|a|} e^{-\frac{itb}{a}} \Phi\left(\frac{t}{a}\right).$$

11. With the same notations, the Fourier transform of $F'(x)$ is $-it\,\Phi(t)$.

12. a) With the preceding notations, the Fourier transform of $x^n F(x)$ is $(-i)^n \Phi^{(n)}(t)$.
  b) If the conditional density function of ξ is x^{2n} $(n = 1, 2, \ldots)$, the (generalized) characteristic function of ξ is $2\pi(-1)^n \, \delta^{(2n)}(t)$, where $\delta(t)$ denotes Dirac's delta (cf. § 9, p. 355).

13. a) Let $\zeta_1, \zeta_2, \ldots, \zeta_n$ be independent random variables having the same normal distribution, $E(\xi_k) = m$, $\bar{\xi} = \frac{1}{n}(\xi_1 + \xi_2 + \ldots + \xi_n)$, and

$$s = \sqrt{\frac{(\xi_1 - \bar{\xi})^2 + (\xi_2 - \bar{\xi})^2 + \ldots + (\xi_n - \bar{\xi})^2}{n-1}}.$$

Show that

$$\tau = \frac{\sqrt{n}\,(\bar{\xi} - m)}{s}$$

has Student's distribution with $n - 1$ degrees of freedom.

  b) Let $\zeta_1, \ldots, \zeta_{n+m}$ be independent random variables having the same normal distribution. Put

$$\bar{\xi}^{(1)} = \frac{1}{n}(\xi_1 + \ldots + \xi_n), \quad \bar{\xi}^{(2)} = \frac{1}{m}(\xi_{n+1} + \ldots + \xi_{n+m}),$$

$$s = \sqrt{\frac{\sum_{k=1}^{n}(\xi_k - \bar{\xi}^{(1)})^2 + \sum_{k=n+1}^{n+m}(\xi_k - \bar{\xi}^{(2)})^2}{n+m-2}}.$$

Show that

$$\tau = \frac{\bar{\xi}^{(1)} - \bar{\xi}^{(2)}}{s} \sqrt{\frac{nm}{n+m}}$$

has Student's distribution with $n + m - 2$ degrees of freedom.

14. Prove that the following property characterizes the normal distribution: If $f(x)$ $(-\infty < x < +\infty)$ is a continuously differentiable positive density function such that for any three real numbers x, y, z the function

$$f(x - t)f(y - t)f(z - t)$$

has its maximum at $t = \dfrac{x + y + z}{3}$, then

$$f(x) = \frac{1}{\sqrt{2\pi}\,\sigma} \exp\left[-\frac{x^2}{2\sigma^2}\right].$$

Hint. By assumption, $f(x)$ is positive. If we put $g(x) = \dfrac{f'(x)}{f(x)}$ and $s = \dfrac{x + y + z}{3}$ we have

$$g(x - s) + g(y - s) + g(z - s) = 0. \tag{2}$$

For $x = y = z$, we obtain $g(0) = 0$. Take now any two x and y and put $z = \dfrac{x + y}{2} = s$; the relation (2) and $g(0) = 0$ lead to

$$g\left(\frac{x - y}{2}\right) + g\left(\frac{y - x}{2}\right) = 0.$$

Hence $g(x)$ is an odd function. If we put $u = x - s$, $v = y - s$ and thus $z - s = -(u + v)$, we can write Formula (2) in the form

$$g(u) + g(v) = g(u + v). \tag{3}$$

Since by assumption, $g(u)$ is continuous, we know that (3) implies

$$g(x) = \frac{f'(x)}{f(x)} = Cx \qquad (C = \text{constant}).$$

Hence by integrating,

$$f(x) = A \exp\left[\frac{Cx^2}{2}\right].$$

As $f(x)$ is a density function, we have $C = -\dfrac{1}{\sigma^2}$ and $A = \dfrac{1}{\sqrt{2\pi}\,\sigma}$, with $\sigma > 0$.

Remark. The result is valid under weaker conditions too.[1]

15. If ξ_1 and ξ_2 are independent random variables having the same nondegenerate distribution with finite variance and if the random variable $a\xi_1 + b\xi_2$ $(0 < a \le b < 1$; $a^2 + b^2 = 1)$ has again the same distribution, then this distribution is normal with expectation zero.

16. The distribution having the density function

$$f(x) = \frac{1}{\sqrt{2\pi}} e^{-\frac{1}{2x}} x^{-\frac{3}{2}} \qquad (x > 0)$$

[1] Cf. Á. Császár [2].

is stable and corresponds to the parameters $\alpha = \dfrac{1}{2}$, $\beta = -1$, $\gamma = 0$, $c = 1$.[1]

17. If $\varphi(t)$ is a characteristic function, then

$$\psi(t) = \frac{1}{t} \int_0^t \varphi(u)\, du$$

is a characteristic function as well.

18. Show that the gamma distributions are infinitely divisible.

19. Let $\zeta(s)$ denote Riemann's zeta-function

$$\zeta(s) = \sum_{n=1}^{\infty} \frac{1}{n^s} \quad \text{with} \quad s = \sigma + it, \ \sigma > 1.$$

Show that the function

$$\varphi(t) = \frac{\zeta(\sigma + it)}{\zeta(\sigma)} \qquad (\sigma > 1)$$

is the characteristic function of an infinitely divisible distribution.

20. We know (Ch. IV, § 10) that the quotient of two independent $N(0, 1)$ random variables has a Cauchy distribution. Show that this property is *not* characteristic for the normal distribution. If ξ and η are independent, have the same distribution of zero expectation and if $\dfrac{\xi}{\eta}$ has a Cauchy distribution, then it does *not* follow that ξ and η are normally distributed.

Hint. Take for density function of ξ and η

$$f(x) = \frac{\sqrt{2}}{\pi} \frac{1}{1 + x^4} \qquad (-\infty < x < +\infty).$$

We obtain for the density function of $\dfrac{\xi}{\eta}$

$$g(x) = \frac{2}{\pi^2} \int_{-\infty}^{\infty} \frac{|y|\, dy}{(1 + y^4)(1 + x^4 y^4)} = \frac{1}{\pi(1 + x^2)}.$$

Remark. This example is due to Laha.[2]

---

[1] $f(x)$ is the density function of ξ^{-2}; where ξ is $N(0, 1)$; this distribution is sometimes called the "inverse normal distribution".

[2] Cf. R. G. Laha [1].

CHAPTER VII

LAWS OF LARGE NUMBERS

§ 1. Chebyshev's and related inequalities

In the present paragraph we shall deal with an inequality due to Chebyshev and with some similar inequalities, all needed in the proofs of the laws of large numbers. First we prove the famous inequality of Chebyshev.[1]

THEOREM 1. *Let ξ be any random variable with expectation $M = E(\xi)$ and with a positive finite standard deviation $D = D(\xi)$. If λ is a real number, $\lambda > 1$, we have*

$$P(|\xi - M| > \lambda D) \leq \frac{1}{\lambda^2}. \tag{1}$$

Remark. If $0 < \lambda \leq 1$, (1) remains valid, but becomes trivial.

PROOF. If we apply Markov's inequality (cf. Ch. IV, § 13, Theorem 1) to the random variable $\eta = (\xi - M)^2$ with λ^2 instead of λ, we obtain immediately (1).

If we apply Markov's inequality to other positive functions of ξ, we obtain other inequalities, related to Chebyshev's inequality. Thus for instance if we put $\eta = |\xi - M|^\alpha$ $(\alpha > 0)$ and

$$M_\alpha = E(|\xi - M|^\alpha) \tag{2}$$

(thus M_α is the α-th absolute central moment of ξ), then we get

$$P(|\xi - M| > \lambda D) \leq \frac{M_\alpha}{(\lambda D)^\alpha}. \tag{3}$$

Of course for $\alpha = 2$ (3) reduces to (1).

In order to get an inequality as sharp as possible, we have to choose α in such a manner that $\dfrac{M_\alpha}{(\lambda D)^\alpha}$ should be as small as possible.

---

[1] Called also the Bienaymé−Chebyshev inequality.

We can also apply Markov's inequality to the random variable $\eta = e^{\varepsilon(\xi - M)}$. If we put

$$E(e^{\varepsilon(\xi-M)}) = \mathscr{M}(\varepsilon), \tag{4}$$

we obtain

$$P(e^{\varepsilon(\xi-M)} > \mathscr{M}(\varepsilon)\,e^{t}) < e^{-t}, \tag{5}$$

where t is a positive number; the exponential function being monotone, it follows for $\varepsilon > 0$ that

$$P\left(\xi > M + \frac{t + \ln \mathscr{M}(\varepsilon)}{\varepsilon}\right) < e^{-t}. \tag{6}$$

In order to get the sharpest possible bound we have to choose ε such that the expression $\dfrac{t + \ln M(\varepsilon)}{\varepsilon}$ is minimal or at least nearly minimal.

In § 4 an improvement of Chebyshev's inequality, which is due to Bernstein, will be deduced from inequality (6).

§ 2. Stochastic convergence

We have mentioned already in Chapter III, § 17 the most elementary case of the laws of large numbers, discovered already by Jacob Bernoulli. In order to prove a more general theorem, we introduce first the concept of stochastic convergence, due to Slutsky.

If $\xi_1, \xi_2, \ldots, \xi_n, \ldots$ is a sequence of random variables for which the relation

$$\lim_{n \to \infty} P(\,|\xi_n| \geq \varepsilon) = 0 \tag{1}$$

holds for every positive ε however small, then the sequence ξ_n $(n = 1, 2, \ldots)$ is said to *converge stochastically* (or *in probability*) to zero. If the random variables ζ_n $(n = 1, 2, \ldots)$ fulfil the relation

$$\lim_{n \to \infty} P(\,|\zeta_n - a| \geq \varepsilon) = 0 \tag{2}$$

for any fixed $\varepsilon > 0$ we shall say that the sequence ζ_n $(n = 1, 2, \ldots)$ *converges in probability* (or *stochastically*) *to the constant a* and indicate this by

$$\lim_{n \to \infty} \operatorname{st} \zeta_n = a \tag{3}$$

or by

$$\zeta_n \xrightarrow{P} a. \tag{4}$$

With this definition Bernoulli's theorem may be formulated as follows:
In a series of independent experiments the relative frequency of the event A tends stochastically to the probability P(A) of A when the number of experiments increases infinitely.

Bernoulli's theorem is an immediate consequence of Chebyshev's inequality, established in the preceding paragraph.

In fact let $\zeta_n = \dfrac{k}{n}$ be the relative frequency of the event A in a series of experiments. The random variable $n\zeta_n$ has a binomial distribution with expectation np and standard deviation $\sqrt{npq}$ $(q = 1 - p)$. Chebyshev's inequality leads to

$$P\left[|\zeta_n - p| > \lambda \sqrt{\frac{pq}{n}}\right] < \frac{1}{\lambda^2} . \tag{5}$$

In particular, if we put $\lambda = \varepsilon \sqrt{\dfrac{n}{pq}}$, (5) becomes

$$P(|\zeta_n - p| > \varepsilon) < \frac{pq}{n\varepsilon^2} ; \tag{6}$$

if now n tends to infinity, the expression on the right of (6) tends to 0, which proves the theorem.

The definition of stochastic convergence can also be given in the following form: the sequence ζ_n $(n = 1, 2, \ldots)$ converges stochastically to the number p when to every pair ε, δ of positive numbers (however small) there can be chosen a number $N = N(\varepsilon, \delta)$ so that for every $n \geq N$

$$P(|\zeta_n - p| > \varepsilon) < \delta. \tag{7}$$

This condition can also be expressed in terms of the distribution function $F_n(x)$ of ζ_n. In fact (7) is equivalent to

$$\lim_{n \to \infty} F_n(x) = \begin{cases} 0 & \text{for } x < p, \\ 1 & \text{for } x > p. \end{cases} \tag{8}$$

If $D_p(x)$ denotes the (degenerate) distribution function of the constant p, (8) is equivalent to

$$\lim_{n \to \infty} F_n(x) = D_p(x) \text{ for } x \neq p. \tag{9}$$

Conversely, it is easy to see that the stochastic convergence of ζ_n to the constant p follows from (8) or (9).

The concept of stochastic convergence can be generalized still further. We say that a sequence of random variables ζ_n $(n = 1, 2, \ldots)$ tends in probability (or stochastically) to the random variable ζ, if for every positive ε one has

$$\lim_{n \to \infty} P(|\zeta_n - \zeta| > \varepsilon) = 0; \tag{10}$$

in this case we write

$$\lim_{n \to \infty} \text{st } \zeta_n = \zeta \tag{11}$$

or

$$\zeta_n \xrightarrow{P} \zeta. \tag{12}$$

It is easy to prove the following

THEOREM 1. *If* $\zeta_n \xrightarrow{P} \zeta$, *the distribution function* $F_n(x)$ *of* ζ_n *tends to the distribution function* $F(x)$ *of* ζ *at every point of continuity of the latter.*

PROOF. If A_n is the event $|\zeta_n - \zeta| < \varepsilon$, we have

$$P(\zeta_n < x) = P(\zeta_n < x \mid A_n) P(A_n) + P(\zeta_n < x \mid A_n) P(\bar{A}_n), \tag{13a}$$

hence

$$P(\zeta_n < x) \leq P(\zeta_n < x \mid A_n) + P(\bar{A}_n). \tag{13b}$$

But

$$P(\zeta_n < x \mid A_n) \leq P(\zeta < x + \varepsilon \mid A_n) \leq \frac{P(\zeta < x + \varepsilon)}{P(A_n)}. \tag{14}$$

(13b) and (14) imply

$$\overline{\lim_{n \to \infty}} P(\zeta_n < x) \leq P(\zeta < x + \varepsilon) \text{ for every } \varepsilon > 0. \tag{15}$$

On the other hand, by (13a)

$$P(\zeta_n < x) \geq P(A_n) P(\zeta_n < x \mid A_n) \geq P(\zeta < x - \varepsilon) - P(\bar{A}_n). \tag{13c}$$

From this follows

$$\underline{\lim_{n \to \infty}} P(\zeta_n < x) \geq P(\zeta < x - \varepsilon) \text{ for every } \varepsilon > 0. \tag{16}$$

Since ε can be chosen arbitrarily small, (15) and (16) imply the statement of Theorem 1.

§ 3. Generalization of Bernoulli's law of large numbers

In the preceding paragraph Chebyshev's inequality was applied to the random variable $n\zeta_n$, where ζ_n denotes the relative frequency of the event A (with $P(A) = p$) in a sequence of n independent experiments. Obviously

$$\zeta_n = \frac{\xi_1 + \xi_2 + \ldots + \xi_n}{n},$$

where ξ_k ($k = 1, 2, \ldots$) is the indicator of the event A in the k-th experiment, that is

$$\xi_k = \begin{cases} 1 & \text{if the event } A \text{ occurs at the } k\text{-th experiment,} \\ 0 & \text{otherwise.} \end{cases}$$

The ξ_k are by assumption identically distributed, independent random variables, assuming the values 0 and 1 only. Their expectation is $E(\xi_k) = p$. Bernoulli's theorem states that

$$\lim_{n\to\infty} \text{st} \ \frac{\xi_1 + \xi_2 + \ldots + \xi_n}{n} = E(\xi_k); \tag{1}$$

i.e. that the empirical mean tends in probability to the common expectation of the ξ_k. It is easy to show that this property remains valid for arbitrary independent identically distributed random variables with finite variance.

THEOREM 1. *Let ξ_n ($n = 1, 2, \ldots$) be pairwise independent and identically distributed random variables with finite expectation $E(\xi_n) = M$ and variance $D^2(\xi_n) = D^2$. Then*

$$\lim_{n\to\infty} \text{st} \left(\frac{1}{n} \sum_{k=1}^{n} \xi_k \right) = M.$$

PROOF. Put

$$\zeta_n = \frac{1}{n} \sum_{k=1}^{n} \xi_k.$$

We have

$$E(\zeta_n) = M \quad \text{and} \quad D(\zeta_n) = \frac{D}{\sqrt{n}}.$$

By applying Chebyshev's inequality we obtain

$$P(|\zeta_n - M| > \varepsilon) < \frac{D^2}{n\varepsilon^2},$$

which proves the statement of Theorem 1.

The suppositions of Theorem 1 can be weakened. It is not necessary to assume that ξ_k are identically distributed, it suffices to assume the existence of the limit

$$\lim_{n\to\infty} \frac{1}{n} \sum_{k=1}^{n} E(\xi_k) = M$$

and the validity of

$$\lim_{n\to\infty} \frac{S_n}{n} = 0,$$

where

$$S_n = \sqrt{\sum_{k=1}^{n} D^2(\xi_k)}.$$

Thus we obtain a form of the law of large numbers which is due to A. A. Markov:

THEOREM 2. *Let ξ_k $(k = 1, 2, \ldots)$ be pairwise independent random variables such that $M_k = E(\xi_k)$ and $D_k = D(\xi_k)$ $(k = 1, 2, \ldots)$ are finite. Suppose further the validity of the following two conditions:*
a) *the limit*

$$\lim_{n\to\infty} \frac{1}{n} \sum_{k=1}^{n} M_k = M \tag{2}$$

exists and is finite;
b) *for*

$$S_n = \sqrt{\sum_{k=1}^{n} D_k^2}$$

we have[1]

$$\lim_{n\to\infty} \frac{S_n}{n} = 0. \tag{3}$$

Then for the random variable

$$\zeta_n = \frac{1}{n} \sum_{k=1}^{n} \xi_k$$

the relation

$$\zeta_n \xrightarrow{P} M$$

is valid.

---

[1] This condition is certainly fulfilled e.g. if the random variables ξ_k (or at least the numbers D_k) are uniformly bounded.

PROOF. Similarly as in proving Theorem 1, we apply Chebyshev's inequality to

$$\zeta_n^* = \frac{1}{n} \sum_{k=1}^{n} (\xi_k - M_k).$$

Taking into account that $E(\zeta_n^*) = 0$ and $D(\zeta_n^*) = \dfrac{S_n}{n}$, we obtain the relation

$$\lim_{n \to \infty} \text{st } \zeta_n^* = 0.$$

Now

$$\zeta_n^* = \zeta_n - \frac{1}{n} \sum_{k=1}^{n} M_k$$

and by assumption

$$\left| \frac{1}{n} \sum_{k=1}^{n} M_k - M \right| < \frac{\varepsilon}{2}$$

if n is large enough. As $|\zeta_n - M| > \varepsilon$ can hold only if $|\zeta_n^*| > \dfrac{\varepsilon}{2}$, it follows

$$\lim_{n \to \infty} \text{st } (\zeta_n - M) = 0. \tag{4}$$

The assumptions of the above theorem can still be weakened. Instead of the pairwise independence of ξ_k it suffices to assume that there does not exist a strong positive correlation between most pairs. More precisely, the following theorem holds, due essentially to S. N. Bernstein:

THEOREM 3. *Let* ξ_k $(k = 1, 2, \ldots)$ *be random variables with finite expectations* $M_k = E(\xi_k)$ *and variances* $D_k^2 = D^2(\xi_k)$ $(k = 1, 2, \ldots)$. *Let* R_{ij} *denote the correlation coefficient of* ξ_i *and* ξ_j. *Assume further the validity of the following three conditions:*

a) *There exists the limit*

$$\lim_{n \to \infty} \frac{1}{n} \sum_{k=1}^{n} M_k = M;$$

b) *For* $S_n^2 = \sum_{k=1}^{n} D_k^2$ *we have* $S_n^2 < Kn$, *where K is a constant independent from n;*

c) $R_{ij} \le R(|i - j|)$, *where R(k) is a nonnegative function of k such that* $R(0) = 1$ *and*

$$\lim_{n \to \infty} \frac{1}{n} \sum_{k=1}^{n} R(k) = 0.$$

Then $\zeta_n = \dfrac{1}{n} \sum\limits_{k=1}^{n} \xi_k$ *converges in probability to* M:

$$\zeta_n \xrightarrow{P} M. \tag{5}$$

PROOF. If we consider the proof of Theorem 2 we see that it suffices to prove the relation

$$\lim_{n \to \infty} D(\zeta_n) = 0; \tag{6}$$

if this is done, the remaining part of the proof can be repeated word by word.

We prove therefore (6). We have

$$D^2(\zeta_n) = \frac{1}{n^2} \sum_{i=1}^{n} \sum_{j=1}^{n} D_i D_j R_{ij} \leq \frac{1}{n^2} \sum_{i=1}^{n} \sum_{j=1}^{n} D_i D_j R(|i-j|),$$

hence

$$D^2(\zeta_n) \leq \frac{1}{n^2} \left(S_n^2 + 2 \sum_{k=1}^{n-1} R(k) \sum_{i=1}^{n-k} D_i D_{i+k} \right). \tag{7}$$

By Cauchy's inequality we have

$$\sum_{i=1}^{n-k} D_i D_{i+k} \leq S_n^2;$$

if we put this into (7) we obtain

$$D^2(\zeta_n) \leq \frac{S_n^2}{n^2} + \frac{2S_n^2}{n} \left(\frac{1}{n} \sum_{k=1}^{n} R(k) \right). \tag{8}$$

Hence by condition b)

$$D^2(\zeta_n) \leq \frac{K}{n} + 2K \left(\frac{1}{n} \sum_{k=1}^{n} R(k) \right).$$

Relation (6) follows and because of condition c) Theorem 3 is herewith proved.

We return now to the case of pairwise independent random variables ξ_k with the same distribution. Let M denote the (finite) expectation of the ξ_k. It was shown by Khintchine that in this case the mean value of

$$\zeta_n = \frac{1}{n} \sum_{k=1}^{n} \xi_k$$

tends in probability to M when n increases, even if $D(\xi_k)$ does not exist.

Thus we have

THEOREM 4. *Let ξ_k be pairwise independent and identically distributed random variables and suppose that the expectation*

$$E(\xi_k) = M \qquad (9)$$

exists. Then for

$$\zeta_n = \frac{1}{n} \sum_{k=1}^{n} \xi_k$$

one has

$$\zeta_n \xrightarrow{P} M. \qquad (10)$$

PROOF. Without restricting generality we may assume $M = 0$. Put

$$\xi_k^* = \begin{cases} \xi_k & \text{for } |\xi_k| \le k, \\ 0 & \text{for } |\xi_k| > k. \end{cases} \qquad (11)$$

If $F(x)$ is the distribution function of the random variables ξ_k, we have

$$\frac{1}{n} \sum_{k=1}^{n} E(\xi_k^*) = \frac{1}{n} \sum_{k=1}^{n} \int_{-k}^{+k} x\, dF(x). \qquad (12)$$

Since by assumption

$$\lim_{k \to \infty} \int_{-k}^{+k} x\, dF(x) = \int_{-\infty}^{+\infty} x\, dF(x) = 0,$$

we can write

$$\lim_{n \to \infty} \frac{1}{n} \sum_{k=1}^{n} E(\xi_k^*) = 0. \qquad (13)$$

On the other hand

$$D^2(\xi_k^*) \le E(\xi_k^{*2}) = \int_{-k}^{+k} x^2\, dF(x),$$

hence

$$\frac{1}{n^2} \sum_{k=1}^{n} D^2(\xi_k^*) \le \frac{1}{n} \int_{-n}^{+n} x^2\, dF(x) \le \frac{1}{\sqrt{n}} \int_{-\sqrt{n}}^{+\sqrt{n}} |x|\, dF(x) + \int_{|x| > \sqrt{n}} |x|\, dF(x) \qquad (14)$$

and consequently

$$\lim_{n \to \infty} \frac{1}{n^2} \sum_{k=1}^{n} D^2(\xi_k^*) = 0. \qquad (15)$$

If we put

$$\zeta_{n,r}^* = \frac{1}{n}\left(\sum_{k=1}^r \xi_k + \sum_{k=r+1}^n \xi_k^*\right),$$

Theorem 2 implies

$$\lim_{n\to\infty} \text{st } \zeta_{n,r}^* = 0.$$

On the other hand we have

$$P(\zeta_{n,r}^* \neq \zeta_n) \leq \sum_{k=r+1}^n P(\xi_k^* \neq \xi_k) = \sum_{k=r+1}^n \int_{|x|>k} dF(x) \leq \int_{|x|>r} |x|\, dF(x). \quad (16)$$

If r is sufficiently large, we have for any $\delta > 0$ the inequality

$$P(\zeta_{n,r}^* \neq \zeta_n) < \delta.$$

Hence for any $\varepsilon > 0$

$$P(|\zeta_n| > \varepsilon) = P(|\zeta_n| > \varepsilon, \zeta_{n,r}^* \neq \zeta_n) + P(|\zeta_n| > \varepsilon, \zeta_{n,r}^* = \zeta_n) \quad (17)$$

and thus

$$\limsup_{n\to\infty} P(|\zeta_n| > \varepsilon) \leq \delta. \quad (18)$$

But since $\delta > 0$ is arbitrary, it follows

$$\lim_{n\to\infty} P(|\zeta_n| > \varepsilon) = 0,$$

which was to be proved.

Remark. When the random variables ξ_k are not only pairwise but completely independent, Theorem 4 can be proved by the method of characteristic functions. We have seen in § 2 that the stochastic convergence of ζ_n to 0 is equivalent to the convergence of the distribution function $F_n(x)$ of ζ_n to the degenerate distribution function $D_0(x)$ of the constant 0 (i.e. to 1 for $x > 0$ and to 0 for $x < 0$). Because of Theorem 3 (Ch. VI), § 4, it suffices thus to show that the characteristic function $\varphi_n(t)$ of ζ_n tends, for every t, to the characteristic function of the constant 0, i.e. to 1. If $\varphi(t)$ is the characteristic function of the random variables ξ_k, then by assumption $\varphi'(0) = 0$. Put

$$\varepsilon(t) = \frac{\varphi(t) - 1}{t}, \quad (19)$$

then

$$\lim_{t\to 0} \varepsilon(t) = 0. \quad (20)$$

Since for the characteristic function $\varphi_n(t)$ of ζ_n we have $\varphi_n(t) = \left[\varphi\left(\dfrac{t}{n}\right)\right]^n$, (20) implies, for every real value of t,

$$\lim_{n\to\infty} \varphi_n(t) = \lim_{n\to\infty}\left(1 + \frac{t}{n}\,\varepsilon\left(\frac{t}{n}\right)\right)^n = 1, \tag{21}$$

which was to be proved.

It was shown by Kolmogorov that in Theorem 4 the assumption of the existence of the expectation of ξ_k can be replaced by the weaker postulate of the existence of the limit

$$\lim_{n\to\infty} \int_{-n}^{+n} x\,dF(x)$$

and of the relation

$$\lim_{x\to\infty} x[F(-x) + (1 - F(x))] = 0$$

and that these conditions are not only sufficient but necessary as well for

$$\zeta_n = \frac{1}{n}\sum_{k=1}^{n} \xi_k$$

to converge in probability to a constant as $n \to \infty$. As regards the proof of this theorem of Kolmogorov, cf. § 14, Exercise 24.

We now give an example to which the law of large numbers does not apply. If the random variables ξ_k are completely independent and if all have the same Cauchy distribution (with the density function $\dfrac{1}{\pi(1 + x^2)}$), then

$$\zeta_n = \frac{1}{n}\sum_{k=1}^{n} \xi_k$$

also has the same Cauchy distribution.

As a matter of fact the characteristic function of the variables ξ_k is equal to $e^{-|t|}$, thus that of ζ_n is equal to $(e^{-\frac{|t|}{n}})^n = e^{-|t|}$. Evidently, $\lim \operatorname{st} \zeta_n = p$ does not hold in this case.

The fact that for independent random variables ξ_k with a common Cauchy distribution the random variable

$$\zeta_n = \frac{1}{n}\sum_{k=1}^{n} \xi_k$$

has the same Cauchy distribution as ξ_k, can be interpreted as follows: When we take a sample from a population with a Cauchy distribution with density function $\dfrac{1}{\pi(1 + (x - m)^2)}$, we do not obtain more information concerning the number m from the mean of a sample, however large, than from a single observation.

§ 4. Bernstein's improvement of Chebyshev's inequality

In the preceding paragraph we applied Chebyshev's inequality to the sum of a large number of independent random variables in order to prove the law of large numbers. It was shown by Bernstein that in this case Chebyshev's inequality can be considerably improved. Put

$$\mathcal{M}(\varepsilon) = E(e^{\varepsilon(\xi - M)}),$$

where $\varepsilon > 0$, and M is the expectation of ξ. We proved in § 1 the inequality

$$P\left(\xi \geq M + \frac{t + \ln \mathcal{M}(\varepsilon)}{\varepsilon}\right) \leq e^{-t} \qquad (t > 0). \tag{1}$$

It was already observed that the choice of ε minimizing $\dfrac{t + \ln \mathcal{M}(\varepsilon)}{\varepsilon}$ makes the inequality (1) as sharp as possible.

We prove now the following

LEMMA. *Let* $\xi_1, \xi_2, \ldots, \xi_n$ *be completely independent bounded random variables with* $E(\xi_k) = 0$, $D(\xi_k) = D_k$ *and suppose* $|\xi_k| \leq K$ $(k = 1, 2, \ldots, n)$. *Put further*

$$\xi = \xi_1 + \xi_2 + \ldots + \xi_n \quad D(\xi) = D, \quad and \quad \mathcal{M}(\varepsilon) = E(e^{\varepsilon \xi}),$$

where ε *is an arbitrary positive number. We have then*

$$\ln \mathcal{M}(\varepsilon) \leq \frac{\varepsilon^2 D^2}{2}\left(1 + \frac{\varepsilon K}{3} e^{\varepsilon K}\right).$$

PROOF. Since   (2)

$$\mathcal{M}(\varepsilon) = \prod_{k=1}^{n} E(e^{\varepsilon \xi_k}),$$

we evaluate first $E(e^{\varepsilon \zeta_k})$. Since

$$e^{\varepsilon \zeta_k} = \sum_{n=0}^{\infty} \frac{\varepsilon^n \zeta_k^n}{n!}$$

and the ζ_k are bounded, the series is uniformly convergent and the expectation of $e^{\varepsilon \zeta_k}$ can be calculated term by term from the power series. Thus we obtain

$$E(e^{\varepsilon \zeta_k}) = 1 + \frac{\varepsilon^2 D_k^2}{2} + \sum_{n=3}^{\infty} \frac{\varepsilon^n E(\zeta_k^n)}{n!} . \tag{3}$$

As $|\zeta_k| \le K$ implies the inequality

$$E(\zeta_k^n) \le D_k^2 K^{n-2},$$

we obtain

$$\sum_{n=3}^{\infty} \frac{\varepsilon^n E(\zeta_k^n)}{n!} \le \varepsilon^2 D_k^2 \sum_{n=3}^{\infty} \frac{(\varepsilon K)^{n-2}}{n!} .$$

As

$$\frac{1}{n!} \le \frac{1}{6} \cdot \frac{1}{(n-3)!}$$

for $n \ge 3$, substitution into (3) gives

$$E(e^{\varepsilon \zeta_k}) \le 1 + \varepsilon^2 D_k^2 \left[\frac{1}{2} + \sum_{n=3}^{\infty} \frac{(\varepsilon K)^{n-2}}{n!} \right] \le 1 + \varepsilon^2 D_k^2 \left[\frac{1}{2} + \frac{\varepsilon K e^{\varepsilon K}}{6} \right],$$

which leads to

$$E(e^{\varepsilon \zeta}) \le \prod_{k=1}^{n} \left[1 + \varepsilon^2 D_k^2 \left(\frac{1}{2} + \frac{\varepsilon K e^{\varepsilon K}}{6} \right) \right].$$

Because of $1 + x < e^x$, we obtain

$$\mathscr{M}(\varepsilon) = E(e^{\varepsilon \zeta}) \le \exp \left[\frac{\varepsilon^2 D^2}{2} \left(1 + \frac{\varepsilon K e^{\varepsilon K}}{3} \right) \right],$$

which is equivalent to (2).

Since by assumption $M = 0$, it follows from (1) that

$$P \left(\zeta \ge \frac{t + \dfrac{\varepsilon^2 D^2}{2} \left[1 + \dfrac{\varepsilon K e^{\varepsilon K}}{3} \right]}{\varepsilon} \right) \le e^{-t}. \tag{4}$$

Put

$$\varepsilon = \frac{\sqrt{2t}}{D}. \tag{5}$$

Then (4) leads to

$$P\left(\xi \geq D\sqrt{2t}\left[1 + \frac{K\sqrt{2t}}{6D}\exp\left(\frac{K\sqrt{2t}}{D}\right)\right]\right) \leq e^{-t}. \tag{6}$$

Substitution of $\lambda = \sqrt{2t}$ gives

$$P\left(\xi \geq \lambda D\left[1 + \frac{\lambda K}{6D}e^{\frac{\lambda K}{D}}\right]\right) \leq e^{-\frac{\lambda^2}{2}}. \tag{7}$$

Thus if λ is large and $\dfrac{\lambda K}{D}$ small, we obtain a much sharper inequality than that of Chebyshev's.

If we apply the obtained result to $-\xi$, we find that

$$P\left(|\xi| \geq \lambda D\left[1 + \frac{\lambda D}{6D}e^{\frac{\lambda K}{D}}\right]\right) \leq 2e^{-\frac{\lambda^2}{2}}. \tag{8}$$

In order to transform (8) into a more convenient form, we restrict ourselves to the case $\dfrac{\lambda K}{D} \leq 1$. We have then

$$e^{\frac{\lambda K}{D}} \leq e < 3.$$

From this, putting $\mu = \lambda\left(1 + \dfrac{\lambda K}{2D}\right)$, we obtain from (8) because of $\lambda \leq$.

$$\leq \mu \leq \lambda\left(1 + \frac{\mu K}{2D}\right)$$

$$P(|\xi| \geq \mu D) \leq 2\exp\left[-\frac{\mu^2}{2\left(1 + \dfrac{\mu K}{2D}\right)^2}\right]. \tag{9}$$

We may free ourselves from the condition $E(\xi_k) = 0$ by applying (9) to the random variable

$$\xi - E(\xi) = \sum_{k=1}^{n}[\xi_k - E(\xi_k)].$$

Thus we have proved the following theorem:

THEOREM 1. *If $\xi_1, \xi_2, \ldots, \xi_n$ are completely independent random variables such that $E(\xi_k) = M_k$, $D(\xi_k) = D_k$ exist and $|\xi_k - M_k| \leq K$ $(k = 1, 2, \ldots, n)$, then for $\xi = \xi_1 + \xi_2 + \ldots + \xi_n$ we have*

$$P(|\xi - M| \geq \mu D) \leq 2 \exp\left[-\frac{\mu^2}{2\left(1 + \frac{\mu K}{2D}\right)^2}\right]. \tag{10}$$

In this formula

$$M = \sum_{k=1}^{n} M_k \quad \text{and} \quad D = \sqrt{\sum_{k=1}^{n} D_k^2},$$

while μ is a positive number such that $\mu \leq \dfrac{D}{K}$.

Let us apply now this result to the case where the ξ_k have a common distribution. Let M_1 be the expectation of the ξ_k and D_1^2 their variance. Then the expectation of the sum $\xi = \sum_{k=1}^{n} \xi_k$ is equal to nM_1 and its variance to $D_1^2 n$. It follows from (10) for $\mu \leq \dfrac{D_1}{K} \sqrt{n}$ that

$$P(|\xi - nM_1| \geq \mu D_1 \sqrt{n}) \leq 2 \exp\left[-\frac{\mu^2}{2\left(1 + \frac{\mu K}{2D_1 \sqrt{n}}\right)^2}\right]. \tag{11}$$

If the ξ_k $(k = 1, 2, \ldots, n)$ are the indicators of an event A in a sequence of n experiments, we get from (11)

THEOREM 2. *Let A be one of the possible results of an experiment, suppose $p = P(A) > 0$ and put $q = 1 - p$. Let the random variable ζ_n denote the relative frequency of A in an experiment consisting of n independent trials. Then for $0 < \varepsilon \leq pq$ we have*

$$P(|\zeta_n - p| \geq \varepsilon) \leq 2 \exp\left[-\frac{n\varepsilon^2}{2pq\left(1 + \frac{\varepsilon}{2pq}\right)^2}\right]. \tag{12}$$

PROOF. (12) follows from (11) when we put $\mu = \varepsilon \sqrt{\dfrac{n}{pq}}$. Chebyshev's inequality [cf. § 2, Formula (6)] leads only to

$$P(|\zeta_n - p| \geq \varepsilon) \leq \frac{pq}{n\varepsilon^2}.$$

Thus e.g. for $p = q = \dfrac{1}{2}$, $\varepsilon = \dfrac{1}{20}$, Chebyshev's inequality guarantees the validity of

$$P\left(\left|\zeta_n - \frac{1}{2}\right| \geq \frac{1}{20}\right) \leq \frac{1}{100} \qquad (13)$$

only for $n \geq 10\,000$, while by using (12) we find that (13) holds already for $n \geq 1283$.

If we take $\varepsilon = \dfrac{1}{50}$, we find, applying Chebyshev's inequality, that

$$P\left(\left|\zeta_n - \frac{1}{2}\right| \geq \frac{1}{50}\right) \leq \frac{1}{100}$$

is valid for $n \geq 62\,500$; while applying (12) we see that it is valid already for $n \geq 7164$.

In these examples $\varepsilon > 0$ and $\delta > 0$ were given and we wanted to estimate the least number $n_0 = n_0(\varepsilon, \delta)$ such that for $n \geq n_0$

$$P(|\zeta_n - p| \geq \varepsilon) \leq \delta$$

holds. This question is answered in general by the following theorem which follows from Theorem 2:

THEOREM 3. *We perform n independent experiments with possible outcomes A and $\bar{A}$. Suppose $p = P(A) > 0$. Let ζ_n be the relative frequency of the event A in the sequence of experiments, and suppose $\delta > 0$, $0 < \varepsilon \leq p(1 - p)$ and*

$$n \geq \frac{9 \ln \dfrac{2}{\delta}}{8\varepsilon^2} = n_0(\varepsilon, \delta).$$

Then

$$P(|\zeta_n - p| \geq \varepsilon) \leq \delta. \qquad (14)$$

PROOF. It follows from Theorem 2 that (14) is fulfilled for

$$n \geq \frac{2pq\left(1 + \dfrac{\varepsilon}{2pq}\right)^2 \ln \dfrac{2}{\delta}}{\varepsilon^2}.$$

Since $2pq \leq \dfrac{1}{2}$ and $\left(1 + \dfrac{\varepsilon}{2pq}\right)^2 \leq \dfrac{9}{4}$ for $0 < \varepsilon \leq pq$, (14) is always fulfilled for $n \geq \dfrac{9}{8\varepsilon^2} \ln \dfrac{2}{\delta}$.

Theorem 2 can be rephrased in the following manner: If we put $x = \varepsilon \sqrt{\dfrac{n}{pq}}$, we have for $0 < x < \sqrt{npq}$

$$P\left(|\zeta_n - p| \geq x \sqrt{\frac{pq}{n}}\right) \leq 2 \exp\left[-\frac{x^2}{2\left(1 + \dfrac{x}{2\sqrt{npq}}\right)^2}\right]; \qquad (15)$$

the sharpness of the inequality can be appreciated by comparing it to the Moivre–Laplace theorem (Ch. III, § 16) concerning the convergence of the binomial distribution to the normal distribution.

Finally, let us make some general remarks about the laws of large numbers. In connection with many random mass phenomena, there are quantities which, though depending on chance, are practically constant. As an example, consider the pressure of a gas enclosed into a vessel. This pressure is the result of the impacts of molecules on the wall of the vessel. The number of the impacts as well as the velocity of the molecules depends on chance; nevertheless the resulting pressure is practically constant, provided that the number of the molecules is very large. Such phenomena can be explained as instances of the law of large numbers.

According to the law of large numbers, by calculating the probability of an event or the expectation of a random variable, one can obtain information about the relative frequency, or of the arithmetic mean, of the results if a large number of independent experiments (observations) were performed. This is the reason why the law of large numbers is basic for so many practical applications of probability theory.

§ 5. The Borel–Cantelli lemma

We now prove the following useful and simple lemma:

LEMMA A. *If $\{A_n\}$ $(n = 1, 2, \ldots)$ is any infinite sequence of events such that*

$$\sum_{n=1}^{\infty} P(A_n) < +\infty, \qquad (1)$$

then with probability 1 at most finitely many events A_n occur simultaneously. Or to put it otherwise, if we put

$$A_\infty = \prod_{n=1}^{\infty} \sum_{k=n}^{\infty} A_k, \qquad (2)$$

then

$$P(A_\infty) = 0. \tag{3}$$

Remark. The right side of (2) is denoted in set theory by

$$\limsup_{n \to +\infty} A_n.$$

PROOF. We have

$$P(A_\infty) \le P(\sum_{k=n}^\infty A_k) \le \sum_{k=n}^\infty P(A_k) \tag{4}$$

for every n. Because of (1), the right hand side of (4) tends to zero as $n \to +\infty$, hence we have (3).

If the A_n are completely independent, (1) is not only sufficient, but also necessary in order that with probability 1 at most finitely many of the A_n should occur. If $\sum_{n=1}^\infty P(A_n) = +\infty$, then $P(A_\infty)$ is not only positive but equal to 1. Thus we have

LEMMA B. *If $A_1, A_2, \ldots, A_n, \ldots$ are completely independent events with* $\sum_{n=1}^\infty P(A_n) = +\infty$, *then*

$$P(A_\infty) = 1. \tag{5}$$

PROOF. Evidently

$$\bar{A}_\infty = \sum_{n=1}^\infty \prod_{k=n}^\infty \bar{A}_k , \tag{6}$$

it suffices thus to show that

$$P(\prod_{k=n}^\infty \bar{A}_k) = 0 \text{ for } n = 1, 2, \ldots; \tag{7}$$

(7) and (6) imply $P(\bar{A}_\infty) = 0$, hence (5). But for $N > n$

$$P(\prod_{k=n}^\infty \bar{A}_k) \le P(\prod_{k=n}^N \bar{A}_k) = \prod_{k=n}^N (1 - P(A_k)) < \exp\left[-\sum_{k=n}^N P(A_k)\right]. \tag{8}$$

The series $\sum_{k=1}^\infty P(A_k)$ being divergent, the right hand side of (8) tends to zero as $N \to \infty$. Lemma B is thus proved. Lemmas A and B together are called the *Borel–Cantelli lemma*.

The hypothesis of Lemma B concerning the complete independence of the A_n-s can be replaced by the weaker condition that $A_1, A_2, \ldots, A_n, \ldots$ be pairwise independent. Even somewhat more is true, namely

LEMMA C. *If $A_1, A_2, \ldots, A_n, \ldots$ are arbitrary events, fulfilling the conditions*

$$\sum_{n=1}^{\infty} P(A_n) = +\infty \tag{9}$$

and

$$\liminf_{n \to \infty} \frac{\sum_{k=1}^{n} \sum_{l=1}^{n} P(A_k A_l)}{\left(\sum_{k=1}^{n} P(A_k)\right)^2} = 1, \tag{10}$$

then (5) holds; thus there occur with probability 1 infinitely many of the events A_n.

PROOF. The proof is based on Chebyshev's inequality. Let α_n ($n = 1, 2, \ldots$) denote the indicator of the event A_n, i.e. put

$$\alpha_n = \begin{cases} 1 & \text{if } A_n \text{ occurs,} \\ 0 & \text{otherwise.} \end{cases}$$

Then $E(\alpha_n) = P(A_n)$ and by Chebyshev's inequality we have

$$P\left(\left|\sum_{k=1}^{n} \alpha_k - \sum_{k=1}^{n} P(A_k)\right| > \varepsilon \sum_{k=1}^{n} P(A_k)\right) \leq \frac{D^2\left(\sum_{k=1}^{n} \alpha_k\right)}{\varepsilon^2 \left(\sum_{k=1}^{n} P(A_k)\right)^2}. \tag{11}$$

Now $E(\alpha_k \alpha_l) = P(A_k A_l)$, hence

$$D^2\left(\sum_{k=1}^{n} \alpha_k\right) = \sum_{k=1}^{n} \sum_{l=1}^{n} P(A_k A_l) - \left(\sum_{k=1}^{n} P(A_k)\right)^2. \tag{12}$$

Thus it follows from (10) that

$$\liminf_{n \to \infty} P\left(\left|\sum_{k=1}^{n} \alpha_k - \sum_{k=1}^{n} P(A_k)\right| > \frac{1}{2} \sum_{k=1}^{n} P(A_k)\right) = 0. \tag{13}$$

If we put

$$d_n = P\left(\sum_{k=1}^{n} \alpha_k < \frac{1}{2} \sum_{k=1}^{n} P(A_k)\right), \tag{14}$$

we have

$$\liminf_{n \to \infty} d_n = 0. \tag{15}$$

It follows from this that one can choose an infinite subsequence of positive integers $n_1 < n_2 < \ldots < n_j < \ldots$, such that

$$\sum_{j=1}^{\infty} d_{n_j} < +\infty \; ; \tag{16}$$

hence by Lemma A we have with probability 1

$$\sum_{k=1}^{n_j} \alpha_k \geq \frac{1}{2} \sum_{k=1}^{n_j} P(A_k),$$

except for a finite number of values of j. Thus by (9) the series $\sum_{k=1}^{\infty} \alpha_k$ is divergent with probability 1, which proves our statement.

The just proved lemmas will serve us well in proofs dealing with improvements of the law of large numbers.

§ 6. Kolmogorov's inequality

In order to establish another group of laws of large numbers we need an improvement of Chebyshev's inequality due to A. N. Kolmogorov.

THEOREM 1. *Let the random variables* $\eta_1, \eta_2, \ldots, \eta_n$ *be completely independent; put further* $E(\eta_k) = M_k$, $D(\eta_k) = D_k$. *If* ε *is an arbitrary positive number, we have*

$$P\left(\max_{1 \leq k \leq n} \left| \sum_{j=1}^{k} (\eta_j - M_j) \right| \geq \varepsilon\right) \leq \frac{\sum_{k=1}^{n} D_k^2}{\varepsilon^2}. \tag{1}$$

PROOF. Put $\eta_k^* = \eta_k - M_k$, $\zeta_k = \sum_{j=1}^{k} \eta_j^*$ $(k = 1, 2, \ldots, n)$. Let further A_k denote the event that ζ_k is the first among the random variables $\zeta_1, \zeta_2, \ldots, \zeta_n$ which is not less than ε, i.e.

$$|\zeta_1| < \varepsilon, \ldots, |\zeta_{k-1}| < \varepsilon \quad \text{and} \quad |\zeta_k| \geq \varepsilon.$$

The events A_k $(k = 1, 2, \ldots, n)$ exclude each other but obviously do not form a complete system of events. Let A_0 denote the event $|\zeta_k| < \varepsilon$ for $k = 1, 2, \ldots, n$, then the events $A_0, A_1, \ldots, A_n$ form already a complete system of events. The probability figuring in (1) can be written in the form

$$P\left(\max_{1 \leq k \leq n} \left| \sum_{j=1}^{k} \eta_j^* \right| \geq \varepsilon\right) = \sum_{k=1}^{n} P(A_k). \tag{2}$$

On the other hand we have

$$\sum_{k=1}^{n} D_k^2 = D^2(\zeta_n) = \sum_{k=0}^{n} P(A_k) E(\zeta_n^2 | A_k). \tag{3}$$

The right hand side of (3) becomes smaller if the term $k = 0$ is omitted from the summation. Hence

$$\sum_{k=1}^{n} D_k^2 \geq \sum_{k=1}^{n} P(A_k) E(\zeta_n^2 | A_k). \tag{4}$$

Let us now consider the conditional expectation $E(\zeta_n^2 | A_k)$. From $\zeta_n = \zeta_k + \sum_{j=k+1}^{n} \eta_j^*$ it follows that

$$\zeta_n^2 = \zeta_k^2 + \sum_{j=k+1}^{n} \eta^{*2} + 2 \sum_{j=k+1}^{n} \zeta_k \eta_j^* + 2 \sum_{k<i<j\leq n} \eta_i^* \eta_j^*. \tag{5}$$

In the definition of the event A_k only occur the random variables $\eta_1^*, \ldots, \eta_k^*$. If α_k denotes the indicator of A_k, then it follows that α_k does not depend on η_j^* $(j > k)$. Hence we have for $k < j \leq n$

$$E(\zeta_k \eta_j^* | A_k) = \frac{E(\zeta_k \eta_j^* \alpha_k)}{P(A_k)} = \frac{E(\eta_j^*) E(\zeta_k \alpha_k)}{P(A_k)} = 0. \tag{6}$$

Similarly, we obtain that

$$E(\eta_i^* \eta_j^* | A_k) = 0 \tag{7}$$

because of the independence of $\eta_i^*, \eta_j^*, \alpha_k$ for $j > i > k$.
(5), (6) and (7) lead to

$$E(\zeta_n^2 | A_k) = E(\zeta_k^2 | A_k) + \sum_{j=k+1}^{n} E(\eta_j^{*2} | A_k) \geq E(\zeta_k^2 | A_k). \tag{8}$$

Now, by the definition of the events A_k, we have $| \zeta_k | \geq \varepsilon$ whenever A_k occurs; hence $E(\zeta_k^2 | A_k) \geq \varepsilon^2$ and thus by (8)

$$E(\zeta_n^2 | A_k) \geq \varepsilon^2.$$

Because of (4) it follows that

$$\sum_{k=1}^{n} D_k^2 \geq \varepsilon^2 \sum_{k=1}^{n} P(A_k),$$

which proves (1).

§ 7. The strong law of large numbers

A sequence η_n of random variables is said to *converge almost surely* (with probability 1) to 0 if

$$P(\lim_{n \to \infty} \eta_n = 0) = 1.$$

Almost sure convergence is a stronger condition than convergence in probability. Indeed we have the following

LEMMA 1. *The condition*

$$P(\lim_{n \to \infty} \eta_n = 0) = 1 \tag{1}$$

is equivalent to the condition

$$\lim_{n \to \infty} \text{st} \; (\sup_{m \geq n} |\eta_m|) = 0. \tag{2}$$

Consequently, (1) *implies*

$$\lim_{n \to \infty} \text{st} \; \eta_n = 0. \tag{3}$$

PROOF. We show first that (2) follows from (1). Let $\varepsilon > 0$; let $A_n(\varepsilon)$ denote the event $\sup_{m \geq n} |\eta_m| \geq \varepsilon$ and C the event $\lim_{n \to \infty} \eta_n = 0$; put further $B_n(\varepsilon) = = CA_n(\varepsilon)$. Then $B_{n+1}(\varepsilon) \subset B_n(\varepsilon)$ and the set $\prod_{n=1}^{\infty} B_n(\varepsilon)$ is obviously empty. It follows from this (cf. Ch. II, § 7, Theorem 3) that

$$\lim_{n \to \infty} P(B_n(\varepsilon)) = 0.$$

Now because of $P(C) = 1$ we have

$$P(B_n(\varepsilon)) = P(A_n(\varepsilon)).$$

Thus (2) holds and, because of (3),

$$|\eta_n| \leq \sup_{m \geq n} |\eta_m|$$

holds as well. Conversely, assume that (2) holds. Let $D(\varepsilon)$ denote the event $\limsup_{n \to \infty} |\eta_n| > \varepsilon$ ($\varepsilon > 0$, arbitrary).

Since obviously $D(\varepsilon) \subset A_n(\varepsilon)$ for $n = 1, 2, \ldots$, (2) implies $P(D(\varepsilon)) = 0$. As $\bar{C} \subset \sum\limits_{k=1}^{\infty} D\left(\dfrac{1}{k}\right)$, we get $P(\bar{C}) = 0$, hence (1). Herewith the lemma is proved.

Remark. Statement (1) concerning the almost sure convergence of η_k to 0 can be rephrased: the probability of the set of the elementary events ω for which $\lim\limits_{n \to \infty} \eta_n(\omega) = 0$ does not hold is equal to zero. In measure theory, this is expressed by saying that the variables $\eta_n(\omega)$ *converge almost everywhere* to 0. Convergence in probability is called in measure theory *convergence in measure.*

In order to emphasize the particular character of the strong law of large numbers we restrict ourselves first to a simple case.

THEOREM 1. *Let* $\xi_1, \xi_2, \ldots, \xi_n, \ldots$ *be (completely) independent and identically distributed random variables with finite expectation* $E(\xi_n) = M$ *and variance* $D^2(\xi_n) = D^2$. *Put*

$$\zeta_n = \frac{1}{n} \sum_{k=1}^{n} \xi_k.$$

Then

$$P(\lim_{n \to \infty} \zeta_n = M) = 1. \tag{4}$$

Remark. According to Lemma 1, (4) is a stronger statement than the law of large numbers: $\lim\limits_{n \to \infty} \text{st } \zeta_n = M$. Therefore Theorem 1 is called the *strong law of large numbers.*

In order to see better the meaning of (4), consider the case where the ξ_k are indicators of an event A in a sequence of experiments. ζ_n is then the relative frequency of A. Put $p = P(A)$. In this case the only thing the law of large numbers says is $\lim\limits_{n \to \infty} \text{st } \zeta_n = p$, while the strong law of large numbers tells us that $\lim\limits_{n \to \infty} \zeta_n = p$ with probability 1. That is, according to Lemma 1, all relations

$$|\zeta_{n+k} - p| < \varepsilon \qquad (k = 1, 2, \ldots)$$

are simultaneously fulfilled with a probability $\geq 1 - \delta$ for $\varepsilon > 0$ and $\delta > 0$ however small, if the index n is larger than a number n_0 depending on ε and δ.

PROOF OF THEOREM 1. We consider

$$\Delta_N = \sup_{n \geq N} |\zeta_n - M| \quad \text{and} \quad \Delta_{a,b} = \max_{a \leq n < b} |\zeta_n - M|.$$

If the inequality $\Delta_N \geq \varepsilon$ is fulfilled for an N such that $2^s \leq N < 2^{s+1}$, then $\Delta_{2^l,2^{l+1}} \geq \varepsilon$ is fulfilled for at least one $l \geq s$. Hence

$$P(\Delta_N \geq \varepsilon) \leq \sum_{l=s}^{\infty} P(\Delta_{2^l,2^{l+1}} \geq \varepsilon) \text{ for } 2^s \leq N < 2^{s+1}. \tag{5}$$

On the other hand we have

$$P(\Delta_{2^l,2^{l+1}} \geq \varepsilon) \leq P(\max_{1 \leq k < 2^{l+1}} k|\zeta_k - M| \geq \varepsilon \cdot 2^l). \tag{6}$$

If we apply to the random variables ξ_k $(k = 1, 2, \ldots, 2^{l+1} - 1)$ Kolmogorov's inequality (proved in the preceding section), we obtain

$$P(\max_{1 \leq k < 2^{l+1}} k|\zeta_k - M| \geq \varepsilon 2^l) \leq \frac{2^{l+1} D^2}{\varepsilon^2 \cdot 2^{2l}} = \frac{2D^2}{\varepsilon^2 \cdot 2^l}. \tag{7}$$

If we substitute this into (6) and consider (5), we have

$$P(\Delta_N \geq \varepsilon) \leq \frac{2D^2}{\varepsilon^2} \sum_{l=s}^{\infty} \frac{1}{2^l} = \frac{4D^2}{\varepsilon^2 \cdot 2^s}. \tag{8}$$

If $N \to \infty$, it follows from (6) and from $2^s > \frac{N}{2}$ that

$$\lim_{N \to \infty} P(\Delta_n \geq \varepsilon) = 0.$$

Hence by Lemma 1 our theorem is proved.

The following two generalizations of the strong law of large numbers are due to A. N. Kolmogorov.

THEOREM 2. *Let $\xi_1, \xi_2, \ldots, \xi_n, \ldots$ be a sequence of (completely) independent random variables for which $E(\xi_k) = M_k$ and $D(\xi_k) = D_k$ exist. Assume further that the series $\sum_{k=1}^{\infty} \frac{D_k^2}{k^2}$ converges. If we put*

$$\zeta_n = \frac{1}{n} \sum_{k=1}^{n} (\xi_k - M_k),$$

then

$$P(\lim_{n \to \infty} \zeta_n = 0) = 1. \tag{9}$$

THEOREM 3. *Let $\xi_1, \xi_2, \ldots, \xi_n, \ldots$ be (completely) independent identically distributed random variables. The random variable*

$$\zeta_n = \frac{1}{n} \sum_{k=1}^{n} \xi_k$$

converges with probability 1 *to a constant* C *iff the expectation* $M = E(\xi_k)$ *exists; in this case* $C = M$ *and, consequently,*

$$P(\lim_{n \to \infty} \zeta_n = M) = 1.$$

The hypothesis of the existence of the variance in Theorem 1 is therefore superfluous.

PROOF OF THEOREM 2. Put, as in the proof of Theorem 1,

$$\Delta_N = \sup_{k \geq N} |\zeta_k| \quad \text{and} \quad \Delta_{a,b} = \max_{a \leq k < b} |\zeta_k|.$$

It follows from $\Delta_N \geq \varepsilon$, $2^s \leq N < 2^{s+1}$ that $\Delta_{2^l, 2^{l+1}} \geq \varepsilon$ for at least one $l \geq s$; hence

$$P(\Delta_N \geq \varepsilon) \leq \sum_{l=s}^{\infty} P(\Delta_{2^l, 2^{l+1}} \geq \varepsilon). \tag{10}$$

By application of Kolmogorov's inequality we obtain

$$P(\Delta_{2^l, 2^{l+1}} \geq \varepsilon) \leq P(\max_{1 \leq k < 2^{l+1}} k|\zeta_k| \geq \varepsilon \cdot 2^l) \leq \frac{1}{2^{2l} \varepsilon^2} \sum_{k=1}^{2^{l+1}-1} D_k^2.$$

Hence by (10),

$$P(\Delta_N \geq \varepsilon) \leq \frac{1}{\varepsilon^2} \sum_{l=s}^{\infty} \frac{1}{2^{2l}} \sum_{k=1}^{2^{l+1}-1} D_k^2.$$

By interchanging the order of summation we find

$$P(\Delta_N \geq \varepsilon) \leq \frac{1}{3 \cdot 4^{s-1} \varepsilon^2} \sum_{k=1}^{2^{s+1}-1} D_k^2 + \frac{16}{3\varepsilon^2} \sum_{k=2^{s+1}}^{\infty} \frac{D_k^2}{k^2}. \tag{11}$$

Now it can be shown that the right hand side of inequality (11) tends to zero as n increases (hence as N increases too) provided that the series $\sum_{k=1}^{\infty} \frac{D_k^2}{k^2}$ is convergent. To show this we need the following lemma due to L. Kronecker.

LEMMA 2. *If the series* $\sum_{k=1}^{\infty} a_k$ *is convergent and if* q_n *is an increasing sequence of positive numbers, tending to* $+\infty$ *for* $n \to \infty$, *then*

$$\lim_{n \to \infty} \frac{1}{q_n} \sum_{k=1}^{n} a_k q_k = 0. \tag{12}$$

PROOF. Put $r_n = \sum\limits_{k=n}^{\infty} a_k$ and choose a number $n_0 = n_0(\varepsilon)$ (ε is an arbitrary small positive number) large enough in order that $n \geq n_0$ should imply $|r_n| < \dfrac{\varepsilon}{3}$. It is easy to see that (with $q_0 = 0$)

$$\frac{1}{q_n} \sum_{k=1}^{n} a_k q_k = \frac{1}{q_n} \sum_{k=1}^{n} r_k(q_k - q_{k-1}) - r_{n+1}. \tag{13}$$

If we put $\max\limits_{k \geq 1} |r_k| = A$, we have for $n \geq n_0$

$$\frac{1}{q_n} \left| \sum_{k=1}^{n} a_k q_k \right| \leq \frac{A q_{n_0}}{q_n} + \frac{2\varepsilon}{3}.$$

Choose now $n_1 > n_0$ such that $\dfrac{A q_{n0}}{q_{n_1}} \leq \dfrac{\varepsilon}{3}$. Then for $n > n_1$

$$\left| \frac{1}{q_n} \sum_{k=1}^{n} a_k q_k \right| < \varepsilon,$$

which proves Lemma 2.

This lemma and the convergence of the series $\sum\limits_{k=1}^{\infty} \dfrac{D_k^2}{k^2}$ imply immediately

$$\lim_{n \to \infty} \frac{\sum\limits_{k=1}^{n} D_k^2}{n^2} = 0;$$

hence the right hand side of (11) tends to 0 as $N \to \infty$. This and Lemma 1 lead to Theorem 2.

PROOF OF THEOREM 3. We show first that the existence of $M = E(\xi_k)$ suffices to imply $P(\lim\limits_{n \to \infty} \zeta_n = M) = 1$. For the sake of simplicity assume $M = 0$. Let the random variables ξ_k^* ($k = 1, 2, \ldots$) be defined by

$$\xi_k^* = \begin{cases} \xi_k & \text{for } |\xi_k| \leq k, \\ 0 & \text{otherwise} \end{cases}$$

and the random variables ξ_k^{**} by

$$\xi_k^{**} = \xi_k - \xi_k^*.$$

Put

$$\zeta_n = \frac{1}{n}\sum_{k=1}^{n}\xi_k, \quad \zeta_n^* = \frac{1}{n}\sum_{k=1}^{n}\xi_k^*, \quad \zeta_n^{**} = \frac{1}{n}\sum_{k=1}^{n}\xi_k^{**}.$$

Let $F(x)$ denote the distribution function of the ξ_k. Since we have assumed $M = 0$, we have

$$\lim_{n\to\infty}\frac{1}{n}\sum_{k=1}^{n}E(\xi_k^*) = \lim_{n\to\infty}\frac{1}{n}\sum_{k=1}^{n}E(\xi_k^{**}) = 0.$$

The expectation of the random variables ξ_k exists by assumption, hence the integral $\int_{-\infty}^{+\infty}|x|\,dF(x)$ is convergent. Since $\zeta_n = \zeta_n^* + \zeta_n^{**}$, it suffices evidently to prove that

$$P(\lim_{n\to\infty}\zeta_n^* = 0) = 1, \tag{14a}$$

$$P(\lim_{n\to\infty}\zeta_n^{**} = 0) = 1. \tag{14b}$$

Theorem 2 applies to the random variables ξ_k^*, since it is easy to show that $D(\xi_k^*)$ exists and that $\sum_{k=1}^{\infty}\frac{D^2(\xi_k^*)}{k^2} < +*\infty$. Indeed

$$D^2(\xi_k^*) \le E(\xi_k^{*2}) = \int_{-k}^{+k} x^2\,dF(x).$$

Hence, because of

$$\sum_{k=j+1}^{\infty}\frac{1}{k^2} \le \sum_{k=j+1}^{\infty}\frac{1}{k(k-1)} = \frac{1}{j},$$

we have

$$\sum_{k=1}^{\infty}\frac{D^2(\xi_k^*)}{k^2} \le \sum_{j=1}^{\infty}\left[\int_{j-1}^{j} x^2\,dF(x) + \int_{-j}^{-j+1} x^2\,dF(x)\right]\sum_{k=j}^{\infty}\frac{1}{k^2} \le$$

$$\le 2\int_{-\infty}^{+\infty}|x|\,dF(x).$$

Hence (14a) must hold. Now consider the random variables ξ_k^{**}. We have

$$P(\xi_k^{**} \ne 0) = \int_{|x|>k} dF(x), \tag{15}$$

hence

$$\sum_{k=1}^{\infty}P(\xi_k^{**} \ne 0) \le \int_{-\infty}^{+\infty}|x|\,dF(x). \tag{16}$$

Lemma A of § 5 permits to state that with probability 1 $\zeta_k^{**} = 0$ holds, for all but a finite number of values of k. This implies (14b) which proves the first half of Theorem 3.

Now we show that the condition is necessary. Assume the validity of $P(\lim_{n\to\infty} \zeta_n = C) = 1$, where C is a constant. Then with probability 1

$$\lim_{n\to\infty} \frac{\zeta_n}{n} = \lim_{n\to\infty} \left(\zeta_n - \frac{n-1}{n}\zeta_{n-1}\right) = C - C = 0.$$

Thus $\left|\frac{\zeta_n}{n}\right| > 1$ holds with probability 1 only for a finite number of values of n. Since the ζ_n are independent, it follows from Lemma B of § 5 that the series $\sum_{n=1}^{\infty} P\left(\left|\frac{\zeta_n}{n}\right| > 1\right)$ is convergent. Now since $P\left(\frac{|\zeta_n|}{n} > 1\right) = \int_{|x|>n} dF(x)$
the series

$$\sum_{n=1}^{\infty} \int_{|x|>n} dF(x) = \sum_{n=1}^{\infty} n\left(\int_n^{n+1} dF(x) + \int_{-(n+1)}^{-n} dF(x)\right)$$

is convergent. But we have

$$\int_{-\infty}^{+\infty} |x|\,dF(x) \le 1 + \sum_{n=1}^{\infty} n\left(\int_n^{n+1} dF(x) + \int_{-(n+1)}^{-n} dF(x)\right)$$

hence $\int_{-\infty}^{+\infty} |x|\,dF(x)$ exists and $M = E(\xi_k)$ exists, too. Hence by the first part of the theorem $C = M$ and thus our theorem is completely proved.

§ 8. The fundamental theorem of mathematical statistics

In the present paragraph we prove a theorem due to Glivenko, which is of fundamental importance in mathematical statistics.

THEOREM 1. *Let the random variables $\xi_1, \xi_2, \ldots, \xi_N$ be the elements of a sample drawn from a population, i.e. let $\xi_1, \xi_2, \ldots, \xi_N$ be identically distributed independent random variables with common distribution function $F(x)$. Let $F_N(x)$ denote the empirical distribution function of the sample, i.e. let $NF_N(x)$ be the number of the indices k for which $\xi_k < x$. Put further*

$$\Delta_N = \sup_{-\infty < x < +\infty} |F_N(x) - F(x)|.$$

Then

$$P(\lim_{N \to \infty} \Delta_N = 0) = 1. \tag{1}$$

Remark. Glivenko's theorem states that the empirical distribution function $F_N(x)$ of a sample of N elements converges, with probability 1, as $N \to \infty$ uniformly in x $(-\infty < x < +\infty)$ to the distribution function $F(x)$ of the population from which the sample was drawn.

PROOF. If $x_{M,k}$ (M a positive integer; $k = 1, 2, \ldots, M$) is the least number x fulfilling $F(x) \le \dfrac{k}{M} \le F(x + 0)$, we have

$$\Delta_N \le \max \ (\Delta_N^{(1)}, \Delta_N^{(2)}) + \frac{1}{M}, \tag{2}$$

where

$$\Delta_N^{(1)} = \max_{1 \le k \le M} |F_N(x_{M,k}) - F(x_{M,k})|,$$

$$\Delta_N^{(2)} = \max_{1 \le k \le M} |F_N(x_{M,k} + 0) - F(x_{M,k} + 0)|.$$

By the strong law of large numbers we have for every fixed x

$$P(\lim_{N \to \infty} F_N(x) = F(x)) = 1$$

and

$$P(\lim_{N \to \infty} F_N(x + 0) = F(x + 0)) = 1.$$

Thus it follows from (2) that

$$P\left(\limsup_{N \to \infty} \ \Delta_N > \frac{1}{M}\right) = 0 \tag{3}$$

for any natural number M, which proves the theorem.

This theorem of prime importance states that a large enough sample gives an almost exact information concerning the distribution of the population.

Glivenko's theorem can also be rephrased in the following manner: If ε and δ are two given positive numbers however small, then there exists an N_0 such that

$$P(\sup_{n \ge N_0} \Delta_n \le \varepsilon) > 1 - \delta. \tag{4}$$

This particular form shows clearly that the strong law of large numbers and Glivenko's theorem have a definite meaning even for the practical case when only finitely many observations are made. In fact, always when a large sample is studied, this theorem is implicitly used; hence it has the right to be called the fundamental theorem of mathematical statistics.

On the other hand it must be noticed that Glivenko's theorem does not give any information how N_0 figuring in (4) depends on ε and δ. This question will be answered by a theorem of Kolmogorov dealt with later on (cf. Ch. VII, § 10).

§ 9. The law of the iterated logarithm

The strong law of large numbers can be still further improved. To get acquainted with the methods needed for this, we give here first a new proof of Theorem 1 of § 7 concerning bounded random variables. The proof rests upon the Borel–Cantelli Lemma A.

Let $\xi_1, \xi_2, \ldots, \xi_n, \ldots$ be independent and identically distributed bounded random variables, and suppose $|\xi_n| \leq K$. Suppose further $E(\xi_n) = 0$, and put $D = D(\xi_n)$ and $\zeta_n = \dfrac{1}{n} \sum\limits_{k=1}^{n} \xi_k$. Then, according to Theorem 1 of § 4, we have the inequality

$$P(|\zeta_n| \geq \varepsilon) \leq 2q^n \qquad (1)$$

for $0 < \varepsilon < \dfrac{D^2}{K}$, where $q = \exp\left[-\dfrac{\varepsilon^2}{2D^2 \left(1 + \dfrac{\varepsilon K}{2D^2}\right)^2} \right]$. From (1) follows

the convergence of the series

$$\sum_{n=1}^{\infty} P(|\zeta_n| \geq \varepsilon) \qquad (2)$$

and because of Lemma A of § 5, we find that the inequality $|\zeta_n| < \varepsilon$ is fulfilled with probability 1 for every sufficiently large n. This implies the strong law of large numbers. Thus we obtained another proof of this law. Notice, however, that a supplementary hypothesis was needed for this proof: the random variables ξ_k were supposed to be bounded. This hypothesis allows to prove a far more precise theorem, called *the law of the iterated logarithm*:

THEOREM 1. *Let* $\xi_1, \xi_2, \ldots, \xi_n, \ldots$ *be uniformly bounded independent random variables with common expectation* $M = E(\xi_n)$ *and standard deviation* $D = D(\xi_n)$. *Put*

$$\eta_n = \frac{\xi_1 + \xi_2 + \ldots + \xi_n - nM}{D} . \qquad (3)$$

Then

$$P\left(\limsup_{n\to\infty} \frac{|\eta_n|}{\sqrt{2n \ln \ln n}} \le 1\right) = 1.\tag{4}$$

In order to prove this we first have to prove a lemma similar to Kolmogorov's inequality of § 6.

LEMMA 1. *Let* $\xi_1, \xi_2, \ldots, \xi_n, \ldots$ *be independent random variables with common expectation* $E(\xi_k) = M$ *and variance* $D^2(\xi_k) = D^2$. *Put*

$$\zeta_k = \xi_1 + \xi_2 + \ldots + \xi_k - kM.$$

Then

$$P(\max_{1\le k\le n} \zeta_k \ge x) \le \frac{4}{3} P(\zeta_n \ge x - 2D\sqrt{n}).\tag{5}$$

PROOF. Let A_k $(k = 1, 2, \ldots, n)$ denote the event that the inequalities

$$\zeta_1 < x,\ \zeta_2 < x, \ldots,\ \zeta_{k-1} < x,\ \zeta_k \ge x$$

are fulfilled. Let B_k denote the event $\zeta_n - \zeta_k > -2\sqrt{n}D$ and A the event $\zeta_n \ge x - 2\sqrt{n}D$. If both A_k and B_k occur, A occurs as well. The events A_k $(k = 1, 2, \ldots, n)$ mutually exclude each other, thus the same holds for the events A_kB_k; the events A_k and B_k are evidently independent since B_k depends only on the random variables $\xi_{k+1}, \ldots, \xi_n$ and A_k depends only on $\xi_1, \ldots, \xi_k$. Since $A_1B_1 + \ldots + A_nB_n \subseteq A$, the independence of A_k and B_k implies

$$\sum_{k=1}^{n} P(A_k)\,P(B_k) = \sum_{k=1}^{n} P(A_k B_k) = P(\sum_{k=1}^{n} A_k B_k) \le P(A).\tag{6}$$

On the other hand:

$$1 - P(B_k) \le P(|\zeta_n - \zeta_k| \ge 2D\sqrt{n}),\tag{7}$$

hence, by Chebyshev's inequality,

$$1 - P(B_k) \le \frac{n-k}{4n} \le \frac{1}{4}.\tag{8}$$

Thus

$$P(B_k) \ge \frac{3}{4}.\tag{9}$$

(6) and (9) lead to

$$\frac{3}{4}\sum_{k=1}^{n} P(A_k) = \frac{3}{4} P(\max_{1\leq k\leq n} \zeta_k \geq x) \leq P(A) = P(\zeta_n \geq x - 2D\sqrt{n}),$$

which was to be proved.

Now we proceed to the proof of the law of the iterated logarithm. We show first that with probability 1 only finitely many of the events $\eta_n >$ $> (1 + \varepsilon)\sqrt{2n \ln \ln n}$ occur if $\varepsilon > 0$ is arbitrarily small. Let γ be $\sqrt{1 + \varepsilon}$ and let N_k be the least positive integer larger than γ^k. Let $A_k(\varepsilon)$ denote the event

$$\max_{N_k\leq n<N_{k+1}} \eta_n > (1 + \varepsilon)\sqrt{2N_k \ln \ln N_k}. \tag{10}$$

In order to show that with probability 1 at most finitely many of the events $A_k(\varepsilon)$ occur, it suffices, in view of the Borel–Cantelli lemma, to show that the series $\sum_{k=1}^{\infty} P(A_k(\varepsilon))$ is convergent for every $\varepsilon > 0$. Now

$$P(A_k(\varepsilon)) \leq P(\max_{1\leq n<N_{k+1}} \eta_n \geq (1 + \varepsilon)\sqrt{2N_k \ln \ln N_k}), \tag{11}$$

hence, because of Lemma 1,

$$P(A_k(\varepsilon)) \leq \frac{4}{3} P(\eta_{N_{k+1}} \geq (1 + \varepsilon)\sqrt{2N_k \ln \ln N_k} - 2\sqrt{N_{k+1}}). \tag{12}$$

If we put

$$\mu = \mu_k = (1 + \varepsilon)^{\frac{7}{8}} \sqrt{\frac{2N_k \ln \ln N_k}{N_{k+1}}}$$

we conclude from (12) by applying Theorem 1 of § 4 that the following relation holds if k is large enough:

$$P(A_k(\varepsilon)) \leq \frac{8}{3} \exp\left[-\frac{\mu_k^2}{2\left(1 + \dfrac{\mu_k K}{2\sqrt{N_{k+1}}}\right)}\right]. \tag{13}$$

Since $\lim_{k\to\infty} \dfrac{\mu_k}{\sqrt{2\ln\ln N_k}} = \dfrac{(1+\varepsilon)^{\frac{7}{8}}}{\sqrt{\gamma}} = (1+\varepsilon)^{\frac{5}{8}}$ and $\lim_{k\to\infty} \dfrac{\mu_k K}{2\sqrt{N_{k+1}}} = 0$ there exists a number k_0 depending only on ε such that for $k \geq k_0$ we have

$$\frac{\mu_k^2}{2\left(1 + \dfrac{\mu_k K}{2\sqrt{N_{k+1}}}\right)} \geq (1 + \varepsilon)\ln\ln N_k \geq (1 + \varepsilon)\ln(k \ln \gamma).$$

Hence for $k \ge k_0$

$$P\big(A_k(\varepsilon)\big) \le \frac{8}{3}\, e^{-(1+\varepsilon)\ln(k\ln\gamma)} = \frac{8}{3\left(\dfrac{1}{2}\ln(1+\varepsilon)\right)^{1+\varepsilon} k^{1+\varepsilon}}. \tag{14}$$

The series $\sum\limits_{k=1}^{\infty} P\big(A_k(\varepsilon)\big)$ is thus convergent for every positive ε and according to the Borel–Cantelli lemma,

$$P\left(\limsup_{n\to\infty} \frac{\eta_n}{\sqrt{2n\ln\ln n}} \le 1\right) = 1. \tag{15a}$$

It can be shown in a similar manner that

$$P\left(\liminf_{n\to\infty} \frac{\eta_n}{\sqrt{2n\ln\ln n}} \ge -1\right) = 1. \tag{15b}$$

This proves Theorem 1.

It should be noticed that (4) cannot be improved; in fact, it can be shown that

$$P\left(\limsup_{n\to\infty} \frac{\eta_n}{\sqrt{2n\ln\ln n}} = 1\right) = 1 \tag{16a}$$

and

$$P\left(\liminf_{n\to\infty} \frac{\eta_n}{\sqrt{2n\ln\ln n}} = -1\right) = 1. \tag{16b}$$

In particular, if the ζ_k are indicators of an event A with probability $P(A) = p$ in a sequence of independent experiments, the conditions of Theorem 1 are fulfilled. Thus we have

THEOREM 2. *If ζ_n represents the relative frequency of an event A in a sequence of independent experiments and if $P(A) = p$ $(0 < p < 1,\ q = 1 - p)$, then*

$$P\left(\limsup_{n\to\infty} \frac{|\zeta_n - p|}{\sqrt{\dfrac{2pq\ln\ln n}{n}}} \le 1\right) = 1. \tag{17}$$

As we have seen, there even holds the more precise relation

$$P\left(\limsup_{n\to\infty} \frac{\zeta_n - p}{\sqrt{\dfrac{2pq \ln \ln n}{n}}} = +1\right) =$$

$$= P\left(\liminf_{n\to\infty} \frac{\zeta_n - p}{\sqrt{\dfrac{2pq \ln \ln n}{n}}} = -1\right) = 1. \tag{18}$$

For the proof of (18) cf. § 16, Exercise 23.

§ 10. Sequences of mixing sets

Let $[\Omega, \mathscr{A}, P]$ be a Kolmogorov probability space and $A_n \in \mathscr{A}$ ($n = 1, 2, \ldots$) a sequence of sets such that for every $B \in \mathscr{A}$ the relation

$$\lim_{n\to\infty} P(A_n B) = dP(B) \tag{1}$$

holds, where d is a number, not depending on B, such that $0 < d < 1$. Then the sequence $\{A_n\}$ is said to be *mixing*; d is called the *density* of the sequence $\{A_n\}$.

THEOREM 1. *If $A_0 = \Omega$, $A_n \in \mathscr{A}$ ($n = 1, 2, \ldots$), $0 < P(A_n) < 1$ and*

$$\lim_{n\to\infty} P(A_n \mid A_k) = d \qquad (k = 0, 1, \ldots), \tag{2}$$

then $\{A_n\}$ is mixing.

Remark. Evidently condition (2) is also necessary, as it is a particular case of (1) for $B = A_k$ ($k = 0, 1, \ldots$).

PROOF. We use elements of the theory of Hilbert spaces. Let $\mathscr{H}$ be the set of all random variables ξ for which $E(\xi^2)$ exists. Put $(\xi, \eta) = E(\xi\eta)$ and $\|\xi\| = (\xi, \xi)^{\frac{1}{2}}$. $\mathscr{H}$ is then a Hilbert space. Let α_n denote the indicator of the event A_n:

$$\alpha_n = \alpha_n(\omega) = \begin{cases} 1 & \text{for } \omega \in A_n \\ 0 & \text{for } \omega \in \bar{A}_n \end{cases} \qquad (n = 0, 1, \ldots).$$

If β is the indicator of B and if $\alpha_n - d = \gamma_n$, we can write (1) in the form

$$\lim_{n\to\infty} (\beta, \gamma_n) = 0, \tag{3}$$

while (2) is equivalent to

$$\lim_{n\to\infty} (\gamma_k, \gamma_n) = 0 \qquad (k = 0, 1, \ldots). \tag{4}$$

We show that (4) implies (3) for every $\beta \in \mathscr{H}$ (hence not merely for the β which are indicators of sets).

Let $\mathscr{H}_1$ denote the set of those elements of $\mathscr{H}$ which are linear combinations of the γ_n or limits of such elements, in the sense of strong convergence, that is $\delta_n \to \delta$ means that $\lim_{n\to\infty} \| \delta_n - \delta \| = 0$. In other words, $\mathscr{H}_1$ is the least subspace of $\mathscr{H}$ containing the elements γ_n $(n = 0,1\ldots)$. Obviously, (4) implies (3) when β is a finite linear combination of the γ_n, and also when $\beta \in \mathscr{H}_1$. In fact, in the latter case there exists for every $\varepsilon > 0$ a $\gamma = \sum_{k=1}^{n} c_k \gamma_k$ with $\| \beta - \gamma \| < \varepsilon$. Because of Schwarz' inequality and of

$$\| \gamma_n \| = E\big((\alpha_n - d)^2\big) = P(A_n)(1 - d)^2 + \big(1 - P(A_n)\big) d^2 \leq 1$$

we have

$$| (\beta, \gamma_n) - (\gamma, \gamma_n) | = | (\beta - \gamma, \gamma_n) | \leq \| \beta - \gamma \| \cdot \| \gamma_n \| \leq \varepsilon.$$

By (4) $\lim_{n\to\infty} (\gamma, \gamma_n) = 0$, thus $\limsup_{n\to\infty} | (\beta, \gamma_n) | \leq \varepsilon$, since for $\varepsilon > 0$ there can be chosen any positive number however small. (3) is therefore proved for every $\beta \in \mathscr{H}_1$.

Let now $\mathscr{H}_2$ be the set of elements δ of $\mathscr{H}$ such that $(\delta, \gamma_n) = 0$ for $n = 0,1,\ldots$. $\mathscr{H}_2$ is then the subspace of $\mathscr{H}$ orthogonal to $\mathscr{H}_1$. For $\beta \in \mathscr{H}_2$ (3) is trivial. Now according to a well-known theorem of the theory of Hilbert spaces[1] $\beta \in \mathscr{H}$ can be written in the form $\beta = \beta_1 + \beta_2$, where $\beta_1 \in \mathscr{H}_1$ and $\beta_2 \in \mathscr{H}_2$. Furthermore,

$$(\beta, \gamma_n) = (\beta_1, \gamma_n) + (\beta_2, \gamma_n) = (\beta_1, \gamma_n)$$

hence (3) holds for every $\beta \in \mathscr{H}$. Theorem 1 is thus proved. As an application we prove now a theorem which shows new aspects of the laws of large numbers.

THEOREM 2. *Let $\xi_1, \xi_2, \ldots, \xi_n, \ldots$ be independent random variables whose arithmetic mean $\zeta_n = \dfrac{\xi_1 + \ldots + \xi_n}{n}$ tends in probability to a random variable ζ. Then ζ is equal with probability 1 to a constant.*

---

[1] Cf. e.g. B. Sz.-Nagy [1] or F. Riesz and B. Sz.-Nagy [1].

PROOF. Choose two numbers a and b such that $P(a \leq \zeta < b) > 0$ and a, b are two points of continuity of the distribution function of ζ. Then $P(a \leq \zeta_n < b) > 0$ for $n > n_0$. Let $A_0 = \Omega$ and let A_k denote the event $a \leq \zeta_{n_0 + k + 1} < b$ $(k = 1, 2, \ldots)$. For $k \geq 1$ we have

$$P(A_n \mid A_k) =$$

$$= P\left(a \leq \frac{(n_0 + k + 1)\zeta_{n_0 + k + 1} + \xi_{n_0 + k + 2} + \cdots + \xi_{n_0 + n + 1}}{n_0 + n + 1} < b \,\middle|\, A_k\right) \leq$$

$$\leq P\left(a - \frac{(n_0 + k + 1)b}{n_0 + n + 1} \leq \frac{\xi_{n_0 + k + 2} + \cdots + \xi_{n_0 + n + 1}}{n_0 + n + 1} < b - \frac{(n_0 + k + 1)a}{n_0 + n + 1}\right)$$

$$\leq P\left(a - \varepsilon \leq \frac{\xi_{n_0 + k + 2} + \cdots + \xi_{n_0 + n + 1}}{n_0 + n + 1} \leq b + \varepsilon\right)$$

for any $\varepsilon > 0$ whenever n is large enough. Similarly, for sufficiently large n,

$$P(A_n \mid A_k) \geq P\left(a + \varepsilon \leq \frac{\xi_{n_0 + k + 2} + \cdots + \xi_{n_0 + n + 1}}{n_0 + n + 1} \leq b - \varepsilon\right).$$

Now by assumption $\zeta_n \xrightarrow{P} \zeta$, hence also $\dfrac{\xi_{n_0 + k + 2} + \cdots + \xi_n}{n} \xrightarrow{P} \zeta$. Now for $\varepsilon > 0$ there can be chosen any positive number however small, thus Theorem 1 of § 2 leads to

$$\lim_{n \to \infty} P(A_n \mid A_k) = P(a \leq \zeta < b),$$

since a and b are points of continuity of the distribution function of ζ. Thus by Theorem 1 the sequence of events $\{A_n\}$ is a mixing sequence with the density $d = P(a \leq \zeta < b)$.

Now let B be the event $a \leq \zeta < b$. We have

$$\lim_{n \to \infty} P(A_n \mid B) = d = P(B).$$

The random variables ζ_n tend in probability to ζ, also when taken on the probability space $[\Omega, \mathcal{A}, P(A \mid B)]$. Thus $\lim\limits_{n \to \infty} P(A_n \mid B) = P(B \mid B) = 1$ by Theorem 1 of § 2. Consequently, $P(B) = 1$ since $P(a \leq \zeta < b) > 0$, $P(a \leq \zeta < b) = 1$. But this means that ζ is a constant with probability 1; in fact if the distribution function of ζ would increase at more than one point, there could be found a pair $a < b$ with $0 < P(a \leq \zeta < b) < 1$ such that a and b are points of continuity of the distribution function of ζ. Theorem 2 is herewith proved. The arithmetic means of independent random variables either tend in probability to a constant or do not converge at all.

§ 11. Stable sequences of events

In the present section we deal with a generalization of the notion of mixing sequences of events, introduced in the preceding section. Let $[\Omega, \mathscr{A}, P]$ be a Kolmogorov probability space; a sequence $\{A_n\}$ of events ($A_n \in \mathscr{A}$; $n = 1, 2, \ldots$) such that for any event $B \in \mathscr{A}$ there exists the limit

$$\lim_{n \to \infty} P(A_n B) = Q(B) \tag{1}$$

will be called a *stable sequence of events*.[1] We shall prove first that the set function $Q(B)$ on the right hand side of (1) is always a measure, i.e. we prove

THEOREM 1. *If* $\{A_n\}$ *is a stable sequence of events, the set function* $Q(B)$ *defined by* (1) *is a measure which is moreover absolutely continuous with respect to the probability measure P.*

PROOF. Obviously, $Q(B)$ is a nonnegative and additive set function and $Q(B) \le P(B)$, hence if $P(B) = 0$, then $Q(B) = 0$. From this the assertion of our theorem follows directly, by Theorem 3 of Chapter II, § 7.

According to the Radon–Nikodym theorem the derivative

$$\frac{dQ}{dP} = \alpha(\omega) \tag{2}$$

exists, furthermore, for every event $B \in \mathscr{A}$ one has

$$Q(B) = \int_B \alpha \, dP. \tag{3}$$

It follows directly from the inequality $Q(B) \le P(B)$ that

$$0 \le \alpha(\omega) \le 1 \tag{4}$$

with probability 1. The random variable $\alpha = \alpha(\omega)$ is called the (local) *density* of the stable sequence of events $\{A_n\}$.

If α is constant almost everywhere, $\alpha = d \, (0 < d < 1)$, then clearly the stable sequence of events $\{A_n\}$ is mixing and has density d. On the other hand, if α is not constant, the stable sequence of events $\{A_n\}$ cannot be a mixing sequence. Hence the notion of the stable sequence of events is a generalization of that of a mixing sequence of events.

Let us now consider an example of a stable but not mixing sequence of events. Let $[\Omega, \mathscr{A}, P]$ be a Kolmogorov probability space, let $\Omega_1 \in \mathscr{A}$, $0 < P(\Omega_1) < 1$ and $\Omega_2 = \overline{\Omega}_1$. Consider further the probability spaces $[\Omega, \mathscr{A}, P_1]$ and $[\Omega, \mathscr{A}, P_2]$, where for every $A \in \mathscr{A}$ $P_1(A) = P(A \mid \Omega_1)$ and $P_2(A) = P(A \mid \Omega_2)$.

[1] Cf. A. Rényi [36].

Let A'_n be a mixing sequence of events in the probability space $[\Omega, \mathscr{A}, P_1]$ with density d_1 and A''_n a mixing sequence of events in the probability space $[\Omega, \mathscr{A}, P_2]$ with density d_2 $(0 < d_1 < d_2 < 1)$; put $A_n = A'_n \Omega_1 + A''_n \Omega_2$. Then clearly we have for every event $B \in \mathscr{A}$

$$P(A_n B) = P(\Omega_1) P_1 (A'_n B) + P(\Omega_2) P_2(A''_n B),$$

hence

$$\lim_{n \to \infty} P(A_n B) = Q(B),$$

where

$$Q(B) = d_1 P(B\Omega_1) + d_2 P(B\Omega_2).$$

Let the random variable $\alpha = \alpha(\omega)$ be defined in the following manner:

$$\alpha(\omega) = \begin{cases} d_1 & \text{if } \omega \in \Omega_1, \\ d_2 & \text{if } \omega \in \Omega_2, \end{cases}$$

then

$$Q(B) = \int_B \alpha dP.$$

Thus the sequence of events $\{A_n\}$ is stable but not mixing, since its density is not constant but assumes two distinct values with positive probabilities. Clearly, there can be constructed in a similar manner stable sequences of events with densities having an arbitrary prescribed discrete distribution.

Now we shall prove the generalization of Theorem 1 of § 10 concerning stable sequences of events.

THEOREM 2. *If* $A_n \in \mathscr{A}$ $(n = 1, 2, \ldots)$, $A_1 = \Omega$ *and if the limits*

$$\lim_{n \to \infty} P(A_n A_k) = Q_k \qquad (k = 1, 2, \ldots)$$

exist, then the sequence of the events $\{A_n\}$ *is a stable sequence of events.*

PROOF. The proof of this theorem corresponds nearly step-by-step to that of Theorem 1 in § 10, hence it will be only sketched. Let $\mathscr{H}$ denote the Hilbert space of all random variables with finite standard deviation on the probability space $[\Omega, \mathscr{A}, P]$; scalar product and norm are (as usual) defined by $(\xi, \eta) = E(\xi\eta)$ and $\| \xi \| = (\xi, \xi)^{\frac{1}{2}}$, respectively. Let α_n be the indicator of the event A_n. Let $\mathscr{H}_1$ be the subspace of the Hilbert space $\mathscr{H}$ spanned by the elements $\alpha_1, \alpha_2, \ldots, \alpha_k, \ldots$; thus $\mathscr{H}_1$ consists of the finite linear combinations (with real coefficients) of the elements of the sequence $\{\alpha_k\}$ and of the (strong) limits of these elements. It is easy to see that if $\xi \in \mathscr{H}_1$, then the limit

$$\lim_{n \to \infty} (\xi, \alpha_n) = L(\xi) \tag{5}$$

exists; in fact if $\xi = c_1\alpha_1 + \ldots + c_k\alpha_k$, then

$$\lim_{n \to \infty} (\xi, \alpha_n) = c_1 Q_1 + \ldots + c_k Q_k,$$

while if ξ is the limit of linear combinations of the α_k, the limit (5) exists again, since

$$|(\xi, \alpha_n) - (\xi', \alpha_n)| = |(\xi - \xi', \alpha_n)| \le \|\xi - \xi'\|.$$

Now in the same way as was done in Section 10 we decompose $\xi \in \mathscr{H}$ as $\xi = \xi_1 + \xi_2$, where $\xi_1 \in \mathscr{H}_1$ and $(\xi_2, \alpha_k) = 0$ $(k = 1, 2, \ldots)$; hence the limit (5) exists for any $\xi \in \mathscr{H}$, and our theorem is proved.

Clearly the functional $L(\xi)$ has the following properties: If $\xi \in \mathscr{H}, \eta \in \mathscr{H}$, furthermore a and b are real constants, then

$$L(a\xi + b\eta) = aL(\xi) + bL(\eta)$$

and

$$|L(\xi)| \le \|\xi\|,$$

To put it otherwise, $L(\xi)$ is a bounded linear functional and thus, according to the well-known theorem of F. Riesz (cf. F. Riesz, and B. Sz.-Nagy, [1]), there exists an $\alpha \in \mathscr{H}$ such that

$$L(\xi) = (\xi, \alpha),$$

i.e. the sequence α_n converges to α, in the sense of weak convergence in the Hilbert space. (A sequence of elements α_n of a Hilbert space is said to converge weakly to α $(\alpha \in \mathscr{H})$, if for any element $\xi \in \mathscr{H}$

$$\lim_{n \to \infty} (\xi, \alpha_n) = (\xi, \alpha).$$

This fact is denoted by $\alpha_n \rightharpoonup \alpha$.)

The preceding discussion contains the proof of the following

THEOREM 3. *Let α_n denote the indicator of the event A_n and $\mathscr{H}$ the Hilbert space formed by the random variables with finite second moments defined on the probability space $[\Omega, \mathscr{A}, P]$. The sequence of events $\{A_n\}$ belonging to the probability space $[\Omega, \mathscr{A}, P]$ is stable, iff α_n converges weakly in $\mathscr{H}$ to an element $\alpha \in \mathscr{H}$. If the sequence of events $\{A_n\}$ is stable and if $\alpha_n \rightharpoonup \alpha$, then α is the density of the sequence of events $\{A_n\}$.*

A stable sequence of events $\{A_n\}$ is mixing, iff there exists a number d $(0 < d < 1)$ such that for every event A

$$Q(A) = dP(A). \tag{6}$$

It is readily seen from Theorem 1 of § 10 that for $A = \Omega$ and $A = A_k$ $(k = 1, 2, \ldots)$ it suffices to assume the validity of (6); from this it follows that $\{A_n\}$ is a mixing sequence of events.

Finally, we prove another theorem showing the great generality of the notion of stable sequences of events.

THEOREM 4. *From any sequence of events one can select a stable subsequence.*

PROOF. Theorem 4 is a direct consequence of a well-known theorem of the Hilbert space theory (cf. F. Riesz, and B. Sz.-Nagy, [1]) stating that from any sequence of elements with bounded norm of the Hilbert space a weakly convergent subsequence can be selected.

As an application of these results we shall discuss in the following section sequences of *exchangeable* events.

§ 12. Sequences of exchangeable events

The notion of a sequence of exchangeable events was already encountered (cf. Ch. II, § 12, Exercises 38–43). Here we shall deal only with *infinite sequences of exchangeable events*. First, let us repeat the definition.

A sequence $\{A_n\}$ $(n = 1, 2, \ldots)$ of events is said to be exchangeable if the probability of the joint occurrence of k distinct events chosen arbitrarily from this sequence depends only on k for every positive value of k, but does not depend on which k events were chosen. Thus there can be given a sequence of numbers p_k such that

$$P(A_{n_1} A_{n_2} \ldots A_{n_k}) = p_k \qquad (k = 1, 2, \ldots), \qquad (1)$$

whenever $n_1 < n_2 < \ldots < n_k$.

First we prove the following theorem:

THEOREM 1. *A sequence of exchangeable events is always stable and is mixing, iff the events are independent and have the same probability.*

PROOF. Let $\{A_n\}$ be a sequence of exchangeable events. Then according to our assumption

$$P(A_n) = p_1,$$

hence

$$\lim_{n \to \infty} P(A_n) = p_1,$$

further by assumption

$$P(A_n A_k) = p_2 \quad \text{if} \quad n > k,$$

hence

$$\lim_{n \to \infty} P(A_n A_k) = p_2 \qquad (k = 1, 2, \ldots).$$

Therefore by Theorem 2 of § 11 $\{A_n\}$ is a stable sequence of events. Now according to § 11 a stable sequence of events $\{A_n\}$ is a mixing sequence, iff α is a constant, $\alpha = d$. In this case $p_1 = d$ and $p_2 = d^2$, furthermore

$$p_3 = P(A_n A_2 A_1), \quad \text{if} \quad n \geq 3,$$

hence

$$p_3 = \lim_{n \to \infty} P(A_n A_2 A_1) = dP(A_2 A_1) = d^3,$$

and, similarly, it can be seen that for every $k \geq 1$ we have

$$p_k = d^k,$$

i.e.

$$P(A_{n_1} A_{n_2} \ldots A_{n_k}) = P(A_{n_1}) P(A_{n_2}) \ldots P(A_{n_k}),$$

whenever $1 \leq n_1 < n_2 < \ldots < n_k$. But this means that the events A_n are independent.

Now we prove a theorem due to B. de Finetti.

THEOREM 2. *If $\{A_n\}$ is a sequence of exchangeable events and the numbers p_k are defined by* (1), *there can be given on the interval* $[0, 1]$ *a distribution function $F(x)$ (fulfilling $F(0) = 0$ and $F(1 + 0) = 1$) such that*

$$p_k = \int_0^1 x^k \, dF(x) \qquad (k = 1, 2, \ldots). \tag{2}$$

PROOF. By Theorem 1 the sequence $\{A_n\}$ is stable. Let α denote the density of this sequence. Then

$$p_1 = P(A_k) = \lim_{n \to \infty} P(A_n) = \int_\Omega \alpha \, dP,$$

furthermore $p_2 = P(A_n A_k)$ if $n > k$ and thus

$$p_2 = \lim_{n \to \infty} P(A_n A_k) = \int_{A_k} \alpha \, dP = \int_\Omega \alpha \, \alpha_k \, dP \qquad (k = 1, 2, \ldots),$$

hence by Theorem 3 of § 11

$$p_2 = \lim_{k \to \infty} \int_\Omega \alpha \alpha_k \, dP = \int_\Omega \alpha^2 \, dP.$$

Similarly,

$$p_3 = P(A_n A_k A_l) \quad \text{if} \quad n > k > l,$$

thus

$$p_3 = \lim_{n \to \infty} P(A_n A_k A_l) = \int_\Omega \alpha \alpha_k \alpha_l \, dP,$$

hence − by taking the limit first for $k \to \infty$, then for $l \to \infty$, − we obtain that

$$p_3 = \int_\Omega \alpha^3 \, dP$$

and, in general, for every positive integral value of k we get

$$p_k = \int_\Omega \alpha^k \, dP. \tag{3}$$

Let now $F(x)$ denote the distribution function of the random variable α. Thus (cf. Theorem 6 of Ch. IV, § 11)

$$p_k = \int_0^1 x^k \, dF(x) \qquad (k = 1, 2, \ldots),$$

which was to be proved. The proof gives, however, somewhat more than stated by Theorem 2; in fact we have proved the more general

THEOREM 3. *Let* $\{A_n\}$ *be an arbitrary exchangeable sequence of events; let the numbers* p_k *be defined by* (1). *Let* α_n *denote the indicator of the event* A_n *and* $\alpha = \alpha(\omega)$ *the density of the sequence* $\{A_n\}$. *Then, for any choice of the positive integers* $1 \le k_1 < k_2 < \ldots < k_r$ *and for any integer* $s \ge 0$ *we have*

$$\int_\Omega \alpha_{k_1} \alpha_{k_2} \ldots \alpha_{k_r} \alpha^s \, dP = p_{r+s}. \tag{4}$$

Remark. Theorem 2 is contained as a particular case in Theorem 3. In fact, if $r = 0$, relation (4) reduces to (3). On the other hand, if $s = 0$, then (4) reduces to (1); i.e. to the definition of the sequence of numbers p_r.

Let $\{A_n\}$ be an exchangeable sequence of events. Let us compute the probability of the joint occurrence of k distinct events selected from this sequence and of the simultaneous nonoccurrence of l events distinct from the former and from each other; i.e. let us compute the probability $P(A_{n_1} A_{n_2} \ldots A_{n_k} \bar{A}_{m_1} \bar{A}_{m_2} \ldots \bar{A}_{m_l})$, where $n_1 < n_2 < \ldots < n_k$, $m_1 < m_2 < \ldots < m_l$ and $n_i \ne m_j$ ($i = 1, 2, \ldots, k$; $j = 1, 2, \ldots, l$). It can be seen by an easy calculation that

$$P(A_{n_1} A_{n_2} \ldots A_{n_k} \bar{A}_{m_1} \bar{A}_{m_2} \ldots \bar{A}_{m_l}) = \sum_{j=0}^{l} p_{k+j} (-1)^j \binom{l}{j}. \tag{5}$$

Equation (5) is valid for the case $k = 0$ too, if we understand by p_0 the value 1.

The expression on the right hand side of (5) can be written in the form

$$\sum_{j=0}^{l} p_{k+j} (-1)^j \binom{l}{j} = (-1)^l \varDelta^l p_{k+l}, \qquad (6)$$

where $\varDelta$ denotes the difference operator, defined by $\varDelta x_k = x_k - x_{k-1}$. Since the probability on the left hand side of (5) is nonnegative, we have obtained that for the sequence of numbers p_k the inequalities

$$(-1)^l \varDelta^l p_{k+l} \geq 0 \qquad (7)$$

hold. Sequences of numbers having property (7) are called *absolutely monotonic* sequences. Hence an absolutely monotonic sequence is nonincreasing, its first differences form a nondecreasing sequence (i.e. the sequence is *convex*), its second differences form a nonincreasing sequence, etc. Note that inequality (7) can be obtained from the representation of the sequence of numbers p_k in (2) or (3), since

$$(-1)^l \varDelta^l p_{k+l} = \int_0^1 x^k (1 - x)^l \, dF(x) = \int_\Omega \alpha^k (1 - \alpha)^l \, dP. \qquad (8)$$

It was shown by F. Hausdorff that every absolutely monotone sequence of numbers p_k $(k = 0, 1, \ldots)$ for which $p_0 = 1$ can be represented in the form (2), where $F(x)$ is a distribution function on the interval $(0, 1)$. This theorem can be deduced from the above Theorem 2. It suffices to show that if p_k $(k = 0, 1, \ldots)$ is an arbitrary absolutely monotone sequence of numbers for which $p_0 = 1$, there can be constructed a sequence of exchangeable events $\{A_n\}$ on a suitable probability space, fulfilling (1). This construction is readily performed, e.g. by the fundamental theorem of Kolmogorov (Ch. IV, § 22). In fact, if α_n denotes the indicator of the event A_n, then

$$P(A_{n_1} \ldots A_{n_k} \bar{A}_{m_1} \ldots \bar{A}_{m_l}) = P(\alpha_{n_1} = 1, \ldots, \alpha_{n_k} = 1, \alpha_{m_1} = 0, \ldots, \alpha_{m_l} = 0).$$

Hence we can see from (5) that given the sequence p_k, the joint distribution function of a finite number of random variables chosen arbitrarily from $\alpha_1, \alpha_2, \ldots, \alpha_n, \ldots$ is given as well; the conditions of compatibility are, obviously, fulfilled and thus the existence of th e(exchangeable) sequence of events with the required properties is ensured by the fundamental theorem of Kolmogorov.

Now we shall prove the following

THEOREM 4. *Let $\{A_n\}$ be a sequence of exchangeable events with density α and let α_n denote the indicator of the event A_n. Then we have with probability 1*

$$\lim_{n \to \infty} \frac{\alpha_1 + \alpha_2 + \ldots + \alpha_n}{n} = \alpha. \tag{9}$$

PROOF. Let

$$\zeta_n = \frac{\alpha_1 + \alpha_2 + \ldots + \alpha_n}{n} - \alpha = \frac{\sum_{k=1}^{n} (\alpha_k - \alpha)}{n}.$$

We calculate the expectation of ζ_n^4. According to (4), if k_1, k_2, k_3, k_4 are distinct positive integers, we have

$$E\big((\alpha_{k_1} - \alpha)(\alpha_{k_2} - \alpha)(\alpha_{k_3} - \alpha)(\alpha_{k_4} - \alpha)\big) = 0. \tag{10}$$

Similarly, it can be seen that if k_1, k_2, k_3 are distinct, then

$$E\big((\alpha_{k_1} - \alpha)^2 (\alpha_{k_2} - \alpha)(\alpha_{k_3} - \alpha)\big) = 0, \tag{11}$$

furthermore, if $k_1 \neq k_2$, then

$$E\big((\alpha_{k_1} - \alpha)^3 (\alpha_{k_2} - \alpha)\big) = 0. \tag{12}$$

On the other hand, if $k_1 \neq k_2$, then

$$E\big((\alpha_{k_1} - \alpha)^2 (\alpha_{k_2} - \alpha)^2\big) = p_2 - 2p_3 + p_4 = A \tag{13}$$

and

$$E\big((\alpha_{k_1} - \alpha)^4\big) = p_1 - 4p_2 + 6p_3 - 3p_4 = B, \tag{14}$$

hence

$$E(\zeta_n^4) = \frac{nB + 3n(n-1)A}{n^4} = O\left(\frac{1}{n^2}\right). \tag{15}$$

Thus the series $\sum_{n=1}^{\infty} E(\zeta_n^4)$ is convergent, hence (by the Beppo Levi theorem) the series $\sum_{n=1}^{\infty} \zeta_n^4$ is convergent with probability 1, i.e. $\lim_{n \to \infty} \zeta_n = 0$ with probability 1, which implies the statement of Theorem 4. Our result may be rephrased as follows: if $\{A_n\}$ is a sequence of exchangeable events and v_n denotes the number among the events $A_1, A_2, \ldots, A_n$ which occur, then the limit $\lim_{n \to \infty} \dfrac{v_n}{n}$ exists with probability 1 and is equal to the density of the sequence of events $\{A_n\}$.

Clearly, Theorem 4 is a generalization of the strong law of large numbers concerning the relative frequency. Notice that the limit of the arithmetic means of the random variables α_k is, in general, not constant — contrary to the case of independent random variables.

Let us now consider as an example Pólya's urn model (cf. Ch. III, § 3, Point 9). Let there be in an urn M red and $N - M$ white balls. Balls are drawn at random, the drawn ball is replaced and simultaneously there are added into the urn $R \geq 1$ balls having the same colour as the one drawn. If A_n denotes the event that the ball drawn at the n-th drawing is a red one, according to Formula (10) of § 3 in Chapter III the sequence of events $\{A_n\}$ is exchangeable and

$$p_k = \prod_{l=0}^{k-1} \frac{M + lR}{N + lR} \quad (k = 1, 2, \ldots).\tag{16}$$

It is readily seen that in this case

$$p_k = \int_0^1 x^k \, dF(x),\tag{17}$$

where $F(x)$ is the distribution function of the beta-distribution with parameters $a = \dfrac{M}{R}$ and $b = \dfrac{N - M}{R}$. That is (cf. Ch. IV, § 10, (10)) we have

$$F(x) = \frac{\Gamma\left(\dfrac{N}{R}\right)}{\Gamma\left(\dfrac{M}{R}\right)\Gamma\left(\dfrac{N-M}{R}\right)} \int_0^x t^{\frac{M}{R}-1} (1-t)^{\frac{N-M}{R}-1} \, dt.\tag{18}$$

Thus by Theorem 4 in case of Pólya's urn model the relative frequency of the drawings yielding a red ball among the first n drawings converges with probability 1 to a random variable having a beta-distribution of order $\left(\dfrac{M}{R}, \dfrac{N-M}{R}\right)$. From this it follows that the *distribution* of this relative frequency converges to the mentioned beta-distribution. In fact, if a sequence η_n of the random variables converges with probability 1 to a random variable η, then (cf. § 7) also η_n tends in probability to η and thus (cf. Theorem 1 of § 2) the distribution of η_n tends to that of η. Hence we have

THEOREM 5. *Let in Pólya's urn scheme* v_n *denote the number of red balls drawn in the course of the first* n *drawings, then*

$$\lim_{n \to \infty} P\left(\frac{v_n}{n} < x\right) = \frac{\Gamma\left(\dfrac{N}{R}\right)}{\Gamma\left(\dfrac{M}{R}\right)\Gamma\left(\dfrac{N-M}{R}\right)} \int_0^x t^{\frac{M}{R}-1} (1-t)^{\frac{N-M}{R}-1} \, dt,$$

i.e. the limit distribution of the relative frequency of the red balls drawn is
a beta-distribution of order $\left(\dfrac{M}{R}, \dfrac{N-M}{R}\right)$.

In particular, if $M = R = 1$ and $N = 2$, the relative frequency of red
balls will be in the limit uniformly distributed on the interval $(0, 1)$.

Furthermore, it is easy to see that Formula (10) of Chapter III, § 3 is a
special case of the present Formula (5).

As is seen from this example, the general theory of stable sequences of
events permits a deeper insight into some particular problems already discussed.

§ 13. The zero-one law

In § 5 we proved the following statement: If the events A_n $(n = 1, 2, \ldots)$
are independent, the probability that there occur infinitely many of them is
either 0 or 1 according to the series $\sum\limits_{n=1}^{\infty} P(A_n)$ being convergent or divergent. Thus the probability in question cannot be equal to any other value than
0 or 1.

This phenomenon is explained by the following general theorem, called
the *"zero-one law"*.

THEOREM 1. *Let* $A_1, A_2, \ldots, A_n, \ldots$ *be a sequence of independent events
and* $\mathscr{B}^{(n)}$ *the least σ-algebra containing all sets* $A_{n+1}, A_{n+2}, \ldots$; *if C is an event
which belongs for every n to the σ-algebra* $\mathscr{B}^{(n)}$, *then either* $P(C) = 0$ *or*
$P(C) = 1$.

PROOF. We need a lemma from set theory. For its formulation the follow-
ing definition has to be introduced: If $\mathscr{M}$ is a system of subsets of Ω having
two properties:

1) $B_n \in \mathscr{M}$ and $B_{n+1} \subseteq B_n$ $(n = 1, 2, \ldots)$ imply $\lim\limits_{n \to \infty} B_n = \prod\limits_{n=1}^{\infty} B_n \in \mathscr{M}$,

2) $B_n \in \mathscr{M}$ and $B_n \subseteq B_{n+1}$ $(n = 1, 2, \ldots)$ imply $\lim\limits_{n \to \infty} B_n = \sum\limits_{n=1}^{\infty} B_n \in \mathscr{M}$,

then $\mathscr{M}$ is said to be a *monotone class*.

LEMMA. *If a monotone class of sets contains an algebra of sets* $\mathscr{A}$, *it contains
also the least σ-algebra* $\sigma(\mathscr{A})$ *containing* $\mathscr{A}$.

PROOF. It suffices to show that if $\mathscr{M}(\mathscr{A})$ is the least monotone class containing $\mathscr{A}$, it is identical with $\sigma(\mathscr{A})$. Let A be any subset of Ω and $\mathscr{M}_A$ the

family of the sets B such that $A - B$, $B - A$, and $A + B$ all belong to $\mathcal{M}(\mathcal{A})$. Clearly, $\mathcal{M}_A$ is a monotone class, since for every (increasing or decreasing) monotone sequence $\{B_n\}$ with $B_n \in \mathcal{M}_A$ we have

$$\lim_{n \to \infty} B_n - A = \lim_{n \to \infty} (B_n - A),$$

$$A - \lim_{n \to \infty} B_n = \lim_{n \to \infty} (A - B_n),$$

$$A + \lim_{n \to \infty} B_n = \lim_{n \to \infty} (A + B_n).$$

By assumption, $\mathcal{A}$ is an algebra of sets, hence $A \in \mathcal{A}$ implies $\mathcal{A} \subseteq \mathcal{M}_A$. Since, by definition, $\mathcal{M}(\mathcal{A})$ is the least monotone class containing $\mathcal{A}$, we have $\mathcal{M}(\mathcal{A}) \subseteq \mathcal{M}_A$ for $A \in \mathcal{A}$. Thus if $C \in \mathcal{M}(\mathcal{A})$ and $A \in \mathcal{A}$, then $C \in \mathcal{M}_A$. Hence $A \in \mathcal{M}_C$ and for every $C \in \mathcal{M}(\mathcal{A})$, $\mathcal{A} \subseteq \mathcal{M}_C$; therefore $\mathcal{M}(\mathcal{A}) \subseteq \mathcal{M}_C$. Consequently $\mathcal{M}(\mathcal{A})$ is an algebra of sets, since for any other D with $D \in \mathcal{M}(\mathcal{A})$, we have $D \in \mathcal{M}_C$, hence $C + D \in \mathcal{M}(\mathcal{A})$ and $C - D \in \mathcal{M}(\mathcal{A})$. But a monotone algebra of sets is obviously a σ-algebra, thus $\sigma(\mathcal{A}) \subseteq \mathcal{M}(\mathcal{A})$. And since $\sigma(\mathcal{A})$ is a monotone class containing $\mathcal{A}$ and $\mathcal{M}(\mathcal{A})$ is the least class of this kind, we have necessarily $\sigma(\mathcal{A}) = \mathcal{M}(\mathcal{A})$, which proves the lemma.

Now we shall prove the theorem. Let $\mathcal{B}_n$ denote the least algebra of sets containing $A_1, A_2, \ldots, A_n$. Let $\mathcal{M}_n$ be the collection of the sets which are independent of all elements of $\mathcal{B}_n$. $\mathcal{M}_n$ is a monotone class. If $\mathcal{A}^{(n)}$ is the least algebra containing $A_{n+1}, A_{n+2}, \ldots$, then $\mathcal{A}^{(n)} \subseteq \mathcal{M}_n$. Hence, because of the lemma, $\sigma(\mathcal{A}^{(n)}) = \mathcal{B}^{(n)} \subseteq \mathcal{M}_n$. Thus if $C \in \prod_{n=1}^{\infty} \mathcal{B}^{(n)}$, then C is independent of every A with $A \in \mathcal{B}_n$ ($n = 1, 2, \ldots$). If $\mathcal{M}(C)$ is the collection of all sets independent of C, we have $\mathcal{B}_n \subseteq \mathcal{M}(C)$, hence $\sum_{n=1}^{\infty} \mathcal{B}_n \subseteq \mathcal{M}(C)$, too.

By applying once more the lemma we find $\sigma \left(\sum_{n=1}^{\infty} \mathcal{B}_n \right) = \mathcal{B}^{(1)} \subseteq \mathcal{M}(C)$. It follows that $C \in \mathcal{M}(C)$, hence $P(CC) = P(C) P(C)$ or $P(C) = P^2(C)$. But this is impossible, unless either $P(C) = 0$ or $P(C) = 1$. Thus Theorem 1 is proved. Finally we mention a generalization of the above theorem.

THEOREM 2. *Let $\xi_1, \xi_2, \ldots, \xi_n, \ldots$ be an infinite sequence of independent random variables and $\mathcal{B}^{(n)}$ the least σ-algebra with respect to which the random variables $\xi_{n+1}, \xi_{n+2}, \ldots$ are measurable. If $C \in \prod_{n=1}^{\infty} \mathcal{B}^{(n)}$, then either $P(C) = 0$ or $P(C) = 1$.*

The proof will only be sketched, because it is similar to that of Theorem 1. Let $\mathcal{F}$ be the collection of sets independent of C. As in the previous proof, we show for the least σ-algebra $\mathscr{B}_n$, relative to which the random variables $\xi_1, \xi_2, \ldots, \xi_n$ are measurable, that $\mathscr{B}_n \subseteq \mathscr{F}$ $(n = 1, 2, \ldots)$. Hence, according to the lemma, $\mathscr{B}^{(1)} \subseteq \mathscr{F}$, therefore $C \in \mathscr{F}$ and, consequently, $P^2(C) = P(C)$. Notice that Theorem 2 of § 10 can also be deduced from this.

§ 14. Kolmogorov's three-series theorem

THEOREM 1. Let $\eta_1, \eta_2, \ldots, \eta_n, \ldots$ be independent random variables and let λ be an arbitrary positive number. Let the random variables η_n^* be defined by

$$\eta_n^* = \begin{cases} \eta_n \ \text{for} \ |\eta_n| < \lambda, \\ \lambda \ \text{otherwise.} \end{cases}$$

The series $\sum_{n=1}^{\infty} \eta_n$ converges with probability 1 iff the following three series converge:

$$\sum_{n=1}^{\infty} P(\eta_n \neq \eta_n^*), \tag{1}$$

$$\sum_{n=1}^{\infty} E(\eta_n^*), \tag{2}$$

$$\sum_{n=1}^{\infty} D^2(\eta_n^*). \tag{3}$$

Remark. It is easy to see from the zero-one law that $\sum_{n=1}^{\infty} \eta_n$ converges either with probability one or with probability zero.

PROOF. We show first that the conditions (1), (2), (3) are sufficient. From (1) and the Borel–Cantelli lemma it follows with probability 1 that $\eta_n = \eta_n^*$ for sufficiently large values of n; hence the series $\sum_{n=1}^{\infty} \eta_n$ and $\sum_{n=1}^{\infty} \eta_n^*$ are, with probability 1, simultaneously convergent or divergent. Thus it suffices to show that $\sum_{n=1}^{\infty} \eta_n^*$ converges with probability 1. Because of (2), it suffices to prove this for the series $\sum_{n=1}^{\infty} \delta_n$, where $\delta_n = \eta_n^* - E(\eta_n^*)$. We know that the random variables δ_n are completely independent, further that $E(\delta_n) = 0$ and $\sum_{n=1}^{\infty} D^2(\delta_n) < +\infty$. Hence for an $\varepsilon > 0$, however small, Kolmogorov's inequality gives

$$P(\max_{n \leq m \leq N} |\sum_{k=n}^{m} \delta_k| \geq \varepsilon) \leq \frac{1}{\varepsilon^2} \sum_{k=n}^{N} D^2(\delta_k), \tag{4}$$

and, if we let tend N to infinity for every fixed $\dot{n}$

$$P(\sup_{n \leq m} |\sum_{k=n}^{m} \delta_k| > \varepsilon) \leq \frac{1}{\varepsilon^2} \sum_{k=n}^{\infty} D^2(\delta_k). \tag{5}$$

Choose now from the sequence of all positive integers a subsequence n_j $(n_1 < n_2 < \ldots)$, such that the series $\sum\limits_{j=1}^{\infty} d_j$ converges, where

$$d_j = \sum_{k=n_j}^{\infty} D^2(\delta_k).$$

Then the series

$$\sum_{j=1}^{\infty} P(\sup_{n_j \leq m} |\sum_{k=n_j}^{m} \delta_k| > \varepsilon) \tag{6}$$

converges as well; by applying the Borel–Cantelli lemma we obtain that the relation

$$|\sum_{k=n_j}^{m} \delta_k| \leq \varepsilon$$

holds with probability 1 for sufficiently large j and $m \geq n_j$. If n is an integer between n_j and m, we have with probability 1

$$|\sum_{k=n}^{m} \delta_k| \leq |\sum_{k=n_j}^{n-1} \delta_k| + |\sum_{k=n_j}^{m} \delta_k| \leq 2\varepsilon. \tag{7}$$

If we replace ε successively by $\dfrac{1}{2M}$ $(M = 1, 2, \ldots)$, we obtain (the union of denumerably many sets of zero measure being a set of zero measure too) with probability 1 the relation

$$|\sum_{k=n}^{m} \delta_k| \leq \frac{1}{M} \quad \text{for} \quad m > n, \tag{8}$$

whenever n is greater than a bound $N(M)$ depending on M. But this means that $\sum\limits_{k=1}^{\infty} \delta_k$ converges with probability 1. Herewith the first part of Theorem 1 is proved.

We show now that conditions (1), (2), (3) are necessary as well for the almost sure convergence of the series $\sum\limits_{n=1}^{\infty} \eta_n$. If this series converges with probability 1, $\eta_n \to 0$ as $n \to \infty$ with probability 1. Hence we must have with probability 1 $|\eta_n| < \lambda$, i.e. $\eta_n = \eta_n^*$, except for finitely many values of n. Thus according to the Borel–Cantelli lemma series (1) converges. Hence it suffices to consider the series $\sum\limits_{n=1}^{\infty} \eta_n^*$, the sum of which, a random variable itself, will be denoted by η^*. In what follows we shall need a lemma due to J. L. Doob.

LEMMA. *If ξ is a bounded random variable $|\xi| \le M$ with variance $D^2 = D^2(\xi)$, its characteristic function $\varphi_\xi(t)$ fulfils for $|t| \le \dfrac{1}{4M}$ the inequality*

$$|\varphi_\xi(t)| \le e^{-\frac{D^2 t^2}{8}}. \tag{9}$$

PROOF. Put $\xi^* = \xi - E(\xi)$. We have $E(\xi^*) = 0$, $|\xi^*| \le 2M$, and $|\varphi_{\xi^*}(t)| = |\varphi_\xi(t)|$. We can write

$$|E(\xi^{*n})| \le (2M)^{n-2} E(\xi^{*2}) = (2M)^{n-2} D^2$$

and

$$\varphi_{\xi^*}(t) - 1 + \frac{D^2 t^2}{2} = \sum_{n=3}^{\infty} \frac{E(\xi^{*n})(it)^n}{n!}.$$

For $|t| \le \dfrac{1}{4M}$ we obtain

$$\left| \varphi_{\xi^*}(t) - 1 + \frac{D^2 t^2}{2} \right| \le D^2 t^2 \sum_{n=3}^{\infty} \frac{(2M|t|)^{n-2}}{n!} \le \frac{D^2 t^2}{6},$$

i.e.

$$|\varphi_{\xi^*}(t)| \le \left| \varphi_{\xi^*}(t) - 1 + \frac{D^2 t^2}{2} \right| + \left(1 - \frac{D^2 t^2}{2} \right) \le 1 - \frac{D^2 t^2}{3} \le e^{-\frac{D^2 t^2}{8}},$$

which proves the lemma.

Now we return to the series $\sum\limits_{n=1}^{\infty} \eta_n^*$. By assumption $|\eta_n^*| \le \lambda$, hence if $\psi_n(t)$ is the characteristic function of η_n^* and D_n^2 its variance, we have, according to the lemma

$$D_n^2 \le \frac{3}{|t|^2} \ln \frac{1}{|\psi_n(t)|} \qquad \text{for} \quad |t| \le \frac{1}{4\lambda}. \tag{10}$$

Since $\sum\limits_{n=1}^{N} \eta_n^*$ converges to η^* with probability 1 as $N \to \infty$, the distribution function of $\sum_{n=1}^{N} \eta_n^*$ converges to the distribution function of η^* at every point of continuity of the latter. Hence

$$\prod_{n=1}^{\infty} \psi_n(t) = \psi(t),$$

where $\psi(t)$ denotes the characteristic function of η^*; furthermore there exists an $\varepsilon > 0$ with $\psi(t) \neq 0$ for $|t| < \varepsilon$. Thus if $|t| < \min\left(\varepsilon, \dfrac{1}{4\lambda}\right)$, then $- \sum_{n=1}^{N} \ln |\psi_n(t)|$ tends to $- \ln |\psi(t)|$. Because of (10) we have

$$\sum_{n=1}^{\infty} D_n^2 \leq - \frac{3}{|t|^2} \ln |\psi(t)|; \tag{11}$$

$\sum_{n=1}^{\infty} D_n^2$ is thus convergent and according to the first part of the theorem the series $\sum_{n=1}^{\infty} \delta_n$ converges with probability 1. Since by hypothesis the same holds for $\sum_{n=1}^{\infty} \eta_n^*$, the series

$$\sum_{n=1}^{\infty} E(\eta_n^*) = \sum_{n=1}^{\infty} (\eta_n^* - \delta_n)$$

converges with probability 1 too. Theorem 1 is thus proved.

It is easy to see from this proof that the following result is also valid:

THEOREM 2. *If η_n $(n = 1, 2, \ldots)$ are independent random variables and if $E(\eta_n)$, $D^2(\eta_n)$ exist, further if the series*

$$\sum_{n=1}^{\infty} E(\eta_n) \tag{12}$$

and

$$\sum_{n=1}^{\infty} D^2(\eta_n) \tag{13}$$

converge, then the series $\sum_{n=1}^{\infty} \eta_n$ converges with probability 1.

The assumptions of Theorem 2 are not necessary for the convergence of $\sum_{n=1}^{\infty} \eta_n$; even the existence of $E(\eta_n)$ and $D^2(\eta_n)$ is not necessary. Theorem 2 can be obtained as the limiting case for $\lambda \to +\infty$ of Theorem 1.

It should be noticed that the hypotheses of Theorems 1 and 2 do not guarantee the absolute convergence of the series $\sum_{n=1}^{\infty} \eta_n$. Thus for instance if $P\left(\eta_n = \pm \dfrac{1}{n}\right) = \dfrac{1}{2}$, $E(\eta_n) = 0$, $D^2(\eta_n) = \dfrac{1}{n^2}$, then the conditions of

Theorem 2 are fulfilled but the series $\sum\limits_{n=1}^{\infty} |\eta_n|$ is divergent. By Theorem 2,

however, the series $\sum\limits_{n=1}^{\infty} \eta_n$ converges with probability 1 for any rearrangements of its terms; the sum and the set of its points of divergence depend of course on the rearrangement in question.

For the almost certain convergence of $\sum\limits_{n=1}^{\infty} |\eta_n|$ it is sufficient, according

to a well-known theorem of Beppo Levi, that $\sum\limits_{n=1}^{\infty} E(|\eta_n|) < +\infty$. On the

other hand, it is sufficient that the series (1) and $\sum\limits_{n=1}^{\infty} E(|\eta_n^*|)$ converge, where

η_n^* is defined as in Theorem 1. This condition is necessary as well, since if Theorem 1 is applied to the sequence $|\eta_n^*|$, it can be seen that the convergence of $\sum\limits_{n=1}^{\infty} |\eta_n^*|$ implies that of $\sum\limits_{n=1}^{\infty} E(|\eta_n^*|)$.

Theorem 1 is stronger than the law of large numbers. Thus for instance Theorem 2 of § 7 can be deduced from the present Theorem 2 as follows: Let $\xi_1, \xi_2, \ldots, \xi_n, \ldots$ be independent random variables with $E(\xi_n) = 0$, $D(\xi_n) = D_n$, and assume $\sum\limits_{n=1}^{\infty} \dfrac{D_n^2}{n^2} < +\infty$. If we put $\eta_n = \dfrac{\xi_n}{n}$, the hypo-

theses of Theorem 2 are fulfilled; hence the series $\sum\limits_{n=1}^{\infty} \eta_n$ converges with probability 1. According to Kronecker's lemma (Lemma 2 of § 7) with $q_n = n$ it follows that with probability 1

$$\lim_{n \to \infty} \frac{\sum\limits_{k=1}^{n} k\eta_k}{n} = \lim_{n \to \infty} \frac{1}{n} \sum_{k=1}^{n} \xi_k = 0,$$

which proves Theorem 2 of § 7.

§ 15. Laws of large numbers on conditional probability spaces

The relation of conditional probability to conditional relative frequency is the same as the relation of ordinary probability to ordinary relative frequency. This is reflected in the following

THEOREM 1. *Let* $\mathscr{F} = [\Omega, \mathscr{A}, \mathscr{B}, P(A \mid B)]$ *be a conditional probability space, C an event,* $C \in \mathscr{B}$, ξ_n $(n = 1, 2, \ldots)$ *a sequence of random variables on* $\mathscr{F}$ *which are independent with respect to C. Let further V be a Borel set on the real axis and* B_n *the set of the elements* $\omega \in C$ *such that* $\xi_n(\omega) \in V$. *Suppose* $B_n \in \mathscr{B}$ $(n = 1, 2, \ldots)$. *Then clearly* $B_n \subseteq C$ $(n = 1, 2, \ldots)$. *Let further the conditional variances*

$$D^2(\xi_n \mid B_n) = D_n^2 \qquad (n = 1, 2, \ldots) \tag{1}$$

exist and assume that the conditional expectations

$$E(\xi_n \mid B_n) = M \qquad (n = 1, 2, \ldots) \tag{2}$$

do not depend on n. Put

$$p_n = P(B_n \mid C) \tag{3}$$

and assume that $p_n > 0$ $(n = 1, 2, \ldots)$. *Suppose further that*

$$\sum_{n=1}^{\infty} p_n = +\infty \tag{4}$$

and

$$\sum_{n=1}^{\infty} \frac{\frac{p_n D_n^2}{n}}{\left(\sum_{k=1}^{n} p_k\right)^2} < +\infty. \tag{5}$$

Define

$$S_n(V) = \sum_{\substack{1 \le k \le n \\ \xi_k \in V}} \xi_k \quad \text{and} \quad N_n(V) = \sum_{\substack{1 \le k \le n \\ \xi_k \in V}} 1. \tag{6}$$

$S_n(V)$ *is thus the sum of the* ξ_k $(1 \le k \le n)$ *whose values belong to V and* $N_n(V)$ *is their number. Then*

$$P\left[\lim_{n \to \infty} \frac{S_n(V)}{N_n(V)} = M \mid C\right] = 1.$$

PROOF. Let us put

$$\varepsilon_k = \varepsilon_k(\omega) = \begin{cases} 1 & \text{for } \xi_k(\omega) \in V \text{ i.e. for } \omega \in B_k, \\ 0 & \text{otherwise} \end{cases}$$

and

$$\delta_k = \frac{\varepsilon_k(\xi_k - M)}{\sum_{j=1}^{k} p_j}.$$

Consider the series $\sum\limits_{k=1}^{\infty} \delta_k$. The δ_k are independent under condition C; further $E(\delta_k \mid C) = 0$ and

$$D^2(\delta_k \mid C) = \frac{p_k D_k^2}{k} \bigg/ \left(\sum\limits_{j=1}^{k} p_j\right)^2 ;$$

thus by hypothesis (5) the series $\sum\limits_{k=1}^{\infty} D^2(\delta_k \mid C)$ is convergent.

Kolmogorov's three-series theorem shows that the series $\sum\limits_{k=1}^{\infty} \delta_k$ converges with probability 1 under condition C. If we apply Kronecker's lemma with $q_n = \sum\limits_{k=1}^{n} p_k$, we obtain

$$P\left(\lim_{n\to\infty} \frac{\sum\limits_{k=1}^{n} \varepsilon_k (\xi_k - M)}{\sum\limits_{k=1}^{n} p_k} = 0 \,\bigg|\, C\right) = 1. \qquad (7)$$

Put now

$$\eta_k = \frac{\varepsilon_k - p_k}{k} \bigg/ \sum\limits_{j=1}^{k} p_j .$$

By repeating the preceding reasoning for the series $\sum\limits_{k=1}^{\infty} \eta_k$ we find that $E(\eta_k \mid C) = 0$ and

$$D^2(\eta_k \mid C) = \frac{p_k (1 - p_k)}{k} \bigg/ \left(\sum\limits_{j=1}^{k} p_j\right)^2 .$$

The series $\sum\limits_{k=1}^{\infty} D^2(\eta_k \mid C)$ converges by the lemma of Abel and Dini;[1] it follows (as in the proof of (7)) that

$$P\left(\lim_{n\to\infty} \frac{\sum\limits_{k=1}^{n} \varepsilon_k}{\sum\limits_{k=1}^{n} p_k} = 1 \,\bigg|\, C\right) = 1. \qquad (8)$$

[1] Cf. K. Knopp [1].

(7) and (8) lead to

$$P\left(\lim_{n\to\infty}\frac{\sum\limits_{k=1}^{n}\varepsilon_k\zeta_k}{\sum\limits_{k=1}^{n}\varepsilon_k} = M \middle| C\right) = 1. \tag{9}$$

Since $\sum\limits_{k=1}^{n}\varepsilon_k\zeta_k = S_n(V)$ and $\sum\limits_{k=1}^{n}\varepsilon_k = N_n(V)$ Theorem 1 is herewith proved.

The quotient $\dfrac{S_n(V)}{N_n(V)}$ can be considered as the empirical conditional average of the random variables ζ_n; indeed it is the arithmetic mean of the ζ_k $(k = 1, 2, \ldots)$ whose values belong to V.

The conditional strong law of large numbers can therefore be stated as follows: If (4) and (5) are valid, the arithmetic mean of those values of ζ_1, $\zeta_2, \ldots, \zeta_n$ which belong to V converges with probability 1 to the common conditional expectation of the ζ_k under condition $\zeta_k \in V$.

Remarks.

1. If D_k is bounded, e.g. independent of k, the condition (5) is a consequence of (4) by the Abel–Dini theorem; hence in this case it suffices that (4) is fulfilled.

2. If instead of (2) we suppose that $M_n = E(\zeta_n \mid B_n)$ fulfils

$$\sum_{n=1}^{\infty}\frac{\sum\limits_{k=1}^{n}p_k |M_k - M|}{\sum\limits_{k=1}^{n}p_k} < +\infty \tag{10}$$

and if we replace (5) by condition

$$\sum_{n=1}^{\infty}\frac{p_n(D_n^2 + M_n^2(1 - p_n))}{\left(\sum\limits_{k=1}^{n}p_k\right)^2} < +\infty, \tag{5'}$$

Theorem 1 remains valid; the proof is similar.

3. If the whole real axis is taken for V, then clearly $B_n = C$ and $p_n = 1$; hence (4) is trivially fulfilled and (5) reduces to the condition $\sum\limits_{n=1}^{\infty}\dfrac{D_n^2}{n^2} < +\infty$.

We obtain thus, as a particular case of Theorem 1, Theorem 2 of § 7 for the ordinary probability space $\mathscr{F}_C = [\Omega, \mathscr{A}, P(A \mid C)]$. Notice that it is possible to state and to prove Theorem 1 without reference to conditional probability spaces; however, in the form given above it shows how the strong law of large numbers can be extended to conditional probability spaces.

4. Consider now the following special case: Let V be the set which consists only of the two elements 0 and 1, let further be $P(\xi = 1 \mid B_n) = p$ and $P(\xi = 0 \mid B_n) = 1 - p = q$. This situation can be described as follows: $\mathscr{F}$ represents an infinite sequence of independent experiments, A and B are the possible outcomes of the individual experiments. If at the k-th experiment both A and B occur, we have $\xi_k = 1$; if $\bar{A}$ and B occur, we have $\xi_k = 0$; finally if B does not occur at the k-th experiment, ξ_k takes on a value distinct from 0 or 1. Then

$$M = E(\xi_n \mid B_n) = P(\xi_n = 1 \mid B_n) = p.$$

The number p can be considered as the conditional probability of A with respect to B; we write therefore $p = P(A \mid B)$. Furthermore, $D_n = \sqrt{pq}$ and (4) implies (5). The quotient $f_n(A \mid B) = \dfrac{S_n(V)}{N_n(V)}$ is thus the conditional relative frequency of A with respect to B in the course of the n first experiments.

Theorem 1 states that $f_n(A \mid B)$ tends to $P(A \mid B)$ with conditional probability 1 for a given condition C. In this interpretation C is some condition concerning the infinite sequence of experiments. $p_n = P(B_n \mid C)$ is the conditional probability of B at the n-th $(n = 1, 2, \ldots)$ experiment.

§ 16. Exercises

1. Prove Theorem 2 of § 3 by the method of characteristic functions in the case where the ξ_k are not merely pairwise but completely independent.

2. Prove the generalization of Theorem 2 in § 3: Let the random variables $\xi_1, \xi_2, \ldots$ satisfy the following conditions:
a) the expectations $M_n = E(\xi_n)$ exist and

$$\lim_{n \to \infty} \frac{1}{n} \sum_{k=1}^{n} M_k = M;$$

b) the variances $D_n^2 = D^2(\xi_n)$ exist and

$$\lim_{n \to \infty} \frac{1}{n^2} \sum_{k=1}^{n} D_k^2 = 0,$$

c) the correlation coefficients $R_{ij} = R(\xi_i, \xi_j)$ fulfil the inequalities

$$\sum_{i=1}^{\infty} \sum_{j=1}^{\infty} R_{ij} x_i x_j \leq C \sum_{i=1}^{\infty} x_i^2$$

for every system of real values x_i such that $\sum_{i=1}^{\infty} x_i^2$ converges; C is a positive constant. Given these conditions,

$$\lim_{n \to \infty} \text{st} \frac{1}{n} \sum_{k=1}^{n} \xi_k = M.$$

3. Prove the following theorem: If $f(x, y)$ is a uniformly continuous function of wo variables, if $\lim_{n \to \infty} \text{st} \; \xi_n = \xi$ and $\lim_{n \to \infty} \text{st} \; \eta_n = \eta$, then

$$\lim_{n \to \infty} \text{st} \; f(\xi_n, \eta_n) = f(\xi, \eta).$$

If $\xi = a$ and $\eta = b$ are constants, continuity of $f(x, y)$ at the point (a, b) is sufficient.

4. Let $\xi_n \; (n = 1, 2, \ldots)$ be independent random variables with $P(\xi = \pm \sqrt{n}) = \frac{1}{2}$, hence $E(\xi_n) = 0$ and $D(\xi_n) = \sqrt{n}$. Therefore the condition $\lim_{n \to \infty} \frac{1}{n^2} \sum_{k=1}^{n} D^2(\xi_k) = 0$ of Theorem 2 in § 3 is not fulfilled. Put $\zeta_n = \dfrac{\xi_1 + \xi_2 + \ldots + \xi_n}{n}$ and show that ζ_n does *not* tend in probability to zero.

Hint. Let $\varphi_n(t)$ be the characteristic function of ζ_n, then

$$\varphi_n(t) = \prod_{k=1}^{n} \cos \frac{t \sqrt{k}}{n} \text{ and } \lim_{n \to \infty} \varphi_n(t) = e^{-\frac{t^2}{4}};$$

$\varphi_n(t)$ does not tend to 1.

Remark. The distribution of ζ_n converges to a normal distribution.

5. Let $\xi_1, \xi_2, \ldots, \xi_n, \ldots$ be pairwise independent random variables with

$$P(\xi_n = \pm n^\delta) = \frac{1}{2}.$$

Show that the law of large numbers holds for the sequence ξ_n if $0 < \delta < \dfrac{1}{2}$.

6. Let the events $A_1, A_2, \ldots, A_n$ be the possible results of an experiment. Let there be performed N such independent experiments. The probability that the event A_k occurs exactly $\nu_k(N)$ times $(k = 1, 2, \ldots, n)$, and in a given order, is equal to

$$\pi_N = \prod_{k=1}^{n} p_k^{\nu_k(N)},$$

where $p_k = P(A_k)$. Since π_N depends on the sequence $\nu_k(N) \; (k = 1, 2, \ldots, n)$ and $\nu_k(N)$ are random variables, π_N is a random variable as well. Obviously

$$E\left(\frac{1}{N} \log_2 \frac{1}{\pi_N}\right) = -\sum_{k=1}^{n} p_k \log_2 p_k.$$

The quantity $H(\mathcal{A}) = -\sum_{k=1}^{n} p_k \log_2 p_k$ is called the *entropy* of the complete system of events $\mathcal{A} = (A_1, A_2, \ldots, A_n)$ (cf. Appendix). Prove the limit relation

$$\lim_{N \to \infty} \operatorname{st} \frac{1}{N} \log_2 \frac{1}{\pi_N} = H(\mathcal{A}).$$

Hint. According to the law of large numbers

$$\lim_{N \to \infty} \operatorname{st} \frac{\nu_k(N)}{N} = p_k \quad \text{for} \quad k = 1, 2, \ldots, n.$$

7. Let an urn contain a_0 white and b_0 red balls. If we draw from the urn a white ball, we put it back and besides we add to the urn a_1 white balls and b_1 red balls. If we draw a red ball we put it back and add to the urn a_2 white and b_2 red balls where $a_1 + b_1 = a_2 + b_2$, $a_2 > 0$. The same procedure is repeated after all subsequent drawings. Let ζ_n denote the number of white balls drawn in the first n drawings. Prove the relation

$$\lim_{n \to \infty} \operatorname{st} \frac{\zeta_n}{n} = \frac{a_2}{b_1 + a_2}.$$

Hint. It is easy to show that $\lim_{n \to \infty} E\left(\frac{\zeta_n}{n}\right) = \frac{a_2}{b_1 + a_2}$; further $\lim_{n \to \infty} D\left(\frac{\zeta_n}{n}\right) = 0$; hence our statement follows from Chebyshev's inequality.

8.a) Let η_n $(n = 1, 2, \ldots)$ be bounded random variables, $|\eta_n| \leq C$. The necessary and sufficient condition that η_n should converge in probability to zero is the fulfilment of the relation

$$\lim_{n \to \infty} E(|\eta_n|) = 0. \tag{1}$$

Hint. Applying Markov's inequality to the random variable $|\eta_n|$, we obtain

$$P(|\eta_n| > \varepsilon) \leq \frac{E(|\eta_n|)}{\varepsilon},$$

hence condition (1) is sufficient for $\eta_n \overset{p}{\to} 0$.

Suppose now that $\lim_{n \to \infty} \operatorname{st} \eta_n = 0$. Let $A_n(\delta)$ be the event $|\eta_n| > \delta$, with an arbitrary $\delta > 0$. We have then

$$E(|\eta_n|) = E(|\eta_n| \,|\, A_n(\delta)) P(A_n(\delta)) + E(|\eta_n| \,|\, \overline{A_n(\delta)}) P(\overline{A_n(\delta)}) \leq CP(A_n(\delta)) + \delta.$$

By assumption $\lim_{n \to \infty} P(A_n(\delta)) = 0$. Hence $\lim_{n \to \infty} \sup E(|\eta_n|) \leq \delta$. Since δ can be arbitrarily small, the necessity of (1) is proved.

b) Suppose that $\lim_{n \to \infty} \operatorname{st} \zeta_n = c$ and that $f(x)$ is a Borel-measurable bounded function which is continuous at the point c. Then $\lim_{n \to \infty} E(f(\zeta_n)) = f(c)$.

Hint. Evidently, $\lim_{n \to \infty} \operatorname{st} (f(\zeta_n) - f(c)) = 0$. Since $f(x)$ is bounded, it follows because of Exercise 8 a) that

$$\lim_{n \to \infty} E(|f(\zeta_n) - f(c)|) = 0,$$

hence

$$\lim_{n \to \infty} E(f(\zeta_n)) = f(c).$$

9. Let $f(x)$ and $g(x)$ be continuous functions on the closed interval $[0, 1]$ which fulfil the relation $0 \leq f(x) < Cg(x)$, where C is a positive constant. Then

$$\lim_{n \to \infty} \int_0^1 \int_0^1 \cdots \int_0^1 \frac{f(x_1) + f(x_2) + \ldots + f(x_n)}{g(x_1) + g(x_2) + \ldots + g(x_n)} \, dx_1 dx_2 \ldots dx_n = \frac{\int_0^1 f(x) \, dx}{\int_0^1 g(x) \, dx}.$$

Hint. Choose in the unit-cube of the n-dimensional space a point P at random (with uniform probability distribution); let $\xi_1, \xi_2, \ldots, \xi_n$ denote its coordinates; ξ_k $(k = 1, 2, \ldots, n)$ are thus independent and uniformly distributed on $[0, 1]$. Put

$$\eta_n = \frac{f(\xi_1) + f(\xi_2) + \ldots + f(\xi_n)}{n}$$

and

$$\zeta_n = \frac{g(\xi_1) + g(\xi_2) + \ldots + g(\xi_n)}{n}.$$

We have thus

$$\lim_{n \to \infty} \text{st } \eta_n = \int_0^1 f(x) \, dx, \quad \lim_{n \to \infty} \text{st } \zeta_n = \int_0^1 g(x) \, dx,$$

and, since $\int_0^1 g(x) \, dx > 0$, we have by the result of Exercise 3,

$$\lim_{n \to \infty} \text{st } \frac{\eta_n}{\zeta_n} = \frac{\int_0^1 f(x) \, dx}{\int_0^1 g(x) \, dx}.$$

Since further $0 \leq \dfrac{\eta_n}{\zeta_n} < C$, we get from the result of Exercise 8. b)

$$\lim E\left(\frac{\eta_n}{\zeta_n}\right) = \frac{\int_0^1 f(x) \, dx}{\int_0^1 g(x) \, dx},$$

q.e.d.

10. Prove that the limit relation

$$\lim_{n \to \infty} \sum_{k=1}^{\infty} f\left(x + \frac{k}{n}\right) e^{-nh} \frac{(nh)^k}{k!} = f(x + h)$$

holds for every $h > 0$ and $x > 0$ if $f(x)$ is a bounded continuous function on $(0, +\infty)$ (Theorem of E. Hille).

Hint. Let $\xi_1, \xi_2, \ldots$ be independent random variables having a common Poisson distribution with $E(\xi_k) = h$. Put $\zeta_n = \dfrac{\xi_1 + \xi_2 + \ldots + \xi_n}{n}$. Then $\lim\limits_{n \to \infty} \text{st } \zeta_n = h$. Since

$f(x)$ is by assumption continuous and bounded, it follows according to Exercise 8.b) that

$$\lim_{n \to \infty} E(f(x + \zeta_n)) = f(x + h),$$

which was to be proved.

11. Let $g(s)$ denote the Laplace transform of a function $f(x)$ which is bounded and continuous in the interval $[0, +\infty)$:

$$g(s) = \int_0^\infty e^{-sx} f(x)\, dx.$$

Prove the Post Widder inversion formula

$$f(x) = \lim_{n \to \infty} \frac{(-1)^{n-1} n^n g^{(n-1)} \left(\dfrac{n}{x} \right)}{x^n (n-1)!}. \quad \text{for } x > 0.$$

Hint. Let $\xi_1, \xi_2, \ldots, \xi_n, \ldots$ be independent random variables having the same exponential distribution with expectation x, i.e. $P(\xi_k < t) = 1 - e^{-\frac{t}{x}}$ for $t \geq 0$. If we put $\zeta_n = \dfrac{1}{n} \sum_{k=1}^{n} \xi_k$, then $\lim_{n \to \infty} \text{st } \zeta_n = x$, hence (see Exercise 8.b))

$$\lim_{n \to \infty} E(f(\zeta_n)) = f(x).$$

Now we have (cf. Formula (24) of Ch. IV, § 9)

$$E(f(\zeta_n)) = \int_0^\infty f(t) \frac{n^n t^{n-1} \exp\left[-\dfrac{nt}{x} \right]}{(n-1)! \, x^n} \, dt = \frac{n^n (-1)^{n-1} g^{(n-1)} \left(\dfrac{n}{x} \right)}{x^n (n-1)!}$$

which leads to our statement.

12. Let η be uniformly distributed on $[0, 1]$. Let $\zeta_n(r)$ denote the number of occurrences of the digit r ($r = 0, 1, \ldots, 9$) among the first n digits of the decimal expansion of η; show that

a) $$P\left(\lim_{n \to \infty} \frac{\zeta_n(r)}{n} = \frac{1}{10} \right) = 1 \qquad (r = 0, 1, \ldots, 9),$$

b) $$P\left(\limsup_{n \to \infty} \frac{\left| \zeta_n(r) - \dfrac{n}{10} \right|}{\sqrt{2n \ln \ln n}} \leq \frac{10}{3} \right) = 1.$$

Hint. Let the random variable $\xi_n(r)$ be equal either to 1 or to 0 according to the n-th digit in the decimal expansion of η being equal to r or distinct from it; then $\xi_n(r)$ ($n = 1, 2, \ldots$) are independent and have the same distribution, $P(\xi_n(r) = 1) = \dfrac{1}{10}$, $P(\xi_n(r) = 0) = \dfrac{9}{10}$ ($r = 0, 1, \ldots, 9$). a) is obtained from the strong law of large numbers, b) from the law of the iterated logarithm.

13. Let $\xi_1, \xi_2, \ldots, \xi_n$ be independent random variables with

$$P(\xi_k = \pm 1) = \frac{1}{2} \qquad (k = 1, 2, \ldots, n).$$

Put $\eta_k = \xi_1 + \xi_2 + \ldots + \xi_k$ and $\zeta_n = \max_{1 \leq k \leq n} \eta_k$.

a) If $q > 0$, then

$$E(q^{\zeta_n}) = \frac{1}{2^{2n}} \left(1 + \frac{1}{q}\right) \left[\sum_{k=0}^{n-1} \binom{2n}{k} q^{2n-2k} + \binom{2n-1}{n-1}\right].$$

b) Show by means of the result of a) that

$$E(q^{\zeta_n}) \leq \left(1 + \frac{1}{q}\right) \left(\frac{q + \frac{1}{q}}{2}\right)^{2n}.$$

c) Using b) show that

$$P\left(\limsup_{n \to \infty} \frac{\zeta_n}{\sqrt{2n \ln \ln n}} \leq 1\right) = 1.$$

d) Show that

$$E(\zeta_n) = \sum_{k=1}^{n-1} \frac{\left(\left[\frac{k-1}{2}\right]\right)}{2^k} \qquad \text{for} \quad n = 2, 3, 4, \ldots$$

and conclude from it that $E(\zeta_n) \approx \sqrt{\dfrac{2n}{\pi}}$.

e) Show finally that

$$\lim_{n \to \infty} P(\zeta_n < x\sqrt{n}) = \begin{cases} \sqrt{\dfrac{2}{\pi}} \displaystyle\int_0^x e^{-\frac{u^2}{2}} \, du & \text{for } x \geq 0, \\ 0 & \text{otherwise.} \end{cases}$$

Hint. Let $p_{n,k} = P(\zeta_n = k)$ $(k = -1, 0, 1, \ldots, n)$. We have the following recursive formulas:

$$p_{n+1,k} = \frac{1}{2}(p_{n,k-1} + p_{n,k+1}) \quad \text{for} \quad k = 2, 3, \ldots, n+1,$$

$$p_{n+1,1} = \frac{1}{2}(p_{n,0} + p_{n,2} + p_{n,-1}),$$

$$p_{n+1,-1} = \frac{1}{2}(p_{n,-1} + p_{n,0})$$

$$p_{n+1,0} = \frac{1}{2} p_{n,1}.$$

a) follows by induction.

14. Let $\xi_1, \xi_2, \ldots, \xi_n, \ldots$ be independent random variables with $E(\xi_n) = 0$, $D(\xi_n) = D_n$ and put $\zeta_n = \xi_1 + \xi_2 + \ldots + \xi_n$; then for any $\varepsilon > 0$ and $n = 1, 2, \ldots$

$$P\left(\sup_{k \geq n} \left| \frac{\zeta_k}{k} \right| \geq \varepsilon\right) \leq \frac{1}{\varepsilon^2}\left(\frac{1}{n^2} \sum_{k=1}^{n} D_k^2 + \sum_{k=n+1}^{\infty} \frac{D_k^2}{k^2}\right). \tag{2}$$

This inequality is due to J. Hájek.[1]

Hint. Put $\eta = \sum_{k=J}^{\infty} \zeta_k^2 \left(\frac{1}{k^2} - \frac{1}{(k+1)^2}\right)$. Then

$$E(\eta) = \frac{1}{n^2} \sum_{k=1}^{n} D_k^2 + \sum_{k=n+1}^{\infty} \frac{D_k^2}{k^2}. \tag{3}$$

Let A_k $(k \geq n)$ denote the event that the inequalities

$$\frac{|\zeta_m|}{m} < \varepsilon \ (m = n, n+1, \ldots, k-1) \text{ and } \frac{|\zeta_k|}{k} \geq \varepsilon$$

hold and put $A = \sum_{k=n}^{\infty} A_k$; then $\bar{A}, A_n, A_{n+1}, \ldots$ is a complete system of events. Hence

$$E(\eta) = E(\eta \mid \bar{A}) P(\bar{A}) + \sum_{k=n}^{\infty} E(\eta \mid A_k) P(A_k).$$

Clearly

$$E(\eta \mid A_k) = \sum_{m=n}^{\infty} E(\zeta_m^2 \mid A_k)\left(\frac{1}{m^2} - \frac{1}{(m+1)^2}\right) \geq$$

$$\geq \sum_{m=k}^{\infty} E(\zeta_m^2 \mid A_k)\left(\frac{1}{m^2} - \frac{1}{(m+1)^2}\right).$$

If $m > k$, it can be shown as in the proof of Kolmogorov's inequality, that

$$E(\zeta_m^2 \mid A_k) \geq E(\zeta_k^2 \mid A_k) \geq k^2 \varepsilon^2.$$

Hence

$$E(\eta \mid A_k) \geq \varepsilon^2,$$

and

$$E(\eta) \geq \varepsilon^2 \sum_{k=n}^{\infty} P(A_k) = \varepsilon^2 P\left(\sup_{k \geq n} \left| \frac{\zeta_k}{k} \right| \geq \varepsilon\right). \tag{4}$$

(3) and (4) imply (2).

15. Deduce Theorem 2 of § 7 from Exercise 14.

16. Prove the following generalization of Exercise 14. If $\xi_1, \xi_2, \ldots$ are completely independent and if $E(\xi_k) = 0$, $D^2(\xi_k) = D_k^2$, further if $0 < B_1 \leq B_2 \leq \ldots$ is a sequence of positive numbers such that $\sum_{k=1}^{\infty} \frac{D_k^2}{B_k} < +\infty$, we have, for $\varepsilon > 0$,

$$P\left(\sup_{k \geq n} \frac{|\zeta_k|}{\sqrt{B_k}} \geq \varepsilon\right) \leq \frac{1}{\varepsilon^2}\left(\frac{1}{B_n} \sum_{k=1}^{n} D_k^2 + \sum_{k=n+1}^{\infty} \frac{D_k^2}{B_k}\right).$$

[1] For this proof see J. Hájek and A. Rényi [1].

Hint. The proof of Exercise 14 can be repeated almost word by word.

17. Deduce from the inequality in Exercise 16 the following theorem: Let $\eta_1, \eta_2, \ldots,$ $\eta_n, \ldots$ be completely independent random variables with expectations $E(\eta_k) = M_k > 0$ and with finite variances $D^2(\eta_k) = D_k^2$. Suppose that

$\alpha) \sum_{k=1}^{\infty} M_k = +\infty,$

$\beta)$ the series $\sum_{k=1}^{\infty} \dfrac{D_k^2}{\left(\sum_{j=1}^{k} M_j\right)^2}$ is convergent. Then with probability 1

$$\lim_{n \to \infty} \frac{\sum_{k=1}^{n} \eta_k}{\sum_{k=1}^{n} M_k} = 1.$$

Hint. This is a generalization of Theorem 2 of § 7; in fact, if $\eta_k = \zeta_k - M_k + 1$, then $E(\eta_k) = 1$, $D(\eta_k) = D_k$; thus if $\sum_{k=1}^{\infty} \dfrac{D_k^2}{k^2} < +\infty$ then with probability 1

$$\lim_{n \to \infty} \frac{1}{n} \sum_{k=1}^{n} \eta_k = 1$$

and thus

$$\lim_{n \to \infty} \frac{1}{n} \sum_{k=1}^{n} (\zeta_k - M_k) = 0.$$

18. Prove Theorem 1 of § 15 by means of Exercise 17.

Hint. Let $\eta_k = \varepsilon_k(\zeta_k - M + 1)$. Then

$$E(\eta_k \mid C) = P(B_k \mid C) = p_k, \quad D(\eta_k \mid C) = D_k.$$

Thus if conditions (4) and (5) of Theorem 1 of § 15 are fulfilled, conditions $\alpha)$ and $\beta)$ of Exercise 17 are fulfilled as well and it follows with conditional probability 1 with respect to the condition C that

$$\lim_{n \to \infty} \frac{\sum_{k=1}^{n} \varepsilon_k (\zeta_k - M + 1)}{\sum_{k=1}^{n} p_k} = 1. \tag{5}$$

If we apply the theorem in Exercise 17 to the sequence $\eta_k = \varepsilon_k$ we have again, with conditional probability 1 under condition C

$$\lim_{n \to \infty} \frac{\sum_{k=1}^{n} \varepsilon_k}{\sum_{k=1}^{n} p_k} = 1. \tag{6}$$

From (5) and (6) it follows that

$$P\left(\lim_{n \to \infty} \frac{\sum_{k=1}^{n} \varepsilon_k \zeta_k}{\sum_{k=1}^{n} \varepsilon_k} = M \;\middle|\; C\right) = 1.$$

19. If the random variables $\xi_1, \xi_2, \ldots, \xi_n, \ldots$ are identically distributed and if the fourth moment of ξ_k exists, then for the validity of the strong law of large numbers it is sufficient that the ξ_k are *four-by-four independent* (instead of being completely independent); thus, if any four of the random variables ξ_k are independent and if $E(\xi_k) = 0$, $D^2(\xi_n) = D^2$, $E(\xi_k^4) = M_4$, further if we put $\zeta_n = \dfrac{1}{n} \sum_{k=1}^{n} \xi_k$, then we have $P(\lim_{n \to \infty} \zeta_n = 0) = 1$.

Hint. If we apply Markov's inequality to the random variables ζ_n^4, we obtain

$$P(\,|\,\zeta_n\,| \geq \varepsilon) \leq \frac{E(\zeta_n^4)}{\varepsilon^4} = \frac{nM_4 + 3n(n-1)\,D^4}{\varepsilon^4\,n^4}.$$

Hence the series $\sum_{n=1}^{\infty} P(|\,\xi_n\,| \geq \varepsilon)$ is convergent and we can use Lemma A of § 5. (The idea of this proof is due to F. P. Cantelli.)

20. If $\xi_1, \xi_2, \ldots, \xi_n$ are identically distributed random variables with finite variance, it suffices for the validity of the strong law of large numbers the still weaker criterion that the ξ_k are *pairwise uncorrelated* (instead of completely independent).

Hint. Let $E(\xi_k) = 0$, $D(\xi_k) = D$ and

$$\zeta_n = \frac{1}{n} \sum_{k=1}^{n} \xi_k.$$

According to Chebyshev's inequality, $P(|\zeta_{n^2}| \geq \varepsilon) \leq \dfrac{D^2}{n^2\,\varepsilon^2}$; hence the series $\sum_{n=1}^{\infty} P(|\zeta_{n^2}| \geq \varepsilon)$ is convergent. By the Borel–Cantelli lemma

$$|\,\zeta_{n^2}\,| \leq \varepsilon \tag{7}$$

with probability 1 for a large enough n. On the other hand

$$P\left(\max_{n^2 < N < (n+1)^2} \left| \sum_{k=n^2+1}^{N} \xi_k \right| \geq \varepsilon n^2 \right) \leq \sum_{N=n^2+1}^{(n+1)^2-1} P\left(\left| \sum_{k=n^2+1}^{N} \xi_k \right| \geq \varepsilon n^2 \right).$$

Hence by Chebyshev's inequality,

$$P\left(\max_{n^2 < N < (n+1)^2} \left| \sum_{k=n^2+1}^{N} \xi_k \right| \geq \varepsilon n^2 \right) \leq \sum_{N=n^2+1}^{(n+1)^2-1} \frac{(N-n^2)\,D^2}{\varepsilon^2\,n^4} \leq \frac{4D^2}{\varepsilon^2\,n^2}.$$

Applying again the Borel–Cantelli lemma, we find that with probability 1,

$$\max_{n^2+1 \leq N < (n+1)^2} \left| \sum_{k=n^2+1}^{N} \xi_k \right| \leq \varepsilon n^2 \tag{8}$$

for a sufficiently large n. (7) and (8) lead with probability 1 for $n^2 \leq N < (n+1)^2$ to the inequality $|\zeta_n| \leq 2\,\varepsilon$ for a large enough n, which proves our statement.

21. If $\xi_1, \xi_2, \ldots$ are pairwise uncorrelated, if $E(\xi_k) = 0$, $D^2(\xi_k) = D_k^2$ and if the series $\sum_{k=1}^{\infty} \dfrac{D_k^2}{k^{3/2}}$ is convergent, then the strong law of large numbers is valid.

Hint. Use the method of Exercise 20.

Remark. By a different method[1] it can be proved that even the convergence of the series $\sum_{k=1}^{\infty} D_k^2 \dfrac{\ln^2 k}{k^2}$ is sufficient.

22. Let us develop the positive number x lying between 0 and 1 into Cantor's series

$$x = \sum_{n=1}^{\infty} \frac{\varepsilon_n(x)}{q_1 q_2 \cdots q_n}$$

belonging to the sequence q_n ($q_n \geq 2$, q_n integer), where the "digits" $\varepsilon_n(x)$ may take on the values $0, 1, \ldots, q_n - 1$ ($n = 1, 2, \ldots$). If η is a random variable uniformly distributed on the interval $(0, 1)$, let $\zeta_n(k)$ denote the number of digits $\varepsilon_j(\eta)$ equal to k ($j = 1, 2, \ldots, n$). Assume that the sequence q_n fulfils the conditions $\lim_{n \to \infty} q_n = +\infty$ and $\sum_{n=1}^{\infty} \dfrac{1}{q_n} = +\infty$. Show that

$$P\left(\lim_{n \to \infty} \frac{\zeta_n(k)}{\sum_{j=1}^{n} \frac{1}{q_j}} = 1 \right) = 1 \quad \text{for } k = 0, 1, \ldots.$$

Hint. Let

$$\xi_{nk} = \begin{cases} 1 & \text{for } \varepsilon_n(\eta) = k, \\ 0 & \text{otherwise.} \end{cases}$$

Hence, for $q_n > k$, $E(\xi_{nk}) = \dfrac{1}{q_n}$ and $D^2(\xi_{nk}) = \dfrac{1}{q_n}\left(1 - \dfrac{1}{q_n}\right)$, the convergence of

$$\sum_{n=1}^{\infty} \frac{D^2(\xi_{nk})}{\left(\sum_{j=1}^{n} \frac{1}{q_j} \right)^2}$$

follows from the Abel–Dini theorem. Thus we can apply the result of Exercise 17. The statement of the present exercise can also be obtained as a particular case of Theorem 1 of § 15.

23. Let η_n be the frequency of the event A in a sequence of n independent experiments, while $P(A) = p$ ($0 < p < 1$; $q = 1 - p$). Prove the complete form of the law of the iterated logarithm, i.e.

$$P\left(\limsup_{n \to \infty} \frac{\eta_n - np}{\sqrt{2npq \ln \ln n}} = 1 \right) = 1 \tag{9}$$

and

$$P\left(\liminf_{n \to \infty} \frac{\eta_n - np}{\sqrt{2npq \ln \ln n}} = -1 \right) = 1. \tag{10}$$

Hint. The proof of the Moivre–Laplace theorem shows that

$$P\left(\frac{\eta_n - np}{\sqrt{npq}} > x \right) = 1 - \Phi(x) + O\left(\frac{x^3}{\sqrt{n}} \right)$$

[1] Cf. J. L. Doob [2], p. 158, Theorem 5.2.

for $x = O(\sqrt{\ln \ln n})$; hence

$$P\left(\frac{\eta_n - np}{\sqrt{2npq \ln \ln n}} \geq 1 - \varepsilon\right) = 1 - \Phi\left((1 - \varepsilon)\sqrt{2\ln \ln n}\right) + O\left(\frac{(\ln \ln n)^{\frac{3}{2}}}{\sqrt{n}}\right).$$

Since we have (cf. Ch. III, § 18, Exercise 18):

$$1 - \Phi(x) > \frac{1}{\sqrt{2\pi x}} \frac{1}{1 + \frac{1}{x^2}} e^{-\frac{x^2}{2}},$$

it follows that

$$P\left(\frac{\eta_n - np}{\sqrt{2npq \ln \ln n}} \geq 1 - \varepsilon\right) \geq \frac{c}{\ln n} \quad \text{for} \quad n \geq n_0,$$

where $c > 0$ is a constant. Let $n_k = 2^k$ and let A_k denote the event

$$\frac{\eta_{n_k} - n_k p}{\sqrt{2n_k pq \ln \ln n_k}} \geq 1 - \varepsilon.$$

Thus the series $\sum_{k=1}^{\infty} P(A_k)$ is divergent. It is easy to show that the sequence A_k fulfils the condition of Lemma C of § 5. Hence there occur with probability 1 infinitely many events A_k. Since $\varepsilon > 0$ is arbitrarily small, (9) is obtained. (10) can be proved in a similar way.

24. Let $\xi_1, \xi_2, \ldots, \xi_n, \ldots$ be pairwise independent random variables with common distribution function $F(x)$. Put $\zeta_n = \frac{\xi_1 + \xi_2 + \ldots + \xi_n}{n}$. Show that in order that $\lim_{n \to \infty} \text{st } \zeta_n = 0$ should hold, the following two conditions are sufficient:

a) $$\lim_{n \to \infty} \int_{-n}^{+n} x \, dF(x) = 0,$$

b) $$\lim_{x \to -\infty} xF(x) = \lim_{x \to +\infty} x(1 - F(x)) = 0,$$

(Theorem of Kolmogorov).

Hint. Let

$$\xi_{nk} = \begin{cases} \xi_k & \text{for } |\xi_k| \leq n, \\ 0 & \text{otherwise} \end{cases}$$

and

$$\zeta_n^* = \frac{1}{n} \sum_{k=1}^{n} \xi_{nk}.$$

Then

$$P(\zeta_n \neq \zeta_n^*) \leq \sum_{k=1}^{n} P(|\xi_k| > n) = n[F(-n) + 1 - F(n)]$$

and, because of b),

$$\lim_{n \to \infty} P(\zeta_n \neq \zeta_n^*) = 0.$$

Thus it suffices to show that $\lim_{n \to \infty} \text{st } \zeta_n^* = 0$; as because of

$$P(\,|\,\zeta_n\,| > \varepsilon) \leq P(\,|\,\zeta_n^*\,| \geq \varepsilon) + P(\zeta_n \neq \zeta_n^*)$$

it follows from this that $\lim_{n \to \infty} \text{st } \zeta_n = 0$. On the other hand, by a)

$$\lim_{n \to \infty} E(\zeta_n^*) = \lim_{n \to \infty} \int_{-n}^{+n} x\,dF(x) = 0.$$

Furthermore

$$D^2(\zeta_n^*) \leq \frac{1}{n} \int_{-n}^{+n} x^2\,dF(x) \leq \frac{1}{n} \sum_{k=1}^{n} k^2 \int_{k-1}^{k} dF(x) + \frac{1}{n} \sum_{k=1}^{n} k^2 \int_{-k}^{-(k-1)} dF(x).$$

Putting $a_k = 1 - F(k)$, $b_k = F(-k)$ we may write

$$D^2(\zeta_n^*) \leq \frac{1}{n} \sum_{k=1}^{n} k^2 \,[(a_{k-1} - a_k) + (b_{k-1} - b_k)] \leq$$

$$\leq \frac{1}{n} \sum_{k=0}^{n-1} (a_k + b_k)(2k + 1).$$

Because of b)

$$\lim_{k \to \infty} (2k + 1)(a_k + b_k) = 0,$$

hence

$$\lim_{n \to \infty} D^2(\zeta_n^*) = 0,$$

and thus $\lim_{n \to \infty} \text{st } \zeta_n^* = 0$.

Remark. In case of completely independent ξ_n the proof can be somewhat simplified by employing characteristic functions; in fact

$$E(e^{it\zeta_n}) = \left(1 + \int_{-\infty}^{+\infty} \left(\exp\frac{itx}{n} - 1\right)dF(x)\right)^n = \left(1 + \frac{it\int_{-n}^{+n} x\,dF(x) + \delta_n}{n}\right)^n,$$

where

$$|\,\delta_n\,| \leq n\big[F(-n) + 1 - F(n)\big] + \frac{t^2}{n} \int_{-n}^{+n} x^2\,dF(x).$$

Hence by a) and b) follows

$$\lim_{n \to \infty} E(e^{it\zeta_n}) = 1$$

for every real t.

CHAPTER VIII

THE LIMIT THEOREMS OF PROBABILITY THEORY

§ 1. The central limit theorems

On the basis of the theorems on characteristic functions established in Chapter VI, we may now pass to the proofs of theorems concerning limit distributions. Most important among these are the so-called "central limit theorems", which express the fact that the distribution of the sum of a large number of independent random variables approaches, under very general conditions, the normal distribution. These theorems disclose the reason why in applications distributions close to normal distributions are so often encountered.

A typical example is the case of errors of measurements; the total error is usually composed of many small errors. The central limit theorems justify the assumption that the errors of measurement are normally distributed; that is, why the normal distribution is sometimes called the *law of errors*.

The simplest case of the central limit theorem, namely the Moivre–Laplace theorem, was already dealt with in Chapter III. First we give here a somewhat different formulation of this theorem.

Let there be performed n independent experiments having for their possible outcomes either the occurrence of an event A or its non-occurrence $\bar{A}$. Put $p = P(A)$, $q = 1 - p = P(\bar{A})$. Let the value of the random variable ξ_k be either 1 or 0, according as the event A occurs or does not occur at the k-th $(k = 1, 2, \ldots, n)$ experiment. Put further

$$\zeta_n = \xi_1 + \xi_2 + \ldots + \xi_n.$$

We know already that $E(\zeta_n) = np$ and $D(\zeta_n) = \sqrt{npq}$. The linear transformation which transforms the random variable ξ into the random variable $\xi^* = \dfrac{\xi - E(\xi)}{D(\xi)}$ having expectation zero and standard deviation one, is called *standardization*. Let $\zeta_n^* = \dfrac{\zeta_n - np}{\sqrt{npq}}$ be the standardized variable corresponding to ζ_n. Since ζ_n is evidently a binomial variable, $P(\zeta_k = k) = \begin{pmatrix} n \\ k \end{pmatrix} p^k q^{n-k}$. Therefore the Moivre–Laplace theorem (Ch. III, § 16,

Theorem 3) can be formulated as follows:

$$\lim_{n \to +\infty} P(\zeta_n^* < x) = \Phi(x), \tag{1}$$

where

$$\Phi(x) = \frac{1}{\sqrt{2\pi}} \int_{-\infty}^{x} e^{-\frac{u^2}{2}} \, du \tag{2}$$

is the normal distribution function. In other words, the distribution of the standardized relative frequency of the event A during n independent experiments tends to the normal distribution as $n \to +\infty$.

The random variables ξ_k are of a very restricted nature: they assume only the values 0 and 1. The central limit theorem in its most simple form, which is an immediate generalization of the Moivre–Laplace theorem, can be stated as follows:

THEOREM 1. *Let* $\xi_1, \xi_2, \ldots, \xi_n, \ldots$ *be independent identically distributed random variables for which* $M = E(\xi_n)$ *and* $D = D(\xi_n) > 0$ *exist; put*

$$\zeta_n = \sum_{k=1}^{n} \xi_k \quad and \quad \zeta_n^* = \frac{\zeta_n - E(\zeta_n)}{D(\zeta_n)}.$$

If $F_n(x)$ *is the distribution function of* ζ_n^*, *we have*

$$\lim_{n \to \infty} F_n(x) = \Phi(x) \quad uniformly \ for \quad -\infty < x < +\infty. \tag{3}$$

PROOF. Let $\varphi(t)$ be the characteristic function of $\eta_k = \xi_k - M$ and $\psi_n(t)$ the characteristic function of ζ_n^*. Since $E(\zeta_n) = nM$ and $D(\zeta_n) = D\sqrt{n}$ it follows from Theorems 3 and 6 of Chapter VI, § 2 that

$$\psi_n(t) = \varphi\left(\frac{t}{D\sqrt{n}}\right)^n. \tag{4}$$

From Theorem 9 of Chapter VI, § 2 it follows, because of $E(\eta_k) = 0$, that

$$\varphi\left(\frac{t}{D\sqrt{n}}\right) = 1 - \frac{t^2}{2n} + o\left(\frac{1}{n}\right). \tag{5}$$

By applying the well-known formula

$$\lim_{n \to +\infty} \left(1 + \frac{x_n}{n}\right)^n = e^x \quad if \quad \lim_{n \to +\infty} x_n = x, \tag{6}$$

we conclude that for all values of t

$$\lim_{n \to +\infty} \psi_n(t) = e^{-\frac{t^2}{2}} \quad (-\infty < t < +\infty) \tag{7}$$

which, because of Theorem 3 of Chapter VI, § 4, proves (1). It is easy to see (in view of the continuity of $\Phi(x)$ for all x) that the convergence is necessarily uniform in x. Evidently Theorem 1 contains the Moivre–Laplace theorem as a particular case.

The statement of Theorem 1 remains valid under much more general conditions. This was shown first by Chebyshev and Markov by an entirely different method, namely by the method of moments (see § 13, Exercise 27).

The method of characteristic functions was first employed by Liapunov. He proved by essentially this method that the central limit theorem can be applied under much more general conditions than those of Chebyshev and Markov.[1] The result of Liapunov can be stated as follows:

THEOREM 2. *Let* $\xi_1, \xi_2, \ldots, \xi_n, \ldots$ *be independent random variables, the first three moments*

$$M(\xi_k) = M, \quad D^2(\xi_k) = D_k^2 > 0, \quad M(|\xi_k - M_k|^3) = H_k^3$$

of which exist $(k = 1, 2, \ldots)$. *Put*

$$S_n = \sqrt{D_1^2 + D_2^2 + \ldots D_n^2}, \tag{8}$$

$$K_n = \sqrt[3]{H_1^3 + H_2^3 + \ldots + H_n^3}, \tag{9}$$

$$\zeta_n = \sum_{k=1}^{n} \xi_k \quad and \quad \zeta_n^* = \frac{\zeta_n - E(\zeta_n)}{D(\zeta_n)}. \tag{10}$$

Let $F_n(x)$ *denote the distribution function of* ζ_n^*. *If Liapunov's condition*

$$\lim_{n \to +\infty} \frac{K_n}{S_n} = 0 \tag{11}$$

is fulfilled, then

$$\lim_{n \to +\infty} F_n(x) = \Phi(x) \quad (-\infty < x < +\infty). \tag{12}$$

Remark. The condition (11) is evidently fulfilled when all ξ_k have the same distribution. In effect, in this case $D_k = D$, $H_k = H$, $S_n = D\sqrt{n}$, $K_n = $

---

[1] Later on it was proved by Markov that Liapunov's theorem can also be proved by the method of moments.

$= H\sqrt[3]{n}$, hence

$$\lim_{n \to +\infty} \frac{K_n}{S_n} = \frac{H}{D} \lim_{n \to +\infty} \frac{1}{\sqrt[6]{n}} = 0.$$

It is again fulfilled when the random variables $\xi_k - M_k$ are uniformly bounded and $\lim_{n \to \infty} S_n = +\infty$, In fact, from $|\xi_k - M_k| \leq C$ follows

$$H_k^3 \leq CD_k^2,$$

hence

$$\frac{K_n}{S_n} \leq \sqrt[3]{\frac{C}{S_n}}$$

and since $S_n \to +\infty$, condition (11) is satisfied.

Liapunov proved the central limit theorem starting from still more general hypotheses. As a matter of fact, it suffices to assume the existence of the moments of order β (for some arbitrary $\beta > 2$) instead of the third moment; in this case instead of (11) one has to suppose

$$\lim_{n \to +\infty} \frac{K_n(\beta)}{S_n} = 0, \tag{13}$$

where

$$K_n(\beta) = \sum_{k=1}^{n} E(|\xi_k - M_k|^\beta)^{\frac{1}{\beta}}. \tag{14}$$

Lindeberg proved the central limit theorem under still more general conditions. His condition is, in a certain sense, necessary as well. It is formulated in the following theorem due to Lindeberg:

THEOREM 3. *Let* $\xi_1, \xi_2, \ldots, \xi_n, \ldots$ *be independent random variables for which the expectations* $M_k = E(\xi_k)$ *and the standard deviations* $D_k = D(\xi_k)$ *exist* $(k = 1, 2, \ldots)$. *Put*

$$S_n = \sum_{k=1}^{n} D_k^{2^{\frac{1}{2}}} \tag{15}$$

and let $F_k(x)$ *be the distribution function of* $\xi_k - M_k$. *If for every positive* ε *the so-called Lindeberg condition*

$$\lim_{n \to +\infty} \frac{1}{S_n^2} \sum_{k=1}^{n} \int\limits_{|x| > \varepsilon S_n} x^2 \, dF_k(x) = 0 \tag{16}$$

is fufillled, then for

$$\zeta_n^* = \frac{\sum\limits_{k=1}^n (\xi_k - M_k)}{S_n} \tag{17}$$

we have

$$\lim_{n \to +\infty} P(\zeta_n^* < x) = \Phi(x). \tag{18}$$

Remark. From Liapunov's condition (11) one can deduce (16); indeed we have

$$\frac{1}{S_n^2} \sum_{k=1}^n \int\limits_{|x| > \varepsilon S_n} x^2 \, dF_k(x) \le \frac{1}{\varepsilon S_n^3} \sum_{k=1}^n \int\limits_{-\infty}^{+\infty} |x|^3 \, dF_k(x) = \frac{1}{\varepsilon} \left(\frac{K_n}{S_n} \right)^3. \tag{19}$$

Similarly, (16) can be deduced from (13), too. Hence it suffices to prove Theorem 3 (Lindeberg's theorem); then Liapunov's theorem (Theorem 2) will also be proved.

PROOF OF THEOREM 3. If $\varphi_k(t)$ is the characteristic function of $\eta_k = \xi_k - M_k$, then

$$\varphi_k \left(\frac{t}{S_n} \right) = \int\limits_{-\infty}^{+\infty} e^{\frac{itx}{S_n}} \, dF_k(x). \tag{20}$$

We need the following elementary lemma:

LEMMA. *For a real u and $k = 1, 2, \ldots$, we have*

$$\left| e^{iu} - \sum_{j=0}^{k-1} \frac{(iu)^j}{j!} \right| \le \frac{|u|^k}{k!}. \tag{21}$$

PROOF OF THE LEMMA. In fact,

$$|e^{iu} - 1|^2 = 2(1 - \cos u) = 2 \int\limits_0^u \sin v \, dv \le 2 \int\limits_0^u v \, dv = u^2, \tag{22}$$

hence (21) holds for $k = 1$; if (21) holds for any k, it follows from

$$e^{iu} - \sum_{j=0}^k \frac{(iu)^j}{j!} = \int\limits_0^u i \left(e^{iv} - \sum_{j=0}^{k-1} \frac{(iv)^j}{j!} \right) dv \tag{23}$$

that (21) holds for $k + 1$ too; hence by induction (21) holds for every k. Thus the lemma is proved.

We have therefore

$$e^{\frac{itx}{S_n}} = 1 + \frac{itx}{S_n} + \theta_1 \frac{x^2 t^2}{2S_n^2} \quad \text{where} \quad |\theta_1| \leq 1 \tag{24}$$

and

$$e^{\frac{itx}{S_n}} = 1 + \frac{itx}{S_n} - \frac{x^2 t^2}{2S_n^2} + \theta_2 \frac{x^3 t^3}{6S_n^3} \quad \text{where} \quad |\theta_2| \leq 1. \tag{25}$$

Now let $\varepsilon > 0$ be given. The integral (20) can be separated into two parts:

$$\varphi_k\left(\frac{t}{S_n}\right) = \int_{-\varepsilon S_n}^{+\varepsilon S_n} e^{\frac{itx}{S_n}} dF_k(x) + \int_{|x| > \varepsilon S_n} e^{\frac{itx}{S_n}} dF_k(x). \tag{26}$$

Consider first the first integral on the right side of (26). Because of (25) we have

$$\int_{-\varepsilon S_n}^{\varepsilon S_n} e^{\frac{itx}{S_n}} dF_k(x) = \int_{-\varepsilon S_n}^{\varepsilon S_n} dF_k(x) + \frac{it}{S_n} \int_{-\varepsilon S_n}^{\varepsilon S_n} x \, dF_k(x) - \frac{t^2}{2S_n^2} \int_{-\varepsilon S_n}^{\varepsilon S_n} x^2 dF_k(x) + R_k^{(1)}, \tag{27}$$

with

$$|R_k^{(1)}| \leq \frac{|t|^3}{6S_n^3} \int_{-\varepsilon S_n}^{\varepsilon S_n} |x|^3 dF_k(x) \leq \frac{\varepsilon |t|^3}{6S_n^2} D_k^2. \tag{28}$$

Apply now Formula (24) to the second integral in (26); we obtain

$$\int_{|x| > \varepsilon S_n} e^{\frac{itx}{S_n}} dF_k(x) = \int_{|x| > \varepsilon S_n} dF_k(x) + \frac{it}{S_n} \int_{|x| > \varepsilon S_n} x \, dF_k(x) + R_k^{(2)}, \tag{29}$$

with

$$|R_k^{(2)}| \leq \frac{|t|^2}{2S_n^2} \int_{|x| > \varepsilon S_n} x^2 dF_k(x). \tag{30}$$

If we add (27) and (29), we obtain by (28), (30) and by taking into account that $E(\eta_k) = 0$,

$$\varphi_k\left(\frac{t}{S_n}\right) = 1 - \frac{t^2 D_k^2}{2S_n^2} + R_k^{(3)}, \tag{31}$$

with

$$R_k^{(3)} \le \frac{|t|^3 \varepsilon D_k^2}{6 S_n} + \frac{t^2}{S_n^2} \int\limits_{|x| > \varepsilon S_n} x^2 \, dF_\iota(x). \tag{32}$$

We show now that (16) implies

$$\lim_{n \to \infty} \frac{\max\limits_{1 \le k \le n} D_k}{S_n} = 0. \tag{33}$$

In fact

$$D_k^2 = \int\limits_{|x| \le \varepsilon S_n} x^2 \, dF_k(x) + \int\limits_{|x| > \varepsilon S_n} x^2 \, dF_k(x) \le \varepsilon^2 S_n^2 + \sum_{k=1}^{n} \int\limits_{|x| > \varepsilon S_n} x^2 \, dF_k(x),$$

where

$$\frac{\max\limits_{1 \le k \le n} D_k^2}{S_n^2} \le \varepsilon^2 + \frac{1}{S_n^2} \sum_{k=1}^{n} \int\limits_{|x| > \varepsilon S_n} x^2 \, dF_k(x). \tag{34}$$

It follows, because of (16), that

$$\limsup_{n \to +\infty} \frac{\max\limits_{1 \le k \le n} D_k}{S_n} \le \varepsilon. \tag{35}$$

Since $\varepsilon > 0$ may be chosen arbitrarily small, (33) is proved.

Choose now $n_0(\varepsilon)$ such that for $n > n_0(\varepsilon)$

$$\frac{1}{S_n^2} \sum_{k=1}^{n} \int\limits_{|x| > \varepsilon S_n} x^2 \, dF_k(x) < \varepsilon \tag{36}$$

and

$$\frac{\max\limits_{1 \le k \le n} D_k}{S_n} < \varepsilon. \tag{37}$$

This can be done because of (16) and (33). Let further be $\varepsilon < \dfrac{1}{|t|}$, thus

$$\frac{1}{2} \le 1 - \frac{t^2 D_k^2}{2 S_n^2} \le 1$$

for $n > n_0(\varepsilon)$ and $1 \le k \le n$. Because of the identity

$$\prod_{k=1}^{n} (a_k + b_k) - \prod_{k=1}^{n} a_k = \sum_{j=1}^{n} b_j \left(\prod_{k<j} a_k \right) \left(\prod_{j<k} (a_k + b_k) \right) \tag{38}$$

(where an empty product is to be replaced by 1), it follows from (32) and (36) that

$$\left| \prod_{k=1}^{n} \varphi_k \left(\frac{t}{S_n} \right) - \prod_{k=1}^{n} \left(1 - \frac{t^2 D_k^2}{2S_n^2} \right) \right| \leq \sum_{k=1}^{n} |R_k^{(3)}| \leq \varepsilon \left(\frac{|t|^3}{6} + t^2 \right). \tag{39}$$

The inequality $|e^{-x} - 1 + x| \leq x^2$ for $|x| \leq \dfrac{1}{2}$ and the identity (38) imply, by considering (37),

$$\left| \prod_{k=1}^{n} \left(1 - \frac{t^2 D_k^2}{2S_n^2} \right) - \prod_{k=1}^{n} e^{-\frac{t^2 D_k^2}{2S_n^2}} \right| \leq \sum_{k=1}^{n} \frac{t^4 D_k^4}{4S_n^4} \leq \frac{t^4 \varepsilon^2}{4}. \tag{40}$$

Hence from (39) and (40) it follows for $n \geq n_0(\varepsilon)$ that

$$\left| \prod_{k=1}^{n} \varphi_k \left(\frac{t}{S_n} \right) - e^{-\frac{t^2}{2}} \right| \leq \varepsilon \left(\frac{|t|^3}{6} + t^2 \right) + \frac{t^4 \varepsilon^2}{4}. \tag{41}$$

Since $\varepsilon > 0$ can be chosen arbitrarily small, (41) implies

$$\lim_{n \to +\infty} \prod_{k=1}^{n} \varphi_k \left(\frac{t}{S_n} \right) = \lim_{n \to +\infty} E(e^{it\zeta_n^*}) = e^{-\frac{t^2}{2}}. \tag{42}$$

Because of Theorem 3 of Chapter VI, § 4, our theorem is thus proved. The convergence in (42) is even uniform on every finite interval $|t| \leq T$.

If the ξ_k are identically distributed and possess a finite standard deviation D, we have

$$F_k(x) = F(x) \quad (k = 1, 2, \ldots, n, \ldots), \quad \int_{-\infty}^{+\infty} x^2 dF(x) = D^2, \text{ and } S_n = D\sqrt{n}.$$

Thus

$$\lim_{n \to +\infty} \frac{1}{S_n^2} \sum_{k=1}^{n} \int_{|x| > \varepsilon S_n} x^2 dF_k(x) = \frac{1}{D^2} \lim_{n \to +\infty} \int_{|x| > \varepsilon D\sqrt{n}} x^2 dF(x) = 0, \tag{43}$$

Theorem 3 therefore contains Theorem 1 as a particular case. Notice that Theorem 1 does not follow from Theorem 2, since it is possible that $\int_{-\infty}^{+\infty} x^2 dF(x)$ exists but $\int_{-\infty}^{+\infty} |x|^3 dF(x) = +\infty$ and even that $\int_{-\infty}^{+\infty} |x|^\beta dF(x) = +\infty$ for every $\beta > 2$.

Let us add that Lindeberg first proved his theorem by a different method, viz. by a direct study of the convolution of the distributions (see § 12).

Lindeberg's condition (16) is, as was shown by W. Feller, necessary as well, in the following sense: If $\xi_1, \xi_2, \ldots, \xi_n, \ldots$ are independent random variables with finite expectation and finite standard deviation, if $F_k(x)$ is the distribution function of $\xi_k - E(\xi_k)$ and if $\zeta_n = \xi_1 + \xi_2 + \ldots + \xi_n$, $A_n = E(\zeta_n)$, $S_n = D(\zeta_n)$ and $\zeta_n^* = \dfrac{\zeta_n - A_n}{S_n}$, then

$$\lim_{n \to +\infty} P(\zeta_n^* < x) = \Phi(x) \qquad (44)$$

and

$$\lim_{n \to +\infty} P(\max_{1 \le k \le n} |\xi_k - E(\xi_k)| > \varepsilon S_n) = 0 \qquad (45)$$

hold iff (16) is fulfilled for every $\varepsilon > 0$.

If (45) is satisfied, the variables $\xi_k - E(\xi_k)$ are said to be "negligible" (or "infinitesimal"). Condition (45) follows from (16) by the inequality

$$P(\max_{1 \le k \le n} |\xi_k - E(\xi_k)| > \varepsilon S_n) \le \sum_{k=1}^{n} P(|\xi_k - E(\xi_k)| > \varepsilon S_n) \le$$

$$\le \frac{1}{\varepsilon^2 S_n^2} \sum_{k=1}^{n} \int_{|x| > \varepsilon S_n} x^2 \, dF_k(x).$$

Lindeberg's condition implies thus that the variables $(\xi_k - E(\xi_k))/S_n$ are, in a certain sense, "uniformly small" with great probability. We do not prove here that (16) is a necessary condition. Neither do we deal with further generalizations of the central limit theorem.[1]

The results of this section can be generalized in the following way: instead of a sequence ξ_k ($k = 1, 2, \ldots$) consider a matrix (ξ_{nk}), ($k = 1, 2, \ldots, k_n$; $n = 1, 2, \ldots$) of random variables such that the variables $\xi_{n1}, \xi_{n2}, \ldots, \xi_{nk_n}$ are independent for every n and put

$$\zeta_n = \sum_{k=1}^{k_n} \xi_{nk}.$$

By the same method which served to prove Theorem 3 we can prove the following, somewhat more general, theorem:

THEOREM 4. *Let ξ_{nk} ($k = 1, 2, \ldots, k_n$) be for every n ($n = 1, 2, \ldots$) independent random variables with finite variance. Put $M_{nk} = E(\xi_{nk})$, $D_{nk} = D(\xi_{nk})$ and let $F_{nk}(x)$ denote the distribution function of $\xi_{nk} - M_{nk}$. We assume that $\sum_{k=1}^{k_n} D_{nk}^2 = 1$. If the Lindeberg condition*

$$\lim_{n \to +\infty} \sum_{k=1}^{k_n} \int_{|x| > \varepsilon} x^2 \, dF_{nk}(x) = 0$$

[1] Cf. the books of B. V. Gnedenko and A. N. Kolmogorov [1] and W. Feller [7], Vol. 2, containing the detailed discussion of many further results in this domain.

is satisfied for any $\varepsilon > 0$, then

$$\lim_{n \to +\infty} P\left(\sum_{k=1}^{n} (\xi_{nk} - M_{nk}) < x \right) = \Phi(x).$$

Theorem 4 evidently contains Theorem 3 as a particular case; it suffices to put $\xi_{nk} = \dfrac{\xi_k}{S_n}$ $(k = 1, 2, \ldots, n)$.

The proof of Theorem 4 is very similar to that of Theorem 3; hence we leave it to the reader.

The central limit theorem can be completed by the evaluation of the remainder, viz. by giving an asymptotic expansion for the distribution function of ζ_n^*, the first term of which is given by $\Phi(x)$ while further terms progress according to powers of $\dfrac{1}{\sqrt{n}}$.

§ 2. The local form of the central limit theorem

In the preceding section we have seen that the distribution function $F_n(x)$ of the standardized sum ζ_n^* of n independent random variables $\xi_1, \xi_2, \ldots,$ $\xi_n, \ldots$ converges, under certain conditions, to the distribution function of the normal distribution as $n \to \infty$. It is therefore natural to ask under which conditions the density function of ζ_n^* (if it exists) tends to the density function of the normal distribution. For this the conditions must certainly be stronger, since it is known that $F_n(x) \to \Phi(x)$ does not necessarily imply $F_n'(x) \to \Phi'(x)$. We prove first in this respect a theorem due to B. V. Gnedenko:

THEOREM 1. *Let* $\xi_1, \xi_2, \ldots, \xi_n, \ldots$ *be independent, identically distributed random variables which have a bounded density function* $f(x)$; *assume further that* $E(\xi_n) = \int\limits_{-\infty}^{+\infty} xf(x)dx = 0$ *and that the integral* $D^2 = \int\limits_{-\infty}^{\infty} x^2 f(x)dx$ *exists; then the density function* $f_n(x)$ *of*

$$\zeta_n^* = \frac{\xi_1 + \xi_2 + \ldots + \xi_n}{D\sqrt{n}} \tag{1}$$

tends to the density function of the normal distribution; hence we have

$$\lim_{n \to \infty} f_n(x) = \frac{1}{\sqrt{2\pi}} e^{-\frac{x^2}{2}}. \tag{2}$$

The convergence is uniform in x.[1]

[1] Fig. 25. The figure represents the case when the random variables ξ_k are uniformly distributed on $(-\sqrt{3}, +\sqrt{3})$.

PROOF. The supposition $D = 1$ does not restrict the generality. Let $\varphi(t)$ be the characteristic function of ξ_k and $\varphi_n(t)$ that of ζ_n^*. We know that

$$\varphi_n(t) = \left[\varphi\left(\frac{t}{\sqrt{n}}\right)\right]^n. \tag{3}$$

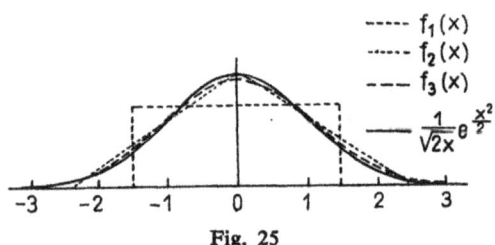

Fig. 25

We first prove the following

LEMMA. *If the density function $g(x)$ is bounded, $g(x) \le K$, and the characteristic function*

$$\psi(t) = \int_{-\infty}^{+\infty} g(x)\, e^{itx}\, dx \tag{4}$$

is nonnegative, then the integral $\int_{-\infty}^{+\infty} \psi(t)\, dt$ exists.

PROOF. (4) implies for $v > 0$

$$\int_{-v}^{+v} \psi(t)\, dt = 2 \int_{-\infty}^{+\infty} g(x)\, \frac{\sin vx}{x}\, dx, \tag{5}$$

hence, for $T > 0$

$$\int_{0}^{2T} \left(\int_{-v}^{+v} \psi(t)\, dt\right) dv = 2 \int_{-\infty}^{+\infty} g(x)\, \frac{1 - \cos 2Tx}{x^2}\, dx. \tag{6}$$

Since $\psi(t) \ge 0$, we have on the other hand

$$\int_{0}^{2T} \left(\int_{-v}^{+v} \psi(t)\, dt\right) dv = \int_{-2T}^{+2T} \psi(t)\,(2T - |t|)\, dt \le T \int_{-T}^{+T} \psi(t)\, dt, \tag{7}$$

hence

$$\int_{-T}^{+T} \psi(t)\, dt \le \frac{2}{T} \int_{-\infty}^{+\infty} g(x)\, \frac{1 - \cos 2Tx}{x^2}\, dx. \tag{8}$$

Because of $g(x) \leq K$ and

$$\int_{-\infty}^{+\infty} \frac{1 - \cos 2Tx}{x^2}\, dx = 2\pi T$$

(cf. Ch. VI, § 4, Formula (40)) we get

$$\int_{-T}^{+T} \psi(t)\, dt \leq 4\pi K. \qquad (9)$$

Since T in (9) can be chosen arbitrarily large and by assumption $\psi(t)$ is nonnegative, the lemma is herewith proved.

Now if the density function of one of two independent random variables is bounded, the density function of their sum is bounded as well (cf. Ch. IV, § 9, Formula (4)). Thus, the density function $f(x)$ being bounded, that of $\xi_1 - \xi_1^*$ is bounded if ξ_1^* is independent of ξ_1 and has the same distribution, and the characteristic function of $\xi_1 - \xi_1^*$ is equal to $|\varphi(t)|^2$. Thus by our lemma $|\varphi(t)|^2$ is integrable. From this it follows by applying Theorem 2 of Chapter VI, § 4 that for $n \geq 2$

$$f_n(x) = \frac{1}{2\pi} \int_{-\infty}^{+\infty} \varphi\left(\frac{t}{\sqrt{n}}\right)^n e^{-ixt}\, dt. \qquad (10)$$

On the other hand, we have

$$\frac{1}{\sqrt{2\pi}} e^{-\frac{x^2}{2}} = \frac{1}{2\pi} \int_{-\infty}^{+\infty} e^{-\frac{t^2}{2}} e^{-ixt}\, dt. \qquad (11)$$

Furthermore, for every $T > 0$, because of the uniform convergence of $\varphi_n(t)$ to $e^{-\frac{t^2}{2}}$ on every finite interval, we have

$$\lim_{n \to +\infty} \frac{1}{2\pi} \int_{-T}^{+T} \varphi_n(t)\, e^{-ixt}\, dt = \frac{1}{2\pi} \int_{-T}^{+T} e^{-\frac{t^2}{2}} e^{-ixt}\, dt \qquad (12)$$

uniformly in x.

We show now that the integral

$$I_n(T) = \int_{|t| > T} \left(\varphi_n(t) - e^{-\frac{t^2}{2}}\right) e^{-ixt}\, dt \qquad (13)$$

can be made arbitrarily small, uniformly in x, by choosing T and n sufficiently large. Because of (12), the theorem will then be proved. In order to show that (13) can be made smaller than any positive number by an appropriate choice of n and T, notice first that

$$|I_n(T)| \leq 2 \int\limits_{T}^{+\infty} \left| \varphi\left(\frac{t}{\sqrt{n}}\right) \right|^n dt + 2 \int\limits_{T}^{+\infty} e^{-\frac{t^2}{2}} dt. \tag{14}$$

The second integral does not depend on n and becomes arbitrarily small by choosing T sufficiently large. It suffices thus to study the first integral. In order to evaluate it, we separate it into two parts. For $u \to 0$ we have

$$\varphi(u) = 1 - \frac{u^2}{2} + o(u^2).$$

For an $\varepsilon > 0$ sufficiently small and $|u| < \varepsilon$ we have thus

$$|\varphi(u)| < 1 - \frac{u^2}{4} \leq e^{-\frac{u^2}{4}}$$

and it follows that

$$\int\limits_{T}^{\varepsilon\sqrt{n}} \left| \varphi\left(\frac{t}{\sqrt{n}}\right) \right|^n dt \leq \int\limits_{T}^{\infty} e^{-\frac{u^2}{4}} du, \tag{15}$$

which tends to zero as $T \to +\infty$, independently of n. It remains to show that the integral

$$\int\limits_{\varepsilon\sqrt{n}}^{+\infty} \left| \varphi\left(\frac{t}{\sqrt{n}}\right) \right|^n dt \tag{16}$$

tends to zero as $n \to +\infty$. First we choose $q = q(\varepsilon)$ with $0 \leq q < 1$, so that $|\varphi(t)| \leq q$ when $|t| \geq \varepsilon > 0$.

In fact, according to Theorem 8 of Chapter VI, § 2,

$$\lim_{t \to +\infty} |\varphi(t)| = 0.$$

Since the ξ_k do not possess a lattice distribution, $|\varphi(t)| \neq 1$ for every $t \neq 0$; therefore if we put

$$\sup_{|t| \geq \varepsilon} |\varphi(t)| = q$$

we have $0 < q < 1$. Then, however,

$$\int\limits_{\varepsilon\sqrt{n}}^{+\infty} \left| \varphi\left(\frac{t}{\sqrt{n}}\right) \right|^n dt = \sqrt{n} \int\limits_{\varepsilon}^{+\infty} |\varphi(u)|^n du \leq \sqrt{n}\, q^{n-2} \int\limits_{-\infty}^{+\infty} |\varphi(u)|^2 du. \tag{17}$$

Since we have already shown that $\int\limits_{-\infty}^{+\infty} |\varphi(u)|^2\, du$ is finite and $\lim\limits_{n \to +\infty} \sqrt{nq^{n-2}} =$
$= 0$, the integral (16) tends also to zero as $n \to +\infty$. All these restrictions are valid uniformly in x. (2) holds thus uniformly for $-\infty < x < +\infty$. Theorem 1 is herewith proved.

When $f(x)$ is not bounded but for any given k $f_k(x)$ is, (2) remains still valid. This can be shown by a slight modification of the above proof. The condition that $f_k(x)$ be bounded for a value of k (and, consequently, also for every $n \geq k$) is evidently necessary for the uniform convergence of $f_n(x)$ to $\dfrac{1}{\sqrt{2\pi}}\, e^{-\frac{x^2}{2}}$.

§ 3. The domain of attraction of the normal distribution

If ξ_k $(k = 1, 2, \ldots)$ are independent, identically distributed variables and if the standard deviation $D = D(\xi_k)$ exists, then, according to Theorem 1 of § 1, the random variables

$$\zeta_n = \sum_{k=1}^{n} \xi_k$$

satisfy the limit relation

$$\lim_{n \to +\infty} P\left(\frac{\zeta_n - A_n}{S_n} < x\right) = \Phi(x) \qquad (-\infty < x < +\infty), \tag{1}$$

where $A_n = E(\zeta_n)$ and $S_n(\zeta_n) = D\sqrt{n}$ $(n = 1, 2, \ldots)$. Now we have to consider, whether the existence of $D(\xi_k)$ is necessary for the validity of (1) with suitably chosen sequences $\{A_n\}$ and $\{S_n\}$. In the present section we show that the existence of the standard deviation $D(\xi_k)$ can be replaced by the weaker assumption (2).

We define the *domain of attraction* of the normal distribution as the set of distribution functions $F(x)$, possessing the following property:

If $\xi_1, \xi_2, \ldots, \xi_n, \ldots$ are independent random variables with the common distribution function $F(x)$, then (1) is fulfilled for suitably chosen sequences of numbers $\{A_n\}$ and $\{S_n\}$.

In the present section we shall determine the domain of attraction of the normal distribution and we shall prove a theorem due to P. Lévy, W. Feller and A. J. Khintchine:

THEOREM 1. *Let $\xi_1, \xi_2, \ldots, \xi_n, \ldots$ be independent, identically distributed random variables with a common distribution function $F(x)$. If for $F(x)$ the* *limit relation*

$$\lim_{y \to +\infty} \frac{y^2 [F(-y) + (1 - F(y))]}{\int_{-y}^{+y} x^2 dF(x)} = 0 \tag{2}$$

holds, then (1) *is valid for every suitably chosen sequence of numbers* $\{A_n\}$ *and* $\{S_n\}$.

Notes

1. Condition (2) is not only sufficient but also necessary for the validity of (1). But this will not be proved here.

2. If the standard deviation of the random variables ξ_k exists, i.e. if $\int_{-\infty}^{+\infty} x^2 dF(x)$ is finite, (2) is evidently true; this follows immediately from the inequality

$$y^2 [F(-y) + (1 - F(y))] \leq \int_{|x| \geq y} x^2 dF(x).$$

Thus we can see that Theorem 1 of the present section comprises Theorem 1 of § 1.

3. If the standard deviation D and the expectation M of ξ_k exist and if $M = 0$, then in (1) $A_n = 0$ and $S_n = D \sqrt{n}$. Conversely, if (1) holds with $A_n = 0$ and $S_n = D\sqrt{n}$, the standard deviation of ξ_k exists and is equal to D. In fact, in this case Theorem 3 of Chapter VI, § 4 permits to state the following: If $\varphi(t)$ is the characteristic function of the random variables ξ_k, we have

$$\lim_{\to +\infty} \varphi \left(\frac{t}{D \sqrt{n}} \right)^n = e^{-\frac{t^2}{2}}$$

hence

$$\lim_{n \to +\infty} \frac{\ln \varphi \left(\frac{t}{D\sqrt{n}} \right) - \ln \varphi(0)}{\left(\frac{t}{D\sqrt{n}} \right)^2} = -\frac{D^2}{2},$$

and from this follows that $D^2 = \int_{-\infty}^{+\infty} x^2 dF(x)$. Thus if $\int_{-\infty}^{+\infty} x^2 dF(x)$ does not exist, the sequence of numbers $\{S_n\}$ for which (1) holds cannot have the order of magnitude $\sqrt{n}$. (Clearly, by the proof of Theorem 1, S_n tends to infinity faster than $\sqrt{n}$.)

4. As an example of a distribution for which the standard deviation does not exist though (2) holds, we mention the distribution with density function

$$f(x) = \begin{cases} \dfrac{1}{|x|^3} & \text{for } |x| > 1, \\ 0 & \text{otherwise.} \end{cases}$$

Now we give the

PROOF OF THEOREM 1.

LEMMA. *If we have*

$$\frac{y^2\left[F(-y) + (1 - F(y))\right]}{\int\limits_{-y}^{+y} x^2 dF(x)} \le \alpha < 1 \quad \text{for } \; y \ge y_0 > 0, \tag{3}$$

then $\int\limits_{-\infty}^{+\infty} |x|\, dF(x)$ *exists.*

If $Y > y \ge y_0 > 0$, we may write:

$$\int\limits_{y}^{Y} x dF(x) = y(1 - F(y)) - Y(1 - F(Y)) + \int\limits_{y}^{Y} (1 - F(x))\, dx$$

and

$$\int\limits_{-Y}^{-y} x dF(x) = YF(-Y) - yF(-y) - \int\limits_{-Y}^{-y} F(x)\, dx,$$

hence

$$\int\limits_{y \le |x| \le Y} |x|\, dF(x) \le y(1 - F(y) + F(-y)) + \int\limits_{y}^{Y} (1 - F(x) + F(-x))\, dx, \tag{4}$$

and by (3)

$$\int\limits_{y \le |x| \le Y} |x|\, dF(x) \le y(1 - F(y) + F(-y)) + \alpha \int\limits_{y}^{Y} \frac{\int\limits_{-x}^{+x} t^2 dF(t)}{x^2}\, dx,$$

thus

$$\int\limits_{y \le x \le Y} |x|\, dF(x) \le y(1 - F(y) + F(-y)) + \alpha \int\limits_{y \le |t| \le Y} |t|\, dF(t) +$$

$$+ \frac{\alpha}{y} \int\limits_{-y}^{+y} t^2\, dF(t).$$

If we subtract now from both sides $\alpha \int\limits_{y \leq |t| \leq Y} |t| \, dF(t)$ and divide by $1 - \alpha$, we get

$$\int\limits_{y \leq |x| \leq Y} |x| \, dF(x) \leq \frac{2\alpha}{(1-\alpha)y} \int\limits_{-y}^{+y} t^2 \, dF(t). \tag{5}$$

Since the right hand side of (5) does not depend on Y, we conclude that $\int\limits_{-\infty}^{+\infty} |x| \, dF(x)$ exists; the lemma is herewith proved.

In what follows, we assume for sake of simplicity that $F(x)$ is continuous and symmetrical with respect to the origin, so that $F(-x) = 1 - F(x)$. In the general case the proof runs similarly, but the calculations are somewhat more complicated. Furthermore, we may assume $F(x) < 1$ for every $x < +\infty$.

The existence of $M = \int\limits_{-\infty}^{+\infty} x \, dF(x)$ follows from the lemma proved above. Because of the symmetry of $F(x)$, we have

$$\int\limits_{-\infty}^{+\infty} x \, dF(x) = 0. \tag{6}$$

(In the general case we consider the random variable $\xi_k' = \xi_k - M$.) We put

$$\delta(y) = \frac{y^2 (1 - F(y))}{\int\limits_0^y x^2 \, dF(x)}. \tag{7}$$

Then by assumption $\lim\limits_{y \to +\infty} \delta(y) = 0$. Put further

$$\Delta(y) = \frac{\delta(y)}{(1 - F(y))^2}. \tag{8}$$

It follows from

$$\Delta(y) = \frac{y^2}{(1 - F(y)) \int\limits_0^y x^2 \, dF(x)} \geq \frac{1}{(1 - F(y)) \left(F(y) - \dfrac{1}{2} \right)} \tag{9}$$

that

$$\lim\limits_{y \to +\infty} \Delta(y) = +\infty. \tag{10}$$

By assumption, $F(x)$ is continuous, hence $\Delta(y)$ is also continuous for $y \geq \geq y_0 > 0$. Let C_n be for $n \geq n_0$ the least positive number $\geq y_0$ such that

$$\Delta(C_n) = n^2. \tag{11}$$

Evidently, $C_n \to +\infty$ and

$$n(1 - F(C_n)) = \sqrt{\delta(C_n)} . \tag{12}$$

Put now

$$S_n^2 = n \int_{-C_n}^{+C_n} x^2 dF(x), \tag{13}$$

and let $\varphi(t)$ be the characteristic function of ξ_k, $\varphi_n(t)$ the characteristic function of ζ_n/S_n. We have

$$\varphi_n(t) = E(e^{\frac{it\zeta_n}{S_n}}) = \left[\varphi\left(\frac{t}{S_n}\right)\right]^n . \tag{14}$$

However, we have

$$\varphi\left(\frac{t}{S_n}\right) = 1 + \int_{-C_n}^{+C_n} (e^{\frac{it\zeta_n}{S_n}} - 1)\, dF(x) + \int_{|x| > C_n} (e^{\frac{itx}{S_n}} - 1)\, dF(x) \tag{15}$$

and

$$\left| \int_{|x| > C_n} (e^{\frac{itx}{S_n}} - 1)\, dF(x) \right| \le 4(1 - F(C_n)) = \frac{4\sqrt{\delta(C_n)}}{n} . \tag{16}$$

By the lemma of §1

$$\int_{-C_n}^{+C_n} (e^{\frac{itx}{S_n}} - 1)\, dF(x) = -\frac{t^2}{2n} + R_n \tag{17}$$

holds with

$$|R_n| \le \int_{-C_n}^{+C_n} \frac{|t|^3 |x|^3}{6S_n^3}\, dF(x) \le \frac{C_n}{S_n} \frac{|t|^3}{6n} = \frac{\sqrt[4]{\delta(C_n)}}{6\sqrt{2}\, n} |t|^3 . \tag{18}$$

Relations (14) through (18) lead to

$$\varphi_n(t) = \left(1 - \frac{t^2}{2n} + \frac{\theta_n \sqrt[4]{\delta(C_n)}}{n}\right)^n$$

with a θ_n which remains in absolute value below a bound not depending on n. Since $\lim_{n \to +\infty} \sqrt[4]{\delta(C_n)} = 0$, we get

$$\lim_{n \to +\infty} \varphi_n(t) = e^{-\frac{t^2}{2}}, \tag{19}$$

which implies the statement of Theorem 1.

As regards the question whether other distributions than the normal also have a domain of attraction, the following example shows that this is possible. Let $\xi_1, \xi_2, \ldots, \xi_n, \ldots$ be completely independent random variables possessing a common stable distribution of order $\alpha(0 < \alpha < 2)$ and characteristic function $e^{-|t|^\alpha}$, then the characteristic function of $\dfrac{1}{n^{1/\alpha}} \sum\limits_{k=1}^{n} \xi_k$ is exactly e^{-t^α}.

Thus any stable distribution has a domain of attraction which contains at least the distribution itself. The domain of attraction of a stable distribution with $0 < \alpha < 2$ is very narrow compared with that of the normal distribution; it contains only distributions very similar to the stable distribution considered. As regards the determination of the domain of attraction in the case of a stable distribution with $0 < \alpha < 2$, we refer to the book of Gnedenko and Kolmogorov [1].

§ 4. Convergence to the Poisson distribution

We have already proved in Chapter III, § 12 that the binomial distribution of order n and parameter p tends, as $n \to +\infty$, to the Poisson distribution with the expectation λ, if p tends to zero in such a way that $np \to \lambda$. This is a particular case of the following, more general theorem:

THEOREM 1. *Let $\xi_{n1}, \xi_{n2}, \ldots, \xi_{nk_n}$ be independent random variables assuming nonnegative integral values only. Put*

$$P(\xi_{nk} = r) = p_{nk}(r) \qquad (r = 0, 1, \ldots; \; k = 1, 2, \ldots, k_n \,; n = 1, 2, \ldots) \quad (1)$$

and

$$R_{nk} = \sum_{r=2}^{\infty} p_{nk}(r). \tag{2}$$

If the following conditions

$$\lim_{n \to +\infty} \sum_{k=1}^{k_n} p_{nk}(1) = \lambda, \tag{A}$$

$$\lim_{n \to +\infty} \max_{1 \le k \le k_n} (1 - p_{nk}(0)) = 0, \tag{B}$$

$$\lim_{n \to +\infty} \sum_{k=1}^{k_n} R_{nk} = 0, \tag{C}$$

are satisfied, then the distribution function of

$$\eta_n = \xi_{n1} + \xi_{n2} + \ldots + \xi_{nk_n} \tag{3}$$

tends, as $n \to +\infty$, to the distribution function of the Poisson distribution with expectation λ.

PROOF. Let $g_{nk}(z)$ denote the generating function of the random variable ξ_{nk}:

$$g_{nk}(z) = \sum_{r=0}^{\infty} p_{nk}(r) z^r \qquad (|z| \le 1). \tag{4}$$

Clearly

$$|g_{nk}(z) - p_{nk}(0) - p_{nk}(1) z| \le R_{nk} \quad \text{for} \quad |z| \le 1. \tag{5}$$

Since

$$p_{nk}(0) - (1 - p_{nk}(1)) = -R_{nk}, \tag{6}$$

we can write

$$|g_{nk}(z) - 1 - p_{nk}(1)(z - 1)| \le 2R_{nk}. \tag{7}$$

The identity (38) of § 1 implies, since $|g_{nk}(z)| \le 1$ and $|1 + p_{nk}(1)(z - 1)| \le 1$,

$$\left| \prod_{k=1}^{k_n} g_{nk}(z) - \prod_{k=1}^{k_n} (1 + p_{nk}(1)(z - 1)) \right| \le 2 \sum_{k=1}^{k_n} R_{nk}. \tag{8}$$

If

$$\max_{1 \le k \le k_n} p_{nk}(1) \le \max_{1 \le k \le k_n} (1 - p_{nk}(0)) \le \frac{1}{4},$$

which is because of (B) fulfilled for $n \ge n_0$, then identity (38) of § 1 leads to

$$\left| \prod_{k=1}^{k_n} (1 + p_{nk}(1)(z - 1)) - \prod_{k=1}^{k_n} \exp (p_{nk}(1)(z - 1)) \right| \le$$

$$\le \max_{1 \le k \le k_n} (1 - p_{nk}(0)) |z - 1|^2 \prod_{k=1}^{k_n} p_{nk}(1). \tag{9}$$

It follows now by our assumptions from (8) and (9) that

$$\lim_{n \to +\infty} \prod_{k=1}^{k_n} g_{nk}(z) = e^{\lambda(z-1)}. \tag{10}$$

Since $\prod_{k=1}^{k_n} g_{nk}(z)$ is the generating function of η_n and $e^{\lambda(z-1)}$ that of the Poisson distribution $\{\lambda^k e^{-\lambda}/k!\}$, our theorem is proved in view of Theorem 4 of Chapter III, § 15.

When $k_n = n$ and the random variables ξ_{nk} take on only the values 0 and 1 with the probabilities $P(\xi_{nk} = 0) = 1 - \dfrac{\lambda}{n}$ and $P(\xi_{nk} = 1) = \dfrac{\lambda}{n}$, then conditions (A), (B), (C) are evidently fulfilled. In this case $\eta_n = \sum\limits_{k=1}^{n} \xi_{nk}$ has a binomial distribution:

$$P(\eta_n = j) = \binom{n}{j}\left(\frac{\lambda}{n}\right)^j \left(1 - \frac{\lambda}{n}\right)^{n-j}.$$

Thus we have as a particular case of Theorem 1:

$$\lim_{n \to +\infty} \binom{n}{j}\left(\frac{\lambda}{n}\right)^j \left(1 - \frac{\lambda}{n}\right)^{n-j} = \frac{\lambda^k e^{-\lambda}}{k!}.$$

Theorem 1 is therefore a generalization of the convergence of the binomial distribution of order n and parameter $p = \dfrac{\lambda}{n}$ to the Poisson distribution of parameter λ, already dealt with in Chapter III.

§ 5. The central limit theorem for samples from a finite population

The statement of the central limit theorem is valid for certain sequences of weakly dependent random variables. In the present and in the following two sections we prove some results in this direction. These results have practical importance, too, since in the applications the independence is often only approximately true. The following theorem[1] refers to samples taken from a finite population, a situation very often encountered in practice.

THEOREM 1. *Let* $a_{N,1}, a_{N,2}, \ldots, a_{N,N}$ *be any real numbers* $(N = 1, 2, \ldots)$,

$$M_N = \sum_{k=1}^{N} a_{N,k} \tag{1}$$

and

$$D_n^2 = \sum_{k=1}^{N} \left(a_{N,k} - \frac{M_N}{N}\right)^2. \tag{2}$$

---

[1] Cf. P. Erdős and A. Rényi [1]. See also J. Hájek [2], where it is shown that Theorem 1 is essentially best possible.

Let further $n = n(N) < N$ be a positive integer-valued function of N and put

$$D_{N,n} = D_N \sqrt{\frac{n}{N}\left(1 - \frac{n}{N}\right)}. \tag{3}$$

From the numbers $a_{N,1}, a_{N,2}, \ldots, a_{N,N}$ we randomly choose n numbers in such a way that all $\binom{N}{n}$ combinations have the same probability. Let the random variable $\zeta_{N,n}$ denote the sum of the so chosen $a_{N,k}$ and put

$$\zeta_{N,n}^* = \zeta_{N,n} - \frac{n}{N} M_N. \tag{4}$$

Put further

$$d_{N,n}(\varepsilon) = \frac{1}{D_N^2} \sum_{\left|a_{N,k} - \frac{M_N}{N}\right| > \varepsilon D_{N,n}} \left(a_{N,k} - \frac{M_N}{N}\right)^2. \tag{5}$$

If the condition

$$\lim_{N \to +\infty} d_{N,n}(\varepsilon) = 0 \tag{6}$$

is satisfied for any $\varepsilon > 0$, then we have for $-\infty < x < +\infty$

$$\lim_{N \to +\infty} P\left(\frac{\zeta_{N,n}^*}{D_{N,n}} < x\right) = \Phi(x). \tag{7}$$

PROOF. Condition (6) implies that $n \to +\infty$ as $N \to +\infty$. Indeed it follows from (6) that there exists for every $\varepsilon > 0$ a number N_0 such that for $N > N_0$ the inequality $d_{N,n}(\varepsilon) < \frac{1}{2}$ holds; but then we have

$$\frac{1}{2} < \frac{1}{D_N^2} \sum_{\left|a_{N,k} - \frac{M_N}{N}\right| \leq \varepsilon D_{N,n}} \left(a_{N,k} - \frac{M_N}{N}\right)^2 \leq N\varepsilon^2 \frac{D_{N,n}^2}{D_N^2} \leq n\varepsilon^2.$$

Hence, for $N > N_0$ we have $n = n(N) \geq \frac{1}{2\varepsilon^2}$ and since $\varepsilon > 0$ can be chosen arbitrarily small, we get

$$\lim_{N \to +\infty} n(N) = +\infty. \tag{8}$$

We may assume

$$M_N = 0. \tag{9}$$

In fact, if (9) is not fulfilled, consider instead of the numbers $a_{N,k}$ the numbers $a'_{N,k} = a_{N,k} - \dfrac{M_N}{N}$; for these (9) is clearly fulfilled and if Theorem 1 holds for $a'_{N,k}$, it remains valid for $a_{N,k}$ too.

Furthermore, we can assume

$$1 \le n \le \frac{N}{2} ; \tag{10}$$

the random variables $\zeta_{N,n}$ and $-\zeta_{N,N-n}$ have indeed the same distribution and if $n > \dfrac{N}{2}$, we may take instead of n the number $N - n$.

We compute now the characteristic function $\varphi_{N,n}(t)$ of $\zeta^*_{N,n}$:

$$\varphi_{N,n}(t) = \frac{1}{\binom{N}{n}} \sum_{1 \le j_1 < j_2 < \ldots < j_n \le N} \exp \left[it(a_{N,j_1} + a_{N,j_2} + \ldots + a_{N,j_n}) \right], \tag{11}$$

where the summation is to be extended over all combinations of order n of the numbers $1, 2, \ldots, N$. In order to prove Theorem 1, it suffices to establish the relation

$$\lim_{N \to +\infty} \varphi_{N,n} \left(\frac{t}{D_{N,n}} \right) = e^{-\frac{t^2}{2}} . \tag{12}$$

We put

$$\lambda = \frac{n}{N} \tag{13}$$

and

$$B_{N,n}(\lambda) = \binom{N}{n} \lambda^n (1 - \lambda)^{N-n}. \tag{14}$$

By using the fact that

$$\frac{1}{2\pi} \int_{-\pi}^{+\pi} e^{ik\varphi} \, d\varphi = \begin{cases} 1 & \text{for } k = 0, \\ 0 & \text{for } k = \pm 1, \pm 2, \ldots \end{cases} \tag{15}$$

we obtain easily the relation

$$\varphi_{N,n}(t) = \frac{1}{2\pi B_{N,n}(\lambda)} \int_{-\pi}^{+\pi} \prod_{k=1}^{N} \left[(1 - \lambda) e^{-i\lambda(\varphi + ta_{N,k})} + \lambda e^{i(1-\lambda)(\varphi + ta_{N,k})} \right] d\varphi. \tag{16}$$

Indeed if we calculate the value of the expression behind the sign of integration, by taking in the product $(N - m)$ times the first and m times the second term, we obtain a term multiplied by the factor $e^{i(m-n)\varphi}$; such a term vanishes therefore when the integration is carried out provided that $m \neq n$. If $N \to +\infty$ and $N - n \to +\infty$, which is certainly fulfilled in our case because of (8) and (10), it follows from Stirling's formula[1] that

$$B_{N,n}(\lambda) \approx \frac{1}{\sqrt{2\pi N\lambda(1 - \lambda)}}. \tag{17}$$

Hence if we introduce a new variable of integration $\psi = \varphi \sqrt{N\lambda(1 - \lambda)}$ and if we replace t by $\dfrac{t}{D_{N,n}}$, we obtain

$$\varphi_{N,n}\left(\frac{t}{D_{N,n}}\right) \approx \frac{1}{\sqrt{2\pi}} \int\limits_{-\pi\sqrt{N\lambda(1-\lambda)}}^{+\pi\sqrt{N\lambda(1-\lambda)}} \prod_{k=1}^{N} \varrho_k(\psi, t)\, d\psi, \tag{18}$$

where we put

$$\varrho_k(\psi, t) = (1 - \lambda) \exp\left[-i\lambda\left(\frac{\psi}{\sqrt{N\lambda(1 - \lambda)}} + \frac{ta_{N,k}}{D_{N,n}}\right)\right] +$$

$$+ \lambda \exp\left[i(1 - \lambda)\left(\frac{\psi}{\sqrt{N\lambda(1 - \lambda)}} + \frac{ta_{N,k}}{D_{N,n}}\right)\right]. \tag{19}$$

According to the lemma of § 1 we have

$$(1 - \lambda) e^{-i\lambda v} + \lambda e^{i(1 - \lambda)v} = 1 - \frac{v^2 \lambda(1 - \lambda)}{2} + R_1 \tag{20a}$$

with

$$|R_1| \leq \frac{\lambda(1 - \lambda)|v|^3}{6} \tag{20b}$$

and

$$(1 - \lambda) e^{-i\lambda v} + \lambda e^{i(1 - \lambda)v} = 1 + R_2 \tag{21a}$$

with

$$|R_2| \leq \frac{\lambda(1 - \lambda) v^2}{2}. \tag{21b}$$

---

[1] If λ is fixed, (17) follows directly from the Moivre-Laplace theorem. In our case λ depends on N, hence the latter theorem cannot be applied. But Stirling's formula leads easily to (17).

On the other hand, an easy calculation shows that

$$|(1 - \lambda)\,e^{-i\lambda v} + \lambda e^{i(1-\lambda)v}| = \sqrt{1 - 2\lambda(1 - \lambda)(1 - \cos v)} \le$$
$$\le 1 - \lambda(1 - \lambda)(1 - \cos v). \tag{22}$$

Now let $\varepsilon > 0$ be given and suppose $|\psi| < 2\varepsilon \sqrt{N\lambda(1 - \lambda)}$. If k is an index such that $\dfrac{|t| \cdot |a_{N,k}|}{D_{N,n}} \le \varepsilon$, then (20) implies

$$\varrho_k(\psi, t) = 1 - \frac{\lambda(1 - \lambda)}{2} \left(\frac{\psi}{\sqrt{N\lambda(1 - \lambda)}} + \frac{ta_{N,k}}{D_{N,n}} \right)^2 (1 + \theta_1 \varepsilon) \tag{23}$$

with $|\theta_1| \le 1$; (21) implies that for every value of k and for $|\psi| < 2\varepsilon\sqrt{N\lambda(1 - \lambda)}$,

$$|\varrho_k(\psi, t) - 1| \le \frac{\lambda(1 - \lambda)}{2} \left(\frac{\psi}{\sqrt{N\lambda(1 - \lambda)}} + \frac{ta_{N,k}}{D_{N,n}} \right)^2. \tag{24}$$

From this follows for $|\psi| < 2\varepsilon \sqrt{N\lambda(1 - \lambda)}$ the relation

$$\prod_{k=1}^{N} \varrho_k(\psi, t) = \exp \left[-\left(\frac{\psi^2 + t^2}{2} \right)(1 + \theta_2\,\varepsilon) \right](1 + \eta_N), \tag{25}$$

where $|\theta_2| \le C_1$ and

$$|\eta_N| \le C_2 \left[d_{N,n}\left(\frac{\varepsilon}{|t|} \right) + \varepsilon^2 \lambda(1 - \lambda) l_N \right] \tag{26}$$

with

$$l_N = \sum_{|a_{N,k}| > \frac{\varepsilon D_{N,n}}{|t|}} 1, \tag{27}$$

C_1 and C_2 being positive constants.

We have already

$$l_N \le \frac{t^2}{\varepsilon^2 \lambda(1 - \lambda)} \, d_{N,n}\left(\frac{\varepsilon}{|t|} \right); \tag{28}$$

it follows from (26), (27) and (28), because of (16), that

$$\lim_{N \to +\infty} \eta_N = 0. \tag{29}$$

For $|\psi| \ge 2\varepsilon \sqrt{N\lambda(1 - \lambda)}$ the following estimates can be used: For $|a_{N,k}| \cdot |t| / D_{N,n} \ge \varepsilon$ the trivial inequality $|\varrho_k(\psi, t)| \le 1$; for

$|a_{N,k}| \, |t|/D_{N,n} < \varepsilon$ the inequality

$$|\varrho_k(\psi,t)| \leq 1 - \lambda(1-\lambda)(1-\cos\varepsilon), \qquad (30)$$

which can be derived from (22).

Thus we obtain

$$\int\limits_{2\varepsilon < \frac{\psi}{\sqrt{N\lambda(1-\lambda)}} < \pi} \left| \prod_{k=1}^{N} \varrho_k(\psi,t) \right| d\psi \leq 2\pi\sqrt{N\lambda} \left(1 - \frac{\lambda(1-\cos\varepsilon)}{2}\right)^{N-l_N}. \quad (31)$$

Since

$$\sqrt{N\lambda}\left(1 - \frac{\lambda(1-\cos\varepsilon)}{2}\right)^{N-l_N} \leq \sqrt{n}\,\exp\left(-\frac{n(1-\cos\varepsilon)}{2} + \lambda l_N\right),$$

the right hand side of (31) tends to zero as $N \to +\infty$ because of (28) and because

$$\lim_{N\to+\infty} \sqrt{n}\,\exp\left(-\frac{n(1-\cos\varepsilon)}{2}\right) = 0.$$

From (18), (25), and (31) we obtain, since ε can be taken arbitrarily small

$$\lim_{N\to+\infty} \varphi_{N,n}\left(\frac{t}{D_{N,n}}\right) = e^{-\frac{t^2}{2}} \frac{1}{\sqrt{2\pi}} \int\limits_{-\infty}^{+\infty} e^{-\frac{\psi^2}{2}}\,d\psi = e^{-\frac{t^2}{2}}, \qquad (32)$$

which concludes the proof of our theorem.

A case of particular importance occurs when M of the numbers $a_{N,k}$ are equal to 1 and $N - M$ are equal to zero. Then

$$P(\zeta_{N,n} = m) = \frac{\binom{M}{m}\binom{N-M}{n-m}}{\binom{N}{n}} \qquad (33)$$

i.e. $\zeta_{N,n}$ has a hypergeometric distribution. Furthermore, $M_N = M$, $D_N = \sqrt{M(N-M)/N}$. Condition (6) is satisfied for $\dfrac{M(N-M)\,n(N-n)}{N^3} \to$ $\to +\infty$ as $N \to +\infty$; in effect, for every $\varepsilon > 0$, $d_{N,n}(\varepsilon) = 0$ as soon as N (depending on ε) is sufficiently large.

If $M \leq \dfrac{N}{2}$ and $n \leq \dfrac{N}{2}$, which can be assumed without restriction of generality, the above condition is equivalent to

$$\frac{nM}{N} \to +\infty. \tag{34}$$

When $\frac{M}{N}$ is constant or remains above a positive bound, this means that n must tend to $+\infty$ with N. From $\frac{nM}{N} \to +\infty$ it follows, because of $M \leq \frac{N}{2}$ and $n \leq \frac{N}{2}$, that $N \to +\infty$, $n \to +\infty$, and $M \to +\infty$. Theorem 1 contains thus as a particular case

THEOREM 2. *If N, M and n are positive integers, $1 \leq M \leq \frac{N}{2}, 1 \leq n \leq \frac{N}{2}$ further if we put $p = \frac{M}{N}$ and $\lambda = \frac{n}{N}$, then*

$$\lim_{\substack{Np\lambda \to +\infty}} \sum_{k \leq np+x\sqrt{np(1-p)(1-\lambda)}} \frac{\binom{M}{k}\binom{N-M}{n-k}}{\binom{N}{n}} = \Phi(x). \tag{35}$$

A particular case of this theorem, when $p = \frac{M}{N}$ = constant, was derived by S. N. Bernstein.

Note further that if $p = \frac{M}{N}$ is constant and n increases more slowly than N (if, for instance $\frac{n^2}{N} \to 0$), (35) can be derived from the Moivre–Laplace theorem by approximating the terms of the hypergeometric distribution by those of the binomial distribution (see Chapter II, § 12, Exercise 18). However, the general case cannot be treated in this way.

Theorem 2 can also be proved directly by merely considering the asymptotic behaviour of the terms of the hypergeometric distribution, but this procedure leads to tiresome calculations.

§ 6. Generalization of the central limit theorem through the application of mixing theorems

A sequence $\eta_1, \eta_2, \ldots, \eta_n, \ldots$ of random variables possessing a limit distribution, i.e. such that

$$\lim_{n \to +\infty} P(\eta_n < x) = F(x)$$

holds at every continuity point x of the distribution function $F(x)$, is said to be *mixing*, if for any event B with positive probability the relation

$$\lim_{n \to +\infty} P(\eta_n < x \mid B) = F(x) \tag{1}$$

holds at every continuity point x of $F(x)$. We prove now

THEOREM 1. *Let* $\xi_1, \xi_2, \ldots, \xi_n, \ldots$ *be independent random variables and put*

$$\zeta_n = \sum_{k=1}^{n} \xi_k.$$

Assume that there exist two sequences $\{C_n\}$ *and* $\{S_n\}$ *with* $S_n \to +\infty$ *and a distribution function* $F(x)$ *such that at every continuity point of* $F(x)$ *the distribution function of*

$$\eta_n = \frac{\zeta_n - C_n}{S_n}$$

tends to $F(x)$:

$$\lim_{n \to +\infty} P(\eta_n < x) = F(x).$$

Then the sequence of the random variables η_n *is mixing.*

PROOF. We shall use the following lemma due to H. Cramér[1]

LEMMA 1. *Let* θ_n *and* ε_n *(n = 1, 2, ...) be two sequences of random variables; assume that the sequence* θ_n *has a limit distribution with distribution function* $F(x)$, *that is, at every continuity point of* $F(x)$ *we have*

$$\lim_{n \to +\infty} P(\theta_n < x) = F(x). \tag{2}$$

Assume further that $\lim_{n \to +\infty} \text{st } \varepsilon_n = 0$. *Then*

$$\lim_{n \to +\infty} P(\theta_n + \varepsilon_n < x) = F(x) \tag{3}$$

at every continuity point x *of* $F(x)$.

PROOF. We have for an arbitrary $\delta > 0$

$$P(\theta_n + \varepsilon_n < x) = P(|\varepsilon_n| \geq \delta) \, P(\theta_n + \varepsilon_n < x \,\|\, |\varepsilon_n| \geq \delta) +$$
$$+ P(\theta_n + \varepsilon_n < x \,\|\, |\varepsilon_n| < \delta). \tag{4}$$

[1] Cf. H. Cramér [3].

By assumption

$$\lim_{n \to +\infty} P(|\varepsilon_n| \geq \delta) = 0.$$

From $\theta_n + \varepsilon_n < x$ and $|\varepsilon_n| < \delta$ follows $\theta_n < x + \delta$; from $\theta_n < x - \delta$ and $|\varepsilon_n| < \delta$ follows $\theta_n + \varepsilon_n < x$. Hence we can conclude from (4)

$$F(x - \delta) - P(|\varepsilon_n| \geq \delta) \leq P(\theta_n + \varepsilon_n < x| |\varepsilon_n| < \delta) \leq F(x + \delta), \qquad (5)$$

$$F(x - \delta) \leq \varliminf_{n \to +\infty} P(\theta_n + \varepsilon_n < x) \leq \varlimsup_{n \to +\infty} P(\theta_n + \varepsilon_n < x) \leq F(x + \delta). \qquad (6)$$

Since x is by assumption a continuity point of $F(x)$ and $\delta > 0$ may be taken arbitrarily small, (3) is proved.

Let now x be a continuity point of $F(x)$ and suppose $F(x) > 0$. Then by assumption we can find an n_0 such that $P(\eta_n < x) > 0$ for $n > n_0$. Put $A_0 = \Omega$ and denote by A_k the event $\eta_{n_0+k} < x \, (k = 1, 2, \ldots)$. Then $P(A_k) > 0$ and, by assumption,

$$\lim_{n \to +\infty} P(A_n | A_0) = \lim_{n \to +\infty} P(A_n) = F(x) > 0.$$

By Theorem 1 of Chapter VII, § 10 it suffices to prove the relations

$$\lim_{n \to +\infty} P(A_n | A_k) = F(x) \qquad (k = 1, 2, \ldots). \qquad (7)$$

Apply Lemma 1 with $\theta_n = \eta_n$ and $\varepsilon_n = -\dfrac{\zeta_{n_0+k}}{S_n}$. This can be done, since the hypotheses of the lemma are fulfilled. We find

$$\lim_{n \to +\infty} P\left(\eta_n - \frac{\zeta_{n_0+k}}{S_n} < x\right) = F(x). \qquad (8)$$

Since

$$\eta_n - \frac{\zeta_{n_0+k}}{S_n} = \frac{\zeta_n - \zeta_{n_0+k} - C_n}{S_n}.$$

does not depend on η_{n_0+k}, we have

$$\lim_{n \to +\infty} P\left(\eta_n - \frac{\zeta_{n_0+k}}{S_n} < x \,\middle|\, A_k\right) = F(x). \qquad (9)$$

If we apply Lemma 1 again, to the random variables

$$\theta_n = \eta_n - \frac{\zeta_{n_0+k}}{S_n}, \qquad \varepsilon_n = \frac{\zeta_{n_0+k}}{S_n}$$

on the probability space $[\Omega, \mathscr{A}, P(A \mid A_k)]$, we get already (7):

$$\lim_{n \to +\infty} P(\eta_n < x \mid A_k) = \lim_{n \to +\infty} P(A_n \mid A_k) = F(x).$$

The conditions of Theorem 1 of Chapter VII, § 10 are thus satisfied and for every B with $P(B) > 0$ the relation

$$\lim_{n \to +\infty} P(\eta_n < x \mid B) = F(x) \tag{10}$$

holds. The theorem is therefore proved for every x such that $F(x) > 0$. If x is a continuity point of $F(x)$ such that $F(x) = 0$, we have

$$\lim_{n \to +\infty} P(\eta_n < x) = 0$$

and, if $P(B) > 0$,

$$\lim_{n \to +\infty} P(\eta_n < x \mid B) \leq \lim_{n \to +\infty} \frac{P(\eta_n < x)}{P(B)} = 0.$$

Theorem 1 is herewith completely proved.

Theorem 1 can also be formulated as follows: If the random variables ξ_k are independent, if $S_n \to +\infty$ as $n \to +\infty$, further if the random variables

$$\eta_n = \frac{\sum_{k=1}^{n} \xi_k - C_n}{S_n}$$

possess a limit distribution, then η_n is in the limit independent of any random variable θ in the following sense: For every y such that $P(\theta < y) > 0$ the relation

$$\lim_{n \to +\infty} P(\eta_n < x, \theta < y) = \lim_{n \to +\infty} P(\eta_n < x) P(\theta < y) \tag{11}$$

holds at every continuity point of the limit distribution of η_n.

The following is an interesting corollary of Theorem 1:

THEOREM 2. *Suppose that the random variables $\xi_1, \xi_2, \ldots, \xi_n, \ldots$ are independent and put $\zeta_n = \sum_{k=1}^{n} \xi_k$. If there exist two sequences C_n and S_n $(n = 1, 2, \ldots)$ with $\lim_{n \to +\infty} S_n = +\infty$ fulfilling the relation*

$$\lim_{n \to +\infty} P(\eta_n < x) = F(x),$$

where $\eta_n = \dfrac{\zeta_n - C_n}{S_n}$ and $F(x)$ is a nondegenerate distribution function, then η_n cannot converge in probability to a random variable η_∞.

PROOF. Assume that there exists such a variable η_∞ with $\eta_a - \eta_\infty \xrightarrow{P} 0$. If we apply the lemma with $\theta_n = \eta_n$ and $\varepsilon_n = \eta_\infty - \eta_n$, we find

$$P(\eta_\infty < x) = F(x). \tag{12}$$

On the other hand, Theorem 1 allows to state the following: If x is a continuity point of $F(x)$ with $0 < F(x) < 1$ ($F(x)$ being nondegenerate; such a point always exists) and if B denotes the event $\eta_\infty < x$, we have

$$\lim_{n \to +\infty} P(\eta_n < x \mid B) = F(x) = P(B). \tag{13}$$

If we apply the lemma to the random variables $\theta_n = \eta_n$ and $\varepsilon_n = \eta_\infty - \eta_n$ on the probability space $[\Omega, \mathscr{A}, P(A \mid B)]$, we get

$$P(\eta_\infty < x \mid B) = P(B \mid B) = P(B),$$

i.e. $F(x) = P(B) = 1$, which contradicts our assumption that $0 < F(x) < 1$. Hence Theorem 2 is proved.

Naturally, it follows from Theorem 2 that, under the conditions of the theorem, the limit of η_n cannot exist almost everywhere. Still more is true: the probability of the existence of the limit $\lim\limits_{n \to +\infty} \eta_n$ is equal to zero. The set C of the elements $\omega \in \Omega$ for which $\lim\limits_{n \to +\infty} \eta_n(\omega) = \eta_\infty(\omega)$ exists is obviously measurable. Suppose we have $P(C) > 0$, then η_n would converge on the probability space $[\Omega, \mathscr{A}, P(A \mid C)]$, with probability 1 and therefore also in probability, which contradicts Theorem 2.

We now prove a lemma.

LEMMA 2. *Let $[\Omega, \mathscr{A}, P]$ be a probability space and $Q(A)$ ($A \in \mathscr{A}$) a second probability measure on the σ-algebra $\mathscr{A}$, absolutely continuous with respect to P. If the sequence of sets $A_n \in \mathscr{A}$ is mixing on $[\Omega, \mathscr{A}, P]$ with the density d, then*

$$\lim_{n \to +\infty} Q(A_n) = d. \tag{14}$$

PROOF. According to the Radon–Nikodym theorem there exists a measurable function $\chi(\omega)$ such that for every $A \in \mathscr{A}$

$$Q(A) = \int_A \chi(\omega)\, dP.$$

If $\chi(\omega)$ is a step function, (14) is clearly fulfilled. According to the definition of the Lebesgue integral there can always be found a step function $\chi_1(\omega)$ such that

$$\int_\Omega |\chi(\omega) - \chi_1(\omega)| \, dP < \varepsilon.$$

Hence (14) is always fulfilled.[1]

Lemma 2 allows to rephrase Theorem 1 in the following stronger form:

THEOREM 3. *If $\xi_1, \xi_2, \ldots, \xi_n, \ldots$ are independent random variables on the probability space $[\Omega, \mathscr{A}, P]$ and if the hypotheses of Theorem 1 are satisfied, then for every probability measure Q absolutely continuous with respect to P the relation*

$$\lim_{n \to +\infty} Q(\eta_n < x) = F(x) \tag{15}$$

holds at every continuity point x of $F(x)$.

Theorem 3 allows to extend limit theorems to sequences of weakly dependent random variables. Assume indeed that the random variables ξ_k are not independent with respect to the probability measure Q, but let there exist a second probability measure P such that Q is absolutely continuous with respect to P while ξ_k are independent with respect to P. If one of the theorems about the limit distributions can be applied to $[\Omega, \mathscr{A}, P]$, Theorem 3 guarantees its applicability to $[\Omega, \mathscr{A}, Q]$ as well.[2]

§ 7. The central limit theorem for sums of a random number of random variables

In the present section $\xi_1, \xi_2, \ldots, \xi_n, \ldots$ denote independent, identically distributed random variables with zero expectation and unit variance. Hence by Theorem 1 of § 1 for the random variables

$$\zeta_n = \frac{\sum\limits_{k=1}^{n} \xi_k}{\sqrt{n}} \tag{1}$$

the relation

$$\lim_{n \to +\infty} P(\zeta_n < x) = \Phi(x) \tag{2}$$

---

[1] The sequence of events $\{A_n\}$ is also mixing with respect to $[\Omega, \mathscr{A}, Q]$ since $Q^*(A) = Q(A \mid B)$ is also absolutely continuous with respect to P. Hence by Lemma 2 $\lim\limits_{n \to +\infty} Q(A_n \mid B) = d$.

[2] Cf. P. Révész [1].

is valid. Let now v_n $(n = 1, 2, \ldots)$ be a sequence of random variables which assume only positive integer values and which are supposed to obey the relation

$$v_n \xrightarrow{P} +\infty \tag{3}$$

i.e. to fulfill for any $N > 0$

$$\lim_{n \to +\infty} P(v_n > N) = 1.$$

We want to find conditions under which the random variables ζ_{v_n} are in the limit normally distributed. It is easy to prove

THEOREM 1. *If the random variables* v_n $(n = 1, 2, \ldots)$ *are independent of the random variables* $\zeta_1, \zeta_2, \ldots, \zeta_k, \ldots$ *and if* (1), (2) *and* (3) *are fulfilled, then*

$$\lim_{n \to +\infty} P(\zeta_{v_n} < x) = \Phi(x). \tag{4}$$

PROOF. Put

$$C_{nk} = P(v_n = k) \qquad (n, k = 1, 2, \ldots). \tag{5}$$

The matrix (C_{nk}) possesses the following properties:

$$C_{nk} \geq 0 \qquad (n, k = 1, 2, \ldots), \tag{6a}$$

$$\sum_{k=1}^{\infty} C_{nk} = 1 \qquad (n = 1, 2, \ldots), \tag{6b}$$

$$\lim_{n \to +\infty} C_{nk} = 0 \qquad (k = 1, 2, \ldots). \tag{6c}$$

(6c) is a consequence of (3); (6a) and (6b) express the fact that $C_{n1}, C_{n2}, \ldots, C_{nk}, \ldots$ is a probability distribution. The three conditions (6) can be expressed by saying that (C_{nk}) is a permanent Toeplitz matrix. A theorem[1] known from the theory of series permits to conclude that, if

$$\lim_{n \to +\infty} S_n = S$$

then

$$\lim_{n \to +\infty} \sum_{k=1}^{\infty} C_{nk} S_k = S.$$

Now

$$P(\zeta_{v_n} < x) = \sum_{k=1}^{\infty} P(\zeta_k < x, v_n = k). \tag{7}$$

[1] Cf. K. Knopp [1].

Since v_n does not depend on ξ_k, we get

$$P(\zeta_{v_n} < x) = \sum_{k=1}^{\infty} C_{nk} P(\zeta_k < x). \qquad (8)$$

From (2) and the above-mentioned theorem from the theory of series we obtain (4) and Theorem 1 is proved.

The situation is somewhat more complicated if we do not suppose that v_n is independent of the variables ξ_k. In this case a stronger condition than (3) must be imposed upon v_n. As an example we prove now a theorem which is a particular case of Anscombe's theorem.[1] The reasoning is inspired by W. Doeblin.

THEOREM 2. *If* (2) *is fulfilled and if*

$$\frac{v_n}{n} \cdot \overset{P}{\to} c, \qquad (9)$$

where c is a positive constant, then (4) *is valid.*

PROOF. Put

$$\eta_n = \sum_{k=1}^{n} \xi_k \text{ and } \lambda_n = [nc]. \qquad (10)$$

Then, because of (9),

$$\frac{v_n}{\lambda_n} \overset{P}{\to} 1. \qquad (11)$$

Furthermore

$$\zeta_{v_n} = \sqrt{\frac{\lambda_n}{v_n}} \left(\frac{\eta_{\lambda_n}}{\sqrt{\lambda_n}} + \frac{\eta_{v_n} - \eta_{\lambda_n}}{\sqrt{\lambda_n}} \right). \qquad (12)$$

Now we need a simple lemma.

LEMMA. *If the sequence of random variables θ_n has a limit distribution and if $\gamma_n \overset{P}{\to} 1$, then the sequence $\theta_n \gamma_n$ has the same limit distribution as the sequence θ_n.*

PROOF. As $\theta_n \gamma_n = \theta_n + \theta_n(\gamma_n - 1)$, it follows that for every $N > 0$

$$P(|\theta_n(\gamma_n - 1)| > \varepsilon) \le P(|\theta_n| > N) + P\left(|\gamma_n - 1| > \frac{\varepsilon}{N}\right).$$

[1] Cf. A. Rényi [22].

Let $\delta > 0$ be arbitrary. Choose N and n_1 so that for $n \geq n_1$ the inequality $P(|\,\theta_n\,| > N) < \delta$ should hold. Choose $n_2 > n_1$ such that for $n \geq n_2$ the inequality $P\left(|\,\gamma_n - 1\,| > \dfrac{\varepsilon}{N}\right) < \delta$ should be valid. So $P(|\,\vartheta_n\,(\gamma_n - 1)\,| > \varepsilon) < 2\delta$ for $n \geq n_2$. Consequently, $\theta_n\,(\gamma_n - 1) \xrightarrow{P} 0$ and the present lemma follows from Lemma 1 of § 6.

According to these two lemmas it suffices for the proof of Theorem 2 to show that

$$\frac{\eta_{\nu_n} - \eta_{\lambda_n}}{\sqrt{\lambda_n}} \xrightarrow{P} 0. \qquad (13)$$

Let $\varepsilon > 0$ and $\delta > 0$ be arbitrary; choose n_1 such that

$$P(|\,v_n - \lambda_n\,| > \delta \varepsilon^2 \lambda_n) < \delta \quad \text{for } n \geq n_1. \qquad (14)$$

Clearly

$$P\left(\left|\frac{\eta_{\nu_n} - \eta_{\lambda_n}}{\sqrt{\lambda_n}}\right| > \varepsilon\right) = \sum_{k=1}^{\infty} P\left(\frac{|\,\eta_k - \eta_{\lambda_n}\,|}{\sqrt{\lambda_n}} > \varepsilon, v_n = k\right) \qquad (15)$$

and because of (14) we obtain the inequality

$$P\left(\frac{|\,\eta_{\nu_n} - \eta_{\lambda_n}\,|}{\sqrt{\lambda_n}} > \varepsilon\right) \leq \delta + P\left(\max_{|k - \lambda_n| \leq \varepsilon^2 \delta \lambda_n} \frac{|\,\eta_k - \eta_{\lambda_n}\,|}{\sqrt{\lambda_n}} > \varepsilon\right). \qquad (16)$$

Now Kolmogorov's inequality (Chapter VII, § 6, Theorem 1) implies

$$P\left(\max_{|k - \lambda_n| < \varepsilon^2 \delta \lambda_n} \frac{|\,\eta_k - \eta_{\lambda_n}\,|}{\sqrt{\lambda_n}} > \varepsilon\right) < 2\delta. \qquad (17)$$

Inequalities (16) and (17) prove (13).

Finally, as an application of Theorem 1 of § 6 we prove a theorem in which v_n fulfills a condition of other type than (9).

THEOREM 3. *Let α be a positive discrete random variable; suppose that* (2) *holds and assume further that*

$$v_n = [n\alpha], \qquad (18)$$

where $[x]$ denotes the integral part of x. Under these conditions (4) *is valid.*

PROOF. Let a_k $(k = 1, 2, \ldots)$ be the values taken on by α with positive probability and A_k the event $\alpha = a_k$; we have

$$P(\zeta_{\nu_n} < x) = \sum_{k=1}^{\infty} P(\zeta_{[na_k]} < x \mid A_k) P(A_k). \qquad (19)$$

When $P(A_k)$ is positive, then because of $a_k > 0$, (2) and Theorem 1 of § 6 we get the relation

$$\lim_{n \to +\infty} P(\zeta_{[na_k]} < x \mid A_k) = \Phi(x). \tag{20}$$

Hence for every fixed m

$$\lim_{n \to +\infty} \sum_{k=1}^{m} P(\zeta_{[na_k]} < x \mid A_k) P(A_k) = \Phi(x) \sum_{k=1}^{m} P(A_k). \tag{21}$$

As there can be found for every $\varepsilon > 0$ an m such that

$$\sum_{k=m+1}^{\infty} P(A_k) < \varepsilon, \tag{22}$$

(4) follows from (21) and (22); thus Theorem 3 is proved.

We can deduce from Theorem 3 the following more general theorem:

THEOREM 4. *Let* α *be a positive discrete random variable; suppose* (2) *and*

$$\frac{v_n}{n} \xrightarrow{P} \alpha. \tag{23}$$

Then (4) *is valid.*

The proof[1] rests upon Theorem 3 and uses the same method as the proof of Theorem 2.

§ 8. Limit distributions for Markov chains

In the present section we shall deal with an important class of sequences of dependent random variables: Markov chains. A sequence of random variables ζ_n $(n = 0, 1, \ldots)$ is called a *Markov chain* if the following conditions are fulfilled; for every n $(n = 0, 1, \ldots)$ and for every system of real numbers $x_0, x_1, \ldots, x_{n+1}$ one has with probability 1

$$P(\zeta_{n+1} < x_{n+1} \mid \zeta_0 = x_0, \ \zeta_1 = x_1, \ldots, \zeta_n = x_n) = P(\zeta_{n+1} < x_{n+1} \mid \zeta_n = x_n). \tag{1}$$

(The conditional probabilities figuring in (1) are defined according to § 2 of Chapter V.) In the present section we shall deal only with Markov chains $\{\zeta_n\}$ such that the values of ζ_n belong to a denumerable set $\mathcal{M}$; without

---

[1] Cf. A. Rényi [31]; later J. Mogyoródi [1], further J. R. Blum, L. Hanson and J. Rosenblatt [1], have proved that in Theorem 4 the restriction that α should have a discrete distribution can be omitted.

any essential restriction of the generality it can be assumed that $\mathcal{M}$ is the set of the nonnegative integers. In this particular case (1) can be written in the following form: If $n, k, j_0, j_1, \ldots, j_n$ are any nonnegative integers, then

$$P(\zeta_{n+1} = k \mid \zeta_0 = j_0, \ \zeta_1 = j_1, \ldots, \zeta_n = j_n) = P(\zeta_{n+1} = k \mid \zeta_n = j_n). \qquad (2a)$$

Markov chains are usually interpreted as follows: Let S be a physical system which can be in the states $A_0, A_1, \ldots, A_k, \ldots$. Let the state of the system change at random in time; consider the states of the system at the time instants $t = 0, 1, \ldots$ and put $\zeta_n = k$ if at time n the system is in the state k.

The hypothesis that the random changes of state of a system form a Markov chain can then be expressed as follows: *The past of the system can influence its future only through its present state.*

If we multiply both sides of (2a) by $P(\zeta_0 = j_0, \ldots, \zeta_n = j_n)$ and add the equations obtained for all values of $j_0, j_1, \ldots, j_{r-1}$ ($1 \leq r \leq n$) and further divide by $P(\zeta_r = j_r, \ldots, \zeta_n = j_n)$, then we obtain

$$P(\zeta_{n+1} = k \mid \zeta_r = j_r, \ldots, \zeta_n = j_n) = P(\zeta_{n+1} = k \mid \zeta_n = j_n). \qquad (2b)$$

Similarly, we can show that for arbitrary integers $0 \leq n_1 < n_2 < \ldots < < n_s \leq n$,

$$P(\zeta_{n+1} = k \mid \zeta_{n_1} = j_1, \ldots, \zeta_{n_s} = j_s) = P(\zeta_{n+1} = k \mid \zeta_{n_s} = j_s). \qquad (2c)$$

The conditional probabilities $P(\zeta_{n+m} = k \mid \zeta_n = j)$ are called the *m-step transition probabilities*, as they give the probability that the system passes from the state A_j to the state A_k during the time interval $(n, n + m)$ ($m = 1, 2, \ldots$), i.e. in m steps. These quantities depend in general on the time n. If not, then the Markov chain is said to be (time-) *homogeneous*. In this Section we deal only with homogeneous Markov chains.

Thus by assumption the probability $P(\zeta_{n+m} = k \mid \zeta_n = j)$ will be independent of n and we may put

$$p_{jk}^{(m)} = P(\zeta_{n+m} = k \mid \zeta_n = j) \qquad (j, k = 0, 1, \ldots). \qquad (3)$$

It is reasonable to consider the numbers $p_{jk}^{(m)}$ ($j, k = 0, 1, \ldots$) as the elements of a matrix

$$\Pi_m = (p_{jk}^{(m)}). \qquad (4)$$

Instead of $p_k^{(1)}$ we write simply p_{jk} and instead of Π_1 simply Π.

Clearly for every positive integer m and for $j \geq 0$ the relation

$$\sum_{k=0}^{\infty} p_{jk}^{(m)} = 1 \tag{5}$$

holds. In fact, the terms of the sum are the probabilities belonging to a complete system of events. Hence the matrix Π_m, which has nonnegative terms only, has the property that the sum of terms in each row is equal to 1. Such matrices with nonnegative elements are called *stochastic matrices*. The matrix Π_m can be computed from Π as follows. According to the theorem of complete probability (cf. Chapter III, § 2, Formula (2)) we have for $1 \leq r < m$

$$P(\zeta_{n+m} = k \mid \zeta_n = j) = \sum_{l=0}^{\infty} P(\zeta_{n+m} = k \mid \zeta_{n+r} = l, \zeta_n = j) P(\zeta_{n+r} = l \mid \zeta_n = j), \tag{6}$$

and it follows from (2c) that

$$p_{jk}^{(m)} = \sum_{l=0}^{\infty} p_{jl}^{(r)} p_{lk}^{(m-r)}. \tag{7}$$

Thus we have

$$\Pi_m = \Pi_r \Pi_{m-r} \qquad (m = 2, 3, \ldots; \ r = 1, 2, \ldots, m-1). \tag{8}$$

Consequently, $\Pi_m = \Pi \Pi_{m-1} = \Pi^2 \Pi_{m-2}$ etc., hence

$$\Pi_m = \Pi^m \qquad (m = 2, 3, \ldots). \tag{9}$$

The matrix of m-step transition probabilities is thus the m-th power of the matrix of one-step transition probabilities.

So far we have only considered transition probabilities, i.e. conditional probabilities. In order to determine from these the probability distribution of ζ_n, we must know the state of the system at the instant $t = 0$ or at least the probabilities of the initial state of the system, i.e. the probability distribution $P(\zeta_0 = k)$ $(k = 0, 1, \ldots)$. With the notation $P(\zeta_n = k) = P_n(k)$ $(n = 0, 1, \ldots)$ one can thus write

$$P_n(k) = \sum_{j=0}^{\infty} P_0(j) p_{jk}^{(n)} \tag{10a}$$

or, more generally,

$$P_n(k) = \sum_{j=0}^{\infty} P_r(j) p_{jk}^{(n-r)} \qquad (r = 0, 1, \ldots, n-1). \tag{10b}$$

If ζ_0 is constant, e.g. equal to j_0, then $P_0(j_0) = 1$ and $P_0(j) = 0$ for $j \neq j_0$.

In this case

$$P_n(k) = p_{j_0k}^{(n)}. \tag{11}$$

As an example of a Markov chain consider a machine in a factory, which is switched on and off as time proceeds. At any instant there are two possibilities: either the machine works (state A_1) or it stands idle (state A_0). Let p_{jk} denote the probability that at the instant $n + 1$ the machine is in the state A_k provided that at the instant n it was in the state A_j $(j, k = 0, 1)$, further put $p_{01} = \lambda, p_{10} = \mu$ $(0 < \lambda < 1, 0 < \mu < 1)$. In this case the matrix of transition probabilities is

$$\Pi = \begin{pmatrix} 1 - \lambda & \lambda \\ \mu & 1 - \mu \end{pmatrix}.$$

A simple calculation gives the n-step transition probabilities. Further we derive from these

$$P_n(1) = \frac{\lambda}{\lambda + \mu} + (1 - \lambda - \mu)^n \left(P_0(1) - \frac{\lambda}{\lambda + \mu} \right) \tag{12a}$$

and

$$P_n(0) = \frac{\mu}{\lambda + \mu} + (1 - \lambda - \mu)^n \left(P_0(0) - \frac{\mu}{\lambda + \mu} \right), \tag{12b}$$

where $P_0(1)$ and $P_0(0)$ are the probabilities that at time 0 the machine works and does not work, respectively. Since $0 < \lambda < 1$, $0 < \mu < 1$, we have always $|1 - \lambda - \mu| < 1$; hence (12a) and (12b) lead to

$$\lim_{n \to +\infty} P_n(1) = \frac{\lambda}{\lambda + \mu}, \quad \lim_{n \to +\infty} P_n(0) = \frac{\mu}{\lambda + \mu}; \tag{13}$$

hence the distribution of ζ_n tends to a limit distribution as $n \to +\infty$.

Notice that the limit values (13) do not depend on the distribution of ζ_0. If $P_0(1) = \dfrac{\lambda}{\lambda + \mu}$, and hence $P_0(0) = \dfrac{\mu}{\lambda + \mu}$, we have $P_n(1) = \dfrac{\lambda}{\lambda + \mu}$ and $P_n(0) = \dfrac{\mu}{\lambda + u}$ for every n and not only in the limit.[1]

If in a Markov chain the distribution of the random variables ζ_n tends to a limit distribution (thus if $\lim_{n \to +\infty} P_n(k) = P(k)$) which does not depend

---

[1] If $\lambda + \mu = 1$, we have $P_n(1) = \lambda$ and $P_n(0) = \mu$, without any assumption on the initial state; in this case the ζ_n-s are independent from each other!

on the initial distribution $P_0(j)$, then the Markov chain is called *ergodic*. An initial distribution such that ζ_n has the same distribution for every value of n, is called a *stationary distribution*. If the Markov chain is ergodic and there exists a stationary distribution, the latter is evidently the limit distribution of ζ_n. It is easy to show that there exists a stationary distribution, iff the system of equations

$$x_k = \sum_{j=0}^{\infty} p_{jk} x_j \qquad (k = 0, 1, \ldots) \qquad (14)$$

admits a solution $x_0, x_1, \ldots$ with $x_k \geq 0$ $(k = 0, 1, \ldots)$ and $\sum_{k=1}^{\infty} x_k = 1$; in this case the numbers x_k constitute a stationary distribution. For the example considered above Equations (14) can be written as

$$x_0 = (1 - \lambda) x_0 + \mu x_1,$$

$$x_1 = \lambda x_0 + (1 - \mu) x_1. \qquad (15)$$

The single solution such that $x_0 + x_1 = 1$ is

$$x_1 = \frac{\lambda}{\lambda + \mu}, \quad x_0 = \frac{\mu}{\lambda + \mu}. \qquad (16)$$

In this example there exists a stationary distribution and the Markov chain is ergodic.

The following theorem, due essentially to A. A. Markov, shows that this holds under rather general conditions.

THEOREM 1. *Let a system possess a finite number of possible states* A_0, $A_1, \ldots, A_N$. *Assume that the changes of state of the system form a homogeneous Markov chain and denote by* $p_{jk}^{(m)}$ *the probability that the system passes from state* A_j *to state* A_k *in m steps. Assume further that there exist integers* $s > 0$ *and* $k_0 \geq 0$ *such that for* $j = 0, 1, \ldots, N$,

$$p_{jk_0}^{(s)} > 0, \qquad (17)$$

i.e. that the matrix Π^s *has at least one column in which all elements are positive. In this case the chain is ergodic; the limits*

$$\lim_{n \to +\infty} p_{jk}^{(n)} = P_k \qquad (j, k = 0, 1, \ldots, N) \qquad (18)$$

exist and do not depend on j. The sequence of numbers $P_0, \ldots, P_N$ *is the unique nonnegative solution of the system of equations*

$$P_k = \sum_{j=0}^{N} P_j p_{jk} \qquad (k = 0, 1, \ldots, N), \tag{19}$$

which satisfies

$$\sum_{j=0}^{N} P_j = 1. \tag{20}$$

The limit distribution $\{P_j\}$ is thus a stationary distribution of the chain.

PROOF. By assumption

$$d = \min_{0 \le j \le N} p_{jk_0}^{(r)} > 0. \tag{21}$$

Put

$$m_k^{(n)} = \min_{0 \le l \le N} p_{lk}^{(n)}, \quad M_k^{(n)} = \max_{0 \le l \le N} p_{lk}^{(n)} \qquad (k = 0, 1, \ldots, N). \tag{22}$$

Clearly

$$p_{jk}^{(n+1)} = \sum_{l=0}^{N} p_{jl} p_{lk}^{(n)}, \tag{23}$$

hence, for $0 \le j \le N$,

$$m_k^{(n)} = m_k^{(n)} \sum_{l=0}^{N} p_{jl} \le \sum_{l=0}^{N} p_{jl} p_{lk}^{(n)} = p_{jk}^{(n+1)}, \tag{24}$$

and

$$m_k^{(n)} \le m_k^{(n+1)}. \tag{25}$$

Similarly, for $0 \le j \le N$,

$$M_k^{(n)} = M_k^{(n)} \sum_{l=0}^{N} p_{jl} \ge \sum_{l=0}^{N} p_{jl} p_{lk}^{(n)} = p_{jk}^{(n+1)}, \tag{26}$$

hence

$$M_k^{(n)} \ge M_k^{(n+1)}. \tag{27}$$

Furthermore, by (22),

$$0 \le m_k^{(n)} \le M_k^{(n)} \le 1, \tag{28}$$

which implies the existence of the limits

$$m_k = \lim_{n \to +\infty} m_k^{(n)} \quad \text{and} \quad M_k = \lim_{n \to +\infty} M_k^{(n)} \tag{29}$$

with

$$m_k \leq M_k. \tag{30}$$

If we can prove that $m_k = M_k$ for $k = 0, 1, \ldots, N$, then (18) will be proved. Now for a suitable l_1 the equality

$$M_k^{(n+s)} = p_{l_1 k}^{(n+s)} = \sum_{j=0}^{N} p_{l_1 j}^{(s)} p_{jk}^{(n)} \tag{31}$$

holds and for a certain l_0;

$$m_k^{(n+s)} = p_{l_0 k}^{(n+s)} = \sum_{j=0}^{N} p_{l_0 j}^{(s)} p_{jk}^{(n)}. \tag{32}$$

Hence

$$M_k^{(n+s)} - m_k^{(n+s)} = \sum_{j=0}^{N} (p_{l_1 j}^{(s)} - p_{l_0 j}^{(s)}) p_{jk}^{(n)}. \tag{33}$$

Let H be the set of all j $(0 \leq j \leq N)$ for which $p_{l_1 j}^{(s)} - p_{l_0 j}^{(s)} \geq 0$ and let $\bar{H}$ be the complementary set of H, i.e. the set of those j $(0 \leq j \leq N)$, for which $p_{l_1 j}^{(s)} - p_{l_0 j}^{(s)} < 0$ holds. Put

$$A = \sum_{j \in H} (p_{l_1 j}^{(s)} - p_{l_0 j}^{(s)}) \quad \text{and} \quad B = \sum_{j \in \bar{H}} (p_{l_1 j}^{(s)} - p_{l_0 j}^{(s)}). \tag{34}$$

Then $A \geq 0$ and

$$A + B = \sum_{j=0}^{N} p_{l_1 j}^{(s)} - \sum_{j=0}^{N} p_{l_0 j}^{(s)} = 1 - 1 = 0,$$

hence $B = -A$ and it follows from (33) that

$$M_k^{(n+s)} - m_k^{(n+s)} \leq (M_k^{(n)} - m_k^{(n)}) A. \tag{35}$$

Two cases are now possible: either $k_0 \in H$ or $k_0 \in \bar{H}$. In the first case we have

$$B \geq - (1 - p_{l_1 k_0}^{(s)}) \geq - (1 - d),$$

and, since $A = -B$,

$$A \leq 1 - d. \tag{36}$$

In the second case we have

$$A \leq 1 - p_{l_0 k_0}^{(s)} \leq 1 - d,$$

hence (36) is valid in both cases. It follows thus from (35) and (36) that

$$M_k^{(n+s)} - m_k^{(n+s)} \leq (1 - d)(M_k^{(n)} - m_k^{(n)}). \tag{37}$$

Furthermore, because of (21)

$$M_k^{(s)} - m_k^{(s)} \le 1 - d. \tag{38}$$

By induction we conclude from (37) and (38) that

$$M_k^{(ns)} - m_k^{(ns)} \le (1 - d)^n \tag{39}$$

and, because of $d > 0$, by passing to the limit $n \to +\infty$ and taking into account (29) and (30), we get $m_k = M_k$. This proves (18) with $P_k = m_k = = M_k$.

The passing to the limit in (7) and (5) shows that $\{P_k\}$ fulfils Equations (19) and (20). It remains merely to prove that the P_k $(k = 0, 1, \ldots, N)$ are uniquely determined by (19) and (20). This can be shown as follows: If $Q_0, \ldots, Q_N$ were a distribution distinct from $P_0, \ldots, P_N$ and satisfying (19), thus if $\sum_{k=0}^{N} Q_k = 1$ and

$$Q_k = \sum_{l=0}^{N} Q_l p_{lk} \qquad (k = 0, 1, \ldots, N) \tag{40}$$

held, then after multiplication of (40) by p_{kt} and summation over k we would obtain

$$Q_t = \sum_{l=0}^{N} Q_l p_{lt}^{(2)}.$$

By repetition of these operations we find for every integer n

$$Q_t = \sum_{l=0}^{N} Q_l p_{lt}^{(n)}.$$

Because of $\lim\limits_{n \to +\infty} p_{lt}^{(n)} = P_t$ and $\sum\limits_{l=0}^{N} Q_l = 1$ there follows $Q_t = P_t$, which was to be proven.

The numbers P_k fulfill the equations

$$P_k = \sum_{l=0}^{N} P_l p_{lk}^{(n)} \tag{41}$$

for every $n = 1, 2, \ldots$; Equation (19) is a particular case of (41).

If the distribution of ζ_0 is known, $P(\zeta_0 = i) = P_0(i)$, then from (10a) one can derive the relation

$$\lim_{n \to +\infty} P_n(k) = P_k \qquad (k = 0, 1, \ldots, N). \tag{42}$$

Finally, let us mention the following particular case: assume that for the matrix of the transition probabilities (p_{jk}) the sum of all columns is equal to 1:

$$\sum_{j=0}^{N} p_{jk} = 1 \quad \text{for} \quad k = 0, 1, \ldots, N.$$

The matrix $\Pi = (p_{jk})$ as well as its transpose $\Pi^* = (p_{kj})$ are stochastic matrices; such a matrix Π is called a *doubly stochastic* matrix. In this case (40) is fulfilled for $Q = \dfrac{1}{N+1}$ $(k = 0, 1, \ldots, N)$; the solution of (19) being unique, there follows $P_k = \dfrac{1}{N+1}$. Thus for a doubly stochastic matrix Π fulfilling the conditions of Theorem 1 the relation

$$\lim_{n \to +\infty} p_{jk}^{(n)} = \frac{1}{N+1} \quad \text{holds for} \quad j, k = 0, 1, \ldots, N.$$

It follows from (42) that the probabilities of the $N + 1$ states are in the limit equal to each other, regardless of the initial distribution.

A particular class of the Markov chains is that of the so-called *additive Markov chains*. If $\xi_0, \xi_1, \ldots, \xi_n, \ldots$ are independent random variables and if we put $\zeta_n = \xi_0 + \xi_1 + \ldots + \xi_n$, the random variables ζ_n $(n = 0, 1, \ldots)$ form a Markov chain, since

$$P(\zeta_{n+1} < x \mid \zeta_0 = x_0, \ldots, \zeta_n = x_n) = P(\xi_{n+1} < x - x_n) =$$

$$= P(\zeta_{n+1} < x \mid \zeta_n = x_n). \tag{43}$$

If ξ_k are identically distributed, the chain is homogeneous. In this case the problem of finding the limit distribution of the chain $\{\zeta_n\}$ can be reduced to the study of sums of independent random variables, already dealt with. If ξ_k take on only integer values, if their expectation is zero and if the greatest common divisor of the values assumed by $\xi_1 - \xi_2$ with positive probabilities is equal to 1, then for every pair k, l of integers the relation (cf. Chapter VI, § 9, Theorem 8)

$$\lim_{n \to +\infty} \frac{P(\zeta_n = k)}{P(\zeta_n = l)} = 1 \tag{44}$$

holds, hence ζ_n has in limit a uniform conditional distribution on the set of all integers. (This may happen also for nonadditive chains.) Further, if the expectation of ξ_k is zero and their variance is equal to 1, then from

Theorem 1 of § 1 there follows the relation

$$\lim_{n \to +\infty} P\left(\frac{\zeta_n}{\sqrt{n}} < x\right) = \Phi(x).$$

For homogeneous Markov chains with a finite number of states the suitably standardized sum

$$\eta_n = \zeta_0 + \zeta_1 + \ldots + \zeta_n$$

is under general conditions in the limit normally distributed. This will be proved here only for the simplest case of a chain with two states.

THEOREM 2. *Let the random variables $\{\zeta_n\}$ form a homogeneous Markov chain with two states. Put $\zeta_n = 0$ or $\zeta_n = 1$, according to whether the system is in state A_0 or in state A_1 at the instant n. Let $\begin{pmatrix} 1-\lambda & \lambda \\ \mu & 1-\mu \end{pmatrix}$ with $0 < \lambda < 1$, $0 < \mu < 1$ be the matrix of transition probabilities. Put*

$$\eta_n = \sum_{k=0}^{n} \zeta_k. \tag{45}$$

Then

$$\lim_{n \to \infty} P\left(\frac{\eta_n - \dfrac{\lambda}{\lambda+\mu}n}{\sqrt{\dfrac{n\lambda\mu(2-\lambda-\mu)}{(\lambda+\mu)^3}}} < x\right) = \Phi(x). \tag{46}$$

Remark. If $\lambda + \mu = 1$, then the variables ζ_n are independent and η_n has a binomial distribution of order n and parameter λ, further in this case

$$\frac{\lambda\mu(2-\lambda-\mu)}{(\lambda+\mu)^3} = \lambda(1-\lambda).$$

Hence (46) reduces to

$$\lim_{n \to +\infty} P\left(\frac{\eta_n - \lambda n}{\sqrt{n\lambda(1-\lambda)}} < x\right) = \Phi(x).$$

Theorem 2 is thus a generalization of the Moivre–Laplace theorem.

PROOF. If τ_n denotes the instant when the system returns for the n-th time to the state A_1, we have $0 \le \tau_1 < \tau_2 < \ldots < \tau_n < \ldots$; $\zeta_{\tau_n} = 1$;

$\zeta_k = 0$ for $k < \tau_1$ or $\tau_n < k < \tau_{n+1}$ $(n = 1, 2, \ldots)$. Put $\delta_1 = \tau_1$, $\delta_n = \tau_n -$ $- \tau_{n-1}$. It is easy to see that δ_n are independent and (δ_1 excepted) have the same distribution. For by the definition of Markov chains, $\tau_n - \tau_{n-1}$ is independent of the random variables $\delta_1, \delta_2, \ldots, \delta_{n-1}$ which depend only on the states of the system at instants $t < \tau_{n-1}$. The fact that the random variables δ_n $(n = 2, 3, \ldots)$ are identically distributed follows from the homogeneity of the Markov chain. Clearly for every $n \geq 2$

$$P(\delta_n = 1) = 1 - \mu$$

and

$$P(\delta_n = k) = \mu\lambda(1 - \lambda)^{k-2} \text{ for } k \geq 2.$$

Hence, for $n \geq 2$

$$E(\delta_n) = \frac{\lambda + \mu}{\lambda}$$

and

$$D^2(\delta_n) = \frac{\mu(2 - \lambda - \mu)}{\lambda^2} .$$

If $\zeta_0 = 1$, δ_1 has the same distribution as the other δ_n $(n \geq 2)$; if $\zeta_0 = 0$, we have $P(\delta_1 = 1) = \lambda$, $P(\delta_1 = k) = (1 - \lambda)^{k-1}\lambda$ for $k \geq 2$ hence $E(\delta_1) =$ $= \frac{1}{\lambda}, D^2(\delta_1) = \frac{1 - \lambda}{\lambda^2} .$

By Theorem 1 of § 1 and the lemma of § 6 follows

$$\lim_{k \to +\infty} P \left(\frac{\tau_k - k \dfrac{\lambda + \mu}{\lambda}}{\dfrac{\sqrt{k\mu(2 - \lambda - \mu)}}{\lambda}} < x \right) = \Phi(x). \tag{47}$$

Now obviously $P(\eta_n < k) = P(\tau_k > n)$; in fact $\eta_n < k$ means that up to the moment n the system was less than k times in the state A_1, thus its k-th entrance into the state A_1 occurs after the moment n, hence $\tau_k > n$ and conversely. If we put

$$k = \left[\frac{\lambda}{\lambda + \mu} n + x \sqrt{\frac{n\lambda\mu(2 - \lambda - \mu)}{(\lambda + \mu)^3}} \right],$$

a simple calculation gives that

$$n = \frac{k(\lambda + \mu)}{\lambda} - \frac{x\sqrt{k\mu(2 - \lambda - \mu)}}{\lambda} + O(1),$$

hence

$$P\left(\frac{\eta_n - \frac{\lambda}{\lambda+\mu}n}{\sqrt{\frac{n\lambda\mu(2-\lambda-\mu)}{(\lambda+\mu)^3}}} < x\right) = P\left(\frac{\tau_k - \frac{\lambda+\mu}{\lambda}k}{\frac{\sqrt{k\mu(2-\lambda-\mu)}}{\lambda}} > -x + O\left(\frac{1}{\sqrt{k}}\right)\right). \quad (48)$$

Thus (47) leads to

$$\lim_{n\to+\infty} P\left(\frac{\eta_n - \frac{\lambda}{\lambda+\mu}n}{\sqrt{\frac{n\lambda\mu(2-\lambda-\mu)}{(\lambda+\mu)^3}}} < x\right) = 1 - \Phi(-x) = \Phi(x). \quad (49)$$

Theorem 2 is thus proved.

§ 9. Limit distributions for "order statistics"

The theory of "order statistics" (i.e. the theory of observations arranged according to their magnitude) is becoming more and more important in mathematical statistics.

In the present section we shall show how the theory of order statistics can be reduced to the study of certain Markov chains.[1] We start from the following particular case: Let $\zeta_1, \zeta_2, \ldots, \zeta_n$ denote the outcomes of n independent observations of a quantity having an exponential distribution; $\zeta_1, \zeta_2, \ldots, \zeta_n$ are thus independent random variables with the same exponential distribution:

$$F(x) = P(\zeta_k < x) = 1 - e^{-\lambda x} \quad \text{for} \quad x \geq 0 \, (\lambda > 0).$$

We use the following property of the exponential distribution: for any x and y one has

$$P(\zeta > x + y \mid \zeta > y) = P(\zeta > x). \quad (1)$$

Property (1) is characteristic of the exponential distribution. In fact, it is equivalent to

$$G(x + y) = G(x) G(y) \quad (2)$$

if $F(x)$ is the distribution function of ζ and $G(x) = 1 - F(x)$; we know already (cf. Chapter III, § 13) that the only nonincreasing solutions of (2),

[1] For the method see A. Rényi [9], [10].

the trivial solutions $G(x) \equiv 0$ and $G(x) \equiv 1$ excepted, are the functions of the form $G(x) = \exp(-\lambda x)$ with $\lambda > 0$.

The meaning of (1) becomes particularly clear if we interpret ζ as the duration of an event which takes a certain lapse of time to occur. In this case (1) expresses the fact that the future duration of an event which is still in course at a moment y does not depend on the time passed already since the beginning of this event.

Arrange the random variables $\zeta_1, \zeta_2, \ldots, \zeta_n$ in increasing order and let

$$\zeta_k^* = R_k(\zeta_1, \zeta_2, \ldots, \zeta_n)$$

be the k-th of the ranked variables ζ_j. Then[1]

$$\zeta_1^* < \zeta_2^* < \ldots < \zeta_n^*.$$

It is easy to determine the distribution of the ζ_k^*.

If the ζ_j are interpreted as durations of independent events beginning at the same time, then ζ_k^* is the duration of that event which is the k-th that ends. We compute now the distribution of the differences $\zeta_{k+1}^* - \zeta_k^*$. Clearly

$$P(\zeta_{k+1}^* - \zeta_k^* > x \mid \zeta_k^* = y) = P(\zeta_{k+1}^* > x + y \mid \zeta_k^* = y). \tag{3}$$

From the $n - k$ events still in course at the moment y none must cease before the moment $x + y$; by (1) the probability of this is

$$[P(\zeta > x)]^{n-k} = e^{-(n-k)\lambda x}.$$

The conditional distribution function of $\zeta_{k+1}^* - \zeta_k^*$ with respect to the condition $\zeta_k^* = y$ is thus

$$P(\zeta_{k+1}^* - \zeta_k^* < x \mid \zeta_k^* = y) = 1 - e^{-(n-k)\lambda x}. \tag{4}$$

The function thus obtained does not depend on y hence it is equal to the unconditional distribution function of $\zeta_{k+1}^* - \zeta_k^*$. This difference has thus itself an exponential distribution and its expectation is $\dfrac{1}{(n-k)\lambda}$ $(k = 1, 2, \ldots, n-1)$. ζ_1^* also has an exponential distribution, with expectation $\dfrac{1}{n\lambda}$. If we put $\zeta_0^* \equiv 0$ and

$$\delta_{k+1} = (n-k)(\zeta_{k+1}^* - \zeta_k^*) \qquad (k = 0, 1, \ldots, n-1), \tag{5}$$

then δ_{k+1} $(k = 0, 1, \ldots, n-1)$ have all the same exponential distribution

---

[1] $F(x)$ being continuous, the probability that two of the ζ_j are equal is zero; this possibility can thus be omitted.

with expectation $\dfrac{1}{\lambda}$. It is easy to see that the δ_k are independent. In fact the conditional probability

$$P(\zeta^*_{k+1} - \zeta^*_k < x \,|\, \zeta^*_1 = y_1, \ldots, \zeta^*_k - \zeta^*_{k-1} = y_k) \tag{6}$$

does not depend on $y_1, y_2, \ldots, y_k$; since the conditions are equivalent to the relations $\zeta^*_j = y_1 + \ldots + y_j$ $(j = 1, 2, \ldots, k)$, they give thus for $j = 1, 2, \ldots, k$ the moment when the j-th of the events starting at $t = 0$ ends. This means that at the moment $t = y_1 + y_2 + \ldots + y_k$ there are exactly $n - k$ events in course. The probability that between t and $t + x$ at least one event comes to an end is $1 - \exp[-(n - k)\lambda x]$. This does not depend on the variables $y_1, \ldots, y_k$; hence the random variables $\delta_1, \delta_2, \ldots, \delta_n$ are independent. Thus ζ^*_k may be written in the form

$$\zeta^*_k = \frac{\delta_1}{n} + \frac{\delta_2}{n - 1} + \ldots + \frac{\delta_k}{n - k + 1}, \tag{7}$$

where δ_j $(j = 1, 2, \ldots, k)$ are independent and possess the same distribution. Equation (7) shows that the variables ζ^*_k form an *additive Markov chain*. By means of (7) the distribution of ζ^*_k can be determined explicitly.

Let the preceding result now be applied to the theory of order statistics. Let $\xi_1, \xi_2, \ldots, \xi_n$ be independent random variables with the same continuous distribution function $F(x)$. As above, put $\xi^*_k = R_k(\xi_1, \xi_2, \ldots, \xi_n)$, hence $\xi^*_1 < \xi^*_2 < \ldots < \xi^*_n$ are the variables ξ_j arranged in increasing order. The theory of order statistics deals with the study of ξ^*_k; ξ^*_k is called the k-th order statistic. This study can be reduced to the case when ξ_k are exponentially distributed and by (7) we then have to consider sums of independent random variables only. In order to show this, put

$$\zeta_k = \ln \frac{1}{F(\xi_k)} \qquad (k = 1, 2, \ldots, n) \tag{8}$$

and

$$\zeta^*_k = R_k(\zeta_1, \ldots, \zeta_n). \tag{9}$$

Since $\ln \dfrac{1}{F(x)}$ is nonincreasing, we have

$$\zeta^*_k = \ln \frac{1}{F(\xi^*_{n+1-k})} \qquad (k = 1, 2, \ldots, n) \tag{10}$$

and as ξ_k are independent, the same is valid for ζ_k.

Consider now the distribution of ζ_k. Let $y = F^{-1}(x)$ $(0 \le x \le 1)$ be the inverse function of $x = F(y)$ $(-\infty < y < +\infty)$. Then the relation

$$P(\zeta_k < x) = P\big(F(\xi_k) > e^{-x}\big) = P\big(\xi_k > F^{-1}(e^{-x})\big)$$

is valid, i.e.

$$P(\zeta_k < x) = 1 - F(F^{-1}(e^{-x})) = 1 - e^{-x} \tag{11}$$

for $0 \leq x < +\infty$. Hence the random variables ζ_k are exponentially distributed with expectation 1. Thus ξ_k^* can be written in the form

$$\xi_k^* = F^{-1}(e^{-\zeta_{n+1-k}^*}) = F^{-1}\left(\exp\left(-\frac{\delta_1}{n} - \frac{\delta_2}{n-1} - \ldots - \frac{\delta_{n+1-k}}{k}\right)\right), \tag{12}$$

where $\delta_1, \delta_2, \ldots, \delta_n$ are independent random variables with expectation 1. Our result implies the theorem of van Dantzig and Malmquist stating that the ratios $\dfrac{F(\xi_{k+1}^*)}{F(\xi_k^*)}$ $(k = 0, 1, \ldots, n)$ are independent of each other (cf. Chapter IV, § 17, Exercise 17). Indeed we have according to (12)

$$\frac{F(\xi_{k+1}^*)}{F(\xi_k^*)} = \exp\left(\frac{\delta_{n+1-k}}{k}\right) \qquad (k = 1, 2, \ldots, n), \tag{13}$$

(We have to put here $F(\xi_{n+1}^*) \equiv 1$.)

The random variables $\xi_1^*, \ldots, \xi_n^*$ form a Markov chain, since because of (12) for $x_1 < x_2 < \ldots < x_k < x$ the relation

$$P(\xi_{k+1}^* < x \mid \xi_1^* = x_1, \ldots, \xi_k^* = x_k) = P\left(\delta_{n+1-k} < k \ln \frac{F(x)}{F(x_k)} \,\middle|\, \delta_{n+1-j} = \right.$$

$$\left. = j \ln \frac{F(x_{j+1})}{F(x_j)} \,;\; 1 \leq j \leq k-1, \xi_k^* = x_k\right) \tag{14}$$

is valid. Because of the independence of δ_j we get

$$P(\xi_{k+1}^* < x \mid \xi_1^* = x_1, \ldots, \xi_k^* = x_k) =$$

$$= P\left(\delta_{n+1-k} < k \ln \frac{F(x)}{F(x_k)} \,\middle|\, \xi_k^* = x_k\right) = P(\xi_{k+1}^* < x \mid \xi_k^* = x_k). \tag{15}$$

Let us notice that as

$$P(F(\xi_k) < x) = P(\xi_k < F^{-1}(x)) = F(F^{-1}(x)) = x \tag{16}$$

holds for $0 < x < 1$, the random variables $F(\xi_k)$ are uniformly distributed in the interval $(0, 1)$. The random variables $F(\xi_k^*)$ are thus the ordered elements of a sample selected from a population uniformly distributed on $(0, 1)$.

Starting from this point of view, many results on order statistics can be derived quite easily. As an example, consider the following problem: What is the limit distribution of the random variables ξ_k^* when both n and k tend to infinity in such a way that $\dfrac{k}{n}$ tends to a limit $q (0 < q < 1)$? In particular we consider the case $k = [nq] + 1$; $\xi_{[nq]+1}^*$ is called the *sample quantile of order q.*

We prove a theorem which implies in particular that the sample quantile of order q is in the limit normally distributed, provided that the distribution function $F(x)$ fulfills certain conditions.

THEOREM. *Let $\xi_1, \xi_2, \ldots, \xi_n$ be independent, identically distributed random variables, with an absolutely continuous distribution function $F(x)$. Suppose that the density function $f(x) = F'(x)$ is continuous and positive on the interval $[a, b]$. If $0 < F(a) < q < F(b) < 1$, and if $k(n)$ is a sequence of integers such that*

$$\lim_{n \to +\infty} \sqrt{n} \left| \frac{k(n)}{n} - q \right| = 0, \tag{17}$$

further if ξ_k^* denotes the k-th order statistic of the sample $\xi_1, \xi_2, \ldots, \xi_n$, then $\xi_{k(n)}^*$ is in the limit normally distributed, viz.

$$\lim_{n \to +\infty} P\left(\frac{\xi_{k(n)}^* - Q}{D} < \frac{x}{\sqrt{n}} \right) = \Phi(x), \tag{18}$$

where

$$Q = F^{-1}(q) \tag{19}$$

and

$$D = \frac{1}{f(Q)} \sqrt{q(1 - q)}. \tag{20}$$

PROOF. We consider first the limit distribution of

$$\zeta_{n+1-k(n)}^* = \ln \frac{1}{F(\xi_{k(n)}^*)} . \tag{21}$$

By (12)

$$\zeta_{n+1-k(n)}^* = \sum_{j=1}^{n+1-k(n)} \frac{\delta_j}{n + 1 - j} , \tag{22}$$

where δ_j are independent and exponentially distributed with density function e^{-x} $(x > 0)$. Hence $E(\delta_j) = D(\delta_j) = 1$ and

$$E(|\delta_j - 1|^3) = \int_0^\infty |x - 1|^3 e^{-x} dx < 3. \tag{23}$$

It follows from (17) and from the known formula

$$\sum_{k=1}^{N} \frac{1}{k} = \ln N + C + O\left(\frac{1}{N}\right),$$

where C is Euler's constant,[1] that

$$M_n = E(\zeta^*_{n+1-k(n)}) = \ln \frac{1}{q} + o\left(\frac{1}{\sqrt{n}}\right). \tag{24}$$

Since

$$\sum_{k=N_1}^{N_2} \frac{1}{k^2} = \frac{1}{N_1} - \frac{1}{N_2} + O\left(\frac{1}{N_1^2}\right)$$

we get

$$S_n^2 = D^2(\zeta^*_{n+1-k(n)}) = \frac{1-q}{nq} + O\left(\frac{1}{n^{\frac{3}{2}}}\right); \tag{25}$$

according to (23) and from

$$\sum_{k=N_1}^{N_2} \frac{1}{k^3} = \frac{1}{2N_1^2} - \frac{1}{2N_2^2} + O\left(\frac{1}{N_1^3}\right)$$

it follows that

$$K_n^3 = \sum_{j=1}^{n+1-k(n)} E\left(\left|\frac{\delta_j - 1}{n+1-j}\right|^3\right) = O\left(\frac{1}{n^2}\right), \tag{26}$$

$$\left(\frac{K_n}{S_n}\right)^3 = O\left(\frac{1}{\sqrt{n}}\right). \tag{27}$$

Thus Liapunov's form of the central limit theorem (Theorem 4 of § 1) can be applied to the sums (22). Taking into account the lemmas of Sections 6 and 7 we get

$$\lim_{n \to +\infty} P\left(\frac{\zeta^*_{n+1-k(n)} - \ln\frac{1}{q}}{\sqrt{\frac{1-q}{q}}} < \frac{x}{\sqrt{n}}\right) = \Phi(x). \tag{28}$$

[1] Cf. K. Knopp [1].

Now because of (21)

$$P\left(\frac{\zeta^*_{n+1-k(n)} - \ln\dfrac{1}{q}}{\sqrt{\dfrac{1-q}{q}}} < \frac{x}{\sqrt{n}}\right) =$$

$$= P\left(\xi^*_{k(n)} > F^{-1}\left(q \exp\left(-x\sqrt{\frac{1-q}{nq}}\right)\right)\right). \qquad (29)$$

The mean value theorem allows us to write

$$F^{-1}\left(q \exp\left(-x\sqrt{\frac{1-q}{nq}}\right)\right) =$$

$$= F^{-1}(q) + \frac{q\left(\exp\left(-x\sqrt{\dfrac{1-q}{nq}}\right) - 1\right)}{f(Q\theta_n)} \qquad (30)$$

where $\lim\limits_{n\to+\infty} \theta_n = 1$; further

$$q\left(\exp\left(-x\sqrt{\frac{1-q}{nq}}\right) - 1\right) = -x\sqrt{\frac{q(1-q)}{n}} + O\left(\frac{1}{n}\right).$$

Now (29), the continuity of $f(x)$, and the lemmas of § 6 and § 7 imply (18), hence the theorem is proved.

The theorem states that the empirical sample quantile of order q of a sample of n elements is for sufficiently large n nearly normally distributed with expectation $Q = F^{-1}(q)$ and standard deviation $\dfrac{1}{f(Q)}\sqrt{\dfrac{q(1-q)}{n}}$.

This fact can be used in practical applications (e.g. in quality control). In case of a symmetric distribution, expectation and median coincide; thus in this case the sample median can be used as an approximation of the expectation.

§ 10. Limit theorems for empirical distribution functions

In the preceding section we have seen how to determine the quantiles of a distribution function $F(x)$ by means of an ordered sample of independent random variables with distribution function $F(x)$. The empirical distribution function of such a sample may help us to get information also about the whole course of the distribution function $F(x)$. Glivenko's fundamental theorem of mathematical statistics (Chapter VII, § 8, Theorem 1) states that the difference between the empirical and theoretical distribution functions tends uniformly to zero with probability 1 as the sample size tends to infinity. Glivenko's theorem, however, says nothing about the "rapidity" of the convergence. But this information is supplied by the theorems of Smirnov and Kolmogorov, which we shall state now without proofs.[1]

Let $\xi_1, \xi_2, \ldots, \xi_n$ be independent, identically distributed random variables with the continuous distribution function $F(x)$. As in the preceding section, let ξ_k^* denote the k-th order statistic. Put

$$F_n(x) = \begin{cases} 0 & \text{for} \quad x \le \xi_1^*, \\[2mm] \dfrac{k}{n} & \text{for} \quad \xi_k^* < x \le \xi_{k+1}^* \quad (k = 1, 2, \ldots, n-1), \\[2mm] 1 & \text{for} \quad \xi_n^* < x; \end{cases} \tag{1}$$

$F_n(x)$ is the empirical distribution function of the sample $\xi_1, \ldots, \xi_n$.

THEOREM 1 (Smirnov).

$$\lim_{n \to +\infty} P\left(\sqrt{n} \sup_{-\infty < x < +\infty} (F_n(x) - F(x)) < y\right) = \begin{cases} 1 - e^{-2y^2} & \text{for } y > 0, \\ 0 & \text{otherwise.} \end{cases}$$

THEOREM 2 (Kolmogorov).

$$\lim_{n \to +\infty} P\left(\sqrt{n} \sup_{-\infty < x < +\infty} |F_n(x) - F(x)| < y\right) = \begin{cases} K(y) & \text{for } y > 0, \\ 0 & \text{otherwise,} \end{cases}$$

where

$$K(y) = \sum_{k=-\infty}^{+\infty} (-1)^k e^{-2k^2 y^2}. \tag{2}$$

Notice that in these two theorems the limit distributions do not depend on $F(x)$. It suffices that $F(x)$ is continuous, this guarantees the validity of these and all further theorems in this section. The values of the function $K(y)$ figuring in Kolmogorov's theorem are given in Table 8 at the end of this book.

The theorems of Smirnov and Kolmogorov may serve to test the hypothesis that a sample of size n was drawn from a population with a given continuous distribution function $F(x)$.

The theorems of Kolmogorov and Smirnov refer to the maximal deviation between $F_n(x)$ and $F(x)$. Often it is more convenient to consider the maximum

---

[1] For the proof of Theorem 1 cf. § 13, Exercise 23.

for $F(x) \geq a > 0$ of the relative deviation $\dfrac{F_n(x) - F(x)}{F(x)}$. The following theorems are concerned with this relative deviation.[1]

THEOREM 3. *We have*

$$\lim_{n \to +\infty} P\left(\sqrt{n} \sup_{x_a \leq x < +\infty} \frac{F_n(x) - F(x)}{F(x)} < y\right) =$$

$$= \begin{cases} \sqrt{\dfrac{2}{\pi}} \displaystyle\int_0^{y\sqrt{\frac{a}{1-a}}} e^{-\frac{x^2}{2}}\, dx & \text{for } y > 0, \\ 0 & \text{otherwise,} \end{cases}$$

where x_a is defined by $F(x_a) = a$, $0 < a < 1$.

THEOREM 4. *We have*

$$\lim_{n \to +\infty} P\left(\sqrt{n} \sup_{x_a \leq x < +\infty} \left|\frac{F_n(x) - F(x)}{F(x)}\right| < y\right) = \begin{cases} L\left(y\sqrt{\dfrac{a}{1-a}}\right) & \text{for } y > 0; \\ 0 & \text{otherwise,} \end{cases}$$

where

$$L(z) = \frac{4}{\pi} \sum_{k=0}^{\infty} (-1)^k \frac{\exp\left[-(2k+1)^2 \dfrac{\pi^2}{8z^2}\right]}{2k+1} \tag{3}$$

and x_a is defined by $F(x_a) = a$, $0 < a < 1$.

The values of the function $L(z)$ defined by (3) are tabulated in Table 9.

We may be interested in the maximum of the relative deviation over an interval (x_a, x_b), where x_a and x_b are defined by $F(x_a) = a$ and $F(x_b) = b$ $(0 < a < b < 1)$. This problem is solved by

THEOREM 5. *If $0 < a < b < 1$, $F(x_a) = a$, $F(x_b) = b$, then the relation*

$$\lim_{n \to +\infty} P\left(\sqrt{n} \sup_{x_a \leq x < x_b} \frac{F_n(x) - F(x)}{F(x)} < y\right) =$$

$$= \frac{1}{\pi} \sqrt{\frac{b}{1-b}} \int_{-\infty}^{y} \exp\left(-\frac{bt^2}{2(1-b)}\right) \int_0^{(y-t)\sqrt{\frac{ab}{b-a}}} e^{-\frac{u^2}{2}}\, du\, dt$$

is valid.

[1] Cf. A. Rényi [9].

First of all, we have to note a surprising corollary of Theorem 5. It follows from the theorem of Smirnov that

$$\lim_{n \to +\infty} P\left(\sup_{-\infty < x < +\infty} \left(F_n(x) - F(x) \right) < 0 \right) = 0,$$

i.e. the probability, that the empirical distribution function remains everywhere under the theoretical distribution function, tends to zero. According to Theorem 3 the same holds if we restrict ourselves to values of x superior to x_a ($a > 0$). However, if we consider an interval $[x_a, x_b]$ with $0 < a < < b < 1$, then by Theorem 5,

$$\lim_{n \to +\infty} P\left(\sup_{x_a \leq x \leq x_b} \left(F_n(x) - F(x) \right) < 0 \right) =$$

$$= \frac{1}{\pi} \sqrt{\frac{b}{1-b}} \int_{+\infty}^{0} \exp\left(-\frac{bt^2}{2(1-b)} \right) \int_{0}^{-t\sqrt{\frac{ab}{b-a}}} e^{-\frac{u^2}{2}} \, du \, dt, \qquad (4)$$

i.e. the probability of the difference $F_n(x) - F(x)$ being in an interval $[x_a, x_b]$ ($0 < a < b < 1$) everywhere negative remains positive even in the limit. Obviously, this result is of practical importance.

One can simplify the right hand side of (4) by a probabilistic consideration, without calculations. In fact, the right hand side of (4) is equal to the probability that a point (x, y) normally distributed on the plane lies in an angular domain $0 < x < +\infty$, $0 < y < x$, where x and y are independent and have the respective standard deviations $\sqrt{\dfrac{a(1-b)}{b-a}}$ and 1. Now this probability is equal to

$$\frac{1}{2\pi} \arctan \sqrt{\frac{a(1-b)}{b-a}} = \frac{1}{2\pi} \arcsin \sqrt{\frac{a(1-b)}{b(1-a)}}. \qquad (5)$$

As a matter of fact for a normal distribution symmetrical in x and y the probability of the random point lying in an angle φ is $\dfrac{\varphi}{2\pi}$; an affine transformation leads from this to the general case. Thus we have

THEOREM 6. *If* $0 \leq a < b \leq 1$, $F(x_a) = a$, $F(x_b) = b$, *then*

$$\lim_{n \to +\infty} P\left(\sup_{x_a \leq x \leq x_b} \left(F_n(x) - F(x) \right) < 0 \right) = \frac{1}{\pi} \arcsin \sqrt{\frac{a(1-b)}{b(1-a)}}. \qquad (6)$$

If we take $a = 0$ or $b = 1$, we see that the right hand side of (6) is equal to zero.

Another fundamental problem of mathematical statistics consists in the decision whether two samples may originate from the same population or not. This question occurs often in problems of medicine, biology, economics and many other domains. Essentially, the question is whether the deviation between the results of two experiments is or is not significant.

The following theorems can serve to decide, provided that the distribution function of the basic population is continuous; the knowledge of the population distribution is, however, not necessary.

Let $\xi_1, \xi_2, \ldots, \xi_n, \eta_1, \eta_2, \ldots, \eta_n$ be independent random variables. Let ξ_k and η_j have continuous distribution functions $F(x)$ and $G(x)$ respectively, which are not necessarily known.

The problem consists of testing the hypothesis $F(x) = G(x)$ by comparing the empirical distributions $F_n(x)$ and $G_n(x)$.

The following two theorems were proved by Smirnov:

THEOREM 7. *If* $F(x) = G(x)$, *then*

$$\lim_{n \to +\infty} P\left(\sqrt{\frac{n}{2}} \sup_{-\infty < x < +\infty} (F_n(x) - G_n(x)) < y\right) = \begin{cases} 1 - e^{-2y^2} & \text{for } y > 0, \\ 0 & \text{otherwise.} \end{cases}$$

THEOREM 8. *If* $F(x) = G(x)$, *then*

$$\lim_{n \to +\infty} P\left(\sqrt{\frac{n}{2}} \sup_{-\infty < x < +\infty} |F_n(x) - G_n(x)| < y\right) = \begin{cases} K(y) & \text{for } y > 0, \\ 0 & \text{otherwise} \end{cases}$$

where $K(y)$ *is defined by* (2).

The theorems of Smirnov can be derived by passage to the limit from the following theorems due to Gnedenko and Koroljuk, which give the exact distributions of the quantities $\sup (F_n(x) - G_n(x))$ and $\sup |F_n(x) - G_n(x)|$ for finite values of n.

THEOREM 9. *If* $\{x\}$ *denotes the least integer* $\geq x$, *if* $c = \{z \sqrt{2n}\}$, *and if* $F(x) \equiv G(x)$, *then*

$$P\left(\sqrt{\frac{n}{2}} \sup_{-\infty < x < +\infty} (F_n(x) - G_n(x)) < z\right) =$$

$$= \begin{cases} 0 & \text{for } z \leq 0, \\ 1 - \dfrac{\dbinom{2n}{n-c}}{\dbinom{2n}{n}} & \text{for } 0 < z \leq \sqrt{\dfrac{n}{2}}, \\ 1 & \text{otherwise.} \end{cases}$$

THEOREM 10. *Under the same conditions as in Theorem 9, one has*

$$P\left(\sqrt{\frac{n}{2}} \sup_{-\infty < x < +\infty} |F_n(x) - G_n(x)| < z\right) =$$

$$= \begin{cases} 0 & \text{for } z \leq \dfrac{1}{\sqrt{2n}}, \\[2ex] \dfrac{1}{\dbinom{2n}{n}} \displaystyle\sum_{k=-\left[\frac{n}{c}\right]}^{+\left[\frac{n}{c}\right]} (-1)^k \binom{2n}{n - kc} & \text{for } \dfrac{1}{\sqrt{2n}} < z \leq \sqrt{\dfrac{n}{2}}, \\[2ex] 1 & \text{otherwise.} \end{cases}$$

The values of

$$\frac{1}{\dbinom{2n}{n}} \sum_{k=-\left[\frac{n}{c}\right]}^{+\left[\frac{n}{c}\right]} (-1)^k \binom{2n}{n - kc}$$

are tabulated in Table 7, for $n \leq 30$; for $n > 30$ Theorem 8 can already be applied.

First we prove Theorems 9 and 10; Theorems 7 and 8 can then be derived by passing to the limit. Collect the random variables $\xi_1, \ldots, \xi_n, \eta_1, \ldots, \eta_n$ into one sequence and arrange these $2n$ numbers in increasing order; let ζ_k^* denote the k-th number in this ordered sequence. One can suppose that $\zeta_1^* < \zeta_2^* < \ldots < \zeta_{2n}^*$. Put

$$\theta_k = \begin{cases} 1 & \text{if } \zeta_k^* \text{ is one of the } \xi_j, \\ -1 & \text{otherwise.} \end{cases}$$

Thus in the sequence $\theta_1, \theta_2, \ldots, \theta_{2n}$, n numbers are equal to 1 and n numbers are equal to -1. Put $S_k = \theta_1 + \theta_2 + \ldots + \theta_k$. We prove first

LEMMA. *The relations*

$$\sup_{-\infty < x < +\infty} (F_n(x) - G_n(x)) = \frac{\max_{1 \leq k \leq 2n} S_k}{n}$$

and

$$\sup_{-\infty < x < +\infty} |F_n(x) - G_n(x)| = \frac{\max_{1 \leq k \leq 2n} |S_k|}{n}$$

are valid.

The number $n[F_n(x) - G_n(x)]$ is the difference between the number of the ξ_j inferior to x $(j = 1, \ldots, n)$ and the number of the η_l inferior to x $(l = 1, \ldots, n)$. If x runs through the real numbers, the quantity $n[F_n(x) - G_n(x)]$ changes only if x passes through one of the values ζ_k^* $(k = 1, 2, \ldots, 2n)$; in this case it changes by θ_k. Hence

$$\sup_{-\infty < x < +\infty} n\big(F_n(x) - G_n(x)\big) = \max_{1 \le k \le 2n} n\big(F_n(\zeta_k^* + 0) - G_n(\zeta_k^* + 0)\big) = \max_{1 \le k \le 2n} S_k$$

and, similarly,

$$\sup_{-\infty < x < +\infty} n\,|F_n(x) - G_n(x)| = \max_{1 \le k \le 2n} n\,|F_n(\zeta_k^* + 0) - G_n(\zeta_k^* + 0)| = \max_{1 \le k \le 2n} |S_k|.$$

The lemma is herewith proved.

PROOF OF THEOREM 9. Clearly, the number of the possible sequences $\theta_1, \theta_2, \ldots, \theta_{2n}$ is equal to the number of the possible arrangements of $2n$ elements among which n are equal to 1 and n to -1; thus this number is $\binom{2n}{n}$. Since $\xi_1, \ldots, \xi_n$, $\eta_1, \ldots, \eta_n$ are independent and identically distributed, all arrangements are equiprobable and each has probability $\dfrac{1}{\binom{2n}{n}}$. In order to determine the probability for $\max_{1 \le k \le 2n} S_k < z\sqrt{2n}$ we must find the number of the sequences $\theta_1, \ldots, \theta_{2n}$ fulfilling this condition and then divide this number by $\binom{2n}{n}$. We arrived thus at a combinatorial problem. Its solution will be facilitated by the following geometrical representation: Assign to every sequence $\theta_1, \ldots, \theta_n$ a broken line in the (x, y) plane starting from the point $(0, 0)$ with the points (S_k, k) $(k = 1, 2, \ldots, 2n)$ as vertices. (Here (a, b) denotes the point with coordinates $x = a$, $y = b$.) There corresponds thus to every sequence $\theta_1, \ldots, \theta_{2n}$ a "path" in the plane; all paths start from $(0, 0)$ and end at $(0, 2n)$; all are composed of segments forming with the x-axis an angle either of $+45°$ or of $-45°$. We have to determine the number of those paths which do not intersect the line $x = z\sqrt{2n}$. Let this number be denoted by $U_n^+(z)$. If a path intersects the line $x = z\sqrt{2n}$, it is clear that it reaches the line $x = \{z\sqrt{2n}\} = c$, too.

Thus we have to count those paths which lie everywhere below the line $x = c$. First we count the paths which intersect the line $x = c$.

If a path intersects the line $x = c$, we uniquely assign to it a path which is identical with the original one up to the first intersection with the line $x = c$ and from this point on is the reflection of the original path with respect to the line $x = c$. The new path ends at the point $(2c, 2n)$. By this procedure, we assign to every path going from $(0, 0)$ to $(0, 2n)$ and intersecting the line $x = c$ in a one-to-one manner a path which goes from $(0, 0)$ to $(2c, 2n)$ and is composed of segments which again form an angle of $\pm 45°$ with the x-axis. The number of paths having one or more points in common with the line $x = c$ is thus equal to the total number of the paths going from $(0, 0)$ to $(2c, 2n)$. This number is $\binom{2n}{n - c}$. Hence

$$U_n^+(z) = \binom{2n}{n} - \binom{2n}{n - c}.$$

Because of the lemma, Theorem 9 is herewith proved.

PROOF OF THEOREM 10. We use a similar argument. The number of paths going from $(0, 0)$ to $(0, 2n)$ and having no point in common either with $x = z\sqrt{2n}$ or with $x = -z\sqrt{2n}$ is equal to the number of paths going from $(0, 0)$ to $(0, 2n)$ and having no point in common with the lines $x = \pm c$. Let this number be denoted by $U_n(z)$.

Let N_+ and N_- denote the number of paths intersecting $x = c$ and $x = -c$, respectively. Let N_{+-} (and N_{-+}) denote the number of the paths which after intersecting $x = c$ (and $x = -c$) intersect also $x = -c$ (and $x = c$), respectively, etc. Let N_0 denote the number of the paths which do not intersect either $x = c$ or $x = -c$. There can be shown as in Chapter II (§ 3, Theorem 9) that

$$N_0 = N - N_+ - N_- + N_{+-} + N_{-+} - N_{+-+} - N_{-+-} + \ldots . \quad (7)$$

We know that $N_+ = \binom{2n}{n + c}$; by reasons of symmetry we have $N_- = N_+$. We calculate now N_{+-} (which is equal to N_{-+}). Let us take the reflection to the line $x = c$ of the section of the path which follows the first common point of the path with the line $x = c$, then let us take another reflection to the line $x = 3c$ of the section of the new path which follows the first common point of the path with the line $x = 3c$; we obtain thus a path which goes from $(0, 0)$ to $(4c, 2n)$. Conversely, there corresponds to every such path exactly one of the original paths intersecting first the line $x = c$, then the line $x = -c$. Hence N_{+-} (and N_{-+} too) is equal to the number of sequences which consist of $n + 2c$ elements equal to 1 and of $n - 2c$ elements equal to -1, viz. to $\binom{2n}{n + 2c}$. Similarly, we obtain

$$N_{\varepsilon_1, \varepsilon_2, \ldots, \varepsilon_k} = \binom{2n}{n + kc} = \binom{2n}{n - kc},$$

where $\varepsilon_1, \ldots, \varepsilon_k$ is a sequence of alternating signs $+$ and $-$ beginning with $+$ or with $-$. By (7) and by the lemma, Theorem 10 is thus proved.

In order to derive Theorem 7 from Theorem 9 it suffices to remark that for $c = \{z \sqrt{2n}\}$ Stirling's formula gives

$$\lim_{n \to +\infty} \frac{\binom{2n}{n+c}}{\binom{2n}{n}} = e^{-2z^2}.$$

Theorem 8 can be derived from Theorem 10 in a similar way.

§ 11. Limit distributions concerning random walk problems

In this section we shall study limit theorems of another type than those encountered so far. As we do not strive at the greatest possible generality but rather wish to present the different types of limit distributions, we shall restrict ourselves mainly to the simplest case, i.e. to the case of the one-dimensional random walk (classical ruin problem). We shall find in the study of this simple problem a lot of surprising laws which contribute to a better understanding of the nature of chance. Theorems 1 and 2 are concerned with the problem of random walk in n-space.

Let the random variables $\xi_1, \xi_2, \ldots, \xi_n, \ldots$ be independent and let each of them assume the values $+1$ and -1 with probability $\dfrac{1}{2}$. The random variable

$$\zeta_n = \sum_{k=1}^{n} \xi \tag{1}$$

may be considered as a gambler's gain in a game of coin tossing after n tosses, provided that the stake is 1 unit of money. The value of ζ_n, which is always an integer, can also be interpreted as the abscissa at the time $t = n$ of a point moving in a random manner on the real axis. This point performs on the real axis a "random walk", in the sense that it moves during the time intervals $(0, 1)$, $(1, 2)$, $\ldots$ either one unit step to the right or one unit step to the left, both with probability $\dfrac{1}{2}$. We shall deal with the laws of this random walk.

Consider first a generalization of the problem to several dimensions. Let G_r denote the set of the points of the r-dimensional Euclidean space which have integer coordinates, i.e. the set of points of the "r-dimensional lattice". Imagine a point which moves "at random" over this lattice. We understand by a "random walk" the following: If the moving point can be found at a time $t = n$ at a certain lattice point, then the probability that at the time $t = n + 1$ it can be found at one of the adjacent points of the lattice is equal to $\dfrac{1}{2r}$ for all adjacent points which have $r - 1$ coordinates equal to those of the preceding point and one coordinate differing by ± 1. If the position of the point at the time $t = n$ is given by the vector $\zeta_n^{(r)}$, then the random vectors $\zeta_n^{(r)}$ ($n = 0, 1, \ldots$) form a homogeneous additive Markov chain, namely

$$\zeta_n^{(r)} = \zeta_0^{(r)} + \sum_{k=1}^{n} \xi_k^{(r)},$$

where the random vector $\xi_k^{(r)}$ represents the displacement of the point during the time interval $(k - 1, k)$; by assumption, the random vectors $\xi_k^{(r)}$ are independent and identically distributed. For $r = 1$ we obtain the one-dimensional random walk problem discussed above; in this case we write simply ζ_n and ξ_k instead of $\zeta_k^{(r)}$ and $\xi_k^{(r)}$.

We prove now first a famous theorem of G. Pólya.[1]

THEOREM 1. *The probability that a point performing a random walk over the lattice G_r returns infinitely often to its initial position is equal to one for $r = 1$ and $r = 2$ and is equal to zero for $r \geq 3$.*

PROOF. Without the restriction of generality we can assume that at the time $t = 0$ the moving point is found at the origin of the coordinate system.

Let $P_n^{(r)}$ denote the probability that at the time $t = n$ the moving point is again at the origin. The moving point returns to the origin after performing in the direction of each of the axes exactly as many steps to the "right" as to the "left". Hence $P_{2n+1}^{(r)} = 0$ and

$$P_{2n}^{(r)} = \frac{1}{(2r)^{2n}} \sum_{n_1 + \ldots + n_r = n} \frac{(2n)!}{(n_1! \ldots n_r!)^2} =$$

$$= \frac{1}{(2r)^{2n}} \binom{2n}{n} \sum_{n_1 + \ldots + n_r = n} \left(\frac{n!}{n_1! \ldots n_r!} \right)^2. \qquad (2)$$

In particular

$$P_{2n}^{(1)} = \frac{\binom{2n}{n}}{2^{2n}},$$

---

[1] Cf. G. Pólya [2].

$$P_{2n}^{(2)} = \frac{\binom{2n}{n}^2}{4^{2n}},$$

$$P_{2n}^{(3)} = \frac{\binom{2n}{n}}{6^{2n}} \sum_{k+l \leq n} \left[\frac{n!}{k!\, l!\, (n-k-l)!} \right]^2.$$

By applying Stirling's formula we obtain

$$P_{2n}^{(1)} \approx \frac{1}{\sqrt{\pi n}} \tag{3a}$$

and

$$P_{2n}^{(2)} \approx \frac{1}{\pi n}. \tag{3b}$$

We give now an estimation of $P_{2n}^{(r)}$ for $r \geq 3$. We know that

$$\sum_{n_1 + \ldots + n_r = n} \frac{n!}{n_1! \ldots n_r!} = r^n.$$

On the other hand, it is easy to see that among the polynomial coefficients the largest are those in which the numbers $n_1, n_2, \ldots, n_r$ differ at most by ± 1 from each other (cf. Chapter III, § 18, Exercise 3). Hence

$$P_{2n}^{(r)} = \frac{\binom{2n}{n}}{(2r)^{2n}} \sum_{n_1 + \ldots + n_r = n} \left(\frac{n!}{n_1! \ldots n_r!} \right)^2 \leq$$

$$\leq \frac{\binom{2n}{n}}{(4r)^n} \max_{\substack{\Sigma n_j = n \\ 1}} \frac{n!}{n_1! \ldots n_r!} = O\left(\frac{1}{n^{\frac{r}{2}}} \right). \tag{3c}$$

On the other hand, it can be proved that $P_{2n}^{(r)}$ can be represented by the following integral:

$$P_{2n}^{(r)} = \frac{\binom{2n}{n}}{(2\pi)^{r-1} (2r)^{2n}} \times$$

$$\times \int_{-\pi}^{+\pi} \ldots \int_{-\pi}^{+\pi} \left[r + 2 \sum_{1 \leq i < j \leq r-1} \cos(\theta_i - \theta_j) + 2 \sum_{i=1}^{r-1} \cos \theta_i \right]^n d\theta_1 \ldots d\theta_{r-1}.$$

Hence we derive the asymptotic relation

$$P_{2n}^{(r)} \approx \frac{1}{2^{r-1}} \left(\frac{r}{n\pi} \right)^{\frac{r}{2}}. \tag{4}$$

From (3a) and (3b) it follows that

$$\sum_{n=1}^{\infty} P_{2n}^{(r)} = +\infty \quad \text{for} \quad r = 1 \text{ and } r = 2,$$

and from (3c) that

$$\sum_{n=1}^{\infty} P_{2n}^{(r)} < +\infty \quad \text{for } r \geq 3.$$

In the latter case the Borel–Cantelli lemma permits to state that for $r \geq 3$ the moving point returns with probability 1 at most finitely many times to its initial position.

For $r = 1$ and $r = 2$ we shall show that with probability 1 the moving point will sooner or later (and therefore infinitely often) return to its initial position. In order to prove this, consider the time interval which passes until the first return of the moving point. Let $Q_n^{(r)}$ denote the probability that the point walking at random on the r-dimensional lattice reaches its initial position for the time after n steps. Obviously,

$$P_{2n}^{(r)} = Q_{2n}^{(r)} + \sum_{k=1}^{n-1} P_{2k}^{(r)} Q_{2n-2k}^{(r)}. \tag{5}$$

Put

$$G_r(x) = \sum_{k=1}^{\infty} P_{2k}^{(r)} x^k \tag{6}$$

and

$$H_r(x) = \sum_{k=1}^{\infty} Q_{2k}^{(r)} x^k, \tag{7}$$

then from (5)

$$G_r(x) = H_r(x) + G_r(x) H_r(x), \tag{8}$$

hence

$$H_r(x) = \frac{G_r(x)}{1 + G_r(x)} \tag{9a}$$

and

$$G_r(x) = \frac{H_r(x)}{1 - H_r(x)}. \tag{9b}$$

Clearly,

$$Q^{(r)} = \sum_{k=1}^{\infty} Q_{2k}^{(r)} = H_r(1) = \lim_{x \to 1-0} \frac{G_r(x)}{1 + G_r(x)} , \tag{10}$$

where $Q^{(r)}$ denotes the probability that the moving point returns at least once to the origin. For $r = 1$ and $r = 2$ the series $\sum_{k=1}^{+\infty} P_{2k}^{(r)}$ is divergent, hence $Q^{(r)} = 1$, while for $r \geq 3$

$$Q^{(r)} = \frac{\sum_{k=1}^{\infty} P_{2k}^{(r)}}{1 + \sum_{k=1}^{\infty} P_{2k}^{(r)}}$$

hence $0 < Q^{(r)} < 1$. (E.g., for $r = 3$, $Q^{(3)} \approx 0.35$.) Thus we have proved Theorem 1; at the same time we have obtained

THEOREM 2. *For $r \geq 3$ a point performing random walk over the lattice G_r has a probability less than 1 to return to its original position.*

It can be shown in a similar manner that for $r = 1$ and $r = 2$ the moving point passes infinitely many times through every point of the lattice with probability 1, while this is not true for $r \geq 3$.

In what follows we shall deal with the case $r = 1$ only. First we give the explicit form of the probability $Q_{2k}^{(1)}$. The generator function (6) is here

$$G_1(x) = \sum_{k=1}^{\infty} \frac{\binom{2k}{k}}{4^k} x^k = \sum_{k=1}^{\infty} \binom{-\frac{1}{2}}{k} (-x)^k = \frac{1}{\sqrt{1-x}} - 1;$$

this and (9a) lead to

$$H_1(x) = \frac{G_1(x)}{1 + G_1(x)} = 1 - \sqrt{1-x} = \sum_{k=1}^{\infty} \binom{\frac{1}{2}}{k} (-1)^{k-1} x^k.$$

Hence

$$Q_{2k}^{(1)} = -\frac{\binom{2k-2}{k-1}}{k\,2^{2k-1}} \approx \frac{1}{2\sqrt{\pi}\,k^{\frac{3}{2}}} . \tag{11}$$

A simple calculation shows that

$$Q_{2k}^{(1)} = P_{2k-2}^{(1)} - P_{2k}^{(1)}. \tag{12}$$

Let v_1 be the number of the steps in which the moving point first returns to its initial position; hence v_1 is a random variable and $P(v_1 = 2k) = Q_{2k}^{(1)}$. It follows from the asymptotic behaviour of the sequence $Q_{2k}^{(1)}$ that the expectation of v_1 is infinite. Let $\varphi(t)$ be the characteristic function of v_1:

$$\varphi(t) = 1 - \sqrt{1 - e^{2it}},$$

hence

$$\lim_{n \to +\infty} \varphi\left(\frac{t}{n^2}\right)^n = \exp(-\sqrt{-2it}). \tag{13}$$

But we have

$$\exp(-\sqrt{-2it}) = \frac{1}{\sqrt{2\pi}} \int_0^{+\infty} \frac{\exp\left(ixt - \frac{1}{2x}\right)}{x^{\frac{3}{2}}} \, dx, \tag{14}$$

hence $\exp(-\sqrt{-2it})$ is the characteristic function of the distribution with the density function

$$f(x) = \begin{cases} \dfrac{e^{-\frac{1}{2x}}}{\sqrt{2\pi}\, x^{\frac{3}{2}}} & \text{for } x > 0, \\ 0 & \text{otherwise.} \end{cases}$$

Because of the identity

$$\int_0^x \frac{e^{-\frac{1}{2u}}}{\sqrt{2\pi}\, u^{\frac{3}{2}}} \, du = 2\left(1 - \Phi\left(\frac{1}{\sqrt{x}}\right)\right) \tag{15}$$

(where $\Phi(x)$ is the distribution function of the normal distribution), we obtain

THEOREM 3. *If* $v_1, v_2, \ldots, v_n, \ldots$ *denotes the moments when the moving point performing a random walk on the line returns to its initial position, i.e. when* $\zeta_{v_k} = 0$, *then for* $x > 0$ *the relation*

$$\lim_{k \to +\infty} P\left(\frac{v_k}{k^2} < x\right) = 2\left(1 - \Phi\left(\frac{1}{\sqrt{x}}\right)\right) \tag{16}$$

is valid.

PROOF. The random variables $v_1, v_2 - v_1, \ldots, v_k - v_{k-1}, \ldots$ are evidently independent and identically distributed. Further

$$v_k = v_1 + (v_2 - v_1) + \ldots + (v_k - v_{k-1}).$$

Hence (16) follows from (13), (14) and (15).

Remark. The distribution figuring in Theorem 3, with the characteristic function $\exp(-\sqrt{2it})$, is a stable distribution corresponding to the parameter $\alpha = \dfrac{1}{2}$; hence the distribution $\{Q_{2k}^{(1)}\}$ belongs to the domain of attraction of this stable distribution.

Theorem 3 can also be formulated in a different way. Let θ_n denote the number of the zeros in the sequence $\zeta_1, \ldots, \zeta_n$; then

$$P(v_k \le n) = P(\theta_n \ge k). \tag{17}$$

From this follows

THEOREM 4.

$$\lim_{n \to +\infty} P\left(\frac{\theta_n}{\sqrt{n}} < y\right) = \begin{cases} 2\Phi(y) - 1 & \text{for } y > 0, \\ 0 & \text{otherwise;} \end{cases} \tag{18}$$

hence $\dfrac{\theta_n}{\sqrt{n}}$ has in limit the same distribution as the absolute value of a random variable with the standard normal distribution.

Now we shall investigate the number of positive and negative terms in the sequence $\zeta_1, \ldots, \zeta_n$. If $\zeta_j = 0$ but $\zeta_{j-1} > 0$, then ζ_j is counted as positive; if $\zeta_j = 0$ but $\zeta_{j-1} < 0$, then it is counted as negative. Let π_n denote the number of the positive terms (in the above-mentioned sense) of the sequence $\zeta_1, \zeta_2, \ldots, \zeta_n$. We prove

THEOREM 5. *For every positive integer n the relation*[1]

$$P(\pi_{2n} = 2k) = \frac{\binom{2k}{k}\binom{2n-2k}{n-k}}{2^{2n}} \qquad (k = 0, 1, \ldots, n) \tag{19}$$

holds.

Remark. Clearly, π_{2n} cannot be an odd number; in fact, π_2 is either 0 or 2, according as $\zeta_1 = 1$ or $\zeta_1 = -1$. Similarly, $\pi_{2n} - \pi_{2n-2}$ is either 0 or 2 since ζ_{2n-2} is an even number: if $\zeta_{2n-2} \ge 0$, we have necessarily $\zeta_{2n-2} \ge 2$, hence $\pi_{2n} - \pi_{2n-2} = 2$; if $\zeta_{2n-2} < 0$, we have necessarily $\zeta_{2n-2} \le -2$, hence $\pi_{2n} - \pi_{2n-2} = 0$; if $\zeta_{2n-2} = 0$, then $\pi_{2n} - \pi_{2n-2}$ is equal to 0 or to 2.

[1] We put $\binom{2k}{k} = 1$ for $k = 0$.

We need the following

LEMMA 1. *For every integer $n \geq 1$ the relation*

$$\frac{1}{2^{2n}} \sum_{k=0}^{n} \binom{2k}{k} \binom{2n-2k}{n-k} = 1 \tag{20}$$

holds.

Remark. Relation (20) is a corollary of (19); in fact if we add the probabilities $P(\pi_{2n} = 2k)$ for $k = 0, 1, \ldots, n$, we obtain 1, remembering that π_{2n} is always even. But since we wish to use (20) for the proof of (19), we have to prove (20) directly.

PROOF. As we have seen, for $|x| < 1$

$$\frac{1}{\sqrt{1-x}} = \sum_{k=0}^{\infty} \frac{\binom{2k}{k}}{4^k} x^k. \tag{21}$$

Let us take the square of both sides of (21); since on the left side we get $\frac{1}{1-x} = \sum_{k=0}^{\infty} x^k$, (20) is obtained by comparing the coefficients of x^n on both sides.

Now we prove (19) by induction. Clearly, (19) is true for $n = 1$; in effect

$$P(\pi_2 = 0) = P(\pi_2 = 2) = \frac{1}{2}.$$

Suppose that (19) is valid for $n < N$ and let ν_1 denote the least index j for which $\zeta_j = 0$; ν_1 is necessarily an even number. Furthermore

$$P(\pi_{2N} = 2k) = \sum_{l=1}^{N} P(\pi_{2N} = 2k, \nu_1 = 2l) + P(\pi_{2N} = 2k, \nu_1 > 2N).$$

But

$$P(\pi_{2N} = 2k, \nu_1 = 2l) =$$
$$= P(\pi_{2N} = 2k, \nu_1 = 2l, \zeta_1 = +1) + P(\pi_{2N} = 2k, \nu_1 = 2l, \zeta_1 = -1),$$

on the one hand

$$P(\pi_{2N} = 2k, \nu_1 = 2l, \zeta_1 = +1) = \frac{1}{2} P\big(\pi_{2(N-l)} = 2(k-l)\big) P(\nu_1 = 2l),$$

and on the other hand

$$P(\pi_{2N} = 2k, v_1 = 2l, \zeta_1 = -1) = \frac{1}{2} P(\pi_{2(N-l)} = 2k) P(v_1 = 2l).$$

According to (12)

$$P(v_1 = 2l) = Q_{2l}^{(1)} = \frac{\binom{2l-2}{l-1}}{2^{2l-2}} - \frac{\binom{2l}{l}}{2^{2l}}$$

which leads to the recursion formula

$$P(\pi_{2N} = 2k) = P(\pi_{2N} = 2k, v_1 > 2N) +$$

$$+ \frac{1}{2} \sum_{l=1}^{N} \left(\frac{\binom{2l-2}{l-1}}{2^{2l-2}} - \frac{\binom{2l}{l}}{2^{2l}} \right) [P(\pi_{2(N-l)} = 2k) + P(\pi_{2(N-l)} = 2(k-l))]. \quad (22)$$

The probability $P(\pi_{2n} = 2k, v_1 > 2N)$ is evidently zero for $0 < k < N$.
If $k = 0$ or $k = N$,

$$P(\pi_{2N} = 2N, v_1 > 2N) = P(\pi_{2N} = 0, v_1 > 2N) = \frac{\binom{2N}{N}}{2^{2N+1}}. \quad (23)$$

Now if the relation (19) is true for $n = 1, 2, \ldots, N - 1$, then it follows by some simple calculations from (20) and (22) that it is also true for $n = N$. Thus Theorem 5 is proved.

This theorem implies the so-called *arc sine law*:

THEOREM 6.

$$\lim_{N \to +\infty} P\left(\frac{\pi_N}{N} < x \right) = \frac{2}{\pi} \text{ arc } \sin \sqrt{x} \text{ for } 0 \le x \le 1. \quad (24)$$

PROOF. According to (3a)

$$\frac{\binom{2k}{k}}{2^{2k}} \approx \frac{1}{\sqrt{\pi k}}. \quad (25)$$

For $0 < x < y < 1$ we obtain from (19)

$$P\left(x \le \frac{\pi_{2n}}{2n} < y \right) \approx \frac{1}{\pi} \sum_{k=[nx]+1}^{[ny]} \frac{1}{\sqrt{\frac{k}{n}\left(1 - \frac{k}{n}\right)}} \frac{1}{n}, \quad (26)$$

hence

$$\lim_{n \to +\infty} P\left(x \le \frac{\pi_{2n}}{2n} < y\right) = \frac{1}{\pi} \int_x^y \frac{dt}{\sqrt{t(1-t)}} =$$

$$= \frac{2}{\pi} (\text{arc sin} \sqrt{y} - \text{arc sin} \sqrt{x}). \tag{27}$$

Now since $\pi_{2n} \le \pi_{2n+1} \le \pi_{2n} + 1$, the limit distribution of $\dfrac{\pi_{2n+1}}{2n+1}$ coincides with that of $\dfrac{\pi_{2n}}{2n}$ which proves Theorem 6.

This theorem can be proved in a more elegant way, which, however, requires more powerful tools. This rests upon the following generalization of Lemma 1:

LEMMA 2. *We have*

$$\frac{1}{2^{2n}} \sum_{k=0}^{n} \binom{2k}{k} \binom{2n-2k}{n-k} e^{it(2k-n)} = P_n(\cos t), \tag{28}$$

where $P_n(x)$ *denotes the n-th Legendre polynomial:*

$$P_n(x) = \frac{1}{2^n n!} \frac{d^n}{dx^n} (x^2 - 1)^n.$$

PROOF. We see that the left side of (28) is the coefficient of x^n in the power series expansion of

$$\frac{1}{\sqrt{(1 - e^{it} x)(1 - e^{-it} x)}} = \frac{1}{\sqrt{1 - 2x \cos t + x^2}}.$$

On the other hand we know that[1]

$$\frac{1}{\sqrt{1 - 2x \cos t + x^2}} = \sum_{n=0}^{+\infty} P_n(\cos t) x^n \tag{29}$$

where $P_n(z)$ is the n-th Legendre polynomial. By comparing coefficients we obtain (28).

Theorem 6 can be derived from (28) as follows: We have

$$E\left(\exp\left(it \frac{\pi_{2n}}{2n}\right)\right) = e^{\frac{it}{2}} P_n\left(\cos \frac{t}{2n}\right).$$

[1] Cf. G. Pólya and G. Szegő [1], Vol. II, p. 291.

By the Laplace formula[1]

$$P_n(x) = \frac{1}{\pi} \int_0^\pi (x + i \cos \varphi \sqrt{1 - x^2})^n \, d\varphi$$

we obtain

$$\lim_{n \to +\infty} e^{\frac{it}{2}} P_n\left(\cos \frac{t}{2n}\right) = \frac{e^{\frac{it}{2}}}{\pi} \int_0^\pi \exp\left(\frac{it \cos \varphi}{2}\right) d\varphi =$$

$$= \frac{1}{\pi} \int_{-1}^{+1} \frac{e^{it(1+u)/2}}{\sqrt{1 - u^2}} \, du = \frac{1}{\pi} \int_0^1 \frac{e^{itx} \, dx}{\sqrt{x(1 - x)}}$$

which implies the statement of Theorem 6 in view of Theorem 3 in Chapter VI, § 4.

Lemma 2 permits an easy calculation of the moments of π_{2n}. For reasons of symmetry $E(\pi_{2n}) = n$. The standard deviation can be derived from (28):

$$D^2(\pi_{2n}) = P'_n(1) = \binom{n+1}{2},$$

hence

$$D(\pi_{2n}) = \sqrt{\frac{n(n+1)}{2}}. \tag{30}$$

Remark. Theorem 6 expresses an interesting paradoxical fact. The derivative of the distribution function $F(x) = \frac{2}{\pi} \arcsin \sqrt{x}$, i.e.

$$F'(x) = \frac{1}{\pi \sqrt{x(1 - x)}}$$

is namely symmetrical with respect to the point $x = \frac{1}{2}$ and has a *minimum* at this point. Consequently, this value is the least probable one for the random variable $\frac{\pi_N}{N}$: the probability that the value of $\frac{\pi_N}{N}$ is in the neighbourhood of a point x $(0 < x < 1)$ is the greater the farther the number x is away from $\frac{1}{2}$. One would expect rather the contrary: indeed it would

---

[1] Cf. G. Pólya and G. Szegő [1], Vol. II, p. 291.

seem quite natural that the moving point would pass approximately half of its time on the positive and the other half on the negative semiaxis. However, Theorem 6 shows that this is not the case. Or, to put it in the terms of coin tossing: One would consider as the most probable that both players are leading during nearly $\frac{1}{2}$ of the whole time. But this is not so; on the contrary, $\frac{1}{2}$ is the least probable value for the fraction of time during which one of the players is leading. However, a little reflexion shows this to be quite natural; indeed ζ_n varies quite slowly, because of $\zeta_{n+1} - \zeta_n = \pm 1$; if ζ_n reaches for a certain n a large positive value, ζ_{n+k} will remain for a long time positive and a similar reasoning holds for large negative values too.

Theorem 6 is due to P. Lévy.[1]

The theorem can be generalized. It was proved by Erdős and Kac[2] that if $\xi_1, \xi_2, \ldots$ are independent random variables with $E(\xi_k) = 0$, $D(\xi_k) = 1$ which satisfy the condition of Lindeberg, further if we put $\zeta_n = \sum_{k=1}^{n} \xi_k$ and if π_N denotes the number of positive terms in the sequence $\zeta_1, \zeta_2, \ldots, \zeta_N$ then π_N fulfils (24). Sparre-Andersen[3] proved that if $\xi_1, \xi_2, \ldots, \xi_n$ are independent random variables with the same symmetrical and continuous distribution, then

$$P(\pi_n = k) = \frac{\binom{2k}{k}\binom{2n-2k}{n-k}}{2^{2n}}. \tag{31}$$

In this case (24) is valid even if the variance does not exist.

We now determine the exact distribution of θ_n (the number of the zeros in the sequence $\zeta_1, \zeta_2, \ldots, \zeta_n$) for even values of n. We prove first

THEOREM 7. *For every positive integer* n

$$P(\theta_{2n} = k) = \frac{2^k}{2^{2n}} \binom{2n-k}{n} \qquad (k = 0, 1, \ldots, 2n) \tag{32}$$

holds.

PROOF. If ν_1 denotes the least positive integer for which $\zeta_{\nu_1} = 0$, then

$$P(\theta_{2n} = k) = \sum_{r=1}^{n-1} P(\theta_{2n} = k, \nu_1 = 2r) + P(\theta_{2n} = k, \nu_1 \geq 2n).$$

[1] Cf. P. Lévy [2].
[2] Cf. P. Erdős and M. Kac [2].
[3] Cf. E. Sparre-Andersen [1].

From this we derive (32) by the method used in the proof of Theorem 5. We obtain from (32) by Stirling's formula

$$E(\theta_n) \approx \sqrt{\frac{2n}{\pi}} \, . \tag{33}$$

Theorem 6 may also be formulated as follows: Put

$$\varepsilon_k = \operatorname{sgn} \zeta_k = \begin{cases} 1 & \text{for } \zeta_k > 0 \text{ or for } \zeta_k = 0 \text{ and } \zeta_{k-1} = 1, \\ -1 & \text{otherwise.} \end{cases}$$

Then for $-1 \le x \le +1$,

$$\lim_{n \to +\infty} P\left(\frac{\varepsilon_1 + \varepsilon_2 + \ldots + \varepsilon_n}{n} < x \right) = \frac{2}{\pi} \arcsin \sqrt{\frac{1+x}{2}} \, . \tag{34}$$

Consequently, the ratio $\dfrac{\varepsilon_1 + \varepsilon_2 + \ldots + \varepsilon_n}{n}$ does not tend to zero as $n \to +\infty$, though this would seem to be quite "plausible". However the following theorem due to Erdős and Hunt[1] is valid:

THEOREM 8.

$$P\left(\lim_{N \to +\infty} \frac{\displaystyle\sum_{k=1}^{N} \frac{\varepsilon_k}{k}}{\displaystyle\sum_{k=1}^{N} \frac{1}{k}} = 0 \right) = 1. \tag{35}$$

PROOF. Clearly $E(\varepsilon_k) = 0$, $E(\varepsilon_k^2) = 1$. We determine $E(\varepsilon_n \varepsilon_m)$ $(m > n)$:

$$E(\varepsilon_n \varepsilon_m) = 2 \sum_{k=1}^{n} P(\zeta_n = k) \big(2P(\zeta_m > 0 \mid \zeta_n = k) - 1 \big) \le$$

$$\le 4 \sum_{k=1}^{n} P(\zeta_n = k) P(-k < \zeta_{m-n} \le 0).$$

The greatest term of the binomial distribution of order n and parameter $\dfrac{1}{2}$ is the central term, asymptotically equal to $\sqrt{\dfrac{2}{n\pi}}$; from this we conclude that

$$P(-k < \zeta_{m-n} \le 0) < \frac{C_1 k}{\sqrt{m-n}} \, ,$$

[1] Cf. P. Erdős and G. A. Hunt [1].

hence

$$|E(\varepsilon_n \varepsilon_m)| \leq C_2 \sqrt{\frac{n}{m-n}} \; ; \tag{36a}$$

here and in what follows C_1, C_2, ... are positive constants. If $m - n \leq n$, we use instead of (36a) the trivial inequality

$$|E(\varepsilon_n \varepsilon_m)| \leq 1. \tag{36b}$$

Thus we obtain

$$E\left(\left(\sum_{n=1}^{N} \frac{\varepsilon_n}{n}\right)^2\right) \leq C_3 \ln N. \tag{37}$$

If we put

$$\varDelta_N = \frac{\displaystyle\sum_{n=1}^{N} \frac{\varepsilon_n}{n}}{\displaystyle\sum_{n=1}^{N} \frac{1}{n}}, \tag{38}$$

we find

$$E(\varDelta_N) = 0 \quad \text{and} \quad E(\varDelta_N^2) \leq \frac{C_4}{\ln N}.$$

Hence, by applying Chebyshev's inequality,

$$P(|\varDelta_N| > \varepsilon) \leq \frac{C_4}{\varepsilon^2 \ln N}. \tag{39}$$

From this we obtain that the series $\sum\limits_{k=1}^{\infty} P(|\varDelta_{2^{k^2}}| > \varepsilon)$ converges for every $\varepsilon > 0$. Thus by the Borel–Cantelli lemma the inequality $|\varDelta_{2^{k^2}}| \leq \varepsilon$ is satisfied with probability 1 for a sufficiently large k. But for $2^{k^2} \leq n < 2^{(k+1)^2}$ we have

$$|\varDelta_n| \leq \varDelta_{2^{k^2}} + \frac{C_5}{k}.$$

Hence the inequality $|\varDelta_n| \leq 2\varepsilon$ is fulfilled with probability 1 for a sufficiently large n. Since $\varepsilon > 0$ can be chosen arbitrarily small, Theorem 8 is proved.

In conclusion we mention some theorems concerning the largest fluctuations of the one-dimensional random walk.

THEOREM 9.

$$\lim_{n \to +\infty} P\left(\max_{1 \leq k \leq n} \zeta_k < x \sqrt{n}\right) = \begin{cases} 2\Phi(x) - 1 & \text{for } x > 0, \\ 0 & \text{otherwise.} \end{cases} \tag{40}$$

THEOREM 10.

$$\lim_{n \to +\infty} P(\max_{1 \le k \le n} |\zeta_k| < x\sqrt{n}) =$$

$$= \begin{cases} \dfrac{4}{\pi} \displaystyle\sum_{k=0}^{\infty} \dfrac{(-1)^k \exp\left[-\dfrac{(2k+1)^2 \pi^2}{x^2}\right]}{2k+1} & \text{for } x > 0, \\[4mm] 0 & \text{otherwise.} \end{cases} \tag{41}$$

Theorem 9 can be derived from the following formula (cf. Chapter III, § 18, Exercise 19):[1]

$$P(\max_{1 \le k \le n} \zeta_k < m) = 1 - \frac{1}{2^m} \sum_{k=0}^{\left[\frac{n-m}{2}\right]} \binom{m+2k}{k} \frac{k}{m+2k} \frac{1}{4^k}. \tag{42}$$

It was shown by Erdős and Kac[2] that Theorems 9 and 10 can be amply generalized. They can be extended to the sums of independent, identically distributed random variables.[3]

It is interesting to compare Theorems 9 and 10 with the results of the preceding section. Those results can be put in the following form:

THEOREM 11.

$$\lim_{n \to +\infty} P(\max_{1 \le k \le 2n} \zeta_k < x\sqrt{2n} \,|\, \zeta_{2n} = 0) = \begin{cases} 1 - e^{-2x^2} & \text{for } x \ge 0, \\ 0 & \text{otherwise.} \end{cases} \tag{43}$$

THEOREM 12.

$$\lim_{n \to +\infty} (\max_{1 \le k \le 2n} |\zeta_k| < x\sqrt{2n} \,|\, \zeta_{2n} = 0) = \begin{cases} \displaystyle\sum_{k=-\infty}^{+\infty} (-1)^k e^{-2k^2x^2} & \text{for } x \ge 0, \\ 0 & \text{otherwise.} \end{cases} \tag{44}$$

Theorems 11 and 12 describe the properties of paths which after $2n$ steps return to the origin. One expects that under this condition the path does not deviate as far from its origin as in the general case. Indeed the expectation of the distribution (43) is

$$\int_0^{\infty} 4x^2 e^{-2x^2} dx = \frac{1}{4} \sqrt{2\pi} = 0.627,$$

while that of the distribution (40) is $\sqrt{\dfrac{2}{\pi}} = 0.798$.

---

[1] For another proof of Theorem 9 see Chapter VII, § 16, Exercise 13. e.
[2] Cf. P. Erdős and M. Kac [1].
[3] For the extension to random variables which are not identically distributed, see A. Rényi [9].

§ 12. Proof of the limit theorems by the operator method

The most important limit theorems of probability calculus can also be proved without the use of characteristic functions, by a direct method. This method will be presented here. It is to be noted that this method, if applicable, is more simple than the method of characteristic functions; but it does not replace the latter. As a matter of fact, the method of characteristic functions has a far wider range of applications and there are many limit theorems which can only be proved by means of characteristic functions, or at least their proof by any other method becomes very complicated indeed.

The method to be dealt with in the present section can be called the *operator method*, since it uses certain functional operators. For the sake of simplicity we shall introduce the method first by proving Liapunov's theorem; then we shall pass to the proof by this method of the more general Lindeberg theorem. Finally, we shall prove a theorem about the convergence to the Poisson distributions (§ 4, Theorem 1) and the theorem of § 3.

We recall some definitions and notations. Let C_3 be the set of all uniformly continuous and bounded real-valued functions defined on the real number axis which are three times differentiable while their first three derivatives are also uniformly continuous and bounded on the whole number axis. If $f = f(x) \in C_3$, put

$$\|f\| = \sup_x |f(x)|. \tag{1}$$

The number $\|f\|$ is called the *norm* of the function $f = f(x)$. Clearly, if $f \in C_3$ and $g \in C_3$, then $f + g \in C_3$, further if $f \in C_3$ and α is a real number, then $\alpha f \in C_3$. It is easy to see that if $f \in C_3$ and $g \in C_3$, then $\|f + g\| \leq \leq \|f\| + \|g\|$ and if $f \in C_3$ and α is a real number, then $\|\alpha f\| = = |\alpha| \|f\|$. An operator A which assigns to every function $f \in C_3$ an element $g = g(x) = Af$ of C_3 is called a *linear operator* if it possesses the following properties:

1) If $f \in C_3$ and $g \in C_3$, then $A(f + g) = Af + Ag$.
2) If $f \in C_3$ and α is a real number, then $A(\alpha f) = \alpha \cdot Af$.
3) There exists a number $K > 0$ such that for any function $f \in C_3$ the inequality $\|Af\| \leq K \cdot \|f\|$ holds.

If 3) is fulfilled for $K = 1$, the operator A is called a *contraction operator*. If A and B are two operators, we define the operator $A + B$ by $(A + B)f = = Af + Bf$. The product of two operators is defined by the consecutive application of the operators, i.e. by $(AB)f = A(Bf)$. The multiplication of operators is associative, but it is usually not commutative. The addition of operators is obviously both associative and commutative. For the multiplication and addition of operators the distributive law is valid:

$$A(B + C) = AB + AC.$$

If A is an operator and α a real number, we understand by αA the operator defined by $(\alpha A)f = \alpha \cdot Af$. Clearly, if α and β are real numbers and A and B operators, then $\alpha(A + B) = \alpha A + \alpha B$, $\alpha(\beta A) = (\alpha\beta)A$ and $(\alpha + \beta)A = \alpha A + \beta A$, further $A + (-A) = O$ and $O \cdot A = O$, where O is the zero operator which assigns to every function $f \in C_3$ the function identically equal to 0. To the reader acquainted with the elements of functional analysis all this is of course familiar.

LEMMA 1. *Let $F(x)$ be an arbitrary distribution function, then the linear operator A_F defined by*

$$A_F f = \int\limits_{-\infty}^{+\infty} f(x + y)\, dF(y) \tag{2}$$

is a contraction operator.

PROOF. It is easy to see that if $f \in C_3$ then $A_F f \in C_3$. Clearly A_F fulfils conditions (1) and (2) in the definition of linear operators, further

$$\|A_F f\| \le \|f\| \int\limits_{-\infty}^{+\infty} dF(y) = \|f\|,$$

hence A_F is indeed a contraction operator.

The operator A_F is called the *operator associated with the distribution function F.*

If A and B are two operators such that for every $f \in C_3$ $ABf = BAf$, then the operators A and B are said to be commutative.

LEMMA 2. *Let $F(x)$ and $G(x)$ be any two distribution functions. The operators A_F and A_G associated with them are commutative and $A_F A_G = A_H$, where $H = H(x)$ is the convolution of the distribution functions $F(x)$ and $G(x)$, i.e.*

$$H(x) = \int\limits_{-\infty}^{+\infty} F(x - y)\, dG(y).$$

PROOF. Clearly

$$A_F A_G f = \int_{-\infty}^{+\infty} \Big(\int_{-\infty}^{+\infty} f(x + y + z)\, dG(z) \Big)\, dF(y) = \int_{-\infty}^{+\infty} f(x + u)\, dH(u).$$

LEMMA 3. *Let A be a contraction operator and B an arbitrary operator, then*

$$\| ABf \| \le \| Bf \|.$$

PROOF. The statement follows from the definition of the contraction operator.

LEMMA 4. *If $U_1, U_2, \ldots, U_n$ and $V_1, V_2, \ldots, V_n$ are operators associated with probability distributions and if $f \in C_3$, then*

$$\| U_1 U_2 \ldots U_n f - V_1 V_2 \ldots V_n f \| \le \sum_{k=1}^{n} \| U_k f - V_k f \|. \tag{3}$$

PROOF. Clearly, for arbitrary linear operators we have the identity

$$U_1 U_2 \ldots U_n - V_1 V_2 \ldots V_n = \sum_{k=1}^{n} U_1 U_2 \ldots U_{k-1} (U_k - V_k) V_{k+1} \ldots V_n. \tag{4}$$

By Lemmas 1, 2 and 3 we get immediately (3).

Now we can begin the proof of the central limit theorem under the Liapunov conditions. Instead of Theorem 2 of § 1 we prove the following somewhat more general theorem:

THEOREM 1. *Let $\xi_{n1}, \xi_{n2}, \ldots, \xi_{nn}$ be independent random variables with finite variances $D_{nk}^2 = D^2(\xi_{nk})$ and third absolute central moments $H_{nk}^3 = E(|\xi_{nk} - E(\xi_{nk})|^3)$. Put $\zeta_n = \sum_{k=1}^{n} \xi_{nk}$ and $\zeta_n^* = \dfrac{\zeta_n - E(\zeta_n)}{D(\zeta_n)}$, further*

$$S_n = D(\zeta_n) = \sqrt{\sum_{k=1}^{n} D_{nk}^2} \quad and \quad K_n = \sqrt[3]{\sum_{k=1}^{n} H_{nk}^3}.$$

Let $F_k(x)$ denote the distribution function of ζ_n^*. If Liapunov's condition

$$\lim_{n \to \infty} \frac{K_n}{S_n} = 0 \tag{5}$$

is fulfilled, then

$$\lim_{n \to \infty} F_n(x) = \Phi(x) = \frac{1}{\sqrt{2\pi}} \int_{-\infty}^{x} e^{-\frac{u^2}{2}}\, du. \tag{6}$$

PROOF. Without restricting generality we may assume $E(\xi_{nk}) = 0$ $(k = 1, 2, \ldots, n)$. Let U_{nk} denote the operator associated with the distribution function $F_{nk}(x)$ of the random variable $\dfrac{\xi_{nk}}{S_n}$. Further let V_{nk} denote the associated operator of the normal distribution with expectation 0 and standard deviation $\dfrac{D_{nk}}{S_n}$. Then $U_{n1} U_{n2} \ldots U_{nn}$ is nothing else than the operator associated with the distribution function $F_n(x)$ and $V_{n1}, V_{n2} \ldots V_{nn}$ is the operator A_Φ associated with the standard normal distribution function $\Phi(x)$ of expectation 0 and standard deviation 1. If we apply Lemma 4, we obtain that for any $f \in C_3$

$$\| A_{F_n} f - A_\Phi f \| \leq \sum_{k=1}^{n} \| U_{nk} f - V_{nk} f \|. \tag{7}$$

Now

$$U_{nk} f = \int_{-\infty}^{+\infty} f(x + y)\, dF_{nk}(y) \quad \text{and} \quad V_{nk} f = \int_{-\infty}^{+\infty} f(x + y)\, d\Phi\left(\frac{yS_n}{D_{nk}}\right).$$

Since $f \in C_3$, $f(x + y)$ can be expanded into a finite Taylor series up to three terms

$$f(x + y) = f(x) + yf'(x) + \frac{y^2}{2} f''(x) + \frac{y^3}{6} f'''(x + \theta y), \tag{8}$$

where $0 < \theta < 1$; of course θ depends on x and y. Thus, taking into account that

$$\int_{-\infty}^{+\infty} dF_{nk}(y) = 1, \quad \int_{-\infty}^{+\infty} y\, dF_{nk}(y) = 0, \quad \text{and} \quad \int_{-\infty}^{+\infty} y^2\, dF_{nk}(y) = \frac{D_{nk}^2}{S_n^2},$$

we obtain

$$U_{nk} f = f(x) + \frac{f''(x)}{2} \cdot \frac{D_{nk}^2}{S_n^2} + \frac{1}{6} \int_{-\infty}^{+\infty} y^3 f'''(x + \theta y)\, dF_{nk}(y) \tag{9}$$

and

$$V_{nk} f = f(x) + \frac{f''(x)}{2} \frac{D_{nk}^2}{S_n^2} + \frac{1}{6} \int_{-\infty}^{+\infty} y^3 f'''(x + \theta y)\, d\Phi\left(\frac{yS_n}{D_{nk}}\right). \tag{10}$$

Hence, if $\sup |f'''(x)| = M$, we get

$$\| U_{nk}f - V_{nk}f \| \le \frac{M}{6} \left(\int_{-\infty}^{+\infty} |y|^3 \, dF_{nk}(y) + \int_{-\infty}^{+\infty} |y|^3 \, d\Phi \left(\frac{yS_n}{D_{nk}} \right) \right). \quad (11)$$

Since

$$\int_{-\infty}^{+\infty} |y|^3 \, dF_{nk}(y) = \left(\frac{H_{nk}}{S_n} \right)^3, \quad (12)$$

and

$$\int_{-\infty}^{+\infty} |y|^3 \, d\Phi \left(\frac{yS_n}{D_{nk}} \right) = \left(\frac{D_{nk}}{S_n} \right)^3 \int_{-\infty}^{+\infty} |y|^3 \, d\Phi(y) \le 2 \left(\frac{D_{nk}}{S_n} \right)^3, \quad (13)$$

there follows

$$\| U_{nk}f - V_{nk}f \| \le \frac{M}{6S_n^3} (H_{nk}^3 + 2D_{nk}^3). \quad (14)$$

Because of the Hölder inequality one has for every random variable

$$E(\xi^2)^{\frac{1}{2}} \le E(|\xi|^3)^{\frac{1}{3}}, \quad (15)$$

hence

$$D_{nk} \le K_{nk}, \quad (16)$$

and thus

$$\sum_{k=1}^{n} D_{nk}^3 \le K_n^3. \quad (17)$$

(7) and (14) lead to

$$\| A_{F_n}f - A_\Phi f \| \le \frac{M}{2} \left(\frac{K_n}{S_n} \right)^3, \quad (18)$$

and thus by (5)

$$\lim_{n \to \infty} \| A_{F_n}f - A_\Phi f \| = 0. \quad (19)$$

Thus we proved that if $f \in C_3$, then for any value of x (and even uniformly in x)

$$\lim_{n \to \infty} \int_{-\infty}^{+\infty} f(x+y) \, dF_n(y) = \int_{-\infty}^{+\infty} f(x+y) \, d\Phi(y). \quad (20)$$

From this follows that (6) holds for every x. Indeed if $\varepsilon > 0$ is arbitrary, let $f_\varepsilon(x)$ be a function belonging to C_3 with the following properties: $f_\varepsilon(x) = 1$ if $x \leq 0, f_\varepsilon(x) = 0$ if $x \geq \varepsilon$ and $f_\varepsilon(x)$ is decreasing if x lies between 0 and ε. Such a function can be given readily, e.g. the following function has all the required properties:

$$f_\varepsilon(x) = \begin{cases} 1 & \text{for } x \leq 0, \\ \left(1 - \left(\dfrac{x}{\varepsilon}\right)^4\right)^4 & \text{for } 0 \leq x \leq \varepsilon, \\ 0 & \text{for } \varepsilon \leq x. \end{cases} \tag{21}$$

Then

$$\Phi(x + \varepsilon) \geq \int_{-\infty}^{+\infty} f_\varepsilon(x + y)\,d\Phi(y) \geq \Phi(x), \tag{22}$$

and

$$F_n(x + \varepsilon) \geq \int_{-\infty}^{+\infty} f_\varepsilon(x + y)\,dF_n(y) \geq F_n(x). \tag{23}$$

Hence

$$\limsup_{n \to \infty} F_n(x) \leq \Phi(x + \varepsilon), \tag{24}$$

and

$$\liminf_{n \to \infty} F_n(x + \varepsilon) \geq \Phi(x). \tag{25}$$

If we apply (25) to $x - \varepsilon$ instead of x, we obtain

$$\liminf_{n \to \infty} F_n(x) \geq \Phi(x - \varepsilon), \tag{26}$$

i.e. we obtain from (24) and (26) that

$$\Phi(x - \varepsilon) \leq \liminf_{n \to \infty} F_n(x) \leq \limsup_{n \to \infty} F_n(x) \leq \Phi(x + \varepsilon). \tag{27}$$

Since (27) is valid for every positive ε, it follows that (6) is fulfilled for every x. Theorem 1 is herewith proved.

Now we pass to the proof of the Lindeberg theorem by the operator method. We prove the theorem in its most general form, i.e. we present the proof of Theorem 4 of § 1.

PROOF OF THEOREM 4, § 1 BY THE OPERATOR METHOD. We may assume without restriction of generality that $M_{nk} = E(\xi_{nk}) = 0$ $(k = 1, 2, \ldots, n)$. Put $\zeta_n = \sum_{k=1}^{n} \xi_{nk}$, and let $F_n(x)$ denote the distribution function of ζ_n.

We prove that for every $f \in C_3$

$$\lim_{n \to \infty} A_{F_n} f = A_\Phi f. \tag{28}$$

As we have seen in the proof of Theorem 1, it then follows that for every real x

$$\lim_{n \to \infty} F_n(x) = \Phi(x).$$

Let U_{nk} denote the operator associated with the distribution function $F_{nk}(x)$ of the random variable ξ_{nk} and V_{nk} the operator associated to the normal distribution with expectation 0 and standard deviation D_{nk}. Then according to our assumptions

$$A_{F_n} = U_{n1} U_{n2} \ldots U_{nn}, \quad \text{and} \quad A_\Phi = V_{n1} V_{n2} \ldots V_{nn}. \tag{29}$$

Further by Lemma 4 for every $f \in C_3$

$$\| A_{F_n} f - A_\Phi f \| \le \sum_{k=1}^{n} \| U_{nk} f - V_{nk}) f \|. \tag{30}$$

Now if we expand $f(x + y)$ into a finite Taylor series up to the second and third term respectively, we get

$$f(x + y) = f(x) + yf'(x) + \frac{y^2}{2} f''(x + \theta_1 y), \tag{31}$$

and

$$f(x + y) = f(x) + yf'(x) + \frac{y^2}{2} f''(x) + \frac{y^3}{6} f'''(x + \theta_2 y), \tag{32}$$

where $0 < \theta_1 < 1$ and $0 < \theta_2 < 1$; of course θ_1 and θ_2 depend on x and y. Let $\varepsilon > 0$. Clearly

$$U_{nk} f = \int_{-\varepsilon}^{+\varepsilon} f(x + y) \, dF_{nk}(x) + \int_{|y| > \varepsilon} f(x + y) \, dF_{nk}(x). \tag{33}$$

Use in the first integral on the right hand side of (33) the equality (32) and in the second integral (31). We obtain

$$U_{nk} f = f(x) + \frac{1}{2} f''(x) D_{nk}^2 + \frac{1}{6} \int_{|y| \le \varepsilon} y^3 f'''(x + \theta_2 y) \, dF_{nk}(y) +$$

$$+ \frac{1}{2} \int_{|y| > \varepsilon} y^2 \left(f''(x + \theta_1 y) - f''(x) \right) dF_{nk}(y). \tag{34}$$

Put

$$\sup |f''(x)| = M_1 \quad \text{and} \quad \sup |f'''(x)| = M_2,$$

then

$$\left| U_{nk}f - f(x) - \frac{1}{2} D_{nk}^2 f''(x) \right| \leq \frac{1}{6} \varepsilon M_2 D_{nk}^2 + M_1 \int_{|y|>\varepsilon} y^2 \, dF_{nk}(x). \tag{35}$$

On the other hand

$$V_{nk}f = f(x) + \frac{1}{2} D_{nk}^2 f''(x) + \frac{1}{6} \int_{-\infty}^{+\infty} y^3 f'''(x + \theta_2 y) \, d\Phi\left(\frac{y}{D_{nk}}\right), \tag{36}$$

hence

$$\left| V_{nk}f - f(x) - \frac{1}{2} D_{nk}^2 f''(x) \right| \leq \frac{1}{6} M_2 D_{nk}^3 \int_{-\infty}^{+\infty} |y|^3 \, d\Phi(y) \leq \frac{M_2 D_{nk}^3}{3}. \tag{37}$$

(35) and (37) lead to

$$\sum_{k=1}^{n} \| U_{nk}f - V_{nk}f \| \leq \frac{1}{6} \varepsilon M_2 + M_1 \sum_{k=1}^{n} \int_{|y|>\varepsilon} y^2 \, dF_{nk}(y) + \frac{M_2}{3} \sum_{k=1}^{n} D_{nk}^3. \tag{38}$$

Since by our assumption the Lindeberg condition is fulfilled, we have

$$\lim_{n \to \infty} \sum_{k=1}^{n} \int_{|y|>\varepsilon} y^2 \, dF_{nk}(y) = 0, \tag{39}$$

furthermore

$$D_{nk}^2 = \int_{|x|\leq\varepsilon} x^2 \, dF_{nk}(x) + \int_{|x|>\varepsilon} x^2 \, dF_{nk}(x) \leq \varepsilon^2 + \int_{|x|>\varepsilon} x^2 \, dF_{nk}(x),$$

therefore

$$\sum_{k=1}^{n} D_{nk}^3 \leq \max_{1\leq k\leq n} D_{nk} \leq \sqrt{\varepsilon^2 + \sum_{k=1}^{n} \int_{|x|>\varepsilon} x^2 \, dF_{nk}(x)},$$

and thus for every positive ε

$$\lim_{n \to \infty} \sup \sum_{k=1}^{n} D_{nk}^3 \leq \varepsilon,$$

that is

$$\lim_{n \to \infty} \sum_{k=1}^{n} D_{nk}^3 = 0, \tag{40}$$

hence (30) and (38) lead to (28). Theorem 4 of § 1 is herewith proved.

As a further illustration of the operator method we prove now a theorem concerning the convergence to the Poisson distribution.

THEOREM 2. *Let ξ_{nk} ($k = 1, 2, \ldots, n$) be independent random variables which assume only the values 0 and 1; put further*

$$P(\xi_{nk} = 1) = p_{nk}. \tag{41}$$

Put

$$\lambda_n = \sum_{k=1}^{n} p_{nk} \tag{42}$$

and suppose that

$$\lim_{n \to \infty} \lambda_n = \lambda \tag{43}$$

and

$$\lim_{n \to \infty} \max_{1 \leq k \leq n} p_{nk} = 0. \tag{44}$$

Then the distribution of

$$\zeta_n = \xi_{n1} + \xi_{n2} + \ldots + \xi_{nn} \tag{45}$$

converges to the Poisson distribution with expectation λ.

Remark. Theorem 2 is a particular case of Theorem 1 of § 4; the latter can be proved in a similar manner. Merely for simplicity's sake we restrict ourselves to the proof of Theorem 2.

PROOF. Let K denote the set of all real-valued bounded functions $f(x)$ ($x = 0, 1, 2, \ldots$) defined on the nonnegative integers. Put $\|f\| = \sup|f(x)|$.

Let there be associated with every probability distribution $\mathscr{P} = \{p_0, p_1, \ldots, p_n, \ldots\}$ an operator defined by

$$A_{\mathscr{P}} f = \sum_{r=0}^{\infty} f(x + r) p_r \tag{46}$$

for every $f \in K$. Clearly, $A_{\mathscr{P}}$ maps the set K into itself, $A_{\mathscr{P}}$ is a linear contraction operator, further if $\mathscr{P}$ and Q are any two distributions defined on the nonnegative integers, then $A_{\mathscr{P}} A_{Q} = A_{R}$ where $\mathscr{R} = \mathscr{P} \cdot Q$, i.e. $\mathscr{R}$ is the convolution of the distributions $\mathscr{P}$ and Q; that is, if $\mathscr{P} = \{p_n\}$ and $Q = \{q_n\}$, then $\mathscr{R} = \{r_n\}$, where

$$r_n = \sum_{k=0}^{n} p_k q_{n-k}.$$

Let U_{nk} denote the operator associated with the distribution $\mathscr{P}_{nk}$ of the random variable ξ_{nk} and V_{nk} the operator associated with the Poisson distribution with parameter p_{nk}. Then $U_{n1} U_{n2} \ldots U_{nn}$ is nothing else than the operator $A_{\mathscr{P}_n}$ associated with the distribution $\mathscr{P}_n$ of the random variable ζ_n, while $V_{n1} V_{n2} \ldots V_{nn}$ is the operator Q_{λ_n} associated with the Poisson distribution with parameter λ_n (taking into account that if Q_λ is the Poisson distribution with parameter λ, then $Q_\lambda Q_\mu = Q_{\lambda+\mu}$).

In order to prove Theorem 2 it suffices to show that for every element f of K the relation

$$\lim_{n \to \infty} \| A\mathscr{P}_n f - AQ_{\lambda_n} f \| = 0 \tag{47}$$

holds. In fact, if (47) holds for every $f \in K$, choose for f the function for which $f(0) = 1$ and $f(x) = 0$ for $x \geq 1$; then it follows from (47) that for every r (and even uniformly in r)

$$\lim_{n \to \infty} \left(P(\zeta_n = r) - \frac{\lambda_n^r e^{-\lambda_n}}{r!} \right) = 0, \tag{48}$$

and since by our assumption $\lambda_n \to \lambda$, it follows from (48) that

$$\lim_{n \to \infty} P(\zeta_n = r) = \frac{\lambda^r e^{-\lambda}}{r!} \qquad (r = 0, 1, \ldots). \tag{49}$$

Now Lemma 4 is valid in this case too, and applying it we obtain

$$\| A\mathscr{P}_n f - AQ_{\lambda_n} f \| \leq \sum_{k=1}^{n} \| U_{nk} f - V_{nk} f \|. \tag{50}$$

On the other hand

$$U_{nk} f - V_{nk} f = f(x)(1 - p_{nk} - e^{-p_{nk}}) +$$

$$+ f(x+1)(p_{nk} - p_{nk} e^{-p_{nk}}) - \sum_{r=2}^{\infty} f(x+r) \frac{p_{nk}^r e^{-p_{nk}}}{r!}, \tag{51}$$

and thus

$$\| U_{nk} f - V_{nk} f \| \leq \|f\| \left(e^{-p_{nk}} - (1 - p_{nk}) + p_{nk}(1 - e^{-p_{nk}}) + \right.$$

$$+ \left[1 - e^{-p_{nk}}(1 + p_{nk}) \right]). \tag{52}$$

Since

$$1 - x \leq e^{-x} \leq 1 - x + x^2 \quad \text{for} \quad x \leq 1, \tag{53}$$

there follows

$$\| U_{nk} f - V_{nk} f \| \leq 3 \|f\| p_{nk}^2. \tag{54}$$

Thus (50) and (54) lead to

$$\| A\mathscr{P}_n f - AQ_{\lambda_n} f \| \leq 3 \|f\| \cdot \sum_{k=1}^{n} p_{nk}^2. \tag{55}$$

Because of

$$\sum_{k=1}^{n} p_{nk}^2 \leq \lambda_n \cdot \max_{1 \leq k \leq n} p_{nk}, \tag{56}$$

by (44) it follows that

$$\lim_{n\to\infty} \|A_{\mathscr{P}_n f} - AQ_{\lambda_n}f\| = 0. \tag{57}$$

As we have seen, the assertion of Theorem 2 follows.

Finally, we give a proof by the operator method of the theorem proved in § 3. Just like there, we assume that the distribution in question is continuous and symmetric with respect to the point 0, i.e. we prove the following

THEOREM 3. *Let* $\xi_1, \xi_2, \ldots, \xi_n, \ldots$ *be independent identically distributed random variables; let their common distribution function be denoted by* $F(x)$. *Suppose that* $F(x)$ *is continuous and the distribution is symmetric with respect to the point 0, i.e.* $F(-x) = 1 - F(x)$ $(x \geq 0)$. *Assume further that*

$$\lim_{y\to\infty} \frac{y^2(1 - F(y))}{\int_0^y x^2 \, dF(x)} = 0. \tag{58}$$

Put $\zeta_n = \xi_1 + \xi_2 + \ldots + \xi_n$. *Then there exists a sequence of numbers* S_n *such that for every* x

$$\lim_{n\to\infty} P\left(\frac{\zeta_n}{S_n} < x\right) = \Phi(x). \tag{59}$$

PROOF. Put

$$\delta(y) = \frac{y^2(1 - F(y))}{\int_0^y x^2 \, dF(x)}, \tag{60}$$

then by assumption

$$\lim_{y\to+\infty} \delta(y) = 0. \tag{61}$$

Put further

$$\Delta(y) = \frac{\delta(y)}{(1 - F(y))^2} = \frac{y^2}{(1 - F(y))\int_0^y x^2 \, dF(x)}, \tag{62}$$

then, as was shown in § 3,

$$\lim_{n\to\infty} \Delta(y) = +\infty. \tag{63}$$

By our assumption $\Delta(y)$ is continuous for $y \geq y_0$. Let C_n denote the least positive number for which

$$\Delta(C_n) = n^2, \tag{64}$$

then $C_n \to \infty$, furthermore

$$n(1 - F(C_n)) = \sqrt{\delta(C_n)}. \tag{65}$$

Put

$$S_n^2 = n \int\limits_{-C_n}^{+C_n} x^2 \, dF(x) = \frac{2C_n^2}{\sqrt{\delta(C_n)}}, \tag{66}$$

then

$$\frac{C_n}{S_n} = \frac{\sqrt[4]{\delta(C_n)}}{\sqrt{2}}. \tag{67}$$

Now let U_{nk} be the operator associated with the distribution of the random variable $\frac{\zeta_n}{S_n}$ and V_{nk} the operator associated with the normal distribution with expectation 0 and standard deviation $\frac{1}{\sqrt{n}}$ ($k = 1, 2, \ldots, n$). Then $U_{n1} U_{n2} \ldots U_{nn}$ is the operator associated with the distribution function $F_n(x)$ of the random variable ζ_n, while $V_{n1} V_{n2} \ldots V_{nn}$ is the operator associated with the standard normal distribution function $\Phi(x)$ (having expectation 0 and standard deviation 1). Thus by Lemma 4 for every $f \in C_3$ we have

$$\| A_{F_n}f - A_\Phi f \| \le \sum_{k=1}^{n} \| U_{nk}f - V_{nk}f \| = n \| U_{n1}f - V_{n1}f \|.$$

Hence it suffices to prove that

$$\lim_{n \to \infty} n \| U_{n1}f - V_{n1}f \| = 0. \tag{68}$$

Now

$$U_{n1}f = \int\limits_{-\frac{C_n}{S_n}}^{+\frac{C_n}{S_n}} f(x + y) \, dF(S_n y) + \int\limits_{|y| > \frac{C_n}{S_n}} f(x + y) \, dF(S_n y). \tag{69}$$

If $\sup |f(x)| = A$, then

$$\left| \int\limits_{|y| > \frac{C_n}{S_n}} f(x + y) \, dF(S_n y) \right| \le A(1 - F(C_n)) = \frac{A\sqrt{\delta(C_n)}}{n}. \tag{70}$$

On the other hand, if in the integral on the right hand side of (69) $f(x + y)$ is expanded into a Taylor series up to the third term and if it is taken into account that by our assumption the distribution with the distribution function $F(y)$ is symmetric with respect to the point 0, then we have

$$\int\limits_{-\frac{C_n}{S_n}}^{+\frac{C_n}{S_n}} f(x + y) \, dF(S_n y) = f(x)\left(1 - 2(1 - F(C_n))\right) + \frac{f''(x)}{2n} + R_n, \tag{71}$$

where

$$R_n = \frac{1}{6S_n^3} \int_{-C_n}^{+C_n} y^3 f'''(x + \theta y) \, dF(y), \tag{72}$$

and $0 < \theta < 1$.

If $\sup |f'''(x)| = B$. then

$$|R_n| \leq \frac{B}{6 S_n^3} \int_{-C_n}^{+C_n} |y|^3 \, dF(y) \leq \frac{BC_n}{3nS_n} = \frac{B}{3n} \frac{\sqrt[4]{\delta(C_n)}}{\sqrt{2}}. \tag{73}$$

On the other hand

$$V_{n1} f = f(x) + \frac{f''(x)}{2n} + O\left(\frac{1}{n^{\frac{3}{2}}}\right), \tag{74}$$

thus (69)–(74) lead to

$$n \| U_{n1} f - V_{n1} f \| \leq 3A \sqrt{\delta(C_n)} + \frac{B}{3\sqrt{2}} \sqrt[4]{\delta(C_n)} + O\left(\frac{1}{\sqrt{n}}\right), \tag{75}$$

hence because of $\delta(C_n) \to 0$ the validity of (68) follows. Herewith Theorem 3 is proved.

Finally we make some remarks concerning the relation between operator and characteristic function methods.

The convergence of a sequence F_n of distribution functions to a distribution function F is proved by the operator method by showing that for every $f \in C_3$ one has $A_{F_n} f \to A_F f$. This implies that the characteristic function φ_n of the distribution function F_n tends to the characteristic function φ of the distribution function F; in fact, if $f(x) = e^{itx}$, then $f \in C_3$ and $A_{F_n} f = e^{itx} \int_{-\infty}^{+\infty} e^{ity} dF_n(y)$, hence $A_{F_n} f = e^{itx} \varphi_n(t)$ and, similarly, $A_F f = e^{itx} \varphi(t)$.

Hence, from the fact that for every $f \in C_3$

$$A_{F_n} f \to A_F f \tag{76}$$

it follows that for every real t $\varphi_n(t) \to \varphi(t)$.

Therefore the operator method proves slightly more than the characteristic function method. In effect, we prove for every $f \in C_3$ the validity of (76) and even that (76) is fulfilled uniformly in x. This makes the proof of the relation $F_n(x) \to F(x)$ simpler, because while the implication of the relation $F_n(x) \to F(x)$ by the relation $\varphi_n(t) \to \varphi(t)$ is a comparatively deep theorem (the so-called continuity theorem of characteristic functions, cf. Theorem 3 of Chapter VI, § 4) it is quite easy to see that (76) implies $F_n(x) \to F(x)$ (for every x which is a continuity point of $F(x)$). On the other hand, the method by which we proved (76) in each of the above discussed cases, can be applied for distributions of sums of *independent* random variables only, while the method of characteristic functions can be applied in other cases too (cf. e.g. § 5 or Exercise 26 of § 13).

§ 13. Exercises

1. Prove Theorem 2' of Chapter VI, § 5 by means of the central limit theorem (Chapter VIII, § 1, Theorem 1).

Hint. If $F(x)$ is a distribution function with expectation 0 and variance 1 such that

$$F\left(\frac{x}{\sigma_1}\right) * F\left(\frac{x}{\sigma_2}\right) = F\left(\frac{x}{\sqrt{\sigma_1^2 + \sigma_2^2}}\right),$$

then $F(x)$ is equal to the n-fold convolution of $F(x\sqrt{n})$. This converges to the normal distribution as $n \to +\infty$.

2. Let $\xi_1, \xi_2, \ldots, \xi_n, \ldots$ be independent random variables and suppose

$$P(\xi_n = a_n) = P(\xi_n = -a_n) = \frac{1}{2} \qquad (n = 1, 2, \ldots).$$

Under what conditions on the positive numbers a_n does Liapunov's condition of the central limit theorem hold for the random variables ξ_n?

Hint. Put

$$S_n = \sqrt{\sum_{k=1}^{n} a_k^2}, \quad K_n = \sqrt[3]{\sum_{k=1}^{n} a_k^3}, \quad \text{and} \quad m_n = \max_{1 \le k \le n} a_k.$$

It follows that

$$\frac{m_n}{S_n} \le \frac{K_n}{S_n} \le \left(\frac{m_n}{S_n}\right)^{\frac{1}{3}},$$

and Liapunov's condition $\lim\limits_{n \to +\infty} \dfrac{K_n}{S_n} = 0$ is fulfilled, iff $\lim\limits_{n \to +\infty} \dfrac{m_n}{S_n} = 0$.

3.a) Let ξ_λ be a random variable having a Poisson distribution with expectation λ. Show by the method of characteristic functions that the distribution function of the random variable $\dfrac{\xi_\lambda - \lambda}{\sqrt{\lambda}}$ tends to the normal distribution function as $\lambda \to +\infty$ (cf. also Ch. III, § 18, Exercise 28).

b) Let ξ_n be a random variable having a gamma distribution of order n with $E(\xi_n) = \dfrac{n}{\lambda}$. Show that the distribution function of $\sqrt{n}\left(\dfrac{\lambda \xi_n}{n} - 1\right)$ tends to the normal distribution function with expectation 0 and standard deviation 1.

c) Let ξ_n be a random variable having a beta distribution of order (np, nq). Show

that the distribution function of the random variable $\sqrt{\dfrac{n}{pq}}\,(p+q)^{\frac{3}{2}}\left(\xi_n - \dfrac{p}{p+q}\right)$
tends to the normal distribution function with expectation 0 and standard deviation 1 as $n \to \infty$.

4. Let $\varepsilon_n(x)$ denote the n-th digit in the decimal expansion of x $(0 \le x \le 1)$; the values of $\varepsilon_n(x)$ are thus the numbers $0, 1, \ldots, 9$. Put $S_n(x) = \sum\limits_{k=1}^{n} \varepsilon_k(x)$. If $E_n(y)$ is the set of the numbers x for which $\dfrac{2S_n(x) - 9n}{\sqrt{33n}} < y$, and if $|E_n(y)|$ denotes the Lebesgue measure of $E_n(y)$, show that

$$\lim_{n \to +\infty} |E_n(y)| = \frac{1}{\sqrt{2\pi}} \int_{-\infty}^{y} e^{-\frac{u^2}{2}}\, du.$$

Hint. We choose a point η at random in $(0, 1)$; i.e. η is a random variable uniformly distributed in the interval $(0, 1)$. The random variables $\xi_n = \varepsilon_n(\eta)$ are then independent and identically distributed; the central limit theorem can be applied. We have:

$$E(\xi_n) = \frac{9}{2}, \quad D(\xi_n) = \frac{\sqrt{33}}{2}.$$

5. Let $q_1, q_2, \ldots, q_n, \ldots$ be a sequence of integers ≥ 2. It is easy to show that every number x $(0 \le x \le 1)$ (a denumerable set of numbers expected) can be represented in one and only one way in the following form:

$$x = \sum_{n=1}^{\infty} \frac{\varepsilon_n(x)}{q_1 q_2 \ldots q_n},$$

where $\varepsilon_n(x)$ may take on the values $0, 1, \ldots, q_n - 1$. As in Exercise 4 put $S_n(x) = \sum\limits_{k=1}^{n} \varepsilon_k(x)$. Now if $E_n(y)$ denotes the set of numbers x $(0 < x < 1)$ such that

$$\frac{S_n(x) - \dfrac{1}{2}\sum\limits_{k=1}^{n}(q_k - 1)}{\sqrt{\dfrac{1}{12}\sum\limits_{k=1}^{n}(q_k^2 - 1)}} < y$$

and $|E_n(y)|$ is the Lebesgue measure of $E_n(y)$, then we have

$$\lim_{n \to +\infty} |E_n(y)| = \frac{1}{\sqrt{2\pi}} \int_{-\infty}^{y} e^{-\frac{u^2}{2}}\, du,$$

provided that the condition

$$\lim_{n \to +\infty} \frac{\max\limits_{1 \le k \le n} q_k}{\sqrt{\sum\limits_{k=1}^{n} q_k^2}} = 0$$

is fulfilled.

Hint. Choose at random a point $\eta \in (0, 1)$ and put $\xi_n = \varepsilon_n(\eta)$. It is easy to see that the random variables ξ_n are independent. Furthermore

$$E(\xi_n) = \frac{q_n - 1}{2}, \quad D^2(\xi_n) = \frac{q_n^2 - 1}{12}, \quad \text{and} \quad E(|\xi_n - E(\xi_n)|^3) < C_1 q_n^3,$$

where C_1 is a positive constant. Liapunov's condition is thus satisfied.

6. Let $\zeta_1, \zeta_2, \ldots, \zeta_n, \ldots$ be independent random variables, with the same normal distribution. Put

$$\bar{\xi}_n = \frac{1}{n} \sum_{k=1}^{n} \xi_k, \quad \sigma_n = \sqrt{\frac{\sum_{k=1}^{n} (\xi_k - \bar{\xi}_n)^2}{n - 1}}, \quad \text{and} \quad \tau_n = \frac{\bar{\xi}_n}{\sigma_n} \sqrt{n}.$$

Show that

$$\lim_{n \to +\infty} P(\tau_n < x) = \frac{1}{\sqrt{2\pi}} \int_{-\infty}^{x} e^{-\frac{u^2}{2}} du.$$

Hint. The distribution of τ_n is Student's distribution with $n - 1$ degrees of freedom (cf. Ch. IV, § 10). Its density function is

$$S_{n-1}(x) = \frac{1}{\sqrt{(n-1)\pi}} \frac{\Gamma\left(\frac{n}{2}\right)}{\Gamma\left(\frac{n-1}{2}\right)} \left(1 + \frac{x^2}{n-1}\right)^{-\frac{n}{2}},$$

and we find that

$$\lim_{n \to +\infty} S_{n-1}(x) = \frac{1}{\sqrt{2\pi}} e^{-\frac{x^2}{2}}.$$

Another proof can be obtained by noticing that the distribution function of $\bar{\xi}_n \sqrt{n}$ tends to the normal distribution function as $n \to +\infty$ and $\lim_{n \to +\infty} \text{st } \sigma_n = 1$; the result follows then from the lemma of § 7.

7. Prove that any subsequence of a Markov chain is also a Markov chain.

8. (Ehrenfest's model of heat conduction.) Consider two urns and N balls labelled from 1 to N. The balls are arbitrarily distributed between the two urns; assume that the first contains M and the second $N - M$ balls. We put into a box N cards labelled also from 1 to N. We draw a card from the box and put the ball bearing the same number from the urn in which it is contained into the other urn. After this the card is replaced into the box and the operation is repeated. Let ζ_n denote the number of the balls in the first urn after the n-th step (i.e. after drawing n cards) ($n = 1, 2, \ldots$; $\zeta_0 = M$). The states of the system consisting of the two urns form a Markov chain. The transition probabilities are

$$P_{k,k+1} = 1 - \frac{k}{N} \qquad (k = 0, 1, \ldots, N - 1),$$

$$P_{k,k-1} = \frac{k}{N} \qquad (k = 1, 2, \ldots, N) \tag{1}$$

$$P_{k,l} = 0 \text{ for } |k - l| \neq 1.$$

Show that

$$E(\zeta_n) = \frac{N}{2} + \left(M - \frac{N}{2}\right)\left(1 - \frac{2}{N}\right)^n.$$

(This example contains the statistical justification of Newton's law of cooling.)

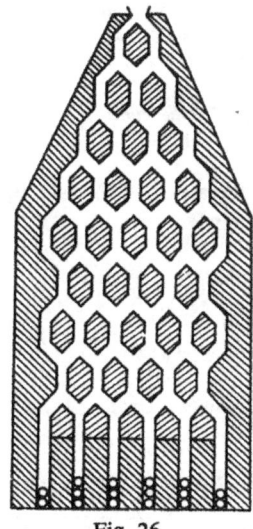

Fig. 26

9. Let Galton's desk be modified in the following manner (cf. Fig. 26): From the N-th row on the number of pegs is alternatingly equal to the number in the $(N - 1)$-th row and in the N-th row. On the whole desk there are $N + n$ rows of pegs. Determine the distribution of the balls in the containers when the number n of balls is large.

10. The random variables $\zeta_0, \zeta_1, \ldots, \zeta_n \ldots$ form a homogeneous Markov chain; all ζ_n take on values in $(0, 1)$; let the conditional distribution of ζ_{n+1} under the condition $\zeta_n = y$ be absolutely continuous for every value of y $(0 \leq y \leq 1)$; let $p(x, y)$ be the corresponding conditional density function. We assume that for $0 \leq x \leq 1$ and $0 \leq y \leq 1$ the function $p(x, y)$ is always positive and that for every x $(0 \leq x \leq 1)$ $\int_0^1 p(x, y) \, dy = 1$ holds, further that $p(x, y)$ is continuous. Let $p_n(x, y)$ be the conditional density function of ζ_n under the condition $\zeta_0 = y$. Show that the relation

$$\lim_{n \to +\infty} p_n(x, y) = 1$$

is valid uniformly in x and y.

11. Let a moving point perform a random walk on a plane regular triangular lattice. If the moving point is at the moment $t = n$ at an arbitrary lattice-point, it may pass at the moment $t = n + 1$ with the same probability to any of the 6 neighbouring lattice points. Show that the moving point will return with probability 1 to its initial position, but that the expectation of the time passing until this return is infinite.

The following Exercises 12 through 18 all deal with homogeneous Markov chains with a finite number of states, fulfilling the conditions of Theorem 1, § 8. The notations are the same. The states are denoted by $A_0, A_1, \ldots, A_N$. The random variable ξ_n is equal to k if the system is in the state A_k at the time n $(k = 0, 1, \ldots, N)$. We put $P(\xi_0 = j) = P_0(j)$, $P_{jk}^{(n)} = P(\xi_{n+m} = k \mid \xi_m = j)$, $p_{jk}^{(1)} = p_{jk}$ and $P_n(k) = P(\xi_n = k)$. We assume that $\min_{l,k} p_{lk} = d > 0$. According to Theorem 1 of § 8 the limits $\lim_{n \to \infty} p_{jk}^{(n)} = P_k$ exist and are independent of j. Furthermore $\sum_{k=0}^{N} P_k = 1$.

12. Let

$$\eta_n^{(k)} = \begin{cases} 1 & \text{if the system is in state } A_k \text{ at the time } t = n, \\ 0 & \text{otherwise.} \end{cases}$$

We put $\zeta_n^{(k)} = \sum_{j=1}^{n} \eta_j^{(k)}$. Show that

$$\lim_{n \to +\infty} \text{st} \, \frac{\zeta_n^{(k)}}{n} = P_k \qquad (k = 0, 1, \ldots, N)$$

i.e. the system passes approximately a fraction P_k of the whole time in the state A_k.

Hint. We have

$$E(\eta_n^{(k)}) = P(\xi_n = k) = P(k),$$

$$D(\eta_n^{(k)}) = \sqrt{P_n(k)\left(1 - P_n(k)\right)}$$

and

$$E(\eta_n^{(k)} \eta_{n+r}^{(k)}) - E(\eta_n^{(k)}) E(\eta_{n+r}^{(k)}) = P_n(k)\left(p_{kk}^{(r)} - P_{n+r}(k)\right).$$

Furthermore, Formula (39) of § 8 implies $|p_{kk}^{(r)} - P_k| \leq (1 - d)^r$. Hence

$$R(\eta_n^{(k)}, \eta_{n+r}^{(k)}) \leq C(1 - d)^r,$$

where $R(\eta_n^{(k)}, \eta_{n+r}^{(k)})$ is the correlation coefficient of $\eta_n^{(k)}$ and $\eta_{n+r}^{(k)}$, and C is positive. Thus the result follows from Theorem 3 of Chapter VII, § 3.

13. Let $\zeta_n = \xi_1 + \xi_2 + \ldots + \xi_n$. Show that

$$\lim_{n \to +\infty} \text{st} \, \frac{\zeta_n}{n} = \sum_{k=1}^{N} k P_k.$$

Hint. ζ_n can be expressed as a function of the variables $\zeta_n^{(k)}$, namely $\zeta_n = \sum_{k=1}^{N} k \zeta_n^{(k)}$. The result follows from that of the preceding exercise.

14. Assume that at $t = 0$ the system is in the state A_k. It returns to it for the first time after a certain number of steps. Let this random number be denoted by $\nu^{(k)}$. Show that $P(\nu^{(k)} > n) < (1 - d)^n$.

Hint. We have $P(\nu^{(k)} > 1) = 1 - p_{kk}$; hence the inequality is true for $n = 1$. Suppose for a proof by induction that the inequality is true for n. Then

$$P(\nu^{(k)} > n + 1) = \sum_{h \neq k} \sum_{l \neq k} P(\nu^{(k)} > n, \zeta_n = j, \zeta_{n+1} = h),$$

hence

$$P(\nu^{(k)} > n + 1) = \sum_{l \neq k} P(\nu^{(k)} > n, \zeta_n = j) \sum_{h \neq k} p_{lh} \leq$$
$$\leq (1 - d) P(\nu^{(k)} > n) \leq (1 - d)^{n+1}.$$

15. Show that:
a) the expectation and the variance of the random variable $\nu^{(k)}$ defined in the previous exercise exist.

b) $E(\nu^{(k)}) = \dfrac{1}{P_k}$.

Hint. a) follows from Exercise 14. Let further $V_k(z)$ denote the generating function of $\nu^{(k)}$; we have

$$V_k(z) = 1 - \frac{1}{U_k(z)},$$

where

$$U_k(z) = 1 + \sum_{n=1}^{\infty} p_{kk}^{(n)} z^n.$$

The relations

$$U_k(z) = \frac{P_k z}{1 - z} + 1 + \sum_{n=1}^{\infty} (p_{kk}^{(n)} - P_k) z^n, \quad |p_{kk}^{(n)} - P_k| \leq (1 - d)^n$$

lead to

$$\lim_{z \to 1} U_k(z)(1 - z) = \lim_{z=1} U_k'(z)(1 - z)^2 = P_k,$$

which implies b).

16. Let the numbers $\mu_r^{(k)}$ $(r = 0, 1, \ldots)$ denote the values of n for which $\eta_n^{(k)} = 1$ $(\mu_0^{(k)} < \mu_1^{(k)} < \ldots)$; $\eta_n^{(k)}$ is defined here as in Exercise 12. Show that the standardized distribution of $\mu_r^{(k)}$ tends to the normal distribution as $r \to +\infty$.

Hint. The random variables $\nu_r^{(k)} = \mu_r^{(k)} - \mu_{r-1}^{(k)}$ are independent and identically distributed. According to the preceding exercise, the expectation and the variance of $\nu_r^{(k)}$ exist and are equal to those of $\nu^{(k)}$. Hence the central limit theorem can be applied.

17. Show that the distribution of the random variables $\zeta_n^{(k)}$ introduced in Exercise 12 tends, after standardization, to the normal distribution as $n \to +\infty$. (Generalization of Theorem 2 in § 8.)

Hint. It is easy to see that $P(\zeta_n^{(k)} < r) = P(\mu_r^{(k)} > n)$; for if the system passes less than r times through the state A_k during the first n steps, then it will return to it for the r-th time after the moment $t = n$, and conversely. Thus we are back to Exercise 16 and find

$$\lim_{n \to +\infty} P\left(\frac{\zeta_n^{(k)} - nP_k}{D_k P_k^{3/2} \sqrt{n}} < x \right) = \Phi(x),$$

where we have put $D_k = D(\nu^{(k)})$.

18. Put

$$P_{lk}^{(-n)} = P(\xi_n = k \mid \xi_{n+1} = j) \qquad (n = 1, 2, \ldots; j, k = 0, 1, \ldots, N).$$

Show that the limits $\lim_{n \to +\infty} P_{lk}^{(-n)} = P_{lk}^*$ exist and form a stochastic matrix.

Hint. We have

$$P_{lk}^{(-n)} = \frac{P(\xi_n = k)\, p_{kl}}{P(\xi_{n+1} = j)}.$$

Hence

$$\lim_{n \to +\infty} P_{lk}^{(-n)} = \frac{P_k p_{kl}}{P_l} = P_{lk}^*$$

and

$$\sum_{k=0}^{N} P_{lk}^* = \frac{1}{P_l} \sum_{k=0}^{N} P_k p_{kl}.$$

But the P_k satisfy the system of equations

$$\sum_{k=0}^{N} P_k p_{kl} = P_l \qquad (j = 0, 1, \ldots, N),$$

hence we find that $\sum_{k=0}^{N} P_{lk}^* = 1$; the transition probabilities P_{lk}^* define thus again a Markov chain.

Remark. A Markov chain is said to be *reversible*, if $P_{lk}^* = p_{kl}$ for $j, k = 0, 1, \ldots, N$. It is necessary and sufficient for this that the matrix of transition probabilities is doubly stochastic.

In Exercises 19 through 23 $\xi_1, \ldots, \xi_n$ are independent, identically distributed random variables with the common continuous distribution function $F(x)$. $\xi_{n,k}^*$ denotes the k-th order statistic $(k = 1, 2, \ldots, n)$.

19. Suppose $F(0) = 0$ and $F'(0) = \lambda > 0$. Show that

$$\lim_{n \to +\infty} P(n \xi_{n,k}^* < x) = \frac{1}{(k-1)!} \int_0^{\lambda x} t^{k-1} e^{-t}\, dt \qquad (k = 1, 2, \ldots).$$

Hint. We have

$$P(n\,\xi_{n,k}^* < x) = \sum_{r=k}^{n} \binom{n}{r} F\left(\frac{x}{n}\right)^r \left(1 - F\left(\frac{x}{n}\right)\right)^{n-r}$$

and, consequently,

$$\lim_{n \to +\infty} P(n\,\xi_{n,k}^* < x) = \sum_{r=k}^{\infty} \frac{(\lambda x)^r\, e^{-\lambda x}}{r!} = \frac{1}{(k-1)!} \int_0^{\lambda x} t^{k-1} e^{-t}\, dt.$$

20. Suppose $F(x) = x$ $(0 \le x \le 1)$. Show that $n\xi_{n,k}^*$ and $n(1 - \xi_{n,n+1-l}^*)$ are independent in the limit as $n \to +\infty$ and have gamma distributions of order k and j, respectively:

$$\lim_{n \to +\infty} P\left(n\,\xi_{n,k}^* < x,\, n(1 - \xi_{n,n+1-l}^*) < y\right) =$$

$$= \int_0^x \frac{u^{k-1} e^{-u}\, du}{(k-1)!} \int_0^y \frac{v^{j-1} e^{-v}\, dv}{(j-1)!}\ .$$

21. Suppose $F(x) = \Phi\left(\dfrac{x-m}{\sigma}\right)$, where $\Phi(x) = \dfrac{1}{\sqrt{2\pi}} \displaystyle\int_{-\infty}^x e^{-\frac{u^2}{2}}\, du$. Show that the

density function of

$$-\frac{\xi_{n,k}^* - m + \sigma \sqrt{2\ln n} - \sigma \dfrac{\ln\ln n + \ln 4\pi}{2\sqrt{2\ln n}}}{\dfrac{\sigma}{\sqrt{2\ln n}}}$$

tends to $\dfrac{1}{(k-1)!} \exp(-kx - e^{-x})$ as $n \to +\infty$.

22. Let $f(x) = F'(x)$ be continuous and positive on the interval $[a, b]$; suppose further that $a < F^{-1}(q_1) = Q_1 < Q_2 = F^{-1}(q_2) < b$. Show that the two-dimensional distribution of $\sqrt{n}\,(\xi_{n,k_1(n)}^* - Q_1)$ and $\sqrt{n}\,(\xi_{n,k_2(n)}^* - Q_2)$ tends to a two-dimensional normal distribution as $n \to +\infty$, if $|k_1(n) - q_1 n|$ and $|k_2(n) - q_2 n|$ remain bounded.

23. Let $F_n(x)$ be the empirical distribution function of the sample $(\xi_1, \xi_2, \ldots, \xi_n)$. Show that

$$P_n(\varepsilon) = P\left(\sup_x (F(x) - F_n(x)) < \varepsilon\right) =$$

$$= 1 - \varepsilon \sum_{k=0}^{[n(1-\varepsilon)]} \binom{n}{k} \left(1 - \varepsilon - \frac{k}{n}\right)^{n-k} \left(\varepsilon + \frac{k}{n}\right)^{k-1}.$$

Hint. We may assume that the variables ξ_k are uniformly distributed in the interval $(0, 1)$. If $m = [n(1 - \varepsilon)]$, the inequality $\sup_x (F(x) - F_n(x)) < \varepsilon$ is equivalent to

$$\xi_{n,j}^* < \frac{j-1}{n} + \varepsilon \quad \text{for} \quad j = 1, 2, \ldots, m+1.$$

It is easy to prove that

$$P_n(\varepsilon) = n! \int \ldots \int_{T_\varepsilon} dx_1 \ldots dx_n,$$

where T_ε is the domain defined by the inequalities

$$0 \le x_1 \le x_2 \le \ldots \le x_n \le 1, \quad x_j \le \frac{j-1}{n} + \varepsilon \quad \text{for} \quad j = 1, 2 \ldots, m+1.$$

The final result can be obtained by induction.

Remark. We can derive from this result the theorem of Smirnov (§ 10, Theorem 1).

24. (Wilcoxon's test for the comparison of two samples.) Let $\xi_1, \ldots, \xi_m$ and $\eta_1, \ldots, \eta_n$ be independent, identically distributed random variables with the common continuous distribution function $F(x)$. Let the numbers ξ_k and η_l be united into a single sequence, let them be arranged in increasing order and investigate the "places" occupied by $\xi_1, \ldots, \xi_m$. Let $\nu_1, \nu_2, \ldots, \nu_m$ denote the ranks of the elements $\xi_1, \ldots, \xi_m$ in this sequence. Put

$$W = \nu_1 + \nu_2 + \ldots + \nu_m - \frac{m(m+1)}{2}.$$

a) Show that W is equal to the number of pairs (ξ_i, η_l) such that $\xi_i > \eta_l$.

b) Show that $E(W) = \frac{nm}{2}$.

c) Let $G_{nm}(z)$ be the generating function of W:

$$G_{nm}(z) = \sum_k P(W = k) z^k.$$

Show that

$$G_{nm}(z) = \frac{C_{n+m}(z)}{C_n(z) C_m(z)},$$

where we have put

$$C_l(z) = \prod_{j=1}^{l} \frac{(1 - z^j)}{j(1 - z)}.$$

d) Show that

$$D(W) = \sqrt{\frac{n\,m(n + m + 1)}{12}}.$$

e) Derive from c) that the distribution of $W^* = \dfrac{W - E(W)}{D(W)}$ tends to the normal distribution as $n \to +\infty$, $m \to +\infty$, if $\dfrac{m}{n}$ tends to a constant. (Cf. Ch. II, § 12, Exercise 46 and Ch. III, § 18, Exercise 45.)

25. Let $\xi_1, \ldots, \xi_n$, $\eta_1, \ldots, \eta_m$ be independent random variables, with the same continuous distribution function $F(x)$ for all ξ_k and $G(x)$ for all η_l. Let L_1 denote the number of triplets (i, j, k) for which $\xi_i > \eta_k$, $\xi_j > \eta_k$ and $i < j$, and L_2 the

number of those triplets (i, j, k) for which $\eta_i > \xi_k$, $\eta_l > \xi_k$ and $i < j$. We put

$$L = \frac{L_1}{m \binom{n}{2}} + \frac{L_2}{n \binom{m}{2}} .$$

Show that if $F(x) \equiv G(x)$, then $E(L) = \dfrac{2}{3}$ and if $F(x) \neq G(x)$, then $E(L) > \dfrac{2}{3}$.

26. Let there be performed N independent experiments. Let the possible outcomes of every experiment be the events $A_1, \ldots, A_r$. Let $p_k = P(A_k)$ $(k = 1, 2, \ldots, r)$; let ν_k denote the number of occurrences of the event A_k, where $\sum\limits_{k=1}^{r} \nu_k = N$. If

$$\chi_N^2 = \sum_{k=1}^{r} \frac{(\nu_k - Np_k)^2}{Np_k} ,$$

then the distribution of χ_N^2 tends as $N \to +\infty$ to the χ^2-distribution with $r - 1$ degrees of freedom:

$$\lim_{N \to +\infty} P(\chi_N^2 < x) = \frac{1}{2^{\frac{r-1}{2}} \Gamma\left(\dfrac{r-1}{2}\right)} \int_0^x t^{\frac{r-3}{2}} e^{-\frac{t}{2}} dt.$$

Hint. The r-dimensional distribution of the random variables $\nu_1, \ldots, \nu_r$ is a multinomial distribution; hence the characteristic function of the joint distribution of the variables $\dfrac{\nu_k - Np_k}{\sqrt{Np_k}}$ tends to

$$\exp\left[-\frac{1}{2} \left(\sum_{k=1}^{r} t_k^2 - \left(\sum_{k=1}^{r} t_k \sqrt{p_k}\right)^2 \right) \right]$$

(Cf. Ch. VI, § 6, (23)).

27. If $\zeta_1, \zeta_2, \ldots, \zeta_n, \ldots$ are random variables such that the k-th order moment of ζ_n tends as $n \to +\infty$ to the k-th order moment of the standard normal distribution, i.e. if

$$\lim_{n \to +\infty} E(\zeta_n^k) = \frac{1}{\sqrt{2\pi}} \int_{-\infty}^{+\infty} t^k e^{-\frac{t^2}{2}} dt = \begin{cases} 1 . 3 \ldots (k - 1) & \text{for } k \text{ even,} \\ 0 & \text{for } k \text{ odd,} \end{cases}$$

then for every real x

$$\lim_{n \to +\infty} P(\zeta_n < x) = \frac{1}{\sqrt{2\pi}} \int_{-\infty}^{x} e^{-\frac{t^2}{2}} dt.$$

Hint. Apply inequality (21) of § 1 to $u = t \zeta_N$ and $k = 2l$. We find

$$\left| e^{it\zeta_N} - \sum_{j=0}^{2l-1} \frac{(it\zeta_N)^j}{j!} \right| \leq \frac{|t\zeta_N|^{2l}}{(2l)!} .$$

From this we get

$$\left| E(e^{it\zeta_N}) - \sum_{j=0}^{2l-1} \frac{(it)^j}{j!} E(\zeta_N^j) \right| \leq \frac{t^{2l}}{(2l)!} E(\zeta_N^{2l}).$$

If we let N tend to $+\infty$, we obtain

$$\limsup_{N \to +\infty} \left| E(e^{it\zeta_N}) - \sum_{j=0}^{l-1} \frac{\left(-\dfrac{t^2}{2} \right)^j}{j!} \right| \leq \frac{t^{2l}}{2^l l!} \, ;$$

hence, by letting l tend to $+\infty$,

$$\lim_{N \to +\infty} E(e^{it\zeta_N}) = e^{-\frac{t^2}{2}},$$

which is equivalent to the desired result.

28. Let $\xi_{n1}, \ldots, \xi_{nn}$ be random variables which assume only the values 0 and 1 and let η_{nk} be the sum of all products of k distinct elements of the sequence $\xi_{n1}, \ldots, \xi_{nn}$:

$$\eta_{nk} = \sum_{1 \leq i_1 < \ldots < i_k \leq n} \xi_{ni_1} \xi_{ni_2} \ldots \xi_{ni_k}.$$

Show that if $E(\eta_{nk})$ tends to $\dfrac{\lambda^k}{k!}$ $(k = 1, 2, \ldots)$ as $n \to +\infty$, then the distribution of the sum

$$\sum_{j=1}^{n} \xi_{nj} = \eta_{n1}$$

tends to the Poisson distribution with parameter λ.

29. Let $\xi_1, \xi_2, \ldots, \xi_n, \ldots$ be independent random variables assuming the values $+1$ and -1 with probability $\dfrac{1}{2}$. Put $\zeta_n = \xi_1 + \xi_2 + \ldots + \xi_n$ and denote by π_n the number of the positive terms in the sequence $\zeta_1, \zeta_2, \ldots, \zeta_n$. (If $\zeta_k = 0$, ζ_k is to be considered as positive if $\zeta_{k-1} = +1$.) Show that

$$P(\pi_{2n} = 2k, \zeta_{2n} = 0) = \frac{\binom{2n}{n}}{(n+1)2^{2n}} \quad \text{for} \quad k = 0, 1, \ldots, n$$

and

$$P(\pi_{2n} = 2k, \zeta_{2n} = -2j) = \frac{1}{2^{2n}} \sum_{1 \leq l \leq n-k} \binom{2n-2l}{n-l} \frac{1}{(n-l+1)l} \binom{2l}{l-j}$$

for $k = 0, 1, \ldots, n$ and $j = 1, 2, \ldots, n$.

30. By using the results of Exercise 29 show the following: If y_n is any sequence of integers such that y_n and n are of the same parity and $\lim\limits_{n \to +\infty} \dfrac{y_n}{\sqrt{n}} = y$ (y is here

any real number), then

$$\lim_{n \to +\infty} P\left(\frac{\pi_n}{n} < x \,\Big|\, \zeta_n = y_n\right) = \int_0^x f(t \mid y)dt,$$

with

$$f(t \mid y) = \frac{2e^{-\frac{y^2}{2}}}{\sqrt{2\pi}} \int_{\frac{y}{\sqrt{t}}}^{+\infty} \frac{e^{-\frac{u^2}{2}} \, du}{\left(1 - \frac{y^2}{u^2}\right)^{\frac{3}{2}}}$$

for $0 \leq t \leq 1$, $y \geq 0$ and $f(t \mid y) = f(1 - t \mid -y)$ for $y < 0$.

Remark. For $y = 0$ the conditional limit distribution of $\frac{\pi_n}{n}$ with respect to the condition $\zeta_n/\sqrt{n} \to 0$ is thus uniform on $(0, 1)$. If we notice that $\zeta_n/\sqrt{n}$ is, in the limit, normally distributed, it follows that

$$\lim_{n \to +\infty} P\left(\frac{\pi_n}{n} < x \mid \zeta_n \geq 0\right) = \frac{2}{\pi} \int_0^x \sqrt{\frac{t}{1-t}} \, dt;$$

$$\lim_{n \to +\infty} P\left(\frac{\pi_n}{n} < x \mid \zeta_n \leq 0\right) = \frac{2}{\pi} \int_0^x \sqrt{\frac{1-t}{t}} \, dt;$$

from these results, from $P(\zeta_n \geq 0) = \frac{1}{2}$ and from

$$\sqrt{\frac{t}{1-t}} + \sqrt{\frac{1-t}{t}} = \frac{1}{\sqrt{t(1-t)}}$$

the arc sin law can be easily derived.

APPENDIX
INTRODUCTION TO INFORMATION THEORY

§ 1. Hartley's formula

Information theory deals with mathematical problems arising in connection with the storage, transformation, and transmission of information.

In our everyday life we receive continuously various types of information (e.g. a telephone number); the informations received are stored (e.g. noted into a note-book), transmitted (told to somebody), etc. In order to use the informations it is often necessary to transform them in various fashions. Thus for instance in telegraphy the letters of the text are replaced by special signs; in television the continuous parts of the image are transformed into successive signals transmitted by electromagnetic waves. In order to treat such problems of communication mathematically, we need first of all a quantitative measure of information.

It is not at all obvious that the amount of information contained in a message can be defined and even measured. If we wish to introduce such a measure, we must abstract from form and content of the message. We have to work like the telegraph office, where only the number of words is counted in order to calculate the price of the telegram.

It is reasonable to measure the amount of information contained in a message by the number of signs necessary to express its content in the most concise possible form. Any system of signs can be used; the informations to be measured must be transformed into the system chosen. Thus, for instance, letters can be replaced by digits, the binary number system can be taken instead of the decimal, and so on. If we add to the 26 letters of the English alphabet the full stop, the comma, the semicolon, the question-mark, the note of exclamation and the space between the words, the 32 signs so obtained can be assigned to the numbers expressible by means of 5 digits in the binary system. (The numbers expressible by 1, 2, 3, or 4 digits are to be completed by zeros to five digits; thus $0 = 00000$, $1 = 00001$.) In this manner, every telegram can be expressed as a sequence of zeros and ones; the number of necessary signs is five times the number of the letters of the text. Every message, every information may thus be encoded into a sequence of zeros and ones.

It seems reasonable to measure the amount of information of a message by the number of signs necessary to express it by zeros and ones. On this basis, messages of different forms and contents become comparable as to the amount of information contained in them.

Since a digit can assume one of the values 0 and 1, the information specifying which of these two possibilities occurred can be taken as the unit of information. Thus the answer to a question which can only be answered by "yes" or "no" contains just one unit of information, the meaning of the particular question being irrelevant. The unit of information is called "bit", which is an abbreviation for "binary digit".

When one receives some information it happens often that only a part of it is really new. Thus for instance, if the telephone numbers in a certain city have all 6 digits, we can be sure in advance that every inhabitant will have a telephone number which is a number having 6 digits. Every information may thus be considered as a distinctive sign of an element of a set. If we know in advance that some object belongs to a certain set E, to give full information on the thing means to specify which of the elements of the set E is the one in question. The amount of information received depends, evidently, on the number of the elements of E. If E contains exactly $N = 2^n$ elements, these can be labelled by binary numbers having n digits; any element will be uniquely characterized by a sequence of length n consisting of zeros and ones, hence by n units of information. From $N = 2^n$ follows $n = \log_2 N$; this gave Hartley the idea to define by $\log_2 N$ the information necessary for the characterization of an element of a set having N elements, even if N is not a power of 2.

At the first glance it would seem that if $2^n < N < 2^{n+1}$, then $\log_2 N$ units of information do not suffice for the characterization of the elements of E, as somewhat more is necessary for this purpose, namely $n + 1$ units of information. This, however, is not the case. If we consider a sequence of symbols the terms of which are elements of E and if we replace each term of this sequence by a sequence of zeros and ones, we need really $n + 1$ binary digits. However, if we take from the elements of E a sequence of k elements (some of which may be equal), there are N^k such sequences and in order to characterize any one of these we need n_k zero or one sign, where

$$2^{n_k-1} < N^k \le 2^{n_k}.$$

In order to transcribe a symbol of our "alphabet" (an element of E) we need therefore on the average $\dfrac{n_k}{k}$ binary numbers, where

$$k \log_2 N \le n_k < k \log_2 N + 1.$$

It follows

$$\lim_{k \to \infty} \frac{n_k}{k} = \log_2 N.$$

Thus for every $\varepsilon > 0$ we can find a number k such that if we take the elements of E by ordered groups of k, then the identification of one element requires on the average less than $\log_2 N + \varepsilon$ binary digits.

The formula

$$I(E_N) = \log_2 N, \tag{1}$$

in which $I(E_N)$ represents the information necessary to characterize the elements of a set E_N of N elements is called *Hartley's formula*.

Formula (1) is a mathematical definition of the amount of information and thus needs no proof at all. Nevertheless, in order to show that this definition is not arbitrary, we postulate some properties which the function $I(E_N)$ should reasonably possess and show that the postulates in question are fulfilled only by the function $\log_2 N$. These postulates are:

A. $I(E_{NM}) = I(E_N) + I(E_M)$ for $N, M = 1, 2, \ldots$;

B. $I(E_N) \leq I(E_{N+1})$;

C. $I(E_2) = 1$.

Postulate C is the definition of the unit; it is not more and not less arbitrary than the choice of the unit of some physical quantity. The meaning of Postulate B is evident: the larger a set, the more information is gained by the characterization of its elements. Postulate A may be justified as follows.

A set E_{NM} of NM element may be decomposed into N subsets each of M elements; let these be denoted by $E_M^{(1)}, \ldots, E_M^{(N)}$. In order to characterize an element of E_{NM} we can proceed in two steps. First we specify that subset to which the element in question belongs. Let this subset be denoted by $E_M^{(j)}$. We need for this specification an information $I(E_N)$, since there are N subsets. Next we identify the element in $E_M^{(j)}$. The amount of information needed for this purpose is equal to $I(E_M)$ since the subset $E_M^{(j)}$ contains M elements. Now these two informations completely characterize an element of E_{NM}; Postulate A expresses thus that *the information is an additive quantity*.

THEOREM 1. *The Postulates* A, B, C *are fulfilled only by the function* $I(E_N) = \log_2 N$.

PROOF. Let P be an integer larger than 2. Define for every integer r the integer $s(r)$ by

$$2^{s(r)} \leq P^r < 2^{s(r)+1}. \tag{2}$$

Taking the logarithms of base 2 on both sides, we get

$$\frac{s(r)}{r} \leq \log_2 P < \frac{s(r)+1}{r}. \tag{3}$$

Hence

$$\lim_{r \to \infty} \frac{s(r)}{r} = \log_2 P. \tag{4}$$

Put $f(n) = I(E_n)$. It follows from B that for $n < m$

$$f(n) < f(m). \tag{5}$$

(2) and (5) lead to

$$f(2^{s(r)}) \leq f(P^r) < f(2^{s(r)+1}). \tag{6}$$

According to A we can write

$$f(a^k) = k f(a) \tag{7}$$

and, by C, $f(2) = 1$; hence it follows from (6) that

$$s(r) \leq r f(P) < s(r) + 1, \tag{8}$$

thus

$$\lim_{n \to \infty} \frac{s(r)}{r} = f(P). \tag{9}$$

From (4) and (9) we conclude that $f(P) = \log_2 P$ for $P > 2$. Since $f(2) = 1$, $f(1) = 0$, the theorem is herewith proved.

Postulate B can be replaced by the following one:

B*. $\qquad\qquad \lim_{N \to \infty} \big(I(E_{N+1}) - I(E_N)\big) = 0;$

and A can be replaced by a weaker postulate, too:

A*. *If N and M are relatively prime numbers, then*

$$I(E_{NM}) = I(E_N) + I(E_M).$$

P. Erdős[1] proved the following

THEOREM 2. $I(E_N) = \log_2 N$ *is the only function which satisfies the postulates* **A*, B*,** *and* **C.**

PROOF. Let $P > 1$ be any power of a prime number and $f(n) = I(E_n)$ a function satisfying **A*, B*, C.** Put

$$g(n) = f(n) - \frac{f(P) \log_2 n}{\log_2 P} \,. \tag{10}$$

Clearly, $g(n)$ fulfills **A*.** Furthermore we have

$$g(n + 1) - g(n) = f(n + 1) - f(n) + \frac{f(P)}{\log_2 P} \log_2 \frac{n}{n + 1} \,.$$

If we put

$$\varepsilon_n = g(n + 1) - g(n), \tag{11}$$

then **B*** implies

$$\lim_{n \to \infty} \varepsilon_n = 0. \tag{12}$$

Hence $g(n)$ fulfills **B*.** Now it is easy to see that

$$g(P) = 0. \tag{13}$$

Define for every integer n an integer n' by

$$n' = \begin{cases} \left[\dfrac{n}{P}\right] & \text{for} \quad \left(\left[\dfrac{n}{P}\right], P\right) = 1, \\[3mm] \left[\dfrac{n}{P}\right] - 1 & \text{for} \quad \left(\left[\dfrac{n}{P}\right], P\right) > 1, \end{cases} \tag{14}$$

where (a, b) denotes the greatest common divisor of the integers a and b.

---

[1] Cf. P. Erdős [2] and the article of D. K. Fadeev, The notion of entropy in a finite probabilistic pattern (Arbeiten zur Informationstheorie, Vol. I). Fadeev found this theorem independently from Erdős. The proof given here (cf. A. Rényi [29], [30], [37]) is considerably simpler than that of the above two authors.

Clearly

$$n' \leq \frac{n}{P} \tag{15}$$

and

$$n = Pn' + l,$$

where $(n', P) = 1$ and $0 \leq l < 2P$. According to (13), $g(Pn') = g(n')$, hence we can write

$$g(n) = g(n') + g(n) - g(Pn') = g(n') + \sum_{k=Pn'}^{n-1} \varepsilon_k, \tag{16a}$$

where ε_k is defined by (11).

Repeat the decomposition (16a) with n' instead of n, then with n'' instead of n', etc. If we put

$$n^{(0)} = n, \qquad n^{(j+1)} = (n^{(j)})' \quad (j = 0, 1, \ldots),$$

we obtain at the k-th step

$$g(n) = g(n^{(k)}) + \sum_{j=1}^{k} \sum_{h=P_n^{(j)}}^{n^{(j-1)}-1} \varepsilon_h. \tag{16b}$$

But by (15)

$$n^{(k)} \leq \frac{n}{P^k},$$

hence we obtain $n^{(k)} = 0$ after at most $\left[\dfrac{\log_2 n}{\log_2 P} \right] + 1$ steps, hence for every n

$$g(n) = \sum_{i=1}^{b_n} \varepsilon_{h_i}, \tag{17}$$

where $h_1 < h_2 < \ldots < h_{b_n}$ and

$$b_n < 2P \left(\frac{\log_2 n}{\log_2 P} + 1 \right).$$

Thus, according to (12),

$$\lim_{n \to \infty} \frac{g(n)}{\log_2 n} = 0, \tag{18}$$

and by (10)

$$\lim_{n \to \infty} \frac{f(n)}{\log_2 n} = \frac{f(P)}{\log_2 P} . \tag{19}$$

Let c denote the limit of the left hand side of (19). We conclude that for every $P > 1$ which is a power of a prime number

$$f(P) = c \, \log_2 P. \tag{20}$$

If the integer $n > 1$ has a decomposition $n = P_1 P_2 \ldots P_r$, where P_i are powers of primes, then we conclude from the additivity of $f(n)$ that

$$f(n) = \sum_{i=1}^{r} f(P_i) = c \sum_{i=1}^{r} \log_2 P_i = c \, \log_2 n. \tag{21}$$

Because of Postulate C, the value of c must be equal to 1. Furthermore, according to A*, $f(1) = 0$. Theorem 2 is herewith proved.

This theorem will be used in the following section.

§ 2. Shannon's formula

Let $E_1, E_2, \ldots, E_n$ be pairwise disjoint finite sets and put

$$E = E_1 + E_2 + \ldots + E_n.$$

Let N_k be the number of elements of the set E_k; E has therefore $N = \sum_{k=1}^{n} N_k$ elements. We put $p_k = \dfrac{N_k}{N}$ $(k = 1, 2, \ldots, n)$. If we know about an element of E that it belongs to a set E_k, for the complete determination of this element we need some further information, the amount of which is equal to $\log_2 N_k$. Thus in order to characterize an element which is known to belong to a subset E_k of E we need on the average the amount of information

$$I_2 = \sum_{k=1}^{n} \frac{N_k}{N} \log_2 N_k = \sum_{k=1}^{n} p_k \log_2 N p_k. \tag{1}$$

The information necessary for the complete characterization of an element of E can therefore be decomposed into two parts. The first part, I_1 determines the set E_k containing the element in question; the second part, I_2, given by Formula (1), identifies the element in E_k. If the information is additive also in this sense, then the relation

$$\log_2 N = I_1 + I_2 = I_1 + \sum_{k=1}^{n} p_k \, \log_2 N p_k \tag{2}$$

must hold. Since $\sum\limits_{k=1}^{n} p_k = 1$, it follows from (2) that in order to know
to which one of the subsets E_k an element of E belongs, we need an amount
of information equal to

$$I_1 = \sum_{k=1}^{n} p_k \log_2 \frac{1}{p_k}. \tag{3}$$

Formula (3) was first established by Shannon and in what follows we
shall call it *Shannon's formula*. Simultaneously with and independently of
Shannon the same formula was also found by N. Wiener.

In particular, if $p_1 = p_2 = \ldots = p_n = \dfrac{1}{n}$, *Shannon's formula* reduces
to *Hartley's formula* (cf. Formula (1) of § 1). Analysing the above heuristic
considerations it is clear that we implicitly used three assumptions, namely

1. The selection of the considered element from the set E depends on
chance; actually, we are dealing with the observed value of a random
variable.

2. All elements of E are equiprobable; the probability that an element
of E belongs to E_k is therefore $p_k = \dfrac{N_k}{N}$.

3. The amounts of information associated with the different possibilities
must be "weighted" by the corresponding probabilities; essentially, we
consider thus the expectation of the information.

Thus, instead of restricting ourselves to the particular case of the random
selection of an element from a set, we are led to the more general question:
how much information is yielded by the outcome of a random experiment?
We shall see that Shannon's Formula (3) remains valid in this more general
case too (hence not only in the case of rational values of p_k).

The general problem can be put in the following form: Let A_1, A_2, ..., A_n
be the possible outcomes of a random experiment $\mathscr{A}$; put $p_k = P(A_k)$
$(k = 1, 2, \ldots, n)$. We wish to know, how much information is furnished
by a single performance of the experiment $\mathscr{A}$. It seems reasonable to start
from the following postulates:

I. *The information obtained depends only on the probability distribution*
$\mathscr{P} = (p_1, p_2, \ldots, p_n)$, *consequently, it will be denoted by* $I(\mathscr{P})$ *or*
$I(p_1, p_2, \ldots, p_n)$. *We suppose further that* $I(p_1, p_2, \ldots, p_n)$ *is a symmetric
function of its variables* $p_1, p_2, \ldots, p_n$.

II. $I(p, 1 - p)$ *is a continuous function of* p $(0 \le p \le 1)$.

III. $I\left(\dfrac{1}{2}, \dfrac{1}{2}\right) = 1.$

Furthermore, we require:

IV. *The following relation holds:*

$$I(p_1, p_2, \ldots, p_n) =$$

$$= I(p_1 + p_2, p_3, \ldots, p_n) + (p_1 + p_2)\, I\left(\frac{p_1}{p_1 + p_2}, \frac{p_2}{p_1 + p_2}\right). \qquad (4)$$

Condition (4) can be worded as follows: Suppose that an outcome A of an experiment with probability $\alpha = P(A)$ can occur in two ways, A' and A'' which mutually exclude each other. Suppose that the probability of A' is α' and that of A'' is α'' ($\alpha' + \alpha'' = \alpha$). Then if we are told in which of the two forms A actually occurred, the amount of information thus obtained is equal to the information associated with the distribution $\left(\dfrac{\alpha'}{\alpha}, \dfrac{\alpha''}{\alpha}\right)$ taken with the weight α, i.e. to $\alpha\, I\left(\dfrac{\alpha'}{\alpha}, \dfrac{\alpha''}{\alpha}\right)$. Postulate IV requires at the same time the additivity of information as well as the "weighting" of the different informations by the corresponding probabilities.

We shall show that Postulates I–IV are fulfilled only by the function defined by Shannon's Formula (3). The above set of Postulates I–IV is due to D. K. Fadeev; it is a simplified form of a system of postulates given by A. I. Khinchin. In § 6 we shall characterize Shannon's information by different postulates which lead also to alternative measures of information.

In the present section we prove

THEOREM 1. *If to every discrete, finite probability distribution there corresponds a number* $I(\mathscr{S}) = I(p_1, p_2, \ldots, p_n)$ *so that the above Postulates* I–IV *are satisfied, then*[1]

$$I(\mathscr{S}) = \sum_{k=1}^{n} p_k \log_2 \frac{1}{p_k}. \qquad (5)$$

PROOF. The proof consists of six steps.

a) We show first that

$$I(1) = 0, \qquad (6)$$

---

[1] In view of $\displaystyle\lim_{x \to 0} x \log_2 \frac{1}{x} = 0$ we put $0 \log_2 \dfrac{1}{0} = 0$.

i.e. that the occurrence of the sure event does not give us any information. In fact, if $n = 2$, $p_1 = 1$, $p_2 = 0$, it follows from IV that

$$I(1, 0) = I(1) + I(1, 0);$$

thus (6) holds. Similarly, we have

$$I(p_1, p_2, \ldots, p_n, 0) = I(p_1, p_2, \ldots, p_n).$$

b) If we put

$$s_m = \sum_{j=1}^{m} p_j,$$

we can deduce from (4) by induction on m the somewhat more general relation

$$I(p_1, \ldots, p_m, p_{m+1}, \ldots, p_{m+n}) =$$

$$= I(s_m, p_{m+1}, \ldots, p_{m+n}) + s_m I\left(\frac{p_1}{s_m}, \ldots, \frac{p_m}{s_m}\right). \tag{4'}$$

According to (4) the formula holds for $m = 2$. Suppose that it is already proved up to $m - 1$. Then

$$I(p_1 + p_2, p_3, \ldots, p_{m+n}) =$$

$$= I(s_m, p_{m+1}, \ldots, p_{m+n}) + s_m I\left(\frac{p_1 + p_2}{s_m}, \ldots, \frac{p_m}{s_m}\right), \tag{7a}$$

furthermore, because of (4),

$$I(p_1, p_2, \ldots, p_{m+n}) = I(p_1 + p_2, p_3, \ldots, p_{m+n}) +$$

$$+ (p_1 + p_2) I\left(\frac{p_1}{p_1 + p_2}, \frac{p_2}{p_1 + p_2}\right) \tag{7b}$$

and

$$s_m I\left(\frac{p_1 + p_2}{s_m}, \ldots, \frac{p_m}{s_m}\right) + (p_1 + p_2) I\left(\frac{p_1}{p_1 + p_2}, \frac{p_2}{p_1 + p_2}\right) =$$

$$= s_m I\left(\frac{p_1}{s_m}, \ldots, \frac{p_m}{s_m}\right); \tag{7c}$$

(4') follows immediately from (7a), (7b) and (7c).

c) We prove now a still more general relation:

$$I(p_{11}, \ldots, p_{1m_1}, \ldots, p_{n1}, \ldots, p_{nm_n}) =$$

$$= I(s_1, \ldots, s_n) + \sum_{j=1}^{n} s_j I\left(\frac{p_{j1}}{s_j}, \ldots, \frac{p_{jm_j}}{s_j}\right), \tag{4''}$$

where we have put

$$s_j = \sum_{l=1}^{m_j} p_{jl} \qquad (j = 1, 2, \ldots, n). \tag{8}$$

By assumption,

$$\sum_{j=1}^{n} s_j = \sum_{j=1}^{n} \sum_{l=1}^{m_j} p_{jl} = 1.$$

Formula (4'') may be considered as a theorem about the information associated with a mixture of distributions. In effect, if $\mathscr{P}_j$ denotes the distribution $\left(\dfrac{p_{j1}}{s_j}, \ldots, \dfrac{p_{jm_j}}{s_j}\right)$, the left hand side of (4'') is the information associated with the mixture of the distributions $\mathscr{P}_j$ with weights s_j. According to (4'') this information is equal to the sum of the average of the informations $I(\mathscr{P}_j)$ with weights s_j and the information associated with the mixing distribution $S = (s_1, \ldots, s_n)$:

$$I\left(\sum_{j=1}^{n} s_j \mathscr{P}_j\right) = \sum_{j=1}^{n} s_j I(\mathscr{P}_j) + I(S). \tag{4'''}$$

(4'') can be obtained immediately by a repeated application of (4'), taking into account the assumption that $I(p_1, \ldots, p_n)$ is a symmetric function of $p_1, \ldots, p_n$.

d) Let $\mathscr{E}_n$ be the distribution $\left(\dfrac{1}{n}, \dfrac{1}{n}, \ldots, \dfrac{1}{n}\right)$ and put $f(n) = I(\mathscr{E}_n)$. From (4'') we deduce the functional equation

$$f(n, m) = f(n) + f(m). \tag{9}$$

In fact, if in (4'') all m_j are equal to m and all p_{jl} are equal to $\dfrac{1}{mn}$, the left hand side is equal to $f(nm)$ and the right hand side to $f(n) + f(m)$, hence we get (9).

e) If we apply (4') to the case when all probabilities are equal and if we unite them all except the first one, we obtain

$$f(n) = I\left(\frac{1}{n}, 1 - \frac{1}{n}\right) + \left(1 - \frac{1}{n}\right) f(n-1). \tag{10}$$

Now we show that

$$\lim_{n \to \infty} [f(n) - f(n - 1)] = 0. \tag{11}$$

Put

$$f(n) - f(n - 1) = d_n \quad \text{and} \quad I\left(\frac{1}{n}, 1 - \frac{1}{n}\right) = \delta_n \qquad (n = 2, 3, \ldots).$$

It follows from our assumptions that

$$\lim_{n \to \infty} \delta_n = 0. \tag{12}$$

Indeed the assumed continuity of $I(p, 1 - p)$ implies

$$\lim_{n \to \infty} \delta_n = I(0, 1) = I(1),$$

and according to (6) $I(1) = 0$. On the other hand

$$f(n - 1) = d_2 + d_3 + \ldots + d_{n-1};$$

(10) is therefore equivalent to

$$\delta_n = d_n + \frac{d_2 + d_3 + \ldots + d_{n-1}}{n}. \tag{13}$$

Multiplying both sides by n and adding the equalities obtained for $n = 2, 3, \ldots, N$, we get

$$\sum_{n=2}^{N} (nd_n + d_2 + \ldots + d_{n-1}) = \sum_{n=2}^{N} n\delta_n, \tag{14}$$

and by a simple transformation

$$\frac{2}{N+1} \sum_{k=2}^{N} d_k = \frac{\displaystyle\sum_{n=2}^{N} n\delta_n}{\displaystyle\sum_{n=2}^{N} n}. \tag{15}$$

Because of (12) the right hand side of (15) tends to zero for $N \to \infty$. Hence we have

$$\lim_{N \to +\infty} \frac{2}{N+1} \sum_{k=2}^{N} d_k = 0. \tag{16}$$

From (12) and (16) it follows because of (13)

$$\lim_{N \to +\infty} d_N = 0, \tag{17}$$

hence we obtain (11).[1]

We have seen that $f(n)$ fulfills conditions A*, B*, and C of the preceding section; hence by Theorem 2 of § 1

$$f(n) = I(\mathscr{E}_n) = \log_2 n. \tag{18}$$

f) We can now finish rapidly the proof of our theorem.

Consider the function $I(p, 1 - p)$. Let first p be rational, $p = \dfrac{a}{b}$ with integers a and b ($a < b$). If we apply (4'') with

$$n = 2, \quad m_1 = a, \quad m_2 = b - a, \quad p_{11} = p_{12} = \ldots = p_{2m_2} = \frac{1}{b},$$

we find

$$\log_2 b = I\left(\frac{a}{b}, 1 - \frac{a}{b}\right) + \frac{a}{b} \log_2 a + \left(1 - \frac{a}{b}\right) \log_2 (b - a). \tag{19}$$

Since by assumption $I(p, 1 - p)$ is continuous, we have for any p between 0 and 1

$$I(p, 1 - p) = p \log_2 \frac{1}{p} + (1 - p) \log_2 \frac{1}{1 - p}, \tag{20}$$

hence (5) is proved for $n = 2$. We show now by induction that (5) holds in the general case too. Suppose that (5) is valid for a certain integer n; let $\mathscr{P} = (p_1, \ldots, p_{n+1})$ be any distribution having $n + 1$ terms. We conclude from (4) and (20) that

$$I(p_1, \ldots, p_{n+1}) = \sum_{k=1}^{n-1} p_k \log_2 \frac{1}{p_k} +$$

$$+ (p_n + p_{n+1})\left[\log_2 \left(\frac{1}{p_n + p_{n+1}}\right) \cdot i\left(\frac{p_n}{p_n + p_{n+1}}, \frac{p_{n+1}}{p_n + p_{n+1}}\right)\right];$$

---

[1] We use here a well-known theorem of the theory of divergent series (Mercer's theorem) which says: If s_n is a sequence, fulfilling

$$\lim_{n \to \infty} \left(as_n + (1 - a) \frac{s_1 + s_2 + \ldots + s_n}{n}\right) = s$$

($0 < a < 1$), then we have also $\lim_{n \to \infty} s_n = s$. We need only the particular case $a = \dfrac{1}{2}$ (cf. G. H. Hardy [1], Ch. V).

hence because of (20),

$$I(p_1, \ldots, p_{n+1}) = \sum_{k=1}^{n+1} p_k \log_2 \frac{1}{p_k}, \qquad (21)$$

and thus the theorem is proved for every integer n.

Remark. It is easy to see that Postulate IV implies the additivity of information. Suppose that the experiments A and B are independent of each other; let A_j ($j = 1, 2, \ldots, m$) be the possible outcomes of A, and B_k ($k = 1, 2, \ldots, n$) those of B. Let $p_j = P(A_j)$, $q_k = P(B_k)$ denote the corresponding probabilities and put $\mathscr{P} = (p_1, \ldots, p_m)$, $Q = (q_1, \ldots, q_n)$. To perform simultaneously A and B means the same as to perform an experiment AB having the events $A_j B_k$ as possible outcomes with the corresponding probabilities $p_j q_k$. The distribution $\{p_j q_k\} = \mathscr{P} * Q$ is called the *direct product* of the distributions $\mathscr{P}$ and Q. If we apply (4″) to $p_1 q_1, \ldots, p_1 q_n$, $p_2 q_1, \ldots, p_m q_n$, we find that

$$I(\mathscr{P} * Q) = I(\mathscr{P}) + I(Q). \qquad (22)$$

However, (4) does not follow from (22). This is most easily demonstrated by the quantity

$$I_2(p_1, \ldots, p_n) = -\log_2(p_1^2 + \ldots + p_n^2), \qquad (23)$$

which fulfills Postulates I–III and Formula (22), without fulfilling (4). (If it fulfilled (4), it would be equal, by the just-proved theorem, to $\sum_{k=1}^{n} p_k \log_2 \frac{1}{p_k}$, which is not the case.) We shall see in § 6 that the quantity (23) too can be considered as a measure of the information associated with the distribution $\mathscr{P} = (p_1, \ldots, p_n)$. In fact, we shall define a class of information measures depending on a parameter α which contains both Shannon's information (for $\alpha = 1$) and the quantity (23) (for $\alpha = 2$).

We add some further remarks.

1. In connection with the notion of information we also have to mention the concept of *uncertainty*. If we receive some information, the previously existing uncertainty will be diminished. The meaning of information *is* precisely this diminishing of uncertainty.

The uncertainty with respect to an outcome of an experiment may be considered as numerically equal to the information furnished by the occurrence of this outcome; thus uncertainty can also be measured. We could have started equally well from the notion of uncertainty; to speak about information or about uncertainty means essentially the same thing: in the first case we consider an experiment which has been performed, in the second case an experiment not yet performed. The two terminologies will be used alternatively in order to obtain the simplest possible formulation of our results.

2. The quantity (5) is frequently called the *entropy* of the distribution $\mathscr{P} = (p_1, \ldots, p_n)$. Indeed, there is a strong connection between the notion of entropy in thermodynamics and the notion of information (or uncertainty). L. Boltzmann was the first to emphasize the probabilistic meaning of the thermodynamical entropy and thus he may be considered as a pioneer of information theory. It would even be proper to call Formula (5) the Boltzmann–Shannon formula. Boltzmann proved that the entropy of a physical system can be considered as a measure of the disorder in the system. In case of a physical system having many degrees of freedom (e.g. a perfect gas) the number measuring the disorder of the system measures also the uncertainty concerning the states of the individual particles.

3. In order to avoid possible misunderstandings it should be emphasized that when we speak about information, what we have in mind is not the subjective "information" possessed by a particular observer. The terminology is really somewhat misleading as it seems to support that the information depends somehow on the observer. In reality the information contained in an observation is a quantity independent of the fact whether it does or does not reach the perception of an observer (be it a man or some registering device or a computer). The notion of uncertainty should also be interpreted in an objective sense; what we have in mind is not the subjective "uncertainty" existing in the mind of the observer concerning the outcomes of an experiment; it is an uncertainty due to the fact that really several possibilities are to be taken into account. The measure of uncertainty does not depend on anything else than these possible events and in this sense it is entirely objective. The above mentioned relation between information and thermodynamical entropy is noteworthy in this respect too.

§ 3. Conditional and relative information

We associated with every discrete finite probability distribution $\mathscr{P} = (p_1, \ldots, p_n)$ the information $I(\mathscr{P})$. If ξ is a random variable assuming the distinct values $x_1, x_2, \ldots, x_n$ with probabilities $p_1, p_2, \ldots, p_n$, we may say that $I(\mathscr{P})$ is *the information contained in the value of* ξ and we may write $I(\xi)$ instead of $I(\mathscr{P})$. It must, however, be remembered that $I(\xi)$ does *not* depend on the values $x_1, x_2, \ldots, x_n$ of ξ. $I(\xi)$ remains invariant, when we replace $x_1, x_2, \ldots, x_n$ by any other system of mutually different numbers

$x'_1, x'_2, \ldots, x'_n$. The observation of the random variable assuming the values $x'_1, x'_2, \ldots, x'_n$ with probabilities $p_1, p_2, \ldots, p_n$ contains the same amount of information as the observation of ξ. Consequently, if $h(x)$ is a function such that $h(x) \neq h(x')$ for $x \neq x'$, we have $I(h(\xi)) = I(\xi)$. However, without the condition $h(x) \neq h(x')$ for $x \neq x'$ we can state only that $I(h(\xi)) \leq \leq I(\xi)$. This follows from the evident inequality

$$(p+q)\log_2 \frac{1}{p+q} \leq p\log_2 \frac{1}{p} + q\log_2 \frac{1}{q} \tag{1}$$

for $p \geq 0$, $q \geq 0$, $p + q \leq 1$.

We shall often need *Jensen's inequality:* If $g(x)$ is a convex function on an interval (a, b), if $x_1, x_2, \ldots, x_n$ are arbitrary real numbers $a < x_k < b$ and if $w_1, w_2, \ldots, w_n$ are positive numbers with $\sum\limits_{k=1}^{n} w_k = 1$, then we have

$$g\left(\sum_{k=1}^{n} w_k x_k\right) \leq \sum_{k=1}^{n} w_k g(x_k). \tag{2}$$

Inequality (2) can readily be proved by a geometrical reasoning. Consider in the plane (x, y) the points $(x_k, g(x_k))$, $k = 1, 2, \ldots, n$. Suppose that masses w_k are situated in these points; the center of gravity of the so formed system will evidently lie in the smallest convex polygon containing the mentioned points. Since all points lie on the convex curve $y = g(x)$, the center of gravity lies above this curve. Let $\bar{x}$ and $\bar{y}$ denote its coordinates, then $g(\bar{x}) \leq \bar{y}$. As clearly $\bar{x} = \sum\limits_{k=1}^{n} w_k x_k$ and $\bar{y} = \sum\limits_{k=1}^{n} w_k g(x_k)$, we get (2). It can be seen immediately that if $g(x)$ is not linear on any subinterval, then in (2) the equality sign can occur only if $x_1 = x_2 = \ldots = x_n$.

If $g(x)$ is *concave*, we have instead of (2) the inequality

$$g\left(\sum_{k=1}^{n} w_k x_k\right) \geq \sum_{k=1}^{n} w_k g(x_k), \tag{2'}$$

since now $-g(x)$ is convex.

From Jensen's inequality (2) we obtain

$$I(p_1, p_2, \ldots, p_n) \leq \log_2 n. \tag{3}$$

It suffices for this to apply (2) to the convex function $y = x\log_2 x$ $(x > 0)$; with $x_k = p_k$, $w_k = \dfrac{1}{n}$ $(k = 1, 2, \ldots, n)$ we get (3). The equality sign holds for $p_1 = p_2 = \ldots = p_n = \dfrac{1}{n}$ only. That is, if there are n possibilities for the outcome of an experiment, the uncertainty will be maximal when all possibilities are equiprobable.

Formula (3) can be generalized as follows: Let $\mathscr{P} = (p_1, p_2, \ldots, p_n)$ be a probability distribution and $W = (w_{jk})$ a stochastic matrix with n rows and n columns; the elements of W are thus nonnegative and the sum of the terms of each row is equal to 1. Put

$$q_k = \sum_{j=1}^{n} p_j w_{jk} \qquad (k = 1, 2, \ldots, n).$$

Then

$$\sum_{k=1}^{n} q_k = \sum_{j=1}^{n} p_j \sum_{k=1}^{n} w_{jk} = \sum_{j=1}^{n} p_j = 1;$$

hence $Q = (q_1, q_2, \ldots, q_n)$ is a probability distribution and we find

$$I(\mathscr{P}) \leq I(Q). \tag{4}$$

In fact, by putting

$$g(x) = x \log_2 x, \qquad x_j = p_j, \quad w_j = w_{jk} \qquad (j = 1, 2, \ldots, n)$$

we can derive from (2) the inequality

$$q_k \log_2 q_k \leq \sum_{j=1}^{n} w_{jk} p_j \log_2 p_j. \tag{5}$$

If in (5) we sum over k, we obtain (4). Inequality (4) expresses that the uncertainty for a distribution is larger if the terms of the distribution are closer to each other.

We introduce now the notion of conditional information. Let ξ and η be two random variables having finite discrete distributions. Let $x_1, x_2, \ldots, x_m$ be the distinct values taken on by ξ with positive probabilities, $y_1, y_2, \ldots, y_n$ those by η. We write:

$$P(\xi = x_j) = p_j \qquad (j = 1, 2, \ldots, m); \qquad \mathscr{P} = (p_1, p_2, \ldots, p_m). \tag{6a}$$

$$P(\eta = y_k) = q_k \qquad (k = 1, 2, \ldots, n); \qquad Q = (q_1, q_2, \ldots, q_n). \tag{6b}$$

$$P(\xi = x_j, \eta = y_k) = r_{jk} \qquad \begin{pmatrix} j = 1, 2, \ldots, m \\ k = 1, 2, \ldots, n \end{pmatrix}; \qquad \mathscr{R} = (r_{11}, \ldots, r_{mn}). \tag{7}$$

$$P(\xi = x_j \mid \eta = y_k) = p_{j|k}; \qquad \mathscr{P}_k = (p_{1|k}, \ldots, p_{m|k}) \quad (k = 1, 2, \ldots, n). \tag{8a}$$

$$P(\eta = y_k \mid \xi = x_j) = q_{k|j}; \qquad Q_j = (q_{1|j}, \ldots, q_{n|j}) \quad (j = 1, 2, \ldots, m). \tag{8b}$$

According to the definition of conditional probability we have

$$r_{jk} = p_j q_{k|j} = q_k p_{j|k} ;$$ (9)

further

$$\sum_{k=1}^{n} r_{jk} = p_j \qquad (j = 1, 2, \ldots, m)$$ (10a)

and

$$\sum_{j=1}^{m} r_{jk} = q_k \qquad (k = 1, 2, \ldots, n).$$ (10b)

We define now the *conditional information* $I(\xi \mid \eta)$ contained in ξ with respect to the condition that η assumes a given value; this will be the expectation of the information associated with the distribution $\mathscr{P}_k$:

$$I(\xi \mid \eta) = \sum_{k=1}^{n} q_k I(\mathscr{P}_k) = \sum_{k=1}^{n} \sum_{j=1}^{m} r_{jk} \log_2 \frac{q_k}{r_{jk}}.$$ (11)

On the other hand, if $I((\xi, \eta))$ denotes the information associated with the two-dimensional distribution of ξ and η:

$$I((\xi, \eta)) = I(\mathscr{R}) = \sum_{j=1}^{m} \sum_{k=1}^{n} r_{jk} \log_2 \frac{1}{r_{jk}},$$ (12)

then we have

$$I((\xi, \eta)) = I(\eta) + I(\xi \mid \eta).$$ (13)

Formula (13) follows from (9), (10b), (11) and (12):

$$I(\xi \mid \eta) = I((\xi, \eta)) + \sum_{j=1}^{m} \sum_{k=1}^{n} r_{jk} \log_2 q_k = I((\xi, \eta)) - I(\eta).$$

It follows from the definition that $I(\xi \mid \eta) = I(\xi)$ when ξ and η are independent, hence (13) reduces in this case to the relation obtained in the preceding section:

$$I((\xi, \eta)) = I(\xi) + I(\eta).$$ (14)

We may consider (13) as a generalization of the theorem on the additivity of the information: the information contained in the pair of values (ξ, η) is the sum of the information contained in the value of η and of the conditional information contained in the value of ξ when we know that η takes on a certain value.

Now we show that in general the relation

$$I((\xi, \eta)) \leq I(\xi) + I(\eta)$$ (15)

holds, where the sign of equality occurs only if ξ and η are independent. According to (13), relation (15) is equivalent to

$$I(\xi \mid \eta) \leq I(\xi) \tag{16}$$

which means that the "conditional" uncertainty of ξ for a known value of η cannot exceed the "unconditional" uncertainty of ξ. By taking (11) into account we can write

$$I(\xi \mid \eta) = -\sum_{j=1}^{m} \sum_{k=1}^{n} q_k p_{j|k} \, \log_2 p_{j|k}. \tag{17}$$

If we apply Jensen's inequality to the function $x \log_2 x$ with $x_k = p_{j|k}$, $w_k = q_k$ $(k = 1, 2, \ldots, n)$, we obtain, in view of (9),

$$p_j \log_2 p_j \leq \sum_{k=1}^{n} q_k p_{j|k} \log_2 p_{j|k}. \tag{18}$$

From (17) and (18) follows immediately (16), and hence (15) too. The sign of equality in (18) can only hold if all $p_{j|k}$ $(k = 1, 2, \ldots, n)$ are equal, i.e. when ξ and η are independent. We conclude from (13) that

$$I(\xi) - I(\xi \mid \eta) = I(\xi) + I(\eta) - I((\xi, \eta)); \tag{19}$$

the right hand side being symmetric in ξ and η, we have

$$I(\xi) - I(\xi \mid \eta) = I(\eta) - I(\eta \mid \xi). \tag{20}$$

The left hand side of (19) may be interpreted as the decrease of uncertainty due to the knowledge of η, or as the information about ξ which can be gained from the value of η. We call this the *relative information* given by η about ξ and denote it by $I(\xi, \eta)$; we have thus

$$I(\xi, \eta) = I(\xi) - I(\xi \mid \eta). \tag{21a}$$

(We must not confuse $I(\xi, \eta)$ with the information $I((\xi, \eta))$ associated with the two-dimensional distribution of ξ and η.) According to (20)

$$I(\xi, \eta) = I(\eta, \xi); \tag{21b}$$

hence the value of η gives the same amount of information about ξ as the value of ξ gives about η.

$I(\xi, \eta)$ can also be defined by the symmetric expression

$$I(\xi, \eta) = I(\xi) + I(\eta) - I((\xi, \eta)) = \sum_{j=1}^{m} \sum_{k=1}^{n} r_{jk} \log_2 \frac{r_{jk}}{p_j q_k}. \tag{22}$$

According to (16) we have

$$I(\xi, \eta) \geq 0, \tag{23}$$

where the equality sign holds only if ξ and η are independent. Hence if ξ and η are not independent, the value of η gives always information about ξ. On the other hand, from (21a) and (21b) follows

$$I(\xi, \eta) \leq \min\left(I(\xi), I(\eta)\right). \tag{24}$$

Here too, it is easy to find the cases in which the equality sign holds. In fact, if $I(\xi, \eta) = I(\xi)$, then $I(\xi \mid \eta) = 0$, which can occur only if the value of ξ is uniquely determined by the value of η, i.e. if $\xi = f(\eta)$. Similarly, $I(\xi, \eta) = I(\eta)$ can occur only if $\eta = g(\xi)$. The quantity $I(\xi, \eta)$ can be considered as a measure of the stochastic dependence between the random variables ξ and η.

The relation $I(\xi, \eta) = I(\eta, \xi)$, expressing that η contains (on the average) just as much information about ξ as ξ about η, seems to be at the first glance surprising, but a deeper consideration shows it to be quite natural.

The following example is enlightening. Let η be a random variable symmetrically distributed with respect to the origin, with $P(\eta = 0) = 0$, and put $\xi = \eta^2$. There corresponds to every value of η one and only one value of ξ, while conversely ξ determines η only up to its sign. In spite of this, ξ gives just as much information on η as η gives on ξ (viz. $I(\xi)$); the difference is that this information suffices for the complete characterization of ξ but does not determine η completely (only the value of $|\eta|$). In fact, $I(\eta) = I(\xi) + 1$ (if we know already the absolute value of η, η can still take on the values $\pm |\eta|$ with probability $\dfrac{1}{2}$, hence one unit of uncertainty must be added).

We prove now the inequality

$$I(\xi, f(\eta)) \leq I(\xi, \eta), \tag{25}$$

which is equivalent to

$$I(\xi \mid f(\eta)) \geq I(\xi \mid \eta). \tag{26}$$

If instead of η we observe a function $f(\eta)$ of η, then we obtain from the value of $f(\eta)$ at most as much information on ξ as from the value of η; the uncertainty of ξ given the value of $f(\eta)$ is thus not less than its uncertainty given the value of η.

PROOF OF (26). If $f(y_k) \neq f(y_l)$ for $k \neq l$, we have equality in (25); if for instance $f(y_k) = f(y_l) \neq f(y_m)$ for $m \neq k$, $m \neq l$ then to the terms $q_k I(\mathscr{P}_k) + q_l I(\mathscr{P}_l)$ (cf. (11)) figuring on the right hand side of (26) there corresponds a single term on the left hand side, viz. $(q_k + q_l) I(\mathscr{P}_{k,l})$, where $\mathscr{P}_{k,l}$ is the conditional distribution of ξ under the condition that η takes on one of the values y_k or y_l. Clearly

$$P(\xi = x_j \mid \eta = y_k \text{ or } \eta = y_l) = \frac{q_k p_{j|k} + q_l p_{j|l}}{q_k + q_l}.$$

If we apply Jensen's inequality to the convex function $x \log_2 x$, we obtain

$$(q_k + q_l) I(\mathscr{P}_{k,l}) \geq q_k I(\mathscr{P}_k) + q_l I(\mathscr{P}_l).$$

The case when several values of $f(y_k)$ are equal to each other can be dealt with similarly. Thus we proved (26), hence (25) too.

§ 4. The gain of information

The same example which served to derive Shannon's formula can be used to get a heuristic idea of the notion of *gain of information*. Let E be a set containing N elements and let $E_1, \ldots, E_n$ be a partition of this set. If N_k is the number of elements of E_k, we have $N = \sum\limits_{k=1}^{n} N_k$ and we put $p_k = \dfrac{N_k}{N}$. Let the elements of E be labelled from 1 to N, $E = \{e_1, e_2, \ldots, e_N\}$ and let the elements of E_k ($k = 1, \ldots, n$) be labelled from 1 to N_k. An element of E chosen at random (all elements having the same probability $\dfrac{1}{N}$ of being chosen) may be characterized in two distinct manners: a) by giving its serial number in E which we denote by ξ; b) by giving the set E_k to which it belongs and its serial number in E_k. The index k of the relevant set E_k is a random variable which we denote by η. The index of the element in question in the set E_η will be denoted by ζ. Then we have

$$I(\xi) = I(\eta) + I(\zeta \mid \eta) \tag{1}$$

where, clearly

$$I(\xi) = \log_2 N, \quad I(\eta) = \sum_{k=1}^{n} p_k \log_2 \frac{1}{p_k}$$

and

$$I(\zeta \mid \eta) = \sum_{k=1}^{n} p_k \log_2 N_k.$$

Now let E' be a nonempty subset of E and let E'_k $(k = 1, 2, \ldots, n)$ denote the intersection of E_k and E'. Let N'_k be the number of elements of E'_k, N' the number of elements of E' and put $q_k = \dfrac{N'_k}{N}$. Then we have $\sum\limits_{k=1}^{n} N'_k =$ $= N'$, hence $\sum\limits_{k=1}^{n} q_k = 1$. Suppose that we know about an element chosen at random that it belongs to E'; what amount of information will be furnished hereby about η? The original (a priori) distribution of η was $\mathscr{P}(p_1, p_2, \ldots, p_n)$; after the information telling us that the chosen element belongs to E', η has the (a posteriori) distribution $Q = (q_1, q_2, \ldots, q_n)$. At the first sight one could think that the information gained is $I(\mathscr{P}) - I(Q)$. This, however, cannot be true, since $I(\mathscr{P}) - I(Q)$ may be negative, while the gain of information must always be positive. The quantity $I(\mathscr{P}) - I(Q)$ is the decrease of uncertainty of η; we are, however, looking for the gain of information with respect to η resulting from the knowledge that e_ξ belongs to E'. Let the quantity looked for be denoted by $I(Q \| \mathscr{P})$[1]; it can be determined by the following reasoning: The statement $e_\xi \in E'$ contains the information $\log_2 \dfrac{N}{N'}$. This information consists of two parts; first the information given by the proposition $e_\xi \in E'$ about the value of η, next the information given by this proposition about the value of ζ if η is already known. The second part is easy to calculate; in fact if $\eta = k$, the information obtained is equal to $\log_2 \dfrac{N_k}{N'_k}$ and since this information presents itself with probability q_k, the information about the value of ζ is

$$\sum_{k=1}^{n} q_k \log_2 \frac{N_k}{N'_k} \, .$$

Hence

$$\log_2 \frac{N}{N'} = I(Q \| \mathscr{P}) + \sum_{k=1}^{n} q_k \log_2 \frac{N_k}{N'_k} \, . \tag{2}$$

Since

$$\sum_{k=1}^{n} q_k = 1 \quad \text{and} \quad \frac{N N'_k}{N' N_k} = \frac{q_k}{p_k} \, ,$$

we find that

$$I(Q \| \mathscr{P}) = \sum_{k=1}^{n} q_k \log_2 \frac{q_k}{p_k} \, . \tag{3}$$

The quantity $I(Q \| \mathscr{P})$ depends only on the distributions $\mathscr{P}$ and Q; it

[1] We use a double bar $\|$ in $I(Q \| \mathscr{P})$ in order to avoid confusion with the conditional information $I(\xi \mid \eta)$.

follows thus from Jensen's inequality that we have always

$$I(Q \| \mathscr{P}) \geq 0. \tag{4}$$

The equality sign occurs in (4) only if the distributions $\mathscr{P}$ and Q are identical. $I(Q \| \mathscr{P})$ is defined only if every p_k is positive and if there exists a one-to-one correspondence between the individual terms of the two distributions. The quantity $I(Q \| \mathscr{P})$, defined by (3), will be called the *gain of information* resulting from the replacement of the (a priori) distribution $\mathscr{P}$ by the (a posteriori) distribution Q.

The gain of information is one of the most important notions in information theory; it may even be considered as the fundamental one, from which all others can be derived. In § 6 we shall build up information theory in this fashion; the gain of information, as a basic concept, will be defined by postulates.

The relative information introduced in the preceding section can be expressed as follows by means of the gain of information. Let ξ and η be random variables assuming the distinct values $x_1, x_2, \ldots, x_m$ and $y_1, y_2, \ldots, y_n$ with positive probabilities $p_j = P(\xi = x_j)$ and $q_k = P(\eta = y_k)$ respectively; put $\mathscr{P} = (p_1, \ldots, p_m)$, $Q = (q_1, q_2, \ldots, q_n)$,

$$P(\xi = x_j, \eta = y_k) = r_{jk}, \quad P(\xi = x_j \mid \eta = y_k) = p_{j|k},$$

$$\mathscr{P}_k = (p_{1|k}, p_{2|k}, \ldots, p_{m|k}).$$

Then we have

$$I(\xi, \eta) = \sum_{k=1}^{n} q_k \, I(\mathscr{P}_k \| \mathscr{P}). \tag{5}$$

Indeed by (3)

$$I(\mathscr{P}_k \| \mathscr{P}) = \sum_{j=1}^{m} p_{j|k} \log_2 \frac{p_{j|k}}{p_j},$$

hence, because of $q_k p_{j|k} = r_{jk}$

$$\sum_{k=1}^{n} q_k \, I(\mathscr{P}_k \| \mathscr{P}) = \sum_{j=1}^{m} \sum_{k=1}^{n} r_{jk} \log_2 \frac{r_{jk}}{p_j q_k}. \tag{6}$$

From this (5) can be derived by Formula (22) of § 3. Formula (5) means that the amount of information on ξ which is contained in the value of η is equal to the expectation of the gain of information obtained by replacing the distribution $\mathscr{P}$ of ξ by the conditional distribution $\mathscr{P}_k$.

If $\mathscr{P} = (p_1, \ldots, p_n)$ is any distribution having n terms and if $\mathscr{E}_n =$

$$= \left(\frac{1}{n}, \frac{1}{n}, \ldots, \frac{1}{n}\right), \quad \text{we have}$$

$$I(\mathscr{P} \| \mathscr{E}_n) = \sum_{k=1}^{n} p_k \log_2 np_k = \log_2 n - I(\mathscr{P}) = I(\mathscr{E}_n) - I(\mathscr{P}). \qquad (7)$$

The gain of information obtained by replacing the uniform distribution by the distribution $\mathscr{P}$ is thus equal in this case to the decrease of uncertainty. But in general the quantities $I(Q \| \mathscr{P})$ and $I(\mathscr{P}) - I(Q)$ are not equal.

Though in general $I(\mathscr{P}_k \| \mathscr{P}) \neq I(\mathscr{P}) - I(\mathscr{P}_k)$, Formula (5) still expresses that the averages of these two quantities are equal. For according to the first definition of relative information,

$$I(\xi, \eta) = I(\xi) - I(\xi \mid \eta) = \sum_{k=1}^{n} q_k \left(I(\mathscr{P}) - I(\mathscr{P}_k)\right),$$

hence, according to (5)

$$\sum_{k=1}^{n} q_k \left(I(\mathscr{P}) - I(\mathscr{P}_k)\right) = \sum_{k=1}^{n} q_k I(\mathscr{P}_k \| \mathscr{P}). \qquad (8)$$

But only the sums on the two sides of (8) are equal; the single terms have not necessarily the same value.

The following symmetric expression is also often considered in information theory:

$$J(\mathscr{P}, Q) = I(Q \| \mathscr{P}) + I(\mathscr{P} \| Q). \qquad (9)$$

This expression was first studied by Jeffreys. A simple calculation shows that

$$J(\mathscr{P}, Q) = \sum_{k=1}^{n} (p_k - q_k) \log_2 \frac{p_k}{q_k}. \qquad (10)$$

Let us remark that while certain terms of the sum (3) defining $I(Q \| \mathscr{P})$ may be negative and we know only that the sum itself is nonnegative, on the contrary, on the right hand side of (10) all terms are nonnegative.

The relative information can be expressed by means of the gain of information in still another way. If $\mathscr{R}$ is the distribution $\{r_{jk}\}$, $\mathscr{P} * Q$ the distribution $\{p_j q_k\}$, then it follows from Formula (22) of § 3 that

$$I(\xi, \eta) = I(\mathscr{R} \| \mathscr{P} * Q). \qquad (11)$$

The information concerning ξ contained in the value of η is thus equal to the gain of information obtained by replacing the direct product of the distributions of ξ and η by their actual joint distribution.

§ 5. The statistical meaning of information

Let the possible outcomes of an experiment $\mathscr{A}$ be denoted by $A_1, A_2, \ldots,$ A_r; let their probabilities be $P(A_k) = p_k$ $(k = 1, 2, \ldots, r)$. Let $\mathscr{P}$ denote the distribution $(p_1, p_2, \ldots, p_r)$; consider n independent repetitions of the experiment $\mathscr{A}$. The probability of an outcome of this sequence of experiments (when we take into account the order of the experiments) is given by $\pi_n = p_1^{\nu_1}, p_2^{\nu_2}, \ldots, p_r^{\nu_r}$, where ν_k means the number of experiments leading to the outcome A_k. Since the ν_k are random variables, π_n is a random variable too. The expectation of ν_k being equal to np_k, we have

$$E\left(\frac{1}{n} \log_2 \frac{1}{\pi_n}\right) = \sum_{k=1}^{r} p_k \log_2 \frac{1}{p_k} = I(\mathscr{P}). \tag{1}$$

The information $I(\mathscr{P})$ may be interpreted as the expectation of $\frac{1}{n} \log_2 \frac{1}{\pi_n}$. According to the law of large numbers

$$\lim_{n \to \infty} \mathrm{st} \frac{\nu_k}{n} = p_k,$$

hence

$$\lim_{n \to \infty} P\left(\left|\frac{1}{n} \log_2 \frac{1}{\pi_n} - I(\mathscr{P})\right| \le \varepsilon\right) = 1 \tag{2}$$

for every $\varepsilon > 0$ (see Ch. VII, § 14, Exercise 6).

If instead of the expectation of $\frac{1}{n} \log_2 \frac{1}{\pi_n}$ we consider the analogous quantity $\frac{1}{n} \log_2 \frac{1}{E(\pi_n)}$, then we obtain

$$\frac{1}{n} \log_2 \frac{1}{E(\pi_n)} = \log_2 \frac{1}{(\sum\limits_{k=1}^{r} p_k^2)}.$$

This quantity was already mentioned in § 2; it can also be considered as a measure of information. We shall return to this question in § 6.

There is still another point of view showing that the definition of information is suitable. The unit of information ("bit") was defined as the amount of information contained in a symbol which can assume only the two values 0 and 1. Such a symbol will be called a 0−1-*symbol*. We shall now consider whether the outcome of an experiment can actually be characterized on the average by $I(\mathscr{P})$ 0−1-symbols. We show that this is really possible, if certain highly improbable events are neglected.

THEOREM 1. *Let $\mathscr{A}$ be an experiment having possible outcomes $A_1, A_2, \ldots,$ A_r occurring with probabilities $p_k = P(A_k) > 0$ $(k = 1, 2, \ldots, r)$. Put $\mathscr{P} =$ $= (p_1, p_2, \ldots, p_r)$. Then for any given $\varepsilon > 0$ and $\delta > 0$ there exists an n_0 depending only on $\mathscr{P}$, ε, and δ such that if there are performed n $(n \geq n_0)$ independent experiments $\mathscr{A}$, then the outcome of this sequence of experiments can with a probability greater than $1 - \delta$ be expressed uniquely by $n(I(\mathscr{P}) + \varepsilon)$ $0-1$-symbols. If $\varrho > 0$ is arbitrarily small, it is impossible to character-ize the outcome of the sequence of experiments by less than $n(I(\mathscr{P}) - \varepsilon)$ $0-1$-symbols with a probability greater than or equal to ϱ whenever $n \geq n_0'$, where n_0' depends only on $\mathscr{P}$, ε, and ϱ.*

Remark. This means that, if the experiment $\mathscr{A}$ is sufficiently often repeat-ed, for the description of an outcome of the experiment one does not need, on the average, more than $I(\mathscr{P}) + \varepsilon$ $0-1$-symbols; hence the statement that the outcome of $\mathscr{A}$ contains the amount of information $I(\mathscr{P})$ has a quite definite meaning.

PROOF. Choose n_1 large enough so that $n \geq n_1$ should imply

$$P\left[\frac{1}{n}\log_2\frac{1}{\pi_n} - I(\mathscr{P}) \leq \frac{\varepsilon}{2}\right] > 1 - \delta. \tag{3}$$

This is, in view of (2), always possible. This means that the sequences of outcomes obtained by repeating n times the experiment $\mathscr{A}$ can be parti-tioned into two classes: the first consists of the sequences for which

$$\frac{1}{n}\log_2\frac{1}{\pi_n} - I(\mathscr{P}) \leq \frac{\varepsilon}{2}, \tag{4}$$

the second of the remaining ones. According to (3) the probability that a sequence belongs to the second class is less than δ. Let C_n denote the number of sequences of the first class, let $q_1, q_2, \ldots, q_{C_n}$ be their probabilities. By (4) we have

$$2^{-n\left(I(\mathscr{P})+\frac{\varepsilon}{2}\right)} \leq q_j \qquad (j = 1, 2, \ldots, C_n) \tag{5}$$

or, by adding these inequalities,

$$C_n \cdot 2^{-n\left(I(\mathscr{P})+\frac{\varepsilon}{2}\right)} \leq \sum_{j=1}^{C_n} q_j. \tag{6}$$

The sum on the right hand side cannot exceed 1, since it represents precisely the probability that a sequence belongs to the first class. Therefore we have

$$C_n \leq 2^{n\left(I(\mathscr{P})+\frac{\varepsilon}{2}\right)}. \tag{7}$$

Now let us number the events of the first class from 1 to C_n and write these numbers in the binary system. For this $\left[n\left(I(\mathscr{S})+\dfrac{\varepsilon}{2}\right)\right]+1$ binary digits are needed. There can be found an n_2 such that for $n \geq n_2$ the inequality

$$\left[n\left(I(\mathscr{S})+\frac{\varepsilon}{2}\right)\right]+1 < n(I(\mathscr{S})+\varepsilon) \tag{8}$$

holds. Put $n_0 = \max(n_1, n_2)$; it is clear that n_0 depends only on ε, δ, and $\mathscr{S}$ and satisfies the requirements of the theorem. It is easy to show that with large probability $n(I(\mathscr{S}) - \varepsilon)\,0-1$-symbols are not sufficient to describe the outcome of the sequence of experiments. To see this, subdivide again the set of the sequences into two classes: let the first class contain the sequences for which

$$I(\mathscr{S}) - \frac{1}{n}\log_2 \frac{1}{\pi_n} \leq \frac{\varepsilon}{2}, \tag{9}$$

and the second the remaining ones. Choose an n_3 such that for $n \geq n_3$ the probability of (9) exceeds $1 - \delta$; this is possible because of (2). Let D_n denote the number of sequences in the first class and let $r_1, r_2, \ldots, r_{D_n}$ be the corresponding probabilities. We have then

$$2^{-n\left(I(\mathscr{S})-\frac{\varepsilon}{2}\right)} \geq r_j \qquad (j = 1, 2, \ldots, D_n). \tag{10}$$

Furthermore, by assumption

$$\sum_{j=1}^{D_n} r_j \geq 1 - \delta. \tag{11}$$

If we select some outcomes and assign to them sequences of zeros and ones of length not exceeding $n(I(\mathscr{S}) - \varepsilon)$, the number of these sequences will be less than $2^{n(I(\mathscr{S})-\varepsilon)}$, hence the total probability of the selected outcomes will be at most

$$2^{n(I(\mathscr{S})-\varepsilon)}\, 2^{-n\left(I(\mathscr{S})-\frac{\varepsilon}{2}\right)} + \delta = 2^{-\frac{n\varepsilon}{2}} + \delta.$$

The total probability of the outcomes not considered is thus at least $1 - \delta - 2^{-\frac{n\varepsilon}{2}} > 1 - 2\delta$, provided that $n \geq n_4$.

Suppose that $n \geq n_0' = \max(n_3, n_4)$ and that, contrary to the statement in the second half of the theorem, it is possible to characterize the outcome of the sequence of experiments by less than $n(I(\mathscr{S}) - \varepsilon)\,0-1$-symbols with a probability $\geq \varrho > 0$. If we choose δ such that $2\delta < \varrho$, then this contradicts what was just proved.

Theorem 1 is therefore completely proved. It can be sharpened in the following manner:

THEOREM 2. *For every $\delta > 0$ there can be given an n_0 such that for $n \geq n_0$ the outcome of n independent experiments $\mathscr{A}$ can be uniquely expressed, with probability $> 1 - \delta$, by at most $nI(\mathscr{P}) + K\sqrt{n}$ $0-1$-symbols; K is here a positive constant which depends only on δ.*

However, there corresponds to every ϱ between 0 and 1 a constant K' and an integer n_0' such that a unique characterization of the outcome of a sequence of experiments becomes impossible (with a probability $\geq \varrho$, for $n \geq n_0'$) by less than $nI(\mathscr{P}) - K'\sqrt{n}$ $0-1$-symbols.

PROOF. It is easy to show[1] that the distribution of the random variable

$$\sqrt{n}\left(\frac{1}{n}\log_2\frac{1}{\pi_n} - I(\mathscr{P})\right) = \sum_{k=1}^{r} \frac{\nu_k - np_k}{\sqrt{n}}\log_2\frac{1}{p_k}$$

tends to the normal distribution as $n \to \infty$. There exists thus a constant K which depends only on δ such that we have for sufficiently large n

$$P\left(\left|\frac{1}{n}\log_2\frac{1}{\pi_n} - I(\mathscr{P})\right| < \frac{K}{\sqrt{n}}\right) > 1 - \delta. \tag{12}$$

The continuation of the proof runs exactly as that of Theorem 1.

Theorem 1 can also be considered as a justification of Shannon's definition of information. The statement of Theorem 1 can be translated into the language of communication theory as follows: Let a message source at the moment t ($t = 1, 2, \ldots$) emit a random signal ξ_t; let $x_1, x_2, \ldots, x_r$ be the possible signals; let $p_k = P(\xi_t = x_k)$ denote the probabilities of the individual signals. These probabilities are supposed to be independent of t. Assume that the signals are independent of each other. Assume further that for transmission the signals must be transformed (encoded) since the "channel" (transmission network) can only transmit two signs.[2] (This is the case e.g. if the channel works with electric current and at every instant only two cases are possible: the current is either on or off.) Let 0 denote one of

---

[1] The proof is the same as that in Exercise 26 of Ch. VIII, § 12.

[2] In information theory the word "channel" has a very general sense: it means any process capable to transmit information.

the signs and 1 the other. The question is then, how many 0 or 1 symbols are necessary for the transmission of the information contained in n signs $\xi_1, \xi_2, \ldots, \xi_n$ furnished by the source. According to Theorem 1 with probability arbitrarily near to 1 less than $n(I(\mathscr{P}) + \varepsilon)$ symbols are required, provided that n is sufficiently large. This shows the importance of the quantity $I(\mathscr{P})$ for communication engineering.

Let us mention an important particular case. If $p_1 = p_2 = \ldots = p_r = \dfrac{1}{r}$, then $I(\mathscr{P}) = \log_2 r$; therefore in order to encode a signal of such a source into $0-1$-symbols, on the average $\log_2 r$ symbols are necessary. (Of course this can be shown directly.) If for instance a number written in the decimal system is transcribed into a binary system, the number of digits increases on the average by the factor $\log_2 10 = 3.3219 \ldots$. This is of importance for computers, which work in the binary system.

If the source emits signals x_k with probabilities p_k ($k = 1, 2, \ldots, r$) and if the channel can transmit s different signs, approximately $\dfrac{nI(\mathscr{P})}{\log_2 s}$ signs are necessary in order to transmit a message of n signs if the most economic coding is applied.

It is to be noticed that optimal or nearly optimal codings are very complicated and are feasible only for long sequences of signals. Hence in practice usually such codes are employed which to some extent take into account the statistical nature of the source, but are more easy to handle than the nearly optimal codings. In particular, the signals are coded one by one or by small groups (as for instance in the encoding of letters into Morse signals).

The message sources encountered in practice are generally much more complicated than those described above. The individual signals are, in general, not independent of each other. E.g. in every natural language the letters have not only different probabilities, but the probability of a letter depends also on the letters preceding it in the text. This can also be taken into account, but we do not deal with these questions here.

The channels actually used in communication theory are also much more complicated than those discussed above. In practice, it is of the great importance to know how to transmit the information through a channel which, with a certain probability, distorts the transmitted signal. Then one cannot be sure that the received signal is identical to the emitted one. (E.g. in broadcasting the distortions caused by the transmission through the atmosphere are perceived as *noise*.) Such channels are called *noisy channels*. Information theory takes this into account, but our brief introduction does not permit to go into these questions.

§ 6. Further measures of information

In the present section we give another characterization of the information; this approach will show what other quantities can be considered as measures of information besides that of Shannon's.

Shannon's information was defined in § 2 by postulates and by means of Shannon's information we introduced the notion of the gain of information. We shall now follow the inverse procedure: we define first the gain of information by a set of postulates; from this then we shall derive a measure of information.

We start from a generalization of the notion of a random variable. Let $[\Omega, \mathscr{A}, P]$ be a Kolmogorov probability space. We define an *incomplete random variable* as a function $\xi = \xi(\omega)$ measurable with respect to the measure on $\mathscr{A}$ and defined on a subset Ω_1 of Ω, where $\Omega_1 \in \mathscr{A}$ and $P(\Omega_1) > 0$. The only difference between an ordinary random variable and an incomplete random variable is thus that the latter is not necessarily defined for every $\omega \in \Omega$. In this sense, ordinary random variables can be considered as particular cases of incomplete random variables.

If ξ is an incomplete random variable assuming values x_k with probabilities p_k ($p_k > 0$; $k = 1, 2, \ldots, n$), we have $\sum\limits_{k=1}^{n} p_k \leq 1$ and not necessarily

$$\sum_{k=1}^{n} p_k = 1.$$

The discrete incomplete random variables ξ and η are said to be independent, if for any two sets A and B the events $\xi \in A$ and $\eta \in B$ are independent.

The distribution of an incomplete random variable will be called an *incomplete probability distribution*; in this sense the ordinary distributions can be considered as a particular case of the latter. Thus if $p_k > 0$ ($k = 1, \ldots, n$) and $\sum\limits_{k=1}^{n} p_k \leq 1$, then $\{p_k\}$ is a finite discrete incomplete distribution. The direct product of two incomplete distributions $\mathscr{P} = \{p_j\}$ ($j = 1, \ldots, n$) and $Q = \{q_k\}$ ($k = 1, 2, \ldots, n$) is defined as the incomplete distribution $\{p_j q_k\}$ ($j = 1, \ldots, m$; $k = 1, \ldots, n$) and will be denoted by $\mathscr{P} * Q$.

To every incomplete distribution $\mathscr{P} = (p_1, \ldots, p_n)$ there can be assigned an ordinary distribution $\mathscr{P}' = (p'_1, \ldots, p'_n)$ by putting

$$p'_k = \frac{p_k}{\sum\limits_{j=1}^{n} p_j}.$$

Let ξ be an incomplete random variable taking on values x_k with probabilities p_k $(k = 1, 2, \ldots, n)$; put $s = \sum_{k=1}^{n} p_k$. If $0 < s < 1$, ξ can be interpreted as a quantity depending on the outcome of an experiment, but not defined for all outcomes of the experiment. For example ξ is only defined if the outcome is observable, which happens with probability s, where $0 < s < 1$. In this case the corresponding distribution $\mathscr{P}'$ may be interpreted as the conditional distribution of ξ with respect to the condition that the outcome of the experiment is observable. Therefore $\mathscr{P}'$ is said to be the *complete conditional distribution* of the incomplete random variable ξ.

We shall now define the mean gain of information obtained if the (incomplete) distribution $\mathscr{P} = (p_1, \ldots, p_n)$ $(p_k > 0$ for $k = 1, \ldots, n)$ of the incomplete random variable ξ is replaced by the incomplete distribution $Q = (q_1, \ldots, q_n)$. Before stating the postulates we make two remarks:

1. The gain of information denoted by $I(Q \| \mathscr{P})$ is defined only if $\mathscr{P}$ and Q have the same number of terms and these are in a one-to-one correspondence defined by their indices.

2. We supposed $p_k > 0$ for all values of k; however, some q_k (but not all) can be equal to 0.

The quantity $I(Q \| \mathscr{P})$ has to satisfy the following postulates:

POSTULATE I. *If* $\mathscr{P} = \mathscr{P}_1 * \mathscr{P}_2$ *and* $Q = Q_1 * Q_2$, *then*

$$I(Q \| \mathscr{P}) = I(Q_1 \| \mathscr{P}_1) + I(Q_2 \| \mathscr{P}_2). \tag{1}$$

Remark. This means that if we put $\mathscr{P}_i = (p_{i1}, \ldots, p_{in})$, $Q_i = (q_{i1}, \ldots, q_{in})$ $(i = 1$ or $2)$, then there corresponds to the element $p_{1j} p_{2k}$ of $\mathscr{P}$ the element $q_{1j} q_{2k}$ of Q. Postulate I is a general formulation of the additivity of information.

POSTULATE II. *If* $p_k \leq q_k$ $(k = 1, 2, \ldots, n)$, *then we have* $I(Q \| \mathscr{P}) \geq 0$; *for* $p_k \geq q_k$ $(k = 1, 2, \ldots, n)$ *we have* $I(Q \| \mathscr{P}) \leq 0$.

Remark. It follows from this that $I(\mathscr{P} \| \mathscr{P}) = 0$. For complete distributions $\mathscr{P}$ and Q Postulate II asserts nothing more than this, as then the inequalities $p_k \leq q_k$ $(k = 1, 2, \ldots, n)$ occur only if $p_k = q_k$ $(k = 1, 2, \ldots, n)$, since $\sum_{k=1}^{n} p_k = \sum_{k=1}^{n} q_k = 1$. In the case of incomplete distributions, however, this postulate leads to important conclusions.

Let $\mathscr{E}_p$ be the distribution consisting of a single term $\{p\}$ $(0 < p \leq 1)$. We require:

POSTULATE III. $I(\mathscr{E}_1 \| \mathscr{E}_{\frac{1}{2}}) = 1$.

This postulate fixes the unit of gain of information.

Before proceeding further, we determine the function

$$g(q,p) = I(\mathscr{E}_q \| \mathscr{E}_p) \qquad (0 < p \le 1, 0 \le q \le 1).$$

It follows from (1) that

$$g(q_1 q_2, p_1, p_2) = g(q_1, p_1) + g(q_2, p_2). \tag{2}$$

If we put $q_1 = q_2 = 1$, we find

$$g(1, p_1 p_2) = g(1, p_1) + g(1, p_2). \tag{3}$$

If we put $q_1 = p_2 = 1$, $p_1 = p$, $q_2 = q$, we obtain

$$g(q, p) = g(1, p) + g(q, 1). \tag{4}$$

Hence, according to Postulate II,

$$g(1, p) + g(p, 1) = 0. \tag{5}$$

We conclude from (4) and (5) that

$$g(q, p) = g(1, p) - g(1, q). \tag{6}$$

Now $g(1, p)$ being, by Postulate II, a decreasing function of p, it follows from (3) by a well known theorem that

$$g(1, p) = c \, \log_2 \frac{1}{p}$$

with $c > 0$. According to Postulate III $c = 1$, thus

$$I(\mathscr{E}_1 \| \mathscr{E}_p) = g(1, p) = \log_2 \frac{1}{p}, \tag{7}$$

and by (6)

$$I(\mathscr{E}_q \| \mathscr{E}_p) = g(q, p) = \log_2 \frac{1}{p} - \log_2 \frac{1}{q} = \log_2 \frac{q}{p}. \tag{8}$$

If we observe the occurrence of an event having probability p, we get the amount of information $\log_2 \dfrac{1}{p}$; if p is replaced by q the gain of information is $\log_2 \dfrac{q}{p}$.

The quantity $\log_2 \dfrac{1}{p}$ can also be considered as measuring the uncertainty

of the occurrence of an event with probability p; the quantity $\log_2 \dfrac{q}{p} =$

$= \log_2 \dfrac{1}{p} - \log_2 \dfrac{1}{q}$ is the decrease of the uncertainty resulting from

the replacement of p by q (note that this "decrease" can be negative as well; indeed if $q < p$, the uncertainty increases).

We introduce now a new notion. If we replace an incomplete distribution $\mathscr{P} = (p_1, \ldots, p_n)$ by an incomplete distribution $Q = (q_1, \ldots, q_n)$, we obtain

with probability q_k the information $\log_2 \dfrac{q_k}{p_k}$ $(k = 1, 2, \ldots, n)$. Put

$$q'_k = \frac{q_k}{\displaystyle\sum_{j=1}^{n} q_j}.$$

The conditional probability that we obtain the information $\log_2 \dfrac{q_k}{p_k}$ under the condition that at least one observation occurs, is equal to q'_k $(k = 1, 2, \ldots, n)$. Put

$$F(Q, \mathscr{P}, x) = \sum_{\log_2 \frac{q_k}{p_k} < x} q'_k; \tag{9}$$

$F(Q, \mathscr{P}, x)$ will be called the *conditional distribution function of the gain of information*.

Now we can formulate our further requirements:

POSTULATE IV. *$I(Q \parallel \mathscr{P})$ depends only on the function $F(Q, \mathscr{P}, x)$.*

Because of this postulate we can also write $I[F(x)]$ instead of $I(Q \parallel \mathscr{P})$, where $F(x) = F(Q, \mathscr{P}, x)$.

Notes. 1. If $Q = \mathscr{E}_q$, $\mathscr{P} = \mathscr{E}_p$, we have

$$F(Q, \mathscr{P}, x) = \begin{cases} 0 \text{ for } x \le \log_2 \dfrac{q}{p}, \\[2mm] 1 \text{ otherwise.} \end{cases}$$

Postulate IV is thus fulfilled and (8) expresses that for a degenerate distribution function

$$F(x) = D_c(x) = \begin{cases} 0 \text{ for } x \leq c, \\ 1 \text{ otherwise,} \end{cases} \qquad (10)$$

(where c can be any real number) we have the relation $I[D_c(x)] = c$.

2. Every distribution function $F(x)$ of a finite discrete distribution can be written in the form $F(x) = F(Q, \mathscr{P}, x)$ where $\mathscr{P}$ and Q are suitably chosen incomplete distributions. Indeed let a_k be the discontinuity points of $F(x)$, with jumps w_k $(k = 1, 2, \ldots, n; \sum_{k=1}^{n} w_k = 1)$, then we have to determine numbers p_k and q_k $(k = 1, 2, \ldots, n)$ such that the relations

$$a_k = \log_2 \frac{q_k}{p_k}, \quad w_k = \frac{q_k}{q_1 + q_2 + \ldots + q_n} \qquad (k = 1, 2, \ldots, n) \qquad (11)$$

hold. This is the case if we take $q_k = tw_k$ and $p_k = tw_k 2^{-a_k}$. If we choose the number t such that

$$0 < t \leq \min\left(1, \frac{1}{\sum_{k=1}^{n} w_k \cdot 2^{-a_k}}\right), \qquad (12)$$

then we obtain a system of solution satisfying all our hypotheses.

$I[F]$ is thus a functional defined on the set $\mathscr{F}$ of all distribution functions of finite discrete distributions. The following postulates concern the properties of this functional.

POSTULATE V. *If $F \in \mathscr{F}$, $G \in \mathscr{F}$, $F \not\equiv G$ and $G(x) \geq F(x)$ $(-\infty < x < +\infty)$, then*

$$I[G(x)] < I[F(x)].$$

Remark. This postulate contains Postulate II. In fact if $p_k \leq q_k$ $(k = 1, 2, \ldots, n)$ we have $\log_2 \frac{q_k}{p_k} \geq 0$, hence $F(Q, \mathscr{P}, x) \leq D_0(x)$. From this follows by Postulate V that $I(Q \| \mathscr{P}) \geq I[D_0(x)] = 0$; and if $p_k \geq q_k$, the inequality is reversed. However, Postulate II is not superfluous, since in order to state Postulate V we used relation (8) resting on Postulate II.

POSTULATE VI. *Let $F_i \in \mathscr{F}$ $(i = 1, 2, 3)$ and $I[F_2] = I[F_3]$. Then for every*

t $(0 \leq t \leq 1)$ *we have*

$$I[tF_1 + (1 - t)F_2] = I[tF_1 + (1 - t)F_3];$$

furthermore, $I[tF_1 + (1 - t)F_2]$ *is a continuous function of* t.[1]

Remark. Postulate VI may be called the postulate of quasi-linearity. Now we can state

THEOREM 1. *If* $I(Q \| \mathscr{S})$ *satisfies Postulates* I *to* VI, *then:*
— *either there exists a real number* $\alpha \neq 1$ *such that* $I(Q \| P) = I_\alpha(Q \| \mathscr{S})$, *defined by*

$$I_\alpha(Q \| \mathscr{S}) = \frac{1}{\alpha - 1} \log_2 \left(\frac{1}{\sum\limits_{k=1}^{n} q_k} \sum_{k=1}^{n} \frac{q_k^\alpha}{p_k^{\alpha-1}} \right), \tag{13a}$$

— *or* $I(Q \| \mathscr{S}) = I_1(Q \| \mathscr{S})$ *with*

$$I_1(Q \| \mathscr{S}) = \frac{1}{\sum\limits_{k=1}^{n} q_k} \sum_{k=1}^{n} q_k \log_2 \frac{q_k}{p_k}. \tag{13b}$$

(13b) *is the limit of* (13a) *for* $\alpha \to 1$:

$$\lim_{\alpha \to 1} I_\alpha(Q \| \mathscr{S}) = I_1(Q \| \mathscr{S}). \tag{14}$$

Remark. If $\mathscr{S}$ and Q are complete distributions, $I_1(Q \| \mathscr{S})$ is identical to Shannon's gain of information defined by Formula (3) of § 4.

The quantity $I_\alpha(Q \| \mathscr{S})$ will be called the *measure of order* α *of the gain of information;* $I_1(Q \| \mathscr{S})$ *will be called Shannon's gain of information or measure of order* 1 *of the gain of information.*

In order to avoid confusions, Shannon's gain of information will from now on always be denoted by $I_1(Q \| \mathscr{S})$ instead of $I(Q \| \mathscr{S})$.

PROOF.[2] Instead of $I_\alpha(Q \| \mathscr{S})$ we use also the notation $I_\alpha(F(Q, \mathscr{S}, x.))$ Then (13a) and (13b) are written as

$$I_\alpha(F) = \frac{1}{\alpha - 1} \log_2 \int_{-\infty}^{+\infty} 2^{(\alpha-1)x} \, dF(x) \quad \text{for } \alpha \neq 1 \tag{15}$$

[1] The assumption of continuity is not indispensable to the proof of Theorem 1; its only purpose is to simplify our proof.
[2] The following proof is a combination of the proofs of two theorems from the theory of functional equations. Cf. G. H. Hardy, J. Littlewood and G. Pólya [1], pp. 215 and 84.

and

$$I_1(F) = \int_{-\infty}^{+\infty} x\,dF(x).\tag{16}$$

From these formulae we see that $I_a(Q \parallel \mathscr{P})$ satisfies for every α Postulates I through VI. It remains still to show that no other functional can satisfy all these Postulates. A simple calculation shows that

$$F(Q_1 * Q_2, \mathscr{P}_1 * \mathscr{P}_2, x) = \int_{-\infty}^{+\infty} F(Q_1, \mathscr{P}_1, x - y)\,dF(Q_2, \mathscr{P}_2, y),\tag{17}$$

which permits to rewrite Postulates I and III in the following form:

POSTULATE I'. *If $F \in \mathscr{F}$, $G \in \mathscr{F}$ and if we put*

$$F * G = \int_{-\infty}^{+\infty} F(x - y)\,dG(y),$$

then we have

$$I[F * G] = I[F] + I[G].\tag{18}$$

POSTULATE III'. $I[D_1(x)] = 1$.

We show now that Postulates I', III', IV, V and VI are satisfied only by the functionals (15) and (16).

Let $\mathscr{F}_A$ be the class of finite discrete distribution functions with $F(-A) = 0$, $F(A) = 1$. We deduce from Postulate VI by induction that from the relations

$$F_i \in \mathscr{F}, \quad F_i' \in \mathscr{F}, \quad I[F_i] = I[F_i'], \quad w_i > 0 \qquad (i = 1, 2, \ldots, r)$$

and $\sum_{i=1}^{r} w_i = 1$ the relation

$$I[\sum_{i=1}^{r} w_i F_i] = I[\sum_{i=1}^{r} w_i F_i']\tag{19}$$

follows. We know already by Postulates I', III', and V that

$$I[D_c(x)] = c\tag{20}$$

holds for every real c, where $D_c(x)$ is the degenerate distribution function of the constant c (see Formula (10)).

Let

$$\psi_A(t) = I[(1 - t)D_{-A}(x) + tD_A(x)].\tag{21}$$

$\psi_A(t)$ is a strictly increasing continuous function of t; further $\psi_A(0) = -A$ and $\psi_A(1) = +A$. Put $t = \varphi_A(u)$ for $u = \psi_A(t)$. As $\varphi_A(u)$ is the inverse function of $\psi_A(t)$, it is continuous and strictly increasing in the interval $(-A, +A)$. From this we derive

$$I[D_u(x)] = u = \psi_A(t) = I[(1 - \varphi_A(u)) D_{-A}(x) + \varphi_A(u) D_A(x)]. \quad (22)$$

Let $F \in \mathscr{F}_A$ be a distribution function which jumps $w_1, w_2, \ldots, w_n$ at the points $a_1, a_2, \ldots, a_n$, we have

$$F = F(x) = \sum_{k=1}^{n} w_k D_{a_k}(x), \quad (23)$$

and, according to (19) and (22),

$$I[F] = I\Big[\sum_{k=1}^{n} w_k \big((1 - \varphi_A(a_k)) D_{-A}(x) + \varphi_A(a_k) D_A(x)\big)\Big], \quad (24)$$

hence

$$I[F] = I\Big[\big(1 - \sum_{k=1}^{n} w_k \varphi_A(a_k)\big) D_{-A}(x) + \big(\sum_{k=1}^{n} w_k \varphi_A(a_k)\big) D_A(x)\Big] \quad (25)$$

or, according to (22) by writing $\varphi_A^{-1}(t)$ instead of $\psi_A(t)$

$$I[F] = \varphi_A^{-1}\Big(\sum_{k=1}^{n} w_k \varphi_A(a_k)\Big). \quad (26)$$

Formula (26) expresses that $I[F]$ is the Kolmogorov–Nagumo quasilinear mean of the numbers a_k with weights w_k $(k = 1, 2, \ldots, n)$. We shall need the following lemma concerning this mean:

LEMMA. *Let $\varphi_1(x)$ and $\varphi_2(x)$ be two continuous and strictly increasing functions in the interval $[J, K]$. Suppose that for arbitrarily chosen numbers $x_1, x_2, \ldots, x_n$ in $[J, K]$ and for positive numbers $w_1, w_2, \ldots, w_n$ with $\sum_{k=1}^{n} w_k = 1$ we always have*

$$\varphi_1^{-1}\Big(\sum_{k=1}^{n} w_k \varphi_1(x_k)\Big) = \varphi_2^{-1}\Big(\sum_{k=1}^{n} w_k \varphi_2(x_k)\Big). \quad (27)$$

This means that the relation

$$\varphi_2(x) = \alpha\varphi_1(x) + \beta \quad (28)$$

holds, where $\alpha > 0$ and β are two constants. (Conversely, (28) implies (27)).

PROOF. It suffices to prove (28) by supposing (27) to hold for $n = 2$, $w_1 = t$, $w_2 = 1 - t$, $0 < t < 1$. Put $\varphi_2(J) = J'$, $\varphi_2(K) = K'$. If x_1 and x_2 describe the interval $[J, K]$, then $y_1 = \varphi_2(x_1)$ and $y_2 = \varphi_2(x_2)$ describe the interval $[J', K']$. Hence if $J' \le y_1 \le K'$, $J' \le y_2 \le K'$ and if we put $\varphi_1(\varphi_2^{-1}(x)) = \varphi_3(x)$ we find

$$\varphi_3(ty_1 + (1 - t)y_2) = t\varphi_3(y_1) + (1 - t)\varphi_3(y_2). \tag{29}$$

$\varphi_3(y)$ is thus a linear function, which proves the first part of the lemma. The converse is trivial.

We show now that there can be found a function $\varphi(x)$, independent of A, such that (26) remains valid if φ_A is replaced by φ. It suffices to prove that

$$\varphi_A(x) = \frac{\varphi_B(x) - \varphi_B(-A)}{\varphi_B(A) - \varphi_B(-A)} \quad \text{for } 0 < A < B. \tag{30}$$

This follows from

$$\varphi_A^{-1}\left(\sum_{k=1}^n w_k \varphi_A(a_k)\right) = \varphi_B^{-1}\left(\sum_{k=1}^n w_k \varphi_B(a_k)\right) \tag{31}$$

or $0 < A < B$, $|a_k| < A$ $(k = 1, 2, \ldots, n)$; Formula (31) itself follows from $\mathscr{F}_A \subset \mathscr{F}_B$ for $A < B$. From (31) and from the lemma we conclude that

$$\varphi_A(x) = \alpha\varphi_B(x) + \beta, \tag{32}$$

and since $\varphi_A(-A) = 0$, $\varphi_A(A) = 1$, we obtain (30).

Thus we proved the existence of a monotone continuous function $\varphi(x)$ which for every $F \in \mathscr{F}$ having jumps w_k at the points a_k $(k = 1, \ldots, n)$ fulfills the relation

$$I[F] = \varphi^{-1}\left(\sum_{k=1}^n w_k \varphi(a_k)\right) = \varphi^{-1}\left(\int_{-\infty}^{+\infty} \varphi(x) \, dF(x)\right). \tag{33}$$

Now we investigate how $\varphi(x)$ can be chosen such that it fulfils also Postulate I'. Put in I'

$$F(x) = tD_a(x) + (1 - t)D_b(x), \quad G(x) = D_y(x).$$

Then we have

$$\varphi^{-1}\big(t\varphi(a + y) + (1 - t)\varphi(b + y)\big) = \varphi^{-1}\big(t\varphi(a) + (1 - t)\varphi(b)\big) + y. \tag{34}$$

Fix y and put $\varphi^*(t) = \varphi(t + y)$. From (34) follows

$$\varphi^{*-1}\big(t\varphi^*(a) + (1 - t)\varphi^*(b)\big) = \varphi^{-1}\big(t\varphi(a) + (1 - t)\varphi(b)\big) \tag{35}$$

for all values of a and b and for $0 \le t \le 1$. It follows from the lemma that

$$\varphi^*(x) = \varphi(x + y) = A(y)\,\varphi(x) + B(y) \quad (A(y) > 0).$$

If $\varphi(0) = 0$, which can be supposed without restriction of generality, then $B(y) = \varphi(y)$, hence

$$\varphi(x + y) = A(y)\,\varphi(x) + \varphi(y). \tag{36}$$

This relation being fulfilled for every y, we may interchange x and y, hence

$$A(y)\,\varphi(x) + \varphi(y) = A(x)\,\varphi(y) + \varphi(x), \tag{37}$$

thus

$$\frac{A(x) - 1}{\varphi(x)} = \frac{A(y) - 1}{\varphi(y)}. \tag{38}$$

From this we obtain $A(y) = k\varphi(y) + 1$, where k is a constant. From (36) and (38) it follows that

$$\varphi(x + y) = k\varphi(x)\,\varphi(y) + \varphi(x) + \varphi(y). \tag{39}$$

We distinguish two cases: $k = 0$ and $k \ne 0$. If $k = 0$,

$$\varphi(x + y) = \varphi(x) + \varphi(y) \tag{40}$$

and, since $\varphi(x)$ is monotone, $\varphi(x) = Bx$, where B is a constant. If $k \ne 0$, put $h(x) = k\varphi(x) + 1$. We conclude from (39)

$$h(x + y) = h(x)\,h(y), \tag{41}$$

and $h(x)$ being monotone,

$$h(x) = 2^{(\alpha - 1)x} \tag{42}$$

with $\alpha \ne 1$, hence

$$\varphi(x) = \frac{2^{(\alpha - 1)x} - 1}{k}. \tag{43}$$

According to the lemma, $\varphi(x)$ can be replaced by $2^{(\alpha - 1)x}$. And, by taking into account (33), we have thus either (15) or (16). The limit relation (14) can be proved e.g. by the rule of l'Hospital. Theorem 1 is herewith proved.

If $\mathscr{P} = \mathscr{E}_n$ is the uniform distribution of n terms and if Q is any incomplete distribution, then for $\alpha \neq 1$ the relation

$$I_\alpha(Q \| \mathscr{E}_n) = \log_2 n - \frac{1}{1-\alpha} \log_2 \left(\frac{\sum\limits_{k=1}^{n} q_k^\alpha}{\sum\limits_{k=1}^{n} q_k} \right) \tag{44}$$

holds. Thus if we put for any incomplete distribution $\mathscr{P} = (p_1, \ldots, p_n)$

$$I_\alpha(\mathscr{P}) = \frac{1}{1-\alpha} \log_2 \left(\frac{\sum\limits_{k=1}^{n} p_k^\alpha}{\sum\limits_{k=1}^{n} p_k} \right), \tag{45a}$$

we find that

$$I_\alpha(Q \| \mathscr{E}_n) = I_\alpha(\mathscr{E}_n) - I_\alpha(Q). \tag{46}$$

(46) shows that the quantity $I_\alpha(\mathscr{P})$ may be considered as a measure of the amount of information corresponding to the distribution $\mathscr{P}$ (or else as a measure of the uncertainty of a random variable with the distribution $\mathscr{P}$). We call $I_\alpha(\mathscr{P})$ the *information of order* α. It is easy to see that

$$I_1(Q \| \mathscr{E}_n) = \log_2 n - \frac{\sum\limits_{k=1}^{n} q_k \log_2 \dfrac{1}{q_k}}{\sum\limits_{k=1}^{n} q_k}. \tag{47}$$

For any incomplete distribution $\mathscr{P} = (p_1, \ldots, p_n)$ we put

$$I_1(\mathscr{P}) = \frac{\sum\limits_{k=1}^{n} p_k \log_2 \dfrac{1}{p_k}}{\sum\limits_{k=1}^{n} p_k}. \tag{45b}$$

and we call this quantity *Shannon's information* or *information of order* 1. If $\mathscr{P}$ is an ordinary distribution, $I_1(\mathscr{P})$ is the entropy or Shannon's information of the distribution $\mathscr{P}$. In what follows, in order to avoid confusions, we shall write always $I_1(\mathscr{P})$ instead of $I(\mathscr{P})$ used in the preceding sections.

For a complete distribution $\mathscr{P}$ the definition of $I_\alpha(\mathscr{P})$ gives

$$I_\alpha(\mathscr{P}) = \frac{1}{1-\alpha} \log_2 \sum_{k=1}^{n} p_k^\alpha \qquad \text{for } \alpha \neq 1. \tag{45c}$$

Clearly, for every distribution function, complete or incomplete,

$$\lim_{\alpha \to 1} I_\alpha(\mathscr{P}) = I_1(\mathscr{P})$$

holds. We study now $I_\alpha(\mathscr{P})$ as a function of α.

THEOREM 2. *Let $\mathscr{P} = (p_1, \ldots, p_n)$ be an incomplete distribution, $\sum_{k=1}^{n} p_k = s \leq 1$. Then $I_\alpha(\mathscr{P})$ is a positive, decreasing function of α. One has $I_0(\mathscr{P}) = \log_2 \dfrac{n}{s}$; in particular, if $\mathscr{P}$ is a complete distribution, then $I_0(\mathscr{P}) = \log_2 n$. Thus for a complete distribution*

$$0 \leq I_\alpha(\mathscr{P}) \leq \log_2 n \qquad (\alpha \geq 0). \tag{48}$$

PROOF. We can write

$$I_\alpha(\mathscr{P}) = \log_2 \left(\frac{\sum\limits_{k=1}^{n} p_k \left(\dfrac{1}{p_k}\right)^{1-\alpha}}{\sum\limits_{k=1}^{n} p_k} \right)^{\frac{1}{1-\alpha}}.$$

We know[1] that the average

$$\left(\sum_{k=1}^{n} w_k x_k^\beta\right)^{\frac{1}{\beta}} \qquad (x_k > 0, w_k > 0, \sum_{k=1}^{n} w_k = 1)$$

s a monotone increasing function of β. Hence Theorem 2 is proved.

Remark. If $p_1 \leq p_2 \leq \ldots \leq p_n$, we have

$$\lim_{\alpha \to -\infty} I_\alpha(\mathscr{P}) = \log_2 \frac{1}{p_1} \text{ and } \lim_{\alpha \to +\infty} I_\alpha(\mathscr{P}) = \log_2 \frac{1}{p_n}.$$

Concerning the gain of information we obtain the following inequality:

THEOREM 3. *If $\mathscr{P} = (p_1, \ldots, p_n)$ and $Q = (q_1, \ldots, q_n)$ are any incomplete distributions ($\sum_{k=1}^{n} p_k = s \leq 1$; $\sum_{k=1}^{n} q_k = t \leq 1$), then $I_\alpha(Q \,\|\, \mathscr{P})$ is an increasing function of α. Since $I_0(Q \,\|\, \mathscr{P}) = \log_2 \dfrac{t}{s}$, for the complete distributions $\mathscr{P}$ and Q there follows the inequality*

$$I_\alpha(Q \,\|\, \mathscr{P}) \geq 0 \text{ for } \alpha \geq 0. \tag{49}$$

---

[1] Cf. G. H. Hardy, J. E. Littlewood and G. Pólya [1], Theorem 16.

PROOF. We have

$$I_\alpha(Q \| \mathscr{P}) = \log_2 \left(\frac{\sum\limits_{k=1}^{n} q_k \left(\dfrac{q_k}{p_k}\right)^{\alpha-1}}{\sum\limits_{k=1}^{n} q_k} \right)^{\frac{1}{\alpha-1}}, \tag{50}$$

from which Theorem 3 follows by the same theorem on mean values (cf. foot-note) as above.

If α is negative or zero, the properties of $I_\alpha(\mathscr{P})$ and $I_\alpha(Q \| \mathscr{P})$ differ essentially from those of *Shannon's information*. As can be seen from Theorem 3, $I_\alpha(Q \| \mathscr{P})$ is for complete distributions only then positive, when α is positive. The following property is particularly undesirable: Let $\alpha < 0$; modify the complete distribution $\mathscr{P} = (p_1, \ldots, p_n)$ by letting p_1 tend to zero, then $I_\alpha(\mathscr{P})$ tends to infinity. On the other hand, $I_0(\mathscr{P})$ is always equal to $\log_2 n$ whenever $\mathscr{P}$ contains n positive terms. $I_0(\mathscr{P})$ is thus very inadequate to measure the information and we consider only $I_\alpha(\mathscr{P})$ *with positive* α as true measures of information.

Let us now consider some distinctive features of Shannon's information among the informations of the family $I_\alpha(\mathscr{P})$, or of $I_1(Q \| \mathscr{P})$ among the informations of the family $I_\alpha(Q \| \mathscr{P})$. One of these properties is given by

THEOREM 4. *If ξ and η are two random variables with the discrete finite distributions $\mathscr{P}$ and Q and if $\mathscr{R}$ denotes the two-dimensional distribution of the pair (ξ, η), then*

$$I_\alpha(\mathscr{R}) \le I_\alpha(\mathscr{P}) + I_\alpha(Q) \tag{51}$$

holds for every $\mathscr{P}$ and Q with the mentioned properties if and only if $\alpha = 1$.

PROOF. We know already that inequality (51) is valid for $\alpha = 1$ (cf. § 3, Formula (15)).

In the case of $\alpha \ne 1$, (51) is not necessarily fulfilled. In fact let 0 and 1 be the possible values of ξ and η; and suppose

$$P(\xi = 0, \, \eta = 0) = pq + \varepsilon,$$

$$P(\xi = 0, \, \eta = 1) = p(1 - q) - \varepsilon,$$

$$P(\xi = 1, \, \eta = 0) = (1 - p)q - \varepsilon,$$

$$P(\xi = 1, \, \eta = 1) = (1 - p)(1 - q) + \varepsilon$$

with

$$0 < p < 1, \ 0 < q < 1, \ p \neq \frac{1}{2}, \quad q \neq \frac{1}{2}$$

and

$$|\varepsilon| \leq \min \left(pq, \ (1-p)q, \ (1-q)p, \ (1-p)(1-q) \right).$$

If (51) were true, the function

$$g(\varepsilon) = (pq + \varepsilon)^\alpha + \left(p(1-q) - \varepsilon \right)^\alpha + \left((1-p)q - \varepsilon \right)^\alpha + \left((1-p)(1-q) + \varepsilon \right)^\alpha$$

would have an extremum for $\varepsilon = 0$. But this is not the case, since $g'(0) \neq 0$.

The quantity $I_1(Q \parallel \mathscr{P})$ is distinguished among the $I_\alpha(Q \parallel \mathscr{P})$ e.g. by the following property:

THEOREM 5. *If* $\mathscr{P} = (p_1, \ldots, p_r)$, $\mathscr{P}' = (p_1', \ldots, p_n')$ *and* $Q = (q_1, \ldots, q_n)$ *are discrete, finite, incomplete distributions fulfilling the relations*

$$q_k = \sqrt{p_k p_k'} \qquad (k = 1, 2, \ldots, n), \tag{52a}$$

i.e. the relations

$$\log_2 \frac{q_k}{p_k} + \log_2 \frac{q_k}{p_k'} = 0 \qquad (k = 1, 2, \ldots, n), \tag{52b}$$

then the relation

$$I_\alpha(Q \parallel \mathscr{P}) + I_\alpha(Q \parallel \mathscr{P}') = 0 \tag{53}$$

holds for every distribution fulfilling (52a) *only if* $\alpha = 1$.

Remark. The distributions $\mathscr{P}$, $\mathscr{P}'$, and Q can only be all three complete if they are identical. In fact, according to Cauchy's inequality

$$\left(\sum_{k=1}^n q_k \right)^2 = \left(\sum_{k=1}^n \sqrt{p_k p_k'} \right)^2 \leq \left(\sum_{k=1}^n p_k \right) \left(\sum_{k=1}^n p_k' \right),$$

where the equality sign can only hold if $\mathscr{P} = \mathscr{P}'$.

PROOF. For $\alpha \neq 1$ we have

$$I_\alpha(Q \parallel \mathscr{P}) + I_\alpha(Q \parallel \mathscr{P}') = \frac{1}{\alpha - 1} \log_2 \frac{\left(\sum\limits_{k=1}^n \dfrac{q_k^\alpha}{p_k^{\alpha-1}} \right) \left(\sum\limits_{k=1}^n \dfrac{q_k^\alpha}{p_k'^{\alpha-1}} \right)}{\left(\sum\limits_{k=1}^n q_k \right)^2}. \tag{54}$$

It is easy to see that the right hand side of (54) is not identically zero; e.g. it is different from 0 if we put $n = 2$, $q_1 = q_2$, $p_1 \neq p_2$.

§ 7. Statistical interpretation of the information of order α

Let $A_1, A_2, \ldots, A_r$ be the possible and mutually exclusive outcomes of an experiment with probabilities $P(A_k) = p_k$, $\sum_{k=1}^{r} p_k = 1$. Put $\mathscr{P} = (p_1, \ldots, p_r)$. Suppose that $p_1 < p_2 < \ldots < p_r$ and perform n independent repetitions of the experiment. Let v_k be the number of experiments leading to the outcome A_k ($k = 1, 2, \ldots, r$). Put $\pi_n = p_1^{v_1} p_2^{v_2} \ldots p_r^{v_r}$. As in § 5, π_n is thus the probability of a sequence of n observations. Consider the function

$$l(\alpha) = \frac{\sum_{k=1}^{r} p_k^\alpha \log_2 \frac{1}{p_k}}{\sum_{k=1}^{r} p_k^\alpha}. \tag{1}$$

Since

$$l'(\alpha) = -\ln 2 \left(\frac{\sum_{k=1}^{r} p_k^\alpha \log_2^2 \frac{1}{p_k}}{\sum_{k=1}^{r} p_k^\alpha} - \left(\frac{\sum_{k=1}^{r} p_k^\alpha \log_2 \frac{1}{p_k}}{\sum_{k=1}^{r} p_k^\alpha} \right)^2 \right), \tag{2}$$

it follows from Cauchy's inequality that $l(\alpha)$ is a strictly decreasing function; further we have

$$l(1) = I_1(\mathscr{P}), \quad \lim_{\alpha \to -\infty} l(\alpha) = \log_2 \frac{1}{p_1}, \quad \lim_{\alpha \to +\infty} l(\alpha) = \log_2 \frac{1}{p_r}.$$

For $\alpha > 1$ we have thus $l(\alpha) < I_1(\mathscr{P})$. If we put

$$p(\alpha) = 2^{-l(\alpha)} \tag{3}$$

we have

$$\log_2 \frac{1}{p(\alpha)} = l(\alpha) = \frac{\sum_{k=1}^{r} p_k^\alpha \log_2 \frac{1}{p_k}}{\sum_{k=1}^{r} p_k^\alpha} \tag{4}$$

and

$$2^{-I_1(\mathscr{P})} < p(\alpha) < p_r \quad \text{for} \quad \alpha > 1$$

Now let $B_n(\alpha)$ be the event $\pi_n \geq p(\alpha)^n$. Consider the conditional information contained in the outcome of the sequence of experiments, under the condition $B_n(\alpha)$. Put for this

$$C_n(\alpha) = \sum_{\substack{\prod\limits_{k=1}^{r} p_x^{n_k} \geq p(\alpha)^n \\ \sum\limits_{k=1}^{r} n_k = n}} \frac{n!}{n_1! \, n_2! \ldots n_r!} \, . \tag{5}$$

Obviously, $C_n(\alpha)$ is the number of outcomes fulfilling the condition $B_n(\alpha)$. The information in question is thus at most equal to $\log_2 C_n(\alpha)$. Further

$$P(B_n(\alpha)) \geq C_n(\alpha) p(\alpha)^n \tag{6}$$

and on the other hand

$$E(\pi_n^{\alpha-1}) = \left(\sum_{k=1}^{r} p_k^\alpha \right)^n. \tag{7}$$

Hence, because of Markov's inequality,

$$P(B_n(\alpha)) \leq p(\alpha)^n \left(\sum_{k=1}^{r} \left(\frac{p_k}{p(\alpha)} \right)^\alpha \right)^n \tag{8}$$

or, according to (6),

$$C_n(\alpha) \leq \left(\sum_{k=1}^{r} \left(\frac{p_k}{p(\alpha)} \right)^\alpha \right)^n. \tag{9}$$

Put

$$q_k(\alpha) = \frac{p_k^\alpha}{\sum\limits_{j=1}^{r} p_j^\alpha} \, . \tag{10}$$

If Q denotes the distribution $(q_1(\alpha), \ldots, q_r(\alpha))$, we get from (4) by a simple calculation that

$$\log_2 \left(\sum_{k=1}^{r} \left(\frac{p_k}{p(\alpha)} \right)^\alpha \right) = I_1(Q_\alpha), \tag{11}$$

hence, according to (9),

$$C_n(\alpha) \leq 2^{n I_1(Q_\alpha)} \, . \tag{12}$$

Furthermore, we have

$$I_1(Q_\alpha) = \alpha \log_2 \frac{1}{p(\alpha)} - (\alpha - 1) I_\alpha(\mathscr{P}) = I_\alpha(\mathscr{P}) + \frac{\alpha}{1-\alpha} I_1(Q_\alpha \| \mathscr{P}). \quad (13)$$

Choose a sufficiently large h for which

$$p_r^{1+\frac{1}{h}} > (p_1 p_2 \ldots p_{r-1})^{\frac{1}{r-1}}; \quad (14)$$

this is possible because of $p_1 < p_2 < \ldots < p_r$.

Put $n_j(\alpha) = [nq_j(\alpha)] - h \ (j = 1, 2, \ldots, r-1)$ and $n_r(\alpha) = n - \sum_{j=1}^{r-1} n_j(\alpha)$.
Then

$$\prod_{k=1}^{r} p_k^{n_k(\alpha)} \geq p(\alpha)^n. \quad (15)$$

When $v_k = n_k(\alpha) \ (k = 1, 2, \ldots, r)$, the event $B_n(\alpha)$ occurs; hence

$$C_n(\alpha) \geq \frac{n!}{\prod_{k=1}^{r} n_k(\alpha)!}. \quad (16)$$

But according to Stirling's formula

$$\frac{n!}{\prod_{k=1}^{r} n_k(\alpha)!} = 2^{nI_1(Q_\alpha) - O(\ln n)}. \quad (17)$$

Relations (12), (16) and (17) lead to

$$I_1(Q_\alpha) - O\left(\frac{\ln n}{n}\right) \leq \frac{1}{n} \log_2 C_n(\alpha) \leq I_1(Q_\alpha). \quad (18)$$

Therewith we proved

THEOREM 1. *Let* $A_1, A_2, \ldots, A_r$ *be the possible outcomes of an experiment,* $P(A_k) = p_k$, $0 < p_1 < p_2 < \ldots < p_r$, $\sum_{k=1}^{r} p_k = 1$ *and* $\mathscr{P} = (p_1, p_2, \ldots, p_r)$. *Let the experiment be repeated* n *times such that the repetitions are independent of each other. Put further*

$$q_k(\alpha) = \frac{p_k^\alpha}{\sum_{j=1}^{r} p_j^\alpha}, \quad Q_\alpha = (q_1(\alpha), \ldots, q_r(\alpha))$$

with $\alpha > 1$ and

$$p(\alpha) = 2^{-\sum\limits_{k=1}^{r} q_k(\alpha) \log_2 \frac{1}{p_k}}. \tag{19}$$

Let v_k be the number of experiments with outcome A_k, let $\pi_n = \prod\limits_{k=1}^{r} p_k^{v_k}$ and let $B_n(\alpha)$ be the event $\pi_n > p(\alpha)_n$. Now if $B_n(\alpha)$ occurs, the outcome of the sequence of experiments may be characterized completely by a sequence of $0 - 1$-symbols of length

$$nI_1(Q_\alpha) = nI_\alpha(\mathscr{P}) + \frac{n\alpha}{1-\alpha} I_1(Q_\alpha \| \mathscr{P}). \tag{20}$$

If, however, $\varrho > 0$ and $\varepsilon > 0$ are arbitrarily small positive numbers and n is large enough, then $n(I_1(Q_\alpha) - \varepsilon)$ $0-1$-symbols are not sufficient with probability $\geq \varrho$.

Remarks. 1. $I_1(Q_\alpha) = I^{(\alpha)}(\mathscr{P})$ may also be considered as an information measure of the distribution $\mathscr{P}$; it has the following properties:

a) $0 \leq I^{(\alpha)}(\mathscr{P}) \leq \log_2 r$,

b) *if $\mathscr{R} = \mathscr{P} * Q$, we have*

$$I^{(\alpha)}(\mathscr{R}) = I^{(\alpha)}(\mathscr{P}) + I^{(\alpha)}(Q).$$

2. It follows from Jensen's inequality that

$$I_\alpha(\mathscr{P}) \geq \log_2 \frac{1}{p(\alpha)}. \tag{21}$$

§ 8. The definition of information for general distributions

If the random variable ξ takes on denumerably many values x_k with probabilities $p_k = P(\xi = x_k)$ $(k = 1, 2, \ldots)$, then we define the *information of order α contained in the value of ξ*, by the formulas

$$I_\alpha(\xi) = \frac{1}{1-\alpha} \log_2 \left(\sum_{k=1}^{\infty} p_k^\alpha \right) \quad \text{for} \quad \alpha \neq 1 \tag{1}$$

and

$$I_1(\xi) = \sum_{k=1}^{\infty} p_k \log_2 \frac{1}{p_k} \tag{2}$$

if the series on the right hand sides of (1) and (2) converge. The series (2) does not always converge. For instance for

$$p_k = \frac{1}{ck \log^2 (k+1)} \qquad (k = 1, 2, \ldots)$$

it is divergent; c is here a "normalizing" factor:

$$c = \sum_{n=1}^{\infty} \frac{1}{n \log^2 (n+1)} \; .$$

However the series (1) converges always for $\alpha > 1$. In case of discrete infinite distributions the measure of order α of the amount of information is thus always defined if $\alpha > 1$.

Let η be a second random variable which takes on the same values as ξ but has a different probability distribution $P(\eta = x_k) = q_k$ $(k = 1, 2, \ldots)$. Let the gain of information of order α, obtained if the distribution $Q = (q_1, q_2, \ldots)$ is replaced by $\mathscr{P} = (p_1, p_2, \ldots)$, be defined by

$$I_\alpha (Q \| \mathscr{P}) = \frac{1}{\alpha - 1} \log_2 \left(\sum_{k=1}^{\infty} \frac{q_k^\alpha}{p_k^{\alpha-1}} \right) \text{ for } \alpha \neq 1, \qquad (3)$$

and by

$$I_1 (Q \| \mathscr{P}) = \sum_{k=1}^{\infty} q_k \log_2 \frac{q_k}{p_k}, \qquad (4)$$

if the series on the right hand side of (3) or (4) converges (which is not always the case). The series (3) converges according to Hölder's inequality always for $0 < \alpha < 1$.

Let now ξ be a random variable having continuous distribution. We want now to extend the definition of the measure of order α of the amount of information, i.e. $I_\alpha(\xi)$, to this case. If we do this in a straightforward way we obtain that this quantity is, in general, infinite. If for instance ξ is uniformly distributed on $(0, 1)$, we know (cf. Ch. VII, § 14, Exercise 12) that the digits of the binary expansion of ξ are completely independent random variables which take on the values 0 and 1 with probability $\frac{1}{2}$. Hence the exact knowledge of the values of ξ furnishes an information $1 + 1 + 1 + \ldots$ which is infinite. Or, to put it more precisely, the amount of information furnished *would be* infinite if the value of ξ *could be* known exactly. Practically, however, a continuous quantity can only be determined up to a finite number of decimal (or binary) digits.

We see thus that if we want to define $I_\alpha(\xi)$ we encounter problems of divergence. It seems reasonable to approach a continuous distribution by a discrete one and to investigate, how the information associated with the discrete distribution increases as the deviation between the two distributions is diminished. Instead of ξ, we can for instance consider

$$\xi_N = \frac{[N\xi]}{N}, \tag{5}$$

where $[x]$ denotes the largest integer not exceeding x. Suppose $\alpha > 0$ and let $I_\alpha(\xi_1)$ be finite (this is only a restriction for $\alpha \leq 1$). It follows from Jensen's inequality that $I_\alpha(\xi_N)$ is finite for every N and the inequality

$$I_\alpha(\xi_N) \leq I_\alpha(\xi_1) + \log_2 N \tag{6}$$

is valid. If $0 < \alpha < 1$ and if we put

$$p_{N,k} = P\left(\xi_N = \frac{k}{N}\right) = P\left(\frac{k}{N} \leq \xi < \frac{k+1}{N}\right)$$

$$(k = 0, \pm 1, \pm 2, \ldots; N = 1, 2, \ldots),$$

then we have the inequality

$$\sum_{k=-\infty}^{+\infty} p_{N,k}^\alpha \leq N^{1-\alpha} \sum_{k=-\infty}^{+\infty} p_{1,k}^\alpha,$$

from which (6) follows; for $\alpha \geq 1$ (6) can be proved in a similar manner.

When the distribution is continuous, the information $I_\alpha(\xi_N)$ tends to infinity as $N \to \infty$; however, in many cases the limit

$$d_\alpha(\xi) = \lim_{N\to\infty} \frac{I_\alpha(\xi_N)}{\log_2 N} \tag{7}$$

exists. The quantity $d_\alpha(\xi)$ will be called the *dimension of order α of ξ*. If not only $d = d_\alpha(\xi)$ exists but also the limit

$$\lim_{N\to\infty} \left(I_\alpha(\xi_N) - d \log_2 N\right) = I_{\alpha,d}(\xi), \tag{8}$$

the quantity $d_{\alpha,d}(\xi)$ will be called the *d-dimensional information of order α contained in the value of the random variable ξ*.

In the important case when the distribution of ξ is absolutely continuous, we have the following

THEOREM 1. *Let ξ be a random variable having an absolutely continuous distribution with density function $f(x)$, which is supposed to be bounded.[1] If*

---

[1] This supposition is superfluous (cf. A. Rényi [27], [34]); we make it merely in order to simplify the proof.

we put $\xi_N = \dfrac{[N\xi]}{N}$ $(N = 1, 2, \ldots)$ and if we suppose that $I_\alpha(\xi_1)$ is finite $(\alpha > 0)$ then

$$\lim_{N \to \infty} \frac{I_\alpha(\xi_N)}{\log_2 N} = 1, \qquad (9)$$

i.e. the dimension of order α of ξ is equal to 1; if the integral $\int\limits_{-\infty}^{+\infty} f(x)^\alpha dx$ $(\alpha \neq 1)$ exists, then

$$\lim_{N \to \infty} \left(I_\alpha(\xi_N) - \log_2 N\right) = I_{\alpha,1}(\xi) = \frac{1}{1-\alpha} \log_2 \Big(\int\limits_{-\infty}^{+\infty} f(x)^\alpha \, dx \Big); \qquad (10)$$

if

$$\int\limits_{-\infty}^{+\infty} f(x) \log_2 \frac{1}{f(x)} \, dx,$$

exists, then

$$\lim_{N \to \infty} \left(I_1(\xi_N) - \log_2 N\right) = I_{1,1}(\xi) = \int\limits_{-\infty}^{+\infty} f(x) \log_2 \frac{1}{f(x)} \, dx. \qquad (11)$$

PROOF. Consider first the case $\alpha = 1$. Put $p_{Nk} = P\left(\xi_N = \dfrac{k}{N}\right)$ and

$$f_N(x) = N p_{Nk} \quad \text{for} \quad \frac{k}{N} \leq x < \frac{k+1}{N} \qquad (k = 0, \pm 1, \ldots).$$

We have then

$$I_1(\xi_N) - \log_2 N = \sum_{k=-\infty}^{\infty} p_{Nk} \log_2 \frac{1}{N p_{Nk}} = \int\limits_{-\infty}^{+\infty} f_N(x) \log_2 \frac{1}{f_N(x)} \, dx. \qquad (12)$$

If

$$F(x) = \int\limits_{-\infty}^{x} f(u) \, du \qquad (13)$$

is the distribution function of ξ, we have

$$f_N(x) = \frac{F\left(\dfrac{k+1}{N}\right) - F\left(\dfrac{k}{N}\right)}{\dfrac{1}{N}} \quad \text{for} \quad \frac{k}{N} \leq x < \frac{k+1}{N}. \qquad (14)$$

According to the well-known theorem of Lebesgue it follows that

$$\lim_{N \to \infty} f_N(x) = f(x)$$

for almost every x. Now we shall use Jensen's inequality in the following form: If $g(x)$ is a concave function and if $p(x)$ and $h(x)$ are measurable functions with $p(x) \geq 0$ and $\int_a^b p(x)dx = 1$, then we have

$$\int_a^b g(h(x))\, p(x)\, dx \leq g\left(\int_a^b h(x)\, p(x)\, dx\right). \tag{15}$$

This inequality can be proved in the same way as the usual form of Jensen's inequality. If we apply (15) with $g(x) = \log_2 x$, $h(x) = \dfrac{1}{f(x)}$ and

$$p(x) = \frac{f(x)}{p_{Nk}} \quad \text{for} \quad \frac{k}{N} \leq x < \frac{k+1}{N},$$

then we get

$$\int_{\frac{k}{N}}^{\frac{k+1}{N}} f(x) \log_2 \frac{1}{f(x)}\, dx \leq p_{Nk} \log_2 \frac{1}{N p_{Nk}} \tag{16}$$

and, by summing over k

$$\int_{-\infty}^{+\infty} f(x) \log_2 \frac{1}{f(x)}\, dx \leq I_1(\xi_N) - \log_2 N, \tag{17}$$

i.e.

$$\int_{-\infty}^{+\infty} f(x) \log_2 \frac{1}{f(x)}\, dx \leq \liminf_{N \to \infty} (I_1(\xi_N) - \log_2 N). \tag{18}$$

We still have to prove the inequality

$$\limsup_{N \to \infty} (I_1(\xi_N) - \log_2 N) \leq \int_{-\infty}^{+\infty} f(x) \log_2 \frac{1}{f(x)}\, dx. \tag{19}$$

If $f(x) \le K$, we have also $f_N(x) \le K$; thus the functions $f_N(x)$ are uniformly bounded. Hence, by the convergence theorem of Lebesgue,

$$\lim_{N \to \infty} \int_{-A}^{+A} f_N(x) \log_2 \frac{1}{f_N(x)} \, dx = \int_{-A}^{+A} f(x) \log_2 \frac{1}{f(x)} \, dx \qquad (20)$$

for every $A > 0$.

According to Jensen's inequality, we have

$$\sum_{k=lN}^{(l+1)N-1} p_{Nk} \log_2 \frac{1}{N p_{Nk}} \le p_{1l} \log_2 \frac{1}{p_{1l}}. \qquad (21)$$

Since we have assumed that $I_1(\xi_1)$ and $\displaystyle\int_{-\infty}^{+\infty} f(x) \log_2 \frac{1}{f(x)} \, dx$ are finite, we can find for every $\varepsilon > 0$ an $A > 0$ such that

$$\left| \int_{|x|>A} f(x) \log_2 \frac{1}{f(x)} \, dx \right| < \varepsilon \qquad (22a)$$

and

$$\sum_{|l|>A} p_{1l} \log_2 \frac{1}{p_{1l}} < \varepsilon. \qquad (22b)$$

(20), (22a) and (22b) show immediately that the theorem is true for $\alpha = 1$.

Consider now the case $\alpha > 1$. We get from Fatou's lemma[1] that

$$\liminf_{N \to \infty} \int_{-\infty}^{+\infty} f_N(x)^\alpha \, dx \ge \int_{-\infty}^{+\infty} f(x)^\alpha \, dx. \qquad (23)$$

On the other hand, according to Jensen's inequality,

$$\int_{-\infty}^{+\infty} f_N(x)^\alpha \, dx \le \int_{-\infty}^{+\infty} f(x)^\alpha \, dx. \qquad (24)$$

It follows from (23) and (24) that

$$\lim_{N \to \infty} \int_{-\infty}^{+\infty} f_N(x)^\alpha \, dx = \int_{-\infty}^{+\infty} f(x)^\alpha \, dx, \qquad (25)$$

hence (10) is proved for $\alpha > 1$. We have still to examine the case $0 < \alpha < 1$.

[1] Cf. F. Riesz and B. Sz.-Nagy [1], p. 30.

According to Jensen's inequality, we have now

$$\int\limits_{-\infty}^{+\infty} f_N(x)^\alpha \, dx \geq \int\limits_{-\infty}^{+\infty} f(x)^\alpha \, dx. \tag{26}$$

On the other hand, according to the convergence theorem of Lebesgue, we have for every $A > 0$

$$\lim_{N \to \infty} \int\limits_{-A}^{+A} f_N(x)^\alpha \, dx = \int\limits_{-A}^{+A} f(x)^\alpha \, dx \leq \int\limits_{-\infty}^{+\infty} f(x)^\alpha \, dx. \tag{27}$$

Jensen's inequality gives

$$\sum_{k=lN}^{(l+1)N-1} N^{\alpha-1} p_{Nk}^\alpha < p_{1l}^\alpha. \tag{28}$$

Since we supposed $I_\alpha(\xi_1)$ to be finite, we can find for every $\varepsilon > 0$ an $A > 0$ such that

$$\sum_{|l|>A} p_{1l}^\alpha < \varepsilon. \tag{29}$$

From (27), (28) and (29) we conclude that (25) remains valid for $0 < \alpha < 1$. Theorem 1 is thus completely proved.

The quantities

$$I_{1,1}(\xi) = \int\limits_{-\infty}^{+\infty} f(x) \log_2 \frac{1}{f(x)} \, dx \tag{30}$$

or

$$I_{\alpha,1}(\xi) = \frac{1}{1-\alpha} \log_2 \int\limits_{-\infty}^{+\infty} f(x)^\alpha \, dx \qquad (\alpha > 0, \alpha \neq 1) \tag{31}$$

are called (one-dimensional) information of order 1 or order α, associated with the random variable ξ. $I_{1,1}(\xi)$ is called also the entropy of the random variable (or of the density function $f(x)$). The properties of these quantities differ in some respect from the properties of the corresponding quantities for discrete distributions. Thus for instance $I_{1,1}(\xi)$ and $I_{\alpha,1}(\xi)$ can be negative. Another difference is that these quantities are not invariant with respect to any one-to-one transformation of the variable. E.g. for $c > 0$ we have

$$I_{\alpha,1}(c\xi) = I_{\alpha,1}(\xi) + \log_2 c. \tag{32}$$

These facts are explained by realizing that $I_{\alpha,1}(\xi)$ is the limit of a *difference* between two informations.

All what we have said can be extended to the case of r-dimensional random vectors $(r = 2, 3, \ldots)$ with an absolutely continuous distribution. Let $f(x_1, \ldots, x_r)$ be the density function of the random vector $(\xi^{(1)}, \ldots, \xi^{(r)})$. Put $\zeta_N^{(k)} = \dfrac{[N\zeta^{(k)}]}{N}$ $(k = 1, 2, \ldots, r)$. If $I_{\alpha}((\xi_1^{(1)}, \ldots, \xi_1^{(r)}))$ is finite,[1] we have

$$\lim_{N \to \infty} \frac{I_{\alpha}((\zeta_N^{(1)}, \ldots, \zeta_N^{(r)}))}{\log_2 N} = r. \tag{33}$$

The dimension of the (absolutely continuous) distribution of a random vector of r components is thus equal to r; the notion of dimension in information theory is thus in accordance with the notion of geometrical dimension. Furthermore, for $\alpha > 0$, $\alpha \neq 1$ we have

$$\lim_{N \to \infty} \left[I_{\alpha}((\zeta_N^{(1)}, \ldots, \zeta_N^{(r)})) - r \log_2 N \right] = I_{\alpha,r}((\xi^{(1)}, \ldots, \xi^{(r)})) \tag{34a}$$

with

$$I_{\alpha,r}((\xi^{(1)}, \ldots, \xi^{(r)})) = \frac{1}{1 - \alpha} \log_2 \int_{-\infty}^{+\infty} \cdots \int_{-\infty}^{+\infty} f(x_1, \ldots, x_r)^{\alpha} dx_1 \ldots dx_r \tag{34b}$$

and

$$\lim_{N \to \infty} \left[I_1((\zeta_N^{(1)}, \ldots, \zeta_N^{(r)})) - r \log_2 N \right] = I_{1,r}((\xi^{(1)}, \ldots, \xi^{(r)})) \tag{35a}$$

with

$$I_{1,r}((\xi^{(1)}, \ldots, \xi^{(r)})) = \int_{-\infty}^{+\infty} \cdots \int_{-\infty}^{+\infty} f(x_1, \ldots, x_r) \log_2 \frac{1}{f(x_1, \ldots, x_r)} dx_1 \ldots dx_r, \tag{35b}$$

provided of course that the integrals exist.

The quantities $I_{\alpha,r}((\xi^{(1)}, \ldots, \xi^{(r)}))$ and $I_{1,r}((\xi^{(1)}, \ldots, \xi^{(r)}))$ defined by (34) and (35) are called *r-dimensional measure of order α, and of order 1, of the amount of information (or entropy) associated with the distribution of the random vector* $(\xi^{(1)}, \ldots, \xi^{(r)})$.

Consider now briefly the notion of gain of information in the case of general distributions. Let $\mathscr{P}$ and Q be any two probability measures on the measurable space $[\Omega, \mathscr{A}]$. Suppose that Q is absolutely continuous with

---

[1] $I_{\alpha}((\zeta_N^{(1)}, \ldots, \zeta_N^{(r)}))$ denotes the entropy of order α of the distribution of the random vector $(\zeta_N^{(1)}, \ldots, \zeta_N^{(r)})$.

respect to $\mathscr{P}$. Then for every set $A \in \mathscr{A}$

$$Q(A) = \int_A h(\omega)\,d\mathscr{P}, \tag{36}$$

where $h(\omega) \geq 0$ is the Radon–Nikodym derivative $\dfrac{dQ}{d\mathscr{P}}$ and

$$\int_\Omega h(\omega)\,d\mathscr{P} = 1.$$

The *gain of information of order* α (or *of order* 1) obtained if $\mathscr{P}$ is replaced by Q is defined[1] by the formulas

$$I_\alpha(Q \,\|\, \mathscr{P}) = \frac{1}{\alpha - 1} \log_2 \int_\Omega h(\omega)^\alpha\,d\mathscr{P} \tag{37a}$$

or

$$I_1(Q \,\|\, \mathscr{P}) = \int_\Omega h(\omega)\,\log_2 h(\omega)\,d\mathscr{P}. \tag{38a}$$

Formulas (37a) and (38a) remain valid in the discrete case too. The (ordinary) discrete distributions $\mathscr{P} = (p_1, \ldots, p_n)$ and $Q = (q_1, \ldots, q_n)$ may indeed be considered as measures defined on an algebra of events of n elements $\omega_1, \omega_2, \ldots, \omega_n$ with $\mathscr{P}(\omega_k) = p_k$ and $Q(\omega_k) = q_k$ $(k = 1, 2, \ldots, n)$. The condition that Q is absolutely continuous with respect to $\mathscr{P}$ is here automatically fulfilled whenever $p_k > 0$ $(k = 1, 2, \ldots, n)$ and we have

$$h(\omega_h) = \frac{q_k}{p_k} \qquad (k = 1, 2, \ldots, n).$$

The formulas

$$I_\alpha(Q \,\|\, \mathscr{P}) = \frac{1}{\alpha - 1} \log_2 \left(\sum_{k=1}^n \frac{q_k^\alpha}{p_k^{\alpha-1}} \right) \quad \text{and} \quad I_1(Q \,\|\, \mathscr{P}) = \sum_{k=1}^n q_k \log_2 \frac{q_k}{p_k}$$

appear thus as particular cases of (37a) and (38a).

If Ω is the set of real numbers, $\mathscr{A}$ the set of the Borel-measurable subsets of Ω and if $\mathscr{P}$ and Q are absolutely continuous with respect to Lebesgue measure, there exist two functions $p(x)$ and $q(x)$ such that

$$\mathscr{P}(A) = \int_A p(x)\,dx, \quad Q(A) = \int_A q(x)\,dx \quad \text{for } A \in \mathscr{A}.$$

[1] One could deduce Formulas (37a) and (38a) from a certain number of postulates as was done in the discrete case (§ 6). This will not be dealt with here.

The measure Q is absolutely continuous with respect to $\mathscr{P}$ if for every x such that $p(x) = 0$ we have $q(x) = 0$. Then

$$h(x) = \frac{q(x)}{p(x)}. \tag{39}$$

In this case we obtain for the gain of information from (37) and (38)

$$I_\alpha(Q \| \mathscr{P}) = \frac{1}{\alpha - 1} \log_2 \left(\int_{-\infty}^{+\infty} \frac{q(x)^\alpha}{p(x)^{\alpha-1}} \, dx \right) \quad \text{for} \quad \alpha \neq 1 \tag{37b}$$

and

$$I_1(Q \| \mathscr{P}) = \int_{-\infty}^{+\infty} q(x) \log_2 \frac{q(x)}{p(x)} \, dx. \tag{38b}$$

The gain of information for absolutely continuous distributions can be obtained from the gain of information for discrete distributions by a limit process:

THEOREM 2. *Let $\mathscr{P}$ and Q be two distributions, absolutely continuous with respect to Lebesgue measure, Q absolutely continuous with respect to $\mathscr{P}$. Let $p(x)$ and $q(x)$ be the respective density functions of $\mathscr{P}$ and Q. We suppose that $p(x)$ and $q(x)$ are bounded.[1] Further if*

$$p_{Nk} = \int_{\frac{k}{N}}^{\frac{k+1}{N}} p(x) \, dx, \qquad q_{Nk} = \int_{\frac{k}{N}}^{\frac{k+1}{N}} q(x) \, dx \qquad (k = 0, \pm 1, \ldots; \quad N = 1, 2, \ldots),$$

$\mathscr{P}_N$ and Q_N denote the distributions $\{p_{Nk}\}$ and $\{q_{Nk}\}$, if α is positive and $I_\alpha(Q_1 \| \mathscr{P}_1)$ is finite (which means a restriction only for $\alpha \geq 1$), then we have

$$\lim_{N \to \infty} I_\alpha(Q_N \| \mathscr{P}_N) = I_\alpha(Q \| \mathscr{P}), \tag{40}$$

where $I_\alpha(Q \| \mathscr{P})$ is defined by (37b) for $\alpha \neq 1$ and by (38b) for $\alpha = 1$, provided that $I_\alpha(Q \| \mathscr{P})$ exists.

PROOF. This is similar to that of Theorem 1. We define $p_N(x)$ and $q_N(x)$ by $p_N(x) = Np_{Nk}$ for $\frac{k}{N} \leq x < \frac{k+1}{N}$ $(k = 0, \pm 1, \ldots)$ and $q_N(x) = Nq_{Nk}$ for $\frac{k}{N} \leq x < \frac{k+1}{N}$ $(k = 0, \pm 1, \ldots)$. Let further $h_N(x) = \frac{p_N(x)}{q_N(x)}$.

---

[1] This supposition is superfluous and serves only to simplify the proof.

Consider first the case $0 < \alpha < 1$. It is clear that $p_N(x) \to p(x)$ and $q_N(x) \to q(x)$ almost everywhere; further

$$I_\alpha(Q_N \| \mathcal{P}_N) = \frac{1}{\alpha - 1} \log_2 \int_{-\infty}^{+\infty} q_N(x)^\alpha p_N(x)^{1-\alpha} dx. \tag{41}$$

According to Lebesgue's theorem we have for every $A > 0$

$$\lim_{N \to \infty} \int_{-A}^{+A} q_N(x)^\alpha p_N(x)^{1-\alpha} dx = \int_{-A}^{+A} q(x)^\alpha p(x)^{1-\alpha} dx. \tag{42}$$

Since

$$\int_{|x| > A} q_N(x)^\alpha p_N(x)^{1-\alpha} dx$$

can be made arbitrarily small for a sufficiently large A, uniformly in N, Theorem 2 is proved for $0 < \alpha < 1$.

Now suppose $\alpha > 1$. We have, according to Jensen's inequality,

$$\int_{-\infty}^{+\infty} \frac{q(x)^\alpha}{p(x)^{\alpha-1}} dx \geq \int_{-\infty}^{+\infty} \frac{q_N(x)^\alpha}{p_N(x)^{\alpha-1}} dx, \tag{43}$$

and on the other hand by Fatou's lemma

$$\liminf_{N \to \infty} \int_{-\infty}^{+\infty} \frac{q_N(x)^\alpha}{p_N(x)^{\alpha-1}} dx \geq \int_{-\infty}^{+\infty} \frac{q(x)^\alpha}{p(x)^{\alpha-1}} dx, \tag{44}$$

which settles the case $\alpha > 1$.

Finally, let $\alpha = 1$. We have

$$I_1(Q_N \| \mathcal{P}_N) = \int_{-\infty}^{+\infty} q_N(x) \log_2 \frac{q_N(x)}{p_N(x)} dx. \tag{45}$$

Since the function $x \log_2 x$ is convex, Jensen's inequality gives

$$\int_{-\infty}^{+\infty} q(x) \log_2 \frac{q(x)}{p(x)} dx \geq \int_{-\infty}^{+\infty} q_N(x) \log_2 \frac{q_N(x)}{p_N(x)} dx. \tag{46}$$

From $x \log_2 x \geq -\dfrac{\log_2 e}{e}$ we deduce

$$q_N(x) \log_2 \frac{q_N(x)}{p_N(x)} + \frac{\log_2 e}{e} p_N(x) \geq 0.$$

Hence, according to Fatou's lemma,

$$\liminf_{N \to \infty} \int_{-\infty}^{+\infty} q_N(x) \log_2 \frac{q_N(x)}{p_N(x)} \, dx \geq \int_{-\infty}^{+\infty} q(x) \log_2 \frac{q(x)}{p(x)} \, dx. \qquad (47)$$

(46) and (47) lead to

$$\lim_{N \to \infty} \int_{-\infty}^{+\infty} q_N(x) \log_2 \frac{q_N(x)}{p_N(x)} \, dx = \int_{-\infty}^{+\infty} q(x) \log_2 \frac{q(x)}{p(x)} \, dx \qquad (48)$$

and Theorem 2 is herewith proved.

§ 9. Information-theoretical proofs of limit theorems

We have seen that for complete discrete distributions the relation $I_\alpha(Q \| \mathscr{P}) \geq 0$ holds, where the equality sign occurs only if $\mathscr{P}$ and Q are identical. We shall now prove the following property: If $\{Q_n\}$ is a sequence of discrete distributions such that $\lim_{n \to \infty} I_\alpha(Q_n \| \mathscr{P}) = 0$, then the distributions Q_n converge to the distribution $\mathscr{P}$. Thus we have the following

THEOREM 1. *If $\mathscr{P} = (p_1, \ldots, p_r)$ and $Q_n = (q_{n1}, \ldots, q_{nr})$ are probability distributions and if*

$$\lim_{N \to \infty} I_\alpha(Q_n \| \mathscr{P}) = 0 \qquad (\alpha > 0), \qquad (1)$$

then we have also

$$\lim_{n \to \infty} q_{nk} = p_k. \qquad (2)$$

PROOF. If (2) does not hold, there exists a subsequence $n_1 < n_2 < \ldots < n_s < \ldots$ of the integers with

$$\lim_{s \to \infty} q_{n_s k} = p_k' \quad \text{and} \quad \sum_{k=1}^{r} (p_k' - p_k)^2 \neq 0. \qquad (3)$$

Obviously, $\sum_{k=1}^{r} p'_k = 1$; further if we put $\mathscr{P}' = (p'_1, \ldots, p'_r)$, it follows from (3) that

$$\lim_{s \to \infty} I_\alpha(Q_{n_s} \| \mathscr{P}) = I_\alpha(\mathscr{P}' \| \mathscr{P}). \tag{4}$$

According to (1), $I_\alpha(\mathscr{P}' \| \mathscr{P}) = 0$, but this is possible only if $\mathscr{P}' \equiv \mathscr{P}$, i.e. if $p'_k = p_k$ for $k = 1, 2, \ldots, r$, which contradicts (3). Thus Theorem 1 is proved.

As an application of this theorem we shall now prove a theorem about ergodicity of homogeneous Markov chains, which, essentially, is contained in Theorem 1 of Chapter VIII, § 8. We give here a new proof of this result, only to show how the methods of information theory may be used to prove theorems on limit distributions.

THEOREM 2. *Let us consider a homogeneous Markov chain with a finite number of states* $A_0, \ldots, A_N$; *let the probability of transition from* A_j *to* A_k *in* n *steps be denoted by* $p_{jk}^{(n)}$ ($n = 1, 2, \ldots$). *For* $p_{jk}^{(1)}$ *we write simply* p_{jk}. *If there exists an integer* $s \geq 1$ *such that* $p_{jk}^{(s)} > 0$ *for* $j, k = 0, 1, \ldots, N$, *then the equations*

$$\sum_{j=0}^{N} x_j p_{jk} = x_k \qquad (k = 0, 1, \ldots, N) \tag{5a}$$

have a system of solutions $x_k = p_k$ ($k = 0, 1, \ldots, N$) *with*

$$p_k > 0 \;\; (k = 0, 1, \ldots, N) \quad and \quad \sum_{k=0}^{N} p_k = 1, \tag{6}$$

and we have

$$\lim_{n \to \infty} p_{jk}^{(n)} = p_k \qquad (j, k = 0, 1, \ldots, N). \tag{7}$$

PROOF. The existence of a solution $x_k = p_k > 0$ ($k = 0, 1, \ldots, N$) of the system of equations (5a) can be proved directly, without probabilistic considerations, in the following manner:[1] The determinant of the system (5a) is zero, since $\sum_{k=1}^{N} p_{jk} = 1$ ($j = 0, 1, \ldots, N$); thus the system has a non-trivial solution $(x_0, x_1, \ldots, x_N)$. If (5a) is fulfilled, we have

$$|x_k| \leq \sum_{j=0}^{N} p_{jk} |x_j|. \tag{5b}$$

---

[1] We have here a particular case of the Perron–Frobenius theorem; cf. F. R. Gantmacher [1], Vol. 2, p. 46.

If we add the inequalities (5b) for $k = 0, 1, \ldots, N$, we obtain

$$\sum_{k=0}^{N} |x_k| \leq \sum_{j=0}^{N} |x_j|.$$

But this inequality is an equality; hence the same must hold for every inequality (5b), i.e.

$$|x_k| = \sum_{j=0}^{N} p_{jk} |x_j|. \tag{5c}$$

Hence (5a) possesses a nontrivial nonnegative system of solutions, say $(p_0, p_1, \ldots, p_N)$. If we multiply (5a) by p_{kl}, add the equations obtained for $k = 0, 1, \ldots, N$ and repeat the whole procedure n times, then we obtain

$$\sum_{j=0}^{N} p_j p_{ji}^{(h+1)} = \sum_{k=0}^{N} p_k p_{kl}^{(h)}.$$

We find then by induction that

$$\sum_{j=0}^{N} p_j p_{jk}^{(h)} = p_k \quad \text{for} \quad h = 1, 2, \ldots. \tag{5d}$$

Since (5d) is valid for $h = s$, it follows that no p_k can be zero. Because of the homogeneity of the equations, (5a) has thus a positive system of solutions $p_0, p_1, \ldots, p_N$ with $\sum_{k=0}^{N} p_k = 1.$[1] Put

$$\mathscr{P} = (p_0, \ldots, p_N) \quad \text{and} \quad \mathscr{P}_j^{(n)} = (p_{j0}^{(n)}, \ldots, p_{jN}^{(n)}).$$

We consider now the quantities $I_\alpha(\mathscr{P}_j^{(n)} \| \mathscr{P})$ and prove the relation

$$\lim_{n \to \infty} I_\alpha(\mathscr{P}_j^{(n)} \| \mathscr{P}) = 0, \tag{8}$$

then Theorem 2 follows because of Theorem 1. The value of α is immaterial for the proof. We can e.g. assume $\alpha > 1$. By assumption

$$\sum_{j=0}^{N} p_j p_{jk} = p_k. \tag{9}$$

If we put $\pi_{jk} = \dfrac{p_j p_{jk}}{p_k}$, we have

$$\sum_{j=0}^{N} \pi_{jk} = 1. \tag{10}$$

---

[1] This solution is unique; this is a corollary of (7) and need not be proved here separately.

Furthermore, by definition

$$p_{jk}^{(n+1)} = \sum_{l=0}^{N} p_{jl}^{(n)} p_{lk},\tag{11}$$

hence

$$I_\alpha(\mathscr{S}_j^{(n+1)} \| \mathscr{S}) = \frac{1}{\alpha-1} \log_2 \left[\sum_{k=0}^{N} p_k \left(\sum_{l=0}^{N} \frac{p_{jl}^{(n)}}{p_l} \pi_{lk} \right)^\alpha \right].\tag{12}$$

Because of (10), Jensen's inequality leads to

$$\left(\sum_{l=0}^{N} \frac{p_{jl}^{(n)}}{p_l} \pi_{lk} \right)^\alpha \leq \sum_{l=0}^{N} \pi_{lk} \left| \frac{p_{jl}^{(n)}}{p_l} \right|^\alpha.\tag{13}$$

Since $\sum\limits_{k=0}^{N} p_{lk} = 1$, it follows that

$$\sum_{k=0}^{N} p_k \pi_{lk} = p_l \sum_{k=0}^{N} p_{lk} = p_l.\tag{14}$$

If we multiply the inequality (13) by p_k and then take the sum over k, we obtain

$$I_\alpha(\mathscr{S}_j^{(n+1)} \| \mathscr{S}) \leq I_\alpha(\mathscr{S}_j^{(n)} \| \mathscr{S}) \qquad (n = 1, 2, \ldots).\tag{15}$$

$I_\alpha(\mathscr{S}_j^{(n)} \| \mathscr{S})$ $(n = 1, 2, \ldots)$ is a monotone decreasing sequence of nonnegative numbers; it has thus a limit

$$\lim_{n \to \infty} I_\alpha(\mathscr{S}_j^{(n)} \| \mathscr{S}) = \gamma.\tag{16}$$

It remains to show that $\gamma = 0$. Choose a subsequence $n_1 < \ldots < n_t < \ldots$ of the integers such that the limits

$$\lim_{t \to \infty} p_{jk}^{(n_t)} = q_{jk} \qquad (k = 0, 1, \ldots, N)$$

exist. Then $\sum\limits_{l=0}^{N} q_{jl} = 1$ and by (11)

$$\lim_{t \to \infty} p_{jk}^{(n_t + s)} = \sum_{l=0}^{N} q_{jl} p_{lk}^{(s)} = q_{jk}'.$$

Obviously

$$\sum_{k=0}^{N} q_{jk}' = \sum_{l=0}^{N} q_{jl} = 1.$$

Let Q_j and Q'_j denote the distributions $(q_{j0}, \ldots, q_{jN})$ and $(q'_{j0}, \ldots, q'_{jN})$, respectively. If we put $\pi^{(s)}_{jk} = p_j p^{(s)}_{jk}/p_k$, Jensen's inequality implies

$$I_\alpha(Q'_j \| \mathscr{S}) = \frac{1}{\alpha - 1} \log_2 \left[\sum_{k=0}^{N} p_k \left(\sum_{l=0}^{N} \frac{q_{jl}}{p_l} \pi^{(s)}_{lk} \right)^\alpha \right] \leq I_\alpha(Q_j \| \mathscr{S}) \qquad (17)$$

by the same argument that led to (15). But, because of (16), the relations

$$I_\alpha(Q'_j \| \mathscr{S}) = \lim_{t \to \infty} I_\alpha(\mathscr{S}_j^{(n_t+s)} \| \mathscr{S}) = \gamma \qquad (18a)$$

and

$$I_\alpha(Q_j \| \mathscr{S}) = \lim_{t \to \infty} I_\alpha(\mathscr{S}_j^{(n_t)} \| \mathscr{S}) = \gamma \qquad (18b)$$

hold; hence there is equality in (17). Since (17) is derived from Jensen's inequality, it follows that equality can hold only if $q_{jl} = \lambda p_l$ ($l = 0, 1, \ldots, N$). Since $\sum_{l=0}^{N} q_{jl} = \sum_{l=0}^{N} p_l = 1$, we must have $\lambda = 1$; consequently, $Q_j = \mathscr{S}$. But then $I_\alpha(Q_j \| \mathscr{S}) = I_\alpha(\mathscr{S} \| \mathscr{S}) = 0$, hence by (18b) $\gamma = 0$. Theorem 2 is herewith proved.

The idea to prove theorems on limit distributions by means of information theory is due to Yu. V. Linnik. He proved in this way the central limit theorem under Lindeberg conditions by using Shannon's entropy; he proved the convergence of the distribution with the density function $p_n(x)$ to the normal distribution by showing that

$$\lim_{n \to \infty} \int_{-\infty}^{+\infty} p_n(x) \log_2 \frac{p_n(x)}{\varphi(x)} \, dx = 0, \qquad (19)$$

where $\varphi(x)$ is the density function

$$\varphi(x) = \frac{1}{\sqrt{2\pi}} e^{-\frac{x^2}{2}}$$

of the normal distribution. Linnik's proof can be simplified if we use the gain of information of order 2 instead of Shannon's entropy; but even so the proof is too intricate to be reproduced here. However, we can briefly indicate the principle of the method.

Let $\xi_1, \xi_2, \ldots, \xi_n, \ldots$ be independent random variables having the same absolutely continuous distribution with zero expectation and unit variance and let $p_n(x)$ be the density function of $\dfrac{\xi_1 + \xi_2 + \ldots + \xi_n}{\sqrt{n}}$. Then

$$\int_{-\infty}^{+\infty} x^2 p_n(x)\, dx = 1.$$

Relation (19) can be written as

$$\lim_{n \to \infty} \int_{-\infty}^{+\infty} p_n(x) \log_2 \frac{1}{p_n(x)}\, dx = \log_2 \sqrt{2\pi e}. \tag{20}$$

It is easy to show that

$$\log_2 \sqrt{2\pi e} = \int_{-\infty}^{+\infty} \varphi(x) \log_2 \frac{1}{\varphi(x)}\, dx, \tag{21}$$

thus (19) is equivalent to

$$\lim_{n \to \infty} \int_{-\infty}^{+\infty} p_n(x) \log_2 \frac{1}{p_n(x)}\, dx = \int_{-\infty}^{+\infty} \varphi(x) \log_2 \frac{1}{\varphi(x)}\, dx. \tag{22}$$

To say that the distribution with the density function $p_n(x)$ tends to the normal distribution, means therefore that the entropy of this distribution tends to the entropy of the normal distribution. But we can prove that for a density function $p(x)$ such that

$$\int_{-\infty}^{+\infty} x^2 p(x)\, dx = 1 \tag{23}$$

the inequality

$$\int_{-\infty}^{+\infty} p(x) \log_2 \frac{1}{p(x)}\, dx \le \int_{-\infty}^{+\infty} \varphi(x) \log_2 \frac{1}{\varphi(x)}\, dx \tag{24}$$

holds, since because of (21) and (23), (24) is equivalent to the well-known inequality

$$\int_{-\infty}^{+\infty} p(x) \log_2 \frac{p(x)}{\varphi(x)}\, dx \ge 0. \tag{25}$$

The statement of the central limit theorem may therefore be expressed as follows: The entropy of the standardized sum of independent random variables tends, as the number of the variables tends to infinity, to the maximum of the entropy of all random variables with unit variance. Thus the central limit theorem of probability theory is closely connected with the second law of thermodynamics.[1]

§ 10. Extension of information theory to conditional probability spaces

In the present section we consider particular conditional probability spaces $[\Omega, \mathscr{A}, \mathscr{B}, P(A \mid B)]$ only: Let $\Omega = \{\omega_1, \omega_2, \ldots, \omega_n, \ldots\}$ be a denumerable set and $\mathscr{A}$ the class of all subsets of Ω. Let further p_k $(k = 1, 2, \ldots)$ be a sequence of nonnegative numbers with $\sum_{k=1}^{\infty} p_k = +\infty$. We put

$$\mu(A) = \sum_{\omega_n \in A} p_n \quad \text{for} \quad A \in \mathscr{A}.$$

Let $\mathscr{B}$ be the set of those subsets B of Ω for which $\mu(B)$ is finite and positive. For $A \in \mathscr{A}$ and $B \in \mathscr{B}$, $P(A \mid B)$ is defined by

$$P(A \mid B) = \frac{\mu(AB)}{\mu(B)} . \tag{1}$$

We shall indicate by some examples how the concepts of information theory can be extended to conditional probability spaces.

If ξ is a discrete random variable defined by $\xi(\omega) = x_n$ for $\omega \in A_n$ $(n = 1, 2, \ldots)$, where $A_n \in \mathscr{A}$, $\sum_{n=1}^{\infty} A_n = \Omega$, $A_n A_m = O$ for $n \neq m$, then the numbers $P(A_n \mid B)$ $(n = 1, 2, \ldots; B \in \mathscr{B})$ form the conditional distribution of ξ with respect to the condition B. Consider the entropy of this distribution, i.e.

$$I_1(\xi \mid B) = \sum_{n=1}^{\infty} P(A_n \mid B) \log_2 \frac{1}{P(A_n \mid B)} . \tag{2}$$

If Ω_N is the set $\{\omega_1, \omega_2, \ldots, \omega_N\}$, then $\Omega_N \in \mathscr{B}$ for $N \geq N_0$. We define the entropy $I_1(\xi)$ of ξ (in other words the information contained in the value of ξ) by

$$I_1(\xi) = \lim_{N \to \infty} I_1(\xi \mid \Omega_N), \tag{3}$$

---

[1] If the distributions considered concern the velocities of the molecules of a gas, the condition

$$\int_{-\infty}^{+\infty} x^2 p_n(x) dx = 1$$

means that the total kinetic energy of the gas is constant.

if this limit exists and is finite. If it does not exist, the information in question will be characterized by the following two quantities:

$$\lim_{N\to\infty} \inf I_1(\xi \mid \Omega_N) = \underline{I}_1(\xi)$$

and

$$\lim_{N\to\infty} \sup I_1(\xi \mid \Omega_N) = \overline{I}_1(\xi).$$

Consider for instance the binary representation of a positive integer. We show that each of the digits of this representation contains exactly one bit of information, just like a digit of the binary expansion of a real number lying in the interval $(0, 1)$. Let Ω be the set of positive integers, $\omega_n = n$, and $p_n = 1$ $(n = 1, 2, \ldots)$. Consider as above the conditional probability space $[\Omega, \mathcal{A}, \mathcal{B}, P(A \mid B)]$. Let $\varepsilon_k(n)$ denote the k-th digit in the binary expansion of n; we have

$$n = \sum_{k=0}^{\infty} \varepsilon_k(n) 2^k, \tag{4}$$

where $\varepsilon_k(n)$ is equal either to 0 or to 1. Obviously

$$\varepsilon_0(n) = \begin{cases} 1 \text{ if } n \text{ is odd}, \\ 0 \text{ if } n \text{ is even}. \end{cases}$$

If $\Omega_N = \{1, 2, \ldots, N\}$, we have

$$I_1(\varepsilon_0(n) \mid \Omega_N) = \frac{\left[\frac{N}{2}\right]}{N} \log_2 \frac{N}{\left[\frac{N}{2}\right]} + \frac{N - \left[\frac{N}{2}\right]}{N} \log_2 \frac{N}{N - \left[\frac{N}{2}\right]},$$

and by (3)

$$I_1(\varepsilon_0(n)) = 1. \tag{5}$$

It follows in the same way that

$$I_1(\varepsilon_k(n)) = 1 \qquad (k = 1, 2, \ldots). \tag{6}$$

Take now an example for which the limit (3) does not exist. Consider again the binary expansions of the positive integers; let $[\Omega, \mathcal{A}, \mathcal{B}, P(A \mid B)]$ be the same conditional probability space as in the previous example. Let $\eta(n)$ be the largest exponent of 2 in the binary expansion of n; hence

$$n = \sum_{k=1}^{\eta(n)} \varepsilon_k(n) 2^k \quad \text{with} \quad \varepsilon_{\eta(n)} = 1.$$

If now $2^r \leq N < 2^{r+1}$, that is if $r = [\log_2 N]$, then

$$P(\eta(n) = j \mid \Omega_N) = \frac{2^j}{N} \quad \text{for} \quad j \leq r - 1$$

and

$$P(\eta(n) = r \mid \Omega_N) = \frac{N - 2^r + 1}{N}.$$

If N tends to infinity through values for which

$$\lim_{N \to \infty} \frac{2^{[\log_2 N]}}{N} = \gamma \qquad \left(\frac{1}{2} \leq \gamma \leq 1\right),$$

then we have

$$\lim_{N \to \infty} I_1(\eta(n) \mid \Omega_N) = 2\gamma + \gamma \, \log_2 \frac{1}{\gamma} + (1 - \gamma) \log_2 \frac{1}{1 - \gamma} = L(\gamma). \qquad (7)$$

Thus the limit $L(\gamma)$ depends on γ. $L(\gamma)$ is a concave function of γ and we have $L\left(\frac{1}{2}\right) = L(1) = 2$. Furthermore, $L(\gamma)$ takes on its maximum for $\gamma = \frac{4}{5}$ and we have

$$\max_{\frac{1}{2} \leq \gamma \leq 1} L(\gamma) = L\left(\frac{4}{5}\right) = \log_2 5.$$

Consequently, $\underline{I}_1(\eta(n)) = 2$ and $\bar{I}_1(\eta(n)) = \log_2 5$. The information $I_1(\eta(n))$ is not defined in this case; nevertheless, it can be stated that the number of digits in the binary representation of an integer contains at least 2 bits and at most $\log_2 5$ bits of information.

§ 11. Exercises

1.[1] a) How much information is contained in the licence number of a car, if this consists of two letters and four decimal digits? (22.6)

b) How much information is needed to express the outcome of a game of "lotto", in which 5 numbers are drawn at random from the first 90 numbers? (25.4)

c) What amount of information is needed to describe the hand of a player in bridge (each player having 13 cards from 52)? (39.2)

d) How much information is contained in a Hollerith punch-card which has 80 columns and in each column one perforation in one of 12 possible positions? (286.8)

---

[1] The numbers in parentheses are the solutions.

e) How much information is contained in a table of values of a function consisting of 50 pages with 40 lines per page and 25 decimal digits on each line (numbers for identification of the lines not counted)? (166 095.5)

f) How much information is contained in a linear macromolecule consisting of 100 000 single molecules, if there can occur one of four different molecules at every place? (200 000)

g) How much information is transmitted per second by a television broadcasting station which emits 25 images per second each of which consists of 520 000 points, each black or white? (13 000 000)

2. a) Let some integer n ($1 \leq n \leq 2\,000$) be divided by 6, 10, 22, and 35 and let the remainders be given, while we assume that the remainders are compatible. How much information is thus given concerning the number n?

Hint. The information is equal to $\log_2 2000 = 10.96$ (i.e. we get full information on n). In fact the remainders mentioned determine n modulo the least common multiple of 6, 10, 22 and 35; which is equal to $2\,310 > 2\,000$, hence n is uniquely determined.

b) Let the number n be expressed in the system of base (-2), i.e. put

$$n = \sum_{k=0}^{r} b_k (-2)^k,$$

where b_k can take on the values 0 or 1 only. How much information on n is contained in b_k?

Hint. Put $N = \sum_{j=0}^{[\frac{r-1}{2}]} 2^{2j+1}$; then

$$\prod_{k=0}^{r} (1 + x^{(-2)^k}) = \sum_{n=-N}^{2^{r+1}-N-1} x^n.$$

It follows that for fixed r the numbers $-N, -N+1, \ldots, 2^{r+1} - N - 1$ can all uniquely be represented in the form $\sum_{k=0}^{r} b_k (-2)^k$ with $b_k = 0$ or 1; there are thus exactly 2^{r+1} numbers which can be expressed in this form and every "digit" b_k contains one bit of information with respect to n.

c) Let $U(n)$ (and $V(n)$) denote the number of different (and of all) prime factors of n. How much information with respect to n is contained in the value of the difference $V(n) - U(n)$?

Hint. As is known,[1] for every positive integer k the asymptotic density of the numbers n with $V(n) - U(n) = k$ exists. Let this density be denoted by d_k, then

$$\sum_{k=0}^{\infty} d_k z^k = \prod_{p} \left(1 - \frac{1}{p}\right) \left(1 + \frac{1}{p - z}\right) \quad \text{for } |z| < 2,$$

where p runs through all primes. Let $N_k(x)$ denote the number of integers n smaller than x with $V(n) - U(n) = k$, then

$$\lim_{x \to +\infty} \frac{N_k(x)}{x} = d_k.$$

[1] Cf. A. Rényi [16].

$(k = 0, 1, \ldots)$; d_0 is the density of square-free numbers and is equal to $\dfrac{6}{\pi^2}$. Thus the amount of information in question is equal to $\sum\limits_{k=0}^{\infty} d_k \log_2 \dfrac{1}{d_k}$.

3. a) Let the real number x $(0 < x < 1)$ be represented in the form

$$x = \sum_{n=1}^{\infty} \frac{\varepsilon_n(x)}{q^n},$$

where q is a positive integer ≥ 2, and $\varepsilon_n(x)$ can take on the values $0, 1, \ldots, q - 1$ $(n = 1, 2, \ldots)$. How much information with respect to x is contained in the value of $\varepsilon_n(x)$?

b) Expand x $(0 < x < 1)$ into the Cantor series

$$x = \sum_{n=1}^{\infty} \frac{\varepsilon_n(x)}{q_1 q_2 \ldots q_n},$$

where $q_1, q_2, \ldots, q_n, \ldots$ are positive integers ≥ 2, and $\varepsilon_n(x)$ can take on the values $0, 1, \ldots, q_n - 1$. How much information with respect to x is contained in the value of $\varepsilon_n(x)$?

c) Expand x $(0 < x < 1)$ into a regular continued fraction

$$x = \cfrac{1}{a_1(x) + \cfrac{1}{a_2(x) + \ldots}},$$

where each $a_n(x)$ can be an arbitrary positive integer. How much information about x is contained in the value of $a_n(x)$?

Hint. Let $m_n(k)$ denote the measure of the set of those x for which $a_n(x) = k$. As is known[1]

$$\lim_{n \to \infty} m_n(k) = \log_2 \left(1 + \frac{1}{k(k+2)} \right) = \pi_k.$$

Hence

$$\lim_{n \to \infty} I_1(a_n(x)) = \sum_{k=1}^{\infty} \pi_k \log_2 \frac{1}{\pi_k}.$$

Let it be remarked that contrary to Exercises 3.a) and 3.b), the random variables $a_n(x)$ in this example are not independent; the total information contained in a sequence of several digits $a_n(x)$ is not equal to the sum of the informations contained in the individual digits.

4. Let a differentiable function $f(x)$ be defined in $[0, A]$ and suppose $f(0) = 0$ and $|f'(x)| \leq B$. Find an upper bound for the information necessary in order to determine the value of $f(x)$ at every point of $[0, A]$ with an error not exceeding $\varepsilon > 0$.

Hint. Put $x_k = \dfrac{k\varepsilon}{B}$ $\left(k = 0, 1, \ldots, \left\lceil \dfrac{AB}{\varepsilon} \right\rceil \right)$, $x_{\left\lceil \frac{AB}{\varepsilon} \right\rceil + 1} = A$. Let the curve of $f(x)$ be approximated by a polygonal line $y = \varphi(x)$ which can have for its slope in

[1] Cf. e.g. A. J. Khinchin [6].

each of the intervals (x_k, x_{k+1}) either $+B$ or $-B$. If $\varphi(x)$ is already defined for $0 \leq x \leq x_k$, then let the slope in (x_{k+1}) be so chosen that $|f(x_{k+1}) - \varphi(x_{k+1})| \leq \varepsilon$. Obviously, this is always possible. Since $f(x) - \varphi(x)$ is in every interval (x_k, x_{k+1}) monotone, the inequality $|f(x) - \varphi(x)| \leq \varepsilon$ holds in the open intervals (x_k, x_{k+1}) $(k = 0, 1, \ldots)$ as well. Clearly, the number of possible functions $\varphi(x)$ is equal to $2^{\left[\frac{AB}{\varepsilon}\right]+1}$. In order to determine $f(x)$ up to an error ε there suffices therefore $\left[\dfrac{AB}{\varepsilon}\right] + 1$ bits of information.

5. We have n apparently identical coins. One of them is false and heavier than the others. We possess a balance with two scales but without weights. How many weighings are necessary to find the false coin?

Hint. The amount of information needed is equal to $\log_2 n$. Only weighings with an equal number of coins in both scales are worth while to be performed. 3 cases are possible: equilibrium, right scale heavier, and left scale heavier. One weighing gives thus at most $\log_2 3$ bits and there must be performed at least $\left\{\dfrac{\log_2 n}{\log_2 3}\right\}$ weighings ($\{x\}$ denotes the smallest integer greater than or equal to x). It is easy to see that this number of weighings is sufficient. In fact, let k be defined by $3^{k-1} < n \leq 3^k$. At the first weighing we put in each of the dishes $\left[\dfrac{n}{3}\right]$ coins. We know then to which of the three sets, each containing at most 3^{k-1} coins, the false coin belongs. Proceeding in this manner, the false coin will be found after at most k weighings.

6. The "Bar–Kochba" game is played as follows: Player A thinks of any object, player B asks questions which can be answered by "yes" or "no", and has to guess the thing on which A thought from the answers. Naturally, A has to answer all questions honestly.

a) The players agree that A thinks of some nonnegative integer $< N$. What is the minimal number of questions permitting B to find out the considered integer? Give an "optimal" sequence of questions.

Hint. Obviously, at least $\{\log_2 N\}$ questions are needed, since each answer provides at most one bit of information and we need $\log_2 N$ bits. An optimal system of questions is to ask, whether in the binary representation of the number x the first, the second, ..., digit is 0? The aim is arrived at by $\{\log_2 N\}$ questions, since the binary representation of an integer is unique.

b) Suppose $N = 2^s$. How many optimal systems of questions do exist? That is: how many systems of exactly s questions determine x whatever it may be?

Hint. The number of the possible sequences of answers to s questions is evidently 2^s. There corresponds thus to every integer x ($x = 0, 1, \ldots, 2^s - 1$) in a one-to-one manner a sequence of s yes-or-no answers. Every question can be put in the following form: Does x belong to a subset A of the sequence $0, 1, \ldots, 2^s - 1$? Thus to an optimal sequence of s questions there correspond s subsets of the set $M = \{0, 1, \ldots, 2^s - 1\}$; let these be denoted by $A_1, A_2, \ldots, A_s$. According to what has been said, A_1 has to contain exactly 2^{s-1} elements. Let $\bar{A}$ always denote the set complementary to A with respect to M. Then $A_1 A_2$ and $\bar{A}_1 A_2$ have to contain both 2^{s-2} elements; $A_1 \bar{A}_2 A_3$, $\bar{A}_1 A_2 A_3$, $\bar{A}_1 \bar{A}_2 A_3$, and $A_1 A_2 A_3$ have to contain 2^{s-3} elements, and so on.

Conversely, if all sets $\bar{A}_1\bar{A}_2 \ldots \bar{A}_{k-1} A_k$ contain exactly 2^{s-k} elements $(k = 1, 2, \ldots, s)$, where $\bar{A}$ means either A or $\acute{A}$, then the system of sets $A_1, A_2, \ldots, A_s$ is optimal. It follows from this that the number of optimal sequences of questions is

$$\binom{2^s}{2^{s-1}} \binom{2^{s-1}}{2^{s-2}}^2 \binom{2^{s-2}}{2^{s-3}}^4 \cdots \binom{2}{1}^{2^{s-1}} = (2^s)!.$$

If we regard the systems of questions which differ only in the order of questions as identical, then the number looked for is $\dfrac{2^s!}{s!}$.

Remark. In the Bar–Kochba game the questions are, in general, formulated while taking into account the answers already obtained. (In the language of set theory: if the first answers have shown that the object belongs to a subset A of the set M of all possible objects, then the next question is whether it belongs to some subset B of the set A.) It follows from what has been said that the questioner suffers no disadvantage by being obliged to put his questions simultaneously.

7. Suppose that in the Bar–Kochba game type players agree that the objects allowed to be thought of are the n elements of a given set M. Suppose that the questions are asked at random, or in other words, all possible questions have the same probability, independently of the answers already obtained.

a) What is the probability that the questioner finds out the object by k questions?

b) Find the limit of the probability obtained in a) as n and k both tend to $+\infty$ such that

$$\lim_{n \to \infty} (k - \log_2 n) = c .$$

Hints. We may suppose that the elements of the set M are the numbers $1, 2, \ldots, n$. Each possible question is equivalent to asking whether the number thought of does belong to a certain subset of M. The number of possible questions is thus equal to the number of subsets of M, i.e. to 2^n. (For sake of simplicity there are included the two trivial questions corresponding to the whole set and the empty set.) Let $A_1, A_2, \ldots, A_k$ be the sets chosen at random by the questioner: i.e. he asks, whether the number thought of does belong to these sets. By assumption, each of the sets A_n is, with probability $\dfrac{1}{2^n}$, equal to an arbitrary subset of M. Put

$$\varepsilon_j(l) = \begin{cases} 1 & \text{if } A_j \text{ contains the number } l, \\ 0 & \text{otherwise.} \end{cases}$$

The random variables $\varepsilon_j(l)$ $(j = 1, 2, \ldots, k; \ l = 1, 2, \ldots, n)$ are independent of each other and each takes on the values 0 and 1 with probability $\dfrac{1}{2}$. The questioner finds the number x, when the sequence of numbers $\varepsilon_1(x), \varepsilon_2(x), \ldots, \varepsilon_k(x)$ is different from all sequences $\varepsilon_1(y), \varepsilon_2(y), \ldots, \varepsilon_k(y)$ with $y \neq x$. The sequences $\varepsilon_1(l), \varepsilon_2(l), \ldots, \varepsilon_k(l)$ are, with probability $\dfrac{1}{2^k}$ and independently of each other, equal to any sequence consisting of k digits 0 or 1; the problem is thus equivalent to the following urn-problem:

n balls are thrown into 2^k urns; each ball has the same probability to fall into any of the urns. One of the balls is red, all the other balls are white. What is the probability

that the red ball is alone in an urn? The answer is evidently $\left(1 - \dfrac{1}{2^k}\right)^{n-1} = P_{n,k}.$
The answer to question b) is therefore

$$\lim_{\substack{k-\log_2 n \to c \\ n \to \infty}} P_{n,k} = \exp\left(-\frac{1}{2^c}\right).$$

Remarks

1. The number of questions needed (if n is s u f f i c i e n t l y   l a r g e) in order to find the number with a probability ≥ 0.99 by means of this random strategy exceeds only by 7 the number of questions needed in the case the optimal strategy is employed. In fact, $\exp\left(-\dfrac{1}{2^7}\right) > 0.99$. This result is surprising, since one would be inclined to guess that the random strategy is much less advantageous than the optimal strategy.

2. When the questions are asked at random it may happen that the same question occurs twice. But the corresponding probability is so small, if n is large, that it is not worth while to exclude this possibility, though of course this would slightly increase the chances of success.

8. Certain players play the Bar–Kochba game in the following manner: There are $r + 1$ players, r players think of some object; the last player asks them questions. The same question is addressed to every player, who answers by "yes" or "no", according to what is true concerning the object he had thought of.

a) Each of the players thinks of one of the numbers $1, 2, \ldots, n$ $(n \geq r)$, but each of a different number. The questions are asked at random, as in the preceding exercise. What is the probability that the questioner finds all numbers by k questions?

b) $n = r$ and the players agree to think each of a different number of the sequence $1, 2, \ldots, n$; hence it is a permutation of numbers which is to be found. What is the probability that the questioner finds the permutation by k questions? Calculate approximately this probability for $k = 2 \log_2 n + c$.

Hints. a) We are led to the following urn problem: we put n balls into 2^k urns, independently of each other, each ball having the same probability $\dfrac{1}{2^k}$ to get into any one of the urns. Among the n balls there are r red balls, the others are white. What is the probability that all the red balls get into different urns? This probability is

$$P_{n,k,r} = \prod_{j=1}^{r-1}\left(1 - \frac{j}{2^k}\right)\left(1 - \frac{r}{2^k}\right)^{n-r}.$$

For $r = 1$ we find as a particular case the result of the preceding exercise.
b) $n = r$, hence

$$P_{n,k,n} = \prod_{j=1}^{n-1}\left(1 - \frac{j}{2^k}\right);$$

thus

$$\lim_{n \to \infty} P_{n,2\log_2 n+c,n} \approx \exp\left(-\frac{1}{2^{c+1}}\right).$$

Remark. It is surprising that in this game to guess a permutation of the numbers $1, 2, \ldots, n$ approximately twice as many questions are necessary as to guess a single one of these numbers. Of course, in the first case one gets to each question n answers.

9. Let $f(n)$ $(n = 1, 2, \ldots)$ be a completely additive number-theoretical function, i.e.

$$f(nm) = f(n) + f(m) \tag{A}$$

for all pairs of integers n and m. Suppose further that the limits

$$\overline{\lim_{n \to \infty}} \, |f(n + d) - f(n)| = l(d) \tag{B}$$

are finite for every integer d and

$$\lim_{d \to \infty} \frac{l(d)}{\log d} = 0.$$

Then $g(n) = c \log n$, where c is a constant.

Hint. Let P be an integer greater than 1; we put

$$g(n) = f(n) - \frac{f(P) \log n}{\log P}.$$

From $g(P) = 0$ and from (B) it follows that

$$\overline{\lim_{n \to \infty}} \, \frac{|g(n)|}{\log n} \leq \frac{\max\limits_{1 \leq d < P} l(d)}{\log P}.$$

From this we conclude that

$$\lim_{P \to \infty} \overline{\lim_{n \to \infty}} \left| \frac{f(n)}{\log n} - \frac{f(P)}{\log P} \right| = 0.$$

This implies the existence of the limit

$$\lim_{n \to \infty} \frac{f(n)}{\log n} = c.$$

If we put now $h(n) = f(n) - c \log n$, then $l(n)$ is a completely additive function for which

$$\lim_{n \to \infty} \frac{h(n)}{\log n} = 0. \tag{1}$$

But this implies $h(n) \equiv 0$ since otherwise there would exist an integer r with $h(r) \neq 0$ and thus, because of the additivity, $h(r^k) = kh(r)$ for $k = 1, 2, \ldots$, which contradicts (1).

Remark. This problem is due to K. L. Chung but his proof differs from ours. If instead of the complete additivity only simple additivity is required, i.e. that $f(n\,m) = f(n) + f(m)$, if $(n, m) = 1$, then the condition (B) does not imply $f(n) = c \log n$. (The last step of the proof cannot be carried out in this case.)

10. Let $P = \{p_k\}$ be any distribution with

$$\sum_{k=1}^{\infty} k p_k = \lambda > 1.$$

Then the entropy

$$\sum_{k=1}^{\infty} p_k \log_2 \frac{1}{p_k} = I_1(\mathscr{P})$$

takes on its maximal value if

$$p_k = \frac{1}{\lambda} \left(1 - \frac{1}{\lambda}\right)^{k-1}.$$

11. Let $\mathscr{P}$ and Q be two distributions, absolutely continuous with respect to Lebesgue measure, with density functions $p(x)$ and $q(x)$ and further let Q be absolutely continuous with respect to $\mathscr{P}$. It follows from Theorem 2 of § 8 that the gain of information is nonnegative in this case too, i.e. we have the inequalities

$$\int_{-\infty}^{+\infty} q(x) \log_2 \frac{q(x)}{p(x)}\, dx \geq 0$$

and

$$\frac{1}{\alpha - 1} \log_2 \int_{-\infty}^{+\infty} \frac{q(x)^{\alpha}}{p(x)^{\alpha-1}}\, dx \geq 0 \quad \text{for} \quad \alpha > 0, \alpha \neq 1.$$

Prove these inequalities directly (without passing to the limit) by Jensen's inequality generalized for functions, i.e. by inequality (15) of § 8.

12. a) Let ξ be a positive random variable having an absolutely continuous distribution, with $E(\xi) = \lambda > 0$. Show that the entropy (of order 1) of ξ is maximal if the distribution of ξ is exponential.

Hint. Let $f(x)$ be a density function in $(0, +\infty)$ with

$$\int_0^{\infty} x f(x)\, dx = \lambda$$

and put

$$g(x) = \frac{1}{\lambda} \exp\left(-\frac{x}{\lambda}\right).$$

We have (cf. Exercise 11)

$$0 \leq \int_0^{\infty} f(x) \log_2 \frac{f(x)}{g(x)}\, dx = \int_0^{\infty} g(x) \log_2 \frac{1}{g(x)}\, dx - \int_0^{\infty} f(x) \log_2 \frac{1}{f(x)}\, dx.$$

b) Let ξ be a random variable distributed in the interval (a, b) with an absolutely continuous distribution function. Show that the entropy (of order 1) of ξ is maximal if ξ is uniformly distributed in (a, b).

Hint. Let $f(x)$ be a density function which vanishes outside (a, b) and put

$$g(x) = \begin{cases} \dfrac{1}{b-a} & \text{for } a < x < b, \\[2mm] 0 & \text{otherwise.} \end{cases}$$

We have then

$$0 \le \int\limits_a^b f(x) \log_2 \frac{f(x)}{g(x)}\, dx = \int\limits_a^b g(x) \log_2 \frac{1}{g(x)}\, dx - \int\limits_a^b f(x) \log_2 \frac{1}{f(x)}\, dx.$$

13. Let $\xi_1, \xi_2, \ldots, \xi_n$ be random variables and $C = (c_{kl})$ a nonsingular n by n matrix. We put

$$\eta_k = \sum_{l=1}^n c_{lk}\xi_l \qquad (k = 1, 2, \ldots, n).$$

Show that we have for $\alpha > 0$

$$I_{\alpha,n}((\eta_1, \eta_2, \ldots, \eta_n)) = I_{\alpha,n}((\xi_1, \xi_2, \ldots, \xi_n)) + \log_2 |\, ||\, C\, ||\, |$$

where $||C||$ denotes the determinant of C.

14. Let $\xi_1, \xi_2, \ldots, \xi_r$ be independent random variables with absolutely continuous distribution. We have

$$I_{\alpha,r}((\xi_1, \xi_2, \ldots, \xi_r)) = I_{\alpha,1}(\xi_1) + I_{\alpha,2}(\xi_2) + \ldots + I_{\alpha,r}(\xi_r).$$

Hint. This follows from Formulas (34b) and (35b) of § 8.

15. The relative information $I(\xi, \eta)$ contained in the value of η concerning ξ (or conversely) can be defined, when the pair (ξ, η) has an absolutely continuous distribution, by

$$I(\xi, \eta) = I_{1,1}(\xi) + I_{1,1}(\eta) - I_{1,2}((\xi,\eta)).$$

Show that if $\mathcal{R}$ is the joint distribution of ξ and η and if $\mathcal{P} * Q$ denotes the direct product of the distributions $\mathcal{P}$ and Q of ξ and η, then Formula (11) of § 4 remains valid; i.e. $I(\xi, \eta)$ is equal to the gain of information obtained by replacing $\mathcal{P} * Q$ by $\mathcal{R}$.

Hint. If $h(x, y)$ is the density function of the pair (ξ, η), $f(x)$ and $g(y)$ are the density functions of ξ and η, then we have

$$I(\xi, \eta) = \int\limits_{-\infty}^{+\infty} \int\limits_{-\infty}^{+\infty} h(x, y) \log_2 \frac{h(x, y)}{f(x) g(y)}\, dx dy.$$

It follows that

$$I(\xi, \eta) = I_1(\mathcal{R} \,||\, \mathcal{P} * Q),$$

because of Formula (38) of § 8.

16. In the following exercises we always use natural (Napier's) logarithms ln.

a) Calculate the entropy (of dimension 1 and of order 1) of the normal distribution; i.e. show that

$$\int\limits_{-\infty}^{+\infty} \varphi(x) \ln \frac{1}{\varphi(x)}\, dx = \ln \sigma \sqrt{2\pi e},$$

where

$$\varphi(x) = \frac{1}{\sqrt{2\pi}\,\sigma} \exp\left(-\frac{(x - m)^2}{2\sigma^2}\right).$$

b) Calculate the entropy (of dimension r and of order 1) of the r-dimensional normal distribution.

Hint. Let the r-dimensional density function of the random variables $\xi_1, \xi_2, \ldots, \xi_r$ be

$$f(x_1, \ldots, x_r) = \frac{\|B\|^{\frac{1}{2}}}{(2\pi)^{\frac{r}{2}}} \exp\left(-\frac{1}{2}\left[\sum_{i=1}^{r} \sum_{j=1}^{r} b_{ij}(x_i - m_i)(x_j - m_j)\right]\right),$$

where $\|B\|$ is the determinant of the positive definite quadratic form

$$\sum_{i=1}^{r} \sum_{j=1}^{r} b_{ij} z_i z_j.$$

By a suitable orthogonal transformation

$$\eta_k = \sum_{j=1}^{r} c_{kj}(\xi_j - m_j)$$

we obtain for the density function of the random variables $\eta_1, \eta_2, \ldots, \eta_r$:

$$\frac{1}{(2\pi)^{\frac{r}{2}} \sigma_1 \sigma_2 \ldots \sigma_r} \exp\left(-\frac{1}{2}\sum_{j=1}^{r} \frac{y_j^2}{\sigma_j^2}\right),$$

with $\sigma_1 \sigma_2 \ldots \sigma_r = \|B\|^{-\frac{1}{2}}$. According to Exercise 13, the entropy is invariant under such a transformation, since the absolute value of the determinant of an orthogonal transformation is equal to 1. Hence, according to Exercise 14,

$$I_{1,r}((\xi_1, \xi_2, \ldots, \xi_r)) = I_{1,r}((\eta_1, \eta_2, \ldots, \eta_r)) = \sum_{k=1}^{r} I_{1,1}(\eta_k) = \ln(2\pi e)^{\frac{r}{2}} \|B\|^{-\frac{1}{2}}.$$

c) Let the joint distribution of the random variables ξ and η be normal with density function

$$f(x, y) = \frac{\sqrt{AC - B^2}}{2\pi} \exp\left(-\frac{1}{2}(Ax^2 + 2Bxy + Cy^2)\right).$$

Calculate the (relative) information contained in the value η concerning ξ.

Hint. We find the desired information by subtracting from the sum of the informations contained in ξ and in η, the information contained in the distribution of the pair (ξ, η). Hence

$$I(\xi, \eta) = \ln\sqrt{2\pi e\,\frac{A}{AC - B^2}} + \ln\sqrt{2\pi e\,\frac{C}{AC - B^2}} - \ln\frac{2\pi e}{\sqrt{AC - B^2}} =$$

$$= \ln\sqrt{\frac{AC}{AC - B^2}}.$$

If $B = 0$, i.e. if ξ and η are independent, we find of course that $I(\xi, \eta) = 0$.

17. Let ξ be a random variable with absolutely continuous distribution and density function $f(x)$. Let the standard deviation σ of ξ be finite and positive.

a) Show that the entropy (of dimension 1 and of order 1) of ξ is maximal if ξ is normally distributed.

b) Show that the entropy $I_{\alpha,1}(\xi)$ (of dimension 1 and of order $\alpha > 1$) is maximal if

$$f_\alpha(x) = \begin{cases} \dfrac{1}{c} \dfrac{\Gamma\left(\dfrac{2\alpha}{\alpha-1}\right)}{\Gamma^2\left(\dfrac{\alpha}{\alpha-1}\right)} 2^{-\frac{\alpha+1}{\alpha-1}} \left(1 - \dfrac{x^2}{c^2}\right)^{\frac{1}{1-\alpha}} & \text{for } |x| < c, \\[4mm] 0 & \text{otherwise,} \end{cases}$$

where we have put $c = \sigma \sqrt{\dfrac{3\alpha-1}{\alpha-1}}$.

Hint. Put

$$m = M(\xi), \quad \sigma^2 = \int_{-\infty}^{+\infty} (x-m)^2 f(x)\, dx, \quad \varphi(x) = \frac{1}{\sqrt{2\pi}\sigma} \exp\left(-\frac{(x-m)^2}{2\sigma^2}\right).$$

We have then

$$0 \le \int_{-\infty}^{+\infty} f(x) \ln \frac{f(x)}{\varphi(x)}\, dx = \int_{-\infty}^{+\infty} \varphi(x) \ln \frac{1}{\varphi(x)}\, dx - \int_{-\infty}^{+\infty} f(x) \ln \frac{1}{f(x)}\, dx,$$

which implies a). b) can be proved in the same fashion. Let it be noticed that $f_\alpha(x)$ tends to

$$\frac{1}{\sqrt{2\pi}\sigma} \exp\left(-\frac{x^2}{2\sigma^2}\right)$$

as $\alpha \to 1$.

18. Let $f(x)$ and $f_n(x)$ be density functions such that $f_n(x) = 0$ $(n = 1, 2, \ldots)$ for every value of x for which $f(x) = 0$; suppose further that all integrals

$$\int_{-\infty}^{+\infty} \frac{f_n^2(x)}{f(x)}\, dx \qquad (n = 1, 2, \ldots)$$

exist and that

$$\lim_{n \to \infty} \int_{-\infty}^{+\infty} \frac{f_n^2(x)}{f(x)}\, dx = 1.$$

Prove that under these conditions

$$\lim_{n \to \infty} \sup_E \left| \int_E f_n(x)\, dx - \int_E f(x)\, dx \right| = 0,$$

where E runs through all measurable subsets of the set of real numbers.

Hint. Applying Schwarz' inequality, we get

$$\left| \int_E f_n(x)\,dx - \int_E f(x)\,dx \right| \leq \int_{-\infty}^{+\infty} \frac{|f_n(x) - f(x)|}{\sqrt{f(x)}} \sqrt{f(x)}\,dx \leq$$

$$\leq \left(\int_{-\infty}^{+\infty} \frac{(f_n(x) - f(x))^2}{f(x)}\,dx \right)^{\frac{1}{2}}$$

and clearly

$$\int_{-\infty}^{+\infty} \frac{(f_n(x) - f(x))^2}{f(x)}\,dx = \int_{-\infty}^{+\infty} \frac{f_n^2(x)}{f(x)}\,dx - 1.$$

TABLES

TABLE 1

Values of $n!$ and of log $n!$ for $n \leq 50$

n	$n!$	log $n!$	n	$n!$	log $n!$
1	1	0.00000000	26	$40329146 \cdot 10^{19}$	26.60561903
2	2	0.30103000	27	$10888869 \cdot 10^{21}$	28.03698279
3	6	0.77815125	28	$30488834 \cdot 10^{22}$	29.48414082
4	24	1.38021124	29	$88417620 \cdot 10^{23}$	30.94653882
5	120	2.07918125	30	$26525286 \cdot 10^{25}$	32.42366007
6	720	2.85733250	31	$82228387 \cdot 10^{26}$	33.91502177
7	5040	3.70243054	32	$26313084 \cdot 10^{28}$	35.42017175
8	40320	4.60552052	33	$86833176 \cdot 10^{29}$	36.93868569
9	362880	5.55976303	34	$29523280 \cdot 10^{31}$	38.47016460
10	3628800	6.55976303	35	$10333148 \cdot 10^{33}$	40.01423265
11	39916800	7.60115572	36	$37199333 \cdot 10^{34}$	41.57053515
12	$47900160 \cdot 10$	8.68033696	37	$13763753 \cdot 10^{36}$	43.13873687
13	$62270208 \cdot 10^2$	9.79428032	38	$52302262 \cdot 10^{37}$	44.71852047
14	$87178291 \cdot 10^3$	10.94040835	39	$20397882 \cdot 10^{39}$	46.30958508
15	$13076744 \cdot 10^5$	12.11649961	40	$81591528 \cdot 10^{40}$	47.91164507
16	$20922790 \cdot 10^6$	13.32061959	41	$33452527 \cdot 10^{42}$	49.52442892
17	$35568743 \cdot 10^7$	14.55106852	42	$14050061 \cdot 10^{44}$	51.14767822
18	$64023737 \cdot 10^8$	15.80634102	43	$60415263 \cdot 10^{45}$	52.78114667
19	$12164510 \cdot 10^{10}$	17.08509462	44	$26582716 \cdot 10^{47}$	54.42459935
20	$24329020 \cdot 10^{11}$	18.38612462	45	$11962222 \cdot 10^{49}$	56.07781186
21	$51090942 \cdot 10^{12}$	19.70834391	46	$55026222 \cdot 10^{50}$	57.74056969
22	$11240007 \cdot 10^{14}$	21.05076659	47	$25862324 \cdot 10^{52}$	59.41266755
23	$25852017 \cdot 10^{15}$	22.41249443	48	$12413916 \cdot 10^{54}$	61.09390879
24	$62044840 \cdot 10^{16}$	23.79270567	49	$60828186 \cdot 10^{55}$	62.78410487
25	$15511210 \cdot 10^{18}$	25.19064568	50	$30414093 \cdot 10^{57}$	64.48307487

TABLE 2

Binomial coefficients $\binom{n}{k}$ for $n \leq 30$ [1]

k \ n	0	1	2	3	4	5	6	7	8
2	1	2	1						
3	1	3	3	1					
4	1	4	6	4	1				
5	1	5	10	10	5	1			
6	1	6	15	20	15	6	1		
7	1	7	21	35	35	21	7	1	
8	1	8	28	56	70	56	28	8	1
9	1	9	36	84	126	126	84	36	9
10	1	10	45	120	210	252	210	120	45
11	1	11	55	165	330	462	462	330	165
12	1	12	66	220	495	792	924	792	495
13	1	13	78	286	715	1287	1716	1716	1287
14	1	14	91	364	1001	2002	3003	3432	3003
15	1	15	105	455	1365	3003	5005	6435	6435
16	1	16	120	560	1820	4368	8008	11440	12870
17	1	17	136	680	2380	6188	12376	19448	24310
18	1	18	153	816	3060	8568	18564	31824	43758
19	1	19	171	969	3876	11628	27132	50388	75582
20	1	20	190	1140	4845	15504	38760	77520	125970
21	1	21	210	1330	5985	20349	54264	116280	203490
22	1	22	231	1540	7315	26334	74613	170544	319770
23	1	23	253	1771	8855	33649	100947	245157	490314
24	1	24	276	2024	10626	42504	134596	346104	735471
25	1	25	300	2300	12650	53130	177100	480700	1081575
26	1	26	325	2600	14950	65780	230230	657800	1562275
27	1	27	351	2925	17550	80730	296010	888030	2220075
28	1	28	378	3276	20475	98280	376740	1184040	3108105
29	1	29	406	3654	23751	118755	475020	1560780	4292145
30	1	30	435	4060	27405	142506	593775	2035800	5852925

[1] For $n \geq 15$ values of $\binom{n}{k}$ are given for $k \leq \dfrac{n}{2}$ only; the further values can be taken from the table by the relation $\binom{n}{n-k} = \binom{n}{k}$.

TABLE 2

(continued)

9	10	11	12	13	14	15	k / n
							2
							3
							4
							5
							6
							7
							8
1							9
10	1						10
55	11	1					11
220	66	12	1				12
715	286	78	13	1			13
2002	1001	364	91	14	1		14
5005	3003	1365	455	105	15	1	15
11440	8008	4368	1820	560	120	16	16
24310	19448	12376	6188	2380	680	136	17
48620	43758	31824	18564	8568	3060	816	18
92378	92378	75582	50388	27132	11628	3876	19
167960	184756	167960	125970	77520	38760	15504	20
293930	352716	352716	293930	203490	116280	54264	21
497420	646646	705432	646646	497420	319770	170544	22
817190	1144066	1352078	1352078	1144066	817190	490314	23
1307504	1961256	2496144	2704156	2496144	1961256	1307504	24
2042975	3268760	4457400	5200300	5200300	4457400	3268760	25
3124550	5311735	7726160	9657700	10400600	9657700	7726160	26
4686825	8436285	13037895	17383860	20058300	20058300	17383860	27
6906900	13123110	21474180	30421755	37442160	40116600	37442160	28
10015005	20030010	34597290	51895935	67863915	77558760	77558760	29
14307150	30045015	54627300	86493225	119759850	145422675	155117520	30

The terms of the Poisson distribution

TABLE 3

k \ λ	0.1	0.2	0.3	0.4	0.5
0	0.90484	0.81873	0.74082	0.67032	0.60653
1	0.09048	0.16375	0.22225	0.26813	0.30327
2	0.00452	0.01637	0.03334	0.05362	0 07581
3	0.00015	0.00109	0.00333	0.00715	0 01263
4		0.00005	0.00025	0.00071	0.00158
5			0.00001	0.00005	0.00016
6					0.00001

k \ λ	0.6	0.7	0.8	0.9	
0	0.54881	0.49659	0.44933	0.40657	
1	0.32929	0.34761	0.35946	0.36591	
2	0.09878	0.12166	0.14379	0.16466	
3	0.01976	0.02838	0.03834	0.04939	
4	0.00296	0.00496	0.00766	0.01111	
5	0.00035	0.00069	0.00123	0.00200	
6	0.00003	0.00008	0.00016	0.00030	
7			0.00001	0.00003	

k \ λ	1	2	3	4	5
0	0.36788	0.13534	0.04978	0.01831	0.00673
1	0.36788	0.27067	0.14936	0.07326	0.03369
2	0.18394	0.27067	0.22404	0.14653	0.08422
3	0.06131	0.18045	0.22404	0.19537	0.14037
4	0.01532	0.09022	0.16803	0.19537	0.17547
5	0.00306	0.03609	0.10082	0.15629	0.17547
6	0.00051	0.01203	0.05040	0.10420	0.14622
7	0.00007	0.00343	0.02160	0.05954	0.10444
8		0.00085	0.00810	0.02977	0.06527
9		0.00019	0.00270	0.01323	0.03626
10		0.00003	0.00081	0.00529	0.01813
11			0.00022	0.00192	0.00824
12			0.00005	0.00064	0.00343
13			0.00001	0.00019	0.00132
14				0.00005	0.00047
15				0.00001	0.00015
16					0.00004
17					0.00001

(continued)   **TABLE 3**

λ / k	6	7	8	9	10
0	0.00247	0.00091	0.00033	0.00012	0.00004
1	0.01487	0.00638	0.00268	0.00111	0.00045
2	0.04461	0.02234	0.01073	0.00499	0.00227
3	0.08923	0.05212	0.02862	0.01499	0.00756
4	0.13385	0.09122	0.05725	0.03373	0.01891
5	0.16062	0.12772	0.09160	0.06072	0.03783
6	0.16062	0.14900	0.12214	0.09109	0.06305
7	0.13768	0.14900	0.13959	0.11712	0.09007
8	0.10326	0.13038	0.13959	0.13176	0.11260
9	0.06883	0.10140	0.12408	0.13176	0.12511
10	0.04130	0 07098	0.09926	0.11858	0.12511
11	0 02252	0.04517	0.07219	0.09702	0.11374
12	0.01126	0.02635	0.04812	0.07276	0.09478
13	0.00519	0.01418	0.02961	0.05037	0.07290
14	0.00222	0.00709	0.01692	0.03238	0.05207
15	0.00089	0 00331	0 00902	0.01943	0.03471
16	0.00033	0.00144	0.00451	0.01093	0.02169
17	0 00011	0 00059	0.00212	0.00578	0.01276
18	0.00003	0 00023	0.00094	0.00289	0.00709
19	0.00001	0 00008	0.00039	0.00137	0.00373
20		0.00003	0.00015	0.00061	0 00186
21			0.00006	0.00026	0.00088
22			0.00002	0.00010	0.00040
23				0.00004	0.00017
24				0.00001	0.00007
25					0.00002
26					0.00001

TABLE 3   (continued)

k \ λ	11	12	13	14	15
0	0.00001				
1	0.00018	0.00007	0.00002	0.00001	
2	0.00101	0.00044	0.00019	0.00008	0.00003
3	0.00370	0.00177	0.00082	0.00038	0.00017
4	0.01018	0.00530	0.00269	0.00133	0.00064
5	0.02241	0.01274	0.00699	0.00373	0.00193
6	0.04109	0.02548	0.01515	0.00869	0.00483
7	0.06457	0.04368	0.02814	0.01739	0.01037
8	0.08879	0.06552	0.04573	0.03043	0.01944
9	0.10853	0.08736	0.06605	0.04734	0.03240
10	0.11938	0.10484	0.08587	0.06628	0.04861
11	0.11938	0.11437	0.10148	0.08435	0.06628
12	0.10943	0.11437	0.10994	0.09841	0.08285
13	0.09259	0.10557	0.10994	0.10599	0.09560
14	0.07275	0.09048	0.10209	0.10599	0.10244
15	0.05335	0.07239	0.08847	0.09892	0.10244
16	0.03668	0.05429	0.07188	0.08655	0.09603
17	0.02373	0.03832	0.05497	0.07128	0.08473
18	0.01450	0.02555	0.03970	0.05544	0.07061
19	0.00839	0.01613	0.02716	0.04085	0.05574
20	0.00461	0.00968	0.01765	0.02859	0.04181
21	0.00241	0.00553	0.01093	0.01906	0.02986
22	0.00121	0.00301	0.00645	0 01213	0.02036
23	0.00057	0.00157	0.00365	0.00738	0.01328
24	0.00026	0.00078	0.00197	0.00430	0.00830
25	0.00011	0.00037	0.00102	0.00241	0.00498
26	0.00004	0.00017	0.00051	0.00129	0.00287
27	0.00002	0.00007	0.00024	0.00067	0.00159
28		0.00003	0.00011	0 00033	0.00085
29		0.00001	0.00005	0.00016	0.00044
30			0.00002	0.00007	0.00022
31				0,00003	0.00010
32				0.00001	0.00005
33					0.00002
34					0.00001

(continued) TABLE 3

k \ λ	16	17	18	19	20
0					
1					
2	0.00001				
3	0.00007	0.00003			
4	0.00030	0.00014	0.00006	0.00003	0.00001
5	0.00098	0.00049	0.00024	0.00011	0.00005
6	0.00262	0 00138	0.00071	0.00036	0.00018
7	0 00599	0.00337	0.00185	0.00099	0.00052
8	0.01198	0.00716	0.00416	0.00236	0.00130
9	0 02131	0.01352	0.00832	0.00498	0.00290
10	0 03409	0.02300	0.01498	0.00946	0.00581
11	0.04959	0.03554	0.02452	0.01635	0.01057
12	0.06612	0.05035	0.03678	0.02588	0.01762
13	0.08138	0.06584	0.05092	0.03783	0.02711
14	0.09301	0.07996	0.06548	0.05135	0 03874
15	0 09921	0.09062	0.07857	0.06504	0.05165
16	0.09921	0.09628	0.08839	0.07724	0.06456
17	0.09338	0.09628	0.09359	0.08632	0.07595
18	0.08300	0.09093	0.09359	0.09112	0.08439
19	0.06989	0.08136	0.08867	0.09112	0.08883
20	0 05592	0.06915	0.07980	0.08656	0 08883
21	0.04260	0.05598	0.06840	0.07832	0.08460
22	0.03098	0.04326	0.05596	0.06764	0.07691
23	0.02155	0.03197	0.04380	0.05587	0.06688
24	0.01437	0.02265	0.03285	0.04423	0.05573
25	0 00919	0.01540	0.02365	0 03362	0 04458
26	0 00566	0.01007	0.01637	0.02456	0 03429
27	0.00335	0.00634	0.01091	0.01728	0.02540
28	0.00191	0.00385	0.00701	0.01173	0.01814
29	0.00105	0.00225	0.00435	0.00768	0.01251
30	0.00056	0.00127	0.00261	0.00486	0.00834
31	0.00029	0.00070	0.00151	0.00298	0.00538
32	0.00014	0.00037	0.00085	0 00177	0 00336
33	0.00007	0.00019	0.00046	0.00102	0.00203
34	0.00003	0.00009	0.00024	0.00057	0.00119
35	0.00001	0.00004	0.00012	0.00030	0.00068
36		0.00002	0.00006	0.00016	0.00938
37		0.00001	0.00003	0.00008	0.00020
38			0.00001	0.00004	0.00010
39				0.00002	0.00005
40					0.00002
41					0.00001

TABLE 4

The incomplete gamma function

$$\Gamma_n(\lambda) = \frac{1}{(n-1)!} \int_0^\lambda t^{n-1} e^{-t}\, dt = \sum_{k=n}^{\infty} \frac{\lambda^k e^{-\lambda}}{k!} \qquad (n = 1, 2, \ldots)$$

n	$\lambda=0.001$	$\lambda=0.002$	$\lambda=0.003$	$\lambda=0.004$
1	0.000999	0.001988	0.002995	0.003992
2	000001	000002	000005	000008

n	$\lambda=0.005$	$\lambda=0.006$	$\lambda=0.007$	$\lambda=0.008$
1	0.004987	0.005982	0.006976	0.007968
2	000013	000018	000024	000032

n	$\lambda=0.009$	$\lambda=0.01$	$\lambda=0.02$	$\lambda=0.03$
1	0.008960	0.009950	0.019801	0.029555
2	000040	000050	000197	000441
3			000002	000004

n	$\lambda=0.04$	$\lambda=0.05$	$\lambda=0.06$	$\lambda=0.07$
1	0.039211	0.048771	0.058236	0.067606
2	000779	001209	001730	002339
3	000010	000020	000034	000054
4				000001

n	$\lambda=0.08$	$\lambda=0.09$	$\lambda=0.10$	$\lambda=0.11$
1	0.076884	0.086069	0.095163	0.104166
2	003034	003815	004679	005624
3	000080	000114	000155	000204
4	000002	000002	000003	000006

n	$\lambda=0.12$	$\lambda=0.13$	$\lambda=0.14$	$\lambda=0.15$
1	0.113080	0.121905	0.130642	0.139292
2	006649	007752	008932	010186
3	000263	000332	000412	000503
4	000008	000011	000014	000018
5				000001

(continued)   TABLE 4

n	$\lambda=0.16$	$\lambda=0.17$	$\lambda=0.18$	$\lambda=0.19$
1	0.147856	0.156335	0.164730	0.173041
2	011513	012912	014381	015919
3	000606	000721	000850	000992
4	000024	000031	000038	000047
5	000001	000001	000001	000001

n	$\lambda=0.20$	$\lambda=0.22$	$\lambda=0.24$	$\lambda=0.26$
1	0.181269	0.197481	0.213372	0.228948
2	017523	020927	024581	028475
3	001149	001506	001927	002414
4	000057	000082	000113	000154
5	000002	000004	000006	000008
6				000001

n	$\lambda=0.28$	$\lambda=0.30$	$\lambda=0.40$	$\lambda=0.50$
1	0.244216	0.259182	0.329680	0.393469
2	032597	036936	061551	090204
3	002970	003600	007926	014388
4	000205	000366	000776	001752
5	000011	000015	000061	000173
6	000001	000001	000004	000014
7			000001	000001

n	$\lambda=1.0$	$\lambda=1.5$	$\lambda=2.0$	$\lambda=2.5$
1	0.63212	0.77687	0.86466	0.91792
2	26424	44218	59399	71270
3	08030	19115	32332	45619
4	01899	06564	14288	24242
5	00366	01858	05265	10882
6	00059	00446	01656	04202
7	00008	00093	00453	01419
8	00001	00017	00110	00425
9	00001	00002	00024	00114
10			00005	00028
11			00001	00006
12				00001

TABLE 4 (continued)

n	λ=3.0	λ=3.5	λ=4.0	λ=4.5
1	0.95021	0.96980	0.98168	0.98889
2	80035	86411	90842	93890
3	57681	67915	76190	82642
4	35277	46338	56653	65770
5	18474	27456	37116	46789
6	08392	14239	21487	29708
7	03351	06529	11067	16895
8	01191	02674	05113	08659
9	00380	00987	02137	04026
10	00110	00331	00813	01709
11	00029	00102	00284	00667
12	00007	00029	00092	00240
13	00002	00008	00027	00081
14		00001	00008	00025
15			00002	00007
16			00001	00002
17				00001

n	λ=5.0	6=5.5	λ=6.0	λ=6.5
1	0.99326	0.99591	0.99752	0.99850
2	95957	97345	98265	98872
3	87535	91162	93804	95696
4	73497	79830	84880	88816
5	55951	64248	71494	77633
6	38404	47108	55433	63096
7	23782	31396	39370	47347
8	13337	19051	25603	32724
9	06809	10564	15276	20843
10	03183	05378	08392	12262
11	01369	02525	04262	06684
12	00545	01099	02009	03389
13	00202	00445	00883	01603
14	00070	00169	00363	00710
15	00023	00060	00140	00296
16	00007	00020	00051	00116
17	00002	00006	00017	00044
18	00001	00002	00006	00015
19		00001	00002	00005
20			00001	00001

(continued) TABLE 4

n	λ=7.0	λ=7.5	λ=8.0	λ=8.5
1	0.99909	0.99945	0.99966	0.99980
2	99271	99530	99698	99807
3	97036	97975	98625	99072
4	91823	94085	95762	96989
5	82701	86794	90037	92564
6	69929	75856	80876	85040
7	55029	62184	68663	74382
8	40129	47536	54704	61440
9	27091	33803	40745	47689
10	16950	22359	28338	34703
11	09852	13776	18412	23664
12	05335	07924	11192	15134
13	02700	04267	06380	09092
14	01281	02157	03418	05141
15	00572	01026	01726	02743
16	00241	00461	00823	01383
17	00096	00196	00372	00661
18	00036	00079	00160	00300
19	00013	00031	00065	00130
20	00005	00011	00025	00054
21		00004	00010	00020
22		00001	00003	00008
23			00001	00003
24				00001

TABLE 4   (continued)

n	$\lambda=9.0$	$\lambda=9.5$	$\lambda=10.0$	
1	0.999 88	0.99993	0.99996	
2	99877	99921	99950	
3	99377	99584	99724	
4	97877	98514	98966	
5	94504	95974	97075	
6	88431	91147	93291	
7	79322	83505	86986	
8	67610	73134	77978	
9	54435	60818	66719	
10	41259	47817	54207	
11	29401	35467	41696	
12	19699	24801	30322	
13	12423	16338	20844	
14	07385	10186	13554	
15	04147	05999	08346	
16	.02204	03347	04874	
17	01111	01773	02704	
18	00533	00893	01428	
19	00243	00428	00719	
20	00106	00196	00345	
21	00044	00086	00159	
22	00018	00036	00070	
23	00006	00015	00030	
24	00003	00006	00012	
25		00002	00004	
26		00001	00001	

TABLE 5

The function $\varphi(x) = \dfrac{1}{\sqrt{2\pi}} e^{-\frac{x^2}{2}}$

x	$\varphi(x)$	x	$\varphi(x)$	x	$\varphi(x)$	x	$\varphi(x)$
0.00	0.3989						
0.01	0.3989	0.41	0.3668	0.81	0.2874	1.21	0.1919
0.02	0.3989	0.42	0.3653	0.82	0.2850	1.22	0.1895
0.03	0.3988	0.43	0.3637	0.83	0.2827	1.23	0.1872
0.04	0.3986	0.44	0.3621	0.84	0.2803	1.24	0.1849
0.05	0.3984	0.45	0.3605	0.85	0.2780	1.25	0.1826
0.06	0.3982	0.46	0.3589	0.86	0.2756	1.26	0.1804
0.07	0.3980	0.47	0.3572	0.87	0.2732	1.27	0.1781
0.08	0.3977	0.48	0.3555	0.88	0.2709	1.28	0.1758
0.09	0.3973	0.49	0.3538	0.89	0.2685	1.29	0.1736
0.10	0.3970	0.50	0.3521	0.90	0.2661	1.30	0.1714
0.11	0.3965	0.51	0.3503	0.91	0.2637	1.31	0.1691
0.12	0.3961	0.52	0.3485	0.92	0.2613	1.32	0.1669
0.13	0.3956	0.53	0.3467	0.93	0.2589	1.33	0.1647
0.14	0.3951	0.54	0.3448	0.94	0.2565	1.34	0.1626
0.15	0.3945	0.55	0.3429	0.95	0.2541	1.35	0.1604
0.16	0.3939	0.56	0.3410	0.96	0.2516	1.36	0.1582
0.17	0.3932	0.57	0.3391	0.97	0.2492	1.37	0.1561
0.18	0.3925	0.58	0.3372	0.98	0.2468	1.38	0.1539
0.19	0.3918	0.59	0.3352	0.99	0.2444	1.39	0.1518
0.20	0.3910	0.60	0.3332	1.00	0.2420	1.40	0.1497
0.21	0.3902	0.61	0.3312	1.01	0.2396	1.41	0.1476
0.22	0.3894	0.62	0.3292	1.02	0.2371	1.42	0.1456
0.23	0.3885	0.63	0.3271	1.03	0.2347	1.43	0.1435
0.24	0.3876	0.64	0.3251	1.04	0.2323	1.44	0.1415
0.25	0.3867	0.65	0.3230	1.05	0.2299	1.45	0.1394
0.26	0.3857	0.66	0.3209	1.06	0.2275	1.46	0.1374
0.27	0.3847	0.67	0.3187	1.07	0.2251	1.47	0.1354
0.28	0.3836	0.68	0.3166	1.08	0.2227	1.48	0.1334
0.29	0.3825	0.69	0.3144	1.09	0.2203	1.49	0.1315
0.30	0.3814	0.70	0.3123	1.10	0.2179	1.50	0.1295
0.31	0.3802	0.71	0.3101	1.11	0.2155	1.51	0.1276
0.32	0.3790	0.72	0.3079	1.12	0.2131	1.52	0.1257
0.33	0.3778	0.73	0.3056	1.13	0.2107	1.53	0.1238
0.34	0.3765	0.74	0.3034	1.14	0.2083	1.54	0.1219
0.35	0.3752	0.75	0.3011	1.15	0.2059	1.55	0.1200
0.36	0.3739	0.76	0.2989	1.16	0.2036	1.56	0.1182
0.37	0.3725	0.77	0.2966	1.17	0.2012	1.57	0.1163
0.38	0.3712	0.78	0.2943	1.18	0.1989	1.58	0.1145
0.39	0.3697	0.79	0.2920	1.19	0.1965	1.59	0.1127
0.40	0.3683	0.80	0.2897	1.20	0.1942	1.60	0.1109

TABLE 5   (continued)

x	$\varphi(x)$	x	$\varphi(x)$	x	$\varphi(x)$	x	$\varphi(x)$
1.61	0.1092	2.01	0.0529	2.41	0.0219	2.81	0.0077
1.62	0.1074	2.02	0.0519	2.42	0.0213	2.82	0.0075
1.63	0.1057	2.03	0.0508	2.43	0.0208	2.83	0.0073
1.64	0.1040	2.04	0.0498	2.44	0.0203	2.84	0.0071
1.65	0.1023	2.05	0.0488	2.45	0.0198	2.85	0.0069
1.66	0.1006	2.06	0.0478	2.46	0.0194	2.86	0.0067
1.67	0.0989	2.07	0.0468	2.47	0.0189	2.87	0.0065
1.68	0.0973	2.08	0.0459	2.48	0.0184	2.88	0.0063
1.69	0.0957	2.09	0.0449	2.49	0.0180	2.89	0.0061
1.70	0.0940	2.10	0.0440	2.50	0.0175	2.90	0.0060
1.71	0.0925	2.11	0.0431	2.51	0.0171	2.91	0.0058
1.72	0.0909	2.12	0.0422	2.52	0.0167	2.92	0.0056
1.73	0.0893	2.13	0.0413	2.53	0.0163	2.93	0.0055
1.74	0.0878	2.14	0.0404	2.54	0.0158	2.94	0.0053
1.75	0.0863	2.15	0.0396	2.55	0.0154	2.95	0.0051
1.76	0.0848	2.16	0.0387	2.56	0.0151	2.96	0.0050
1.77	0.0833	2.17	0.0379	2.57	0.0147	2.97	0.0048
1.78	0.0818	2.18	0.0371	2.58	0.0143	2.98	0.0047
1.79	0.0804	2.19	0.0363	2.59	0.0139	2.99	0.0046
1.80	0.0790	2.20	0.0355	2.60	0.0136	3.00	0.0044
1.81	0.0775	2.21	0.0347	2.61	0.0132	3.10	0.0033
1.82	0.0761	2.22	0.0339	2.62	0.0129	3.20	0.0024
1.83	0.0748	2.23	0.0332	2.63	0.0126	3.30	0.0017
1.84	0.0734	2.24	0.0325	2.64	0.0122	3.40	0.0012
1.85	0.0721	2.25	0.0317	2.65	0.0119	3.50	0.0009
1.86	0.0707	2.26	0.0310	2.66	0.0116	3.60	0.0006
1.87	0.0694	2.27	0.0303	2.67	0.0113	3.70	0.0004
1.88	0.0681	2.28	0.0297	2.68	0.0110	3.80	0.0003
1.89	0.0669	2.29	0.0290	2.69	0.0107	3.90	0.0002
1.90	0.0656	2.30	0.0283	2.70	0.0104	4.00	0.0001
1.91	0.0644	2.31	0.0277	2.71	0.0101	4.10	0.0001
1.92	0.0632	2.32	0.0270	2.72	0.0099	4.20	0.0001
1.93	0.0620	2.33	0.0264	2.73	0.0096		
1.94	0.0608	2.34	0.0258	2.74	0.0093		
1.95	0.0596	2.35	0.0252	2.75	0.0091		
1.96	0.0584	2.36	0.0246	2.76	0.0088		
1.97	0.0573	2.37	0.0241	2.77	0.0086		
1.98	0.0562	2.38	0.0235	2.78	0.0084		
1.99	0.0551	2.39	0.0229	2.79	0.0081		
2.00	0.0540	2.40	0.0224	2.80	0.0079		

TABLE 6

The normal distribution function $\Phi(x) = \dfrac{1}{\sqrt{2\pi}} \displaystyle\int_{-\infty}^{x} e^{-\frac{u^2}{2}}\, du$

x	$\Phi(x)$	x	$\Phi(x)$	x	$\Phi(x)$	x	$\Phi(x)$
0.00	0.5000						
0.01	0.5040	0.41	0.6591	0.81	0.7910	1.21	0.8869
0.02	0.5080	0.42	0.6628	0.82	0.7939	1.22	0.8888
0.03	0.5120	0.43	0.6664	0.83	0.7967	1.23	0.8907
0.04	0.5160	0.44	0.6700	0.84	0.7995	1.24	0.8925
0.05	0.5199	0.45	0.6736	0.85	0.8023	1.25	0.8944
0.06	0.5239	0.46	0.6772	0.86	0.8051	1.26	0.8962
0.07	0.5279	0.47	0.6808	0.87	0.8078	1.27	0.8980
0.08	0.5319	0.48	0.6844	0.88	0.8106	1.28	0.8997
0.09	0.5359	0.49	0.6879	0.89	0.8133	1.29	0.9015
0.10	0.5398	0.50	0.6915	0.90	0.8159	1.30	0.9032
0.11	0.5438	0.51	0.6950	0.91	0.8186	1.31	0.9049
0.12	0.5478	0.52	0.6985	0.92	0.8212	1.32	0.9066
0.13	0.5517	0.53	0.7019	0.93	0.8238	1.33	0.9082
0.14	0.5557	0.54	0.7054	0.94	0.8264	1.34	0.9099
0.15	0.5596	0.55	0.7088	0.95	0.8289	1.35	0.9115
0.16	0.5636	0.56	0.7123	0.96	0.8315	1.36	0.9131
0.17	0.5675	0.57	0.7157	0.97	0.8340	1.37	0.9147
0.18	0.5714	0.58	0.7190	0.98	0.8365	1.38	0.9162
0.19	0.5753	0.59	0.7224	0.99	0.8389	1.39	0.9177
0.20	0.5793	0.60	0.7257	1.00	0.8413	1.40	0.9192
0.21	0.5832	0.61	0.7291	1.01	0.8438	1.41	0.9207
0.22	0.5871	0.62	0.7324	1.02	0.8461	1.42	0.9222
0.23	0.5910	0.63	0.7357	1.03	0.8485	1.53	0.9236
0.24	0.5948	0.64	0.7389	1.04	0.8508	1.44	0.9251
0.25	0.5987	0.65	0.7422	1.05	0.8531	1.45	0.9265
0.26	0.6026	0.66	0.7454	1.06	0.8554	1.46	0.9279
0.27	0.6064	0.67	0.7486	1.07	0.8577	1.47	0.9292
0.28	0.6103	0.68	0.7517	1.08	0.8599	1.48	0.9306
0.29	0.6141	0.69	0.7549	1.09	0.8621	1.49	0.9319
0 30	0.6179	0.70	0.7580	1.10	0.8643	1.50	0.9332
0.31	0.6217	0.71	0.7611	1.11	0.8665	1.51	0.9345
0.32	0.6255	0.72	0.7642	1.12	0.8686	1.52	0.9357
0.33	0.6293	0.73	0.7673	1.13	0.8708	1.53	0.9370
0.34	0.6331	0.74	0.7703	1.14	0.8729	1.54	0.9382
0.35	0.6368	0.75	0.7734	1.15	0.8749	1.55	0.9394
0.36	0.6406	0.76	0.7764	1.16	0.8770	1.56	0.9406
0.37	0.6443	0.77	0.7794	1.17	0.8790	1.57	0.9418
0.38	0.6480	0.78	0.7823	1.18	0.8810	1.58	0.9429
0.39	0.6517	0.79	0.7853	1.19	0.8830	1.59	0.9441
0.40	0.6554	0.80	0.7881	1.20	0.8849	1.60	0.9452

(continued)

TABLE 6

x	$\Phi(x)$	x	$\Phi(x)$	x	$\Phi(x)$	x	$\Phi(x)$
1.61	0.9463	1.86	0.9686	2.22	0.9868	2.72	0.9967
1.62	0.9474	1.87	0.9693	2.24	0.9875	2.74	0.9969
1.63	0.9484	1.88	0.9699	2.26	0.9881	2.76	0.9971
1.64	0.9495	1.89	0.9706	2 28	0.9887	2.78	0.9973
1.65	0.9505	1.90	0.9713	2.30	0.9893	2.80	0.9974
1.66	0.9515	1.91	0.9719	2.32	0.9898	2.82	0.9976
1.67	0.9525	1.92	0.9726	2.34	0.9904	2.84	0.9977
1.68	0.9535	1.93	0.9732	2.36	0.9909	2.86	0.9979
1.69	0.9545	1.94	0.9738	2.38	0.9913	2.88	0.9880
1.70	0.9554	1.95	0.9744	2.40	0.9918	2.90	0.9981
1.71	0.9564	1.96	0.9750	2.42	0.9922	2.92	0.9982
1.72	0.9572	1.97	0.9756	2.44	0.9927	2.94	0.9984
1.73	0.9582	1.98	0.9761	2.46	0.9931	2.96	0.9985
1.74	0.9591	1.99	0.9767	2.48	0.9934	2.98	0.9986
1.75	0.9599	2.00	0.9772	2.50	0.9938	3.00	0.9986
1.76	0.9608	2.02	0.9783	2.52	0.9941	3.20	0.9993
1.77	0.9616	2.04	0.9793	2.54	0.9945	3.40	0.9996
1.78	0.9625	2.06	0.9803	2.56	0.9948	3.60	0.9998
1.79	0.9633	2.08	0.9812	2.58	0.9951	3.80	0.9999
1.80	0.9641	2.10	0.9821	2.60	0.9953		
1.8!	0.9649	2.12	0.9830	2.62	0.9956		
1.82	0.9656	2.14	0.9838	2.64	0.9959		
1.83	0.9664	2.16	0.9846	2.66	0.9961		
1.84	0.9671	2.18	0.9854	2.68	0.9963		
1.85	0.9678	2.20	0.9861	2.70	0.9965		

TABLES

The values of 100 $P_n(c)$,

where $P_n(c) = P\left(\sqrt{\dfrac{n}{2}} \sup_{-\infty < x < +\infty} |F_n(x) - G_n(x)| > z\right) = 1 - \dfrac{1}{\binom{2n}{n}} \sum_{k=-\left[\frac{n}{c}\right]}^{\left[\frac{n}{c}\right]} (-1)^k \binom{2n}{n+kc} \qquad (c = [z\sqrt{2n}+1])$

n \ c	1	2	3	4	5	6	7	8	9	10	11	12	13	14	15	16	17	18
5	100.00	87.30	35.71	7.94	0.79													
6	100.00	93.07	47.40	14.29	2.60	0.22												
7	100.00	96.27	57.52	21.21	5.30	0.82	0.06											
8	100.00	98.01	66.01	28.27	8.70	1.87	0.25	0.02										
9	100.00	98.95	73.01	35.17	12.59	3.36	0.63	0.07	0.00									
10	100.00	99.45	78.69	41.75	16.78	5.24	1.23	0.21	0.02	0.00								
11	100.00	99.71	83.26	47.92	21.15	7.65	1.70	0.38	0.06	0.01	0.00							
12	100.00	99.85	86.90	53.61	25.58	9.95	3.14	0.79	0.15	0.02	0.00							
13	100.00	99.92	89.78	58.82	29.99	12.65	4.43	1.26	0.29	0.05	0.01	0.00						
14	100.00	99.96	92.06	63.55	34.33	15.49	5.90	1.88	0.49	0.10	0.02	0.00						
15	100.00	99.98	93.83	67.81	38.55	18.44	7.55	2.62	0.77	0.18	0.04	0.01	0.00					
16	100.00	99.99	95.23	71.64	42.63	21.45	9.33	3.50	1.12	0.30	0.07	0.01	0.00					
17	100.00	99.99	96.31	75.06	46.54	24.50	11.24	4.50	1.56	0.46	0.12	0.02	0.00					
18	100.00	100.00	97.15	78.10	50.26	27.54	13.24	5.60	2.07	0.67	0.18	0.04	0.01	0.00				
19	100.00	100.00	97.81	80.81	53.79	30.57	15.32	6.81	2.67	0.92	0.28	0.07	0.02	0.00				
20	100.00	100.00	98.31	83.20	57.13	33.56	17.45	8.11	3.35	1.23	0.40	0.11	0.03	0.01	0.00			
21	100.00	100.00	98.70	85.31	60.28	36.50	19.63	9.48	4.11	1.59	0.55	0.17	0.04	0.01	0.00			
22	100.00	100.00	99.01	87.17	63.24	39.37	21.84	10.93	4.93	2.00	0.73	0.24	0.07	0.02	0.00			
23	100.00	100.00	99.24	88.80	66.01	42.18	24.06	12.43	5.83	2.47	0.95	0.32	0.10	0.03	0.01	0.00		
24	100.00	100.00	99.42	90.24	68.60	44.90	26.28	13.98	6.78	2.99	1.20	0.43	0.14	0.04	0.01	0.00		
25	100.00	100.00	99.55	91.50	71.02	47.55	28.50	15.58	7.79	3.56	1.48	0.56	0.19	0.06	0.02	0.00		
26	100.00	100.00	99.66	92.60	73.27	50.10	30.71	17.20	8.85	4.18	1.81	0.71	0.26	0.08	0.02	0.01	0.00	
27	100.00	100.00	99.74	93.57	75.37	52.56	32.90	18.86	9.96	4.84	2.17	0.98	0.33	0.11	0.03	0.01	0.00	
28	100.00	100.00	99.80	94.41	77.32	54.94	35.06	20.53	11.10	5.55	2.56	1.09	0.42	0.15	0.05	0.01	0.00	
29	100.00	100.00	99.85	95.14	79.12	57.22	37.20	22.21	12.29	6.30	2.99	1.31	0.53	0.20	0.07	0.02	0.01	0.00
30	100.00	100.00	99.88	95.78	80.80	59.41	39.29	23.91	13.50	7.09	3.46	1.56	0.65	0.25	0.09	0.03	0.01	0.00

TABLE 8

The function $K(z) = \sum\limits_{k=-\infty}^{+\infty} (-1)^k e^{-2k^2z^2}$

z	$K(z)$	z	$K(z)$	z	$K(z)$
0.28	0.000001	0.71	0.305471	1.14	0.851394
0.29	0.000004	0.72	0.322265	1.15	0.858038
0.30	0.000009	0.73	0.339113	1.16	0.864442
0.31	0.000021	0.74	0.355981	1.17	0.870612
0.32	0.000046	0.75	0.372833	1.18	0.876548
0.33	0.000091	0.76	0.389640	1.19	0.882258
0.34	0.000171	0.77	0.406372	1.20	0.887750
0.35	0.000303	0.78	0.423002	1.21	0.893030
0.36	0.000511	0.79	0.439505	1.22	0.898104
0.37	0.000826	0.80	0.455857	1.23	0.902972
0.38	0.001285	0.81	0.472041	1.24	0.907648
0.39	0.001929	0.82	0.488030	1.25	0.912132
0.40	0.002808	0.83	0.503808	1.26	0.916432
0.41	0.003972	0.84	0.519366	1.27	0.920556
0.42	0.005476	0.85	0.534682	1.28	0.924505
0.43	0.007377	0.86	0.549744	1.29	0.928288
0.44	0.009730	0.87	0.564546	1.30	0.931908
0.45	0.012590	0.88	0.579070	1.31	0.935370
0.46	0.016005	0.89	0.593316	1.32	0.938682
0.47	0.020022	0.90	0.607270	1.33	0.941848
0.48	0.024683	0.91	0.620928	1.34	0.944872
0.49	0.030017	0.92	0.634286	1.35	0.947756
0.50	0.036055	0.93	0.647338	1.36	0.950512
0.51	0.042814	0.94	0.660082	1.37	0.953142
0.52	0.050306	0.95	0.672516	1.38	0.955650
0.53	0.058534	0.96	0.684636	1.39	0.958040
0.54	0.067497	0.97	0.696444	1.40	0.960318
0.55	0.077183	0.98	0.707940	1.41	0.962486
0.56	0.087577	1.99	0.719126	1.42	0.964552
0.57	0.098656	1.00	0.730000	1.43	0.966516
0.58	0.110395	1.01	0.740566	1.44	0.968382
0.59	0 122760	1.02	0.750826	1.45	0.970158
0.60	0 135718	1.03	0.760780	1.46	0.971846
0.61	0.149223	1.04	0.770434	1.47	0.973448
0.62	0.163225	1.05	0.779794	1.48	0.974970
0.63	0.177753	1.06	0.788860	1.49	0.976412
0.64	0.192677	1.07	0.797636	1.50	0.977782
0.65	0.207987	1.08	0.806128	1.51	0.979080
0.66	0.223637	1.09	0.814342	1.52	0.980310
0.67	0.239582	1.10	0.822282	1.53	0.981476
0.68	0.255780	1.11	0.829950	1.54	0.982578
0.69	0.272189	1.12	0.837356	1.55	0.983622
0.70	0.288765	1.13	0.844502	1.56	0.984610

z	$K(z)$	z	$K(z)$	z	$K(z)$
1.57	0.985544	1.93	0.998837	2.29	0.999944
1.58	0.986426	1.94	0.998924	2.30	0.999949
1.59	0.987260	1.95	0.999004	2.31	0.999954
1.60	0.988048	1.96	0.999079	2.32	0.999958
1.61	0.988791	1.97	0.999149	2.33	0.999962
1.62	0.989492	1.98	0.999213	2.34	0.999965
1.63	0.990154	1.99	0.999273	2.35	0.999968
1.64	0.990777	2.00	0.999329	2.36	0.999970
1.65	0.991364	2.01	0.999380	2.37	0.999973
1.66	0.991917	2.02	0.999428	2.38	0.999976
1.67	0.992438	2.03	0.999474	2.39	0.999978
1.68	0.992928	2.04	0.999516	2.40	0.999980
1.69	0.993389	2.05	0.999552	2.41	0.999982
1.70	0.993828	2.06	0.999588	2.42	0.999984
1.71	0.994230	2.07	0.999620	2.43	0.999986
1.72	0.994612	2.08	0.999650	2.44	0.999987
1.73	0.994972	2.09	0.999680	2.45	0.999988
1.74	0.995309	2.10	0.999705	2.46	0.999989
1.75	0.995625	2.11	0.999723	2.47	0.999990
1.76	0.995922	2.12	0.999750	2.48	0.999991
1.77	0.996200	2.13	0.999770	2.49	0.999992
1.78	0.996460	2.14	0.999790	2.50	0.9999925
1.79	0.996704	2.15	0.999806	2.55	0.9999956
1.80	0.996932	2.16	0.999822	2.60	0.9999974
1.81	0.997146	2.17	0.999838	2.65	0.9999984
1.82	0.997316	2.18	0.999852	2.70	0.9999990
1.83	0.997533	2.19	0.999864	2.75	0.9999994
1.84	0.997707	2.20	0.999874	2.80	0.9999997
1.85	0.997870	2.21	0.999886	2.85	0.99999982
1.86	0.998023	2.22	0.999896	2.90	0.99999990
1.87	0.998145	2.23	0.999904	2.95	0.99999994
1.88	0.998297	2.24	0.999912	3.00	0.99999997
1.89	0.998421	2.25	0.999920		
1.90	0.998536	2.26	0.999926		
1.91	0.998644	2.27	0.999934		
1.92	0.998744	2.28	0.999940		

TABLE 9

The function $L\left(y\sqrt{\dfrac{a}{1-a}}\cdot\right)$ $\left(L(z) = \dfrac{4}{\pi}\sum_{k\,0}^{\infty}(-1)^k\,\dfrac{e^{-\frac{(2k+1)^2\pi^2}{8z^2}}}{2k-1}\right)$

a / y	0.01	0.02	0.03	0.04	0.05	0.06	0.07
0.1							
0.5							
1.0							0.0000	0.0000
1.5			0.0000	0.0000	0.0000	0.0002	0.0009
2.0		0.0000	0.0001	0.0008	0.0036	0.0101	0.0212
2.5		0.0001	0.0022	0.0112	0.0299	0.0578	0.0925
3.0	0.0000	0.0015	0.0151	0.0474	0.0941	0.1487	0.2061
3.5	0.0001	0.0092	0.0491	0.1136	0.1879	0.2628	0.3341
4.0	0.0006	0.0291	0.1052	0.2001	0.2942	0.3804	0.4571
4.5	0.0031	0.0643	0.1776	0.2951	0.4001	0.4901	0.5665
5.0	0.0096	0.1134	0 2582	0.3895	0.4985	0.5873	0.6598
5.5	0.0225	0.1726	0.3406	0.4784	0.5863	0.6707	0.7374
6.0	0.0428	0.2375	0.4204	0.5591	0.6627	0.7409	0.8005
6.5	0.0707	0.3045	0.4952	0.6310	0.7282	0.7989	0.8509
7.0	0.1053	0.3708	0.5638	0.6940	0.7834	0.8461	0.8904
7.5	0.1452	0.4347	0.6258	0.7484	0.8294	0.8838	0.9207
8.0	0.1889	0.4959	0.6811	0.7951	0.8671	0.9135	0.9436
8.5	0 2348	0.5513	0.7301	0.8345	0.8977	0.9365	0.9606
9.0	0 2819	0.6031	0.7731	0.8676	0.9221	0.9540	0.9729
9.5	0 3290	0.6506	0.8104	0.8950	0.9414	0.9672	0.9817
10.0	0.3754	0.6938	0.8427	0 9175	0.9564	0.9770	0.9878
11.0	0.4640	0.7678	0.8939	0.9505	0.9768	0.9891	0.9949
12 0	0.5450	0.8270	0.9303	0.9714	0.9882	0.9951	0.9980
13.0	0.6174	0.8734	0.9555	0.9841	0.9943	0.9980	0.9993
14.0	0 6812	0.9090	0.9724	0.9915	0.9974	0.9992	0.9998
15.0	0.7367	0.9358	0.9833	0.9956	0.9988	0.9997	0 9999
16.0	0.7844	0.9555	0.9902	0.9978	0.9995	0.9999	1.0000
17.0	0.8249	0.9697	0.9944	0.9990	0.9998	1.0000	
18.0	0.8591	0.9797	0.9969	0.9995	0.9999		
19.0	0.8876	0.9867	0.9983	0.9998	1.0000		
20.0	0.9112	0.9915	0.9991	0.9999			
21.0	0.9304	0.9946	0.9996	1.0000			
22 0	0.9459	0.9967	0.9998				
23.0	0.9584	0.9980	0.9999				
24.0	0.9683	0.9988	1.0000				
25.0	0.9760	0.9993					
30.0	0.9949	1.0000					
35.0	0.9991						
40.0	0.9999						
43.0	1.0000						

(continued) TABLE 9

y \ a	0.08	0.09	0.1	0.2	0.3	0.4	0.5
0.1						0.0000	0.0000
0.5				0.0000	0.0000	0.0008	0.0092
1.0	0.0000	0.0000	0.0000	0.0092	0.0716	0.2001	0.3708
1.5	0.0023	0.0050	0.0092	0.1420	0.3542	0.5591	0.7328
2.0	0.0367	0.0563	0.0793	0.3708	0.6193	0.7951	0.9090
2.5	0.1315	0.1730	0.2155	0.5778	0.7966	0.9175	0.9752
3 0	0.2632	0.3184	0.3708	0.7328	0.9009	0.9714	0.9946
3.5	0.3999	0.4599	0.5142	0.8398	0.9561	0.9915	0.9991
4.0	0.5244	0.5835	0.6353	0.9090	0.9823	0.9978	0.9999
4.5	0.6311	0.6860	0.7328	0.9511	0.9936	0.9995	1.0000
5.0	0.7193	0.7683	0.8088	0.9752	0.9979	0.9999	
5.5	0.7903	0.8326	0.8665	0.9881	0.9994	1.0000	
6.0	0.8463	0.8817	0.9090	0.9946	0.9998		
6.5	0.8895	0.9181	0.9395	0.9977	1.0000		
7.0	0.9220	0.9446	0.9607	0.9991			
7.5	0.9460	0.9633	0.9752	0.9996			
8.0	0.9634	0.9763	0.9847	0.9999			
8.5	0.9756	0.9850	0.9908	1.0000			
9.0	0.9841	0.9907	0.9946				
9.5	0.9898	0.9944	0.9969				
10.0	0.9936	0.9967	0.9983				
11.0	0.9976	0.9989	0.9995				
12.0	0.9992	0.9997	0.9999				
13.0	0.9997	0.9999	1.0000				
14.0	0.9999	1.0000					
15.0	1.0000						
16.0							
17.0							
18.0							
19.0							
20.0							
21.0							
22.0							
23.0							
24.0							
25.0							
30.0							
35.0							
40.0							
43.0							

REMARKS AND BIBLIOGRAPHICAL NOTES

These notes wish to call attention to books and papers which may be useful to the reader for further study of subjects dealt with in the present textbook, including books and papers to which reference was made in the text. For topics which are treated in detail in some current textbook, we mention only such books, where the reader can find further references.

As regards topics not discussed in standard textbooks, the sources of the material contained in this book are given in greater detail. These bibliographic notes contain often some remarks on the historical development of the problems dealt with, but to give a full account of the history of probability theory was of course impossible. For the history of Probability Calculus up to Laplace see Todhunter [1].

Concerning less-known theorems or methods from other branches of mathematics, we refer to some current textbook readily accessible to the reader.

The notes are restricted to the most important methodical problems. On several occasions the method of exposition chosen in the present book is compared in the notes to that in other textbooks.

Chapter I

Glivenko was the first to stress in his textbook (Glivenko [3]; cf. also Kolmogorov [9]) the advantage of discussing the algebra of events as a Boolean algebra before the introduction of the notion of probability. It seems to us that the understanding of Kolmogorov's axiomatic theory is hereby facilitated. On the general theory of measure and integration over a Boolean algebra instead of over a field of sets see Carathéodory [1]. Recent results on probability as a measure on a Boolean algebra are summarized in Kappos [1].

§§ 1–4. On Boolean algebras in general see Birkhoff [1], Glivenko [2]. We did not give a system of independent axioms for Boolean algebras, since it seemed to us of much more importance to present the rules of Boolean algebra in a way which makes clear the duality of the two basic operations.

§ 5. See Stone [1]. We follow here Frink [1]; as to the Lemma see Hausdorff [1] and Frink [2].

§ 6. The unsolved problem mentioned in Exercise 7 was first formulated by Dedekind (cf. Birkhoff [1], p. 147). Concerning Exercise 11 see e.g. Gavrilov [1].

Chapter II

From Chapter II on, probability theory is developed on the basis of Kolmogorov's axiomatics, first published in Kolmogorov [5]. The idea to consider probability as an additive set function has — like every important mathematical idea — many forerunners, cf. e.g. Borel [1], Lomnicki [1], Lévy [1], Steinhaus [1], Jordan [1], [2]. The merit of Kolmogorov was to formulate for the first time this idea consequently and in its whole generality and to show how from this idea probability theory can be developed as a strictly axiomatic branch of modern mathematics. Herewith he solved one of the famous problems of Hilbert. Nearly all modern textbooks of probability theory and mathematical statistics (cf. e.g. Blanc-Lapierre and Fortet [1], Cramér [2], Doob [1], Feller [7], Fisz [1], Fréchet [1], Gnedenko [3], Kac [3], Lévy [3], [4], Loève [1], Neyman [1], [2], Onicescu, Mihoc and Ionescu-Tulcea [1], Parzen [1], Richter, [1], Schmetterer [1], van der Waerden [1]) and nearly the whole recent literature of probability theory and mathematical statistics are based on Kolmogorov's axiomatics. Earlier theories and discussions of the concept of probability may be found in Laplace [1], [2], Bernstein [2], von Mises [1], [2], Wald [1]. § 11 and § 12 present a generalized system of axioms which contains that of Kolmogorov as a particular case (cf. Rényi [15]).

§ 3. Concerning Theorem 10 cf. Ch. Jordan [3]. A large number of general identities and inequalities between probabilities of events can be found in Fréchet [2]. With respect to Theorems 11 and 12 see Rényi [23]. The method based on these theorems is closely related to the method of indicator functions of Loève (cf. Loève [1]). About the relation of the two methods to each other cf. Rényi [35].

§ 7. On Measure Theory see Halmos [1], Aumann [1]. The proof of the σ-additivity of Lebesgue–Stieltjes measure (by means of Kolmogorov's Theorem 3) differs from those given usually in textbooks and is due to Catherine Rényi and A. Rényi.

§ 10. On integral geometry cf. Blaschke [1].

§ 11. Some authors (cf. e.g. Reichenbach [1], Popper [1], [2]) have long ago emphasized that conditional probability may and should be chosen as the basic concept of probability theory. But the starting point of these authors was essentially philosophical. They did not try to give a corresponding generalization of Kolmogorov's axiomatic theory. The idea, that unbounded measures may serve as legitimate (conditional) probability distributions, does not appear in these early works. On the other hand, unbounded measures were long ago used in statistics as Bayesian *a priori* distributions (cf. e.g. Jeffreys [1], Dumas [1], [2], Baticle [1], [2]), without an exact mathematical foundation. In the paper [15] of Rényi, these two points of view are, in a certain sense, united and connected to Kolmogorov's axiomatic theory. Concerning the theory of conditional probability algebras cf. further Rényi [14], [18], [19], Császár [1], Kappos [1]. Somewhat later, but independently, Luce [1] constructed a similar system of axioms for finite conditional probability algebras, by using an entirely different reasoning (starting from an investigation of the psychological laws of choice and preference).

§ 12. Concerning exchangeable events (Exercises 38–41) cf. de Finetti [1], Khinchin [2]. For Exercise 46, see Wilcoxon [1], Rényi [12]. On Exercise 47 and in general on applications of probability theory to number theory see e.g. Kac [4], Rényi [25].

Chapter III

§ **2.** Cf. Bayes [1].

§ **3.** On the Pólya-distribution see Eggenberger–Pólya [1], for further generalization see Onicescu–Mihoc [1].

§ **11.** Concerning Theorem 6 see Kantorovich [1].

§ **13.** Poisson [1], Bortkiewicz [1]. For a general theory of simple and composed Poisson processes see Aczél–Jánossy–Rényi [1], Rényi [5], [6], Aczél [2], Prékopa [1], Marczewski [1], Florek, Marczewski and Ryll-Nardzewski [1], Saxer [1].

§ **14.** The idea that in number theory the application of Dirichlet's series can be replaced by a formal algebraic calculus is — according to Hardy — due to Harald Bohr (Hardy and Wright [1], p. 259). Instead of Dirichlet's series, this idea is applied here to power series and hereby to the convolutions of probability distributions. On related problems see Rényi [4].

§ **16.** On Euler's summation formula see Knopp [1].

§ **17.** Concerning Exercise 8, see Bernstein [4]; for a generalization Erdős and Rényi [4]. For Exercise 32, see Bernstein [1] and also Arató–Rényi [1]. For Exercise 35, see Feldheim [2] and Rényi [20].

Chapter IV

§ **7.** On projections of probability distributions see Cramér–Wold [1], Rényi [8].

§ **8.** On the lognormal distribution see Kolmogorov [8], Rényi [32].

§ **9.** Concerning Example 1, see Lobatchewski [1]. On the χ^2-distribution, see Helmert [1], K. Pearson [1]; on the beta-function, Lösch–Schoblik [1].

§ **10.** Example 1: Student [1], Example 2: Fisher [1].

§ **17.** Concerning Exercises 17–18, cf. Malmquist [1], van Dantzig [1], Hajós–Rényi [1]. Exercise 26: Bateman [1], Bharucha-Reid [1]. Exercise 45: see, for instance, Veksler–Grochev–Isaev [1].

Chapter V

§ **1.** The content of this section appeared first in the present book (German edition, 1962).

§ **2.** We follow here Kolmogorov [5]. For the Radon–Nikodym theorem, see e.g. Halmos [1].

§ **3.** Concerning the new deduction of the Maxwell distribution given here, see Rényi [19].

§ **4.** We follow here Kolmogorov [5].

§ **6** and § **7.** See Gebelein [1], Rényi [26], [28], Csáki–Fischer [1], [2]. On the Lemma of Theorem 3 of § 7, see Boas [1]. On Theorem 3, see Rényi [26]. On the applications of the large sieve of Linnik in number theory, see Linnik [1], Rényi [2], Bateman–Chowla–Erdős [1].

§ **7.** Exercises 1–4 treat problems of integral geometry from the point of view of probability theory (see Blaschke [1]). Exercise 6: cf. Hajós–Rényi [1]; Exercises 28–30: Rényi [28].

Chapter VI

The method of characteristic functions, as mentioned by Lévy [4], goes essentially back to Cauchy. As to its application in probability theory, the merit is due to Liapunov [1], Pólya [1] and Lévy [2]. Detailed expositions of the theory may be found in Dugué [1], Esseen [1], Ky Fan [1], Linnik [3], Lukács [4]. On Fourier series and integrals, see Zygmu nd [1], [2].

§ 4. On the theorem of Helly, cf. Lukács [4].

§ 5. Theorem 1: Bernstein [4]; Theorem 3: Cramér [1]; Theorem 4: Linnik–Singer [1]. Concerning the theorem on the singularities of a power series with positive coefficients, which was applied in point 4, see Titchmarsh [1]. On the theorem of H. A. Schwarz, cf. Hurwitz–Courant [1]. For the formula of Faà di Bruno, cf. Jordan [4], Lukács [2]. Theorem 5: cf. Darmois [1] and Skitovitch [1], Theorem 6: Lukács [1]; see also Geary [1], Kawata–Sakamoto [1], Singer [1] and Lukács [3].

§§ 7–8. On the theory of infinitely divisible and stable distributions see Gnedenko–Kolmogorov [1], Lévy [4], Khinchin–Lévy [1], Feldheim [1].

§ 7. Concerning the theory of distributions (theory of generalized functions) we follow the book of Lighthill [1], which develops the theory established by J. Mikusinski. The application of the theory of distributions to probability theory was published for the first time in the German edition of the present book. For Theorem 7, see Robbins [1]. On the method of Laplace, see de Bruijn [1]. On the theory of theta functions, cf. Hurwitz and Courant [1]. Chung and Erdős proved Theorem 8 in a stronger form.

§ 10. Exercise 4: see Shannon [1]; Exercise 6: Hardy–Littlewood–Pólya [1]; Exercise 14 (for a more general theorem): Császár [2]; Exercise 20: Laha [1].

Chapter VII

§ 1. Chebyshev [1], Bienaymé [1]. The first Chebyshev-type inequality is due to Gauss [1].

§ 2. Bernoulli [1], Slutsky [1].

§ 3. Markov [1], Khinchin [4].

§ 4. Bernstein [4], Uspenski [1].

§ 5. Borel [1], Cantelli [1]. On Lemma C, see Erdős–Rényi [2].

§ 6. Kolmogorov [1], [5].

§ 7. Kolmogorov [2]. On Lemma 2: Knopp [1].

§ 8. Glivenko [1].

§ 9. Khinchin [1], Kolmogorov [1], Erdős [1], Feller [4], Rényi [1].

§ 10. Rényi [24], Rényi–Révész [1].

§ 11. Rényi [38].

§ 12. Rényi–Révész [2], Rényi [38].

§ 13. Kolmogorov [5].

§ 14. Kolmogorov [5], Khinchin–Kolmogorov [1], Steinhaus [2] and also Steinhaus, Kac and Ryll-Nardzewski [1]; for the lemma, Doob [2].

§ 15. Rényi [15], [17]. For the theorem of Abel-Dini, cf. Knopp [1], p. 173.

§ 16. Exercise 10, cf. Hille [1]; Exercise 11: Widder [1]; Exercise 13: Rényi [1]; Exercise 14: Hájek–Rényi [1]; Exercise 21: Doob [2] (the theorem mention ed in the remark is due to Menshov); Exercise 24: Kolmogorov [5].

Chapter VIII

§ 1. Chebyshev [1], Markov [1], Liapunov [1], Lindeberg [1], Pólya [1], Feller [1], Khinchin [3], Gnedenko–Kolmogorov [1], Kolmogorov [5], [11], Prékopa–Rényi–Urbanik [1].

§ 2. Gnedenko [2].

§ 3. Lévy [4], Feller [1], Khinchin [5], and also Gnedenko–Kolmogorov [1].

§ 5. Erdős–Rényi [1]; for the particular case $p = $ const, cf. Bernstein [4].

§ 6. For the lemma, cf. Cramér [3]. Theorem 3 was first, under certain restrictions, proved by a different method (cf. Rényi [3]). This result was generalized by Kolmogorov [10]. For the more simple proof given here, cf. Rényi [24]. Theorem 3 may be applied to prove limit theorems for dependent random variables; cf. Révész [1]. On the central limit theorem for dependent random variables, see the fundamental paper of Bernstein [3].

§ 7. Anscombe [1], Doeblin [2], Rényi [22], [31].

§ 8. On the theory and applications of Markov chains and Markov processes see Markov [1], Kolmogorov [3], [7]; Doeblin [1], [2], Feller [2], [3], [6], [7], Doob [2], Chung [1], Bartlett [1], Bharucha-Reid [1], Wiener [2], Chandrasekhar [1], Einstein [1], Hostinsky [1], Lévy [3], Rényi [11].

§ 9. Rényi [9], [10], van Dantzig [1], Malmquist [1]; further references are to be found in Wilks [1], Wang [1].

§ 10. Kolmogorov [6], N. V. Smirnov [1], [2], Gnedenko [1], Gnedenko–Koroljuk [1], Doob [1], Feller [5], Donsker [1].

§ 11. For Theorem 1: Pólya [2]. See also Dvoretzky–Erdős [1]. On the arc sine law (Theorem 6) cf. Lévy [2], Erdős–Kac [2]; Sparre–Andersen [1], [2]; Chung–Feller [1], Rényi [33]. On Lemma 2, Rényi [36]; for other generalizations, Spitzer [1]. For Theorem 8, see Erdős–Hunt [1]; for Theorem 9, Erdős–Kac [1]; for a generalization of it, Rényi [9].

§ 12. Lindeberg [1], Krickeberg [1].

§ 13. Exercise 5: Rényi [17]; for a similar general system of independent functions, see Steinhaus, Kac and Ryll-Nardzewski [1]–[10], Rényi [7]. Exercise 8: Kac [1]; Exercise 24: Wilcoxon [1] and Rényi [12]; Exercise 25: Lehmann [1] and Rényi [12]; the equivalence of the problems considered in these two papers is proved in E. Csáki [1]. Exercise 28: Erdős–Rényi [3]. The result of Exercise 30 is due to Chung and Feller [1]; as regards the presentation given here, cf. Rényi [33].

Appendix

On the concepts of the entropy and information see Boltzmann [1], Hartley [1], Shannon [1], Wiener [1], Shannon–Weaver [1], Woodward [1], Barnard [1], Jeffreys [1], and the papers of Khinchin. Fadeev, Kolmogorov, Gelfand, Jaglom, etc., in Arbeiten zur Informationstheorie I–III .On the role of the notion of information in statistics, see the works of Fisher [1]–[3], and of Kullback [1]. The notion of the dimension of a probability distribution and that of the entropy of the corresponding dimension were introduced in a paper of Balatoni and Rényi (Arbeiten zur Informationstheorie I) and were further developed in Rényi [27], [30]. Measures of information differing from the Shannon-measure were already considered earlier, e.g. by Bhattacharyya [1] and Schützenberger [1]; the theory of entropy and information of order α is developed in Rényi [34], [37].

Part of the material appeared for the first time in the German edition of this book. This appendix covers merely the basic notions of information theory; their application to the transmission of information through a noisy channel, coding theory, etc. are not dealt with here. Besides the already mentioned works of Shannon and Khinchin let there be indicated those of Feinstein [1], [2], McMillan [1] and Wolfowitz [1], [2], [3].

§ 1. Concerning the theorem of Erdős on additive number-theoretical functions, which was rediscovered by Fadeev, see Erdős [2]; the simple proof given in the text is due to Rényi [29].

§ 2. For the theorem of Mercer, see Knopp [1].

§ 6. On the mean value theorem, see de Finetti [2], Kolmogorov [4], Nagumo [1], Aczél [1], Hardy–Littlewood–Pólya [1] (where further references can be found; this book contains also all other inequalities used in the Appendix, e.g. the inequalities of Jensen and of Hölder).

§ 9. The idea that quantities of information theory may be used for the proof of limit theorems is due to Linnik [2].

On the theorem of Perron–Frobenius, see Gantmacher [1].

§ 11. For Exercise 2c, see Rényi [16] and Kac [2]; Exercise 3c: Khinchin [6], for the generalizations: Rényi [21]. Exercise 4: Kolmogorov–Tikhomirov (Arbeiten zur Informationstheorie III). The content of Exercise 9 is due to Chung (unpublished communication), the proof given here differs from that of Chung. Exercise 17b: cf. Moriguti [1].

Tables

Further tables and graphic representations useful in probability calculus may be found in Fisher–Yates [1], E. S. Pearson–H. O. Hartley [1], Molina [1], Koller [1].

REFERENCES

Aczél, J.
[1] On mean values, Bull. Amer. Math. Soc. **54**, 393–400 (1948).
[2] On composed Poisson distributions, III, Acta Math. Acad. Sci. Hung. **3**, 219–224 (1952).
Aczél, J., L. Jánossy and A. Rényi
[1] On composed Poisson distributions, I, Acta Math. Acad. Sci. Hung. **1**, 209–224 (1950).
Alexandrov, P. S. (Александров, C.) [1] Введение в общую теорию множеств и функций (Introduction to the theory of sets and functions), OGIZ, Moscow-Leningrad 1948.
Anscombe, F. J.
[1] Large sample theory of sequential estimation, Proc. Cambridge Phil. Soc. **48**, 600 (1952).
Arató, M. and A. Rényi
[1] Probabilistic proof of a theorem on the approximation of continuous functions by means of generalized Bernstein polynomials, Acta Math. Acad. Sci. Hung. **8**, 91–98 (1957).
ARBEITEN ZUR INFORMATIONSTHEORIE I–III (Teil von A. J. Chintschin, D. K. Faddejew, A. N. Kolmogoroff, A. Rényi und J. Balatoni; Teil II von I. M. Gelfand, A. M. Jaglom, A. N. Kolmogoroff, Chiang Tse-Pei, I. P. Zaregradski; Teil III von A. N. Kolmogoroff und W. M. Tichomirow), VEB Deutscher Verlag der Wissenschaften, Berlin 1957 bzw. 1960.
Aumann, G.
[1] Reelle Funktionen, Springer-Verlag, Berlin–Göttingen–Heidelberg 1954.

Barban, M. B. (Барбан, M. Б.)
[1] Об одной теореме P. Bateman, S. Chowla, P. Erdős, Publications of the Mathematical Institute of the Hungarian Academy of Sciences, 9 A, 429–435 (1964).
Barnard, G. A.
[1] The theory of information, J. Royal Stat. ʌɔc. (B) **13**, 46–69 (1951).
Bartlett, M. S.
[1] An introduction to stochastic processes with special reference to methods and applications, Cambridge Univ. Press, Cambridge 1955.
Bateman, H.
[1] The solution of a system of differential equations occurring in the theory of radioactive transformations, Proc. Cambridge Phil. Soc. **15**, 423–427 (1910).
Bateman, P. T., S. Chowla and P. Erdős
[1] Remarks on the size of $L(1, \chi)$, Publ. Math. Debrecen **1**, 165–182 (1950).
Baticle, E.
[1] Sur une loi de probabilité a priori paramètres d'une loi laplacienne, C. R. Acad. Sci. Paris **226**, 55–57 (1948).

BATICLE, E.
[2] Sur une loi de probabilité a priori pour l'interprétation des résultats de tirages dans une urne, C. R. Acad. Sci. Paris **228**, 902–904 (1949).

BAUER, H.
[1] Wahrscheinlichkeitstheorie und Grundzüge der Masstheorie, Sammlung Göschen 1216/1216a, de Gruyter, Berlin 1964.

BAYES, TH.
[1] Essay towards solving a problem in the doctrine of chances, "Ostwald's Klassiker der Exakten Wissenschaften", Nr. 169, W. Engelmann, Leipzig 1908.

BERNOULLI, J.
[1] Ars Coniectandi (1713) I–II, III–IV. "Ostwald's Klassiker der Exakten Wissenschaften", Nr. 108, W. Engelmann, Leipzig 1899.

BERNSTEIN, S. N. (Бернштейн, С. Н.)
[1] Démonstration du théorème de Weierstrass fondée sur la calcul des probabilités, Soobshchs. Charkovskovo Mat. Obshch. (2) **13**, 1–2 (1912).
[2] Опыт аксиоматического обоснования теории вероятностей (On a tentative axiomatisation probability theory), Charkovskovo Zap. Mat. ot-va 15, 209–274 (1917).
[3] Sur l'extension du théorème limite du calcul des probabilités aux sommes de quantités dépendantes. Math. Ann. **97**, 1–59 (1926).
[4] Теория вероятностей (Probability theory), 4. ed., Gqztehizdat, Moscow 1946.

BHARUCHA-REID, A. T.
[1] Elements of the theory of Markov processes and their applications, McGraw-Hill, New York 1960.

BHATTACHARYYA, A.
[1] On some analogues of the amount of information and their use in statistical estimation, Sankhya **8**, 1–14 (1946).

BIENAYMÉ, M.
[1] Considérations à l'appui de la découverte de Laplace sur la loi des probabilités dans la méthode des moindres carrés, C. R. Acad. Sci. Paris 37, 309–324 (1853).

BIRKHOFF, G.
[1] Lattice theory, 3. ed., American Mathematical Society Colloquium Publications 25. AMS, Providence 1967.

BLANC-LAPIERRE, A., et R. FORTET
[1] Théorie des fonctions aléatoires, Masson et Cie., Paris 1953.

BLASCHKE, W.
[1] Vorlesungen über Integralgeometrie, 3. Aufl., VEB Deutscher Verlag der Wissenschaften, Berlin 1955.

BLUM, J. R., D. L. HANSON and L. H. KOOPMANS
[1] On the strong law of large numbers for a class of stochastic processes, Zeitschrift für Wahrscheinlichkeitstheorie 2, 1–11 (1963).

BOAS, R. P. JR.
[1] A general moment problem, Amer. J. Math. **63**, 361–370 (1941).

BOCHNER, S. and S. CHANDRASEKHARAN
[1] Fourier transforms, Princeton Univ. Press, Princeton 1949.

BOLTZMANN, L.
[1] Vorlesungen über Gastheorie, Johann Ambrosius Barth, Leipzig 1896.

BOREL, É.
[1] Sur les probabilités dénombrables et leurs applications arithmétiques, Rend. Circ. Mat. Palermo **26**, 247–271 (1909).
[2] Éléments de la théorie des probabilités, Hermann et Fils, Paris 1909.

VON BORTKIEWICZ, L.
[1] Das Gesetz der kleinen Zahlen, B. G. Teubner, Leipzig 1898.
DE BRUIJN, N. G.
[1] Asymptotic methods in analysis, North Holland Publ. Comp. Inc., Amsterdam 1958.

CANTELLI, F. P.
[1] La tendenza ad un limite nel senzo del calcolo delle probabilita, Rend. Circ. Mat. Palermo 16, 191–201 (1916).
CARATHÉODORY, C.
[1] Entwurf einer Algebraisierung des Integralbegriffes, Sitzungsber. Math.-Naturwiss. Klasse Bayer. Akad. Wiss., München 1938, S. 24–28.
CHANDRASEKHAR, S.
[1] Stochastic problems in physics and astronomy, Rev. Mod. Phys. 15, 1–89 (1943).
CHEBYSHEV, P. L. (Чебышев, П. Л.)
[1] Теория вероятностей (Theory of probability), Akad. izd., Moscow 1936.
CHUNG, K. L.
[1] Markov chains with stationary transition probabilities, Springer-Verlag, Berlin–Göttingen–Heidelberg 1960.
CHUNG, K. L., and P. ERDŐS
[1] Probability limit theorems assuming only the first moment, I, Mem. Amer. Math. Soc. 6, 1–19 (1950).
CHUNG, K. L. and W. FELLER
[1] On fluctuations in coin-tossing, Proc. Acad. Sci. USA 35, 605–608 (1949).
CRAMÉR, H.
[1] Über eine Eigenschaft der normalen Verteilungsfunktion, Math. Z. 41, 405–414 (1936).
[2] Random variables and probability distributions, Cambridge Univ. Press, Cambridge 1937.
[3] Mathematical methods of statistics, Princeton Univ. Press, Princeton 1946.
CRAMÉR, H. and H. WOLD
[1] Some theorems on distribution functions, J. London Math. Soc. 11, 290–294 (1936).
CSÁKI, E.
[1] On two modifications of the Wilcoxon-test, Publ. Math. Inst. Hung. Acad. Sci. 4, 313–319 (1959).
CSÁKI, P. and J. FISCHER
[1] On bivariate stochastic connection, Publ. Math. Inst. Hung. Acad. Sci. 5, 311–323 (1960).
[2] Contributions to the problem of maximal correlation, Publ. Math. Inst. Hung. Acad. Sci. 5, 325–337 (1950).
CSÁSZÁR, Á.
[1] Sur la structure des espaces de probabilité conditionelle, Acta Math. Acad. Sci. Hung. 6, 337–361 (1955).
[2] Sur une caractérisation de la répartition normale de probabilités, Acta Math. Acad. Sci. Hung. 7, 359–382 (1956).

VAN DANTZIG, D.
[1] Mathematische Statistiek, "Kadercursus Statistiek, 1947–1948", Mathematisch Centrum, Amsterdam 1948.

DARMOIS, G.
[1] Analyse générale des liaisons stochastiques, Revue Inst. Internat. Stat. **21**, 2–8 (1953).

DOEBLIN, W.
[1] Sur les propriétés asymptotiques de mouvements régis par certains types de chaînes simples, Bull. Soc. Math. Roumaine Sci. **39I**, 57–115 (1937); **39II**, 3–61 (1937).
[2] Éléments d'une théorie générale des chaînes simples constantes de Markov, Ann. Sci. École Norm. Sup. (3) **57**, 61–111 (1940).

DONSKER, M. D.
[1] Justification and extension of Doob's heuristic approach to the Kolmogorov–Smirnov theorems, Ann. Math. Stat. **23**, 277–281 (1952).

DOOB, J. L.
[1] Heuristic approach to the Kolmogorov–Smirnov theorems, Ann. Math. Stat. **20**, 393 (1949).
[2] Stochastic processes, Wiley–Chapman, New York–London 1953.

DUGUÉ, D.
[1] Arithmétique de lois de probabilités, Mém. Sci. Math., No. 137, Gauthier-Villars, Paris 1957.

DUMAS, M.
[1] Sur les lois de probabilités divergentes et la formule de Fisher, Interméd. Rech. Math. **9** (1947), Supplement 127–130.
[2] Interprétation de resultats de tirages exhaustifs, C. R. Acad. Sci. Paris **288**, 904–906 (1949).
(See the note from E. Borel after Dumas' article, too.)

DVORETZKY, A. and P. ERDŐS
[1] Some problems on random walk in space, Proc. 2nd Berkeley Symp. Math. Stat. Prob. 1950, Univ. California Press, Berkeley–Los Angeles 1951, 353–367.

EGGENBERGER, F., und G. PÓLYA
[1] Über die Statistik verketteter Vorgänge, Z. angew. Math. Mech. **3**, 279–289 (1923).

EINSTEIN, A.
[1] Zur Theorie der Brownschen Bewegung, Ann. Physik **19**, 371–381 (1906).

ERDŐS, P.
[1] On the law of the iterated logarithm, Ann. Math. **43**, 419–436 (1942).
[2] On the distribution function of additive functions, Ann. Math. **47**, 1–20 (1946).

ERDŐS, P. and G. A. HUNT
[1] Changes of signs of sums of random variables, Pacific J. Math. **3**, 678–679 (1953).

ERDŐS, P. and M. KAC
[1] On certain limit theorems of the theory of probability, Bull. Amer. Math. Soc. **52**, 292–302 (1946).
[2] On the number of positive sums of independent random variables, Bull. Amer. Math. Soc. **53**, 1011–1020 (1947).

ERDŐS, P. and A. RÉNYI
[1] On the central limit theorem for samples from a finite population, Publ. Math. Inst. Hung. Acad. Sci. **4**, 49–61 (1959).
[2] On Cantor's series with convergent $\sum \dfrac{1}{q_n}$, Ann. Univ. Sci. Budapest, Roland o Eötvös nom., Sect. Math. **2**, 93–109 (1959).

ERDŐS, P. and A. RÉNYI
[3] On the evolution of random graphs, Publ. Math. Inst. Hung. Acad. Sci. 5, 17–61 (1960).
[4] On a classical problem of probability theory, Publ. Math. Inst. Hung. Acad. Sci. 6, 215–220 (1961).
ESSEEN, C. G.
[1] Fourier analysis of distribution functions, A mathematical study of the Laplace–Gaussian law, Acta Math. 77, 1–125 (1945).

FEINSTEIN, A.
[1] A new basic theorem of information theory, Trans. Inst. Radio Eng., 2–22 (1954).
[2] Foundations of information theory, McGraw-Hill, New York 1958.
FELDHEIM, E.
[1] Étude de la stabilité des lois de probabilité, Dissertation, Univ. Paris, Paris 1937.
[2] Neuere Beweise und Verallgemeinerung der wahrscheinlichkeitstheoretischen Sätze von Simmons, Mat. Fiz. Lapok 45, 99–114 (1938).
FEILER, W.
[1] Über den zentralen Grenzwertsatz der Wahrscheinlichkeitsrechnung, Math. Z. 40, 521–559 (1935); 42, 301–312 (1947).
[2] Zur Theorie der stochastischen Prozesse, Existenz- und Eindeutigkeitssätze, Math. Ann. 113, 113–160 (1936).
[3] On the integro-differential equations of purely discontinuous Markov processes, Trans. Amer. Math. Soc. 48, 488–515 (1940). Errata: ibidem 58, 474 (1945).
[4] The law of the iterated logarithm for identically distributed random variables, Ann. Math. 47, 631–638 (1946).
[5] On the Kolmogorov–Smirnov limit theorems for empirical distributions, Ann. Math. Stat. 19, 177–189 (1948).
[6] On the theory of stochastic processes, with particular reference to applications, Proc. Berkeley Symp. Math. Stat. Prob. 1945, 1946, Univ. California Press, Berkeley–Los Angeles 1949, 403–432.
[7] An introduction to probability theory and its applications, Vols 1–2, Wiley, New York 1950–1966.
DE FINETTI, B.
[1] Funzione caratteristica di un fenomeno aleatorio, Mem. R. Accad. Lincei (6) 4, 85–133 (1930).
[2] Sul concetto di media, Giorn. Ist. Ital. Att. 2, 369–396 (1931).
FISHER, R. A.
[1] Statistical methods for research workers, 10th edition, Oliver–Boyd Ltd., Edinburgh–London 1948.
[2] The design of experiments, Oliver–Boyd Ltd., London–Edinburgh 1949.
[3] Contributions to mathematical statistics, Wiley–Chapman, New York–London 1950.
FISHER, R. A. and F. YATES
[1] Statistical tables for biological, agricultural and medical research, Oliver–Boyd Ltd., London–Edinburgh 1949.
FISZ, M.
[1] Probability theory and mathematical statistics, 3. ed. Wiley, New York 1963.
FLOREK, K., E. MARCZEWSKI and C. RYLL-NARDZEWSKI
[1] Remarks on the Poisson stochastic process, I, Studia Math. 13, 122–129 (1953).

FRÉCHET, M.
[1] Récherches théoriques modernes, Fascicule 3 du Tome 1 du Traité du calcul des probabilités par É. Borel et divers auteurs, Gauthier–Villars, Paris 1937.
[2] Les probabilités associées à un système d'événements compatibles et dépendants, I–II, Hermann et Cie., Paris 1940 and 1943.
FRINK, O.
[1] Representations of Boolean algebras, Bull. Amer. Math. Soc. 47, 755–756 (1941).
[2] A proof of the maximal chain theorem, Amer. J. Math. 74, 676–678 (1952).

GANTMACHER, F. R. (Гантмахер, Ф. Р.)
[1] Matrize rechnung, I–II, VEB Deutscher Verlag der Wissenschaften, Berlin 1958 bzw. 1959 (Übersetzung aus dem Russischen).
GAUSS, C. F.
[1] Theoria combinationis observationum erroribus minimus obnoxiae, Göttingen 1821.
GAVRILOV, M. A. (Гаврилов, М. А.)
[1] Теория релейно-контактных схем (Theory of relay-contact schemes), Moscow–Leningrad 1950.
GEARY, R. C.
[1] Distribution of Student's ratio for nonnormal samples, J. Royal Stat. Soc. Supplement 3, 178–184 (1936).
GEBELEIN, H.
[1] Das statistische Problem der Korrelation als Variations- und Eigenwertproblem und sein Zusammenhang mit der Ausgleichungsrechnung, Z. angew. Math. Mech. 21, 364–379 (1941).
GLIVENKO, V. I. (Гливенко, В. И.)
[1] Sulla determinazione empirica di una legge probabilita, Gior. Ist. Ital. Att. 4, 1–10 (1933).
[2] Théorie générale des structures, Act. Sci. Industr. Nr. 652, Hermann et Cie., Paris 1938.
[3] Курс тории в ероятностей (A course of probability theory), GONTI, Moscow–Leningrad 1939.
GNEDENKO, B. V. (Гнеденко, Б. В.)
[1] Sur la distribution limite du terme maximum d'une série aléatoire, Ann. Math. 44, 423–453 (1943).
[2] Локальная предельная теорема для плотностей (A local limit theorem for probability densities), Dokl. Akad. Nauk. SSSR 95, 5–7 (1954).
[3] The theory of probability (Transl. from the Russian), Chelsea, New York 1962.
GNEDENKO, B. V. and A. N. KOLMOGOROV (Гнеденко, Б. В. и А. Н. Колмогоров)
[1] Limit distributions for sums of independent random variables, Addison–Wesley. Cambridge (Mass.) 1954.
GNEDENKO, B. V. and V. S. KOROLJUK (Гнеденко, Б. В. и В. С. Королюк)
[1] О максимальном расхождении двух эмпирических распределений (On the maximal divergence of two empirical distributions), Dokl. Akad. Nauk. SSSR 80, 525 (1951).

HÁJEK, J. and A. RÉNYI
[1] Generalization of an inequality of Kolmogorov, Acta Math. Acad. Sci. Hung. 6, 281–283 (1955).

HAJÓS, G. and A. RÉNYI
[1] Elementary proofs of some basic facts concerning order statistics, Acta Math. Acad. Sci. Hung. **5**, 1–6 (1954).
HALMOS, P. R.
[1] Measure Theory, van Nostrand, New York 1950.
HARDY, G. H.
[1] Divergent series, Clarendon Press, Oxford 1949.
HARDY, G. H., J. E. LITTLEWOOD and G. PÓLYA
[1] Inequalities, 2nd edition, Cambridge Univ. Press, Cambridge 1952.
HARDY, G. H. and W. W. ROGOSINSKI
[1] Fourier series, 3rd edition, Cambridge Univ. Press, Cambridge 1956.
HARDY, G. H. and E. M. WRIGHT
[1] An introduction to the theory of numbers, 4th edition, Clarendon Press, Oxford 1960.
HARRIS, T. E.
[1] The theory of branching processes, Springer Verlag, Berlin–Heidelberg–New York 1963.
HARTLEY, R. V.
[1] Transmission of information, Bell Syst. Techn. J. **7**, 535–563 (1928).
HAUSDORFF, F.
[1] Grundzüge der Mengenlehre, B. G. Teubner, Leipzig 1914.
HELMERT, R.
[1] Über die Wahrscheinlichkeit der Potenzsummen der Beobachtungsfehler und über einige damit im Zusammenhang stehende Fragen, Z. Math. Phys. **21**, 192–219 (1876).
HILLE, E.
[1] Functional analysis and semi-groups, Amer. Math. Soc. Coll. Publ., Vol. 31, New York 1948.
HOSTINSKY, B.
[1] Méthodes générales du calcul des probabilités, Mém. Sci. Math. Nr. 52, Gauthier–Villars, Paris 1931.
HURWITZ, A. und R. COURANT
[1] Funktionentheorie, Springer, Berlin 1929.

JEFFREYS, H.
[1] Theory of probability, 2nd edition, Clarendon Press, Oxford 1948.
JORDAN, CH.
[1] On probability, Proc. Phys. Math. Soc. Japan **7**, 96–109 (1925).
[2] Statistique mathématique, Gauthier–Villars, Paris 1927.
[3] Le théorème de probabilité de Poincaré, généralisé au cas de plusieurs variables indépendantes, Acta Sci. Math. Szeged **7**, 103–111 (1934).
[4] Calculus of finite differences, 2nd edition, Chelsea Publ. Comp., New York 1950.
[5] Fejezetek a klasszikus valószínűségszámításból (Chapters from the classical calculus of probabilities), Akadémiai Kiadó, Budapest 1956.

KAC, M.
[1] Random walk and the theory of Brownian motion, Amer. Math. Monthly **54**, 369–391 (1947).

KAC, M.

[2] A remark on the proceeding paper by A. Rényi, Publ. Inst. Math. Beograd 8, 163–165 (1955).

[3] Probability and related topics in physical sciences, Lectures in applied mathematics, Vol. I, Intersci. Publ., London–New York 1959.

[4] Statistical independence in probability, analysis and number theory, Math. Assoc. America 1959.

KANTOROVITCH, L. V. (Канторович, Л. В.)

[1] Sur une problème de M. Steinhaus, Fund. Math. 14, 266–270 (1929).

KAPPOS, D. A.

[1] Strukturtheorie der Wahrscheinlichkeitsfelder und -räume, Springer-Verlag, Berlin–Göttingen–Heidelberg 1960.

KAWATA, I. and H. SAKAMOTO

[1] On the characterization of the normal population by the independence of the sample mean and the sample variance, J. Math. Soc. Japan 1, 111–115 (1949).

KHINCHIN, A. J. (Хинчин, А. Я.)

[1] Über dyadische Brüche, Math. Z. 18, 109-118 (1923).

[2] Sur les classes d'événements équivalents, Mat. Sbornik 39:3, 40–43 (1932).

[3] Asymptotische Gesetze der Wahrscheinlichkeitsrechnung, Springer, Berlin 1933.

[4] Korrelationstheorie der stationärer stochastischer Prozesse, Math. Ann. 109, 604–615 (1934).

[5] Sul dominio di attrazione della legge di Gauss, Giorn. Ist. Ital. Att. 6, 378–393 (1935).

[6] Kettenbrüche, B. G. Teubner, Leipzig 1956.

[7] О классах эквивалентных событии (On classes of equivalent events), Dokladi Akad. Nauk. SSSR 85, 713–714 (1952).

KHINCHIN, A. J. und A. N. KOLMOGOROV (Хинчин, А. Я. и А. Н. Колмогоров)

[1] Über Konvergenz von Reihen, deren Glieder durch den Zufall bestimmt werden, Mat. Sbornik 32, 668–677 (1925).

(KHINCHIN) CHINTSCHIN, A. J. et P. LÉVY

[1] Sur les lois stables, C. R. Acad. Sci. Paris 202, 374–376 (1936).

KNOPP, K.

[1] Theorie und Anwendung der unendlichen Reihen, Springer, Berlin 1924.

KOLLER, S.

[1] Graphische Tafeln zur Beurteilung statistischer Zahlen, Steinkopff, Dresden–Leipzig 1943.

KOLMOGOROV, A. N. (Колмогоров, А. Н.)

[1] Über das Gesetz des iterierten Logarithmus, Math. Ann. 101, 126–136 (1929).

[2] Sur la loi forte des grandes nombres, C. R. Acad. Sci. Paris 191, 910–912 (1930).

[3] Über die analytischen Methoden in der Wahrscheinlichkeitsrechnung, Math. Ann. 104, 415–458 (1930).

[4] Sur la notion de la moyenne, Atti R. Accad. Naz. Lincei 12, 388–391 (1930).

[5] Foundations of the theory of probability, Chelsea, New York 1956.

[6] Sulla determinazione empirica di una legge di distribuzione, Giorn. Ist. Ital. Att. 4, 83–91 (1933).

[7] Цепы Маркова с счетным множеством возможных состояний (Markov chains with denumerably infinite possible states), Bull. Mosk. Univ. 1, 1 (1937).

[8] О логарифмически нормальном законе распределения размеров частиц при дроблении (On the lognormal distribution of the sizes of particles in chopping), Dokl. Akad. Nauk. SSSR 31, 99–101 (1941).

[9] Algèbres de Boole métriques complètes, VI. Zjad Matematykow Polskich, Warsaw 20–23. IX. 1948, Inst. Math. Univ. Krakow. 1950. 22–30.

KOLMOGOROV, A. N. (Колмогоров, А. Н.)
[10] Ein Satz über die Konvergenz der bedingten Erwartungswerte und deren Anwendungen, I. Magyar Matematikai Kongresszus Közleményei, Budapest 1950, 377–386.
[11] Некоторые работы последных лет в области предельних теорем теории вероятностей (On some recent works concerning the limit theorems of probability theory), Vestnik Univ. Moscow **8** (10), 29–38 (1953). *See also* "Arbeiten zur Informationstheorie III."
KRICKEBERG, K.
[1] Wahrscheinlichkeitstheorie, Teubner, Stuttgart 1963.
KULLBACK, S.
[1] Information theory and statistics, Wiley, New York 1959.
KY FAN
[1] Les fonctions définies-positives et les fonctions complètement monotones, leurs applications au calcul des probabilités et à la théorie des espaces distanciées, Mém. Sci. Math. **114**, Gauthier–Villars, Paris 1950.

LAHA, R. G.
[1] An example of a non-normal distribution where the quotient follows the Cauchy law, Proc. Nat. Acad. Sci. USA **44**, 222–223 (1958).
LAPLACE, P. S.
[1] Théorie analytique des probabilités, 1795. Oeuvres Complètes de Laplace, t. 7, Gauthier–Villars, Paris 1886.
[2] Essai philosophique sur les probabilités, I–II, Gauthier–Villars, Paris 1921.
LEHMANN, E. L.
[1] Consistency and unbiasedness of certain nonparametric tests, Ann. Math. Stat. **22**, 165–180 (1951).
LÉVY, P.
[1] Calcul des Probabilités, Gauthier–Villars, Paris 1925.
[2] Sur certains processes stochastiques homogènes, Comp. Math. **7**, 283–339. (1939).
[3] Processus stochastiques et mouvement brownien, Gauthier–Villars, Paris 1948.
[4] Théorie de l'addition des variables aléatoires, 2e ed. Gauthier–Villars, Paris 1954.
LIGHTHILL, M. J.
[1] An introduction to Fourier analysis and generalised functions, Cambridge Univ. Press, Cambridge 1959.
LINDEBERG, J. W.
[1] Eine neue Herleitung des Exponentialgesetzes in der Wahrscheinlichkeitsrechnung, Math. Z. **15**, 211–225 (1922).
LINNIK, Yu. V. (Линник, Ю. В.)
[1] The large sieve, Dokl. Akad. Nauk. SSSR **30**, 292–294 (1941).
[2] Теоретико-информационное доказательство центральной предельной теоремы в условиях Линдеберга (An information theoretic proof of the central limit theorem on Lindeberg conditions), Teor. Verojatn. Prim. **4**, 311–321 (1959).
[3] Разложения вероятностных законов (Decomposition of probability functions), Izd. Univ. Leningrad 1960.
LINNIK, Yu. V. and A. A. SINGER (Линник, Ю. В. и А. А. Зингер)
[1] Об одном аналитическом обобщении теоремы Крамера (On an analytic extension of Cramér's theorem), Vestnik Leningr. Univ. 11, 51–56 (1955).
LJAPUNOV, A. M. (Ляпунов, А. М.)
[1] Избранные труды (Selected works), Akad. izd. Moscow 1948, pp. 179–250.

LOBACHEVSKY, N. I. (Лобачевский, Н. И.)
[1] Sur la probabilité des résultats moyens, tirés des observations répétées, J. reine angew. Math. **24**, 164–170 (1842).

LOÈVE, M.
[1] Probability theory, van Nostrand, New York 1955.

ŁOMNICKI, A.
[1] Nouveaux fondements du calcul des probabilités, Fund. Math. **4**, 34–41 (1923).

LÖSCH, F. und F. SCHOBLIK
[1] Die Fakultät, B. G. Teubner, Leipzig 1951.

LUCE, R. D.
[1] Individual choice behaviour. A theoretical analysis, Wiley, New York 1959.

LUKÁCS, E.
[1] A characterization of the normal distribution, Ann. Math. Stat. **13**, 91–93 (1942).
[2] Application of Faà di Bruno's formula in mathematical statistics, Amer. Math. Monthly **62**, 340–348 (1955).
[3] Characterisation of populations by properties of suitable statistics, Proc. 3rd Berkeley Symp. Math. Stat. Prob. 1954–1955, Vol. II, Univ. California Press, Berkeley–Los Angeles 1956, 215–229.
[4] Characteristic functions, Griffin, London 1960.

LUKÁCS, E. and R. G. LAHA
[1] Applications of characteristic functions, Griffin, London 1964.

MALMQUIST, S.
[1] On a property of order statistics from a rectangular distribution, Skand. Aktuarietidskrift **33**, 214–222 (1950).

MARCZEWSKI, E.
[1] Remarks on the Poisson stochastic process, II, Studia Math. **13**, 130–136 (1953).

MARKOV, A. A. (Марков, А. А.)
[1] Wahrscheinlichkeitsrechnung, B. G. Teubner, Leipzig 1912.

McMILLAN, B.
[1] The basic theorems of information theory, Ann. Math. Stat. **24**, 196–219 (1953).

MEDGYESSY, P.
[1] Decomposition of superpositions of distribution functions, Akad. Kiadó, Budapest 1961.

VON MISES, R.
[1] Wahrscheinlichkeitsrechnung und ihre Anwendung in der Statistik und theoretischen Physik, Deuticke, Leipzig–Wien 1931.
[2] Wahrscheinlichkeit, Statistik und Wahrheit, Springer-Verlag, Berlin 1952.

MOGYORÓDI, J.
[1] On a consequence of a mixing theorem of A. Rényi, MTA Mat. Kut. Int. Közl., **9**, 263–267 (1964).

MOLINA, F. C.
[1] Poisson's exponential binomial limit, van Nostrand, New York 1942.

MORIGUTI, S.
[1] A lower bound for a probability moment of an absolutely continuous distribution with finite variance, Ann. Math. Stat. **23**, 286–289 (1952).

NAGUMO, M.
[1] Über eine Klasse von Mittelwerten, Japan. J. Math. **7**, 71–79 (1930).

NEVEU, J.
[1] Mathematical foundations of the calculus of probability, Holden-Day Inc., San Francisco 1965.
NEYMAN, J.
[1] L'estimation statistique traité comme un problème classique de probabilité, Act. Sci. Industr., Nr. 739, Gauthier-Villars, Paris 1938.
[2] First course in probability and statistics, H. Holt et Co., New York 1950.

ONICESCU, O. et G. MIHOC
[1] La dépendance statistique. Chaînes et familles de chaînes discontinues, Act. Sci. Industr., Nr. 503, Gauthier-Villars, Paris 1937.
ONICESCU, O., G. MIHOC si C. T. IONESCU-TULCEA
[1] Calculul probabilitatilor si applicatii, Bucureşti 1956.

PARZEN, E.
[1] Modern probability theory and its applications, Wiley, New York 1960.
PEARSON, E. S. and H. O. HARTLEY
[1] Biometrical tables for statisticians, Cambridge Univ. Press, Cambridge 1954.
PEARSON, K.
[1] Early statistical papers, Cambridge Univ. Press, Cambridge 1948.
POINCARÉ, H.
[1] Calcul des probabilités, Carré-Naud, Paris 1912.
POISSON, S. D.
[1] Recherches sur la probabilité de judgements, Bachelier, Paris 1837.
PÓLYA, G.
[1] Über den zentralen Grenzwertsatz der Wahrscheinlichkeitsrechnung und das Momentproblem, Math. Z. 8, 171–181 (1920).
[2] Über eine Aufgabe der Wahrscheinlichkeitsrechnung betreffend die Irrfahrt im Straßennetz, Math. Ann. 84, 149–160 (1921).
PÓLYA, G. und G. SZEGŐ
[1] Aufgaben und Lehrsätze aus der Analysis, I–II, Springer, Berlin 1925.
POPPER, K.
[1] Philosophy of science: A personal report, British Philosophy in the Mid-Century, ed. by C. A. MACE, 1956, p. 191.
[2] The logic of scientific discovery, Hutchinson, London 1959.
PRÉKOPA, A.
[1] On composed Poisson distributions, IV, Acta Math. Acad. Sci. Hung. 3, 317–326 (1952).
[2] Valószínűségelmélet műszaki alkalmazásokkal (Probability theory and its applications in technology), Műszaki Könyvkiadó, Budapest 1962.
PRÉKOPA, A., A. RÉNYI and K. URBANIK
[1] О предельном расределении для сумм независимых случайных величин на бикомпактных коммутативных топологических группах (On limit distribution of sums of independent random variables over bicompact commutative topological groups), Acta Math. Acad. Sci. Hung. 7, 11–16 (1956).

REICHENBACH, H.
[1] Wahrscheinlichkeitslehre, Sijthoff, Leiden 1935.

RÉNYI, A.

[1] Simple proof of a theorem of Borel and of the law of the iterated logarithm, Mat. Tidsskrift B, 41–48 (1948).

[2] О представлении четных чисел в виде суммы простого и почти простого числа (On the representation of even numbers as sums of a prime and an almost prime number), Izvestia Akad. Nauk. SSSR, Ser. Mat. 12, 57–78 (1948).

[3] К теории предельных теорем для сумм независимых случайных величин (On limit theorems of sums of independent random variables), Acta Math. Acad. Sci. Hung. 1, 99–108 (1950).

[4] On the algebra of distributions, Publ. Math. Debrecen 1, 135–149 (1950).

[5] On composed Poisson distributions, II, Acta Math. Acad. Sci. Hung. 2, 83–98 (1951).

[6] On some problems concerning Poisson processes, Publ. Math. Debrecen 2, 66–73 (1951).

[7] On a conjecture of H. Steinhaus, Ann. Soc. Polon. Math. 25, 279–287 (1952).

[8] On projections of probability distributions, Acta Math. Acad. Sci. Hung. 3, 131–142 (1952).

[9] On the theory of order statistics, Acta Math. Acad. Sci. Hung. 4, 191–232 (1953).

[10] Eine neue Methode in der Theorie der geordneten Stichproben, Bericht über die Mathematiker-Tagung Berlin 1953, VEB Deutscher Verlag der Wissenschaften, Berlin 1953, 203–213.

[11] Kémiai reakciók tárgyalása a sztochasztikus folyamatok elmélete segítségével (On describing chemical reactions by means of stochastic processes), A Magyar Tudományos Akadémia Alkalmazott Matematikai Intézetének Közleményei 2, 596–600 (1953) (In Hungarian).

[12] Újabb kritériumok két minta összehasonlítására (Some new criteria for comparison of two samples), A Magyar Tudományos Akadémia Alkalmazott Matematikai Intézetének Közleményei 2, 243–265 (1953) (In Hungarian).

[13] Valószínűségszámítás (Probability theory), Tankönyvkiadó, Budapest 1954 (In Hungarian).

[14] Axiomatischer Aufbau der Wahrscheinlichkeitsrechnung, Bericht über die Tagung Wahrscheinlichkeitsrechnung und Mathematische Statistik, VEB Deutscher Verlag der Wissenschaften, Berlin 1954, 7–15.

[15] On a new axiomatic theory of probability, Acta Math. Acad. Sci. Hung. 6, 285–335 (1955).

[16] On the density of sequences of integers, Publ. Inst. Math. Beograd 8, 157–162 (1955).

[17] A számjegyek eloszlása valós számok Cantor-féle előállításaiban (The distribution of the digits in Cantor's representation of the real numbers), Mat. Lapok 7, 77–100 (1956) (In Hungarian).

[18] On conditional probability spaces generated by a dimensionally ordered set of measures, Teor. Verojatn. prim. 1, 61–71 (1956).

[19] A new deduction of Maxwell's law of velocity distribution, Isv. Mat. Inst. Sofia 2, 45–53 (1957).

[20] A remark on the theorem of Simmons, Acta Sci. Math. Szeged. 18, 21–22 (1957).

[21] Representations for real numbers and their ergodic properties, Acta Math. Acad. Sci. Hung. 8, 477–493 (1957).

[22] On the asymptotic distribution of the sum of a random number of independent random variables, Acta Math. Acad. Sci. Hung. 8, 193–199 (1957).

RÉNYI, A.

[23] Quelques remarques sur les probabilités des événements dépendantes, J. Math. pures appl. **37**, 393-398 (1958).

[24] On mixing sequences of sets, Acta Math. Acad. Sci. Hung. **9**, 215-228 (1958).

[25] Probabilistic methods in number theory, Proceedings of the International Congress of Mathematicians, Edinburgh 1958, 529-539.

[26] New version of the probabilistic generalization of the large sieve, Acta Math. Acad. Sci. Hung. **10**, 217-226 (1959).

[27] On the dimension and entropy of probability distributions, Acta Math. Acad. Sci. Hung. **10**, 193-215 (1959).

[28] On measures of dependence, Acta Math. Acad. Sci. Hung. **10**, 441-451 (1959).

[29] On a theorem of P. Erdős and its applications in information theory, Mathematica Cluj **1** (24), 341-344 (1959).

[30] Dimension, entropy and information, Transactions of the II. Prague Conference on Information theory, statistical decision functions, random processes, Praha 1960, 545-556.

[31] On the central limit theorem for the sum of a random number of independent random variables, Acta Math. Acad. Sci. Hung. **11**, 97-102 (1960).

[32] Az aprítás matematikai elméletéről (On the mathematical theory of chopping), Építőanyag 1-8 (1960) (In Hungarian).

[33] Bolyongási problémákra vonatkozó határeloszlástételek (Limit theorems in random walk problems), A Magyar Tudományos Akadémia III (Matematikai és Fizikai) Osztályának Közleményei **10**, 149-170 (1960) (In Hungarian).

[34] Az információelmélet néhány alapvető kérdése (Some fundamental problems of the information theory), A Magyar Tudományos Akadémia III (Matematikai és Fizikai) Osztályának Közleményei **10**, 251-282 (1960) (In Hungarian).

[35] Egy általános módszer valószínűségszámítási tételek bizonyítására (A general method for proving theorems in probability theory), A Magyar Tudományos Akadémia III (Matematikai és Fizikai) Osztályának Közleményei **11**, 79-105 (1961) (In Hungarian).

[36] Legendre polynomials and probability theory, Ann. Univ. Sci. Budapest, R. Eötvös nom., Sect. Math. **3·4**, 247-251 (1961).

[37] On measures of entropy and informations. Proc. Fourth Berkeley Symposium on Math. Stat. Prob. 1960, Vol. I, Univ. California Press, Berkeley-Los Angeles 1961, 547-561.

[38] On stable sequences of events, Sankhya A **25**, 293-302 (1963).

[39] On certain representations of real numbers and on equivalent events, Acta Sci. Math. Szeged **26**, 63-74 (1965).

[40] Új módszerek és eredmények a kombinatórikus analízisben (New methods and results in combinatorial analysis), A Magyar Tudományos Akadémia III (Matematikai és Fizikai) Osztályának Közleményei **16**, 75-105, 159-177 (1966) (In Hungarian).

[41] Sur les espaces simples des probabilités conditionnelles, Ann. Inst. H. Poincaré B **1**, 3-19 (1964).

[42] On the foundations of information theory, Review of the International Statistical Institute **33**, 1-14 (1965).

RÉNYI, A. and P. RÉVÉSZ

[1] On mixing sequences of random variables, Acta Math. Acad. Sci. Hung. **9**, 389-393 (1958).

[2] A study of sequences of equivalent events as special stable sequences, Publicationes Mathematicae Debrecen **10**, 319-325 (1963).

RÉNYI, A. and R. SULANKE
  [1] Über die konvexe Hülle von *n* zufällig gewählten Punkten,· I—II, Zeitschrift für Wahrscheinlichkeitstheorie **2**, 75–84 (1963); **3**, 138–147 (1964).
RÉVÉSZ, P.
  [1] A limit distribution theorem for sums of dependent random variables, Acta Math. Acad. Sci. Hung. **10**, 125–131 (1959).
  [2] The laws of large numbers, Akad. Kiadó, Budapest 1967.
RICHTER, H.
  [1] Wahrscheinlichkeitstheorie, Springer-Verlag, Berlin 1956.
RIESZ, F. and B. SZ.-NAGY
  [1] Functional analysis, Blackie, London–Glasgow 1956.
ROBBINS, H.
  [1] On the equidistribution of sums of independent random variables, Proc. Amer. Math. Soc. **4**, 786–799 (1953).
ROTA, G. C.
  [1] The number of partitions of a set, Amer. Math. Monthly, **71**, 498–504 (1964).

SAXER, W.
  [1] Versicherungsmathematik, II, Springer-Verlag, Berlin–Göttingen–Heidelberg 1958.
SCHMETTERER, L.
  [1] Einführung in die mathematische Statistik, Springer-Verlag, Wien 1956.
SCHÜTZENBERGER, M. P.
  [1] Contributions aux applications statistiques de la théorie de l'information, Inst. Stat. Univ. Paris (A) **2575**, 1–115 (1953).
SHANNON, C. E.
  [1] A mathematical theory of communication, Bell Syst. Techn. J. **27**, 379–423, 623–653 (1948).
SHANNON, C. E. and W. WEAVER
  [1] The mathematical theory of communication, Univ. Illinois Press, Urbana 1949.
SINGER, A. A. (Зингер, А. А.)
  [1] О независимых выборках из нормальной совокупности (On independent samples from a population), Uspehi Mat. Nauk **6**, 172–175 (1951).
SKITOVICH, V. R. (Скитович, В. Р.)
  [1] Об одном свойстве нормального распределения (On a property of the normal distribution), Dokl. Akad. Nauk. SSSR **89**, 217–219 (1953).
SLUTSKY, E.
  [1] Über stochastische Asymptoten und Grenzwerte, Metron **5**, 1–90 (1925).
SMIRNOV, N. V. (Смирнов, Н. В.)
  [1] Über die Verteilung allgemeiner Glieder in der Variationsreihe, Metron **12**, 59–81 (1935).
  [2] Приближение законов распределения случайных величин по эмпирическим данным (Approximation of the laws of distribution of random variables by means of empirical data), Uspehi Mat. Nauk. **10**, 179–206 (1944).
SMIRNOV, V. I. (Смирнов, В. И.)
  [1] Lehrgang der höheren Mathematik, Teil III, 3. Aufl., VEB Deutscher Verlag der Wissenschaften, Berlin 1961.
VON SMOLUCHOWSKI, M.
  [1] Drei Vorträge über Diffusion, Brownsche Molekularbewegung und Koagulation von Kolloidteilchen, Phys. Z. **17**, 557–571, 585–599 (1916).

SPARRE-ANDERSEN, E.
[1] On the number of positive sums of random variables, Skand. Aktuarietidskrift, 1949, 27–36.
[2] On the fluctuations of sums of random variables, I–II, Math. Scand. 1, 263–285 (1953); 2, 193–223 (1954).

SPITZER, F.
[1] A combinatorial lemma and its application to probability theory, Trans. Amer. Math. Soc. 82, 323–339 (1956).

STEINHAUS, H.
[1] Les probabilités dénombrables et leur rapport à la théorie de la mesure, Fund. Math. 286–310 (1923).
[2] Sur la probabilité de la convergence des séries, Studia Math. 2, 21–39 (1951).

STEINHAUS, H., M. KAC et C. RYLL-NARDZEWSKI
[1]–[10] Sur les fonctions indépendantes, I, Studia Mathematica 6, 46–58 (1936); II, ibidem 6, 59–66 (1936); III, ibidem 6, 89–97 (1936); IV, ibidem 7, 1–15 (1938); V, ibidem 7, 96–100 (1938); VI, ibidem 9, 121–132 (1940); VII, ibidem 10, 1–20 (1948); VIII, ibidem 11, 133–144 (1949); IX, ibidem 12, 102–107 (1951); X, ibidem 13, 1–17 (1953).

STONE, M. H.
[1] The theory of representation for Boolean algebras, Trans. Amer. Math. Soc. 4, 31–111 (1936).

STUDENT
[1] —'s Collected papers, Edited by E. S. Pearson and J. Wishart, London 1942.

SZÁSZ, G.
[1] Introduction to lattice theory (transl. from the Hungarian), Akad. Kiadó, Budapest 1963.

SZŐKEFALVI-NAGY, B.
[1] Spektraldarstellung linearer Transformationen des Hilbertschen Raumes, Springer, Berlin 1942.

TITCHMARSH, E. C.
[1] Theory of functions, Clarendon Press, Oxford 1952.

TODHUNTER, L.
[1] History of the mathematical theory of probability, MacMillan, Cambridge–London 1865.

USPENSKI, J. W. (Успенский, Ю. В.)
[1] Introduction to mathematical probability, McGraw-Hill, New York–London 1937.

VEKSLER, V., L. GROSHEV and B. ISAEV (Векслер, В., Л. Грошев и Б. Исаев)
[1] Ионизационные методы исследования излучений (Ionisation methods in the study of radiations), Gostehizdat, Moscow 1949.

WAERDEN, VAN DER, B. L.
[1] Mathematische Statistik, Springer-Verlag, Berlin–Göttingen–Heidelberg 1957.

WALD, A.
[1] Die Widerspruchsfreiheit des Kollektivbegriffes der Wahrscheinlichkeitsrechnung, Erg. Math. Koll. 8, Wien 1935–1936.

WANG, SHOU YEN
[1] On the limiting distribution of the ratio of two empirical distributions, Acta Math. Sinica 5, 253 (1955).

WIDDER, D. V.
[1] The Laplace-transform, Princeton Univ. Press, Princeton 1946.
WIENER, N.
[1] Cybernetics or control and communication in the animal and the machine, Act. Sci. Indust., Nr. 1053, Hermann et Cie, Paris 1948.
[2] Extrapolation, interpolation and smoothing of stationary time series, Wiley, New York 1949.
WILCOXON, F.
[1] Individual comparisons by ranking methods, Biometrics Bull. 1, 80–83 (1945).
WILKS, S. S.
[1] Order statistics, Bull. Amer. Math. Soc. 54, 6–50 (1948).
WOLFOWITZ, J.
[1] The coding of messages subject to chance errors, Illinois J. Math. 1, 591–606 (1957).
[2] Information theory for mathematicians, Ann. Math. Stat. 29, 351–356 (1958).
[3] Coding theorems of information theory, Springer-Verlag, Berlin–Göttingen–Heidelberg 1961.
WOODWARD, P. M.
[1] Probability and information theory with applications to radar, Pergamon Press, London 1953.

ZYGMUND, A.
[1] Trigonometrical series, Warsaw 1935; Dover–New York 1955.
[2] Trigonometric series, I–II, Cambridge Univ. Press, Cambridge 1959.

AUTHOR AND SUBJECT INDEX